Mastering**BIOLOGY**

Your gateway to the most widely-used online biology self-study materials and more...

With your purchase of a new copy of Freeman's *Biological Science,* Third Edition, you should have received a **Student Access Kit for MasteringBiology™**. This kit contains instructions and a code for accessing this dynamic website. Your Student Access Kit looks like this:

DON'T THROW YOUR ACCESS KIT AWAY!

If you did not purchase a new textbook or cannot locate the Student Access Kit, you can purchase your subscription to MasteringBiology™ online with a major credit card by selecting Buy Now. To do so, go to www.masteringbio.com. Identify the exact title and edition of your textbook when prompted.

INSTRUCTORS:
Look for your "Instructor Access Kit" that you received with your book. Contact your Pearson Benjamin Cummings Sales Representative if you need assistance.

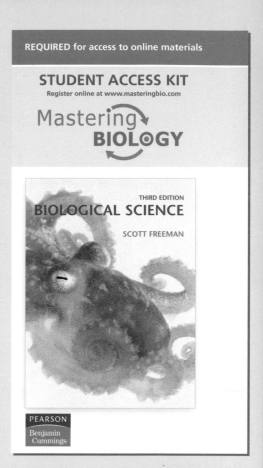

Mastering**BIOLOGY** What is MasteringBiology™?

MasteringBiology™ offers two valuable learning systems:

1. The **Study Area** for studying on your own or in a study group
2. **MasteringBiology™ Assignments** that your instructor may require

The **Study Area** provides state-of-the-art BioFlix™ 3-D Animations, Discovery videos, Web Animations, Cell Biology videos, interactive eBook, answers to textbook questions and much more (see other side).

MasteringBiology™ Assignments may include Tutorials and other Multiple Choice Questions required by your instructor. MasteringBiology™ tutorials coach you with feedback specific to your needs and offer hints when you get stuck, optimizing your study time and maximizing your learning.

(MB) Features of the MasteringBiology™ Study Area for Freeman's *Biological Science,* Third Edition

CHAPTER GUIDE

Prepare for the test by reviewing important information outlined in the Chapter Guide. The Chapter Guide provides a quick overview of key concepts, practice quizzes, and review questions to help students study for the exam. Link directly to the BioFlix™ 3-D animations, Discovery Videos, BioSkills, Cell Biology Animations, and interactive eBook to help students master biology's most difficult topics.

BIOFLIX™ 3-D ANIMATIONS

 BioFlix™ are 3-D movie-quality animations with carefully constructed student tutorials, labeled slide shows, study sheets, and quizzes which help bring biology to life. Topics include tour of an animal cell, tour of a plant cell, cellular respiration, photosynthesis, mitosis, meiosis, protein synthesis, water transport in plants, how neurons work, and muscle contraction.

VIDEOS

Watch brief Discovery Channel™ video clips on 29 different topics including antibiotic resistance, fighting cancer, and introduced species.

WEB ANIMATIONS

Learn about biological concepts in a visual format with online web animations. Animations include pre-quizzes and post-quizzes to test students' understanding of dynamic concepts and processes. Additional Cell Biology Animations provide vivid images of the functions and processes of the cell.

GRAPHIT!

Graphing tutorials show students how to plot, interpret, and critically evaluate real data.

CHAPTER QUIZZES

Students can assess their comprehension of material with 20 multiple-choice quiz questions for each chapter.

ANSWER KEY

Find answers to the book's figure and table caption questions and exercises, "you should be able to…" activities, and end-of-chapter questions.

ART

View and print artwork from the textbook to follow along in class or use as a study tool in review sessions.

AUDIO GLOSSARY

Build your biology vocabulary with our audio glossary. Hear the correct pronunciation and learn the meaning of all key terms.

E-BOOK

Refer to a convenient online version of the book while you study.

BIOLOGICAL SCIENCE

VOLUME **1** **THE CELL, GENETICS, AND DEVELOPMENT**

|← Actual Size: 4 cm →|

ABOUT THE PHOTO:
He'e, *Octopus* sp. (juvenile), photographed at Midway Atoll National
Wildlife Refuge, located in the Northwestern Hawaiian Islands, 29 March
2003, by David Liittschwager and Susan Middleton, authors of
Archipelago: Portraits of Life in the World's Most Remote Island Sanctuary
(National Geographic Society, 2005).

BIOLOGICAL SCIENCE

THIRD EDITION

SCOTT FREEMAN

University of Washington

PEARSON

Benjamin
Cummings

San Francisco Boston New York
Cape Town Hong Kong London Madrid Mexico City Montreal Munich Paris Singapore Sydney Tokyo Toronto

Editor-in-Chief, Biology: Beth Wilbur
Sponsoring Editors: Andrew Gilfillan, Susan Winslow,
  Becky Ruden
Editorial Project Managers: Ann Heath, Sonia DiVittorio
Development Editors: Erin Mulligan,
  Mary Catherine Hager, Susan Weisberg
Senior Production Supervisor: Shannon Tozier
Associate Editor: Mercedes Grandin
Assistant Editor: Anna Amato
Editorial Assistant: Lisa Tarabokjia
Media Producer: Ericka O'Benar
Executive Directors of Development: Carol Trueheart,
  Deborah Gale
Executive Marketing Manager: Lauren Harp
Director of Marketing: Christy Lawrence
Managing Editor: Mike Early

Executive Managing Editor: Erin Gregg
Production Service and Composition: Pre-Press PMG
Illustrations: Kim Quillin, Imagineering Media Services, Inc.,
  Greg Williams
Interior Design: Marilyn Perry
Cover Design: Kim Quillin, Marilyn Perry
Cover Production: Laura Wieglab and Side by Side Studios
Manufacturing Buyer: Mike Early and Michael Pene
Director, Image Resource Center: Melinda Patelli
Manager, Rights and Permissions: Zina Arabia
Visual Research Coordinator: Elaine Soares
Image Permission Coordinator: Debbie Latronica
Photo Researcher: Yvonne Gerin
Cover Printer: Phoenix Color
Printer and Binder: R.R. Donnelley, Willard

Cover Image: Octopus–he'e, *Octopus* sp. (juvenile): David Liittschwager/Susan Middleton

Library of Congress Cataloging-in-Publication Data

Freeman, Scott
  Biological science / Scott Freeman.—3rd ed.
    p. cm.
  Includes index.
  ISBN 0-13-224950-2 (student edition)—ISBN 0-13-224985-5 (professional
copy)—ISBN 0-13-225308-9 (volume 1)—ISBN 0-13-232543-8 (volume 2)
—ISBN 0-13-156816-7 (volume 3)
  1. Biology. I. Title.
  QH308. 2. F73 2007
  570—dc22

2007036866

ISBN 10-digit 0-13-224950-2; 13-digit 978-0-13-224950-8 (Student edition)
ISBN 10-digit 0-13-224985-5; 13-digit 978-0-13-224985-0 (Professional copy)
ISBN 10-digit 0-13-225308-9; 13-digit 978-0-13-225308-6 (Volume 1)
ISBN 10-digit 0-13-232543-8; 13-digit 978-0-13-232543-1 (Volume 2)
ISBN 10-digit 0-13-156816-7; 13-digit 978-0-13-156816-7 (Volume 3)

PEARSON
Benjamin
Cummings

2 3 4 5 6 7 8 9 10—DOW—11 10 09 08
www.aw-bc.com

Brief Contents: Volume 1

About the Author

Scott Freeman received his PhD. in Zoology from the University of Washington and was subsequently awarded an Alfred P. Sloan Postdoctoral Fellowship in Molecular Evolution at Princeton University. His current research focuses on the scholarship of teaching and learning—specifically, (1) how active learning and peer teaching techniques increase student learning and improve performance in introductory biology, and (2) how the levels of exam questions vary among introductory biology courses, standardized post-graduate entrance exams, and professional school courses. He has also done research in evolutionary biology on topics ranging from nest parasitism to the molecular systematics of the blackbird family. Scott teaches introductory biology for majors at the University of Washington and is co-author, with Jon Herron, of the standard-setting undergraduate text *Evolutionary Analysis*.

UNIT ADVISORS

An elite group of eleven content experts and star teachers worked with Scott and Kim on every aspect of the third edition. These advisors came to be treasured as they read and interpreted reviews, provided recommendations on outstanding recent papers to check, answered questions, and provided advice on an array of specific issues. The quality and accuracy of this book are a tribute to their efforts and skills.

Ross Feldberg, *Tufts University* (Unit 1)

David Wilson, *Parkland College* (Unit 1)

Paula Lemons, *Duke University* (Unit 2)

Greg Podgorski, *Utah State University* (Units 3 and 4)

George Gilchrist, *College of William and Mary* (Unit 5)

Brianna Timmerman, *University of South Carolina* (Unit 6)

Marc Perkins, *Orange Coast College* (Unit 6)

Michael Black, *California Polytechnic State University* (Units 6 and 8)

Diane Marshall, *University of New Mexico* (Unit 7)

James M. Ryan, *Hobart and William Smith Colleges* (Unit 8)

Alan Molumby, *University of Illinois, Chicago* (Unit 9)

ILLUSTRATOR

Kim Quillin combines expertise in biology and information design to create pedagogically effective and scientifically accurate visual representations of biological principles. She received her B.A. in Biology at Oberlin College and her Ph.D. in Integrative Biology (as a National Science Foundation Graduate Fellow) from the University of California, Berkeley, and has taught undergraduate biology at both schools. Students and instructors alike have praised Kim's illustration programs for *Biological Science*, as well as *Biology: A Guide to the Natural World*, by David Krogh, and *Biology: Science for Life*, by Colleen Belk and Virginia Borden, for their success at applying core principles of information design to convey complex biological ideas in a visually appealing manner.

Preface to Instructors

☞STUDENTS, There is also a preface for you, located right before Chapter 1. It's called "Using this Book as a Tool for Learning." Please read it—it should help you get organized and be successful in this course. *Scott Freeman*

This book is for instructors who want to help their students learn how to think like a biologist. The content knowledge, problem-solving ability, and analytical skills that this requires can help students become better human beings, in addition to preparing them for success in clinical medicine, scientific research, conservation, law, teaching, journalism, and other careers.

A course goal that focuses on thinking and learning, rather than simply memorizing, is important because today's biology students are going to be tomorrow's problem solvers. This is biology's century, not only because of the breathtaking pace of research but also because many of the most profound challenges we face today—resource shortages, overpopulation, species extinctions, drug resistance, global warming—are biological in nature. The world needs our students.

A Student-Centered Textbook

The first edition of *Biological Science* focused on offering a new approach to teaching biology—one that emphasized higher-order thinking skills over an encyclopedic grasp of what is known about biology. The second edition stayed true to this vision, but added topics and features that made the book easier for professors to use. The third edition also went through several rounds of revision driven by expert reviewers and advisors. Based on input from hundreds of professors around the world, I made thousands of changes to make the book even more accurate, current, and easy to use. But fundamentally, this edition is all about students. For the past three years, the book team and I have used insights from research on student learning—as well as direct feedback from students—as a way to create a better teaching tool.

I plunged into the literature on student learning, and members of the book team and I conducted two dozen focus groups with over 130 students. We asked students the same questions, over and over: What topics are most difficult in this chapter? What helped you "get it"? What tripped you up? How can this figure, or table, or passage of text teach better? Our goal was to create a text that is innovative, engaging, filled with the excitement that drives research, and inspired by data on how students learn.

Fairly quickly, our reading and research identified two fundamental problems:

1. *Novice learners have trouble picking out important information.* This is one of the most striking findings emerging from research on how people learn. It's also an issue that we hear about all the time as instructors. "Do we have to know X?" "Will Y be on the test?" We also see it when students come to office hours and open up their text—they've highlighted everything. Students need help figuring out which material is really important and which offers supporting details. This is crucial to their success, because if we're doing our job right, we're going to test them on the important stuff—not details that they'll forget five minutes after the exam.

2. *Novice learners have a terrible time with self-assessment.* How many times have students told you, "I understood the concepts *so well*, but just did badly on the exam"? Novice learners struggle to understand when they don't understand something. They'll sit in class or read the text and tell themselves, "Yeah, I get this, I get this," but then they crash on the exam. Experts are much more skeptical—they make themselves *use* information and ideas before they're confident they understand what's going on.

Fundamentally, our task is simple: We need to help our students become better students. Textbooks should help learners acquire the skills they need to make the novice-to-expert transition.

Supporting Novice Learners

The Gold Thread—"Learn It"

Our response to the "can't pick out the important points" problem is a battery of tools, highlighted in gold throughout the text.

- **Key Concepts** are listed at the start of the chapter. When material related to these key concepts is presented in the chapter itself, it is flagged with a gold bullet.

- **Chapter Summaries** revisit each of the key concepts. In this way, the big ideas in each chapter are laid out at the start, developed in detail, and then summarized.

- **Check Your Understanding** boxes appear at the ends of key sections within each chapter. Each box briefly summarizes one or two fundamental points—the key ideas that students ought to have mastered before they move on.

- **Highlighted Passages** help students focus on particularly important information throughout the text.

- **Summary Tables** pull information together into a compact format that is easy to review.

In effect, the gold thread offers students expert guidance on picking out and focusing on the important, unifying points in an information-laden discipline. In addition, and in the same spirit,

- **"Pointing Hands"** in illustrations act like your hand at the whiteboard, guiding students' attention to a figure's central teaching points.

The Blue Thread—"Practice It"

Our response to the self-assessment problem is an array of questions and exercises, highlighted in blue throughout the text.

- **In-text "You should be able to's"** offer exercises on topics that professors and students have identified as the most difficult concepts in each chapter.

- **Caption Questions and Exercises** challenge students to critically examine the information in a figure or table—not just absorb it.

- **Check Your Understanding** boxes present two to three tasks that students should be able to complete in order to demonstrate a mastery of specified key ideas.

- **Chapter Summaries** include "You should be able to" problems or exercises related to each of the key concepts declared in the gold, "Learn It" thread.

- **End-of-Chapter Questions** are organized around Bloom's taxonomy of learning, so students can test their understanding at the knowledge, comprehension, and application levels.

The fundamental idea is that if students really understand a piece of information or a concept, they should be able to do something with it.

Supporting Visual Learners

Kim Quillin—this book's illustrator—is a 1-in-6.6-billion talent. She combines superb academic training in biology (a Ph.D. from the University of California, Berkeley), insights from Edward Tufte's research on information architecture, impressive artistic talent, and a teacher's sensitivity to students.

Kim's goals in the first edition were to build a visual narrative that was (1) a direct extension of the text narrative and (2) guided by rigorous attention to the latest advances in information design. For example, figures that illustrate stepwise processes in biology were presented in a standardized, linear, and intuitive format, to spare students the "speed bumps" of convoluted compositions. In all of the figures, the drawings, colors, and labels were placed with surgical precision to maximize cleanliness and present a clear hierarchy of information. And to help bring a narrative voice to the figures, Kim integrated the "pointing hand" feature—inspired by National Academy of Sciences member M.A.R. Koehl—as a quiet and friendly way of highlighting important or challenging information such as data trends in graphs, or implications of phylogenetic trees.

The second edition stayed faithful to the original art philosophy, adding details requested by professors, formalizing experiments into boxes that made the experimental process consistent and explicit, and increasing the vibrancy and color of the drawings. Refinements in the third edition have focused on bringing the content of many individual figures up-to-date and honing the clarity, impact, and student-friendliness of every figure.

Supporting Skill Building

When instructors write learning objectives for introductory biology courses, they are paying close attention to skill building, in addition to the traditional emphasis on mastering content and concepts and the newer focus on developing higher-order thinking skills.

To aid skill building in introductory biology, I've added a set of nine new appendixes, called BioSkills, to the end matter of the book—just before the glossary. The BioSkills are meant to provide background on skills and techniques that are used throughout biological science. In addition, some of the BioSkills should help shore up problems with student preparation in mathematics and chemistry. They are called out at relevant points in the text. Please point them out to students who are having trouble with specific skills.

1. Reading Graphs

2. Reading a Phylogenetic Tree

3. Using Statistical Tests and Interpreting Standard Error Bars

4. Reading Chemical Structures

5. Using Logarithms

6. Making Concept Maps

7. Using Electrophoresis to Separate Molecules

8. Observing Microscopic Structures and Processes

9. Combining Probabilities

Serving a Community of Teachers

There is nothing that inspires me more than watching a passionate teacher work with motivated students. Teaching and learning is the essence of humanity, of who we are.

I write *Biological Science* because it's my way of being part of that work. As a textbook author, my greatest reward comes from interacting with inspired instructors and students from around the world—through e-mails, phone calls, focus groups, reviews, seminars, and teaching workshops.

Introductory biology courses are undergoing a slow and steady, yet remarkable, change: from a memorization-driven, largely passive exercise to a dynamic, active interchange that emphasizes higher-order thinking skills. The change is good. Instructors are stimulated and having more fun. They're approaching teaching problems in a hypothesis-testing framework, collecting data, and changing course designs based on empirical evidence. Students are performing at a higher level and should be better prepared for graduate school, professional school, and careers related to biology. My hope is that *Biological Science* can support this change.

Thank you for considering this text, for your passion about biology, and for your work on behalf of your students. What you do is *important*.

Scott Freeman
University of Washington

This book is dedicated to the world's greatest profession: teaching. Teachers are like Johnny Appleseed—they sow seeds, but seldom get to see the trees or fruit. This edition reflects in particular four teachers who had an exceptional impact on me: Vern Bailey, Owen Jenkins, Sievert Rohwer, and Barbara Wakimoto. Thank you.

Detailed Contents

UNIT 5   EVOLUTIONARY PROCESSES AND PATTERNS   481

Acknowledgments

Expert Feedback

Work on this edition was organized around two pillars: feedback from our advisory board and peer reviewers, and feedback from students.

Advisory Board

A cadre of highly select advisors analyzed reviews, provided citations for particularly important recent papers, and worked with me and Kim Quillin to brainstorm ideas for improving each chapter and unit. It was a joy to work with them, and their expertise and perspective were vital to this edition.

Unit 1   Ross Feldberg, *Tufts University*
          David Wilson, *Parkland College*
Unit 2   Paula Lemons, *Duke University*
Unit 3   Greg Podgorski, *Utah State University*
Unit 4   Greg Podgorski, *Utah State University*
Unit 5   George Gilchrist, *William & Mary College*
Unit 6   Brianna Timmerman, *University of South Carolina*
          Marc Perkins, *Orange Coast College*
          Michael Black, *California Polytechnic State University, San Luis Obispo*
Unit 7   Diane Marshall, *University of New Mexico*
Unit 8   Jim Ryan, *Hobart and William Smith Colleges*
          Michael Black, *California Polytechnic State University, San Luis Obispo*
Unit 9   Alan Molumby, *University of Illinois, Chicago*

Reviewers

Peer review is the backbone of scientific publication. Acting as a reviewer is a fundamental component of our service as professional biologists, and reviewing chapters from an introductory text is one of the most important things we can do to improve the training of the next generation of biologists. I am deeply grateful for the insights offered by the following reviewers, who drew on their extensive content expertise, teaching experience, and research practice.

Marc Albrecht, *University of Nebraska, Kearney*
David Asch, *Youngstown State University*
Mariette Baxendale, *University of Missouri, St. Louis*
Greg Beaulieu, *University of Victoria*
Christopher Beck, *Emory University*
Peter Berget, *Carnegie Mellon University*
Janet Bester-Meredith, *Seattle Pacific University*
Cynthia Bishop, *Seattle Pacific University*
Michael Black, *California Polytechnic State University, San Luis Obispo*

Anthony Bledsoe, *University of Pittsburgh*
Patrice Boily, *University of New Orleans*
Scott Bowling, *Auburn University*
Maureen Brandon, *Idaho State University*
John Briggs, *Arizona State University*
Art Buikema, *Virginia Tech University*
Kim Caldwell, *University of Alabama*
Jeff Carmichael, *University of North Dakota*
Patrick Carter, *Washington State University*
John Caruso, *University of New Orleans*
Mary Lynn Casem, *California State University, Fullerton*
Cynthia Church, *Metropolitan State College*
Alison Cleveland, *University of South Florida*
Anita Davelos Baines, *University of Texas, Pan-American*
Jeff Demuth, *Indiana University*
Todd Duncan, *University of Colorado, Denver*
Johnny El Rady, *University of South Florida*
Peter Facchini, *University of Calgary*
Zen Faulkes, *University of Texas, Pan-American*
Ross Feldberg, *Tufts University*
Lewis Feldman, *University of California, Berkeley*
Jonathan Fisher, *St. Louis University*
Steve Frankel, *Northeastern University*
Amy Frary, *Mount Holyoke College*
Jed Fuhrman, *University of Southern California*
Caitlin Gabor, *Texas State University, San Marcos*
Michael Gaines, *University of Miami*
John R. Geiser, *Western Washington University*
D. Timothy Gerber, *University of Wisconsin, La Crosse*
Lisa Gerheart, *University of California, Davis*
Kathy Gillen, *Kenyon College*
Florence K. Gleason, *University of Minnesota, Twin Cities*
John Godwin, *North Carolina State University*
Reuben Goforth, *Michigan State University*
Linda Green, *University of Virginia*
Joe Harsh, *University of North Carolina, Charlotte*
Clare Hays, *Metropolitan State University*
Kerry Heafner, *University of Louisiana, Monroe*
Harold Heatwole, *North Carolina State University*
Brian Helmuth, *University of Southern California*
Susan Hengeveld, *Indiana University*
Mark Hens, *University of North Carolina, Greensboro*
Albert Herrera, *University of Southern California*
Malcolm Hill, *University of Richmond*
Ron Hoham, *Colgate University*
Kelly Howe, *University of New Mexico*
Cindy Johnson-Groh, *Gustavus Adolphus College*

Walter Judd, *University of Florida*
Nancy Kaufmann, *University of Pittsburgh*
Loren Knapp, *University of South Carolina*
Scott Knight, *Montclair State University*
Paul Lagos, *University of Mississippi*
Paula Lemons, *Duke University*
Vicky Lentz, *SUNY, College at Oneonta*
Georgia Lind, *Kingsborough Community College*
Chris Little, *University of Texas, Pan-American*
Andrea Lloyd, *Middlebury College*
Christopher Loretz, *University of Buffalo*
Cindy Martinez Wedig, *University of Texas, Pan-American*
Andrew McCubbin, *Washington State University*
Kelly McLaughlin, *Tufts University*
Victoria McMillan, *Colgate University*
Jennifer Miskowski, *University of Wisconsin, La Crosse*
Alan Molumby, *University of Illinois, Chicago*
Daniel Moon, *University of North Florida*
Mike Muller, *University of Illinois, Chicago*
Dana Nayduch, *Georgia Southern University*
Jacalyn S. Newman, *University of Pittsburgh*
Harry Nickla, *Creighton University*
Mary Jane Niles, *University of San Francisco*
Shawn Nordell, *St. Louis University*
Celia Norman, *Arapahoe Community College*
Nicole Obert, *University of Illinois, Urbana-Champaign*
John Osterman, *University of Nebraska, Lincoln*
John Palisano, *University of the South, Sewanee*
Glenn Parsons, *University of Mississippi, Oxford*
Andrew Pease, *Villa Julie College*
Deborah Pelli, *University of North Carolina, Greensboro*
Shelley A. Phelan, *Fairfield University*
Debra Pires, *University of California, Los Angeles*
Peggy Pollak, *Northern Arizona University*
Harvey Pough, *Rochester Institute of Technology*
Colin Purrington, *Swarthmore College*
Margaret Qazi, *Gustavus Adolphus College*
Rajinder Ranu, *Colorado State University*
Pamela C. Rasmussen, *Michigan State University*
Ann E. Rushing, *Baylor University*
James Ryan, *Hobart and William Smith Colleges*
Adam Ryburn, *SUNY, College at Oneonta*
Margaret Saha, *College of William and Mary*
Mark Sandheinrich, *University of Wisconsin, La Crosse*
Glenn Sauer, *Fairfield University*
Stephen G. Saupe, *St. John's University*
Andrew Scala, *Dutchess Community College*
Richard Showman, *University of South Carolina, Columbia*
Walter Shriner, *Mt. Hood Community College*
Sue Simon-Westendorf, *Ohio University*
Mark Spiro, *Bucknell University*
Paul Stapp, *California State University, Fullerton*

Scott Steinmaus, *California Polytechnic State University, San Luis Obispo*
John Stiller, *Eastern Carolina University*
John Stolz, *Duquesne University*
Kirk A. Stowe, *University of South Carolina*
Brianna Timmerman, *University of South Carolina*
Martin Tracey, *Florida International University*
Ashok Updhyaya, *University of South Florida*
Ann Vogel, *Illinois State University*
Fred Wasserman, *Boston University*
Elizabeth Weiss, *University of Texas, Austin*
Susan Whittemore, *Keene State College*
Ted Zerucha, *Appalachian State University*

Accuracy Reviewers

Once several rounds of peer review are completed and revised chapters and figures are produced, we rely on accuracy reviewers to check every page for errors. This work is demanding intellectually and has to be done under time pressure. The accuracy reviewers for this edition were exceptionally talented and timely.

Unit 1	Wayne Becker, *University of Wisconsin, Madison*
Unit 2	James Manser, *Harvey Mudd College (formerly)*
Unit 3	Peter Berget, *Carnegie Mellon University*
	Mary Rose Lamb, *University of Puget Sound*
Unit 4	James Manser, *Harvey Mudd College (formerly)*
Unit 5	Jeffrey Feder, *University of Notre Dame*
	Andrew Forbes, *University of Notre Dame*
	Andrew Michel, *University of Notre Dame*
	Tom Powell, *University of Notre Dame*
Unit 6	Laura Baumgartner, *University of Colorado, Boulder*
	Michael Black, *California Polytechnic State University, San Luis Obispo*
	Kimberly Erickson, *University of Colorado, Boulder*
	Steve Trudell, *University of Washington, Seattle*
Unit 7	Susan Waaland, *University of Washington, Seattle*
Unit 8	Warren Burggren, *University of North Texas*
	Susan Whittemore, *Keene State College*
Unit 9	Mark Johnston, *Dalhousie University*
BioSkills	Julie Aires, *Florida Community College at Jacksonville*; Ross Feldberg, *Tufts University*; George Gilchrist, *William and Mary College*; Doug Luckie, *Michigan State University*; Greg Podgorski, *Utah State University*

Correspondents

I am grateful to colleagues who take the initiative to contact me directly or through my publisher to make suggestions on how to improve the text and figures. Please never hesitate to do this—I take your comments to heart, in the spirit of a shared commitment to improved student learning. This list also includes friends and colleagues who were kind enough to respond to emails or calls from me, asking for ideas on how to clarify specific topics.

Julie Aires, *Florida Community College, Jacksonville*
Gerald Borgia, *University of Maryland*
Scott Bowling, *Auburn University*
Elizabeth Cowles, *Rice University*
Fred Delcomyn, *University of Illinois, Urbana-Champaign*
Leslie Dendy, *University of New Mexico, Los Alamos*
John Dudley, *University of Illinois, Urbana-Champaign*
Larry Forney, *University of Idaho*
Arthur Gibson, *University of California, Los Angeles*
Matt Gilg, *University of Northern Florida*
Jean Heitz, *University of Wisconsin, Madison*
Jack Hogg, *University of Montana*
Johnathan Kupferer, *University of Illinois, Chicago*
Hans Landel, *Edmonds Community College*
Frederick Lanni, *Carnegie Mellon University*
Andi Lloyd, *Middlebury College*
Carmen Mannella, *Wadsworth Center, SUNY Albany*
Andrew McCubbin, *Washington State University*
Tim Nelson, *Seattle Pacific University*
Shawn Nordell and students, *University of St. Louis*
Carol Pollock, *University of British Columbia*
Joelle Presson, *University of Maryland*
William Saunders, *LaGuardia Community College*
David Senseman, *University of Texas, San Antonio*
Bryan Spohn, *Florida Community College, Jacksonville*
Scott Steinmaus, *California Polytechnic State University, San Luis Obispo*
Judy Stone, *Colby College*
Dean Wendt, *California Polytechnic State University, San Luis Obispo*

Student Feedback

The second pillar of this edition—in addition to the role played by advisors and reviewers—was an extensive series of focus groups with students who were currently taking introductory biology or who had just completed the course.

Student Focus Group Coordinators

Planning and implementing the student focus groups would have been impossible without the support of key faculty members, who went out of their way to provide opportunities for their students to be heard.

Julie Aires, *Florida Community College, Jacksonville*
Frank Cantelmo, *St. Johns University, New York*
Matt Gilg, *University of Northern Florida*
Bill Hoese, *California State University, Fullerton*
John Nagey, *Scottsdale Community College*
Debra Pires, *University of California, Los Angeles*
Emily Taylor, *California Polytechnic State University, San Luis Obispo*
John Weser, *Scottsdale Community College*

Student Focus Group Participants

The students who attended focus groups were asked three questions about chapters they were assigned to read: (1) what were the most difficult concepts, (2) why were they hard, and (3) what helped you finally get it? We usually had them work in groups, and when they reported back to us, we never failed to be deeply impressed by the quality of their ideas and their ability to articulate them. It is not possible to overstate how important student feedback was to this edition. Combined with the superb advice I was getting from advisors, reviewers, and other colleagues, I had a wealth of ideas on how to make each chapter work better for both instructors and learners. These students were *inspiring*.

California Polytechnic State University, San Luis Obispo
Jenna Arruda, Katie Camfield, Benjamin Capper, Mandsa Chandra, Rebekah Clarke, Annalisa Conn, Marisa Crawford, Katie Duffield, T.J. Eames, Megan Fay, Margaret Hackney, Steffani Hall, Gemma Hill, John Kong, Taylor Lindblom, Adam Marre, Vik Mediratta, Serena Moen, Sunil Patel, Corinne Ross, Teresa Sais, Jessie Singer, Stephanie Szeto, Kelsey Tallon, Gregory Thurston, Greg Vidovic, Melody Wilkinson, Taiga Young

California State University, Fullerton
Redieat Assefa, Josemari Feliciano, Civon Gewelber, Sarah Harpst, Jeff Kuhnlein, Linda Ong, T. Richard Parenteau, Robert Tran, Nicole Bournival

Fairfield University
Sally Casper, Tamika Dickens, Pryce Gaynor, Cindi Munden

Florida Community College, Jacksonville
Algen Albritten III, Danielle Boss, Chantel Callier, Eugenia Cruz, Lauren Faulkner, Jonathan Hopkins, Chantae Knight, David Lambert, Amber McCurdy, Tara Pladsen, Lauren Spruiell, Courtney Torgeon, Theresa Tran

LaGuardia Community College
Kristine Azzoli, Felicita Gonzalez, Pedro Granados, Mike Levine, Kris Ragoonath, Maria Reyes

Scottsdale Community College
Tatum Arthur, Shadi Asayesh, Angela Bikl, Abrey Britt, Drew Bryck, Jason Butler, Cindy Clifton, Dean Doty, Tannaz Farahani, Bethany Garcia, Jeff Godfrey, Troy Graziadei, Dina Habhab, Loreley Hall, Crista Jackson, Paul Krueger, David Levine, Chad Massena, Jessica Massena, Brian Martinez, Sam Mohmand, Esther Morantz, Jill Patel, Staci Puckett, Rebecca Rees, David Rosenbaum, Samantha Schrepel, Chris Schroeder, Kelsey Thomsen, Chris Volpe, Jamie Wagner, Lianne Wharton

St. Johns University, New York

Diana Carroccia, Milea Emmons, Tunc Ersoy, Blayre Linker, Zain Mirza, Mohammed Sheikh, Richardson Talarera, Michael Weinberg, Win Aung Yeni

University of California, Los Angeles

Farhan Banani, Stephanie Davis, Krystal De La Rosa, Samantha Hammer, Jennifer Okuda Hein, Neha Jashi, Marissa Lee, Calvin Leung, Venkat Mocherla, David Nguyen, Isabella Niu, Aya Obara

University of Maryland

Megan Berg, Lauren Fitzgerald, Megan Janssen, Avita Jones, Deidre Robinson

University of Northern Florida

Elysia Brennan, Christopher Ferrara, Lindsay Googe, Samantha Grogan, Marie Haagensen, Crystal Harris, Madeline Parhalo, Sherline Pierre, Stacy Pohlman, Nichole Polito, Sarah Lynn Redding, Megan Richardson, Megan Smart, Frank Snyder

Supplements Contributors

Our goal for the supplements package to accompany the Third Edition was to create learning tools that incorporate principles of active learning. Research shows that students do better in class when they are asked to use the material they are learning about. From a new workbook that encourages students to practice biology, to interactive web animations that test their knowledge, our supplements ask students to work with information, not just memorize it. My sincere thanks to the following people for their important contribution to the book's core teaching values.

Media Supplements

Marc Albrecht, *University of Nebraska, Kearney*
John Bell, *Brigham Young University*
Michael Black, *California Polytechnic State University, San Luis Obispo*
Warren Burggren, *University of North Texas*
Fannie Chen
Carol Chihara, *University of San Francisco*
Clarissa Dirks, *University of Washington, Seattle*
Kimberly Erickson, *University of Colorado, Boulder*
Zen Faulkes, *University of Texas, Pan-American*
Kathy Gillen, *Kenyon College*
Mary Catherine Hager
Susan Hengeveld, *Indiana University, Bloomington*
Loren Knapp, *University of South Carolina*
Jonathan Lochamy, *Georgia Perimeter College*
James Manser, *Harvey Mudd College (formerly)*
Cynthia Martinez-Wedig, *University of Texas, Pan-American*
Victoria McMillan, *Colgate University*
Andrew Pease, *Villa Julie College*

Debra Pires, *University of California, Los Angeles*
Pamela Rasmussen, *Michigan State University*
Susan Rouse, *Brenau University*
Christina Russin, *Northwestern University*
William Russin, *Northwestern University*
Cheryl Ingram Smith, *Clemson University*
Ellen M. Smith
Mark Spiro, *Bucknell University*
Eric Stavney, *DeVry University*
Michael Wenzel, *California State University, Sacramento*

Print Supplements

Marc Albrecht, *University of Nebraska, Kearney*
Charles Austerberry, *Creighton University*
Brian Bagatto, *University of Akron*
Jay Brewster, *Pepperdine University*
Warren Burggren, *University of North Texas*
Cynthia Giffen, *University of Wisconsin, Madison*
Jean Heitz, *University of Wisconsin, Madison*
Laurel Hester, *Cornell University*
Cynthia Martinez-Wedig, *University of Texas, Pan-American*
Jenny McFarland, *Edmonds Community College*
Greg Podgorski, *Utah State University*
Carol Pollock, *University of British Columbia*
Susan Rouse, *Southern Wesleyan University*
Elena Shpak, *University of Tennessee*
Sally Sommers Smith, *Boston University*
Briana Timmerman, *University of South Carolina*
David Wilson, *Parkland College*

Book Team

Finally, this edition would not have been published without the encouragement and support of our publishing partners at Pearson Arts & Science. I would like to acknowledge those individuals in the Pearson Science Group who helped make the Third Edition possible.

Prentice Hall

This edition was launched by Prentice Hall and then transferred—along with all other Prentice Hall biology titles—to their sister company, Benjamin Cummings. The editorial team at Prentice Hall was responsible for establishing the vision that directed this edition. In addition, they recruited the first advisors and media and supplement contributors, and implemented the initial set of student focus groups, prior to turning the project over to their Benjamin Cummings colleagues. I am grateful for their talent, energy, and friendship, and for the extraordinary efforts they made to make the management transition as smooth as possible. These people are Andrew Gilfillan (Sponsoring Editor), Ann Heath (Executive Project Manager), Erin Mulligan (Development Editor), Lisa Tarabokjia (Editorial Assistant), and Carol Trueheart (VP, Executive Director of Development).

Benjamin Cummings

The Benjamin Cummings editorial and production team welcomed *Biological Science* into their publishing house and moved the project forward through the final critical stages of development and into production. The team was initially led by Sponsoring Editor Susan Winslow, who brought a fresh perspective to the project. Then, Market Development Manager Becky Ruden took over the Sponsoring Editor reins with focused energy and verve. She is a bright, young star. Special thanks go to Project Manager Sonia DiVittorio, who has proven herself one of the sharpest talents in textbook publishing. Sonia's tireless pursuit of quality is evident on every page. In tribute, I have given her a new title at the company: Goddess of Bookmaking.

Thanks also go to Senior Production Supervisor Shannon Tozier, who led the production team with enormous skill and perseverance, and to Design Manager Marilyn Perry, who helped create the striking interior and cover designs. Others on the book team deserving acknowledgement include: Mary Catherine Hager and Susan Weisberg (Development Editors), Anna Amato (Assistant Editor), Mercedes Grandin (Associate Editor), Ericka O'Benar (Media Producer), Yvonne Gerin, Elaine Soares, and Debbie Latronica (photo research team), Christy Lawrence, Lauren Harp, Lillian Carr, and Mansour Bethany (marketing team), Josh Frost (market development), and Deborah Gale (Executive Director of Development).

Additionally, I'd like to thank Chris Thillen (copyeditor); Frank Purcell, Ellen Sanders, Pete Shanks (proofreaders); production editor Katy Faria and her production colleagues at Pre-Press PMG; and the illustration team at Imagineering for all the hours and energy put into making this edition the best it can be.

Finally, I'd like to extend my appreciation to the people whose vantage point allows them to assess the bigger picture. I am grateful to Linda Davis, President of Pearson Arts and Sciences, for helping me understand Benjamin Cummings' overarching publishing goals and how *Biological Science* fits into them. Thank you, Beth Wilbur, Editor-in-Chief of Biology, for providing editorial leadership and helping the book take root in its new home and allowing room for its personality to flourish. A very special thank-you goes to Paul Corey, President of the Pearson Science Group. Paul has been an advocate for the core values of *Biological Science* from its beginning over a decade ago. I am grateful for his continuing friendship and professional guidance.

Supplements

The Freeman *Biological Science* supplements package offers a robust suite of print and electronic tools designed to help instructors make the most of their limited time and to help students study efficiently.

INSTRUCTOR RESOURCES

- The entire textbook illustration program is available in JPEG format. All tables, photos, and line drawings with (and without) labels are individually enhanced for optimal in-class projection and are pre-loaded into chapter-correlated PowerPoint presentations.

- A second set of PowerPoint presentations consists of lecture outlines for each chapter, augmented by key text illustrations and hyperlinks to Web Animations.

- A third set of PowerPoint presentations allows select key figures to be presented in a step-by-step manner. In-text figure caption exercises with illustrated answers are included in this step-edited set.

- In-class active lecture questions correlated by chapter can be used with any classroom response system and are available in PowerPoint format.

- The Instructor Guide includes lecture outlines, active learning lecture activities, answers to end-of-chapter questions, and innovative material to help motivate and engage students.

- The Printed Test Bank and Computerized Test Bank have been peer reviewed and student tested. Test questions are ranked according to Bloom's taxonomy and the improved TestGen® software makes assembling tests much easier. The Test Bank is also available in Course Management systems and in Word® format on the Instructor Resources DVD.

- Four-color Transparency Acetates with every illustration from the text are available. Labels and images have been enlarged and modified to ensure optimal projection in a large lecture hall.

STUDENT RESOURCES

- **NEW!** *Practicing Biology: A Student Workbook* offers a variety of activities such as modeling, mapping, and graphing to help students with different learning styles visualize and understand biological processes.

- **NEW!** The eBook addresses the changing needs of students and instructors in the majors biology course by offering an electronic version of the text that links directly to animations, quizzes, and videos.

- **NEW!** Complimentary access to Pearson Tutor Services provides highly interactive one-on-one biology tutoring by qualified instructors seven nights a week during peak study hours. Students will be able to "drop in" for live online help, submit questions to an e-structor anytime, or pre-schedule a tutoring session with an e-structor to receive help at their convenience.

- The Study Guide presents a breakdown of key biological concepts, and helps students focus on the fundamentals of each chapter. It is designed in two parts to help students study more effectively. Part I is intended as a "survival guide," and Part II explores the material in the textbook, chapter by chapter.

MULTIMEDIA RESOURCES

- **NEW!** MasteringBIOLOGY MasteringBiology™ offers in-depth online tutorials on biology's toughest topics. These tutorials provide hints and feedback specific to each student's misconceptions. MasteringBiology also includes diagnostic test questions and is useful for "just-in-time" teaching.

- **NEW!** *BioFlix* BioFlix cover the most difficult biology topics with 3-D, movie-quality animations, labeled slide shows, carefully constructed student tutorials, study sheets, and quizzes that support all types of learners. Topics include Tour of an Animal Cell, Tour of a Plant Cell, Membrane Transport, Cellular Respiration, Photosynthesis, Mitosis, Meiosis, DNA Replication, Protein Synthesis, Water Transport in Plants, How Neurons Work, Synapses, and Muscle Contraction.

- **Web Animation** Web Animations add depth and visual clarity to the most important topics and processes described in the text. Animations include pre-quizzes and post-quizzes to help students prepare for exams.

- **NEW!** Discovery Channel video clips on 29 different topics include antibiotic resistance, fighting cancer, and introduced species. Additional Cell Biology animations provide vivid images of the functions and processes of the cell.

- **NEW!** Access to BioForum, a new online community forum created just for biology educators. Learn from your peers about how they're teaching difficult topics or what resources they use to help them teach. Post activities, handouts, Web links, or other tools you would like to share with forum participants.

- **NEW!** Video demonstration shows biology educators using active learning techniques you can incorporate in your classroom immediately.

- Course Management content for *Biological Science*, 3rd Edition, is available for institutions using **WebCT** or **Blackboard** and is also available in our nationally hosted **Course Compass** course management system. If desired, WebCT and Blackboard cartridges containing only the Test Bank are available for download.

FOCUS ON THE GOLD THREAD ...

The gold thread helps you pick out important information.

KEY CONCEPTS

- Proteins are made of amino acids vary in structure and fu their side chains vary in com

- Proteins vary widely in str structure of a protein ca four levels that form ino acid seq

Start with *Key Concepts* on the first page of every chapter. Read these gold bullet points first to familiarize yourself with the chapter's big ideas.

strate and enzyn en the transition state ctually the transition state.

- Interactions with R-group transition state and thus lower for the reaction to proceed. At t line the active site may form sho ist with the transfer of atoms ant to another. More co R-groups allows

Watch for material related to Key Concepts inside chapters; it will also be flagged with a gold bullet. Slow down and pay close attention to these highlighted passages.

Other particularly important information is highlighted in gold. Gold highlighting is always a signal to slow down and pay special attention.

le tip of the iceberg ity. The next level of orga ondary structure, which is crea ng. Secondary structures are dis f proteins that are stabilized largel occurs between the carboxyl oxyge nd the hydrogen on the amino g atom in the carboxyl group to its high electronegativi group has a pa

together, an

Check Your Unders

If you understand that...

- Amino acids are small mol bonded to a carboxyl grou atom, and a side chain ca

- Each amino acid has di because each has

The gold half of *Check Your Understanding* boxes summarizes important information from the section you just read. Stop and ask yourself, Do I *really* understand every bullet point?

Summary of Key Concepts at the ends of chapters is a good place to start reviewing when it's time to study for an exam. Key Concepts are revisited here in detail.

SUMMARY OF KEY

- **Proteins are made of amino and function because their si**

Once monomers such as an event in chemical evolutio molecules. Researchers from the polymerizati amino acid

The best way to succeed on exams is to *practice*.
If you really understand a piece of information or a
concept, **you should be able to** *do something with it*.
If you can't, you haven't mastered the material.

Practice It activities are always flagged with blue
bullets and blue type. Answers to all *Practice It*
activities are available at **www.masteringbio.com**
under the Study Area button.

Text passages flagged with a
blue bullet, blue type, and the
words *"you should be able to"*
offer exercises on topics that
professors and students have
identified as most difficult.
These are the topics most students struggle with on exams.

process works.

● If you understand the
state the purpose of a Pun
not necessary to create a 2
the two *r* alleles that segreg
umns and the two *R* alleles
ther as rows.
   As an example of the
Punnett square i

aracteristics o

FIGURE 24.13 A Natural Expe
Finch Population in Response t
Drought). The histograms in the
beak depth in medium ground fin
after the drought of 1977. *N* is the
● **QUESTION** Why was the sam
● **EXERCISE** Fill in the predic

Many figures and some
tables include blue *Questions*
or *Exercises* to help you check
your understanding of the
material they present.

The blue half of *Check Your*
Understanding boxes asks
you to do something with the
information in the gold half.
If you can't complete these
exercises, go back and re-read
that section of the chapter.

The cell cycle consi

You should be able to...
Diagram the cell cycle and inc
1) the locations of cell-cycle c
2) the level of Cdk at each sta
3) the level of cyclin at each
4) the levels of MPF activit

es. Researc
m the polymerization
teins, amino acids are joined
boxyl group of one amino aci
amino acid.

You should be able to draw a
peptide bond. You should also
R-groups on amino acids are s
and why some R-groups are re

MB Web Animation at w
densation and Hydroly

Blue *"you should be able to"*
exercises also let you test your
understanding of each item in
the *Summary of Key Concepts*
at the ends of chapters.

Half of this book is text. The other half is figures and tables. Figures always focus on important concepts. Tables present the raw material of science; they are data rich.

Look for pointing hands inside figures. They flag important ideas—the ones your own instructor might point out in lecture.

When ATP binds here, the reaction rate slows dramatically

Fructose-1

Some figures and tables ask you to to analyze data and fill in information. Take the time to do these exercises; they will help you master important concepts.

Experiment Boxes will help you understand how experiments are designed and give you practice interpreting data. Some leave space open for you to fill in null hypotheses, predicted outcomes, or the conclusion.

Prediction of natural selection hyp

Prediction of null hypothesis (no s

Results:
80

Average

Summary Tables pull important information together in a format that's easy to review. A complete list of Summary Tables is provided inside the back cover.

SUMMARY TABLE **14.1 Proteins**

Name	Structure
Opening the helix	
Helicase	

www.masteringbio.com

Part I: *The Study Area* includes Web Animations and BioFlix.

Web Animations Web Animation help you review important textbook topics. Pre- and post-quizzes let you test your proficiency.

BioFlix *BioFlix* cover the most difficult biology topics with 3-D, movie-quality animations, labeled slide shows, carefully constructed tutorials, study sheets, and quizzes that support all types of learners. Topics include: Tour of an Animal Cell, Tour of a Plant Cell, Membrane Transport, Cellular Respiration, Photosynthesis, Mitosis, Meiosis, DNA Replication, Protein Synthesis, Water Transport in Plants, How Neurons Work, Synapses, and Muscle Contraction.

Part 2: *MasteringBiology Instructor Assignments* include in-depth tutorials on biology's toughest topics.

If you are working on a tutorial and get stuck, . . .

You can also ask for hints . . .

. . . you will receive instant feedback specific to your error.

. . . to receive background information and Socratic questions and exercises designed to guide you to mastery of the topic.

Biology and the Tree of Life

A young reef octopus, from Hawaii, also appears on the cover of this book. It is one of perhaps 10 million species alive today.

KEY CONCEPTS

- Biological science was founded with the development of (1) the cell theory, which proposes that all organisms are made of cells and that all cells come from preexisting cells, and (2) the theory of evolution by natural selection, which maintains that species change through time because individuals with certain heritable traits produce more offspring than other individuals do.

- A phylogenetic tree is a graphical representation of the evolutionary relationships among species. These relationships can be estimated by analyzing similarities and differences in traits. Species that share distinctive traits are closely related and are placed close to each other on the tree of life.

- Biologists ask questions, generate hypotheses to answer them, and design experiments that test the predictions made by competing hypotheses.

In essence, biology is a search for ideas and observations that unify our understanding of the diversity of life, from bacteria living in rocks a mile underground to octopuses and humans. Chapter 1 is an introduction to this search.

The goals of this chapter are to introduce the amazing variety of life-forms alive today, consider some fundamental traits shared by all organisms, and explore how biologists go about answering questions about life. The chapter also introduces themes that will resonate throughout this book: (1) analyzing how organisms work at the molecular level, (2) understanding why organisms have the traits they do in terms of their evolutionary history, and (3) helping you learn how to think like a biologist.

We begin by examining two of the greatest unifying ideas in all of science: the cell theory and the theory of evolution by natural selection. When these concepts emerged in the mid-1800s,

they revolutionized the way that biologists understand the world. The cell theory proposed that all organisms are made of cells and that all cells come from preexisting cells. The theory of evolution by natural selection maintained that species have changed through time and that all species are related to one another through common ancestry. The theory of evolution by natural selection established that bacteria, mushrooms, roses, robins, and humans are all part of a family tree, similar to the genealogies or family trees that connect individual people.

A **theory** is an explanation for a very general class of phenomena or observations. The cell theory and the theory of evolution provide a foundation for the development of modern biology because they focus on two of the most general questions possible: What are organisms made of? Where did they come from? Let's begin by tackling the first of these two questions.

1.1 The Cell Theory

The initial conceptual breakthrough in biology—the cell theory—emerged after some 200 years of work. In 1665 Robert Hooke used a crude microscope to examine the structure of cork (a bark tissue) from an oak tree. The instrument magnified objects to just 30 times (30×) their normal size, but it allowed Hooke to see something extraordinary. In the cork he observed small, pore-like compartments that were invisible to the naked eye (**Figure 1.1a**). These structures came to be called cells.

Soon after Hooke published his results, Anton van Leeuwenhoek succeeded in developing much more powerful microscopes, some capable of magnifications up to 300×. With these instruments, Leeuwenhoek inspected samples of pond water and made the first observations of single-celled organisms like the *Paramecium* in **Figure 1.1b**. He also observed and described the structure of human blood cells and sperm cells.

In the 1670s a researcher who was studying the leaves and stems of plants with a microscope concluded that these large, complex structures are composed of many individual cells. By the early 1800s enough data had accumulated for a biologist to claim that *all* organisms consist of cells. But between then and now, biologists have developed microscopes that are tens of thousands of times more powerful than Leeuwenhoek's and have described well over a million new species. Did the biologist's claim hold up?

Are *All* Organisms Made of Cells?

The smallest organisms known today are bacteria that are barely 200 nanometers wide, or 200 *billionths* of a meter. (See the endpapers of this book to review the metric system and its prefixes.) It would take 5000 of these organisms lined up end to end to span a millimeter. This is the distance between the smallest hash marks on a metric ruler. In contrast, sequoia trees can be over 100 meters tall. This is the equivalent of a 20-story building. Bacteria and sequoias are composed of the same fundamental building block, however—the cell. Bacteria consist of a single cell; sequoias are made up of many cells.

Biologists have become increasingly dazzled by the diversity and complexity of cells as advances in microscopy have made it possible to examine cells at higher magnifications. The basic conclusion made in the 1800s is intact, however: As far as is known, all organisms are made of cells. Today, a **cell** is defined as a highly organized compartment that is bounded by a thin, flexible structure called a plasma membrane and that contains concentrated chemicals in an aqueous (watery) solution. The chemical reactions that sustain life take place inside cells. Most cells are also capable of reproducing by dividing—in effect, by making a copy of themselves.

The realization that all organisms are made of cells was fundamentally important, but it formed only the first part of the cell theory. In addition to understanding what organisms are made of, scientists wanted to understand how cells come to be.

Where Do Cells Come From?

Most scientific theories have two components: The first describes a pattern in the natural world, while the second identifies a mechanism or process that is responsible for creating that pattern. Hooke and his fellow scientists had articulated the pattern component of the cell theory. In 1858 Rudolph Virchow added the process component by stating that all cells arise from preexisting cells. ● The complete cell theory, then, can be stated as follows: All organisms are made of cells, and all cells come from preexisting cells.

This claim was a direct challenge to the prevailing explanation, called spontaneous generation. At the time, most biologists believed that organisms arise spontaneously under certain conditions. For example, the bacteria and fungi that spoil foods such as milk and wine were thought to appear in these nutrient-rich media of their own accord—they spring to life from nonliving materials. Spontaneous generation was a **hypothesis**: a proposed explanation. The all-cells-from-cells

(a) The first view of cells: Robert Hooke's drawing from 1665

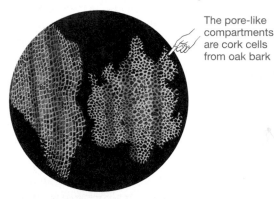

The pore-like compartments are cork cells from oak bark

(b) Anton van Leeuwenhoek was the first to view single-celled "animalcules" in pond water.

A *Paramecium*

FIGURE 1.1 The Discovery of Cells.

hypothesis, in contrast, maintained that cells do not spring to life spontaneously but are produced only when preexisting cells grow and divide. Biologists usually use *theory* to refer to proposed explanations for broad patterns in nature and *hypothesis* to refer to explanations for more tightly focused questions.

Soon after the all-cells-from-cells hypothesis appeared in print, Louis Pasteur set out to test its predictions experimentally. A **prediction** is something that can be measured and that must be correct if a hypothesis is valid. Pasteur wanted to determine whether microorganisms could arise spontaneously in a nutrient broth or whether they appear only when a broth is exposed to a source of preexisting cells. To address the question, he created two treatment groups: a broth that was not exposed to a source of preexisting cells and a broth that was.

The spontaneous generation hypothesis predicted that cells would appear in both treatments. The all-cells-from-cells hypothesis predicted that cells would appear only in the treatment exposed to a source of preexisting cells.

Figure 1.2 shows Pasteur's experimental setup. Note that the two treatments are identical in every respect but one. Both used glass flasks filled with the same amount of the same nutrient broth. Both were boiled for the same amount of time to kill any existing organisms such as bacteria or fungi. But because the flask pictured in **Figure 1.2a** had a straight neck, it was exposed to preexisting cells after sterilization by the heat treatment. These preexisting cells are the bacteria and fungi that cling to dust particles in the air. They could drop into the nutrient broth because the neck of the flask was straight. In contrast, the flask

FIGURE 1.2 The Spontaneous Generation Hypothesis Was Tested Experimentally.

drawn in **Figure 1.2b** had a long swan neck. Pasteur knew that water would condense in the crook of the swan neck after the boiling treatment and that this pool of water would trap any bacteria or fungi that entered on dust particles. Thus, the swan-necked flask was isolated from any source of preexisting cells even though it was still open to the air.

Pasteur's experimental setup was effective because there was only one difference between the two treatments and because that difference was the factor being tested—in this case, a broth's exposure to preexisting cells. ● If you understand this concept, you should be able to identify problems that would arise if he had put different types of broth in the two treatments, heated them for different lengths of time, or used a ceramic flask for one treatment and a glass flask for the other.

And Pasteur's results? As Figure 1.2 shows, the treatment exposed to preexisting cells quickly filled with bacteria and fungi. This observation was important because it showed that the heat sterilization step had not altered the nutrient broth's capacity to support growth, and because it supported the hypothesis that growth started with preexisting cells. But the treatment in the swan-necked flask remained sterile. Even when the broth was left standing for months, no organisms appeared in it.

Because Pasteur's data were in direct opposition to the predictions made by the spontaneous generation hypothesis, the results persuaded most biologists that the all-cells-from-cells hypothesis was correct.

The success of the cell theory's process component had an important implication: If all cells come from preexisting cells, it follows that all individuals in a population of single-celled organisms are related by common ancestry. Similarly, in a multicellular individual such as you, all of the cells present are descended from preexisting cells, tracing back to a fertilized egg. A fertilized egg is a cell created by the fusion of sperm and egg—cells that formed in individuals of the previous generation. In this way, all of the cells in a multicellular organism are connected by common ancestry.

The second great founding idea in biology is similar, in spirit, to the cell theory. It also happened to be published the same year as the all-cells-from-cells hypothesis. This was the realization, made independently by Charles Darwin and Alfred Russel Wallace, that all *species*—all distinct, identifiable types of organisms—are connected by common ancestry.

1.2   The Theory of Evolution by Natural Selection

In 1858 short papers written separately by Darwin and Wallace were read to a small group of scientists attending a meeting of the Linnean Society of London. A year later, Darwin published a book that expanded on the idea summarized in those brief papers. The book was called *The Origin of Species*. The first edition sold out in a day.

What Is Evolution?

Like the cell theory, the theory of evolution by natural selection has a pattern and a process component. Darwin and Wallace's theory made two important claims concerning patterns that exist in the natural world. The first claim was that species are related by common ancestry. This contrasted with the prevailing view in science at the time, which was that species represent independent entities created separately by a divine being. The second claim was equally novel. Instead of accepting the popular hypothesis that species remain unchanged through time, Darwin and Wallace proposed that the characteristics of species can be modified from generation to generation. Darwin called this process "descent with modification."

Evolution, then, means that species are not independent and unchanging entities, but are related to one another and can change through time. This part of the theory of evolution—the pattern component—was actually not original to Darwin and Wallace. Several scientists had already come to the same conclusions about the relationships among species. The great insight by Darwin and Wallace was in proposing a process, called **natural selection**, that explains *how* evolution occurs.

What Is Natural Selection?

Natural selection occurs whenever two conditions are met. The first is that individuals within a population vary in characteristics that are **heritable**—meaning, traits that can be passed on to offspring. A **population** is defined as a group of individuals of the same species living in the same area at the same time. Darwin and Wallace had studied natural populations long enough to realize that variation among individuals is almost universal. In wheat, for example, some individuals are taller than others. As a result of work by wheat breeders, Darwin and Wallace knew that short parents tend to have short offspring. Subsequent research has shown that heritable variation exists in most traits and populations. The second condition of natural selection is that in a particular environment, certain versions of these heritable traits help individuals survive better or reproduce more than do other versions. For example, if tall wheat plants are easily blown down by wind, then in windy environments shorter plants will tend to survive better and leave more offspring than tall plants will.

If certain heritable traits lead to increased success in producing offspring, then those traits become more common in the population over time. In this way, the population's characteristics change as a result of natural selection acting on individuals. This is a key insight: Natural selection acts on individuals, but evolutionary change affects only populations. In this example, populations of wheat that grow in windy environments tend to become shorter from generation to generation. But in any given generation, none of the individual wheat plants get taller or shorter as a result of natural selection. This sort of change in the characteristics of a population, over time, is evolution. Evolution occurs when heritable variation leads to differential success in reproduction. ● If you understand this concept, you should be able to graph how average stem height will change over time in a population that occupies a windy environment versus a non-windy environment—one where tall plants have an advantage due to better access to light. (For help with reading and making graphs, see **BioSkills 1** in the back of this book.)

Darwin also introduced some new terminology to identify what is happening during natural selection. For example, in everyday English the word *fitness* means health and well-being. But in biology, **fitness** means the ability of an individual to produce offspring. Individuals with high fitness produce many offspring. Similarly, the word *adaptation* in everyday English means that an individual is adjusting and changing to function in new circumstances. But in biology, an **adaptation** is a trait that increases the fitness of an individual in a particular environment. Once again, consider wheat: In windswept habitats, wheat plants with short stalks have higher fitness than do individuals with long stalks. Short stalks are an adaptation to windy environments.

To clarify further how natural selection works, consider the origin of the vegetables called the cabbage family plants.

Broccoli, cauliflower, Brussels sprouts, cabbage, kale, savoy, and collard greens descended from the same species—the wild plant in the mustard family pictured in **Figure 1.3a**. To create the plant called broccoli, horticulturists selected individuals of the wild mustard species with particularly large and compact flowering stalks. In mustards, the size and shape of the flowering stalk is a heritable trait. When the selected individuals were mated with one another, their offspring turned out to have larger and more compact flowering stalks, on average, than the original population (**Figure 1.3b**). By repeating this process over many generations, horticulturists produced a population with extraordinarily large and compact flowering stalks. The derived population has been artificially selected for the size and shape of the flowering stalk; as **Figure 1.3c** shows, it barely resembles the ancestral form. Note that during this process, the size and shape of the flowering stalk in each individual plant did not change within its lifetime—the change occurred in the characteristics of the population over time. Darwin's great insight was that natural selection changes the characteristics of a wild population over time, just as the deliberate manipulation of "artificial selection" changes the characteristics of a domesticated population over time.

(a) Wild *Brassica oleracea*: the parental generation

(c) Broccoli: one descendant of wild *Brassica oleracea*

Large, compact flowering stalks

(b) ARTIFICIAL SELECTION

Select these individuals for next generation

Average

Number of individuals

Average

Average

Width of flowering stalk (cm)

These charts, called histograms, document how the width of the flowering stalk changed through time in response to selection

1. Parental generation: Select individuals with the largest and most compact flowering stalks, and breed them.

2. Generation 2: Of the offspring, select the individuals with the largest and most compact flowering stalks, and breed them.

3. Generation 3: Of the offspring, select the individuals with the largest and most compact flowering stalks, and breed them.

4. Repeat dozens of times.

5. Descendants: After many generations, average flowering stalks are much larger and more compact.

FIGURE 1.3 Artificial Selection Can Produce Dramatic Changes in Organisms.

Since Darwin and Wallace published their work, biologists have succeeded in measuring hundreds of examples of natural selection in wild populations. They have accumulated a massive body of evidence documenting that species have changed through time.

Together, the cell theory and the theory of evolution provided the young science of biology with two central, unifying ideas:

1. The cell is the fundamental structural unit in all organisms.

2. All species are related by common ancestry and have changed over time in response to natural selection.

Check Your Understanding

If you understand that...

- Natural selection occurs when heritable variation in certain traits leads to improved success in reproduction. Because individuals with these traits produce many offspring with the same traits, the traits increase in frequency and evolution occurs.

- Evolution is simply a change in the characteristics of a population over time.

You should be able to...

Explain why each of the following common misconceptions about evolution by natural selection is incorrect, using the example of selection on the height of wheat stalks.

1) evolution is progressive, meaning that species always get larger, more complex, or "better" in some sense;

2) individuals as well as populations change when natural selection occurs; or

3) individuals with high levels of fitness are stronger or bigger or "more dominant."

(MB) Web Animation at www.masteringbio.com
Artificial Selection

1.3 The Tree of Life

Section 1.2 focused on how individual populations change through time in response to natural selection. But over the past several decades, biologists have also documented dozens of cases in which natural selection has caused populations of one species to diverge and form new species. This divergence process is called **speciation**. In several instances, biologists are documenting the formation of new species right before our eyes (**Figure 1.4**).

Research on speciation supports a claim that Darwin and Wallace made over a century ago—that natural selection can

(a) *Tragopogon mirus* (left) evolved from *Tragopogon dubius* (right).

(b) Different soapberry bug species feed on native (left) or introduced (right) plants.

(c) Different maggot fly species feed on hawthorn (left) versus apple (right) fruits.

FIGURE 1.4 Speciation in Action. The pairs of organisms shown here are in the process of becoming independent species (see Chapter 26).

lead to change *between* species as well as within species. The broader conclusions are that all species come from preexisting species and that all species, past and present, trace their ancestry back to a single common ancestor. If the theory of evolution by natural selection is valid, biologists should be able to reconstruct a **tree of life**—a family tree of organisms. If life on Earth arose just once, then such a diagram would describe the genealogical relationships among species with a single, ancestral species at its base.

Has this task been accomplished? If the tree of life exists, what does it look like? To answer these questions, we need to step back in time and review how biologists organized the diversity of organisms *before* the development of the cell theory and the theory of evolution.

Linnaean Taxonomy

In science, the effort to name and classify organisms is called **taxonomy**. This branch of biology began to flourish in 1735 when a botanist named Carolus Linnaeus set out to bring order to the bewildering diversity of organisms that were then being discovered.

The building block of Linnaeus' system is a two-part name unique to each type of organism. The first part indicates the organism's **genus** (plural: **genera**). A genus is made up of a closely related group of species. For example, Linnaeus put humans in the genus *Homo*. Although humans are the only living species in this genus, several extinct organisms, all of which walked upright and made extensive use of tools, were later also assigned to *Homo*. The second term in the two-part name identifies the organism's species. Section 1.1 defined a species as a distinct, identifiable type of organism. More formally, a **species** is made up of individuals that regularly breed together or have characteristics that are distinct from those of other species. Linnaeus gave humans the species name *sapiens*.

An organism's genus and species designation is called its scientific name or Latin name. Scientific names are always italicized. Genus names are always capitalized, but species names are not—for instance, *Homo sapiens*. Scientific names are based on Latin or Greek word roots or on words "Latinized" from other languages (see **Box 1.1**). Linnaeus gave a scientific name to every species then known. (He also Latinized his own name—from Karl von Linné to Carolus Linnaeus.)

Linnaeus maintained that different types of organisms should not be given the same genus and species names. Other species may be assigned to the genus *Homo*, and members of other genera may be named *sapiens*, but only humans are named *Homo sapiens*. Each scientific name is unique.

Linnaeus' system has stood the test of time. His two-part naming system, or **binomial nomenclature**, is still the standard in biological science.

Taxonomic Levels   To organize and classify the tremendous diversity of species being discovered in the 1700s, Linnaeus created a hierarchy of taxonomic groups: From the most specific grouping to the least specific, the levels are **species**, **genus**, **family**, **order**, **class**, **phylum** (plural: **phyla**), and **kingdom**. **Figure 1.5** shows how this nested, or hierarchical, classification scheme works, using humans as an example. Although our species is the sole living member of the genus *Homo*, humans are now grouped with the orangutan, gorilla, common chimpanzee, and bonobo in a family called Hominidae. Linnaeus grouped members of this family with gibbons, monkeys, and lemurs in an order called Primates. The Primates are grouped in the class Mammalia with rodents, bison, and other organisms that have fur and produce milk. Mammals, in turn, join other animals with structures called notochords in the phylum

FIGURE 1.5 Linnaeus Defined Taxonomic Levels. In the Linnaean system, each animal species is placed in a taxonomic hierarchy with seven levels. Lower levels are nested within higher levels.

Chordata, and they join all other animals in the kingdom Animalia. Each of these named groups—primates, mammals, or *Homo sapiens*—can be referred to as a **taxon** (plural: **taxa**). The essence of Linnaeus's system is that lower-level taxa are nested within higher-level taxa.

Aspects of this hierarchical scheme are still in use. As biological science matured, however, several problems with Linnaeus's original proposal emerged.

How Many Kingdoms Are There?   Linnaeus proposed that species could be organized into two kingdoms—plants and animals. According to Linnaeus, organisms that do not move and that produce their own food are plants; organisms that move and acquire food by eating other organisms are animals.

Not all organisms fall neatly into these categories, however. Molds, mushrooms, and other fungi survive by absorbing nutrients from dead or living plants and animals. Even though they do not make their own food, they were placed in the kingdom Plantae because they do not move. The tiny, single-celled organisms called bacteria also presented problems. Some bacteria can move, and many can make their own food. But initially they, too, were thought to be plants.

In addition, a different fundamental division emerged when advances in microscopy allowed biologists to study the contents of individual cells in detail. In plants, animals, and an array of other organisms, cells contain a prominent component called a nucleus (**Figure 1.6a**). But in bacteria, cells lack this kernel-like structure (**Figure 1.6b**). Organisms with a nucleus are called **eukaryotes** ("true-kernel"); organisms without a nucleus are called **prokaryotes** ("before-kernel"). The vast majority of prokaryotes are unicellular ("one-celled"); many eukaryotes are multicellular ("many-celled"). These findings suggested that the most fundamental division in life was between prokaryotes and eukaryotes.

In response to new data about the diversity of life, biologists proposed alternative classification schemes. In the late 1960s one researcher suggested that a system of five kingdoms best reflects the patterns observed in nature. This five-kingdom system is depicted in **Figure 1.7**. Although the scheme was widely used, it represents just one proposal out of many. Other biologists suggested that organisms are organized into three, four, six, or eight kingdoms.

Using Molecules to Understand the Tree of Life

About the time that the five-kingdom proposal was published, Carl Woese (pronounced *woes*) and colleagues began working on the problem from a radically different angle. Instead of assigning organisms to kingdoms based on characteristics such as the presence of a nucleus or the ability to move or to manufacture food, they attempted to understand the relationships among organisms by analyzing their chemical components. Their goal

(a) Eukaryotic cells have a membrane-bound nucleus.

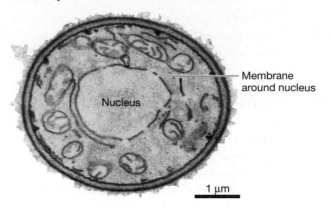

(b) Prokaryotic cells do *not* have a membrane-bound nucleus.

FIGURE 1.6 Eukaryotes and Prokaryotes.

● **EXERCISE** Study the scale bars; then draw two ovals that accurately represent the relative sizes of a eukaryotic cell and a prokaryotic cell.

was to understand the **phylogeny** (meaning "tribe-source") of all organisms—their actual genealogical relationships.

To understand which organisms are closely versus distantly related, Woese and co-workers needed to study a molecule that is found in all organisms. The molecule they selected is called small subunit rRNA. It is an essential part of the machinery that all cells use to grow and reproduce.

Although rRNA is a large and complex molecule, its underlying structure is simple. The rRNA molecule is made up of sequences of four smaller chemical components called ribonucleotides. These ribonucleotides are symbolized by the letters

---

BOX 1.1 Scientific Names and Terms

Scientific names and terms are often based on Latin or Greek word roots that are descriptive. For example, *Homo sapiens* is derived from the Latin *homo* for "man" and *sapiens* for "wise" or "knowing." The yeast that bakers use to produce bread and that brewers use to brew beer is called *Saccharomyces cerevisiae*. The Greek root *saccharo* means "sugar," and *myces* refers to a fungus. *Saccharomyces* is aptly named "sugar fungus" because yeast is a fungus and because the domesticated strains of yeast used in commercial baking and brewing are often fed sugar. The specific name of this organism, *cerevisiae*, is Latin for "beer." Loosely translated, then, the scientific name of brewer's yeast means "sugar fungus for beer."

Most biologists find it extremely helpful to memorize some of the common Latin and Greek roots. To aid you in this process, new terms in this text are often accompanied by a reference to their Latin or Greek word roots in parentheses.

KINGDOM MONERA
(includes all
prokaryotes)

5 µm

KINGDOM PROTISTA
(includes several
groups of unicellular
eukaryotes)

20 µm

KINGDOM PLANTAE

3 cm

KINGDOM FUNGI

1 cm

KINGDOM ANIMALIA

1 mm

FIGURE 1.7 The Five-Kingdom Scheme. For decades, most biologists accepted the hypothesis that organisms naturally fall into the five kingdoms depicted here.

● **QUESTION** How many times bigger is the fruit fly in the bottom panel than one of the prokaryotic cells in the top panel?

A, U, C, and G. In rRNA, ribonucleotides are connected to one another linearly, like boxcars of a freight train (**Figure 1.8**).

Why might rRNA be useful for understanding the relationships among organisms? The answer is that the ribonucleotide sequence in rRNA is a trait, similar to the height of wheat stalks or the size of flowering stalks of broccoli, that can change during the course of evolution. Although rRNA performs the same function in all organisms, the sequence of ribonucleotide building blocks in this molecule is not identical among species. In land plants, for example, the molecule might start with the sequence A-U-A-U-C-G-A-G. In green algae, which are closely related to land plants, the same section of the molecule might contain A-U-A-U-G-G-A-G. But in brown algae, which are not closely related to green algae or to land plants, the same part of the molecule might consist of A-*A*-A-U-*G*-G-A-G.

The research program that Woese and co-workers pursued was based on a simple premise: If the theory of evolution is correct, then rRNA sequences should be very similar in closely related organisms but less similar in organisms that are less closely related. Groups that are closely related, like the plants, should share certain changes in rRNA that no other species have.

To test this premise, the researchers determined the sequence of ribonucleotides in the rRNA of a wide array of species. Then they considered what the similarities and differences in the sequences implied about relationships among the species. The goal was to produce a diagram that described the phylogeny of the organisms in the study. ● A diagram that depicts evolutionary history in this way is called a **phylogenetic tree**. Just as a family tree shows relationships among individuals, a phylogenetic tree shows relationships among species. On a phylogenetic tree, branches that are close to one another represent species that are closely related; branches that are farther apart represent species that are more distantly related.

Compare the rRNA nucleotide
sequence observed in land plants...

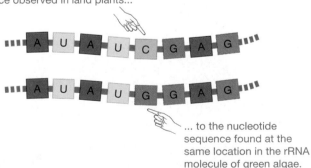

... to the nucleotide
sequence found at the
same location in the rRNA
molecule of green algae.

FIGURE 1.8 RNA Molecules Are Made Up of Smaller Molecules.
The complete small subunit rRNA molecule contains about 2000 ribonucleotides; just 8 are shown in this comparison.

● **QUESTION** Suppose that in the same portion of rRNA, molds and other fungi have the sequence A-U-A-U-G-G-A-C. According to these data, are fungi more closely related to green algae or to land plants? Explain your logic.

The Tree of Life Estimated from an Array of Genes To construct a phylogenetic tree, researchers use a computer to find the arrangement of branches that is most consistent with the similarities and differences observed in the data. Although the initial work was based only on the sequences of ribonucleotides observed in rRNA, biologists now use data sets that include sequences from a wide array of genes. A recent tree produced by comparing these sequences is shown in **Figure 1.9**. Because this tree includes species from many different kingdoms and phyla, it is often called the universal tree, or the tree of life. For help in learning how to read a phylogenetic tree, see **BioSkills 2**.

The tree of life implied by rRNA and other genetic data astonished biologists. For example:

● The fundamental division in organisms is not between plants and animals or even between prokaryotes and eukaryotes. Rather, *three* major groups occur: (1) the Bacteria; (2) another group of prokaryotic, single-celled organisms called the Archaea; and (3) the eukaryotes. To accommodate this new perspective on the diversity of organisms, Woese created a new taxonomic level called the **domain**. As Figure 1.9 indicates, the three domains of life are now called the Bacteria, Archaea, and Eukarya.

● Some of the kingdoms that had been defined earlier do not reflect how evolution actually occurred. For example, recall that Linnaeus grouped the multicellular eukaryotes known as fungi with plants. But the genetic data indicate that fungi are much more closely related to animals than they are to plants.

● Bacteria and Archaea are much more diverse than anyone had imagined. If the differences among animals, fungi, and plants warrant placing them in separate kingdoms, then dozens of kingdoms exist among the prokaryotes.

The Tree of Life Is a Work in Progress Just as researching your family tree can help you understand who you are and where you came from, so the tree of life helps biologists understand the relationships among organisms and the history of species. For example, the discovery of the Archaea and the placement of lineages such as the fungi qualify as exciting breakthroughs in our understanding of life's diversity. Work on the tree of life continues at a furious pace, however, and the location of certain branches on the tree is hotly debated. As databases expand and as techniques for analyzing data improve, the shape of the tree of life presented in Figure 1.9 will undoubtedly change.

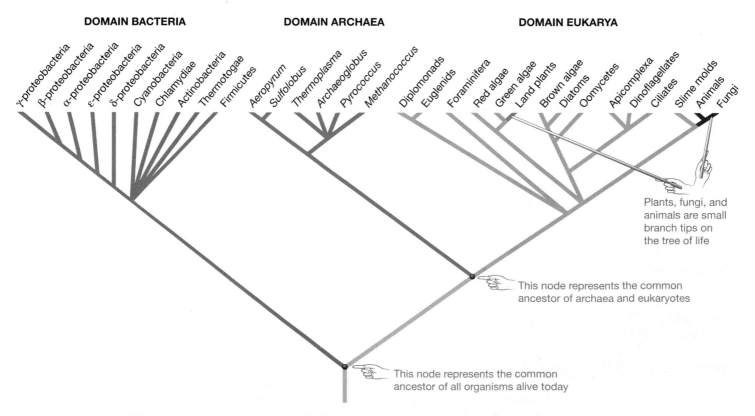

FIGURE 1.9 The Tree of Life. "Universal tree" estimated from a large amount of gene sequence data. The three domains of life revealed by the analysis are labeled. Common names are given for most lineages in the domains Bacteria and Eukarya. Genus names are given for members of the domain Archaea, because most of these organisms have no common names.

Check Your Understanding

If you understand that...

- A phylogenetic tree shows the evolutionary relationships among species.
- To infer where species belong on a phylogenetic tree, biologists examine the characteristics of the species involved. Closely related species should have similar characteristics, while less closely related species should be less similar.

You should be able to...

Examine the following sequences and draw a phylogenetic tree showing the relationships among Species A, B, and C that these data imply:

Species A:	A	A	C	T	A	G	C	G	C	G	A	T
Species B:	A	A	C	T	A	G	C	G	C	C	A	T
Species C:	T	T	C	T	A	G	C	G	G	T	A	T

1.4 Doing Biology

This chapter has introduced some of the great ideas in biology. The development of the cell theory and the theory of evolution by natural selection provided cornerstones when the science was young; the tree of life is a relatively recent insight that has revolutionized the way researchers understand the diversity of life on Earth.

These theories are considered great because they explain fundamental aspects of nature, and because they have consistently been shown to be correct. They are considered correct because they have withstood extensive testing. How do biologists test ideas about the way the natural world works? The answer is that they test the predictions made by alternative hypotheses, often by setting up carefully designed experiments. To illustrate how this approach works, let's consider two questions currently being addressed by researchers.

Why Do Giraffes Have Long Necks? An Introduction to Hypothesis Testing

If you were asked why giraffes have long necks, you might say that long necks enable giraffes to reach food that is unavailable to other mammals. This hypothesis is expressed in African folktales and has traditionally been accepted by many biologists. The food competition hypothesis is so plausible, in fact, that for decades no one thought to test it. Recently, however, Robert Simmons and Lue Scheepers assembled data suggesting that the food competition hypothesis is only part of the story. Their analysis supports an alternative hypothesis—that long necks allow giraffes to use their heads as effective weapons for battering their opponents.

How did biologists test the food competition hypothesis? What data support their alternative explanation? Before attempting to answer these questions, it's important to recognize that hypothesis testing is a two-step process. The first step is to state the hypothesis as precisely as possible and list the predictions it makes. The second step is to design an observational or experimental study that is capable of testing those predictions. If the predictions are accurate, then the hypothesis is supported. If the predictions are not met, then researchers do further tests, modify the original hypothesis, or search for alternative explanations.

The Food Competition Hypothesis: Predictions and Tests Stated precisely, the food competition hypothesis claims that giraffes compete for food with other species of mammals. When food is scarce, as it is during the dry season, giraffes with longer necks can reach food that is unavailable to other species and to giraffes with shorter necks. As a result, the longest-necked individuals in a giraffe population survive better and produce more young than do shorter-necked individuals, and average neck length of the population increases with each generation. To use the terms introduced earlier, long necks are adaptations that increase the fitness of individual giraffes during competition for food. This type of natural selection has gone on so long that the population has become extremely long necked.

The food competition hypothesis makes several explicit predictions. For example, the food competition hypothesis predicts that (1) neck length is variable among giraffes; (2) neck length in giraffes is heritable; and (3) giraffes feed high in trees, especially during the dry season, when food is scarce and the threat of starvation is high.

The first prediction is correct. Studies in zoos and natural populations confirm that neck length is variable among individuals.

The researchers were unable to test the second prediction, however, because they studied giraffes in a natural population and were unable to do breeding experiments. As a result, they simply had to accept this prediction as an assumption. In general, though, biologists prefer to test every assumption behind a hypothesis.

What about the prediction regarding feeding high in trees? According to Simmons and Scheepers, this is where the food competition hypothesis breaks down. Consider, for example, data collected by a different research team about the amount of time that giraffes spend feeding in vegetation of different heights. **Figure 1.10a** shows that in a population from Kenya, both male and female giraffes spend most of their feeding time eating vegetation that averages just 60 percent of their full height. Studies on other populations of giraffes, during

(a) Most feeding is done below neck height.

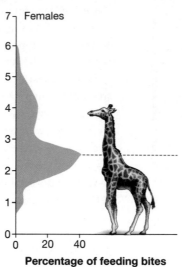

(b) Typical feeding posture in giraffes

FIGURE 1.10 Giraffes Do Not Usually Extend Their Necks to Feed.

both the wet and dry seasons, are consistent with these data. Giraffes usually feed with their necks bent **(Figure 1.10b)**.

These data cast doubt on the food competition hypothesis, because one of its predictions does not appear to hold. Biologists have not abandoned this hypothesis completely, though, because feeding high in trees may be particularly valuable during extreme droughts, when a giraffe's ability to reach leaves far above the ground could mean the difference between life and death. Still, Simmons and Scheepers have offered an alternative explanation for why giraffes have long necks. The new hypothesis is based on the mating system of giraffes.

The Sexual Competition Hypothesis: Predictions and Tests Giraffes have an unusual mating system. Breeding occurs year round rather than seasonally. To determine when females are coming into estrus (or "heat") and are thus receptive to mating, the males nuzzle the rumps of females. In response, the females urinate into the males' mouths. The males then tip their heads back and pull their lips to and fro, as if tasting the liquid. Biologists who have witnessed this behavior have proposed that the males taste the females' urine to detect whether estrus has begun.

Once a female giraffe enters estrus, males fight among themselves for the opportunity to mate. Combat is spectacular. The bulls stand next to one another, swing their necks, and strike thunderous blows with their heads. Researchers have seen males knocked unconscious for 20 minutes after being hit and have cataloged numerous instances in which the loser died. Giraffes are the only animals known to fight in this way.

These observations inspired a new explanation for why giraffes have long necks. The sexual competition hypothesis is based on the idea that longer-necked giraffes are able to strike harder blows during combat than can shorter-necked giraffes. In engineering terms, longer necks provide a longer moment arm. A long moment arm increases the force of the impact. (Think about the type of sledge hammer you'd use to bash down a concrete wall—one with a short handle or one with a long handle?) Thus, longer-necked males should win more fights and, as a result, father more offspring than do shorter-necked males. If neck length in giraffes is inherited, then the average neck length in the population should increase over time. Under the sexual competition hypothesis, long necks are adaptations that increase the fitness of males during competition for females.

Although several studies have shown that long-necked males are more successful in fighting and that the winners of fights gain access to estrous females, the question of why giraffes have long necks is not closed. With the data collected to date, most biologists would probably concede that the food competition hypothesis needs further testing and refinement and that the sexual selection hypothesis appears promising. It could also be true that both hypotheses are correct. For our purposes, the important take-home message is that all hypotheses must be tested rigorously.

In many cases in biological science, testing hypotheses rigorously involves experimentation. Experimenting on giraffes is difficult. But in the case study considered next, biologists were able to test an interesting hypothesis experimentally.

Why Are Chili Peppers Hot? An Introduction to Experimental Design

Experiments are a powerful scientific tool because they allow researchers to test the effect of a single, well-defined factor on a particular phenomenon. Because experiments testing the effect of neck length on food and sexual competition in giraffes have yet to be done, let's consider a different question: Why do chili peppers taste so spicy?

The jalapeño, Anaheim, and cayenne peppers used in cooking descended, via artificial selection, from a wild shrub that is native to the deserts of the American Southwest. As **Figure 1.11a** shows, wild chilies produce fleshy fruits with seeds inside, just like their domesticated descendants. In both wild chilies and the cultivated varieties, the "heat" or pungent flavor of the fruit and seeds is due to a molecule called capsaicin. In humans and other mammals, capsaicin binds to heat-sensitive cells in the tongue and mouth. In response to this binding, signals are sent to the brain that produce the sensation of burning. Similar signals would be transmitted if you drank boiling water. Asking why chilies are hot, then, is the same as asking why chilies contain capsaicin.

Josh Tewksbury and Gary Nabhan proposed that the presence of capsaicin is an adaptation that protects chili fruits from being eaten by animals that destroy the seeds inside. To understand this hypothesis, it's important to realize that the seeds inside a fruit have one of two fates when the fruit is eaten. If the seeds are destroyed in the animal's mouth or digestive system, then they never germinate (sprout). In this case, "seed predation" has occurred. But if seeds can travel undamaged through the animal, then they are eventually "planted" in a new location along with a valuable supply of fertilizer. In this case, seeds are dispersed. Here's the key idea: Natural selection should favor fruits that taste bad to animal species that act as seed predators. But these same fruits should not deter species that act as seed dispersers. This proposal is called the directed dispersal hypothesis.

Does capsaicin deter seed predators, as the directed dispersal hypothesis predicts? To answer this question, the researchers captured some cactus mice (**Figure 1.11b**) and birds called curve-billed thrashers (**Figure 1.11c**). These species are among the most important fruit- and seed-eating animals in habitats where chilies grow. Based on earlier observations, the biologists predicted that cactus mice destroy chili seeds but that curve-billed thrashers disperse them effectively.

To test the directed dispersal hypothesis, the biologists offered both cactus mice and curve-billed thrashers three kinds of fruit: hackberries, fruits from a strain of chilies that can't synthesize capsaicin, and pungent chilies that have lots of capsaicin. The non-pungent chilies are about the same size and color as normal chilies and have similar nutritional value. The hackberries look like chilies, except that they are not quite as red and contain no capsaicin. The three fruits were present in equal amounts. For each animal tested, the researchers recorded the percentage of hackberry, non-pungent chili, and pungent chili that was eaten during a specific time interval. Then they calculated the average amount of each fruit that was eaten by five test individuals from each species.

The directed dispersal hypothesis predicts that seed dispersers will eat the pungent chilies readily but that seed predators won't. Recall that a prediction specifies what we should observe if a hypothesis is correct. Good scientific hypotheses make testable predictions—predictions that can be supported or rejected by collecting and analyzing data. If the directed dispersal hypothesis is wrong, however, then there shouldn't be any difference in what various animals eat. This latter possibility is called a **null hypothesis**. A null hypothesis specifies what we should observe when the hypothesis being tested

(a) Wild chilies produce fruits that contain seeds.

(b) Cactus mice are seed eaters.

(c) Curve-billed thrashers are fruit eaters.

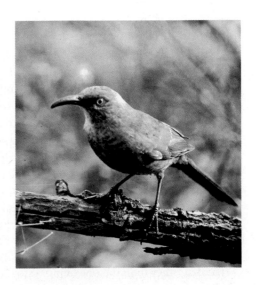

FIGURE 1.11 Chilies … and Chili Eaters?

Experiment

Question: Does the presence of capsaicin in chilies deter some predators but not others?

Dispersal hypothesis: Capsaicin deters cactus mice (seed predators) but not birds (seed dispersers).

Null hypothesis: Cactus mice and birds respond to capsaicin in the same way.

Experimental setup:

Cactus mouse

Allow mice and thrashers to select fruit to eat

Curve-billed thrasher

Hackberry (H) Non-pungent chilies (NP) Pungent chilies (P)

Hackberry (H) Non-pungent chilies (NP) Pungent chilies (P)

Prediction: Both will eat hackberry, but only thrashers will eat pungent chilies.

Prediction of null hypothesis: No difference between thrashers and mice in fruit consumed.

Results:

Cactus mice don't eat chilies that contain capsaicin

Cactus mouse

Thrashers do eat chilies that contain capsaicin

Thrasher

Conclusion: The presence of capsaicin deters cactus mice but not thrashers.

FIGURE 1.12 An Experimental Test: Does Capsaicin Deter Some Fruit Eaters? The graphs in the Results section indicate the average percentage of fruit consumed by the animals that were tested. The thin vertical lines indicate the standard error associated with each average. The standard error is a measure of variability and uncertainty in the data. For more details, see **BioSkills 3**.

doesn't hold. These predictions are listed in **Figure 1.12**. Do the predictions of the directed dispersal hypothesis hold? To answer this question, look at the results plotted in Figure 1.12. View the data for each type of fruit to see if different amounts were eaten by different types of animals, and ask yourself whether the two animals ate the same or different amounts of each type of fruit. Based on your analysis of the data, decide whether the results support the directed dispersal hypothesis or the null hypothesis. Use the conclusion stated in Figure 1.12 to check your answer.

In relation to designing effective experiments, this study illustrates several important points:

- It is critical to include control groups. A control checks for factors, other than the one being tested, that might influence the experiment's outcome. For example, if hackberries had

not been included as a control, it would have been possible to claim that the cactus mice in the experiment didn't eat pungent chilies simply because they weren't hungry. But the not-hungry hypothesis can be rejected because all of the animals ate hackberries.

- The experimental conditions must be carefully controlled. The investigators used the same feeding choice setup, the same time interval, and the same definition of what "fruit consumed" meant in each test. Controlling all of the variables except one—the types of fruits presented—is crucial because it eliminates alternative explanations for the results. For example, what types of problems could arise if the cactus mice were given less time to eat than the thrashers, or if the test animals were always presented with hackberries first and pungent chilies last?

- Repeating the test is essential. It is almost universally true that larger sample sizes in experiments are better. For example, suppose that the experimenters had used just one cactus mouse instead of five, and that this mouse was unlike other cactus mice because it ate almost anything. If so, the resulting data would be badly distorted. By testing many individuals, the amount of distortion or "noise" in the data caused by unusual individuals or circumstances is reduced.

(MB) **Web Animation** at www.masteringbio.com
Introduction to Experimental Design

⊙ If you understand these concepts, you should be able to design an experiment that tests the directed dispersal hypothesis by adding capsaicin to small pieces of apple—a fruit that both mice and thrashers eat readily.

To test the assumption that cactus mice are seed predators and curve-billed thrashers are seed dispersers, the researchers did a follow-up experiment. They fed fruits of the non-pungent chili to each type of predator. When the seeds had passed through the animals' digestive systems and were excreted, the researchers collected and planted the seeds—along with 14 uneaten seeds. Planting uneaten seeds served as a control treatment, because it tested the hypothesis that the seeds were viable and would germinate if they were not eaten. About 50 percent of the uneaten seeds germinated, and almost 60 percent of the seeds eaten by a thrasher germinated. But none of the seeds eaten by cactus mice germinated. The data indicate that seeds pass through curve-billed thrashers unharmed but are destroyed when eaten by cactus mice.

Based on the outcomes of these two experiments, the researchers concluded that curve-billed thrashers are efficient seed dispersers and are not deterred by capsaicin. The cactus mice, in contrast, refuse to eat chilies. If they ate chilies, the mice would kill the seeds. These are exactly the results predicted by the directed dispersal hypothesis. The biologists concluded that the presence of capsaicin in chilies is an adaptation that keeps their seeds from being destroyed by mice. In habitats that contain cactus mice, the production of capsaicin increases the fitness of individual chili plants.

These experiments are a taste of things to come. In this text you will encounter hypotheses and experiments on questions ranging from how water gets to the top of 100-meter-tall sequoia trees to why the bacterium that causes tuberculosis has become resistant to antibiotics. A commitment to tough-minded hypothesis testing and sound experimental design is a hallmark of biological science. Understanding their value is an important first step in becoming a biologist.

Check Your Understanding

If you understand that... ⊙

- Hypotheses are proposed explanations that make testable predictions.
- Predictions are observable outcomes of particular conditions.
- Well-designed experiments alter just one condition—a condition relevant to the hypothesis being tested.

You should be able to... ⊙

1) Design an experiment to test the hypothesis that the use of capsaicin as cooking spice is an adaptation—specifically, that the presence of capsaicin in food kills disease-causing bacteria.

2) State the predictions of the adaptation hypothesis and the null hypothesis in your experiment.

3) Answer the following questions about your experimental design:

    How does the presence of a control group in your experiment allow you to test the null hypothesis?

    Why isn't the experiment valid without the presence of a control group?

    How are experimental conditions controlled or standardized in a way that precludes alternative explanations of the data?

    Why do you propose to repeat the experiment many times?

Chapter Review

SUMMARY OF KEY CONCEPTS

For over two hundred years, biologists have been discovering traits that unify the spectacular diversity of living organisms.

⊙ **Biological science was founded with the development of (1) the cell theory, which proposes that all organisms are made of cells and that all cells come from preexisting cells, and (2) the theory of evolution by natural selection, which maintains that the** characteristics of species change through time—primarily because individuals with certain heritable traits produce more offspring than do individuals without those traits.

The cell theory is an important unifying principle in biology, because it identified the fundamental structural unit common to all life. The theory of evolution by natural selection is another key unifying principle, because it states that all organisms are related by common ancestry. It also offered a robust explanation for why species change through time and why they are so well adapted to their habitats.

You should be able to describe the evidence that supported the cell theory. You should also be able to explain why a population of wild *Brassica oleracea* will evolve by natural selection, if in response to global warming individuals with large leaves begin producing the most offspring. ●

(MB) **Web Animation** at www.masteringbio.com
Artificial Selection

● **A phylogenetic tree is a graphical representation of the evolutionary relationships among species. Phylogenies can be established by analyzing similarities and differences in traits. Species that share many traits are closely related and are placed close to each other on the tree of life.**

The cell theory and the theory of evolution predict that all organisms are part of a genealogy of species, and that all species trace their ancestry back to a single common ancestor. To reconstruct this phylogeny, biologists have analyzed the sequence of components in rRNA and other molecules found in all cells. A tree of

life, based on similarities and differences in these molecules has three major lineages: the Bacteria, Archaea, and Eukarya.

You should be able to explain why biologists can determine whether newly discovered species are members of the Bacteria, Archaea, or Eukarya by analyzing their rRNA or other molecules. ●

● **Biologists ask questions, generate hypotheses to answer them, and design experiments that test the predictions made by competing hypotheses.**

Another unifying theme in biology is a commitment to hypothesis testing and to sound experimental design. Analyses of neck length in giraffes and the capsaicin found in chilies are case studies in the value of testing alternative hypotheses and conducting experiments. Biology is a hypothesis-driven, experimental science.

You should be able to explain (1) the relationship between a hypothesis and a prediction, and (2) why experiments are convincing ways to test predictions. ●

(MB) **Web Animation** at www.masteringbio.com
Introduction to Experimental Design

QUESTIONS

● Test Your Knowledge

1. Anton van Leeuwenhoek made an important contribution to the development of the cell theory. How?
   a. He articulated the pattern component of the theory—that all organisms are made of cells.
   b. He articulated the process component of the theory—that all cells come from preexisting cells.
   c. He invented the first microscope and saw the first cell.
   d. He invented more powerful microscopes and was the first to describe the diversity of cells.

2. Suppose that a proponent of the spontaneous generation hypothesis claimed that cells would appear in Pasteur's swan-necked flask eventually. According to this view, Pasteur did not allow enough time to pass before concluding that life does not originate spontaneously. Which of the following is the best response?
   a. The spontaneous generation proponent is correct: Spontaneous generation would probably happen eventually.
   b. Both the all-cells-from-cells hypothesis and the spontaneous generation hypothesis could be correct.
   c. If spontaneous generation happens only rarely, it is not important.
   d. If spontaneous generation did not occur after weeks or months, it is not reasonable to claim that it would occur later.

3. What does the term *evolution* mean?
   a. The strongest individuals produce the most offspring.
   b. The characteristics of an individual change through the course of its life, in response to natural selection.

c. The characteristics of populations change through time.
d. The characteristics of species become more complex over time.

4. What does it mean to say that a characteristic of an organism is heritable?
   a. The characteristic evolves.
   b. The characteristic can be passed on to offspring.
   c. The characteristic is advantageous to the organism.
   d. The characteristic does not vary in the population.

5. In biology, what does the term *fitness* mean?
   a. how well trained and muscular an individual is, relative to others in the same population
   b. how slim an individual is, relative to others in the same population
   c. how long a particular individual lives
   d. the ability to survive and reproduce

6. Could *both* the food competition hypothesis and the sexual selection hypothesis explain why giraffes have long necks? Why or why not?
   a. No. In science, only one hypothesis can be correct.
   b. No. Observations have shown that the food competition hypothesis cannot be correct.
   c. Yes. Long necks could be advantageous for more than one reason.
   d. Yes. All giraffes have been shown to feed at the highest possible height and fight for mates.

Test Your Knowledge answers: 1. d; 2. d; 3. c; 4. b; 5. d; 6. c

● Test Your Understanding   *Answers are available at www.masteringbio.com*

1. The Greek roots of the term taxonomy can be translated as "arranging rules." Explain why these roots were an appropriate choice for this term.

2. It was once thought that the deepest split among life-forms was between two groups: prokaryotes and eukaryotes. Draw and label a phylogenetic tree that represents this hypothesis. Then draw and

label a phylogenetic tree that shows the actual relationships among the three domains of organisms.

3. Why was it important for Linnaeus to establish the rule that only one type of organism can have a particular genus and species name?

4. What does it mean to say that an organism is adapted to a particular habitat?

5. Compare and contrast natural selection with the process that led to the divergence of a wild mustard plant into cabbage, broccoli, and Brussels sprouts.

6. The following two statements explain the logic behind the use of molecular sequence data to estimate evolutionary relationships:

"If the theory of evolution is true, then rRNA sequences should be very similar in closely related organisms but less similar in organisms that are less closely related."

"On a phylogenetic tree, branches that are close to one another represent species that are closely related; branches that are farther apart represent species that are more distantly related."

Is the logic of these statements sound? Why or why not?

● Applying Concepts to New Situations

Answers are available at www.masteringbio.com

1. A scientific theory is a set of propositions that defines and explains some aspect of the world. This definition contrasts sharply with the everyday usage of the word theory, which often carries meanings such as "speculation" or "guess." Explain the difference between the two definitions, using the cell theory and the theory of evolution by natural selection as examples.

2. Turn back to the tree of life shown in Figure 1.9. Note that Bacteria and Archaea are prokaryotes, while Eukarya are eukaryotes. On the simplified tree below, draw an arrow that points to the branch where the structure called the nucleus originated. Explain your reasoning.

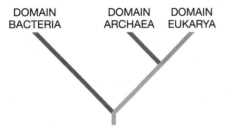

DOMAIN BACTERIA   DOMAIN ARCHAEA   DOMAIN EUKARYA

3. The proponents of the cell theory could not "prove" that it was correct in the sense of providing incontrovertible evidence that all organisms are made up of cells. They could state only that all organisms examined to date were made of cells. Why was it reasonable for them to conclude that the theory was valid?

4. How do the tree of life and the taxonomic categories created by Linnaeus (kingdom, phylum, class, order, family, genus, and species) relate to one another?

www.masteringbio.com is also your resource for • Answers to text, table, and figure caption questions and exercises • Answers to *Check Your Understanding* boxes • Online study guides and quizzes • Additional study tools including the *E-Book for Biological Science* 3rd ed., textbook art, animations, and videos.

2 Water and Carbon: The Chemical Basis of Life

KEY CONCEPTS

- ● Molecules form when atoms bond to each other. Chemical bonds are based on electron sharing. The degree of electron sharing varies from nonpolar covalent bonds, to polar covalent bonds, to ionic bonds.

- ● Of all small molecules, water is the most important for life. Water is highly polar and readily forms hydrogen bonds. Hydrogen bonding makes water an extremely efficient solvent with a high capacity for absorbing energy.

- ● Chemical reactions tend to be spontaneous if they lead to lower potential energy and higher entropy (more disorder). An input of energy is required for nonspontaneous reactions to occur.

- ● Most of the important compounds in organisms contain carbon. Early in Earth's history, the energy in sunlight triggered nonspontaneous reactions that led to the formation of key carbon-containing molecules.

Life began in the ocean, over 3 billion years ago.

Chapter 1 introduced experiments on the hypothesis of spontaneous generation, which tested the idea that life arises from nonliving materials. This work helped build a consensus that spontaneous generation does not occur. But for life to exist now, spontaneous generation must have occurred at least once, early in Earth's history. How did life begin? This simple query has been called "the mother of all questions."

This chapter examines a theory, called chemical evolution, that attempts to answer this question. Like the theories introduced in Chapter 1, the theory of chemical evolution has a pattern component that makes a claim about the natural world and a process component that explains that pattern. **Chemical evolution** is the proposition that early in Earth's history, simple chemical compounds in the atmosphere and ocean combined to form larger, more complex substances. As a result, the chem-

istry of the oceans and atmosphere changed over time. The theory's name is appropriate because in its simplest sense, the term *evolution* means change through time. According to the theory, the process responsible for this pattern was the conversion of the kinetic energy in sunlight and heat to chemical energy in the form of bonds between atoms that formed large, complex compounds.

The theory also maintains that larger, more complex substances began to accumulate and then reacted with one another to produce even more complicated compounds, and that continued chemical evolution eventually led to the origin of life. More specifically, the hypothesis is that one of these complex compounds was able to make a copy of itself, or self-replicate. As this molecule multiplied, the process of evolution by natural selection took over—there was a switch from chemical evolution

● Key Concept   Important Information   ● Practice It

to biological evolution. Eventually, a self-replicating molecule became surrounded by a membrane, and cellular life began.

At first glance, the theory of chemical evolution seems wildly implausible. But is it? What evidence do biologists have that chemical evolution occurred? Let's start with the fundamentals—the atoms and molecules that would have combined to get chemical evolution started.

2.1 The Building Blocks of Chemical Evolution

Just four types of atoms—hydrogen, carbon, nitrogen, and oxygen—make up 96 percent of all matter found in organisms today. Many of the molecules found in living cells contain thousands, or even millions, of these atoms bonded together. But early in Earth's history, it is likely that these elements existed only in simple substances such as water and carbon dioxide, which contain just three atoms apiece.

The theory of chemical evolution maintains that simple compounds in the ancient atmosphere and ocean combined to form the larger, more complex substances found in living cells. To understand how this process could have begun, we need to consider the following questions:

- What is the physical structure of hydrogen, carbon, nitrogen, oxygen, and other atoms found in living cells?

- What is the structure of water, carbon dioxide, and the other simple molecules that acted as the building blocks of chemical evolution?

These questions focus on one of the most central themes in biology: Function follows from structure. To understand how a molecule affects your body or the role it played in chemical evolution, you have to understand how it is put together.

What Atoms Are Found in Organisms?

Figure 2.1 shows a simple way of depicting the structure of an atom, using hydrogen and carbon as examples. Extremely small particles called electrons orbit an atomic nucleus made up of large particles called protons and neutrons. Protons have a positive electric charge, neutrons are electrically neutral, and electrons have a negative electric charge. Opposite charges attract; like charges repel. When the number of protons and the number of electrons in an atom (or molecule) are the same, the charges balance and the atom is electrically neutral.

Figure 2.2 shows a segment of the periodic table of the elements. Notice that each of the **elements** contains a characteristic number of protons, called its **atomic number**. The atomic number is given as a subscript of each element symbol in the table. The number of neutrons present in an element can vary, however. Forms of an element with different numbers of neutrons are known as **isotopes** ("equal-places"). For example, all

(a) Diagrams of atoms

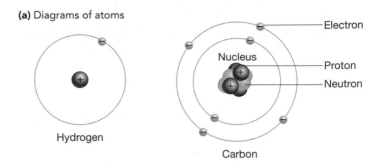

Hydrogen

Carbon

(b) Most of an atom's volume is empty space.

If an atom occupied the same volume as this stadium, the nucleus would be about the size of a pea

FIGURE 2.1 Parts of an Atom. These diagrams of the hydrogen and carbon atoms show how the nucleus, made up of protons and neutrons, is surrounded by orbiting electrons. In reality, electrons do not orbit the nucleus in circles; their actual orbits are complex. As the note on this photograph shows, the diagrams are not to scale.

atoms of the element carbon have 6 protons. But naturally occurring isotopes of carbon can have either 6 or 7 neutrons, giving them a total of 12 or 13 protons and neutrons, respectively. The sum of the protons and neutrons in an atom is called its **mass number**, given as a superscript of each symbol in Figure 2.2. Although the masses of protons, neutrons, and

FIGURE 2.2 A Portion of the Periodic Table. Each element has a unique atomic number and is represented by a unique one- or two-letter symbol.

electrons can be measured in grams, the numbers involved are so small that chemists and physicists prefer to use a special unit called the **atomic mass unit (amu)**, or the dalton. The masses of protons and neutrons are virtually identical and are routinely rounded to 1 amu. A carbon atom that contains 6 protons and 6 neutrons has a mass of 12 amu and a mass number of 12, while a carbon atom with 7 neutrons would have a mass number of 13. These isotopes would be written as ^{12}C and ^{13}C, respectively.

To understand how the atoms involved in chemical evolution behave, though, focus on how electrons are arranged around the nucleus. Electrons move around atomic nuclei in specific regions called **orbitals**. Each electron orbital has a distinctive shape, and each orbital can hold up to two electrons. Orbitals, in turn, are grouped into levels called **electron shells**. These are numbered 1, 2, 3, and so on, to indicate their relative distance from the nucleus. The electrons of an atom fill the innermost shells first, before filling outer shells.

To understand how the structures of atoms differ, take a moment to study **Figure 2.3**. This chart highlights the atoms that are most abundant in living cells. The diagrams in each box of the table show how electrons are distributed in the shells of carbon and other key elements. The outermost shell of an atom is called the atom's valence shell, and the electrons found in that shell are referred to as **valence electrons**. Two observations are important:

1. In each of the highlighted elements, the outermost electron shell is not full. The highlighted elements have at least one unpaired valence electron.

2. The number of unpaired electrons in the valence shell varies among elements. Carbon has four unpaired electrons in its outermost shell; hydrogen has one. The number of unpaired electrons found in an atom is called its **valence**. Carbon's valence is four; hydrogen's valence is one.

These observations are significant because an atom is most stable when its valence shell is filled. One way that shells can be filled is through the formation of **chemical bonds**—strong attractions that bind atoms together.

How Does Covalent Bonding Hold Molecules Together?

To understand how atoms can become more stable by sharing electrons, consider hydrogen. The hydrogen atom has just one electron, which resides in a shell that can hold two electrons. When two atoms of hydrogen approach each other, the two electrons present become shared by the two nuclei (**Figure 2.4**).

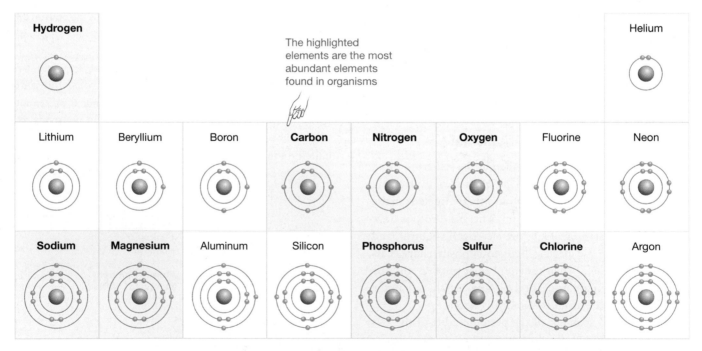

FIGURE 2.3 The Structure of Atoms Found in Organisms. This diagram represents the atomic nucleus as a closed sphere. The first (inner), second, and third electron shells are shown as rings. The dots on the rings represent electrons. Electrons are drawn in pairs if they occupy filled orbitals within the same electron shell; they are drawn singly if they occupy unfilled orbitals.

● **EXERCISE** Unpaired electrons in an unfilled shell can form chemical bonds. Write down the number of bonds each highlighted atom is capable of forming.

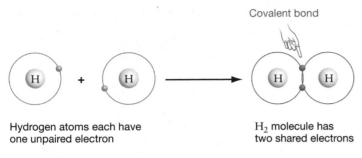

FIGURE 2.4 **Covalent Bonds Result from Electron Sharing.** When two hydrogen atoms come into contact, their electrons are attracted to the positive charge in each nucleus. As a result, their orbitals overlap, the electrons are shared by each nucleus, and a covalent bond forms.

FIGURE 2.5 **Electron Sharing and Bond Polarity. (a)** The electrons in bonds are shared equally if the two atoms involved have equal electronegativity, but **(b)** unequally if the atoms have unequal electronegativity. Delta (δ) symbols in polar covalent bonds refer to partial positive and negative charges that arise due to unequal electron sharing.

Both atoms now have a completely filled shell, and they have formed a chemical bond. Together, the bonded hydrogen atoms are more stable than the two individual hydrogen atoms. The shared electrons "glue" the atoms together in a **covalent bond**. Substances that are held together by covalent bonds are called **molecules**. In this case of two hydrogen atoms, the bonded atoms form a single molecule of hydrogen, written as H–H or H_2.

It's often helpful to think about covalent bonding as electrical attraction and repulsion. As two hydrogen atoms move closer together, their positively charged nuclei repel each other and their negatively charged electrons repel each other. But each proton attracts both electrons, and each electron attracts both protons. Covalent bonds form when the attractive forces overcome the repulsive forces. This is the case when hydrogen atoms interact to form the hydrogen molecule (H_2).

To visualize what's happening when a covalent bond forms, imagine two children who want the same toy. If both children grasp the toy tightly, through it they become attached to each other. For as long as they hold onto the toy, neither child can leave the other's side. ● If you understand this concept, you should be able to explain what the children, the toy, and the children's hold on the toy actually represent, and why the "children" want the toy.

To summarize, covalent bonding is based on electron sharing. In a hydrogen molecule, the electrons involved in the covalent bond are shared between the two nuclei. This situation is illustrated in **Figure 2.5a**. The covalent bond between hydrogen atoms is represented by a dash, and the electrons are drawn as dots between the two nuclei. An equal sharing of electrons, as illustrated in Figure 2.5a, results in a **nonpolar covalent bond**—a covalent bond that is symmetrical.

It's important to note, though, that the electrons participating in covalent bonds are not always shared equally between the atoms involved. Asymmetric sharing of electrons results in polar covalent bonds. Some atoms hold electrons in covalent bonds much more tightly than do other atoms, so the degree of sharing varies by element. Chemists call this property an atom's **electronegativity**. If you understand this concept, you should be able to extend the analogy of a covalent bond being represented by two children clutching the same toy. What would this situation look like if one of the children wanted the toy more than the other? In this case, which of the two bonds would be weaker—meaning, easier to break?

Oxygen is among the most electronegative of all elements: It attracts covalently bonded electrons more strongly than does any other atom commonly found in organisms. Nitrogen's electronegativity is somewhat lower than oxygen's. Carbon and hydrogen, in turn, have relatively low and approximately equal electronegativities. The electronegativities of the four most abundant elements in organisms are related as follows: O >> N > C ≅ H.

To put these ideas into practice, consider the water molecule. Water consists of oxygen bonded to two hydrogen atoms and is written H_2O. As **Figure 2.5b** illustrates, the electrons involved in the covalent bonds in water are not shared equally but are held much more tightly by the oxygen nucleus than by the hydrogen nuclei. Hence, water has two polar covalent bonds. Because electrons are shared unequally, the oxygen atom has a partial negative charge and the hydrogen atoms have a partial positive charge. Notice that the partial charges on the molecule are symbolized by the lowercase Greek letter delta, δ. As Section 2.2 will show, the partial charges on water molecules—due simply to the difference in electronegativity between oxygen and hydrogen—are one of the primary reasons life exists.

How Does Ionic Bonding Hold Ionic Compounds Together?

Ionic bonds are similar in principle to covalent bonds, but instead of electrons being shared between two atoms, the electrons in

(a) A sodium ion being formed

Loss of electron
Cation formation

Sodium ion
has positive
charge

(b) A chloride ion being formed

Gain of electron
Anion formation

Chloride ion
has negative
charge

(c) Table salt is a crystal composed of two ions.

Cl⁻

Na⁺

FIGURE 2.6 Ion Formation and Ionic Bonding. (a) When an atom of sodium (Na) loses an electron, it forms a sodium ion (Na^+). The sodium ion is stable because it has a full valence shell. **(b)** When an atom of chlorine (Cl) gains an electron, it forms a chloride ion (Cl^-), which also has a full valence shell. **(c)** In table salt (NaCl), sodium and chloride ions pack into a crystal structure held together by electrical attraction.

ionic bonds are completely transferred from one atom to the other. The electron transfer occurs because it gives the resulting atoms a full outermost shell. Sodium atoms (Na), for example, tend to lose an electron, leaving them with a full second shell. This is a much more stable arrangement, energetically, than having a lone electron in their third shell (**Figure 2.6a**). The atom that results has a net electric charge of +1, because it has one more proton than it has electrons. An atom or molecule that carries a charge is called an **ion**. The sodium ion is written Na^+ and, like other positively charged ions, is called a **cation**. Chlorine atoms (Cl), in contrast, tend to gain an electron, filling their outermost shell (**Figure 2.6b**). The atom has a net charge of −1, because it has one more electron than protons. This negatively charged ion, or **anion**, is written Cl^- and is called chloride. When sodium and chlorine combine to form

table salt (sodium chloride, NaCl), the atoms pack into a crystal structure consisting of sodium cations and chloride anions (**Figure 2.6c**). The electrical attraction between the ions is so strong that salt crystals are difficult to break apart.

This discussion of covalent and ionic bonding supports an important general observation: ● The degree to which electrons are shared in chemical bonds forms a continuum. As the left-hand side of **Figure 2.7** shows, covalent bonds between atoms with exactly the same electronegativity—for example, between the atoms of hydrogen in H_2—represent one end of the continuum. The electrons in these nonpolar bonds are shared equally. In the middle of the continuum are bonds where one atom is much more electronegative than the other. In these asymmetric bonds, substantial partial charges exist on each of the atoms. These types of polar covalent bonds occur

Equal sharing of electrons ◄——————————————————————► Transfer of electrons

| Nonpolar covalent bonds (atoms have no charge) | Polar covalent bonds (atoms have partial charge) | Ionic bonds (atoms have full charge) |

$H \overset{\circ}{\underset{\circ}{-}} H$

$H \overset{\circ}{\underset{\circ}{-}} \overset{\overset{H}{|}}{\underset{\overset{|}{H}}{C}} \overset{\circ}{\underset{\circ}{-}} H$

$\overset{\delta^+}{H} \overset{\delta^-}{\underset{\circ}{-}} \overset{}{N} \overset{\delta^+}{\underset{\circ}{-}} H$ $\overset{}{\underset{H\ \delta^+}{}}$

$\overset{\delta^-}{\underset{\delta^+ H \qquad H\ \delta^+}{O}}$

$Na^+\ Cl^-$

Hydrogen   Methane   Ammonia   Water   Sodium chloride

FIGURE 2.7 The Electron-Sharing Continuum. The degree of electron sharing in chemical bonds can be thought of as a continuum, from equal sharing in nonpolar covalent bonds to no sharing in ionic bonds.

● **QUESTION** Why do most polar covalent bonds involve nitrogen or oxygen?

when a highly electronegative atom such as oxygen or nitrogen is bound to an atom with a lower affinity for electrons, such as carbon or hydrogen. Water (H_2O) and ammonia (NH_3) contain polar covalent bonds. At the right-hand side of the continuum are molecules made up of atoms with extreme differences in their electronegativities. In this case, electrons are transferred rather than shared, the atoms have full charges, and the bonding is ionic. Common table salt, NaCl, is a familiar example. In biology, ionic bonding is rare. The chemical bonds that occur in biological molecules are on the left-hand side and the middle of this continuum.

Some Simple Molecules Formed from H, C, N, and O

Look again at Figure 2.3 and count the number of unpaired electrons in the valence shells of carbon, nitrogen, and oxygen atoms: Carbon has four, nitrogen has three, and oxygen has two. Each unpaired electron can make up half of a covalent bond. As a result, a carbon atom can form a total of four covalent bonds; nitrogen can form three; and oxygen, two. When each of the four unpaired electrons of a carbon atom covalently bonds with a hydrogen atom, the molecule that results is written CH_4 and is called methane (**Figure 2.8a**). This is the most common molecule found in natural gas. When a nitrogen atom's three unpaired electrons bond with three hydrogen atoms, the result is NH_3, or ammonia. Similarly, an atom of oxygen can form covalent bonds with two atoms of hydrogen, resulting in a water molecule (H_2O).

Double and Triple Bonds Atoms with more than one unpaired electron in the valence shell can form multiple single bonds and sometimes double or triple bonds. **Figure 2.8b** shows how carbon forms double bonds with oxygen atoms to produce carbon dioxide (CO_2). Triple bonds result when three pairs of electrons are shared. **Figure 2.8c** shows the structure of molecular nitrogen (N_2), which forms when two nitrogen atoms establish a triple bond.

The Geometry of Some Simple Molecules Proponents of the theory of chemical evolution argue that simple molecules such as CH_4, NH_3, H_2O, CO_2, and N_2 were important components of Earth's ancient atmosphere and ocean. This reasoning is based on the observation that these molecules are found in volcanic gases on Earth and in the atmospheres of nearby planets. If so, then such molecules provided the building blocks for chemical evolution. Before analyzing how these molecules could have combined to form more complex compounds, however, we need to consider their shape and how scientists represent their composition and geometry. Recall that this is a crucial point—one that will resonate throughout this text. Function follows from structure. For example, the overall shape of a molecule often dictates its behavior and function.

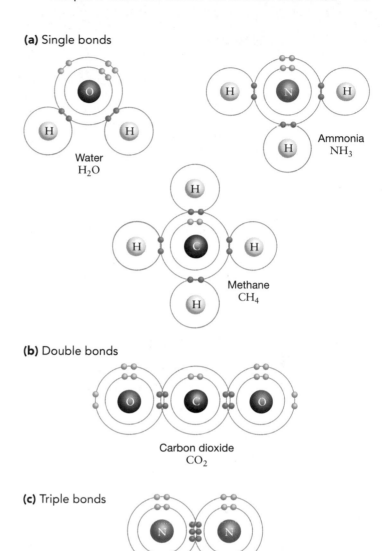

(a) Single bonds

Water
H_2O

Ammonia
NH_3

Methane
CH_4

(b) Double bonds

Carbon dioxide
CO_2

(c) Triple bonds

Molecular nitrogen
N_2

FIGURE 2.8 Unpaired Electrons in the Valence Shell Participate in Covalent Bonds. Covalent bonding is based on sharing of electrons in the outermost shell. Covalent bonds can be **(a)** single, **(b)** double, or **(c)** triple.

● **QUESTION** Why can carbon participate in four single bonds, while oxygen can participate in only two and hydrogen in only one?

The shape of a simple molecule is governed by the geometry of its bonds. Bond angles, in turn, are determined by electrical repulsion that occurs between pairs of electrons. For example, **Figure 2.9a** shows the shape of methane (CH_4).

- The four pairs of electrons in carbon's outer shell form a tetrahedron when bonded to hydrogen because repulsion among the shared electron pairs orients them at an angle of 109.5° to one another.

- The same outer shells and bond geometry occur in ammonia (NH_3), but one of the four orbitals in the nitrogen atom's outer shell is already filled with a pair of electrons—so just three orbitals with unpaired electrons bond with hydrogen.

(a) Methane (CH₄)   **(b)** Ammonia (NH₃)   **(c)** Water (H₂O)

Unshared electrons

Shared electrons

FIGURE 2.9 The Geometry of Methane, Ammonia, and Water. Diagrams showing the shapes of the orbitals in the outermost shells of **(a)** methane, **(b)** ammonia, and **(c)** water. Though some of the same orbitals are involved in bonding, the overall shape of each molecule is radically different.

● **EXERCISE** Label which molecule is bent and planar, which forms a tetrahedron (having four identical faces), and which forms a pyramid.

In this case, the geometry of the orbitals and the geometry of the molecule are different (**Figure 2.9b**).

● In the oxygen molecule, two of the four orbitals present in the outermost shell are filled with electron pairs. When hydrogen atoms bond to the remaining two electrons to form water (H₂O), the resulting molecule is bent and two-dimensional, or planar (**Figure 2.9c**). Section 2.2 will explore how water's shape, in combination with the differing electronegativities between oxygen and hydrogen, make it the most important molecule on Earth.

Representing Molecules   Molecules can be represented in a variety of ways. The simplest representation is a **molecular formula**, which indicates the numbers and types of atoms in a molecule. Water has the molecular formula H₂O; methane has the molecular formula CH₄ (**Figure 2.10a**).

Molecular formulas are a compact way of stating a molecule's composition, but they contain no information about how the molecule is put together. **Structural formulas** indicate which atoms are bonded together. In a structural formula, single, double, and triple bonds are represented by single, double, and triple dashes, respectively. **Figure 2.10b** gives structural formulas for molecules that were important components of ancient Earth's atmosphere and ocean. Structural formulas also provide some information on the molecule's geometry. Water's structural formula indicates that it is bent, while the oxygen molecule is drawn as a linear molecule.

The limitation of structural formulas is that they are two-dimensional, while molecules are three-dimensional. More complex diagrams, such as ball-and-stick models and space-filling models, are based on the three-dimensional structure of molecules. **Figures 2.10c** and **2.10d** show how ball-and-stick and space-filling models give a more sophisticated view of a molecule's geometry as well as the relative sizes of the atoms involved. For more information on interpreting chemical structures, see **BioSkills 4** in the back of the book.

Quantifying the Concentration of Key Molecules   When a **chemical reaction** occurs, one substance is combined with others or broken down into another substance. Chemicals react in simple whole-number combinations. For example, 1 molecule of acetic acid reacts with 1 molecule of ethanol to form 1 molecule of ethyl acetate. But if you wanted to set up this reaction in an experiment, how would you know how much acetic acid and ethanol to add? The problem is that there is no simple way of counting the numbers of molecules present in a solution. Researchers solve this problem using the mole concept.

	Methane	Ammonia	Water	Oxygen
(a) Molecular formulas:	CH₄	NH₃	H₂O	O₂
(b) Structural formulas:	H–C–H with H above and below (on C)	H–N–H with H below (on N)	O with H, H (bent)	O=O
(c) Ball-and-stick models:				
(d) Space-filling models:				

FIGURE 2.10 Molecules Can Be Represented Several Ways. Each method of representing a molecule has particular advantages. **(a)** Molecular formulas are compact and indicate the number and identity of atoms involved. **(b)** Structural formulas show which atoms are bonded to one another and indicate geometry in two dimensions. **(c)** Ball-and-stick models take up more space than do structural formulas but include information about three-dimensional shape. **(d)** Space-filling models are not as easy to read as ball-and-stick models but more accurately depict the spatial relationships between atoms.

A **mole** refers to the number 6.022×10^{23}—just as the unit called the dozen refers to the number 12 or the unit million refers to the number 1×10^6. The mole is a useful unit because the mass of one mole of any molecule is the same as its molecular weight expressed in grams. **Molecular weight** is the sum of the mass numbers of all the atoms in a molecule. For example, summing the mass numbers of 2 atoms of hydrogen and 1 atom of oxygen gives $1 + 1 + 16$, or 18, which is water's molecular weight. It follows that if you weighed a sample of 18 grams of water, it would contain 6.022×10^{23} water molecules, or 1 mole of water molecules.

Now let's get back to your experiment with acetic acid and ethanol. To obtain an equal number of each type of molecule, you use a balance to measure out the two substances and add an equal number of moles of each. For example, if you wanted to react one mole of acetic acid and one mole of ethanol, you'd weigh out a number of grams of acetic acid and ethanol equal to their respective molecular weights and make up a **solution**—a homogenous (uniform) mixture of one or more substances dissolved in a liquid. When substances are dissolved in a liquid, their concentration is expressed in terms of molarity (symbolized by "M"). **Molarity** is the number of moles of the substance present per liter of solution. A 1-molar solution of acetic acid in water, for instance, means that 1 mole of acetic acid—that is, 6.022×10^{23} acetic acid molecules—is contained in 1 liter of water.

Check Your Understanding

If you understand that...

- Covalent bonds are based on electron sharing. Ionic bonds are based on electrical attraction between ions with opposite charges.

- Covalent bonds can be polar or nonpolar, depending on whether the electronegativities of the two atoms involved are the same or different.

You should be able to...

1) Draw the structural formulas of methane (CH_4) and ammonia (NH_3) and add dots to indicate the relative locations of the covalently bonded electrons, based on the relative electronegativities of C, H, and N.

2) Draw the electron shells around sodium ions (Na^+) and chloride ions (Cl^-) and explain why table salt (NaCl) exists.

2.2 The Early Oceans and the Properties of Water

Life is based on water. In a typical living cell, over 75 percent of the volume consists of this molecule. Nearly 70 percent of

FIGURE 2.11 Water Is the Most Abundant Molecule in Organisms. Fruits shrink when they are dried because they consist primarily of water.

your body, by weight, is water (**Figure 2.11**). You can survive for weeks without eating, but you aren't likely to live more than 3–4 days without drinking. In humans, a water loss of just 2 percent leads to weakness, dizziness, irritability, headache, trouble urinating, and a general feeling of being unwell.

Water is essential for life today, and virtually all researchers agree that most of the important steps in chemical evolution, including the origin of life itself, occurred in water. Water is vital for a simple reason: As a **solvent**—an agent for getting substances into solution—water can dissolve more types of substances than any other molecule known. The chemical reactions occurring inside your body right now and the reactions that caused chemical evolution some 3.5 billion years ago depend on direct, physical interaction between the reactants. Substances are most likely to collide and react when they are dissolved. Life is based on water primarily because of water's solvent properties. The formation of Earth's first ocean, about 3.8 billion years ago, was a turning point in chemical evolution, because it gave chemical evolution a place to happen.

Why Is Water Such an Efficient Solvent?

To understand why water is such an effective solvent, recall from Figure 2.5b that the molecule contains two hydrogen atoms bonded to an oxygen atom, and that oxygen and hydrogen differ in their electronegativity—meaning, their ability to attract the electrons involved in covalent bonds. Because oxygen is among the most electronegative of all elements, it attracts covalently bonded electrons much more strongly than does the hydrogen nucleus. As a result, the electrons that participate in the bonds are not shared equally. The covalent bonds in water are polar.

If water were a linear molecule like carbon dioxide, the polarity of its covalent bonds would not matter very much. But because the water molecule is bent, the molecule as a whole is **polar,** meaning that the overall distribution of charge is

(a) Water is polar.

Electrons are
pulled toward
oxygen

(b) Hydrogen bonds form
between water molecules.

FIGURE 2.12 Water Is Polar and Participates in Hydrogen Bonds.
(a) Because of oxygen's high electronegativity, the electrons that are
shared between hydrogen and oxygen spend more time close to the
oxygen nucleus—giving the oxygen atom a partial negative charge and
the hydrogen atom a partial positive charge. **(b)** The electrical attraction
that occurs between the partial positive and negative charges on water
molecules forms a hydrogen bond.

● **EXERCISE** Label the hydrogen bond in part (b).

● **EXERCISE** Water molecules routinely form multiple hydrogen bonds.
Add 2 hydrogen bonds to the H_2O molecule on the right side of part (b).

asymmetrical. As **Figure 2.12a** shows, the side of the molecule
containing the oxygen atom is slightly more negative, and the
side with the hydrogen atoms is slightly more positive. As before, the partial charges on the molecule are symbolized by the
lowercase Greek letter delta, δ.

Figure 2.12b illustrates how the polarity of water affects its interactions with other water molecules. When two liquid water
molecules approach one another, the partial positive charge on
hydrogen attracts the partial negative charge on oxygen. This
weak electrical attraction forms a hydrogen bond between the
molecules. ● If you understand how water's polarity makes
hydrogen bonding possible, you should be able to (1) draw a version of Figure 2.12b that shows water as a linear (not bent) molecule, with partial charges on the oxygen and hydrogen atoms;
and (2) explain why electrostatic attractions between water molecules would be much weaker or even nonexistent, as a result.

In a water-based, or aqueous, solution, hydrogen bonds
also form between water molecules and other polar molecules. Similar interactions occur between water and ions. Ions
and polar molecules stay in solution because of their interactions with water's partial charges (**Figure 2.13a**). Substances
that interact with water in this way are said to be **hydrophilic**
("water-loving").

In contrast, compounds that are uncharged and nonpolar do
not interact with water through hydrogen bonding and do not
dissolve in water. Instead, they interact with each other (**Figure
2.13b**). Substances that do not interact with water are said to
be **hydrophobic** ("water-hating"). The interactions between hydrophobic atoms and molecules are much weaker than interactions based on hydrogen bonding, however.

● To summarize, hydrogen bonding makes it possible for almost any charged or polar molecule to dissolve in water. Although individual hydrogen bonds are not nearly as strong as
covalent or ionic bonds, they are much more common. Hydrogen bonding is extremely important in biology due to the sheer
number of hydrogen bonds that form between water and other
molecules that are polar or carry a charge.

How Does Water's Structure Correlate with Its Properties?

Water's structure is unusual. Its small size, bent shape, highly
polar covalent bonds, and overall polarity are unique among
molecules. Because the structure of molecules routinely correlates with their function, it's not surprising that water has some
remarkable properties. We've already reviewed the most
important of water's chemical attributes—its ability to form
hydrogen bonds with ions and polar compounds and its capacity to act as a solvent.

(a) Polar molecules and ions dissolve readily in water.

Salt in absence of water

Salt dissolved in water

(b) Nonpolar molecules do not dissolve readily in water.

Octane interacts with
itself, but not with water

FIGURE 2.13 Polar Molecules and Ions Dissolve Readily in Water; Nonpolar Molecules Do Not.
(a) Water's polarity makes it a superb solvent for polar molecules and ions. **(b)** In aqueous solution,
nonpolar molecules and compounds interact with each other but not with water.

● **QUESTION** Explain the physical basis of the expression, "Oil and water don't mix."

Water has several other striking physical properties. Each of these is the direct result of hydrogen bonding. In relation to chemical evolution, the most important of these properties are that water expands as it changes from a liquid to a solid and that it has an extraordinarily large capacity for absorbing heat. Let's analyze why hydrogen bonding explains these properties, along with the property called surface tension.

Cohesion, Adhesion, and Surface Tension   Binding between like molecules is called **cohesion**. Water is cohesive—meaning that it stays together—because of the hydrogen bonds that form between individual molecules. Binding between unlike molecules, in contrast, is called **adhesion**. Adhesion is usually analyzed in regard to interactions between a liquid and a solid surface. Water adheres to surfaces that have any polar or charged components.

Cohesion and adhesion are important in explaining how water moves from the roots of plants to their leaves, but you can also see them in action in the concave surface, or **meniscus**, that forms in a glass of water (**Figure 2.14a**). Because water molecules at the surface hydrogen-bond with water molecules below them, they experience a net downward pull. But water molecules at the surface also adhere to the glass, allowing them to resist the downward pull. The result of these two forces is the formation of a meniscus.

Cohesion is also instrumental in the phenomenon known as surface tension. Because hydrogen bonding exerts a pulling force, or tension, at the surface of any body of water, water molecules are not stable at the surface—they are constantly being pulled away. It follows that a body of water as a whole is most stable when its total surface area is minimized. This fact has an important consequence: Water resists any force that increases its surface area. More specifically, any force that depresses a water surface meets with resistance. This resistance makes a water surface act as if it had an elastic membrane (**Figure 2.14b**)—a property called **surface tension**.

All liquids have a surface tension. Water's surface tension is extraordinarily high because of the extensive hydrogen bonding that occurs between molecules. In water, the "elastic membrane" at the surface is stronger than it is in other liquids.

Water Is Denser as a Liquid than as a Solid   When factory workers pour molten metal or plastic into a mold and allow it to cool to the solid state, the material shrinks. When molten lava pours out of a volcano and cools to solid rock, it shrinks. But when you fill an ice tray with water and put it in the freezer to make ice, the water expands.

Unlike most substances, water is denser as a liquid than it is as a solid. In other words, there are more molecules of water in a given volume of liquid water than there are in the same

(a) Cohesion and adhesion explain why a meniscus forms where water meets a solid surface.

The **meniscus** (concave surface) is a result of two forces:

Adhesion: Water molecules at the surface adhere to the glass, allowing them to resist the downward pull of cohesion

Cohesion: Because water molecules at the surface hydrogen-bond with water molecules below them, they experience a net downward pull

(b) Water has high suface tension.

Because of surface tension, light objects do not fall through the water's surface

FIGURE 2.14 Cohesion, Adhesion, and Surface Tension. (a) A combination of adhesion and cohesion causes a meniscus to form when water is bounded by a surface. **(b)** Water resists forces—like the weight of an insect—that increase its surface area. The resistance is great enough that light objects do not break through the surface.

(a) In ice, water molecules form a crystal lattice.

(b) In liquid water, no crystal lattice forms.

(c) Liquid water is denser than ice. As a result, ice floats.

FIGURE 2.15 Hydrogen Bonding Forms the Crystal Structure of Ice.

● **EXERCISE** In ice, each molecule can form four hydrogen bonds at one time. (Each oxygen atom can form two; each hydrogen atom can form one.) Choose two molecules in part (a) and circle the four H bonds. If only three bonds are shown, draw in the fourth—where another water molecule would be added to the crystal.

volume of solid water. **Figure 2.15a** illustrates why this is so. Note that in ice, each water molecule participates in four hydrogen bonds. These hydrogen bonds cause the water molecules to form a regular and repeating structure, or crystal. The crystal structure of ice is fairly open, meaning that there is a relatively large amount of space between molecules.

Now compare the extent of hydrogen bonding and the density of ice with that of liquid water, illustrated in **Figure 2.15b**. Although the exact structure of liquid water is not fully understood, it is clearly dynamic. Hydrogen bonds are constantly being made and broken. Overall, the extent of hydrogen bonding in liquid water is much less than that found in ice. As a result, molecules in the liquid phase are packed much more closely together than are molecules of solid water, even though their temperature is higher. Normally, heating a substance causes it to expand because molecules begin moving faster and colliding more often and with greater force. But heating ice causes hydrogen bonds to break and the open crystal structure to collapse to a more-dense configuration in liquid water. In this way, hydrogen bonding explains why water is denser as a liquid than as a solid.

This property of water has an important result: Ice floats (**Figure 2.15c**). If it did not, ice would sink to the bottom of lakes, ponds, and oceans soon after it formed. The ice would stay frozen in the cold depths. If water weren't so unusual, it is almost certain that Earth's oceans would have frozen solid before life had a chance to start.

Water Has a High Capacity for Absorbing Energy

Hydrogen bonding is also responsible for another of water's remarkable physical properties: Water has a high capacity for absorbing energy. For example, water has an extraordinarily high specific heat. **Specific heat** is the amount of energy required to raise the temperature of 1 gram of a substance by 1°C. Water's specific heat is high because, when a source of energy such as sunlight or a flame strikes it, hydrogen bonds must be broken before heat can be transferred and the water molecules begin moving faster. As a result, it takes an extraordinarily large amount of energy to change the temperature of water and other molecules where extensive hydrogen bonding occurs (**Table 2.1**).

Similarly, it takes a large amount of energy to break the hydrogen bonds in liquid water and change the molecules from the

TABLE **2.1 Specific Heats of Some Liquids**

The specific heats reported in this table were measured at 25°C and are given in units of joules per gram of substance per degree Celsius. (The **joule** is a unit of energy.)	
Liquids with extensive hydrogen bonding	**Specific Heat**
Ammonia	4.70
Water	4.18
Liquids with some hydrogen bonding	
Ethanol (CH_3CH_2OH)	2.44
Ethylene glycol ($HOCH_2CH_2OH$; used in antifreeze)	2.22
Liquids with little or no hydrogen bonding	
Benzene (C_6H_6)	1.80
Xylene (C_8H_{10})	1.72
Sulfuric acid (H_2SO_4)	1.40

liquid phase to the gas phase. Water's **heat of vaporization**—the energy required to change 1 gram of it from a liquid to gas—is higher than that of most molecules that are liquid at room temperature. As a result, water has to absorb a great deal of energy to evaporate. Water's high heat of vaporization is the reason that sweating or dousing yourself with water is an effective way to cool off on a hot day. Because water molecules have to absorb a great deal of energy from your body in order to evaporate, you lose heat.

These properties of water are important to the theory of chemical evolution. As Section 2.3 will show, the energy in sunlight can drive the formation of complex molecules from simple building-block compounds. But energy inputs can also break complex molecules apart. Water's high specific heat insulates dissolved substances from sources of energy like intense sunlight. Because compounds that were important in chemical evolution dissolve readily in water, they would have formed in the ocean or rained out of the atmosphere into the ocean. Once these compounds were in an environment where they were better protected, they would have persisted and slowly increased in concentration over time. Similarly, if chemical evolution took place at the edges of ponds or on beaches, the evaporation of water would have kept the surfaces relatively cool. Hydrogen bonding has a moderating influence on water temperature.

(MB) **Web Animation** at www.masteringbio.com
The Properties of Water

Acid–Base Reactions and pH

One other aspect of water's chemistry is important to understanding chemical evolution and how organisms work today. Water is not a completely stable molecule. In reality, water molecules continually undergo a chemical reaction with themselves. This "dissociation" reaction can be written as follows:

$$H_2O \rightleftharpoons H^+ + OH^-$$

The double arrow indicates that the reaction proceeds in both directions.

The molecules on the right-hand side of the expression are the **hydrogen ion** (H^+) and the **hydroxide ion** (OH^-). A hydrogen ion is simply a proton. In reality, however, protons never exist by themselves. In water, for example, protons associate with water molecules to form the hydronium ion, H_3O^+. Thus, the dissociation of water is more accurately written

$$H_2O + H_2O \rightleftharpoons H_3O^+ + OH^-$$

One of the water molecules on the left-hand side of the expression has given up a proton, while the other water molecule has accepted a proton. Substances that give up protons during chemical reactions and raise the hydrogen ion concentration of water are called **acids**; molecules or ions that acquire protons during chemical reactions and lower the hydrogen ion concentration of water are called **bases**. A chemical reaction that involves a trans-

fer of protons is called an acid–base reaction. Every acid–base reaction requires a proton donor and a proton acceptor—an acid and a base. Most acids act only as acids, and most bases act only as bases; but water can act as both an acid and a base.

Water is an extremely weak acid, though—very few water molecules dissociate to form hydronium ions and hydroxide ions. In contrast, strong acids like the hydrochloric acid in your stomach readily give up a proton when they react with water.

$$HCl + H_2O \rightleftharpoons H_3O^+ + Cl^-$$

Strong bases readily acquire a proton when they react with water. For example, sodium hydroxide (NaOH, commonly called lye), dissociates completely in water to form Na^+ and OH^-. The hydroxide ions then accept a proton from hydronium ions in the water, forming more water molecules.

$$NaOH(aq) \rightarrow Na^+ + OH^-$$

$$OH^- + H_3O^+ \rightleftharpoons 2H_2O$$

To summarize, adding an acid to a solution increases the concentration of protons; adding a base to a solution lowers the concentration of protons.

These points are important because in a solution, the tendency for acid–base reactions to occur is largely a function of the number of protons present. When an acid–base reaction occurs, the transfer of a proton changes the charge of the proton donor and the proton acceptor. When a compound's charge changes, its tendency to participate in hydrogen bonding and other types of electrostatic interactions also changes. This is critical. Even if the concentration of protons in your blood doubles, it is still extremely low. Yet the acid–base reactions triggered by the change in proton concentration are likely to kill you within a few minutes (**Box 2.1** introduces important molecules called buffers, which prevent dramatic pH changes in blood and elsewhere.)

Chemists can measure the concentration of protons in a solution directly. In a sample of pure water at 25°C, the concentration of H^+ is 1.0×10^{-7} M. (Recall that M represents molarity, or moles per liter.) Because this is such a small number (1 ten-millionth), the exponential notation is cumbersome. So chemists and biologists prefer to express the concentration of protons in a solution with a logarithmic notation called **pH**. (The term *pH* is derived from the French *puissance d'hydrogène*, or "power of hydrogen.") By definition, the pH of a solution is the negative of the base-10 logarithm, or log, of the hydrogen ion concentration:

$$pH = -\log[H^+]$$

(The square brackets are a standard notation for indicating "concentration of" a substance in solution.) Taking antilogs gives

$$[H^+] = \text{antilog}(-pH) = 10^{-pH}$$

BOX 2.1 What Is a Buffer?

Acid–base reactions add or remove a proton from ions or molecules. The addition or removal of a proton changes the charge on a compound and thus how it interacts with other polar or charged molecules. Cells are extremely sensitive to changes in pH because they contain many polar or charged substances. In addition, many of the key reactions that occur in cells involve the addition or removal of a proton, and pH affects how likely these reactions are to occur.

Based on these observations, it shouldn't be surprising to learn that minimizing changes in pH inside cells is an important part of maintaining optimal conditions for life. Compounds that minimize variation in pH are called **buffers** because they buffer a solution against the damaging effects of pH change.

Most buffers are weak acids, meaning that they are somewhat likely to give up a proton in solution. To see how they work, consider the disassociation of acetic acid in water to form acetate ions and protons.

$$CH_3COOH \rightleftharpoons CH_3COO^- + H^+$$

When acetic acid and acetate are present in about equal concentrations in a solution, they function as a buffering system. If the concentration of protons increases slightly, the protons react with acetate ions to form acetic acid and pH does not change. If the concentration of protons decreases slightly, acetic acid gives up protons and pH does not change.

In cells, a wide array of naturally occurring molecules act as buffers. Biologists have also developed an array of synthetic compounds to buffer experimental solutions that contain pH-sensitive molecules.

---

pH, then, is a convenient way to indicate the concentration of protons in a solution. To review logarithms, see **BioSkills 5**.

Figure 2.16 shows the pH scale and reports the pH of some common solutions. The pH of pure water at 25°C is 7. Pure water is used as a standard, or point of reference, on the pH scale. Solutions that contain acidic molecules (molecules that act as acids) have a proton concentration larger than 1×10^{-7} M and thus a pH < 7, because acidic molecules tend to release protons into solution. In contrast, solutions that contain basic molecules (molecules that act as bases) have a proton concentration less than 1×10^{-7} M and thus a pH > 7, because basic molecules tend to accept protons from solution. Solutions with a pH of 7 are considered neutral solutions, neither acidic nor basic. The solution inside living cells is about 7. ◉ If you understand this concept, you should be able to explain the difference in hydrogen ion concentration between a slightly basic solution with pH 8 and a slightly acidic solution with pH 6.

Although the pH of today's oceans is about 8, the early oceans probably were much like today's lakes in their pH and chemical composition—simply because it would have taken hundreds of millions of years for rain and wind to slowly erode rocks, and for rivers to deliver dissolved ions that increased the ocean's salinity and increased its pH.

As chemical evolution began, then, water provided the physical environment for key reactions to take place. In some cases, water also acted as an important reactant. Although acid–base reactions were not critical to the initial stages of chemical evolution, they became extremely important once the process was under way.

Table 2.2 summarizes some of the key properties of water for you, as a reference. Now, before looking more closely at the role of carbon in life, let's move on to consider the reactions that created the building blocks of chemical evolution.

FIGURE 2.16 The pH Scale. Because the pH scale is logarithmic, a change in one unit of pH represents a change in the concentration of hydrogen ions equal to a factor of 10. Coffee has a hundred times more H^+ than pure water has.

◉ **QUESTION** How is the pH of black coffee affected if you add milk?

SUMMARY TABLE **2.2 Properties of Water**

Property	Cause	Biological consequences
Solvent for charged or polar compounds		Most chemical reactions important for life take place in aqueous solution.
High **cohesion**	Hydrogen bonds form between water molecules.	Creates surface tension; also important in water transport in plants.
High **adhesion**		Plays a role in water transport in plants.
High **surface tension**	High cohesion makes a water surface resist forces that increase surface area.	Important in water transport in plants; small organisms can walk on water.
Denser as a liquid than a solid	Hydrogen bonding leads to formation of low-density crystal structure in ice.	
High **specific heat**	Water molecules must absorb lots of heat energy to break hydrogen bonds and experience increased movement (and thus temperature).	Oceans absorb and release heat slowly, moderating coastal climates.
High **heat of vaporization**		Evaporation of water from an organism cools the body.

● **EXERCISE** You should be able to fill in the missing cells in this table. You should also be able to make a concept map relating water's structure to the properties listed here. (For an introduction to concept mapping, see **BioSkills 6**.)

2.3 Chemical Reactions, Chemical Evolution, and Chemical Energy

Proponents of the theory of chemical evolution contend that simple molecules present in the atmosphere and ocean of ancient Earth participated in chemical reactions that produced larger, more complex molecules. According to one hypothesis, the early atmosphere was made up of gases ejected from volcanoes. Carbon dioxide, water vapor, and nitrogen are the dominant gases ejected from volcanoes today; a small amount of molecular hydrogen (H_2) and methane (CH_4) may also be present. But when these molecules are placed in a glass tube together and allowed to interact, very little happens. Water vapor condenses to liquid water as the mixture cools, but the simple molecules do not suddenly link together to create large, complex substances like those found in living cells. Instead, their bonds remain intact. How, then, did chemical evolution occur?

To answer this question we need to explore two topics: (1) how chemical reactions occur, and (2) how conditions on ancient Earth made certain reactions possible.

How Do Chemical Reactions Happen?

Chemical reactions are written in a format similar to mathematical equations, with the initial, or **reactant**, atoms or molecules shown on the left and the resulting reaction **product**(s) shown on the right. For example, the most common reaction in the mix of gases and water that emerge from volcanoes is

$$CO_2(g) + H_2O(l) \rightleftharpoons H_2CO_3(aq)$$

This expression indicates that carbon dioxide (CO_2) reacts with water (H_2O), forming carbonic acid (H_2CO_3). The state of each reactant and product is indicated as gas (g), liquid (l), in aqueous solution (aq), or solid (s).

Note that the expression is balanced; that is, 1 carbon, 3 oxygen, and 2 hydrogen atoms are present on each side of the expression. Note also that the expression contains a double arrow, meaning that the reaction is reversible. When the forward and reverse reactions proceed at the same rate, the quantities of reactants and products remain constant, although not necessarily equal. A dynamic but stable state such as this is termed a **chemical equilibrium**. A chemical equilibrium can be disturbed by changing the concentration of reactants or products. For example, adding CO_2 to the mixture would drive the reaction to the right, creating more H_2CO_3 until the equilibrium proportions of reactants and products are reestablished. Adding H_2CO_3 instead would drive the reaction to the left. Removing CO_2 would also drive the reaction to the left; removing H_2CO_3 would drive it to the right.

A chemical equilibrium can also be altered by changes in temperature. For example, the water molecules in this set of interacting elements, or **system**, would be present as a combination of liquid water and water vapor:

$$H_2O(l) \rightleftharpoons H_2O(g)$$

If liquid water molecules absorb enough heat, they transform to the gaseous state. This is called an **endothermic** ("within heating") process because heat is absorbed during the process. In contrast, the transformation of water vapor to liquid water releases heat and is called **exothermic** ("outside heating"). Raising the temperature of this system drives the equilibrium to the right; cooling the system drives it to the left.

In relation to chemical evolution, though, these reactions and changes of state are not particularly interesting. Carbonic acid is not an important intermediate in the formation of more complex molecules. According to models developed by a series of researchers, however, interesting things do begin to happen when the system being studied is the ions and molecules present in the oceans and atmosphere of early Earth, and when large amounts of energy are added.

What Is Energy?

Energy can be defined as the capacity to do work or to supply heat. This capacity exists in one of two ways—as a stored potential or as an active motion.

Stored energy is called **potential energy**. An object gains or loses its ability to store energy because of its position. An electron that resides in an outer electron shell will, if the opportunity arises, fall into a lower electron shell closer to the positive charges on the protons in the nucleus. Because of its position farther from the positive charges in the nucleus, an electron in an outer electron shell has more potential energy than does an electron in an inner shell (**Figure 2.17**).

Kinetic energy is the energy of motion. Molecules have kinetic energy because they are constantly in motion. This form of kinetic energy—the kinetic energy of molecular motion—is called **thermal energy**. The **temperature** of an object is a measure of how much thermal energy its molecules possess. If an object has a low temperature, its molecules are moving slowly. (We perceive this as "cold".) If an object has a high temperature, its molecules are moving rapidly. (We perceive this as "hot".) When two objects with different temperatures come

into contact, thermal energy is transferred between them. We call this transferred energy **heat**.

There are many forms of potential and kinetic energies, and energy can change from one form into another. To drive this point home, consider a water molecule sitting at the top of a waterfall, as in **Figure 2.18a**. This molecule has potential energy because of its position. If the molecule passes over the waterfall, its potential energy is converted to the kinetic energy of motion. When the molecule has reached the rocks below, it has experienced a change in potential energy because it has changed position. The change in potential energy is transformed into an equal amount of energy in other forms: mechanical energy, which tends to break up the rocks; heat (thermal energy), which raises the temperature of the rocks and the water itself; and sound.

An electron in an outer electron shell (**Figure 2.18b**) is analogous to the water molecule at the top of a waterfall. If the electron falls to a lower shell, its potential energy is converted to the kinetic energy of motion. After the electron occupies the lower electron shell, it has less potential energy. As panel 3 in Figure 2.18b shows, the change in potential energy is transformed into an equal amount of energy in other forms—usually thermal energy, but sometimes light.

These examples illustrate the **first law of thermodynamics**, which states that energy is conserved. Energy cannot be created or destroyed, but only transferred and transformed.

Energy transformation is the heart of chemical evolution. According to the best data available, molecules that were part of the young Earth were exposed to massive inputs of energy. Part of this energy was in the form of heat, from the molten mass that initially formed the planet. Part of the energy was in the form of high-energy radiation from the Sun, which today is screened out by molecules in the atmosphere. How would the application of large amounts of heat and radiation affect the course of chemical evolution?

Chemical Evolution: A Model System

To assess the impact of energy inputs on the simple molecules present in the early oceans and atmosphere, researchers have constructed computer models to simulate the reactions that can occur among carbon dioxide, water, ammonia, and hydrogen molecules. The goal of one such study was quite specific: The researchers wanted to determine whether a molecule called formaldehyde (H_2CO) could be produced. Along with hydrogen cyanide (HCN), formaldehyde is a key intermediate in the creation of the larger, more complex molecules found in cells. Forming formaldehyde and hydrogen cyanide is the critical first step in chemical evolution—a trigger that could set the process in motion.

The research group began by proposing that the following reaction could take place:

$$CO_2(g) + 2H_2(g) \rightarrow H_2CO(g) + H_2O(g)$$

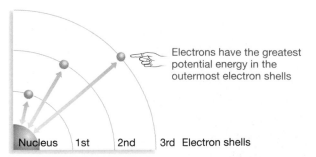

Electrons have the greatest potential energy in the outermost electron shells

Nucleus   1st   2nd   3rd Electron shells

FIGURE 2.17 Potential Energy as a Function of Electron Shells. Electrons in outer shells have more potential energy than do electrons in inner shells, because negative charges in outer shells are farther from the positive charges in the nucleus. Each shell represents a distinct level of potential energy.

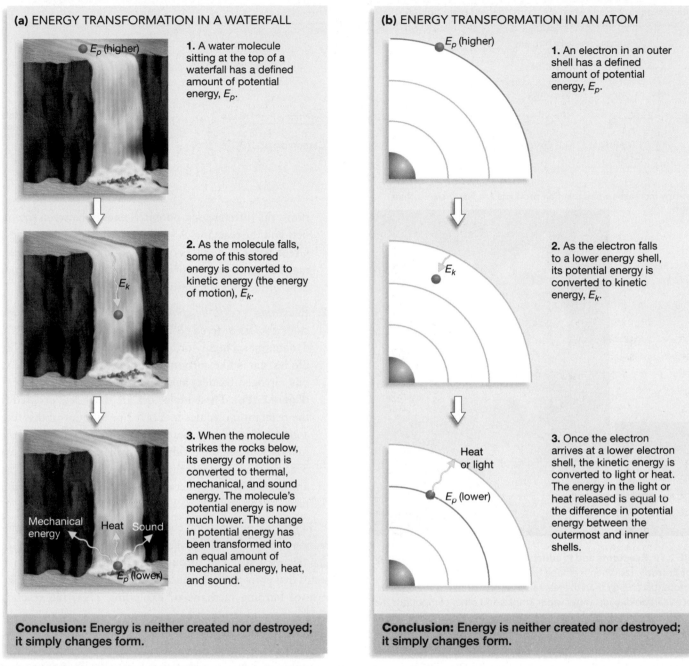

(a) ENERGY TRANSFORMATION IN A WATERFALL

E_p (higher)

1. A water molecule sitting at the top of a waterfall has a defined amount of potential energy, E_p.

E_k

2. As the molecule falls, some of this stored energy is converted to kinetic energy (the energy of motion), E_k.

Mechanical energy   Heat   Sound

E_p (lower)

3. When the molecule strikes the rocks below, its energy of motion is converted to thermal, mechanical, and sound energy. The molecule's potential energy is now much lower. The change in potential energy has been transformed into an equal amount of mechanical energy, heat, and sound.

Conclusion: Energy is neither created nor destroyed; it simply changes form.

(b) ENERGY TRANSFORMATION IN AN ATOM

E_p (higher)

1. An electron in an outer shell has a defined amount of potential energy, E_p.

E_k

2. As the electron falls to a lower energy shell, its potential energy is converted to kinetic energy, E_k.

Heat or light

E_p (lower)

3. Once the electron arrives at a lower electron shell, the kinetic energy is converted to light or heat. The energy in the light or heat released is equal to the difference in potential energy between the outermost and inner shells.

Conclusion: Energy is neither created nor destroyed; it simply changes form.

FIGURE 2.18 Energy Transformations. During an energy transformation, the total amount of energy in the system remains constant.

Before we explore how they tested this hypothesis, however, it will be helpful to know why this reaction doesn't occur spontaneously—that is, why doesn't it occur *without* an input of energy?

What Makes a Chemical Reaction Spontaneous? When chemists say that a reaction is spontaneous, they have a precise meaning in mind: ● Chemical reactions are spontaneous if they proceed on their own, without any continuous external in-

fluence such as added energy. Two factors determine whether a reaction is spontaneous or nonspontaneous:

1. *Reactions tend to be spontaneous if the products have lower potential energy than the reactants.* Reaction products have lower potential energy if their electrons are held more tightly than are the electrons of the reactants. Recall that highly electronegative atoms such as oxygen and nitrogen hold electrons much more tightly than do atoms with a lower electronegativity, such as carbon and hydrogen. For

(a) Potential energy drops when electrons are held more tightly in reaction products than in reactants.

(b) Entropy increases when reaction products are less ordered than reactants.

FIGURE 2.19 Reactants and Products May Differ in Potential Energy and/or Entropy. (a) When methane burns, the products have much lower potential energy than the reactants have. **(b)** When TNT explodes, it produces carbon dioxide, water vapor, smoke, and other compounds that are much less ordered than the original system. The reaction results in an increase in entropy.

● **EXERCISE** In part (a), label which electrons have relatively low potential energy and which have relatively high potential energy.

example, when natural gas burns, methane reacts with oxygen gas to produce carbon dioxide and water:

$$CH_4(g) + 2O_2(g) \rightarrow CO_2(g) + 2H_2O(g)$$

The electrons involved in the C=O and H–O bonds of carbon dioxide and water are held much more tightly than they were in the C–H and O=O bonds of methane and oxygen (**Figure 2.19a**). As a result, the products have much lower potential energy than the reactants. The difference in potential energy between reactants and products is given off as heat, so the reaction is exothermic. In chemical reac-

tions, the difference in potential energy between the products and the reactants is symbolized by ΔH. (The uppercase Greek letter Δ, delta, is often used in chemical and mathematical notation to represent change.) When a reaction is exothermic, ΔH is negative.

2. *Reactions tend to be spontaneous when the product molecules are less ordered than the reactant molecules.* TNT, or dynamite, is a highly ordered structure. But when TNT explodes, gases like carbon dioxide, carbon monoxide, various nitrogen oxides, and small particulates are given off (**Figure 2.19b**). These molecules are much less ordered than the reactant molecules in TNT. The amount of disorder in a group of molecules is called its **entropy**, which is symbolized by S. When the products of a chemical reaction are less ordered than the reactant molecules are, entropy increases and ΔS is positive. Reactions tend to be spontaneous if they increase entropy.

In general, physical and chemical processes proceed in the direction that results in lower potential energy and increased disorder (**Figure 2.20**). The **second law of thermodynamics**, in fact, states that entropy always increases in an isolated system. In the case of burning methane or exploding TNT, the reaction is exothermic *and* results in higher entropy—less-ordered products.

Because reactions tend to be spontaneous when ΔH is negative and ΔS is positive, it's necessary to assess the *combined* contributions of changes in heat and disorder to determine whether a chemical reaction is spontaneous. To do this, chemists define a quantity called the **Gibbs free-energy change**, symbolized by ΔG.

$$\Delta G = \Delta H - T\Delta S$$

Here, T stands for temperature measured on the Kelvin scale. The $T\Delta S$ term simply means that entropy becomes more important in determining free-energy change as the temperature of the molecules increases. The faster molecules are moving, the more important entropy becomes in determining the overall free-energy change.

Chemical reactions are spontaneous when ΔG is less than zero. Such reactions are said to be **exergonic**. Reactions are nonspontaneous when ΔG is greater than zero. Such reactions are termed **endergonic**. When ΔG is zero, reactions are at equilibrium. ⬤ If you understand these concepts, you should be able to explain why the same reaction can be nonspontaneous at low temperature but spontaneous at high temperature. You should also be able to explain why some exothermic reactions are nonspontaneous.

Here are the important things to remember about free energy: Free energy changes when the potential energy and/or entropy of substances changes. Chemical reactions run in the direction that lowers the free energy of the system. Exergonic reactions are spontaneous and release energy; endergonic reactions are nonspontaneous and require an input of energy to proceed.

Energy Inputs and the Start of Chemical Evolution The reaction between carbon dioxide (CO_2) and hydrogen gas (H_2) that forms formaldehyde (H_2CO) and water is endergonic. Formaldehyde and water have more potential energy and are more highly ordered than CO_2 and H_2. The reaction is nonspontaneous because ΔG is positive. For the reaction to occur, a large input of energy is required.

To explore how formaldehyde formation and chemical evolution got started, a research group constructed a computer model of the ancient atmosphere. The model consisted of a list of all possible chemical reactions that can occur among CO_2, H_2O, N_2, NH_3, CH_4, and H_2 molecules. In addition to the spontaneous reactions, they included reactions that occur when these molecules are struck by sunlight. This was crucial because sunlight represents a source of energy.

The sunlight that strikes Earth is made up of packets of light energy called photons. The amount of light energy contained in a photon can vary widely. Today, most of the higher-energy photons in sunlight never reach Earth's lower atmosphere. Instead, they are absorbed by a molecule called ozone (O_3) in the upper atmosphere. But if Earth's early atmosphere was filled with volcanic gases released as the molten planet cooled, it is extremely unlikely that appreciable quantities of ozone existed. Based on this logic, researchers infer that when chemical evolution was occurring, large quantities of high-energy photons bombarded the planet.

To understand why this energy source was so important, recall that the atoms in hydrogen and carbon dioxide have full

(a) Water has a higher potential energy at the top of a waterfall than at the bottom.

High potential energy

Low potential energy

(b) A sugar molecule has higher potential energy and more order (lower entropy) than carbon dioxide and water.

High potential energy, more order

Sugar (glucose)   +   6 O_2

This reaction occurs in your cells and when wood burns

Low potential energy, less order

6 CO_2 +   6 H_2O

(not all product molecules are shown)

FIGURE 2.20 Spontaneous Processes Result in Lower Potential Energy or Increased Disorder, or Both.

outermost shells. This arrangement makes these molecules largely unreactive. But energy from photons can break molecules apart by knocking electrons away from the outer shells of atoms. The atoms that result, called **free radicals**, have unpaired electrons and are extremely unstable (**Figure 2.21**). To mimic the conditions on early Earth more accurately, the computer model included several reactions that produce highly reactive free radicals.

To understand which of the long list of possible reactions would actually occur, and to estimate how much formaldehyde could have been produced in the ancient atmosphere, the researchers needed to consider the effects of two additional factors: temperature and concentration.

The Roles of Temperature and Concentration in Chemical Reactions   Even if a chemical reaction occurs spontaneously, it may not happen quickly. The reactions that convert iron to rust or sugar molecules to carbon dioxide and water are spontaneous, but at room temperature they occur very slowly, if at all. For most reactions to proceed, one chemi-

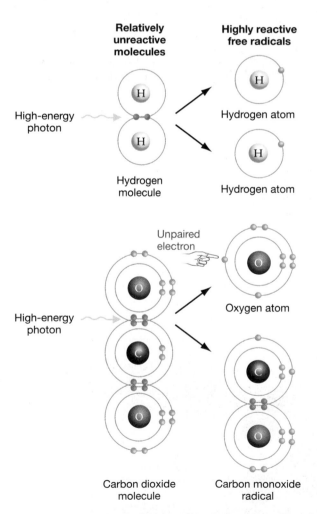

FIGURE 2.21 Free Radicals Are Extremely Reactive. When a high-energy photon strikes a hydrogen or carbon dioxide molecule, free radicals can be created. Formation of free radicals is thought to be responsible for some key reactions in chemical evolution.

cal bond has to break and another one has to form. For this to happen, the substances involved must collide in a specific orientation that brings the electrons involved near each other.

The number of collisions occurring among the substances in a mixture depends on the temperature and the concentrations of the reactants. When the concentration of reactants is high, more collisions should occur and reactions should proceed more quickly. When their temperature is high, reactants should move faster and collide more frequently. Higher concentrations and higher temperatures should tend to speed up chemical reactions. **Figure 2.22** provides data from a set of experiments performed by students from Parkland College in Champaign, Illinois, to test these predictions. As the data in the Results section show, the reaction rate speeded up significantly when the concentration of reactants was increased and when the temperature of the reaction mixture was higher.

To model the behavior of simple molecules in the ancient atmosphere, then, the researchers who were working on the formaldehyde synthesis reaction needed to specify the concentration of each molecule as well as the temperature. Then they were able to assign a rate to each of the reactions listed in their model based on the actual reaction rates observed in experiments conducted at controlled temperatures and concentrations, like those in Figure 2.22.

Their result? They calculated that, under temperature and concentration conditions accepted as reasonable approximations of early Earth conditions by most atmospheric scientists, appreciable quantities of formaldehyde would have been produced. Using a similar model, other researchers have shown that significant amounts of hydrogen cyanide (HCN) could also have been produced in the ancient atmosphere. According to this research, large quantities of the critical intermediates in chemical evolution would have formed in the ancient atmosphere.

How Did Chemical Energy Change during Chemical Evolution?

The initial products of chemical evolution are important, for a simple reason: They have more potential energy than the reactant molecules have. When formaldehyde is produced, an increase in potential energy occurs because the electrons that bond CO_2 and H_2 together are held more tightly than they are in H_2CO or H_2O. This form of potential energy—the potential energy stored in chemical bonds—is called **chemical energy**.

This observation gets right to the heart of chemical evolution: The energy in sunlight was converted to chemical energy—potential energy in chemical bonds. This energy transformation explains how chemical evolution was possible. When small, simple molecules absorb energy, chemical reactions can occur that transform the external energy into potential energy stored in chemical bonds. More specifically, the energy in sunlight was converted to chemical energy in the form of formalde-

Experiment

Question: Do chemical reaction rates increase with increased temperature and concentration?

Rate increase hypothesis: Chemical reaction rates increase with increased temperature. They also increase with increased concentration of reactants.

Null hypothesis: Chemical reaction rates are not affected by increases in temperature or concentration of reactants.

Experimental setup:

Experimental reaction: $3\ HSO_3^-(aq) + IO_3^-(aq) \rightleftharpoons 3\ HSO_4^-(aq) + I^-(aq)$

Reactant concentrations constant
Temperature increases

Almost continuous variation in temperature

−1°C 0°C 9°C 12°C 21°C 22°C 35°C 38°C 50°C

Reactant concentrations vary
Temperature constant

Many replicates at each concentration

Trial 1 Trial 2 Trial 3

Concentration of $3\ HSO_3^-$ (M):	0.0083 → 0.0083		0.0083	0.0083	0.017
Concentration of IO_3^- (M):	0.010 → 0.010		0.005	0.010	0.010
Temperature (°C):	−1 → 50		23	23	23

Prediction: Reaction rate, measured as 1/(time for reaction to go to completion), will increase with increased concentrations of reactants and increased temperature of reaction mix.

Prediction of null hypothesis: No difference in reaction rates among treatments in each set-up.

Results:

Conclusion: Chemical reaction rates increase with increased temperature and concentration.

FIGURE 2.22 Testing the Hypothesis That Reaction Rates Are Sensitive to Changes in Concentration and Temperature.

hyde and hydrogen cyanide. The complete reaction that results in the formation of formaldehyde is written as

$$CO_2(g) + 2H_2(g) + sunlight \rightarrow H_2CO(g) + H_2O(g)$$

Notice that the reaction is balanced in terms of the atoms *and* the energy involved. An increase in chemical energy occurred

that made the production of larger, more complex molecules possible.

This result makes sense intuitively. Energy is the capacity to do work, and it seems logical that building larger, more complex molecules requires work to be done. More specifically, the reactions involved in chemical evolution are endergonic, so inputs of energy were required. Now the question is, what happened to these first building blocks of chemical evolution?

Check Your Understanding

If you understand that...

- Chemical reactions tend to be spontaneous if they lead to lower potential energy and higher entropy (more disorder).
- The combined effects of potential energy and entropy changes are summarized in the equation for the Gibbs free-energy change.

You should be able to...

1) Write out the Gibbs equation and define each of the components.
2) Explain why potential energy might decrease as the result of a chemical reaction and why entropy might increase.

2.4   The Importance of Carbon

Life has been called a carbon-based phenomenon, and with good reason. Except for water, almost all of the molecules found in organisms contain this atom.

Carbon has great importance in biology because it is the most versatile atom on Earth. Due to its four valence electrons, it can form many covalent bonds. With different combinations of single and double bonds, an almost limitless array of molecular shapes are possible. You have already examined the tetrahedral structure of methane and the linear shape of carbon dioxide. When molecules contain more than one carbon atom, these atoms can be bonded to one another in long chains, as in the component of the gasoline called octane

(C_8H_{18}; **Figure 2.23a**), or in a ring, as in the sugar glucose ($C_6H_{12}O_6$; **Figure 2.23b**). Molecules that contain carbon are called **organic** molecules. (Other types of molecules are referred to as *inorganic* compounds.) Carbon atoms provide the structural framework for virtually all of the important compounds associated with life.

Linking Carbon Atoms Together

The formation of carbon-carbon bonds was an important event in chemical evolution. It represented a crucial step toward the production of the types of molecules found in living organisms. Once organic compounds such as formaldehyde and hydrogen cyanide had formed, continued chemical evolution could occur by the addition of heat alone. For example, when molecules of formaldehyde are heated, they react with one another to form a molecule called acetaldehyde. Acetaldehyde contains a carbon-carbon bond. With continued heating, reactions between formaldehyde and acetaldehyde molecules can produce the larger carbon-containing compounds called sugars.

Figure 2.24 summarizes these steps. In studying the figure, be sure to note two key messages: (1) Chemical evolution got under way because abundant sources of outside energy were available to trigger endergonic reactions; and (2) some of these molecules produced by chemical evolution are found in organisms living today. Once chemical evolution was under way, a wide variety of small organic molecules would have accumulated in the ancient Earth.

Functional Groups

In general, the carbon atoms in an organic molecule furnish a skeleton that gives the molecule its overall shape. But the chemical behavior of the compound—meaning the types of reactions

(a) Carbons linked in a linear molecule

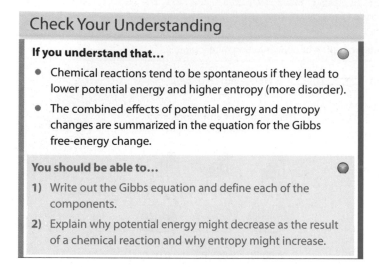

C_8H_{18} Octane

(b) Carbons linked in a ring

$C_6H_{12}O_6$ Glucose

FIGURE 2.23 The Shapes of Carbon-Containing Molecules. (a) Octane is one of the primary ingredients in gasoline. It is a linear molecule. **(b)** Glucose is a sugar that can form the ring-like structure illustrated here.

CHEMICAL EVOLUTION HYPOTHESIS

1. Simple molecules were present in the atmosphere of ancient Earth, including carbon monoxide (CO), carbon dioxide (CO_2), hydrogen (H_2), ammonia (NH_3), water (H_2O), and nitrogen (N_2).

2. The energy in sunlight drove reactions among the simple molecules, resulting in compounds such as formaldehyde (H_2CO) and hydrogen cyanide (HCN).

3. When heated, compounds containing single carbon atoms reacted to form more complex molecules containing carbon-carbon bonds, including acetaldehyde, glycine, and ribose (a sugar).

FIGURE 2.24 The Start of Chemical Evolution—An Overview. Chemical evolution is a process whereby simple molecules containing C, H, O, and N react to form organic compounds with higher potential energy in the form of carbon-carbon bonds. The process is triggered by an energy source such as sunlight or the heat released in a volcanic eruption.

that it participates in—is dictated by groups of H, N, or O atoms that are bonded to one of the carbon atoms in a specific way. These critically important H-, N-, and O-containing groups found in organic compounds are called **functional groups**. The composition and properties of six prominent functional groups recognized by organic chemists are summarized in **Table 2.3**. To understand the role that organic compounds play in organisms, it is important to analyze how these functional groups behave.

- In solution, the amino and carboxyl functional groups tend to attract or drop a proton, respectively. Amino groups function as bases, while carboxyl groups act as acids. During chemical evolution and in organisms today, the most important types of amino- and carboxyl-containing molecules are the amino acids, which Chapter 3 analyzes in detail. Amino acids contain both an amino group and a carboxyl group. These molecules can be linked together by covalent bonds that form between amino- and carboxyl-containing groups on different amino acids.

- The carbonyl group is found on aldehyde and ketone molecules such as formaldehyde, acetaldehyde, and acetone. This functional group is the site of reactions that link these molecules into larger, more complex compounds, like the sugar ribose illustrated in Figure 2.24.

- Hydroxyl groups are important because they act as weak acids. In many cases, the protons involved in acid–base reactions that occur in cells come from hydroxyl groups on organic compounds. Because hydroxyl groups are polar, molecules containing a number of hydroxyl groups will form hydrogen bonds and be highly soluble in water.

- Phosphate groups carry two negative charges. When phosphate groups are transferred from one organic compound to another, the change in charge often dramatically affects the recipient molecule.

- Sulfhydryl groups consist of a sulfur atom bonded to a hydrogen atom. They are important because sulfhydryl groups on different molecules can link together via disulfide (S–S) bonds.

To summarize, functional groups make things happen. The number and type of functional groups attached to a framework of carbon atoms imply a great deal about how that molecule is going to behave. When you encounter an organic compound that is new to you, it's important to do two things: Examine its overall size and shape, and locate any functional groups. Understanding these two features will help you understand the molecule's role in chemical evolution and today's cells.

SUMMARY TABLE **2.3** **Six Functional Groups Commonly Attached to Carbon Atoms**

Functional Group	*Formula	Family of Molecules	Example	Properties of Functional Group
Amino	$R-N\begin{smallmatrix}H\\H\end{smallmatrix}$	Amines	Glycine (an amino acid)	Acts as a base—tends to attract a proton to form: $R-{}^+N\begin{smallmatrix}H\\H\\H\end{smallmatrix}$
Carbonyl	$R-C\begin{smallmatrix}O\\H\end{smallmatrix}$ (Aldehydes) $R-\underset{O}{\overset{}{C}}-R$	Aldehydes / Ketones	Acetaldehyde / Acetone	Aldehydes, especially, react with compounds of form HR_2 to produce larger molecules with form: R group from aldehyde → $R_1-\underset{R_2}{\overset{OH}{C}}-H$
Carboxyl	$R-C\begin{smallmatrix}O\\OH\end{smallmatrix}$	Carboxylic acids	Acetic acid	Acts as an acid—tends to lose a proton to form: $R-C\begin{smallmatrix}O\\O^-\end{smallmatrix}$
Hydroxyl	$R-OH$	Alcohols	Ethanol	Highly polar, so makes compounds more soluble through hydrogen bonding with water; may also act as a weak acid and drop a proton
Phosphate	$R-O-\underset{O^-}{\overset{O}{\underset{\|}{\overset{\|}{P}}}}-O^-$	Organic phosphates	3–Phosphoglyceric acid	When several phosphate groups are linked together, breaking O–P bonds between them releases large amounts of energy
Sulfhydryl	$R-SH$	Thiols	Cysteine	When present in proteins, can form disulfide (S–S) bonds that contribute to protein structure

● **EXERCISE** Predict whether each functional group is polar or nonpolar, based on the electronegativities of the atoms involved.

Once carbon-containing molecules with functional groups had appeared early in Earth history, what happened next? For chemical evolution to continue, two things had to happen. First, reactions among relatively small and simple organic compounds had to produce the building blocks of the large molecules found in living cells. Second, these building blocks had to link together to form proteins, nucleic acids, and carbohydrates—the large, complex compounds found in organisms.

Our task in the next three chapters is to analyze how these events occurred and how proteins, nucleic acids, and carbohydrates function in organisms today.

As far as we know, the jump from nonlife to life has occurred only once in the history of the universe. According to the theory of chemical evolution, it happened as molecules featured in the next three chapters began to accumulate in the waters of the ancient Earth.

Chapter Review

SUMMARY OF KEY CONCEPTS

The theory of chemical evolution maintains that early in Earth's history, complex molecules formed from simple precursor compounds in the ancient atmosphere and ocean, as chemical reactions converted the energy in sunlight and other sources into chemical energy. Chemical energy is a form of potential energy and can be stored in bonds between atoms.

● **Molecules form when atoms bond to each other. Chemical bonds are based on electron sharing. The degree of electron sharing varies from nonpolar covalent bonds, to polar covalent bonds, to ionic bonds.**

When atoms participate in chemical bonds to form molecules, the shared electrons give the atoms full valence shells and thus contribute to the atoms' stability. The electrons may be shared equally or unequally, depending on the relative electronegativities of the two atoms involved. Nonpolar covalent bonds result from equal sharing; polar covalent bonds are due to unequal sharing. Ionic bonds form when an electron is completely transferred from one atom to another.

You should be able to draw the electron-sharing continuum and place molecular oxygen (O_2), carbon dioxide (CO_2), and calcium chloride ($CaCl_2$) on it. ●

● **Of all small molecules, water is the most important for life. Water is highly polar and readily forms hydrogen bonds. Hydrogen bonding makes water an extremely efficient solvent with a high capacity for absorbing energy.**

Water is the most effective solvent known because it is polar, meaning that it has partial positive and negative charges. Water is polar because it is bent and has two polar covalent bonds. Polar molecules and charged substances, including ions, interact with water and stay in solution via hydrogen bonding and electrostatic attraction. Water's ability to participate in hydrogen bonding also gives it an extraordinarily high capacity to absorb heat and cohere to molecules.

You should be able to draw how water interacts with ammonia in solution. ●

(MB) **Web Animation** at www.masteringbio.com
The Properties of Water

● **Chemical reactions tend to be spontaneous if they lead to lower potential energy and higher entropy (more disorder). An input of energy is required for nonspontaneous reactions to occur.**

The reactions involved in chemical evolution resulted in products that had higher potential energy and lower entropy than the reactants had. As a result, these reactions were nonspontaneous. They could occur only because there was an input of kinetic energy in the form of sunlight and heat.

You should be able to explain why cells cannot stay alive without constant inputs of energy. ●

● **Most of the important compounds in organisms contain carbon. Early in Earth history, the energy in sunlight triggered nonspontaneous reactions that led to the formation of key carbon-containing molecules.**

Organic molecules are critical to life because they have complex shapes provided by a framework of carbon atoms, along with complex chemical behavior due to the presence of functional groups. The first step in chemical evolution was the formation of the small organic compounds called formaldehyde and hydrogen cyanide from molecules such as ammonia (NH_3), methane (CH_4), molecular hydrogen (H_2), and carbon dioxide (CO_2.) These reactions occur readily when a source of intense energy, such as the radiation in sunlight, is present.

You should be able to explain why molecules with carbon-carbon bonds have more potential energy and lower entropy than carbon dioxide. ●

QUESTIONS

● **Test Your Knowledge**

1. Which of the following occurs when a covalent bond forms?
   a. The potential energy of electrons drops.
   b. Electrons in valence shells are shared between nuclei.
   c. Ions of opposite charge interact.
   d. Polar molecules interact.

2. If a reaction is exothermic, then which of the following statements is true?
   a. The products have lower potential energy than the reactants.
   b. Energy must be added for the reaction to proceed.
   c. The products have lower entropy (are more ordered) than the reactants.
   d. It occurs extremely quickly.

3. If a reaction is exergonic, then which of the following statements is true?
   a. The products have lower free energy than the reactants.
   b. Energy must be added for the reaction to proceed.
   c. The products have lower entropy (are more ordered) than the reactants.
   d. It occurs extremely quickly.

4. What is thermal energy?
   a. a form of potential energy
   b. the temperature increase that occurs when any form of energy is added to a system
   c. mechanical energy
   d. the kinetic energy of molecular motion, measured as heat

5. What determines whether a chemical reaction is spontaneous?
   a. if it increases the disorder, or entropy, of the substances involved
   b. if it decreases the potential energy of the substances involved
   c. the temperature only—reactions are spontaneous at high temperatures and nonspontaneous at low temperatures
   d. the combined effect of changes in potential energy and entropy

6. Which of the following is *not* an example of an energy transformation?
   a. A shoe drops, converting potential energy to kinetic energy.
   b. A chemical reaction converts the energy in sunlight into the chemical energy in formaldehyde.
   c. The electrical energy flowing through a light bulb's filament is converted into light and heat.
   d. Sunlight strikes a prism and separates into distinct wavelengths.

Test Your Knowledge answers: 1. b; 2. a; 3. a; 4. d; 5. d; 6. d

Test Your Understanding

Answers are available at www.masteringbio.com

1. Consider the reaction between carbon dioxide and water, which forms carbonic acid:

$$CO_2(g) + H_2O(l) \rightleftharpoons H_2CO(aq)$$

In aqueous solution, carbonic acid immediately dissociates to form a proton and the bicarbonate ion, as follows:

$$H_2CO_3(aq) \rightleftharpoons H^+(aq) + HCO_3^-(aq)$$

Does this second reaction raise or lower the pH of the solution? Does the bicarbonate ion act as an acid or a base? If an underwater volcano bubbled additional CO_2 into the ocean, would this sequence of reactions be driven to the left or the right? How would this affect the pH of the ocean?

2. When chemistry texts introduce the concept of electron shells, they emphasize that shells represent distinct potential energy levels. In introducing electron shells, this chapter also emphasized that they represent distinct distances from the positive charges in the nucleus. Are these two points of view in conflict? Why or why not?

3. Draw a ball-and-stick model of the water molecule, and explain why this molecule is bent. Indicate the location of the partial electric charges on it. Why do these partial charges exist?

4. Hydrogen bonds form because the opposite, partial electric charges on polar molecules attract. Covalent bonds form as a result of the electrical attraction between electrons and protons. Covalent bonds are much stronger than hydrogen bonds. Explain why, in terms of the electrical attractions involved.

5. Explain why extensive hydrogen bonding gives water an extraordinarily high specific heat.

6. Explain the relationship between the carbon atoms in an organic molecule and the functional groups on the same molecule.

Applying Concepts to New Situations

Answers are available at www.masteringbio.com

1. Why isn't CO_2 bent and polar, like H_2O? Why is H_2O much more likely to participate in chemical reactions than CO_2?

2. Oxygen is extremely electronegative, meaning that its nucleus pulls in electrons shared in covalent bonds. Because these electrons are close to the oxygen nucleus, they have lower potential energy. Explain the changes in electron position that are illustrated in Figure 2.19a based on oxygen's electronegativity.

3. When nuclear reactions take place, some of the mass in the atoms involved is converted to energy. The energy in sunlight is created during nuclear fusion reactions on the Sun. Explain what astronomers mean when they say that the Sun is burning down and that it will eventually burn out.

4. Why do coastal regions tend to have climates with moderate temperatures and lower annual variation in temperature than do inland areas at the same latitude?

www.masteringbio.com is also your resource for • Answers to text, table, and figure caption questions and exercises • Answers to *Check Your Understanding* boxes • Online study guides and quizzes • Additional study tools including the *E-Book for Biological Science* 3rd ed., textbook art, animations, and videos.

Protein Structure and Function

3

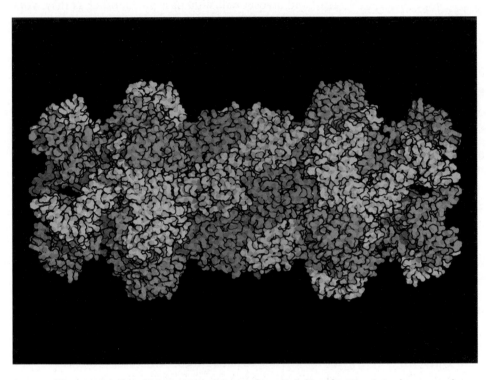

A space-filling model of an AAA+ protease, a type of protein whose function is to chop up and dispose of old or damaged proteins in cells.

KEY CONCEPTS

● Proteins are made of amino acids. Amino acids vary in structure and function because their side chains vary in composition.

● Proteins vary widely in structure. The structure of a protein can be analyzed at four levels that form a hierarchy—the amino acid sequence, substructures called α-helices and β-pleated sheets, interactions between amino acids that dictate a protein's overall shape, and combinations of individual proteins that make up larger, multiunit molecules.

● Proteins vary widely in function. In cells, most proteins are enzymes that function as catalysts. Chemical reactions occur much faster when they are catalyzed by enzymes. During enzyme catalysis, the reactants bind to an enzyme's active site in a way that allows the reaction to proceed efficiently.

Chapter 2 introduced the hypothesis that chemical reactions in the atmosphere and ocean of ancient Earth led to the formation of the first complex carbon-containing compounds. This idea, called chemical evolution, was first proposed in 1923 by Alexander I. Oparin. The hypothesis was published again—independently and six years later—by J. B. S. Haldane. Today, the Oparin–Haldane proposal can best be understood as a formal scientific theory. As Chapter 1 noted, scientific theories typically have two components: a statement about a pattern that exists in the natural world and a proposed mechanism or process that explains the pattern. In the case of chemical evolution, the pattern is that increasingly complex carbon-containing molecules formed in the atmosphere and ocean of ancient Earth. The process responsible for this pattern

was the conversion of energy, from sunlight and other sources, into chemical energy in the bonds of large, complex molecules.

Scientific theories are continuously refined as new information comes to light, and many of Oparin and Haldane's original ideas about how chemical evolution occurred have been extensively revised. In its current form, the theory can be broken into four steps, each requiring an input of energy:

1. Chemical evolution began with the production of small organic compounds such as formaldehyde (H_2CO) and hydrogen cyanide (HCN), from reactants such as H_2, CO_2, and CH_4, and NH_3.

2. Formaldehyde, hydrogen cyanide, and other simple organic compounds reacted to form the mid-sized molecules called

● Key Concept   Important Information   ● Practice It

amino acids, sugars, and nitrogenous bases. These building-block molecules accumulated in the shallow waters of the ancient ocean, forming a complex solution called the **prebiotic soup**.

3. Mid-sized, building-block molecules linked to form the types of large molecules found in cells today, including proteins, nucleic acids, and complex carbohydrates. These large molecules are each made up of distinctive chemical subunits that join together: Proteins are composed of amino acids; nucleic acids are composed of nucleotides; and complex carbohydrates are composed of sugars.

4. Life became possible when one of these large, complex molecules acquired the ability to make a copy of itself. This self-replicating molecule began to multiply by means of chemical reactions that it controlled. At that point, life had begun. Chemical evolution gave way to biological evolution.

Analyzing how the final three steps in chemical evolution occurred is the subject of this chapter and the next three chapters. Each of these chapters focuses on one of the four primary classes of biological molecules found in cells living today: proteins, nucleic acids, carbohydrates, and lipids. What do the subunits of these large molecules look like, and how could they have been created by chemical evolution? How are the subunits linked to form a larger molecule—a protein, nucleic acid, a complex carbohydrate, or lipid? What do these large molecules look like, and what do they do in living cells? Finally, which type of molecule was responsible for the origin of life? This last question is particularly intriguing, because researchers around the world are racing to synthesize a self-replicating molecule—to produce life in a test tube. Which types of molecule are they working with, and why? Let's begin our analysis with the workhorse molecules called proteins.

3.1 What Do Proteins Do?

As a group, proteins serve more types of functions in cells than any other type of molecule does. Consider the red blood cells that are moving through your arteries, capillaries, and veins right now. Each of these cells contains about 300 million copies of the protein called hemoglobin. Hemoglobin carries oxygen from your lungs to cells throughout the body. But every red blood cell also has thousands of copies of a protein called carbonic anhydrase, which is important for moving carbon dioxide from cells back to the lungs, where it can be breathed out. Other proteins protect the body by attacking disease-causing bacteria and viruses; still others make movement possible or form structures that give cells their shape.

Proteins are crucial to most tasks required for cells to exist. The diverse functions of proteins include the following:

- *Defense.*   Proteins called antibodies and complement proteins attack and destroy viruses and bacteria that cause disease.

- *Movement.*   Motor proteins and contractile proteins are responsible for moving the cell itself, or for moving large molecules and other types of cargo inside the cell. As you turn this page, for example, specialized proteins called actin and myosin will slide past one another as they work to flex or extend muscle cells in your fingers and arm.

- *Catalysis.*   Many proteins are specialized to **catalyze**, or speed up, chemical reactions, most of which would not otherwise be able to proceed. A protein that functions as a catalyst is called an **enzyme**. The carbonic anhydrase molecules in red blood cells are an example, and so is the protein called salivary amylase that is present in your mouth. Salivary amylase helps begin the digestion of starch and other complex carbohydrates into simple sugars.

- *Signaling.*   Proteins called peptide hormones bind to receptor proteins on particular cells. In response, the activity of the cell with the receptor protein changes. In this way, proteins are involved in carrying and receiving signals from cell to cell inside the body.

- *Structure.*   Structural proteins give cells mechanical support. They also create structures such as fingernails and hair. The membranes of red blood cells have structural proteins that connect to structural proteins inside the cell itself. Working together, these structural proteins keep the cells flexible and in their normal disc-like shape.

SUMMARY TABLE **3.1 Protein Functions**

Protein Type	Role in Cell or Organism
Antibodies and complement proteins	Defense—destruction of disease-causing viruses and bacteria
Contractile proteins and motor proteins	Movement
Enzymes	Catalyze chemical reactions
Hormones	Act as signals that help coordinate the activities of many cells
Receptor proteins	Receive chemical signals from outside cell and initiate response
Structural proteins	Provide support for cells and tissues; form structures such as hair, feathers, cocoons, and spider webs
Transport proteins	Move substances across cell membrane and throughout body

- *Transport.* Proteins are responsible for allowing particular molecules to enter or exit cells, and for carrying specific compounds throughout the body. Hemoglobin is a particularly well-studied transport protein, but virtually every cell is studded with membrane proteins that control the passage of specific molecules and ions.

These aspects of protein function are summarized in **Table 3.1**; each will be explored in much greater detail later in the text. In many if not all cases, the function of the protein involved is closely correlated with its structure. Before analyzing how proteins are put together, let's briefly consider how the building blocks of proteins could have been synthesized before life began.

3.2 Early Origin-of-Life Experiments

In 1953 a graduate student named Stanley Miller performed a breakthrough experiment in the study of chemical evolution. Miller wanted to answer a simple question: Can complex organic compounds be synthesized from the simple molecules present in Earth's early atmosphere and ocean? In other words, is it possible to recreate the first steps in chemical evolution by simulating ancient Earth conditions in the laboratory?

Miller's experimental setup (**Figure 3.1**) was designed to produce a microcosm of ancient Earth. The large glass flask represented the atmosphere and contained the gases methane (CH_4), ammonia (NH_3), and hydrogen (H_2), all of which have high free energy. This large flask was connected to a smaller flask by glass tubing. The small flask held a tiny ocean—200 milliliters (mL) of liquid water. Miller boiled this water constantly so that water vapor was added to the mix of gases in the large flask. As the vapor cooled and condensed, it flowed back into the smaller flask, where it boiled again. In this way, water vapor circulated continuously through the system. This was important: If the molecules in the simulated atmosphere reacted with one another, the "rain" would carry them into the simulated ocean, forming a simulated version of the prebiotic soup.

Had Miller stopped there, however, little or nothing would have happened. Even at the boiling point of water (100°C), the molecules involved in the experiment are stable. They do not undergo spontaneous chemical reactions, even at high temperatures.

Something did start to happen in the apparatus, however, when Miller sent electrical discharges across the electrodes he'd inserted into the atmosphere. These miniature lightning bolts added a crucial element to the reaction mix—pulses of intense electrical energy. After a day of continuous boiling and sparking, the solution in the boiling flask began to turn pink. After a week, it was deep red and cloudy. When Miller analyzed the molecules dissolved in the solution, he found that several complex carbon-containing compounds were present, including several with carbon-carbon bonds. The experiment, driven by

Experiment

Question: Can simple molecules and kinetic energy lead to chemical evolution?

Hypothesis: If kinetic energy is added to a mix of simple molecules with high free energy, reactions will occur that produce more complex molecules, perhaps including some with carbon-carbon bonds.

Null hypothesis: Chemical evolution will not occur, even with an input of energy.

Experimental setup:

Prediction: Complex organic compounds will be found in the liquid water.

Prediction of null hypothesis: Only the starting molecules will be found in the liquid water.

Results:

Samples taken from the liquid water contain formaldehyde, hydrogen cyanide, and several complex compounds with carbon-carbon bonds, including amino acids

Conclusion: Chemical evolution occurs readily if simple molecules with high free energy are exposed to a source of kinetic energy.

FIGURE 3.1 Miller's Spark-Discharge Experiment. The arrows in the "Experimental setup" diagram indicate the flow of water vapor or liquid. The large glass flask can contain any mixture of gases desired. The condenser consists of a jacket with cold water flowing through it.

● **EXERCISE** Label the parts of the apparatus that mimic the ocean, the atmosphere, rain, and lightning.

heat energy and the energy in the electrical discharges, had recreated the start of chemical evolution.

To find out exactly which products resulted from the initial reactions in the simulated atmosphere, Miller drew samples from the apparatus at intervals. In these samples he found large quantities of hydrogen cyanide (HCN) and formaldehyde (H_2CO). The data were exciting because HCN and (H_2CO) are required for reactions that lead to the synthesis of more complex organic molecules. Indeed, some of these more complex compounds were actually present in the miniature ocean. The sparks and heating had led to the synthesis of compounds that are fundamental to life: amino acids.

3.3   Amino Acids and Polymerization

Based on the presence of amino acids, Miller claimed that his experiment simulated the second stage in chemical evolution—the formation of a prebiotic soup. Although the assumptions behind his experiments, and hence the results, eventually came under fire, follow-up studies have confirmed that the second step in chemical evolution occurred early in Earth's history. For example, organic compounds like methane (CH_4), formaldehyde, and acetaldehyde form when water vapor, carbon monoxide, carbon dioxide, hydrogen (H_2), and other volcanic gases are put into a glass flask and exposed to the types of high-energy radiation found in sunlight. These are realistic simulations of conditions on early Earth (see Chapter 2). Based on results like these, there is now a strong consensus that amino acids and other components of the prebiotic soup are readily produced under conditions that accurately simulate the atmosphere and oceans of ancient Earth.

Now let's look at the molecules themselves. What are amino acids, and how are they linked to form proteins?

The Structure of Amino Acids

The bacterial cells that live on your skin contain several thousand different proteins; as a group, the cells in your body produce tens of thousands of distinct proteins. But most of these proteins are composed of just 21 different building blocks, called **amino acids**. All 21 amino acids have a common backbone structure.

To understand the structure of an amino acid, recall that carbon atoms have four unpaired valence electrons and form four covalent bonds. Every amino acid has a carbon that makes three similar bonds. The first bond attaches this carbon to NH_2—the amino functional group (**Figure 3.2a**). The second bond links this carbon to COOH—the carboxyl functional group. You might recall from Chapter 2 that the carboxyl group is acidic because its two oxygen atoms are highly electronegative. They pull electrons away from the hydrogen atom, which means that it is relatively easy for this group to lose a proton. The combination of the amino and carboxyl groups on these

(a) Non-ionized form of amino acid

(b) Ionized form of amino acid

FIGURE 3.2   Amino Acid Structure. All amino acids have the same general structure: a central carbon, shown in red, bonded to an amino functional group, a carboxyl functional group, a hydrogen atom, and a side chain, or R-group.

molecules inspired the name amino acid. A third bond links the highlighted carbon to a hydrogen atom. In all amino acids, then, a carbon atom is bonded to an amino group, a carboxyl group, and hydrogen. As we'll see in a moment, the nature of the fourth attachment is what makes each amino acid unique.

The presence of an amino group and a carboxyl group in amino acids is important. **Figure 3.2b** shows what happens to these functional groups in solution. In water at pH 7, the concentration of protons causes the amino group to act as a base. It attracts a proton to form NH_3^+. The carboxyl group, in contrast, loses a proton to form COO^-. The charges on these functional groups (1) help amino acids stay in solution where they can interact with each other and with other solutes, and (2) add to their chemical reactivity.

The Nature of Side Chains   In Figure 3.2, the highlighted carbon atom in the amino acid forms a fourth bond with an atom or a group of atoms, abbreviated as "R." Chemists use this symbol to indicate additional atoms called a side chain. In every amino acid, a carbon atom is linked to a hydrogen atom, an amino group, a carboxyl group, and an R-group. The 21 amino acids found in organisms are different because their R-groups are different. These R-groups vary from a single hydrogen atom to large structures containing carbon atoms linked into rings. Several of the side chains found in amino acids contain functional groups highlighted in Chapter 2, Table 2.3. Specifically, some amino acids have R-groups with carboxyl, sulfhydryl, hydroxyl, or amino functional groups.

Figure 3.3 shows the most common 20 amino acids found in cells and sorts them based on whether their side chain is nonpolar, polar, or electrically charged. (The amino acid that is not shown—selenocysteine—is similar in structure to cysteine but has a selenium atom in place of sulfur.) This property of side chains is important because amino acids with nonpolar side chains do not have charged or electronegative

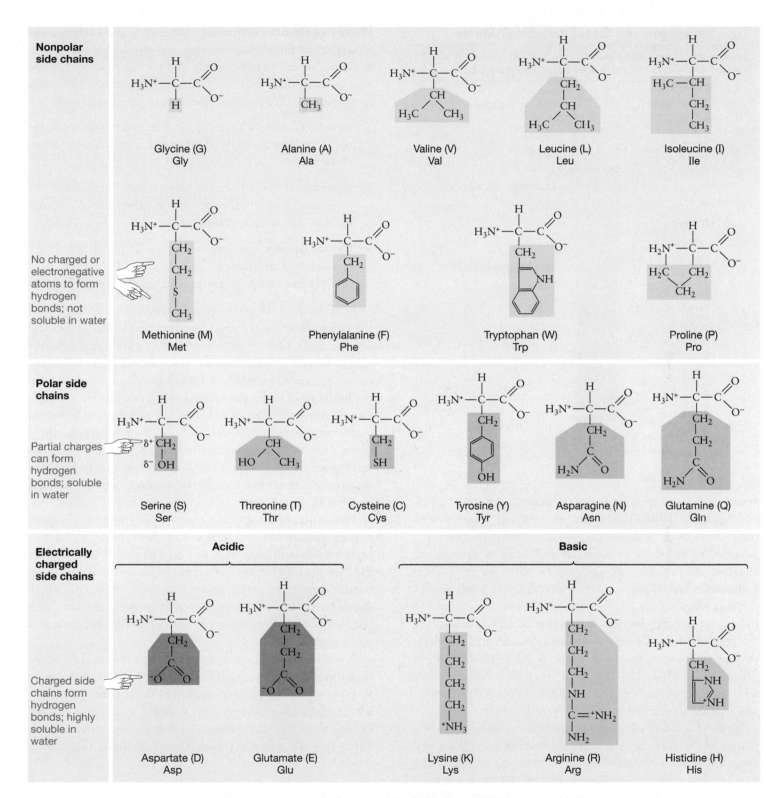

FIGURE 3.3 The 20 Major Amino Acids Found in Organisms. At the pH (about 7.0) found in cells, the 20 major amino acids found in organisms have the structural formulas shown here. The side chains are highlighted, and standard single-letter and three-letter abbreviations for each amino acid are given. For clarity, the carbon atoms in the ring structures of phenylalanine, tyrosine, tryptophan, and histidine are not shown; each bend in a ring is the site of a carbon atom. The hydrogen atoms in these structures are also not shown. A double line inside a ring indicates a double bond.

● **EXERCISE** Explain why the green R-groups are nonpolar and why the pink R-groups are polar, based on the relative electronegativities of O, N, C, and H. Note that sulfur (S) has an electronegativity almost equal that of carbon, and slightly higher than that of hydrogen.

TABLE 3.2 How Amino Acids Interact with Water

20 amino acids are ranked according to how likely they are to interact with water. Color codes are based on Figure 3.3.

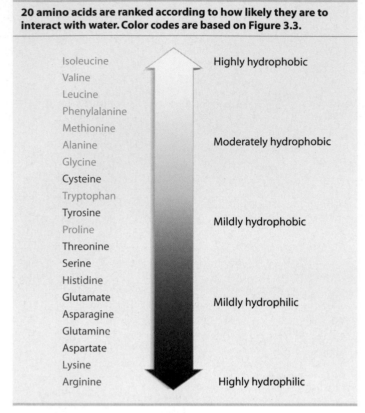

Isoleucine — Highly hydrophobic
Valine
Leucine
Phenylalanine
Methionine
Alanine — Moderately hydrophobic
Glycine
Cysteine
Tryptophan
Tyrosine
Proline — Mildly hydrophobic
Threonine
Serine
Histidine
Glutamate
Asparagine — Mildly hydrophilic
Glutamine
Aspartate
Lysine
Arginine — Highly hydrophilic

atoms capable of forming hydrogen bonds with water. These R-groups are **hydrophobic**—water does not interact with them. As a result, hydrophobic side chains tend to coalesce in aqueous solution. In contrast, amino acids with polar or charged side chains interact readily with water and are **hydrophilic**. Hydrophilic amino acids dissolve in water easily.

Table 3.2 ranks the 20 most common amino acids according to how readily they interact with water. In almost every case, the polarity of the R-group found on an amino acid correlates with its ability to interact with water. ● You should be able to predict which amino acids are most and least soluble in water, and—using the structures in Figure 3.3—explain why cysteine and tyrosine interact relatively poorly with water.

● In addition to affecting the solubility of amino acids, the nature of the side chain influences their chemical reactivity. Some amino acids contain side chains consisting entirely of carbon and hydrogen atoms. These R-groups rarely participate in chemical reactions. As a result, the chemical behavior of these amino acids depends primarily on their size and shape rather than reactivity. In contrast, amino acids that have hydroxyl, amino, carboxyl, or sulfhydryl functional groups in their side chains are more reactive. For example, amino acids with sulfur atoms (S) in their side chains can form bonds that help link different parts of large proteins. The key point here is that variation in the structure of the R-groups in amino acids explains variation in their properties and functions.

What Are Optical Isomers?   When you are analyzing the structure and function of an organic compound like an amino acid, it is important to recognize that molecules with the same molecular formula may have different structures. Molecules like these are called **isomers**. There are three types of isomers:

1. **Structural isomers** have the same atoms but differ in the order in which covalently bonded atoms are attached (**Figure 3.4a**).

2. **Geometric isomers** have the same atoms but differ in the arrangement of atoms or groups on either side of a double bond or ring structure (**Figure 3.4b**).

3. **Optical isomers** have the same atoms but differ in the arrangement of atoms or groups around a carbon atom that has four different groups attached (**Figure 3.4c**).

Most amino acids have optical isomers. Figure 3.4c shows the arrangements in the two optical isomers of the amino acid alanine. Note that the two forms of the molecule are mirror images of one another, just as your left and right hands are mirror images of each other. Like your left and right hands, the left-handed and right-handed forms of alanine cannot be exactly superimposed. They also do not have a plane of symmetry—meaning, you cannot bisect them anywhere with a plane and create identical halves. Carbon atoms with this feature exist in every amino acid except glycine. In fact, any carbon atom that has four different atoms or groups attached to it has an optical isomer.

The existence of optical isomers is an important issue in biology. Because the structures of each optical isomer of a molecule are different, they have different functions. In cells, only the "left-handed" forms of amino acids exist. If the right-handed form of an amino acid is introduced into cells experimentally, it does not function normally. As is the case with most other molecules, an amino acid's function is determined by its structure.

These observations are a challenge to the chemical evolution hypothesis, because no plausible mechanism has been proposed to explain how the process resulted in the production of only left-handed optical isomers. Was it simply a matter of chance? Or was there something unusual about the chemistry of early Earth that we still don't understand? To date, the issue is unresolved.

How Do Amino Acids Link to Form Proteins?

Amino acids link to each other to form proteins. Similarly, the molecular building blocks called nucleotides attach to one another to form nucleic acids, and simple sugars connect to form complex carbohydrates. In general, a molecular subunit such as an amino acid, a nucleotide, or a sugar is called a **monomer** ("one-part"). When monomers bond together, the resulting structure is called a **polymer** ("many-parts"). The process of

(a) Structural isomers differ in the order which their atoms are attached.

Ethanol (C₂H₆O)   Dimethyl ether (C₂H₆O)

(b) Geometric isomers differ in the arrangement of atoms around a double bond.

trans-2-butene (C₄H₈)   *cis*-2-butene (C₄H₈)

(c) Optical isomers are mirror images of one another— they cannot be exactly superimposed.

Hands are mirror images, just as optical isomers are.

Alanine has two forms that cannot be superimposed

Your left and right hands cannot be superimposed (that is, with both thumbs facing the same direction and both palms facing the same direction)

FIGURE 3.4 Structural, Geometric, and Optical Isomers. (a) Ethanol is the active ingredient in alcoholic beverages; dimethyl ether, a gas at room temperature, is used in refrigeration. **(b)** The molecules *trans*-2-butene and *cis*-2-butene are used in the production of gasoline, synthetic rubber, and solvents. **(c)** All amino acids except glycine have optical isomers.

linking monomers is called **polymerization** (**Figure 3.5**). Thus, amino acids polymerize to form proteins. Biologists also use the word **macromolecule** to denote a very large molecule that is made up of smaller molecules joined together. A **protein** is a macromolecule—a polymer—that consists of linked amino acid monomers.

The theory of chemical evolution states that monomers in the prebiotic soup polymerized to form proteins and other types of macromolecules found in organisms. This is a difficult step, because monomers such as amino acids do not spontaneously self-assemble into macromolecules such as proteins. According to the second law of thermodynamics reviewed in Chapter 2, this is not

surprising. Complex and highly organized molecules are not expected to form spontaneously from simpler constituents, because polymerization organizes the molecules involved into a more complex, ordered structure. Stated another way, polymerization decreases the disorder, or entropy, of the molecules involved. In addition, polymers are energetically much less stable than their component monomers. Because the ΔH term of the Gibbs free-energy equation is positive and the $T\Delta S$ term is negative, ΔG is positive at all temperatures. Polymerization reactions are endergonic and nonspontaneous. Monomers must absorb energy in order to link together. How could this have happened during chemical evolution?

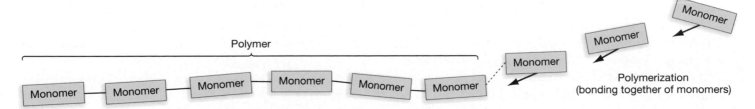

FIGURE 3.5 Monomers Are the Building Blocks of Polymers. Monomers polymerize to form polymers.

(a) Condensation reaction: monomer in, water out

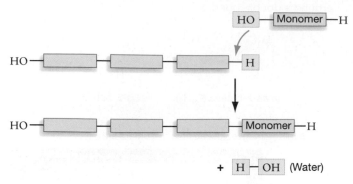

(b) Hydrolysis: water in, monomer out

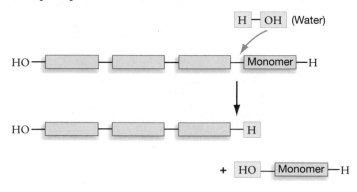

FIGURE 3.6 Polymers Can Be Extended or Broken Apart. (a) In a condensation reaction, a monomer is added to a polymer to make a longer polymer. The new bond that forms results in the formation of a water molecule. **(b)** In hydrolysis, a water molecule reacts with the bond linking the monomers. A monomer is broken off the chain, resulting in a shorter polymer.

Could Polymerization Occur in the Energy-Rich Environment of Early Earth? Researchers who added heat or electrical discharges to solutions of either amino acids or nucleic acids found that polymerization reactions proceed slowly, if at all. Monomers polymerize through **condensation reactions**, also known as **dehydration reactions**. These reactions are aptly named because the newly formed bond results in the loss of a water molecule (**Figure 3.6a**). The reverse reaction, called **hydrolysis**, breaks polymers apart by adding a water molecule (**Figure 3.6b**). The water molecule reacts with the bond linking the monomers, separating one monomer from the polymer chain.

In a solution, such as the prebiotic soup, condensation and hydrolysis represent the forward and reverse reactions of a chemical equilibrium. Hydrolysis dominates because it increases entropy and because it is energetically favorable—it lowers the potential energy of the electrons involved. This is why polymerization reactions are expected to proceed slowly or not at all.

According to recent experiments, the key to overcoming hydrolysis during chemical evolution was, quite literally, as common as mud. Researchers have been able to create stable polymers by mixing monomers with a source of chemical energy and tiny mineral particles—the size found in clay or mud. These experiments are based on the hypothesis that growing macromolecules would be protected from hydrolysis if they clung, or adsorbed, to the mineral surfaces.

More specifically, the experiments were designed to simulate events that could have occurred in the prebiotic soup. In one experiment, researchers put amino acids in a solution with a source of chemical energy and tiny mineral particles and allowed them to react. After a day, the researchers separated the mineral particles from the solution. They then put the particles into a fresh solution containing amino acids and a source of chemical energy. After repeating this procedure for several days, they analyzed the mineral particles and found polymers up to 55 amino acids long. These results support the hypothesis that adsorption to mineral particles protects polymers from hydrolysis. Because the experimental procedure was designed to mimic coastal environments where beaches are repeatedly washed with fresh waves or tidal flows, the results make it reasonable to claim that at least some muddy tide pools and beaches became covered with small proteins early in Earth's history.

The Peptide Bond Exactly how do amino acids polymerize? As **Figure 3.7** shows, amino acids polymerize when a bond forms between the carboxyl group of one amino acid and the amino group of another. The C–N bond that results from this condensation reaction is called a **peptide bond**. Peptide bonds are particularly stable because electrons are partially shared between the neighboring carbonyl functional group and the

FIGURE 3.7 Peptide Bond Formation. When the carboxyl group of one amino acid reacts with the amino group of a second amino acid, a peptide bond forms.

● **QUESTION** Is a peptide bond a hydrogen bond, a nonpolar covalent bond, a polar covalent bond, or an ionic bond?

(a) Polypeptide chain

(b) Numbering system

FIGURE 3.8 Amino Acids Polymerize to Form Polypeptides. (a) Amino acids can be linked into long chains, called polypeptides, by peptide bonds. **(b)** By convention, the sequence of amino acids in a polypeptide chain is numbered from the N-terminus to the C-terminus.

peptide bond. The degree of electron sharing is great enough that peptide bonds actually have some of the characteristics of a double bond. For example, the peptide bond is planar.

When amino acids are linked by peptide bonds into a chain, the amino acids are referred to as **residues** and the resulting molecule is called a **polypeptide**. **Figure 3.8a** shows how the chain of peptide bonds in a polypeptide gives the molecule a structural framework, or a "backbone." Three points are important to note about this backbone: (1) The side chains present in each residue extend out from it, (2) it has directionality, and (3) it is flexible. The backbone is directional because there is an amino group ($-NH_3^+$) on one end of every polypeptide chain and a carboxyl group ($-COO^-$) on the other. By convention, biologists always write amino acid sequences in the same

direction. The end of the sequence that has the free amino group is placed on the left and is called the N-terminus, or amino-terminus, and the end with the free carboxyl group appears on the right-hand side of the sequence and is called the C-terminus, or carboxy-terminus. The amino acids in the chain are always numbered starting from the N-terminus (**Figure 3.8b**), because the N-terminus is the start of the chain when proteins are synthesized in cells. Although the peptide bond itself cannot rotate because of its double-bond nature, the single bonds on either side of the peptide bond can rotate. As a result, the structure as a whole is flexible (**Figure 3.9**).

When fewer than 50 amino acids are linked together in this way, the resulting polypeptide is called an **oligopeptide** ("few peptides") or simply a **peptide**. Polypeptides that contain 50 or

FIGURE 3.9 Proteins Are Flexible. Polypeptide chains are flexible because the groups on either side of each peptide bond can rotate about their single bonds.

● **EXERCISE** Draw an arrow to each peptide bond in this molecule. Circle each amino acid.

more amino acids are formally called **proteins**. Proteins may consist of single polypeptides or multiple polypeptides that are bonded to each other.

Proteins are the stuff of life. Let's take a look at how they are put together, and then at what they do.

Check Your Understanding

If you understand that...

- Amino acids are small molecules with a carbon atom bonded to a carboxyl group, an amino group, a hydrogen atom, and a side chain called an R-group.

- Each amino acid has distinctive chemical properties because each has a unique R-group.

- When the carboxyl group of one amino acid reacts with the amino group of another amino acid, a strong covalent bond called a peptide bond forms.

- Polypeptides are polymers made up of peptide-bonded amino acids. Small polypeptides are called oligopeptides, and large polypeptides are called proteins.

You should be able to ...

1) Draw the general form of an amino acid.

2) Draw and label two amino acids linked by a peptide bond.

(MB) Web Animation at www.masteringbio.com
Condensation and Hydrolysis Reactions

3.4 What Do Proteins Look Like?

With respect to their structure, proteins may be the most diverse class of molecules known. This is important, because diversity in structure and in the chemical reactivity of R-groups explains the incredible functional diversity of proteins.

Figure 3.10 illustrates some of the diverse sizes, shapes, and functions observed in proteins. Note that these proteins range from butterfly-shaped to donut-shaped to globular. In the case of the TATA-box binding protein in Figure 3.10a and the protein called porin in Figure 3.10b, the shape of the molecule has an obvious correlation with its function. The TATA-box binding protein has a groove where DNA molecules fit; porin has a hole that forms a pore. The groove in the TATA-box binding protein interacts with specific regions of DNA inside cells, while porin fits in cell membranes and allows certain hydrophilic molecules to pass through. But most of the proteins found in cells function as enzymes and are globular (Figure 3.10c).

The unparalleled diversity of proteins—in size, shape, and other aspects of structure—is important because function follows from structure. Proteins can serve diverse functions in cells because they are diverse in size and shape as well as in the chemical properties of their amino acids.

How can biologists make sense of this diversity of protein size and shape? Initially, the amount of variation seems overwhelming. Fortunately, it is not. No matter how large or complex a protein may be, its underlying structure can be broken down into just four basic levels of organization.

Primary Structure

Every protein has a unique sequence of amino acids. That simple revelation was the culmination of 12 years of study by Frederick Sanger and co-workers during the 1940s and 1950s. Sanger's group worked out the first techniques for determining the amino acid sequence of a protein and published the completed sequence of the hormone insulin, a protein that helps regulate sugar concentrations in the blood of humans and other mammals. When other proteins were analyzed, it rapidly became clear that each protein has a definite and distinct amino acid sequence.

(a) TATA-box binding protein

Butterfly shaped

(b) Porin

Doughnut shaped

(c) Pyrophosphatase

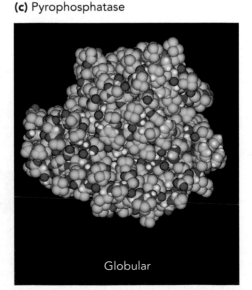

Globular

FIGURE 3.10 In Size and Shape, Proteins Are the Most Diverse Class of Molecules Known.

Biochemists call the unique sequence of amino acids in a protein the **primary structure** of that protein. The sequence of amino acids in Figure 3.8, for example, defines that polypeptide's primary structure. With 21 types of amino acids available and size ranging from two amino acid residues to tens of thousands, the number of primary structures that are possible is practically limitless. There may, in fact, be 21^n different polypeptides of length n. For a polypeptide that is just 10 amino acids long, 21^{10} primary sequences are possible. This is over 16,500 billion.

Recall that the R-groups present on each amino acid affect its solubility and chemical reactivity. Based on this observation, it is reasonable to predict that the R-groups present in a polypeptide will affect the polypeptide's properties and function. In some cases, even a single change in the sequence of amino acids can cause radical changes in the way the macromolecule as a whole behaves. As an example, consider the hemoglobin protein of humans. In some individuals, hemoglobin has a valine instead of a glutamate at the amino acid numbered 6 in a strand of 146 amino acids (**Figure 3.11a**). Valine's side chain is very different from the R-group in glutamate. The change results in a protein that tends to crystallize instead of staying in solution when oxygen concentrations in the blood are low. When hemoglobin crystallizes, the red blood cells that carry the protein adopt a sickled shape (**Figure 3.11b**) and can no longer efficiently pass through the blood vessels called capillaries. In people whose hemoglobin contains this single amino acid change, cells downstream of blocked capillaries become starved for oxygen. A debilitating illness called sickle-cell dis-

ease results. A protein's primary structure is fundamental to its function and to determining higher levels of protein structure: secondary, tertiary, and quaternary.

Secondary Structure

Even though variation in the amino acid sequence of a protein is virtually limitless, it is only the tip of the iceberg in terms of generating structural diversity. The next level of organization in proteins is known as **secondary structure**, which is created in part by hydrogen bonding. Secondary structures are distinctively shaped sections of proteins that are stabilized largely by hydrogen bonding that occurs between the carboxyl oxygen of one amino acid residue and the hydrogen on the amino group of another. The oxygen atom in the carboxyl group has a partial negative charge due to its high electronegativity, while the hydrogen atom in the amino group has a partial positive charge because it is bonded to nitrogen, which has high electronegativity (**Figure 3.12a**).

Note that secondary structure is not created by interactions among side chains, but instead by interactions between atoms that are part of a protein's peptide-bonded backbone. This is a key point, because hydrogen bonding between sections of the backbone is possible only when different parts of the same polypeptide bend in a way that puts carboxyl and amino groups close together. The bending that aligns parts of the backbone and allows these bonds to form occurs in several distinct ways. **Figure 3.12b** shows two of the most important configurations that allow formation of hydrogen bonds: (1) an

(a) Normal amino acid sequence

| Thr | Pro | Glu | Glu |
| 4 | 5 | 6 | 7 |

Normal red blood cells

(b) Single change in amino acid sequence

| Thr | Pro | Val | Glu |
| 4 | 5 | 6 | 7 |

Sickled red blood cells

FIGURE 3.11 Changes in Primary Structure Affect Protein Function. Compare the primary structure of **(a)** normal hemoglobin with that of **(b)** hemoglobin molecules of people with sickle-cell disease. The single amino acid change causes red blood cells to change from their normal disc shape in (a) to a sickled shape in (b) when oxygen concentrations are low. Each red blood cell contains about 300 million hemoglobin molecules.

(a) Hydrogen bonds form between peptide chains.

(b) Secondary structures of proteins result.

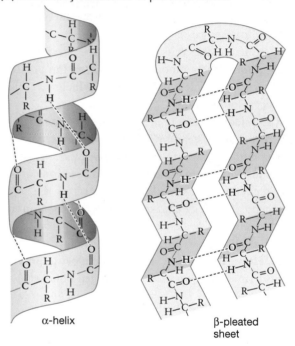

α-helix   β-pleated
sheet

(c) Ribbon diagrams of secondary structure.

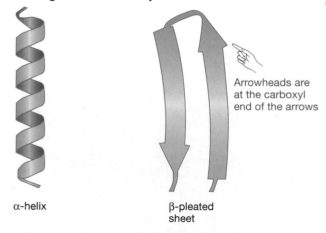

Arrowheads are
at the carboxyl
end of the arrows

α-helix   β-pleated
sheet

FIGURE 3.12 Secondary Structures of Proteins. The peptide-bonded
backbone of a protein can coil or fold in on itself when hydrogen bonds
form between amino groups and carboxyl groups.

● **EXERCISE** β-pleated sheets often involve three or more strands. Add
a third strand to the β-pleated sheets in part (c).

α-helix (alpha helix), in which the polypeptide's backbone is
coiled, and (2) a **β-pleated sheet** (beta-pleated sheet), in which
segments of a peptide chain bend 180° and then fold in the
same plane. When biologists use illustrations called ribbon
diagrams to represent the shape of a protein, α-helices are
shown as coils while β-pleated sheets are shown by groups of
arrows in a plane (**Figure 3.12c**).

In many cases, secondary structure consists exclusively of
α-helices and β-pleated sheets. Which one forms, if either,
depends on the molecule's primary structure—specifically, the
geometry of the amino acids in the sequence. Methionine and
glutamic acid, for example, are much more likely to be
involved in α-helices than in β-pleated sheets. The opposite is
true for valine and isoleucine. Proline, in contrast, is unlikely
to be involved in either type of secondary structure. This is a
key point: Which secondary structures form depends on the
protein's primary structure.

Although each of the hydrogen bonds in an α-helix or a
β-pleated sheet is very weak relative to a covalent bond, the
large number of hydrogen bonds in these structures makes
them highly stable. As a result, they increase the stability of the
molecule as a whole and help define its shape. For overall
shape and stability, though, a protein's tertiary structure is even
more important.

Tertiary Structure

Alpha helices and β-pleated sheets form due to interactions between components of a protein's peptide-bonded backbone. In
contrast, most of the overall shape, or **tertiary structure**, of a
polypeptide results from interactions between R-groups or between R-groups and the peptide backbone. As **Figure 3.13a**
shows, side chains can be involved in a wide variety of bonds
and interactions. Because each contact between R-groups
causes the peptide-bonded backbone to bend and fold, each
contributes to the distinctive three-dimensional shape of a
polypeptide.

Four types of interactions that involve side chains are
particularly important:

1. Hydrogen bonds form in several ways: between hydrogen
   atoms and the carboxyl group in the peptide-bonded
   backbone, and between hydrogen and atoms with partial
   negative charges in side chains.

2. In an aqueous solution, water molecules interact with
   hydrophilic side chains and force hydrophobic side chains
   to coalesce. Once the hydrophobic side chains are close to
   each other, they are stabilized by electrical attractions
   known as **van der Waals interactions**. These weak attractions occur because the constant motion of electrons gives
   molecules a tiny asymmetry in charge that changes with
   time. If molecules get extremely close to each other, the
   minute partial charge on one molecule induces an opposite

(a) Interactions that determine the tertiary structure of proteins

$-CH_2-OH$ ---------- $O=C$

Hydrogen bond between side chain and carboxyl oxygen

$-CH_2-OH$ ---------------- $\overset{H}{\underset{H}{N}}-\overset{O}{C}CH_2-$

Hydrogen bond between two side chains

$CH_3 \quad H_3C$
$-CH_2CH \quad CHCH_2-$
$CH_3 \quad H_3C$

Hydrophobic interactions (van der Waals interactions)

$-(CH_2)_4-NH_3^+ \quad {}^-O-\overset{O}{C}CH_2CH_2-$

Ionic bond

$-CH_2-S-\!\!\!-\!\!\!-S-CH_2-$

Disulfide bond

(b) Tertiary structures are diverse.

A tertiary structure composed mostly of α-helices

A tertiary structure composed mostly of β-pleated sheets

A tertiary structure rich in disulfide bonds

FIGURE 3.13 Tertiary Structure of Proteins Results from Interactions Involving R-Groups (a) Each protein has a unique overall shape called its tertiary structure. Tertiary structure is created by bonds and other interactions that cause proteins to fold in a precise way. **(b)** The tertiary structure of these proteins includes interactions between α-helices and β-pleated sheets. The polypeptide chains are color-coded so that you can follow the chain from the amino-end (red) to the carboxy-end (dark blue).

● **EXERCISE** In part (b): Label the four a-helices in the left-hand protein, the α-helix and one of the β-pleated sheets in the center protein, and the yellow disulfide bonds in the right-hand protein.

partial charge in the nearby molecule and causes an attraction. Although the attraction is very weak relative to covalent bonds or even hydrogen bonds, a large number of van der Waals interactions can occur in a polypeptide when many hydrophobic residues congregate. The result is a significant increase in stability. It is important to recognize how important this is. Hydrophobic interactions are responsible for folding that results in the globular shape of many proteins.

3. Covalent bonds can form between sulfur atoms when a reaction occurs between sulfur-containing R-groups, such as those found in cysteine. These **disulfide** ("two-sulfur") **bonds** are frequently referred to as bridges because they

create strong links between distinct regions of the same polypeptide.

4. Ionic bonds form between groups that have full and opposing charges, such as the ionized amino and carboxyl functional groups highlighted on the right in Figure 3.13a.

It's also important to note that the overall shape of many proteins depends in part on the presence of secondary structures like α-helices and β-pleated sheets. Tertiary structure depends on primary structure and secondary structure.

With so many interactions possible between side chains and peptide-bonded backbones, it's not surprising that polypeptides vary in shape from rod-like filaments to ball-like masses. (See **Figure 3.13b,** and look again at Figure 3.10.)

Quaternary Structure

The first three levels of protein structure involve individual polypeptides. But many proteins contain several distinct polypeptides that interact to form a single structure. The combination of polypeptide subunits gives proteins a **quaternary structure**. The individual polypeptides may be held together by bonds or other interactions among R-groups or sections of their peptide backbones.

In the simplest case, a protein with quaternary structure can consist of just two subunits that are identical. The Cro protein found in a virus called bacteriophage λ (pronounced *LAMB-da*) is an example (**Figure 3.14a**). Proteins with two polypeptide subunits are called dimers ("two-parts"). More than two polypeptides can be linked into a single protein, however, and the polypeptides involved may be distinct in primary, secondary, and tertiary structure. Hemoglobin, for example, is a tetramer ("four-parts"). As **Figure 3.14b** shows, the completed protein consists of two copies of each of two different polypeptides. Proteins that consist of a single polypeptide lack quaternary structure.

Table 3.3 summarizes the four levels of protein structure, using hemoglobin as an example. ◯ The key thing to note is that protein structure is hierarchical. Quaternary structure is based on tertiary structure, which is based in part on secondary structure. All three of the higher-level structures are based on primary structure. ◯ You should be able to describe the elements of the primary, secondary, tertiary, and quaternary structure of the Cro protein shown in Figure 3.14a. (Check your work by using the information about hemoglobin given in Table 3.3.)

The summary table and preceding discussion have two important messages. First, the combination of primary, secondary, tertiary, and quaternary levels of structure is responsible for the fantastic diversity of sizes and shapes observed in proteins. Second, most elements of protein structure are based on folding of polypeptide chains. Does this folding occur spontaneously? What happens if normal folding is disrupted?

Folding and Function

If you were able to synthesize one of the polypeptides in hemoglobin from individual amino acids, and if you placed the resulting chain in water, it would spontaneously fold into the shape of the tertiary structure shown in Table 3.3. This result seems counterintuitive. Because an unfolded protein has many more ways to move about, it has much higher entropy than the folded version. Folding is spontaneous in some cases, however, because the bonds and van der Waals interactions that occur make the folded molecule more stable energetically than the unfolded molecule. Thus, folding may release enough free energy to be exergonic and occur spontaneously.

Folding is also crucial to the function of a completed protein. This point was hammered home in a set of classic experiments by Christian Anfinson and colleagues during the 1950s. Anfinson studied a protein called ribonuclease that is found in many organisms. Ribonuclease is an enzyme that breaks ribonucleic acid polymers apart. Anfinson found that the ribonuclease could be unfolded, or **denatured**, by treating it with compounds that break hydrogen bonds and disulfide bonds (**Figure 3.15**). The denatured ribonuclease was unable to function normally—it could no longer break apart nucleic acids. This is not surprising, given that the function of a protein depends on its structure.

When Anfinson removed the denaturing agents, however, the molecule refolded and began to function normally again. These

(a) Cro protein, a dimer

(b) Hemoglobin, a tetramer

FIGURE 3.14 Quaternary Structures of Proteins Are Created by Multiple Polypeptides. These are ribbon diagrams, which represent the primary sequence as a ribbon. **(a)** The Cro protein is a dimer—it consists of two polypeptides. The polypeptides are identical in this case but are colored yellow and green here. **(b)** Hemoglobin is a tetramer—it consists of four polypeptides. Hemoglobin is composed of two copies of two polypeptides. The yellow and purple polypeptides are identical; so are the blue and green polypeptides.

◯ **EXERCISE** Label the β-pleated sheet that holds the Cro dimer together.

SUMMARY TABLE **3.3** **Protein Structure**

Level	Description	Stabilized by	Example: Hemoglobin
Primary	The sequence of amino acids in a polypeptide	Peptide bonds	Gly — Ser — Asp — Cys
Secondary	Formation of α-helices and β-pleated sheets in a polypeptide	Hydrogen bonding between groups along the peptide-bonded backbone; thus, depends on primary structure.	
Tertiary	Overall three-dimensional shape of a polypeptide (includes contribution from secondary structures)	Bonds and other interactions between R-groups, or between R-groups and the peptide-bonded backbone; thus, depends on primary structure.	One of hemoglobin's subunits
Quaternary	Shape produced by combinations of polypeptides (thus, combinations of tertiary structures)	Bonds and other interactions between R-groups, and between peptide backbones of different polypeptides; thus, depends on primary structure.	Hemoglobin, which consists of four polypeptide subunits

● **EXERCISE** Label the secondary structures within the tertiary structure, and the tertiary structures within the quaternary structure.

experiments confirmed that ribonuclease folds spontaneously and that folding is essential for normal function.

More recent work has shown that in cells, folding is often facilitated by specific proteins called **molecular chaperones**. Many molecular chaperones belong to a family of molecules called the heat-shock proteins. These compounds are produced in large quantities after cells experience high temperatures or other treatments that make proteins lose their tertiary structure. Heat-shock proteins speed the refolding of other proteins into their normal shape after denaturation has occurred.

To summarize, a protein's function depends on its shape. In most cases, the final shape of a protein depends on folding. To drive this point home, **Box 3.1** investigates the way a change in the shape of the prion (pronounced *PREE-on*) protein of mammals converts normal molecules into abnormally folded proteins. Misfolded prions cause the brains of humans and other mammals to disintegrate.

FIGURE 3.15 Proteins Fold into Their Normal, Active Shape. (a) The tertiary structure of ribonuclease is defined primarily by four disulfide bonds, shown in yellow. The protein's primary structure is represented by the green ribbon. **(b)** When the disulfide bonds and various non-covalent bonds are broken, the protein denatures (unfolds).

BOX 3.1 Prions

Over the past several decades, evidence has accumulated that certain proteins can act as infectious, disease-causing agents. The proteins involved are called **prions**, or proteinaceous infectious particles. Where do these molecules come from, and how do they work?

According to the prion hypothesis developed by Stanley Prusiner, infectious proteins are improperly folded forms of normal proteins that are present in healthy individuals. The infectious and normal forms do not necessarily differ in amino acid sequence, however. Instead, their shapes are radically different. Further, the infectious form of a protein can induce other normal protein molecules to change their shape to the altered form.

Figure 3.16 illustrates the shape differences observed in the normal and infectious forms of the first prion ever described. The molecule in Figure 3.16a is called the prion protein (PrP) and is a normal component of mammalian cells. Mutant versions of this protein, like the one in Figure 3.16b, are found in a wide variety of species and cause a family of diseases known as the spongiform encephalopathies—literally, "sponge-brain-illnesses." Hamsters, cows, goats, and humans afflicted with these diseases undergo massive degeneration of the brain. Cattle suffer from "mad cow disease"; sheep and goats acquire scrapie (so called because the animals itch so badly that they scratch off their wool or hair); humans develop kuru or Creutzfeldt-Jakob disease. Although some spongiform encephalopathies can be inherited, in many cases the disease is transmitted when individuals eat tissues containing the infectious form of PrP. All of the prion illnesses are fatal.

(a) Normal prion protein

(b) Misfolded prion protein

FIGURE 3.16 Prions Are Improperly Folded Proteins. Ribbon model of **(a)** a normal prion protein and **(b)** the misfolded form that causes mad cow disease in cattle. Secondary structure is represented by coils α-helices and arrows (β-pleated sheets).

Given how proteins are put together and how they are folded, let's take a detailed look at how they function as catalysts. Diversity in structure and chemical reactivity explain why proteins are easily the most diverse catalysts known.

Check Your Understanding

If you understand that...

- Proteins have up to four levels of structure.
- Primary structure consists of the sequence of amino acids.
- Secondary structure results from interactions between atoms in the peptide-bonded backbone of the same polypeptide, and yields structures such as α-helices and β-pleated sheets.
- Tertiary structure is a consequence of bonds or other interactions between R-groups or between R-groups and the peptide-bonded backbone of a polypeptide, and produces distinctive folds.
- Quaternary structure occurs when multiple polypeptides interact to form a single protein.

You should be able to ...

1) Draw a protein that you make up.
2) Describe elements of its primary through quaternary structure.

3.5 Enzymes: An Introduction to Catalysis

Of all the functions that proteins perform in cells, catalysis may be the most important. To understand why, recall from Chapter 2 that changes in potential energy and entropy determine whether a reaction is spontaneous. Spontaneous chemical reactions are not necessarily fast, however. The reaction that makes iron rust is exothermic and spontaneous, but it occurs slowly. Most of the chemical reactions in cells don't occur fast enough to support life unless a catalyst is present.

To appreciate how enzymes work, it is essential to get a deeper understanding of the factors that limit the speed at which chemical reactions occur. Reaction rates depend on how the chemical bonds involved are broken and re-formed. Recall from Chapter 2 that reactions take place when reactants (1) collide in a precise orientation and (2) have enough kinetic energy to overcome repulsion between electrons that come into contact as a bond forms. Part of the reason enzymes are such effective catalysts is that they bring reactant molecules—called **substrates**—together in a precise orientation, so that the electrons involved in the reaction can interact. But enzymes can also affect the amount of kinetic energy reactants must have for a reaction to proceed. To understand why, it's critical to realize that a collision between reactants creates a combination of old and new bonds called a **transition state**. The amount of free

FIGURE 3.17 **Atoms and Molecules Have More Kinetic Energy at High Temperatures than at Low Temperatures.**

⦿ **EXERCISE** Add a vertical line two-thirds of the way along the horizontal axis. Label this line "Activation energy for reaction." At low and high temperatures, which molecules have enough kinetic energy for this reaction to proceed?

energy required to reach the transition state is called the **activation energy** of the reaction. Reactions happen when reactants have enough kinetic energy to reach the transition state. The kinetic energy of molecules, in turn, is a function of their temperature (**Figure 3.17**). This is why chemical reactions tend to proceed faster at higher temperatures, as noted in Chapter 2.

Figure 3.18 graphs the changes in free energy that take place during the course of a chemical reaction. In this graph, ΔG indicates the overall change in free energy in the reaction—that is, the energy of the products minus the energy of the reactants. In this case the products have lower potential energy than the reactants, meaning that the reaction is exothermic. But because

the activation energy for this reaction, symbolized by E_a is high, the reaction would proceed slowly. The more unstable the transition state, the higher the activation energy and the less likely a reaction is to proceed quickly.

Reaction rates, then, depend on both the kinetic energy of the reactants and the activation energy of the particular reaction—meaning the free energy of the transition state. If the kinetic energy of the participating molecules is high, then molecular collisions are likely to result in completed reactions. But if the activation energy of a particular reaction is also high, then collisions are less likely to result in completed reactions.

In many cases, the electrons in the transition-state molecule can be stabilized when they interact with another ion, atom, or molecule. When this occurs, the activation energy required for the reaction drops and the reaction rate increases. A substance that lowers the activation energy of a reaction and increases the rate of the reaction is called a **catalyst.** It's important to note that a catalyst is not consumed in a chemical reaction, even though it participates in the reaction. The composition of a catalyst is exactly the same after the reaction as it was before.

Figure 3.19 diagrams how catalysts lower the activation energy for a reaction by lowering the free energy of the transition state. Note that the presence of a catalyst does not affect the overall energy change, ΔG, or change the energy of the reactants or the products. A catalyst changes only the free energy of the transition state.

Most enzymes are quite specific in their activity—they catalyze just a single reaction by lowering the activation energy that is required—and many are astonishingly efficient. Most of the important reactions in biology do not occur at all, or else

FIGURE 3.18 **Changes in Free Energy during a Chemical Reaction.** The changes in free energy that occur over the course of a hypothetical reaction between an atom A and a molecule containing atoms B and C. The overall reaction would be written as $A + BC \rightarrow AB + C$. E_a is the activation energy of the reaction, and ΔG is the overall change in free energy.

⦿ **EXERCISE** This graph illustrates an exergonic reaction. Draw the same type of graph for an endergonic reaction.

FIGURE 3.19 **A Catalyst Changes the Activation Energy of a Reaction.** The energy profile for the same reaction diagrammed in Figure 3.18, but with a catalyst present. Even though the energy barrier to the reaction, E_a is much lower, ΔG does not change.

⦿ **QUESTION** Can a catalyst make a nonspontaneous reaction occur spontaneously?

⦿ **EXERCISE** Go back to Figure 3.17 and add another vertical line one-third of the way along the horizontal axis. Label this line "Activation energy for reaction with catalyst." Label the molecules that have enough energy to undergo a reaction with a catalyst but not without a catalyst.

proceed at imperceptible rates, without a catalyst. In contrast, a single molecule of the enzyme carbonic anhydrase can catalyze over 1,000,000 reactions *per second*. It's not unusual for enzymes to speed up reactions by a factor of a million; some enzymes make reactions go many *trillions* of times faster than they would without a catalyst.

To summarize, enzymes bring reactants together in precise orientations and stabilize transition states. Protein catalysts are important because they speed up the chemical reactions that are required for life. Now the question is, how do enzymes do what they do?

How Do Enzymes Work?

The initial hypothesis for how enzymes work was proposed by Emil Fischer in 1894. According to Fischer's lock-and-key model, enzymes are rigid structures analogous to a lock. The keys are substrates that fit into the lock and then react.

Several important ideas in this model have stood the test of time. For example, Fischer was correct in proposing that enzymes bring substrates together in a precise orientation that makes reactions more likely. His model also accurately explained why most enzymes can catalyze only one specific reaction. Enzyme specificity is a product of the geometry and chemical properties of the sites where substrates bind.

As researchers began to test and extend Fischer's model, the location where substrates bind and react became known as the enzyme's **active site**. The active site is where catalysis actually occurs. When techniques for solving the three-dimensional structure of enzymes became available, it turned out that enzymes tend to be very large relative to substrates and roughly globular, and that the active site is in a cleft or cavity within the globular shape. The enzyme hexokinase, which is at work in most cells of your body now, is a good example. (Many enzymes have names that end with –*ase*.) As the left-hand side of

Figure 3.20 shows, the active site in hexokinase is a small notch in an otherwise large, crescent-shaped enzyme.

As knowledge of enzyme action grew, however, Fischer's model was modified. Perhaps the most important change was based on the realization that enzymes are not rigid and static, but flexible and dynamic. In fact, many enzymes undergo a significant change in shape, or conformation, when reactant molecules bind to the active site. This conformational change, called an **induced fit**, is noticeable in the hexokinase molecule on the right of Figure 3.20. As hexokinase binds its substrate—the sugar glucose—the enzyme rocks forward over the active site.

In addition, recent research has clarified the nature of Fischer's key. When one or more substrate molecules enter the active site, they are held in place through hydrogen bonding or other electrical interactions with amino acids in the active site. Once the substrate is bound, one or more R-groups in the active site come into play. The degree of interaction between the substrate and enzyme increases and reaches a maximum when the transition state is formed. Thus, Fischer's key is actually the transition state.

● Interactions with R-groups at the active site stabilize the transition state and thus lower the activation energy required for the reaction to proceed. At the atomic level, R-groups that line the active site may form short-lived covalent bonds that assist with the transfer of atoms or groups of atoms from one reactant to another. More commonly, the presence of acidic or basic R-groups allows the reactants to lose or gain a proton more readily.

Figure 3.21 provides a specific example of how substrates interact with an active site. The panels show a substrate fitting into the active site of the enzyme called ribonuclease (the same enzyme that denatured in Figure 3.15). Notice that once the substrate has bound, R-groups in the active site interact with the substrate to form the transition state and get the reaction under

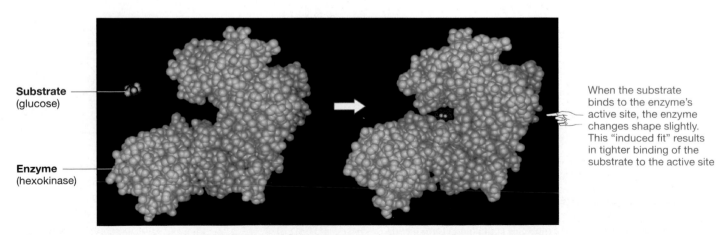

Substrate
(glucose)

Enzyme
(hexokinase)

When the substrate binds to the enzyme's active site, the enzyme changes shape slightly. This "induced fit" results in tighter binding of the substrate to the active site

FIGURE 3.20 Reactant Molecules Bind to Specific Locations in an Enzyme. The reactant molecule, shown in red, fits into a precise location, called the active site, in the green enzyme. In the enzyme shown here and in many others, the binding event causes the protein to change its shape.

● **EXERCISE** Label the active site in the enzyme on the left.

R-GROUPS IN AN ENZYME'S ACTIVE SITE STABILIZE THE TRANSITION STATE OF A SUBSTRATE.

1. Enzyme's active site is empty (three of the R-groups in the active site are shown).

2. Substrate (RNA molecule) fits into active site. Proton from substrate is transferred to R-group 1 of active site.

3. Transition state is stabilized by R-group 2 (an interaction between O⁻ and N⁺H₃); proton from R-group 3 is transferred to substrate, splitting it in two.

4. After several more steps, including a reaction between a water molecule and the phosphate group, the product molecules leave the active site.

FIGURE 3.21 R-groups in an Enzyme's Active Site Stabilize the Transition State. Part of the reaction sequence that occurs when the enzyme ribonuclease catalyzes a reaction that cuts a substrate in two. Note how precisely the substrate fits into the active site and how a few R-groups in the enzyme are involved in key steps in the reaction.

way. ● You should be able to analyze Figure 3.21 and identify two acid–base reactions that occur inside the active site and stabilize the transition state.

Whatever the mechanism, enzymes increase reaction rates by bringing substrates together and lowering the energy required to form the transition state. The products of the reaction have a much lower affinity for the active site than do either the reactants or the transition state, however. Therefore, they are released from the enzyme once they form.

Figure 3.22 summarizes these principles. Enzyme catalysis can be analyzed as a three-step process:

1. **Initiation**. Instead of reactants occasionally colliding in a random fashion, enzymes orient reactants precisely as they bind at specific locations within the active site.

2. **Transition state facilitation**. The act of binding induces the formation of the transition state. In some cases the transition state is stabilized by a change in the enzyme's shape. The interaction between the substrate and R-groups in the enzyme's active site lowers the activation energy required for the reaction. As a result, inside a catalyst's active site, more reactant molecules have sufficient kinetic energy to reach this lowered activation energy. Thus, the catalyzed reaction proceeds much more rapidly than the uncatalyzed reaction.

3. **Termination**. The reaction products have considerably less affinity for the active site than does the transition state. Thus binding ends, the enzyme returns to its original conformation, and the products are released.

Fischer's model inspired intensive and fruitful research into the mechanism of enzyme action.

● You should be able to complete the following sentences:

1. Enzymes speed reaction rates by _____ and lowering activation energy.

2. Activation energies drop because enzymes destabilize bonds in the reactant, stabilize the _____, make acid–base reactions more favorable, and/or change the reaction mechanism through a covalent bonding interaction.

3. Enzyme specificity is a function of the active site's shape and the chemical properties of the _____ at the active site.

4. In enzymes, as in many molecules, function follows from _____.

Do Enzymes Act Alone? Atoms or molecules that are not part of the enzyme's primary structure are often required for the enzyme to function normally. These enzyme **cofactors** can be either (1) metal ions such as Zn^{2+} (zinc) or Mg^{2+} (magnesium), or (2) small organic **molecules** called **coenzymes**. In

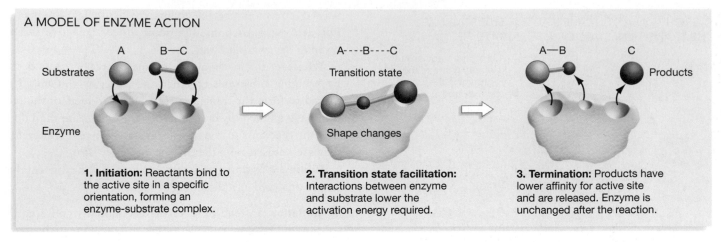

A MODEL OF ENZYME ACTION

Substrates   A   B—C

Enzyme

1. Initiation: Reactants bind to the active site in a specific orientation, forming an enzyme-substrate complex.

A---B---C

Transition state

Shape changes

2. Transition state facilitation: Interactions between enzyme and substrate lower the activation energy required.

A—B   C   Products

3. Termination: Products have lower affinity for active site and are released. Enzyme is unchanged after the reaction.

FIGURE 3.22 Enzyme Action Can Be Analyzed as a Three-Step Process.

many cases, the cofactor binds to the active site and is thought to play a key role in stabilizing the transition state during the reaction. The presence of the cofactor is therefore essential for catalysis. To drive this point home, consider that many of the vitamins in your diet are required for the production of enzyme cofactors. Vitamin deficiencies cause enzyme-cofactor deficiencies. Lack of cofactors, in turn, disrupts normal enzyme function and causes disease. For example, thiamine (vitamin B_1) is required for the production of an enzyme cofactor called thiamine pyrophosphate, which is required by three different enzymes. Lack of thiamine in the diet dramatically reduces the activity of these enzymes and causes an array of nervous system and heart disorders collectively known as beriberi.

In addition, most enzymes are regulated by molecules that are not part of the enzyme itself. An enzyme is active or inactive, depending on the presence or absence of molecules that change the protein's structure in some way. In some cases, catalysis is inhibited when a molecule that is similar in size and shape to the substrate(s) binds to the active site. This event is called **competitive inhibition,** because the molecule involved competes with the substrate(s) for access to the enzyme's active site (Figure 3.23a). In other cases, a regulatory molecule binds at a location other than the active site. This type of regulation is called **allosteric** ("different-structure") **regulation**, because the molecule involved does not affect the active site directly. Instead, the binding event changes the shape of the enzyme in a way that makes the active site accessible or inaccessible (**Figure 3.23b**). Later chapters will provide detailed examples of how regulatory molecules change the activity of specific enzymes.

(MB) Web Animation at www.masteringbio.com
Activation Energy and Enzymes

What Limits the Rate of Catalysis? For several decades after Fischer's model was published, most research on enzymes focused on rates of enzyme action, or what biologists call enzyme kinetics. Researchers observed that, when the amount of product produced per second—indicating the speed of the reac-

tion—is plotted as a function of substrate concentration, a graph like that shown in **Figure 3.24** results. When substrate concentrations are low, the speed of an enzyme-catalyzed reaction increases in a linear fashion. As substrate concentrations increase, however, the increase in speed begins to slow. Eventually the reaction rate plateaus at a maximum speed. This pattern is in striking contrast to the situation for uncatalyzed reactions, in which reaction speed tends to show a continuing linear increase with substrate concentration. The non-linear, "saturation kinetics" of enzyme-catalyzed reactions were taken as strong evidence that the enzyme-substrate complex proposed by Fischer actually exists. The idea was that at some point, active sites cannot accept substrates any faster, no matter how large the concentration of substrates gets. Stated another way, reaction rates level off because all available enzyme molecules are being used.

It's important to recognize, however, that even though the general shape of the curve is similar for all enzymes, its position varies. In enzymes that have a high affinity for the substrate, the curve is shifted to the left. This means that substrate concentrations do not have to be very high for these enzymes to function at maximum speed. In enzymes with a lower affinity for the substrate, though, the curve is shifted to the right. This means that substrate concentrations have to be much higher before an enzyme is able to catalyze reactions at its maximal rate.

Thus, an enzyme's speed is partly a function of the concentration of its substrates (there can be one or more) and partly a function of its intrinsic affinity for the substrates. The processing of ethanol—the active ingredient in alcoholic beverages—in the human liver furnishes an example of these concepts. Ethanol is a poison broken down in two enzyme-catalyzed steps. First ethanol is converted to acetaldehyde (CH_3CHO). Then acetaldehyde is converted to acetic acid (CH_3COOH)—a harmless molecule that gives vinegar its bite. Humans have two versions of the enzyme that catalyzes the second reaction. One version of the enzyme works quickly and the other much more slowly. In some people, the fast-working version of the enzyme has a single amino acid change in its primary sequence that

(a) Competitive inhibition directly blocks the active site.

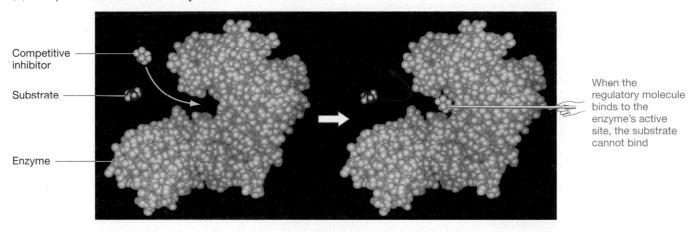

Competitive inhibitor

Substrate

Enzyme

When the regulatory molecule binds to the enzyme's active site, the substrate cannot bind

(b) Allosteric regulation occurs when a regulatory molecule binds somewhere other than the active site.

Substrate

Enzyme

Regulatory molecule

or

Activating the enzyme

Inactivating the enzyme

When the regulatory molecule binds to a different site on the enzyme, it induces a shape change that makes the active site either available to the substrate (left) or unavailable (right)

FIGURE 3.23 An Enzyme's Activity Is Precisely Regulated. Enzymes are turned on or off when specific molecules bind to them.

Maximum speed of reaction

Rate of product formation

Substrate concentration

FIGURE 3.24 Kinetics of an Enzyme-Catalyzed Reaction. The general shape of this curve is characteristic of enzyme-catalyzed reactions.

● **EXERCISE** Label the part of the graph where (1) reaction is most sensitive to changes in substrate concentration, and (2) most or all of the active sites present are occupied.

affects the enzyme's active site. In these individuals, the fast-working version of the enzyme is completely inactivated, so only the slow-working version functions normally. Acetaldehyde must be at a high concentration before the slow-working version of the enzyme works at maximum speed. In these individuals, acetaldehyde builds up to high enough concentrations in the bloodstream that uncomfortable symptoms result, including a rapid heartbeat and pronounced flushing of the skin. People who lack the fast-acting version of the enzyme have virtually zero tolerance for alcoholic beverages. Imbibing even a small amount of ethanol makes them feel ill.

How Do Physical Conditions Affect Enzyme Function?

Given that an enzyme's structure is critical to its function, it's not surprising that an enzyme's activity is sensitive to conditions that alter its structure. In particular, the activity of an enzyme often changes drastically as a function of temperature and pH. Temperature affects the movement of the enzyme as well as the kinetic energy of the substrates; pH affects the makeup and charge of amino acid side chains with carboxyl or amino groups, and the active site's ability to participate in proton-transfer or electron-transfer reactions.

Do data support these assertions? **Figure 3.25a** shows how the activity of an enzyme changes as a function of temperature. Data for the enzyme glucose-6-phosphatase, which is helping to produce usable energy in your cells right now, are shown for two species of bacteria. Note that in both of the bacterial species illustrated, the enzyme has a distinct optimum—a temperature at which it functions best. One of the bacterial species lives inside your gut, where the temperature is about 40°C, while the other lives in hot springs, where temperatures can be close to 100°C. The temperature optimum for the enzyme that reflects these environments. The two types of bacteria have different versions of the enzyme that differ in primary structure. Natural selection—the process introduced in Chapter 1—has favored different structures that have different functions. The

enzymes are adaptations that allow each species to thrive at different temperatures.

Figure 3.25b makes the same point for pH. The enzyme in this graph, called chitinase, protects bacterial cells by digesting a molecule found in the cell walls of fungi that use bacteria as food. The data come from a species of bacterium that lives in acidic pools and a bacterial species that lives in the soil under palm trees. The organism that thrives in an acidic environment has a version of the enzyme that performs best at low pH; the organism that lives near palms has a version of the enzyme that functions best near neutral pH. Each enzyme is sensitive to changes in pH, but each species' version of the enzyme has a structure that allows it to function best at the pH of its environment.

To summarize, the rate of an enzyme-catalyzed reaction depends not only on substrate concentration and the enzyme's intrinsic affinity for the substrate but also on temperature and pH. Temperature affects the movement of the substrates and enzyme; pH affects the enzyme's shape and reactivity.

(a) Enzymes from different organisms may function best at different temperatures.

(b) Enzymes from different organisms may function best at different pHs.

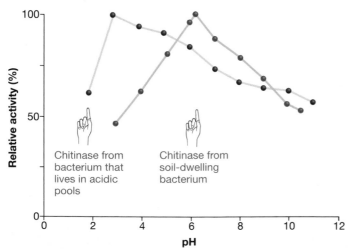

FIGURE 3.25 Enzymes Have an Optimal Temperature and pH. Enzymes are sensitive to changes in temperature and pH. Further, the structures of the enzymes found in a particular organism allow that organism to function well at the **(a)** temperature and **(b)** pH of its environment.

Check Your Understanding

If you understand that...

- Most proteins are enzymes that make specific chemical reactions occur rapidly.

- Enzymes catalyze reactions by means of a three-step mechanism:

  Step 1: Binding of reactants in a precise orientation inside the active site.

  Step 2: Facilitation of the transition state, thus lowering the activation energy required for the reaction. This step often involves a shape change in the enzyme, resulting in an "induced fit" between active site and substrate. Enzyme cofactors are often required at this step.

  Step 3: Release of products, which do not bind tightly to the active site.

- The activity of enzymes is controlled through allosteric regulation or competitive inhibition. Different enzymes work at different rates, and enzymes are sensitive to changes in temperature and pH.

You should be able to...

1) Diagram the three steps in enzyme catalysis.

2) Add drawings and notes indicating how competitive inhibition and allosteric regulation affect these steps.

Was the First Living Entity a Protein?

The theory of chemical evolution holds that life began with a molecule that could make a copy of itself. This self-replicating entity increased in number and formed a population of individuals. The population then began to evolve by natural selection.

For such a molecule to make a copy of itself quickly, it must have been able to catalyze the reactions required for the copying process. This chapter's examination of protein structure and function raises a key question: Was the first self-replicating molecule on Earth a protein?

Several observations argue that the answer to this question is yes. Experimental studies have shown that amino acids were likely to be abundant in the prebiotic soup and that they could have polymerized to form small proteins. In addition, proteins are the most efficient catalysts known, and a self-replicating

molecule had to act as a catalyst during the assembly and polymerization of its copy. These observations support the hypothesis that the self-replicating molecule was a polypeptide. Indeed, several laboratories that are currently working to create life have focused on synthesizing a self-replicating protein.

To date, however, attempts to simulate the origin of life with proteins have not been successful. Most origin-of-life researchers are increasingly skeptical about the hypothesis that life began with a protein. Their reasoning is that to make a copy of something, a mold or template is required. Proteins cannot furnish this information. Nucleic acids, in contrast, can. How they do so is the subject of Chapter 4.

Chapter Review

SUMMARY OF KEY CONCEPTS

- **Proteins are made of amino acids. Amino acids vary in structure and function because their side chains vary in composition.**

  Once monomers such as amino acids have formed, the next key event in chemical evolution is their polymerization into macromolecules. Researchers have observed polypeptide formation from the polymerization of amino acids on clay particles. In proteins, amino acids are joined by a peptide bond between the carboxyl group of one amino acid and the amino group of a second amino acid.

  You should be able to draw a generalized amino acid and a peptide bond. You should also be able to explain why some R-groups on amino acids are soluble versus insoluble in water, and why some R-groups are reactive versus nonreactive. ●

  (MB) Web Animation at www.masteringbio.com
  Condensation and Hydrolysis Reactions

- **Proteins vary widely in structure. The structure of a protein can be analyzed at four levels that form a hierarchy. These are the amino acid sequence, substructures called α-helices and β-pleated sheets, interactions between amino acids that dictate a protein's overall shape, and combinations of individual proteins that make up larger, multiunit molecules.**

  A protein's primary structure, or sequence of amino acids, is responsible for most of its chemical properties. Interactions that take place between carboxyl and amino groups in the same peptide-bonded backbone create secondary structures, which are stabilized primarily by hydrogen bonding. Interactions between R-groups found in the same polypeptide and between R-groups and the peptide-bonded backbone allow the protein to fold into a characteristic overall shape. A complete protein frequently consists of several different polypeptides, bonded together.

  You should be able to make a concept map (see BioSkills 6) that relates the four levels of protein structure to each other and to how proteins function as catalysts. You should be able to describe

how variation in each of these four levels of protein structure can affect a protein's function. ●

- **Proteins vary widely in function. In cells, most proteins are enzymes that function as catalysts. Chemical reactions occur much faster when they are catalyzed by enzymes. During enzyme catalysis, the reactants bind to an enzyme's active site in a way that allows the reaction to proceed efficiently.**

  In organisms, proteins function in defense, movement, signaling, structural support, and transport of materials. In addition, they are far and away the most diverse and efficient class of catalysts known. A catalyst speeds the rate of that reaction by lowering the activation energy required, even though the catalyst itself is unchanged by the reaction. Enzymes are protein catalysts that lower activation energy by stabilizing the transition state of the reaction.

  Enzyme-catalyzed reactions take place at a location called the active site of the enzyme. The active site in each type of enzyme has unique chemical properties and a distinctive size and shape. As a result, most enzymes are able to catalyze only one specific reaction. As a group, enzymes are able to catalyze many types of reactions because the chemical and physical structures of their active sites are so diverse. This diversity is due to the variety of amino acids and the four levels of protein structure.

  Many enzymes function only with the help of cofactors. In addition, virtually all enzyme activity in cells is regulated by molecules that bind at the active site or at locations on the protein that induce a change in the size or shape of the active site. The rate at which enzymes work depends on substrate concentration, their affinity for the substrate, pH, and temperature.

  You should be able to draw an energy diagram for a chemical reaction with and without a catalyst, and relate the change in activation energy to events at an enzyme's active site. ●

  (MB) Web Animation at www.masteringbio.com
  Activation Energy and Enzymes

QUESTIONS

● Test Your Knowledge

1. What two functional groups are present on every amino acid?
   a. a carbonyl group and a carboxyl group
   b. an amino group and a carbonyl group
   c. an amino group and a hydroxyl group
   d. an amino group and a carboxyl group

2. Twenty-one different amino acids are found in the proteins of cells. What distinguishes these molecules?
   a. the location of their carboxyl group
   b. the location of their amino group
   c. the composition of their side chains, or R-groups
   d. their ability to form peptide bonds

3. What determines the primary structure of a polypeptide?
   a. its sequence of amino acids
   b. hydrogen bonds that form between carboxyl and amino groups on different residues
   c. hydrogen bonds and other interactions between side chains
   d. the number, identity, and arrangement of polypeptides that make up a protein

4. In a polypeptide, what is most responsible for the secondary structure called an α-helix?
   a. the sequence of amino acids

   b. hydrogen bonds that form between carboxyl and amino groups on different residues
   c. hydrogen bonds and other interactions between side chains
   d. the number, identity, and arrangement of polypeptides that make up a protein

5. What is a transition state?
   a. the complex formed as covalent bonds are being broken and reformed during a reaction
   b. the place where an allosteric regulatory molecule binds to an enzyme
   c. an interaction between reactants with high kinetic energy, due to high temperature
   d. the shape adopted by an enzyme that has an inhibitory molecule bound at its active site

6. By convention, biologists write the sequence of amino acids in a polypeptide in which direction?
   a. carboxy- to amino-terminus
   b. amino- to carboxy-terminus
   c. polar residues to nonpolar residues
   d. charged residues to uncharged residues

Test Your Knowledge answers: 1. d; 2. c; 3. a; 4. b; 5. a; 6. b

● Test Your Understanding

Answers are available at www.masteringbio.com

1. Explain the lock-and-key model of enzyme activity. Be sure to comment on what role the active site of an enzyme plays.

2. Isoleucine, valine, leucine, phenylalanine, and methionine are amino acids with highly hydrophobic side chains. Suppose a section of a protein contains a long series of these hydrophobic residues. How would you expect this portion of the protein to behave when the molecule is in aqueous solution?

3. Compare and contrast competitive inhibition and allosteric regulation.

4. Does it take energy for polymerization reactions to proceed, or do they occur spontaneously? Why or why not?

5. A major theme in this chapter is that the structure of molecules correlates with their function. Use this theme to explain why proteins can perform so many different functions in organisms and why proteins as a group are such effective catalysts.

6. Explain why temperature and pH affect enzyme function.

● Applying Concepts to New Situations

Answers are available at www.masteringbio.com

1. Researchers are searching for life on other planets. Is it reasonable to expect that protein-based life-forms could exist in extraterrestrial environments that are very hot or highly acidic? Why or why not?

2. Recently, researchers were able to measure movement that occurred in a single amino acid in an enzyme as reactions were taking place in its active site. The amino acid that moved was located in the active site, and the rate of movement correlated closely with the rate at which the reaction was taking place. Discuss the significance of these findings, using the information in Figures 3.20 and 3.22.

3. Researchers can analyze the atomic structure of enzymes during catalysis. In one recent study, investigators found that the transition state included the formation of a free radical, and that a coenzyme bound to the active site donated an electron to help stabilize the free radical. How would the reaction rate and the stability of the transition state change if the coenzyme were not available?

4. In a recent experiment, researchers found that if they took a mixture containing two optical isomers of a monomer and put a thin layer of the mixture on water, a crystal-like structure formed. Only one of the optical isomers—not both—was included in the crystal. The researchers suggested that a similar phenomenon might occur if both optical isomers of a monomer formed a thin coating on a clay mineral particle. How might this observation relate to the fact that proteins in cells contain only one isomer of each amino acid?

www.masteringbio.com is also your resource for • Answers to text, table, and figure caption questions and exercises • Answers to *Check Your Understanding* boxes • Online study guides and quizzes • Additional study tools including the *E-Book for Biological Science* 3rd ed., textbook art, animations, and videos.

Nucleic Acids and the RNA World

4

A space-filling model of deoxyribonucleic acid, or DNA. In today's cells, DNA functions as an information repository. But the nucleic acid called ribonucleic acid, or RNA, was probably responsible for the origin of life.

KEY CONCEPTS

◯ Nucleotides are monomers that consist of a sugar, a phosphate group, and a nitrogen-containing base. Ribonucleotides polymerize to form RNA. Deoxyribonucleotides polymerize to form DNA.

◯ DNA's primary structure consists of a sequence of nitrogen-containing bases, which contain information in the form of a molecular code. DNA's secondary structure consists of two DNA strands running in opposite directions. The strands are held together by complementary base pairing and are twisted into a double helix.

◯ RNA's primary structure also consists of a sequence of nitrogen-containing bases that contain information in the form of a molecular code. Its secondary structure includes short double helices and structures called hairpins. RNA molecules called ribozymes catalyze important chemical reactions.

Chapter 3 began with experimental evidence that chemical evolution produced the monomers called amino acids and their polymers called proteins. But the chapter ended by stating that even though proteins are the workhorse molecules of today's cells, few researchers still favor the hypothesis that life began as a protein molecule. Instead, the vast majority of biologists now contend that life began as a polymer called a nucleic acid—specifically, a molecule of ribonucleic acid (RNA). This proposal is called the RNA world hypothesis. Just what is a nucleic acid? How could these types of molecules have been produced by chemical evolution? What do the nucleic acids RNA and deoxyribonucleic acid (DNA) do in cells today, and why isn't it plausible to propose that life

began with DNA? Before attempting to answer these questions, let's consider an even more basic issue: How would we know when a molecule became alive? Stated another way, What *is* life?

Like many simple questions, the issue of what defines life is not easy to answer. Although there is no precise definition of what constitutes life, most biologists point to two attributes to distinguish life from nonlife. The first attribute is **reproduction**, the ability of something to make a copy of itself. (In organisms that reproduce asexually, the copy is exact. In the case of sexually reproducing organisms, the copy is not exact. Instead, traits from male and female individuals are combined to produce an offspring.) In essence, this chapter is about the first molecule that could make a copy of itself.

◯ Key Concept   Important Information   ● Practice It

For something to be alive, however, most biologists insist that it have a second attribute called **metabolism**: the ability to acquire particular molecules and use them in controlled chemical reactions that maintain conditions suitable for life and that contribute to growth. In today's organisms, chemical reactions are precisely controlled because enzymes and reactants are contained inside a cell by a plasma membrane. To many biologists, then, the presence of a plasma membrane is also required for life.

According to the theory of chemical evolution, these two attributes of life did not emerge simultaneously. Instead, the theory predicts that chemical evolution first led to the existence of a molecule that could make copies of itself. Some researchers argue that this self-replicating molecule was also capable of metabolism, because it had to be able to catalyze the polymerization reactions that allowed copies of itself to form. Later, the theory holds, a descendant of this molecule became enclosed in a membrane. This event created the first cell. The formation of the cell allowed for reactants to be enclosed. As a result, it made precise control over reactions possible.

The key idea is that life started with replication and later became cellular. To distinguish a naked (membrane-less) self-replicator from later, cellular forms of "true" life, this chapter refers to the self-replicator as the first living entity or the first life-form—but *not* as the first organism. The term **organism** is reserved for cellular-based life. This distinction isn't trivial. Naked, self-replicating molecules undoubtedly existed early in Earth's history, and researchers are almost certain to produce one in the laboratory within your lifetime. But it is much less likely that biologists will be able to create cellular life in the foreseeable future. Will humans be able to create life in a test tube? The answer depends, in part, on how you define life.

To understand how life began, it's helpful to delve into more of the chemistry that underlies biology. Let's begin with an analysis of nucleic acids as monomers and end with current research efforts to create a self-replicating molecule.

4.1   What Is a Nucleic Acid?

Nucleic acids are polymers, just as proteins are polymers. But instead of being made up of monomers called amino acids, **nucleic acids** are made up of monomers called **nucleotides**. **Figure 4.1a** diagrams the three components of a nucleotide: (1) a phosphate group, (2) a sugar, and (3) a nitrogenous (nitrogen-containing) base. The phosphate is bonded to the sugar molecule, which in turn is bonded to the nitrogenous base. A **sugar** is an organic compound with a carbonyl group ($\diagdown$C=O) and several hydroxyl (OH) groups. Notice that the "prime" symbols (′) in

FIGURE 4.1 The General Structure of a Nucleotide. (a) The relationship between the phosphate group, the sugar, and the nitrogenous base of a nucleotide. The numbers indicate the position of each carbon in the sugar's ring. The nitrogenous base is bonded to carbon number 1 in the ring, while the phosphate is bonded to carbon number 5. The bond between the phosphate group and the sugar is called a 5′ linkage. Although hydrogen atoms are bonded to the carbon atoms in the ring (see part b), biologists routinely omit them in representations, for clarity. **(b)** Ribose and deoxyribose are similar sugars that are found in nucleotides. **(c)** Purines and pyrimidines are nitrogenous bases. A C–N bond links them to the sugar in a nucleotide. This bond forms at the nitrogen atom that is highlighted in each base. Purines are substantially larger than pyrimidines.

Figure 4.1 indicate that the carbon being referred to is part of the sugar and not of the attached nitrogenous base.

Although a wide variety of nucleotides are found in living cells, origin-of-life researchers concentrate on two types: **ribonucleotides** and **deoxyribonucleotides**. In ribonucleotides, the sugar is ribose; in deoxyribonucleotides, it is deoxyribose. As **Figure 4.1b** shows, these two sugars differ by a single atom. Ribose has an –OH group bonded to the second carbon in the ring. Deoxyribose has an H instead at the same location. Note that *deoxy* means "lacking oxygen."

Cells today have four different ribonucleotides, each of which contains a different nitrogenous base. These bases, diagrammed in **Figure 4.1c**, belong to structural groups called **purines** and **pyrimidines**. Ribonucleotides include the purines adenine (A) and guanine (G), and the pyrimidines cytosine (C) and uracil (U).

Similarly, four different deoxyribonucleotides are found in cells today and are distinguished by the structure of their nitrogenous base. Like ribonucleotides, deoxyribonucleotides include adenine, guanine, and cytosine. But instead of uracil, a closely related pyrimidine called thymine (T) occurs in deoxyribonucleotides (**Figure 4.1c**). You should be able to diagram a ribonucleotide and a deoxyribonucleotide, using a ball for the phosphate group, a pentagon to represent the sugar subunit, and a hexagon to represent the nitrogenous base. Label the 2′, 3′, and 5′ carbons on the sugar molecule, and add the atoms that are bonded to the 2′ carbon.

Could Chemical Evolution Result in the Production of Nucleotides?

Based on data presented in Chapter 3, most researchers contend that amino acids were abundant early in Earth's history. As yet, however, no one has observed the formation of a nucleotide via chemical evolution. The problem lies with mechanisms for synthesizing the sugar and nitrogenous base components of these molecules. Let's consider each issue in turn.

Laboratory simulations have shown that many sugars can be synthesized readily under conditions that mimic the prebiotic soup. Specifically, when formaldehyde (H_2CO) molecules are heated in solution, they react with one another to form almost all of the sugars that have five or six carbons. (These are called pentoses and hexoses, respectively.) Ironically, the ease of forming these sugars creates a problem. In modern experiments the various pentoses and hexoses are produced in approximately equal amounts, but it seems logical to predict that ribose would have had to be particularly abundant for RNA or DNA to form in the prebiotic soup. How ribose came to be the dominant sugar during chemical evolution is still a mystery. Origin-of-life researchers refer to this issue as "the ribose problem."

The origin of the pyrimidines is equally challenging. Simply put, origin-of-life researchers have yet to discover a plausible

mechanism for the synthesis of cytosine, uracil, and thymine molecules prior to the origin of life. Purines, in contrast, are readily synthesized by reactions among hydrogen cyanide (HCN) molecules. For example, both adenine and guanine have been found in the solutions recovered after spark-discharge experiments.

The ribose problem and the questions about the origin of pyrimidine bases are two of the most serious challenges that remain for the theory of chemical evolution. Research on how large quantities of ribose could have formed in the prebiotic soup, along with the pyrimidines, continues. But once formed, how did nucleic acids polymerize to form RNA and DNA? This question has an answer.

How Do Nucleotides Polymerize to Form Nucleic Acids?

Nucleic acids form when nucleotides polymerize. As **Figure 4.2** shows, the polymerization reaction involves the formation of a bond between the phosphate group of one nucleotide and the hydroxyl group of the sugar component of another nucleotide. The result of this condensation reaction is called a **phosphodiester linkage**, or a phosphodiester bond. In Figure 4.2, a phosphodiester linkage joins the 5′ carbon on the ribose of one nucleotide to the 3′ carbon on the ribose of the other. When the nucleotides involved contain the sugar ribose, the polymer that is produced is called **ribonucleic acid**, or simply **RNA**. If the nucleotides contain the sugar deoxyribose instead, then the resulting polymer is **deoxyribonucleic acid**, or **DNA**.

FIGURE 4.2 Nucleotides Polymerize via Phosphodiester Linkages. Ribonucleotides can polymerize via condensation reactions. The linkage that results, between the 3′ carbon of one ribonucleotide and the 5′ carbon of another ribonucleotide, is called a phosphodiester linkage.

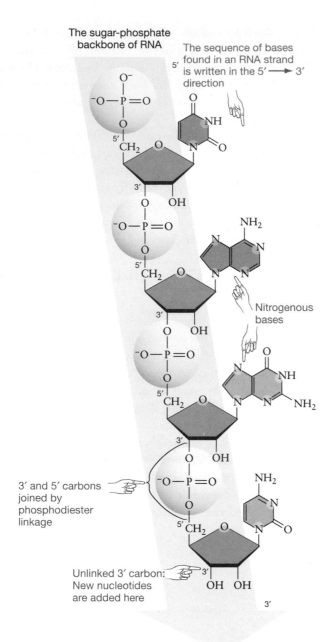

The sugar-phosphate backbone of RNA

The sequence of bases found in an RNA strand is written in the 5′ ⟶ 3′ direction

Nitrogenous bases

3′ and 5′ carbons joined by phosphodiester linkage

Unlinked 3′ carbon: New nucleotides are added here

FIGURE 4.3 RNA Has a Sugar-Phosphate Backbone.

● **EXERCISE** Identify the four bases in this RNA strand, using Figure 4.1c as a key. Then write down the base sequence, starting at the 5′ end.

Figure 4.3 shows how the chain of phosphodiester linkages in a nucleic acid acts as a backbone, analogous to the peptide-bonded backbone found in proteins. The sugar-phosphate spine of a nucleic acid is directional, as is the peptide-bonded spine of a polypeptide. In a strand of RNA or DNA, one end has an unlinked 5′ carbon while the other end has an un-linked 3′ carbon—meaning a carbon that is not linked to an-other nucleotide. By convention, the sequence of bases found in an RNA or DNA strand is always written in the 5′ → 3′ di-rection. (The system is logical because in cells, RNA and

DNA are always synthesized in this direction. Bases are added at the 3′ end of the growing molecule.) The sequence of ni-trogenous bases forms the primary structure of the molecule, analogous to the sequence of amino acids in a polypeptide. When biologists write the primary structure of a stretch of DNA, they simply list the sequence of nucleotides using their single-letter abbreviations. For example, a six-base-long DNA sequence might be ATTAGC.

In cells, the polymerization reactions that form nucleotides are catalyzed by enzymes. Like other polymerization reactions, the process is endergonic. Polymerization can take place in cells because the free energy of the nucleotide monomers is first raised by reactions that add two phosphate groups to the ri-bonucleotides or deoxyribonuclcotides, creating nucleotide triphosphates (**Figure 4.4**). Molecules that have phosphate groups attached in this way are said to be **phosphorylated**. This is a key point, and one that you will encounter again and again in this text: The addition of one or more phosphate groups raises the potential energy of the substrate molecules enough to make an endergonic reaction possible. (Chapter 9 explains how this happens.) In the case of nucleic acid poly-merization, researchers refer to the phosphorylated nucleotides as "activated."

During chemical evolution, activated nucleotides probably polymerized on the surfaces of clay-sized (very fine-grained) mineral particles. In a suite of experiments analogous to those discussed in Chapter 3 for protein synthesis, researchers have produced RNA molecules by incubating activated ribonu-cleotides with tiny mineral particles. The hypothesis was that polymerization could occur without an enzyme if the ribonu-cleotides and growing RNA strands adhered to the clay particles.

The addition of phosphate groups raises the potential energy of the monomer

FIGURE 4.4 Activated Monomers Drive Endergonic Polymerization Reactions. Polymerization reactions involving ribonucleotides are endergonic. But polymerization reactions involving ribonucleotide triphosphates are exergonic.

● **EXERCISE** The reaction shown in Figure 4.2 does not actually occur—only nucleoside triphosphates like the one shown here can form phosphodiester bonds with other nucleotides or with nucleic acids. Add two phosphate groups to the nucleotide in Figure 4.2 to make the reaction realistic. On the right side of the reaction, the products should include the two phosphate groups.

In one experiment, researchers isolated the clay particles after a day of incubation and then added a fresh batch of activated nucleotides. They repeated this reaction-isolation-reaction sequence for a total of 14 days, or 14 additions of fresh ribonucleotides. At the end of the two-week experiment, they analyzed the mineral particles, using techniques called **gel electrophoresis** and **autoradiography** (see **BioSkills 7** at the end of this book), and found RNA molecules up to 40 nucleotides long. Based on these results, there is a strong consensus that if ribonucleotides and deoxyribonucleotides were able to form during chemical evolution, they would be able to polymerize and form RNA and DNA. Now, what would these nucleic acids look like, and what could they do?

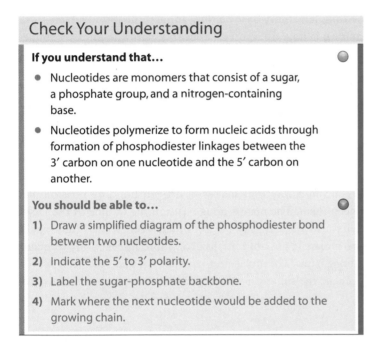

Check Your Understanding

If you understand that...

- Nucleotides are monomers that consist of a sugar, a phosphate group, and a nitrogen-containing base.
- Nucleotides polymerize to form nucleic acids through formation of phosphodiester linkages between the 3′ carbon on one nucleotide and the 5′ carbon on another.

You should be able to...

1) Draw a simplified diagram of the phosphodiester bond between two nucleotides.
2) Indicate the 5′ to 3′ polarity.
3) Label the sugar-phosphate backbone.
4) Mark where the next nucleotide would be added to the growing chain.

4.2 DNA Structure and Function

Nucleic acids have a primary structure that is somewhat similar to the primary structure of proteins. Proteins have a peptide-bonded backbone with a series of R-groups that extend from it; DNA and RNA molecules have a sugar-phosphate backbone, created by phosphodiester linkages, and a sequence of four nitrogenous bases that extend from it. Recall that the primary structures of DNA and RNA contrast in two fundamental aspects: The sugar in DNA's backbone is deoxyribose instead of the ribose in RNA, and RNA contains the nitrogenous base uracil instead of the thymine found in DNA.

DNA and RNA also have secondary structure, analogous to the secondary structure of proteins. But the secondary structures that form in DNA and RNA differ from each other and contrast with those observed in proteins. Distinctive secondary structure, combined with differences in primary structure, dictate why DNA and RNA have different functions in organisms. Let's analyze the secondary structure and function of DNA first, and then dig into the secondary structure and function of RNA.

What Is the Nature of DNA's Secondary Structure?

The solution to DNA's secondary structure, announced in 1953, ranks among the great scientific breakthroughs of the twentieth century. James Watson and Francis Crick presented a model for the secondary structure of DNA in a one-page paper published in *Nature*. At the time, Watson was a 25-year-old postdoctoral fellow and Crick was a 37-year-old graduate student.

Watson and Crick's finding was a hypothesis based on a series of results from other laboratories. They were trying to propose a secondary structure that could explain several important observations about the DNA found in cells:

- Chemists had worked out the structure of nucleotides and knew that DNA polymerized through the formation of phosphodiester linkages. Thus, Watson and Crick knew that the molecule had a sugar-phosphate backbone.

- By analyzing the nitrogenous bases in DNA samples from different organisms, Erwin Chargaff had established two empirical rules: (1) The total number of purines and pyrimidines in DNA is the same; and (2) the numbers of T's and A's in DNA are equal, and the numbers of C's and G's in DNA are equal.

- By bombarding DNA with X-rays and analyzing how it scattered the radiation, Rosalind Franklin and Maurice Wilkins had calculated the distances between groups of atoms in the molecule (see **Bioskills 8** for an introduction to this techique, called **X-ray crystallography**). The scattering patterns showed that three distances were repeated many times: 0.34 nanometer (nm), 2.0 nm, and 3.4 nm. Because the measurements repeated, the researchers inferred that DNA molecules had a regular and repeating structure. The pattern of X-ray scattering suggested that the molecule was helical or spiral in nature.

Based on this work, understanding DNA's structure boiled down to understanding the nature of the helix involved. What type of helix would have a sugar-phosphate backbone and explain both Chargaff's rules and the Franklin–Wilkins measurements?

To answer this question, Watson and Crick began by analyzing the size and geometry of deoxyribose, phosphate groups, and the nitrogenous bases. The bond angles and measurements suggested that the distance of 2.0 nm probably represented the width of the helix and that 0.34 nm was likely to be the distance between bases stacked in a spiral. Now they needed to make sense of Chargaff's rules and the 3.4-nm distance, which

FIGURE 4.5 Building a Physical Model of DNA Structure. Watson (left) and Crick (right) represented the arrangement of the four deoxyribonucleotides in a double-helical arrangement, using wires with precise lengths and geometries.

appeared to be exactly 10 times the distance between a single pair of bases.

To solve this problem, Watson and Crick constructed a series of physical models such as the one pictured in **Figure 4.5**. Building these models allowed them to tinker with different types of helical configurations. After many false starts, they hit on an idea that looked promising. ⬤ They arranged two strands of DNA side by side and running in opposite directions—meaning that one strand ran in the 5′ → 3′ direction while the other was oriented 3′ → 5′. Strands with this orientation are said to be antiparallel. Watson and Crick found that if the antiparallel strands were twisted together to form a double helix, the coiled sugar-phosphate backbones ended up on the outside of the spiral and the nitrogenous bases on the inside. But for the bases from each backbone to fit in the interior of the 2.0-nm-wide structure, they had to form purine-pyrimidine pairs (see **Figure 4.6a**). With that came a fundamental insight: Inside the double helix, the bases lined up in a way that allowed hydrogen bonds to form between certain purines and pyrimidines. More specifically, adenine could form hydrogen bonds with thymine, and guanine could form hydrogen bonds with cytosine (**Figure 4.6b**). Because of this specificity, the A-T and G-C bases were said to be complementary. Two hydrogen bonds form when A and T pair, but three hydrogen bonds form when G and C pair. As a result, the G-C interaction is slightly stronger than the A-T bond.

Watson and Crick had discovered the phenomenon known as **complementary base pairing**. In fact, the term **Watson–Crick pairing** is now used interchangeably with the phrase complementary base pairing.

Figure 4.7 shows how antiparallel strands of DNA form when complementary bases line up and form hydrogen bonds. As you study the figure, notice that DNA is put together like a ladder whose ends have been twisted in opposite directions. The sugar-phosphate backbone forms the supports of the ladder; the base pairs represent the rungs of the ladder. The twisting occurs because it allows the nitrogenous bases to line up in a way that makes hydrogen bonding possible between them. The nitrogenous bases in the middle of the DNA helix also stack tightly on top of each other. This tight packing makes it possible for hydrophobic interactions to occur among bases. The molecule as a whole is water soluble, however, because the sugar-phosphate backbones, which face the exte-

(a) Only purine-pyrimidine pairs fit inside the double helix.

Purine-pyrimidine pair
JUST RIGHT

Purine-purine pair
NOT ENOUGH SPACE

Pyrimidine-pyrimidine pair
TOO MUCH SPACE

Space inside sugar-phosphate backbones

FIGURE 4.6 Complementary Base Pairing Is Based on Hydrogen Bonding. (a) Only purine-pyrimidine pairs fit inside the double helix effectively. **(b)** Hydrogen bonds form when guanine and cytosine pair, and when adenine and thymine pair.

(b) Hydrogen bonds form between G-C pairs and A-T pairs.

DNA contains thymine, whereas RNA contains uracil

FIGURE 4.7 The Secondary Structure of DNA Is a Double Helix. Complementary base pairing is responsible for twisting DNA into a double helix.

rior of the molecule, are negatively charged and thus are hydrophilic.

It is also important to note that the outside of the helical DNA molecule forms two types of grooves and that the types differ in size. The larger of the two is known as the major groove, and the smaller one is known as the minor groove. **Figure 4.8** highlights these grooves and illustrates how DNA's secondary structure explains the measurements observed by Franklin and Wilkins (see **Box 4.1**).

Since the model of the double helix was published, experimental tests have shown that the hypothesis is correct in almost every detail. DNA's secondary structure consists of two antiparallel strands twisted into a **double helix**. The molecule is stabilized by hydrophobic interactions between the bases in its interior and by hydrogen bonding between the complementary base pairs A-T and G-C. ◯ If you understand DNA's secondary structure, you should be able to explain why the major and minor grooves exist, and why the strands are antiparallel to each other rather than parallel. Now the question is, How does this secondary structure affect the molecule's function?

DNA Is an Information-Containing Molecule

Watson and Crick's model created a sensation because it revealed how DNA could store and transmit biological information. In literature, information consists of letters on a page.

FIGURE 4.8 Dimensions of DNA Secondary Structure. The double-helix hypothesis explains the measurements inferred from X-ray analysis of DNA molecules.

◯ **QUESTION** What do the red pentagons and the yellow balls represent? What do the lines of green and purple dots represent?

In music, information is composed of the notes on a staff. But inside cells, information consists of a sequence of monomers in a nucleic acid. The four nitrogenous bases function like letters of the alphabet. A particular sequence of bases is like the sequence of letters in a word—it has meaning. In all cells that have been examined to date, from tiny bacteria to gigantic redwood trees, DNA carries the information required for the organism's growth and reproduction. Exploring how this information is encoded and translated into action is the heart of Chapters 15 through 18.

Here, however, our focus is on how life began. The theory of chemical evolution holds that life began as a self-replicating molecule—a molecule that could make a copy of itself. **Figure 4.9** shows how a copy of DNA can be made by complementary base pairing. In step 1 of this diagram, heating or enzyme-catalyzed reactions cause the double helix to separate. In step 2, free deoxyribonucleotides form hydrogen bonds with complementary bases on the original strand of DNA—also called a **template strand**. As they do, their sugar-phosphate groups form phosphodiester linkages to create a new strand—also called a **complementary strand**. Note that the 5′ → 3′ directionality of the complementary strand is opposite that of the template strand. In this way, complementary base pairing allows each strand of a DNA double helix to be copied exactly, producing the two daughter molecules shown in step 3. Thus, DNA contains the information required for a copy of itself to be made.

Watson and Crick ended their paper on the double helix with one of the classic understatements in the scientific literature: "It has not escaped our notice that the specific pairing we have postulated immediately suggests a possible copying mechanism." The central idea is that DNA's primary structure serves as a mold or template for the synthesis of a complementary strand.

(MB) **Web Animation** at www.masteringbio.com
Nucleic Acid Structure

In today's cells, however, DNA does not self-replicate spontaneously. Instead, the molecule is copied through a complicated series of reactions catalyzed by a large suite of enzymes. Consistent with these data, researchers consider it extremely unlikely that a DNA molecule began copying itself early in Earth's history. Let's explore why.

Is DNA a Catalytic Molecule?

The DNA double helix is a highly structured molecule that is much more stable than RNA and most proteins. DNA is regular and symmetric, with few chemical groups exposed that can participate in chemical reactions. For example, the lack of a 2′ hydroxl group on deoxyribonucleotides (see Figure 4.1b) makes the polymer much less reactive than RNA and highly resistant to chemical degradation. All of these features increase DNA's stability and thus its effectiveness as a reliable information-bearing molecule. Intact stretches of DNA have been recovered from fossils that are tens of thousands of years old. Despite death and exposure to a wide array of pH,

DNA FORMS A TEMPLATE FOR ITS OWN SYNTHESIS.

1. If the hydrogen bonds between complementary base pairs are broken, the DNA helix can separate.

2. Each strand of DNA can serve as a template for the formation of a new strand. Free nucleotides attach according to complementary base pairing.

3. When the new strands polymerize to form a sugar-phosphate backbone, secondary structure is restored.

Conclusion: The original molecule has been copied. Each copy has one strand from the original DNA molecule and one new strand.

FIGURE 4.9 Making a Copy of DNA. If new bases are added to each of the two strands of DNA via complementary base pairing, a copy of the DNA molecule can be produced.

● **QUESTION** When double-stranded DNA is heated to 95°C, the bonds between complementary base pairs break and single-stranded DNA results. Based on this observation, is the reaction shown in step 1 endergonic or exergonic?

BOX 4.1 The Human Side of Research

The discovery of the double helix was a critically important advance in the early 1950s, because experiments conducted in the 1940s—and described in detail in Chapter 14—had established that DNA is the hereditary material. Stated another way, scientists had recently discovered that genes are made of DNA. Immediately thereafter, understanding DNA's structure became *the* question in biology. Researchers who were working on the problem knew that the greatest honor in science—the Nobel Prize—would almost certainly be awarded to the person or persons who were first to discover the correct structure. The race was on: Many of the brightest and most famous scientists of the day were involved, working in different laboratories around the world.

Arguably the most important event in the research was Wilkins's and Franklin's ability to produce X-ray pictures of DNA.

Recall that the photographs made it possible to make crucial measurements and inferences about the molecule's structure. Although Wilkins and Franklin are justifiably given joint credit for this work, they actually did not cooperate closely. It was Franklin who took the photographs that led to the realization that DNA is helical, but it was Wilkins who showed her photographs to Watson and Crick. Wilkins did so without Franklin's knowledge or permission. Today, such an act would be considered a serious form of professional misconduct. Another interesting wrinkle to this story is that Linus Pauling—undoubtedly the century's greatest biochemist—was also working on the problem at the time, but he was denied permission to travel from the United States to England to see the data. The United States was then in the throes of the Cold War, and the U.S. State Department had revoked Pauling's passport

because his political views were considered too liberal.

As expected, Watson and Crick, along with Wilkins, received the Nobel Prize. The award was made in 1962, after experimental work had confirmed that the double-helix model was correct. Tragically, Franklin died of ovarian cancer in 1958, when she was just 37. She never received the Nobel, because the honor is not awarded posthumously.

The moral of the story, if there is one, may be simply that scientific research is a very human enterprise. In addition, times have changed. Although competition can still be intense, biology is an increasingly cooperative and international enterprise; misconduct is usually detected and addressed promptly. Rosalind Franklin's professional life was made difficult at times because she was a woman. Today most of the graduate students in molecular biology are female.

---

temperature, and chemical conditions, the molecules have the same sequence of bases as the organisms had when they were alive.

The orderliness and stability that make DNA such a dependable information repository make it extraordinarily inept at catalysis, however. Recall from Chapter 3 that enzyme function is based on a specific binding event between a substrate and a protein catalyst. Thanks to the enormous diversity of primary through quaternary structures found in proteins, a wide array of binding events can occur. In comparison, DNA's primary and secondary structures are simple. It is not surprising, then, that DNA has never been observed to catalyze any reaction in any organism. Even though researchers have been able to construct single-stranded DNA molecules that can catalyze a few simple reactions in the laboratory, the number and diversity of reactions involved is a minute fraction of the activity catalyzed by proteins.

In short, DNA furnishes an extraordinarily stable template for copying itself. But due to its inability to act as an effective catalyst, virtually no researchers support the hypothesis that the first life-form consisted of DNA. Instead, most biologists who are working on the origin of life support the hypothesis that life began with RNA.

Check Your Understanding

If you understand that...

- DNA's primary structure consists of a sequence of deoxyribonucleotides.

- DNA's secondary structure consists of two DNA molecules that run in opposite $5' \rightarrow 3'$ orientations to each other. The two strands are twisted into a double helix, and they are held together through hydrogen bonds between A-T and G-C pairs and hydrophobic interactions among atoms inside the helix.

- The sequence of deoxyribonucleotides in DNA contains information, much like the sequence of letters and punctuation on this page. Due to complementary base pairing, each DNA strand also contains the information required to form the complementary strand.

You should be able to...

Make a sketch of a DNA molecule that

1) Indicates at least four complementary base pairs.

2) Labels the molecule's sugar-phosphate backbone, the hydrogen bonds between complementary bases, and the location of hydrophobic interactions that stabilize the helix.

3) Indicates the $5' \rightarrow 3'$ orientation of each strand.

4.3   RNA Structure and Function

Like DNA, RNA has a primary structure consisting of a sugar-phosphate backbone formed by phosphodiester linkages and, extending from that backbone, a sequence of four types of nitrogenous bases. But it's important to recall two significant differences between these nucleic acids:

1. The pyrimidine base thymine does not exist in RNA. Instead, RNA contains the closely related pyrimidine base uracil.

2. The sugar in the sugar-phosphate backbone of RNA is ribose, not deoxyribose as in DNA.

The second point is critical because the hydroxyl (–OH) group on the 2′-carbon of ribose is much more reactive than the hydrogen atom on the 2′-carbon of deoxyribose. This functional group can participate in reactions that tear the polymer apart. The presence of this –OH group makes RNA much less stable than DNA.

A comparison of secondary structure in RNA and DNA is just as instructive. Like DNA molecules, most RNA molecules have secondary structure that results from complementary base pairing between purine and pyrimidine bases. In RNA, adenine forms hydrogen bonds only with uracil, and guanine again forms hydrogen bonds with cytosine. (Guanine can also bond with uracil, but it does so much less effectively than with cytosine.) Thus, the complementary base pairs in RNA are A-U and G-C. Three hydrogen bonds form between guanine and cytosine, but only two form between adenine and uracil.

How do the secondary structures of RNA and DNA differ? In the vast majority of cases, the purine and pyrimidine bases in RNA undergo hydrogen bonding with complementary bases on the same strand, rather than forming hydrogen bonds with complementary bases on a different strand, as in DNA. **Figure 4.10** shows how this works. The key is that when bases on one part of an RNA strand fold over and align with ribonucleotides on another segment of the same strand, the two sugar-phosphate strands are antiparallel. In this orientation, hydrogen bonding between complementary bases results in a stable double helix.

If the section where the fold occurs includes a large number of unbonded bases, then the stem-and-loop configuration shown in Figure 4.10 results. This type of secondary structure is called a **hairpin**. Several other types of RNA secondary structures are possible, each involving a different length and arrangement of base-paired segments.

Like the α–helices and β–pleated sheets observed in many proteins, RNA secondary structures are stabilized by hydrogen bonding and occur spontaneously. Even though hairpins and other types of secondary structures reduce the entropy of RNA molecules, they form spontaneously because the energy released in formation of H-bonds makes the overall process favorable. Hydrogen bond formation is exothermic; and overall, the formation of hairpins is exergonic.

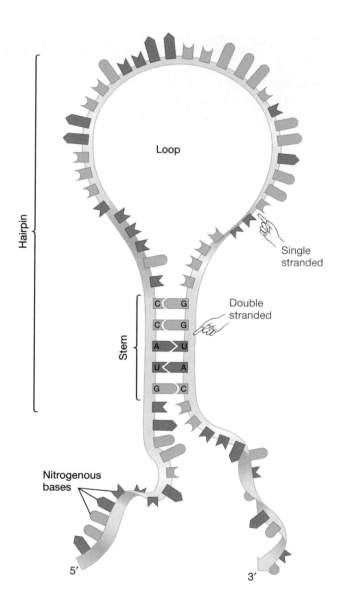

FIGURE 4.10 Complementary Base Pairing and Secondary Structure in RNA: Stem-and-Loop Structures. This RNA molecule has secondary structure. The double-stranded "stem" and single-stranded "loop" form a hairpin. The bonded bases in the stem are antiparallel, meaning that they are oriented in opposite directions.

● RNA molecules can also have tertiary structure and quaternary structure, owing to interactions that fold secondary structures into complex shapes or that hold different RNA strands together. As a result, RNA molecules with different base sequences can have very different overall shapes and chemical properties. Although RNA molecules do not begin to rival proteins in overall structural diversity or the presence of various functional groups, RNA molecules are much more diverse in size, shape, and reactivity than DNA molecules are. Structurally and chemically, RNA is intermediate between the complexity of proteins and the simplicity of DNA.

Based on these observations, it is logical to note that RNA is also intermediate in terms of its function in cells. RNA molecules cannot archive information nearly as efficiently as DNA molecules do; but RNAs do perform key functions in information processing. Likewise, they cannot catalyze as many reactions as proteins do, but certain RNAs catalyze particularly important reactions. RNA molecules are also involved in defending cells, providing structure, and regulating the cell's response to changes in the environment. In cells, RNA molecules function like a jack-knife or a pocket tool with an array of attachments: They perform a wide variety of tasks reasonably well.

The diversity of roles that RNA plays in cells will be a prominent theme in later chapters. Here let's focus on the roles that RNA could have played in the origin of life—as an information-containing entity and as a catalyst.

RNA as an Information-Containing Molecule

Because RNA contains a sequence of bases analogous to the letters in a word, it can function as an information-containing molecule. And because hydrogen bonding occurs specifically between A-U pairs and G-C pairs in RNA, it is possible for RNA to furnish the information required to make a copy of itself. The process occurs as shown for DNA in Figure 4.9, except that the template RNA exists as a single strand instead of a double strand. When RNA is copied, free ribonucleotides form hydrogen bonds with complementary bases on the original strand of RNA—the template strand. As they do, their sugar-phosphate groups form phosphodiester linkages to produce the complementary strand. Finally, the hydrogen bonds between the strands are broken by heating or by a catalyzed reaction. The new RNA molecule now exists independently of the original strand. If these steps were repeated with the new strand used as a template, the resulting molecule would be a copy of the original. In this way, an RNA's primary sequence serves as a mold. ● You should be able to draw the steps involved in synthesizing a complementary strand of RNA, analogous to the steps described for DNA in Figure 4.9.

Although complementary base pairing allows RNA to carry the information required for the molecule to be copied, RNA is not nearly as stable a repository for particular sequences as is DNA. In the prebiotic soup, could an RNA molecule have made a copy of itself before it would have degraded?

Is RNA a Catalytic Molecule?

In terms of diversity in chemical reactivity and overall shape, RNA molecules do not begin to match proteins. The primary structure of RNA molecules is much more restricted because there are only four types of nitrogenous bases in RNA versus the 21 types of amino acids found in proteins, and RNA molecules cannot form the variety of bonds that give proteins their extensive tertiary structure. But because RNA has a degree of structural and chemical complexity, it should be capable of stabilizing a few transition states and catalyzing at least a limited number of chemical reactions. Indeed, Sidney Altman and Thomas Cech shared the 1989 Nobel Prize in chemistry for

showing that catalytic, enzyme-like RNAs, or **ribozymes**, exist in organisms. The ribozymes they isolated, from a single-celled organism called *Tetrahymena*, catalyzed both the hydrolysis and condensation of phosphodiester linkages. Building on that accomplishment, researchers have discovered ribozymes that catalyze dozens of different reactions in cells. For example, ribozymes catalyze the formation of peptide bonds when amino acids polymerize to form polypeptides. Ribozymes are at work in your cells right now.

The discovery of ribozymes was a watershed event in origin-of-life research. Before Altman and Cech published their discovery, biologists thought that proteins were the only type of molecule capable of catalyzing chemical reactions in organisms. But if a ribozyme in *Tetrahymena* catalyzes polymerization reactions similar to those diagrammed in steps 1 and 2 of Figure 4.9, it raises the possibility that an RNA molecule could make a copy of itself. Such a molecule could copy itself and qualify as the first living entity. Does any experimental evidence support this hypothesis?

4.4 The First Life-Form

The theory of chemical evolution maintains that life began as a naked self-replicator—a molecule that existed by itself in solution, without being enclosed in a membrane. To make a copy of itself, that first living molecule had to provide a template that could be copied. It also had to catalyze polymerization reactions that would link monomers into a copy of that template. Because RNA is capable of both processes, most origin-of-life researchers propose that the first life-form was made of RNA. As noted in the introduction to this chapter, that proposal is called the RNA world hypothesis.

Because no self-replicating RNA molecules exist today, researchers test the hypothesis by trying to simulate the RNA world in the laboratory. The eventual goal is to create an RNA molecule that can catalyze its own replication. When this goal is achieved, a life-form will have been created in the laboratory.

To understand how researchers go about this work, consider a recent set of experiments by Wendy Johnston and others working in David Bartel's laboratory. This research group's goal was ambitious; they wanted to create, from scratch, a ribozyme that was capable of catalyzing the addition of ribonucleotide triphosphates to an existing strand, via complementary base pairing. Such a ribozyme would be called RNA replicase. This research program is creating considerable excitement among biologists interested in the origin of life, because adding ribonucleotides to a growing strand is a key attribute of an RNA replicase.

The strategy Bartel's group pursued is outlined in **Figure 4.11**. As you study this figure, note that their procedure essentially mimics the process of natural selection introduced in

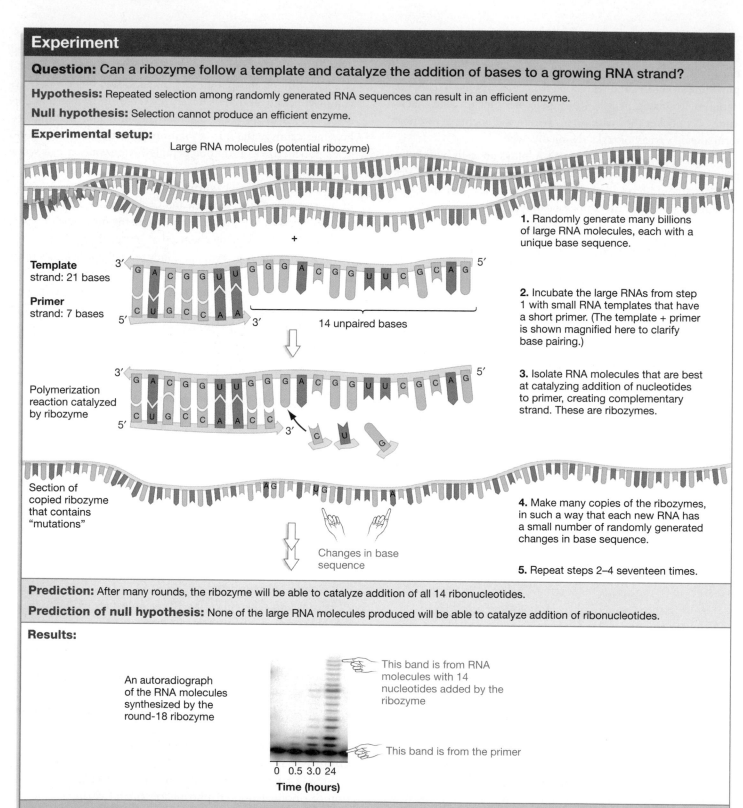

Experiment

Question: Can a ribozyme follow a template and catalyze the addition of bases to a growing RNA strand?

Hypothesis: Repeated selection among randomly generated RNA sequences can result in an efficient enzyme.

Null hypothesis: Selection cannot produce an efficient enzyme.

Experimental setup:

Large RNA molecules (potential ribozyme)

1. Randomly generate many billions of large RNA molecules, each with a unique base sequence.

+

Template strand: 21 bases

Primer strand: 7 bases

14 unpaired bases

2. Incubate the large RNAs from step 1 with small RNA templates that have a short primer. (The template + primer is shown magnified here to clarify base pairing.)

Polymerization reaction catalyzed by ribozyme

3. Isolate RNA molecules that are best at catalyzing addition of nucleotides to primer, creating complementary strand. These are ribozymes.

Section of copied ribozyme that contains "mutations"

Changes in base sequence

4. Make many copies of the ribozymes, in such a way that each new RNA has a small number of randomly generated changes in base sequence.

5. Repeat steps 2–4 seventeen times.

Prediction: After many rounds, the ribozyme will be able to catalyze addition of all 14 ribonucleotides.

Prediction of null hypothesis: None of the large RNA molecules produced will be able to catalyze addition of ribonucleotides.

Results:

An autoradiograph of the RNA molecules synthesized by the round-18 ribozyme

This band is from RNA molecules with 14 nucleotides added by the ribozyme

This band is from the primer

0   0.5 3.0 24
Time (hours)

Conclusion: If allowed to react for 24 hours, the round-18 ribozyme can add all 14 bases to the primer, producing a strand complementary to the template.

FIGURE 4.11 Selection of a Novel Ribozyme. Researchers use variations of this protocol to produce ribozymes that can catalyze a wide variety of reactions.

Chapter 1. For example, the researchers began their experiments by establishing a large population of RNA molecules that were variable in structure and function, analogous to a large population of organisms with a variety of features. In this case, the researchers created millions of large RNA molecules that contained randomly generated primary sequences. They incubated these newly synthesized, large RNAs with a small RNA "primer," similar to the short RNAs that bind to template strands when RNA is being synthesized in cells. When the large RNAs and the primers were allowed to react with a large supply of free ribonucleotides, the researchers found that some of the large molecules could catalyze the addition of a few ribonucleotides that were complementary to the bases in the template. In this "round-1" experiment, the polymerization didn't happen very quickly or accurately, but it did occur. From their initial pool of millions of randomly generated molecules, Bartel's team had discovered a ribozyme that could catalyze the synthesis of RNA.

To continue the experiment, the researchers isolated the round-1 ribozymes. This step was analogous to natural selection, because only certain variants survived to produce "offspring." To create these offspring, which would represent the next generation of ribozymes, Bartel's team copied the selected molecules in a way that introduced a few random changes to their sequence of bases. These changes were analogous to the types of mutations that occur each generation in natural populations. Mutation is a random process that produces new traits. When the researchers incubated the modified ribozymes with a new template RNA, they found that most of the modified ribozymes worked worse than the original ones. Some worked better, however.

The researchers selected the best of these round-2 ribozymes, copied them in a way that introduced a few additional, random changes in their base sequences, and allowed them to react with yet another template. The best ribozymes in this round 3 were isolated, and then the copying, reaction, and selection process was repeated . . . and repeated . . . and repeated. By round 18, the group had found a ribozyme that was a much better catalyst than the original molecule. Evolution had occurred. They had created a ribozyme that was reasonably proficient at adding ribonucleotides to a growing strand.

As this book goes to press, the round-18 ribozyme is the closest biologists have come to creating life. Thanks to similar efforts at other laboratories around the world, researchers have produced an increasingly impressive set of ribozymes—an array of molecules capable of catalyzing many of the key reactions responsible for replication and metabolism. Each result provides support for the RNA world hypothesis. Each result also brings research teams closer to the creation of an RNA replicase. If this goal is met, human beings will have created a living entity in a test tube.

Chapter Review

SUMMARY OF KEY CONCEPTS

● Nucleotides are monomers that consist of a sugar, a phosphate group, and a nitrogen-containing base. Ribonucleotides polymerize to form RNA. Deoxyribonucleotides polymerize to form DNA.

Ribonucleotides have a hydroxyl (–OH) group on their 2′ carbon, while deoxyribonucleotides do not. Both types of nucleotides polymerize via phosphodiester bonds to form nucleic acids, which have a sugar-phosphate backbone with nitrogenous bases attached.

You should be able to diagram the difference between a ribonucleotide and a deoxyribonucleotide. You should also be able to draw a generalized nucleic acid and label the sugar-phosphate backbone, phosphodiester linkages, 5′ to 3′ polarity, and nitrogenous bases. ●

● DNA's primary structure consists of a sequence of nitrogen-containing bases, which contain information in the form of a molecular code. DNA's secondary structure consists of two DNA strands running in opposite directions. The strands are held together by complementary base pairing and are twisted into a double helix.

DNA is an extremely stable molecule that serves as a superb archive for information in the form of base sequences. DNA is stable because deoxyribonucleotides lack a reactive 2′ hydroxyl group and because antiparallel DNA strands form a secondary structure called a double helix. The DNA double helix is stabilized by hydrogen bonds that form between complementary purine and pyrimidine bases and by hydrophobic interactions between bases stacked on the inside of the spiral. This same structural stability and regularity make DNA ineffective at catalysis, however.

In addition to being stable, DNA is readily copied via complementary base pairing. Complementary base pairing occurs between A-T and G-C pairs in DNA.

You should be able to explain why DNA molecules with a high percentage of guanine and cytosine are particularly stable. You should also be able to draw a DNA double helix where there is a purine–purine pair or a pyrimidine–pyrimidine pair. ●

(MB) **Web Animation** at www.masteringbio.com
Nucleic Acid Structure

● RNA's primary structure also consists of a sequence of nitrogen-containing bases that contain information in the form of a molecular code. Its secondary structure includes short double helices and structures called hairpins. RNA molecules called ribozymes catalyze important chemical reactions.

RNA molecules can have secondary structure because complementary base pairing occurs between A-U and G-C pairs on the same strand. Some tertiary and quarternary structure also occurs, because RNA molecules fold in precise ways and may interact with each other.

Compared with proteins and DNA, RNA is exceptionally versatile. The primary function of proteins is to catalyze chemical reactions, and the primary function of DNA is to carry information. But RNA is an "all-purpose" macromolecule that can do both.

You should be able to explain why RNA molecules can have tertiary and quaternary structure, while DNA molecules do not. ●

QUESTIONS

● Test Your Knowledge

1. What are the four nitrogenous bases found in RNA?
   a. uracil, guanine, cytosine, thymine (U, G, C, T)
   b. adenine, guanine, cytosine, thymine (A, G, C, T)
   c. adenine, uracil, guanine, cytosine (A, U, G, C)
   d. alanine, threonine, glycine, cysteine (A, T, G, C)

2. What determines the primary structure of an RNA molecule?
   a. the sugar-phosphate backbone
   b. complementary base pairing and the formation of hairpins
   c. the sequence of deoxyribonucleotides
   d. the sequence of ribonucleotides

3. DNA attains a secondary structure when hydrogen bonds form between the nitrogenous bases called purines and pyrimidines. What are the complementary base pairs that form in DNA?
   a. A-T and G-C
   b. A-U and G-C
   c. A-G and T-C
   d. A-C and T-G

4. By convention, biologists write the sequence of bases in RNA and DNA in which direction?
   a. $3' \rightarrow 5'$
   b. $5' \rightarrow 3'$
   c. N-terminal to C-terminal
   d. C-terminal to N-terminal

5. The secondary structure of DNA is called a double helix. Why?
   a. Two strands wind around one another in a helical, or spiral, arrangement.
   b. A single strand winds around itself in a helical, or spiral, arrangement.
   c. It is shaped like a ladder.
   d. It stabilizes the molecule.

6. In RNA, when does the secondary structure called a hairpin form?
   a. when hydrophobic residues coalesce
   b. when hydrophilic residues interact with water
   c. when complementary base pairing between ribonucleotides on the same strand creates a stem-and-loop structure
   d. when complementary base pairing forms a double helix

Test Your Knowledge answers: 1. c; 2. d; 3. a; 4. b; 5. a; 6. c

● Test Your Understanding

Answers are available at www.masteringbio.com

1. Make a concept map (see **BioSkills 6**) that relates DNA's primary structure to its secondary structure. Your diagram should include deoxyribonucleotides, purines, pyrimidines, phosphodiester bonds, DNA, complementary base pairing, and antiparallel strands.

2. Make a generalized diagram of monomers undergoing condensation reactions to form a polymer. Label the type of bond that forms when nucleotides polymerize into RNA or DNA. Does it take energy for polymerization reactions to proceed, or do they occur spontaneously? Why or why not?

3. Growing strands of nucleic acids are always extended in the $5' \rightarrow 3'$ direction. What do the $5'$ and $3'$ refer to? Now recall that only activated nucleotides (nucleotide triphosphates) can be added to a growing strand. Draw an activated nucleotide being added to an RNA strand.

4. Why is DNA such a stable molecule compared with proteins or RNA?

5. Explain how the secondary structures called hairpins form in RNA. Include a labeled diagram.

6. A major theme in this chapter is that the structure of molecules correlates with their function. Explain why DNA's secondary structure limits its catalytic abilities compared with that of RNA. Why is it logical that RNA molecules can catalyze a modest but significant array of reactions? Why are proteins the most effective catalysts?

● Applying Concepts to New Situations

Answers are available at www.masteringbio.com

1. Do you agree with the "definition" of life given at the start of this chapter? Why or why not? If you were looking for life on Jupiter's moon Europa or on Mars, how would you know when you found it?

2. Suppose that experiments like those reviewed in Section 4.4 succeeded in producing a molecule that could make a copy of itself. Outline a one-page opinion piece for your local newspaper that explains the nature of the research and discusses the ethical and philosophical implications of the discovery.

3. Before Watson and Crick published their model of the DNA double helix, Linus Pauling offered a model based on a triple helix. Draw your conception of what a DNA triple helix might look like. What interactions would keep such a secondary structure together? How could such a molecule be copied?

4. Origin-of-life researcher Robert Crabtree maintains that experiments simulating early Earth conditions are a valid way to test the theory of chemical evolution. Crabtree claims that if scientists working in the field agree that an experiment is a plausible reproduction of early Earth conditions, it is valid to infer that its results are probably correct—that the simulation effectively represents events that occurred some 3.5 billion years ago. Do you agree? Do you find the models and experiments presented in this chapter and previous chapters to be convincing tests of the theory? Explain your answers.

5

An Introduction to Carbohydrates

KEY CONCEPTS

- Sugars and other carbohydrates are highly variable in structure.

- Monosaccharides are monomers that polymerize to form polymers called polysaccharides, via different types of glycosidic linkages.

- Carbohydrates perform a wide variety of functions in cells, ranging from energy storage to formation of tough structural fibers.

This scanning electron micrograph shows plant cell walls (yellow honeycombed structures), which consist primarily of cellulose. The orange granules are stored starch. Both cellulose and starch are carbohydrates. Cellulose provides structural support for these cells. Starch supplies chemical energy.

This unit highlights the four types of macromolecules that are prominent in today's cells: proteins, nucleic acids, carbohydrates, and lipids. Understanding the structure and function of each of these macromolecules is a basic requirement for exploring how life began and how organisms work. Chapters 3 and 4 analyzed the way proteins and nucleic acids are put together and what they do. This chapter focuses on carbohydrates; Chapter 6 will introduce lipids.

The term **carbohydrate** encompasses the monomers called monosaccharides ("one-sugar"), small polymers called oligosaccharides ("few-sugars"), and the large polymers called polysaccharides ("many-sugars"). The name carbohydrate is logical because the chemical formula of many of these molecules is $(CH_2O)_n$, where the n refers to the number of "carbon-hydrate" groups. The name can also be misleading, though, because carbohydrates do not consist of carbon atoms bonded to water molecules. Instead, they are molecules with a carbonyl ($C=O$) and several hydroxyl (–OH) functional groups, along with several to many carbon-hydrogen (C–H) bonds.

Let's begin by considering the structure of monosaccharides and continue by analyzing how these building blocks polymerize to form polysaccharides. The chapter closes with an analysis of the role of carbohydrates in the origin of life and a look at what carbohydrates do in cells today.

● Key Concept   Important Information   ● Practice It

5.1 Sugars as Monomers

Sugars are fundamental to life. They provide chemical energy in cells and furnish some of the molecular building blocks required for the synthesis of larger, more complex compounds. Monosaccharides were also important during chemical evolution, early in Earth's history. For example, the sugar called ribose is required for the formation of nucleotides. Laboratory simulations have shown that ribose and many other monosaccharides could have been produced in the prebiotic soup. What are these compounds, and how do they differ from one another?

Figure 5.1 illustrates the structure of the monomer called a **monosaccharide**, or simple sugar. The carbonyl group that serves as one of monosaccharides' distinguishing features can be found either at the end of the molecule, forming an aldehyde sugar (an aldose), or within the carbon chain, forming a ketone sugar (a ketose). The presence of a carbonyl group along with multiple hydroxyl groups provides an array of functional groups in sugars. Based on this observation, it's not surprising that sugars are able to participate in a large number of chemical reactions.

The number of carbon atoms present also varies in monosaccharides. By convention, the carbons in a monosaccharide are numbered consecutively, starting with the end nearest the carbonyl group. Figure 5.1 features three-carbon sugars, or **trioses**. Ribose, which acts as a building block for nucleotides, has five carbons and is called **pentose**; the glucose that is coursing through your bloodstream and being used by your cells right now is a six-carbon sugar, or **hexose**.

Besides varying in the location of the carbonyl group and the total number of carbon atoms present, monosaccharides can vary in the spatial arrangement of their atoms. There is, for example, a wide array of pentoses and hexoses. Each is distinguished by the configuration of its hydroxyl functional groups. **Figure 5.2** illustrates glucose and galactose, which are six-carbon sugars that are optical isomers—they have the same chemical formula ($C_6H_{12}O_6$) but not the same structure. Although both are aldose sugars with six carbons, they differ in the spatial arrangement of the hydroxyl group at the carbon highlighted in Figure 5.2. Because their structures differ, their functions differ. In cells, glucose is used as a source of chemical energy that sustains life. But for galactose to be used as a source of energy, it first has to be converted to glucose via an enzyme-catalyzed reaction. A total of eight different hexoses exist, due to the different ways of configuring the hydroxyl groups in space. In addition, each hexose comes in two forms, each a mirror image of the other. Thus, 16 distinct structures with the molecular formula $C_6H_{12}O_6$ are possible.

Sugars do not usually exist in the form of linear chains as illustrated in Figure 5.1 and Figure 5.2, however. Instead, in aqueous solution they tend to form ring structures. The chain and ring forms exist in equilibrium when simple sugars are in solution, although almost all are in the ring shape. Glucose serves as the example in **Figure 5.3**. When the cyclic structure forms in glucose, the carbon that is numbered 1 in the linear chain forms a bond with an oxygen atom and with a hydroxyl group that can be oriented in two distinct ways. As the right-hand side of Figure 5.3 shows, the different configurations result in the molecules α-glucose and β-glucose.

FIGURE 5.1 The Carbonyl Group in a Sugar Occurs in One of Two Configurations.

FIGURE 5.2 Sugars May Vary in the Configuration of Their Hydroxyl Groups. The two six-carbon sugars shown here vary only in the spatial orientation of their hydroxyl groups. These molecules are isomers, but they are not mirror images of each other.

● **EXERCISE** Mannose is a six-carbon sugar that is identical to glucose, except that the hydroxyl (OH) group on carbon number 2 is switched in orientation. Next to these structures, draw the structural formula of mannose.

● **QUESTION** What kind of isomers are mannose and glucose?

(a) Linear form of glucose   **(b)** Ring forms of glucose

FIGURE 5.3 **Sugars Exist in Linear and Ring Forms. (a)** The linear form of glucose is rare. **(b)** In solution, almost all glucose molecules spontaneously bend into one of two ring structures, called the α and β forms of glucose. The difference between the two forms lies in whether the hydroxyl group on carbon number 1 is above or below the plane of the ring. The two forms exist in equilibrium, but the β form is more common because it is slightly more stable than the α form.

● To summarize, many distinct monosaccharides exist because so many aspects of their structure are variable: alternative ring forms, optical isomers with different arrangements of hydroxyl groups in space and different mirror-image forms, variation in carbon number, and aldose or ketose placement of the carbonyl group. Each monosaccharide has a unique structure and function.

Laboratory simulations have shown that most monosaccharides are readily synthesized under conditions that mimic the prebiotic soup. For example, when formaldehyde (H_2CO) molecules are heated in solution, they react with one another to form almost all of the pentoses and hexoses, as well as some seven-carbon sugars. In addition, researchers recently announced the discovery of the three-carbon ketose illustrated in Figure 5.1, along with a wide array of compounds closely related to sugars, on the Murchison meteorite that struck Australia in 1969. Based on these observations, investigators suspect that sugars are synthesized on dust particles and other debris in interstellar space and could have rained down onto Earth as the planet was forming. Most researchers interested in chemical evolution maintain that a wide diversity of monosaccharides existed in the prebiotic soup. But it remains a mystery why ribose might have predominated and made the synthesis of nucleotides possible. It also appears highly unlikely that monosaccharides were able to polymerize to form the polysaccharides found in today's cells. Let's explore why.

Check Your Understanding

If you understand that...   ●

- Simple sugars differ in three respects:

    (1) the location of their carbonyl group,

    (2) the number of carbon atoms present, and

    (3) the spatial arrangement of their atoms—particularly the relative positions of hydroxyl (OH) groups.

You should be able to...   ●

Draw the structural formula of a monosaccharide in linear form and then draw other sugars that differ from this one in each of the three aspects listed.

5.2   The Structure of Polysaccharides

Polysaccharides are polymers that form when monosaccharides are linked together. They are also known as complex carbohydrates. The simplest polysaccharides consist of two sugars and are known as **disaccharides**. The two monomers involved may be identical, as in the two α-glucose molecules that link to form maltose. Or they may be different, as in the combination of a glucose molecule and a galactose molecule

(a) Monosaccharides polymerize when hydroxyl groups react to form glycosidic linkages...

The hydroxyl groups from the 1-carbon and 4-carbon react to produce an α-1,4-glycosidic linkage and water

(b) ...between various carbons and with various geometries.

In this case, the hydroxyl groups from the 1-carbon and 4-carbon react (with glucose flipped upside down) to produce a β-1,4-glycosidic linkage

FIGURE 5.4 Monosaccharides Polymerize through Formation of Glycosidic Linkages. A glycosidic linkage occurs when hydroxyl groups on two monosaccharides undergo a condensation reaction to form a bond.

that forms lactose—the most important sugar in milk. (**Box 5.1** on page 88 explains how problems with metabolizing lactose or galactose can lead to illness in humans.)

Simple sugars polymerize when a condensation reaction occurs between two hydroxyl groups, resulting in a covalent bond called a **glycosidic linkage** (**Figure 5.4a**). Glycosidic linkages are analogous to the peptide bonds that hold proteins together and to the phosphodiester bonds that connect the nucleotides in nucleic acids. There is an important difference, however. Peptide and phosphodiester bonds always form at the same location in their monomers. But because glycosidic linkages form between hydroxyl groups, and because every monosaccharide contains at least two hydroxyl groups, the location

and geometry of glycosidic linkages can vary widely among polysaccharides (**Figure 5.4b**). To drive this point home, consider the structures of the most common polysaccharides found in organisms today: starch, glycogen, cellulose, and chitin, along with a modified polysaccharide called peptidoglycan. Each of these macromolecules can consist of a few hundred to many thousands of monomers (**Figure 5.5**), joined by glycosidic linkages at different locations.

Starch: A Storage Polysaccharide in Plants

In plant cells, monosaccharides are stored for later use in the form of starch. **Starch** consists entirely of α-glucose monomers

FIGURE 5.5 Polysaccharides Are Polymers of Monosaccharides. Note that sugars are often drawn in simplified form for legibility.

EXERCISE Circle the glycosidic linkages in this polysaccharide.

that are joined by glycosidic linkages. As the top panel in **Table 5.1** shows, the angle of the linkages between carbons 1 and 4 causes the chain of glucose subunits to coil into a helix. Starch is actually a mixture of two such polysaccharides, however. One is an unbranched molecule called amylose, which contains only α-1,4-glycosidic linkages. The other is a branched molecule called amylopectin. The branching in amylopectin occurs when glycosidic linkages form between carbon 1 of a glucose monomer on one strand and carbon 6 of a glucose monomer on another strand. In amylopectin, branches occur in about one out of every 30 monomers.

Glycogen: A Highly Branched Storage Polysaccharide in Animals

Glycogen performs the same storage role in animals that starch performs in plants. In humans, for example, glycogen is stored in the liver and in muscles. When you start exercising, enzymes begin breaking glycogen into glucose monomers, which are then processed in muscle cells to supply energy. Glycogen is a polymer of α-glucose and is nearly identical to the branched form of starch. But instead of an α-1,6-glycosidic linkage occurring in about one out of every 30 monomers, a branch occurs in about one out of every 10 glucose subunits (see Table 5.1).

Cellulose: A Structural Polysaccharide in Plants

As Chapter 1 noted, all cells are enclosed by a membrane. In most unicellular and most multicellular organisms living today, the cell is also surrounded by a layer of material called a wall. A **cell wall** is a protective sheet that occurs outside the membrane. In algae, plants, bacteria, fungi, and many other groups, the cell wall is composed primarily of one or more polysaccharides.

In plants, cellulose is the major component of the cell wall. **Cellulose** is a polymer of β-glucose monomers, joined by β-1,4-glycosidic linkages. As Table 5.1 shows, the geometry of the bond is such that each glucose monomer in the chain is flipped in relation to the adjacent monomer. This arrangement increases the stability of cellulose strands, because the flipped orientation makes it possible for multiple hydrogen bonds to form between adjacent and parallel strands of cellulose. As Table 5.1 also shows, cellulose often occurs in long, parallel strands that are joined by these hydrogen bonds. The linked cellulose fibers are strong and provide the cell with structural support.

Chitin: A Structural Polysaccharide in Fungi and Animals

Chitin is a polysaccharide that stiffens the cell walls of fungi. It is also found in a few algae and in many animals; it is, for example, the most important component of the external skeletons of insects and crustaceans.

Chitin is similar to cellulose, but instead of consisting of glucose monomers, the monosaccharide involved is one called N-acetylglucosamine. These monomers are abbreviated "NAc." NAc are joined by β-1,4-glycosidic linkages (see Table 5.1). As in cellulose, the geometry of these bonds results in every other residue being flipped in orientation.

Like the glucose monomers in cellulose, the subunits of N-acetylglucosamine in chitin form hydrogen bonds between adjacent strands. The result is a tough sheet that provides stiffness and protection.

Peptidoglycan: A Structural Polysaccharide in Bacteria

Bacteria, like plants, have cell walls. But unlike plants, most bacteria lack the ability to produce cellulose. Instead, a polysaccharide called **peptidoglycan** gives bacterial cell walls strength and firmness.

Peptidoglycan is the most complex of the polysaccharides discussed thus far. It has a long backbone formed by two types of monosaccharides that alternate with each other and are linked by β-1,4-glycosidic linkages. In addition, a chain of amino acids is attached to one of the two sugar types. When molecules of peptidoglycan align, peptide bonds link the amino acid chains on adjacent strands. **Box 5.2** on page 88 describes how certain antibiotics kill bacteria by disrupting the enzymes that catalyze the formation of the peptide-bonded cross-links.

Although the presence of amino acids gives peptidoglycan a more complex structure than either cellulose or chitin, it is important to note that all three of these polysaccharides have an important feature in common. Structural polysaccharides usually exist as sets of long, parallel strands that are linked to each other. This design gives materials that are made of these molecules the ability to withstand forces that pull or push on them—what an engineer would call tension and compression. In this way, the structure and function of structural polysaccharides are correlated.

(MB) **Web Animation** at www.masteringbio.com
Carbohydrate Structure and Function

Polysaccharides and Chemical Evolution

Cellulose is the most abundant organic compound on Earth today, and chitin is probably the second most abundant by weight. Virtually all organisms manufacture glycogen or starch. But despite their current importance to organisms, polysaccharides probably played little to no role in the origin of life. This conclusion is supported by several observations:

SUMMARY TABLE 5.1 Polysaccharides Differ in Structure

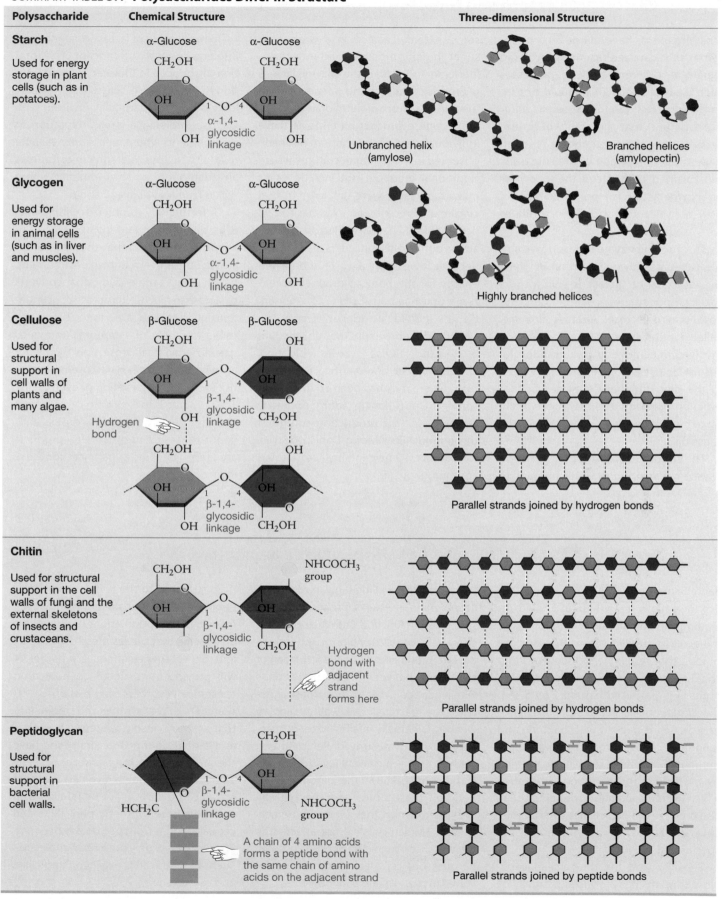

Polysaccharide	Chemical Structure	Three-dimensional Structure
Starch — Used for energy storage in plant cells (such as in potatoes).	α-Glucose, α-Glucose; CH_2OH; OH; 1 4; α-1,4-glycosidic linkage; OH	Unbranched helix (amylose) Branched helices (amylopectin)
Glycogen — Used for energy storage in animal cells (such as in liver and muscles).	α-Glucose, α-Glucose; CH_2OH; OH; 1 4; α-1,4-glycosidic linkage; OH	Highly branched helices
Cellulose — Used for structural support in cell walls of plants and many algae.	β-Glucose, β-Glucose; CH_2OH; OH; 1 4; β-1,4-glycosidic linkage; CH_2OH; Hydrogen bond; OH; β-1,4-glycosidic linkage; CH_2OH	Parallel strands joined by hydrogen bonds
Chitin — Used for structural support in the cell walls of fungi and the external skeletons of insects and crustaceans.	CH_2OH; $NHCOCH_3$ group; OH; 1 4; β-1,4-glycosidic linkage; CH_2OH; Hydrogen bond with adjacent strand forms here	Parallel strands joined by hydrogen bonds
Peptidoglycan — Used for structural support in bacterial cell walls.	CH_2OH; OH; 1 4; β-1,4-glycosidic linkage; HCH_2C; $NHCOCH_3$ group; A chain of 4 amino acids forms a peptide bond with the same chain of amino acids on the adjacent strand	Parallel strands joined by peptide bonds

BOX 5.1 Lactose Intolerance and Galactosemia

Most organisms have a large array of enzymes that catalyze reactions involving sugars. A different enzyme processes each type of sugar. This should not be surprising, considering that sugars differ in shape and that the ability of an enzyme's active site to catalyze a reaction depends on the shape of the substrate. As explained in Chapter 3, the specificity of enzyme action is a prime example of how a molecule's structure correlates with its function.

The human enzyme lactase, for example, catalyzes only one reaction, in which the disaccharide lactose is split into its two component monosaccharides, glucose and galactose. Another enzyme, called 1-phosphate uridyl transferase, is involved in converting galactose to glucose. If lactose and galactose are present in the diet but these enzymes do not function normally, the two sugars build up to high concentration in the body and cause disease.

The condition known as lactose intolerance occurs in most adult humans to some extent. Lactose is a prominent sugar in dairy products; and in lactose-intolerant individuals, the enzyme lactase is produced only during infancy—when the primary source of nutrition is mother's milk. Lactase production ends as these individuals mature. This pattern has been observed in most human cultures where traditional diets did not include cow's milk. Dairy cattle were unknown in Thai culture until recently, for example, and 97 percent of Thai people are lactose intolerant as adults. This pattern is also logical, because it would be a waste of time and energy for the body to produce an enzyme that has no substrate to work on. If these individuals drink milk once lactase production ends, though, the unprocessed lactose causes bloating, abdominal pain, and diarrhea. But in cultures where milk has been an important source of nutrition in adults for many centuries, lactase production continues throughout life. Among Danish people, for example, 97 percent are lactose tolerant throughout life.

To make sense of lactose tolerance and intolerance, recall the concept of adaptation introduced in Chapter 1. An adaptation is a trait that increases the fitness of individuals in a particular environment. Lactose intolerance is an adaptation to environments where milk is not available to adults. Lactose tolerance, in contrast, is an adaptation to environments where milk is readily available to adults.

Unfortunately, problems with galactose metabolism are completely nonadaptive—meaning that they do not increase fitness in any known environment. If a child lacks the enzyme that converts galactose to glucose, galactose concentrations build to high levels in the blood and cause the array of symptoms known as galactosemia ("galactose-sign"). Individuals who suffer from galactosemia are at risk for mental retardation or even death. The only cure is to exclude galactose from the diet. In large enough doses, milk, yogurt, cheese, and other dairy products are life-threatening substances for these individuals.

BOX 5.2 How Do the Penicillins and Cephalosporins Kill Bacteria?

Bacteria are single-celled organisms that live in virtually every habitat known, including the human body. **Antibiotics** are molecules that kill bacteria. The first antibiotic that researchers discovered and widely used to cure bacterial infections in humans is produced naturally, by a soil-dwelling fungus called *Penicillium chrysogenum*. Soil is packed with bacteria and fungi that compete for space, water, and nutrients. Some fungi reduce competition by producing and secreting antibiotics, which kill the bacteria surrounding them. The drug we call penicillin is such a molecule, and it was named in honor of the species that produces it.

Penicillin and the drug called cephalosporin, which is produced naturally by species of soil-dwelling fungi in the genus *Cephalosporium*, are closely related in structure and function. Both molecules are effective because they bind very tightly to the enzymes that catalyze the formation of cross-links between individual strands within peptidoglycan. Without these cross-links, the bacterial cell wall begins to weaken. Eventually it fails and tears open, and the cell is destroyed. Penicillin and cephalosporin cause very few side effects in humans, because the binding event between these drugs and the cell-wall enzyme to which they bind is extremely specific. Human cells are not affected by the presence of penicillin or cephalosporin, because our cells lack the enzymes involved in cross-linking peptidoglycan.

In response to the widespread use of penicillin and cephalosporin, however, many bacterial populations have evolved an enzyme that breaks these drugs apart and renders them ineffective. (Chapter 24 will provide more detail on how drug resistance evolves.) In response, researchers have synthesized molecules that are very closely related to penicillin and cephalosporin structurally and have the same function, but are not as severely affected by the newly evolved bacterial enzyme. You may, in fact, be familiar with some of these synthetic penicillins and cephalosporins from personal experience. They include widely prescribed molecules such as methicillin, oxacillin, ampicillin, and ceftriaxone.

- No plausible mechanism exists for the polymerization of monosaccharides under conditions that prevailed early in Earth's history. In cells and in laboratory experiments, the glycosidic linkages illustrated in Figure 5.4 and Table 5.1 form only with the aid of specialized enzymes.

- It is highly likely that life began in the form of an RNA molecule. However, researchers have yet to discover a ribozyme that can join simple sugars by catalyzing the formation of glycosidic linkages. Thus, it appears extremely unlikely that polysaccharides were present in significant quantities during the RNA world.

- Even though monosaccharides contain large numbers of hydroxyl and carbonyl groups, they lack the diversity of functional groups found in amino acids. Polysaccharides also have simple secondary structures, consisting of linkages between adjacent strands. Thus, they lack the structural and chemical complexity that makes proteins, and to a lesser extent RNA, effective catalysts. To date, no reactions have been discovered that are catalyzed by polysaccharides.

- The monomers in polysaccharides are not capable of complementary base pairing. Like proteins, but unlike DNA and RNA, polysaccharides cannot provide the information required for themselves to be copied. As far as is known, no polysaccharides store information in cells. Thus, no one has proposed that the first living entity might have been a polysaccharide.

Even though polysaccharides probably did not play a significant role in the earliest forms of life, they became enormously important once cellular life evolved. Let's take a detailed look at how they function in today's cells.

Check Your Understanding

If you understand that...

- Polysaccharides form when enzymes catalyze the formation of glycosidic linkages between monosaccharides that are in the α or β form.

- Most polysaccharides are long, linear molecules, but some branch extensively. Among linear forms, it is common for adjacent strands to be linked by hydrogen bonding or other types of linkages.

You should be able to...

1) Make a rough sketch that shows the structures of glycogen and cellulose.

2) Label the glycosidic linkages in each structure.

3) Indicate where hydrogen bonds form between adjacent strands of cellulose.

4) Check your work by using Table 5.1.

5.3 What Do Carbohydrates Do?

Chapter 4 introduced one of the four basic functions that carbohydrates perform in organisms: furnishing the building blocks of molecules needed by cells. Recall that both RNA and DNA contain sugars—the five-carbon sugars ribose and deoxyribose, respectively. In nucleotides, which consist of a sugar, a phosphate group, and a nitrogenous base, the sugar itself acts as a subunit of the larger molecule. But sugars frequently furnish only the raw "carbon skeletons" that are used in the synthesis of important molecules. Amino acids are being synthesized by your cells right now, for example, using sugars as a starting point.

Although the details of how sugars are used in synthesizing amino acids and other complex molecules are beyond the scope of this book, it will be productive to delve into the other three major roles of carbohydrates. ● Carbohydrates have diverse functions in cells: They provide fibrous structural materials, indicate cell identity, and store chemical energy.

Carbohydrates as Structural Molecules

Cellulose and chitin, along with the modified polysaccharide peptidoglycan, are key structural compounds. They form fibers that give cells and organisms strength and elasticity.

To appreciate why cellulose, chitin, and peptidoglycan are effective as structural molecules, recall that they form long strands and that bonds can form between adjacent strands. Table 5.1 detailed how these bonds form. In the cell walls of plants, a collection of about 80 cellulose molecules are cross-linked by hydrogen bonding to create a tough fiber. These cellulose fibers, in turn, crisscross to form a tough sheet (**Figure 5.6a**). Groups of chitin molecules are also organized into fibers held together by hydrogen bonding. Although each individual hydrogen bond in cellulose and chitin is relatively weak, the combination of many weak bonds allows these structural polysaccharides to form large fibers that are strong, yet flexible. Chitin fibers stiffen the cell walls of fungi, and they overlap in the external skeletons of insects to form a dense, waterproof sheet (**Figure 5.6b**). In the cell walls of bacteria, the presence of amino acid residues in peptidoglycan allows adjacent fibers to cross-link, via peptide bonds, into durable sheets (**Figure 5.6c**).

Besides being stiff and strong, the structural carbohydrates are durable. Almost all organisms have the enzymes required to break the α-1,4- and α-1,6-glycosidic linkages that hold starch and glycogen molecules together, but only a few organisms have enzymes capable of hydrolyzing the β-1,4-glycosidic linkages in cellulose, chitin, and peptidoglycan. The shape and orientation of β-1,4-glycosidic linkages make them difficult to break, and few enzymes have active sites with the correct geometry and reactive groups to do so. As a result, the structural polysaccharides are resistant to degradation and decay.

(a) Cellulose in plant cell wall

1 µm

(b) Chitin in insect exoskeleton

5 µm

(c) Peptidoglycan in bacterial cell wall

0.1 µm

FIGURE 5.6 Cellulose, Chitin, and Peptidoglycan Form Tough Fibers or Sheets.

Ironically, the durability of cellulose is important for digestion—even though the cellulose that you ingest when you eat plant cells passes through your gut undigested. The cellulose that you eat is referred to as dietary fiber. The cellulose in dietary fiber, like the cellulose in a paper towel, absorbs water. The presence of dietary fiber adds moisture to the feces. In addition, cellulose adds bulk that helps fecal material move through the intestinal tract more quickly, preventing colon cancer, constipation, and other problems.

The Role of Carbohydrates in Cell Identity

Polysaccharides do not store information in cells, but they do *display* important information—specifically, on the outer surface of a cell. **Figure 5.7** shows how this information display happens. Certain molecules project outward from the cell surface into the surrounding environment. These molecules are called glycoproteins. A **glycoprotein** is a protein that is covalently bonded to a carbohydrate—usually the relatively short chains of sugars called oligosaccharides.

Glycoproteins are key molecules in what biologists call cell–cell recognition and cell–cell signaling. Each cell in your body has glycoproteins on its surface that identify it as part of your body. Immune system cells use these glycoproteins to distinguish your body's cells from foreign cells, such as bacteria. In addition, each distinct type of cell in a multicellular organism—for example, the nerve cells and muscle cells in your body—displays a different set of glycoproteins on its surface. In cells, glycoproteins form a "sugar coating" that acts like the magnetic stripe on a credit card or the personal identification number (PIN) you use to access a bank account—it immediately identifies the individual that bears it.

● If you understand this concept, you should be able to draw two cells, and then add and label (1) glycoproteins that identify each of them as distinct cell types and (2) glycoproteins that identify them as part of your body.

The identification information displayed by glycoproteins helps cells recognize and communicate with each other. The key point here is to recognize that the enormous number of structurally distinct monosaccharides makes it possible for an enormous number of unique oligosaccharides to exist. As a result, each cell type and each species can display a unique identity.

Glycoprotein

Outside of cell

Inside of cell

FIGURE 5.7 Carbohydrates Are an Identification Badge for Cells. Glycoproteins contain sugar groups that project from the surface of the plasma membrane. These sugar groups have distinctive structures that identify the type or species of the cell.

These electrons are shared equally by atoms with similar electronegativity

Carbon dioxide   Water $(CH_2O)_n$ Oxygen
                                                  (Carbohydrate)

FIGURE 5.8 Carbohydrates Have High Free Energy. In these diagrams, the horizontal lines indicate covalent bonds. The dots represent the relative positions of electrons in those bonds.

● **EXERCISE** Circle the bonds in this diagram that have the highest free energy.

The Role of Carbohydrates in Energy Storage

Candy-bar wrappers promise a quick energy boost, and ads for sports drinks claim that their products provide the "carbs" needed for peak activity. If you were to ask friends or family members what carbohydrates do in your body, they would probably say something like "They give you energy." And after pointing out that carbohydrates are also used to establish cell identity, as a structural material, and as a source of carbon skeletons for the synthesis of other complex molecules, you'd have to agree. Carbohydrates store and provide chemical energy in cells. What is it about carbohydrates that makes this possible?

Carbohydrates and Free Energy   Recall from earlier chapters that the essence of chemical evolution was the conversion of kinetic energy in sunlight and heat into chemical energy stored in the bonds of molecules such as formaldehyde (H_2CO) and hydrogen cyanide (HCN). Today, the kinetic energy in sunlight is converted to chemical energy stored in the bonds of carbohydrates, via the process known as **photosynthesis**. Photosynthesis in plants, algae, and some bacteria is a complex set of reactions that can be summarized most simply as follows:

$$CO_2 + H_2O + sunlight \rightarrow (CH_2O)_n + O_2$$

where $(CH_2O)_n$ represents a carbohydrate. **Figure 5.8** shows the structural formulas of the molecules involved and represents the relative positions of their covalently bonded electrons. The keys to understanding this figure are to compare the positions of the electrons in the reactants and the products, and note that the covalently bonded electrons in the C–H bonds of carbohydrates are shared more equally, and thus held less tightly, than they are in the C–O bonds of carbon dioxide. The same is true for the carbon-carbon (C–C) bonds of carbohydrates: Electrons are shared more equally than they are in the C–O bonds of carbon dioxide.

Now, recall from Chapter 2 that when electrons are held tightly by a highly electronegative atom like oxygen, they have

low potential energy. But when they are bonded to a less-electronegative atom like carbon or hydrogen, they are farther from the nucleus and have high potential energy. C–C and C–H bonds have much higher free energy than C–O bonds have. As a result, carbohydrates have much more free energy than carbon dioxide has.

The essence of photosynthesis, then, is that energy in sunlight is transformed into chemical energy that is stored in the C–H and C–C bonds of carbohydrates. The C–C and C–H bonds have high free energy. When a cell needs to tap this stored energy, carbohydrates participate in exergonic reactions that synthesize a molecule called **adenosine triphosphate (ATP)**. More specifically, the energy that is released when sugars are processed is used to synthesize ATP from a precursor called adenosine diphosphate (ADP). The overall reaction can be written as follows:

$$CH_2O + O_2 + ADP + P_i \rightarrow CO_2 + H_2O + ATP$$

To put this in words, the chemical energy stored in the C–H and C–C bonds of carbohydrate is transferred to chemical energy in the form of the third phosphate group in ATP. The free energy in ATP drives what would otherwise be endergonic reactions like polymerization, moves your muscles, and performs other types of work in cells. Carbohydrates are like the water that piles up behind a dam; ATP is like the electricity, generated at a dam, that lights up your home. Carbohydrates store chemical energy; ATP "spends" it.

Later chapters will analyze in detail how sugars and other carbohydrates are made in organisms, and how these carbohydrates are then broken down to provide cells with usable chemical energy in the form of ATP. For now, the important thing is to recognize that carbohydrates store and provide chemical energy to cells because they contain a large number of C–H bonds and because C–H bonds have high free energy.

Figure 5.9 summarizes and extends these points. For example, compare the structures of carbon dioxide and the carbohydrate shown in Figure 5.9a and Figure 5.9b, and you'll see that hydrogen atoms—an electron plus a proton—have been added to the

(a) Carbon dioxide

O═C═O

(b) A carbohydrate

(c) A fatty acid (a component of fat molecules)

FIGURE 5.9 In Organisms, Free Energy is Stored in C–H and C–C Bonds. (a) In carbon dioxide, the electrons involved in covalent bonds are held tightly by oxygen atoms. **(b)** In carbohydrates such as the sugar shown here, many of the covalently bonded electrons are held loosely between C and H atoms. **(c)** The fatty acids found in fat molecules have more C–H bonds and fewer C–O bonds than carbohydrates do.

● **EXERCISE** Label the molecule with the highest amount of free energy, and the molecule with the lowest amount.

● **QUESTION** Which molecule has the highest amount of potential energy? Of chemical energy?

carbohydrate. As Figure 5.9c shows, the fatty-acid subunits found in fats have even more free energy than carbohydrates have. Fatty acids are made up largely of long chains of carbon atoms bonded to hydrogen atoms. Compared with carbohydrates, fats contain many more C–C and C–H bonds and many fewer C–O bonds. To restate briefly, C–C and C–H bonds have high free energy because the electrons are shared equally by atoms with low electronegativities, while C–O bonds have low free energy because the highly electronegative oxygen atom holds the electrons so tightly. Both carbohydrates and fats are used as fuel in cells. Fats will be discussed in more detail in Chapter 6.

How Do Carbohydrates Store Energy? Starch and glycogen are efficient energy-storage molecules because they

polymerize via α-glycosidic linkages instead of the β-glycosidic linkages observed in the structural polysaccharides. The α-linkages in storage polysaccharides are readily hydrolyzed to release sugars, while the β-linkages in structural polysaccharides resist enzymatic degradation. The glucose subunits that are hydrolyzed from starch and glycogen are then processed in reactions that result in the production of ATP. Starch and glycogen are like a candy bar that has segments, so you can break off chunks whenever you need a boost.

The most important enzyme involved in catalyzing the hydrolysis of α-glycosidic linkages in glycogen is called **phosphorylase**. The enzymes involved in breaking these linkages in starch are called **amylases**. Most of your cells contain phosphorylase, so they can break down glycogen to provide glucose on demand. Your salivary glands and pancreas also produce amylases that are secreted into your mouth and small intestine, respectively. These amylases are responsible for digesting the starch that you eat.

Check Your Understanding

If you understand that...

- Carbohydrates have several important functions in cells in addition to providing building blocks for the synthesis of more complex compounds.
- Polysaccharides such as cellulose, chitin, and peptidoglycan form cell walls, which give cells structural strength.
- Glycoproteins project from the surface of cells. They provide a molecular PIN that identifies the cell's type or species.
- Starch and glycogen store sugars for later use in reactions that produce ATP. Sugars contain large amounts of chemical energy because they contain carbon atoms that are bonded to hydrogen atoms or other carbon atoms. These bonds have high free energy because the electrons are shared equally by atoms with low electronegativity.

You should be able to...

1) Identify two aspects of the structures of cellulose, chitin, and peptidoglycan that correlate with their function as structural molecules.
2) Describe how the carbohydrates you ate during breakfast today are functioning in your body right now.

Chapter Review

SUMMARY OF KEY CONCEPTS

● **Sugars and other carbohydrates are highly variable in structure.**

Carbohydrates are organic compounds that have a carbonyl group and several to many hydroxyl groups. They can occur as the simple sugars called monosaccharides or as complex polysaccharides. Many types of monosaccharide exist, each distinguished by one or more of the following features: (1) the location of their carbonyl group—either at the end of the molecule or within it; (2) the number of carbon atoms they contain—from three to seven, usually; and (3) the orientation of their hydroxyl groups in the linear chain and/or the ring form. In addition, most monosaccharides have optical isomers.

You should be able to draw two monosaccharides and explain at least two aspects of their structure that differ. ●

● **Monosaccharides are monomers that polymerize to form polymers called polysaccharides, via different types of glycosidic linkages.**

Monosaccharides can be linked together by covalent bonds, called glycosidic linkages, that join hydroxyl groups on adjacent molecules. Unlike the peptide bonds that form proteins and the phosphodiester bonds of nucleic acids, glycosidic linkages may form at several locations on a monosaccharide. The different disaccharides and polysaccharides are distinguished by the type of monomers involved and the location of the glycosidic linkages between them. The most common polysaccharides in organisms today are starch, glycogen, cellulose, and chitin; peptidoglycan is an abundant polysaccharide that has short chains of amino acids attached.

You should be able to draw two disaccharides and explain at least two aspects of their structure that differ. ●

● **Carbohydrates perform a wide variety of functions in cells, ranging from energy storage to formation of tough structural fibers.**

Starch and glycogen are made up of glucose molecules that are in the α ring form and that are joined by glycosidic linkages between their first and fourth carbons. In some forms of starch, individual chains are occasionally joined to other chains; in glycogen these links are abundant enough to form a highly branched molecule. Both starch and glycogen function as energy-storage molecules. Because sugars contain many C–C and C–H bonds, they contain a significant amount of chemical energy. When cells need energy, enzymes hydrolyze the α-1,4-glycosidic linkages in starch or glycogen. The reaction releases individual glucose molecules, which then are processed in reactions that lead to the production of ATP. Chemical energy in the form of ATP is readily usable by cells.

Cellulose, chitin, and peptidoglycan are made up of monosaccharides joined by β-1,4-glycosidic linkages. When individual molecules of these polysaccharides align side by side, bonds form between them. In cellulose and chitin, the individual molecules are joined by hydrogen bonds; in peptidoglycan, the intermolecular links consist of peptide bonds between amino acid chains that extend from the polysaccharide. Hydrogen bonding produces bundles of cellulose and chitin molecules that form strong, elastic fibers; when peptide bonding occurs between many peptidoglycan strands, the result is a tough fibrous or sheetlike substance. Cellulose, chitin, and peptidoglycan provide cells and organisms with support. In addition, few organisms have enzymes that can degrade β-1,4-glycosidic linkages, which makes cell walls made of cellulose, chitin, and peptidoglycan resistant to attack. In carbohydrates, as in proteins and nucleic acids, structure correlates with function.

You should be able to describe two key differences in the structure of polysaccharides that function in energy storage versus structural support. ●

 at www.masteringbio.com

Carbohydrate Structure and Function

QUESTIONS

● **Test Your Knowledge**

1. What is the difference between a monosaccharide, a disaccharide, and a polysaccharide?
   a. the number of carbon atoms in the molecule
   b. the type of glycosidic linkage between monomers
   c. the spatial arrangement of the various hydroxyl residues in the molecule
   d. the number of monomers in the molecule

2. What type of bond allows sugars to polymerize?
   a. glycosidic linkage
   b. phosphodiester bond
   c. peptide bond
   d. hydrogen bonds

3. What holds cellulose molecules together in bundles large enough to form fibers?
   a. the cell wall
   b. peptide bonds
   c. hydrogen bonds
   d. hydrophobic interactions between different residues in the cellulose helix

4. What are the primary functions of carbohydrates in cells?
   a. energy storage, cell identity, structure, and building blocks for synthesis
   b. catalysis, structure, and transport
   c. information storage and catalysis
   d. signal reception, signal transport, and signal response

5. Why is it unlikely that carbohydrates played a large role in the origin of life?
   a. They cannot be produced by chemical evolution.
   b. They have optical isomers.
   c. More types of glycosidic linkages are possible than are actually observed in organisms.
   d. They do not polymerize without the aid of enzymes.

6. What is a "quick and dirty" way to assess how much free energy an organic molecule has?
   a. Count the number of oxygen atoms it contains.
   b. Count the number of C–H bonds it contains.
   c. Count the number of C–O bonds it contains.
   d. Determine whether it contains a carbonyl group.

Test Your Knowledge answers: 1. d; 2. a; 3. c; 4. a; 5. d; 6. b

---

● Test Your Understanding

Answers are available at www.masteringbio.com

1. Explain why the structure of carbohydrates supports their function in signaling the identity of a cell.

2. Draw the ring structure of glucose in the α form. Now add another glucose molecule with an α-1,4-glycosidic linkage, and then a third glucose molecule with an α-1,6-glycosidic linkage.

3. Compare and contrast the structures and functions of starch and glycogen. How are these molecules similar? How are they different?

4. Why do the bonds in a carbohydrate store a large amount of chemical energy compared to the chemical energy stored in the bonds of carbon dioxide?

5. What aspects of the structure of cellulose and chitin supports their function in protecting and stiffening cells and organisms?

6. Both glycogen and cellulose consist of glucose monomers that are linked end to end. How do the structures of these polysaccharides differ? How do their functions differ?

● Applying Concepts to New Situations

Answers are available at www.masteringbio.com

1. A weight-loss program for humans that emphasizes minimal consumption of carbohydrates has become popular in recent years. What is the logic behind this diet? (Note: This diet plan has caused controversy and is not endorsed by some physicians and researchers.)

2. To treat galactosemia, physicians exclude the monosaccharide galactose from the diet. Why does the disaccharide lactose also have to be excluded from the diet?

3. Amylase, an enzmye found in human saliva, catalyzes the hydrolysis of the α-1,4-glycosidic linkages in starch. If you hold a salty cracker in your mouth long enough, it will begin to taste sweet. Why?

4. Lysozyme, an enzyme found in human saliva, tears, and other secretions, catalyzes the hydrolysis of the β-1,4-glycosidic linkages in peptidoglycan. What effect does contact with this enzyme have on bacteria?

www.masteringbio.com is also your resource for • Answers to text, table, and figure caption questions and exercises • Answers to *Check Your Understanding* boxes • Online study guides and quizzes • Additional study tools including the *E-Book for Biological Science* 3rd ed., textbook art, animations, and videos.

Lipids, Membranes, and the First Cells

These bacterial cells have been stained with a red compound that inserts itself into the plasma membrane. The plasma membrane defines the cell—the basic unit of life. In single-celled organisms like those shown here, the membrane creates a physical separation between life on the inside and nonlife on the outside.

KEY CONCEPTS

- Phospholipids are amphipathic molecules—they have a hydrophilic region and a hydrophobic region. In solution, they spontaneously form bilayers that are selectively permeable—meaning that only certain substances cross them readily.

- Ions and molecules diffuse spontaneously from regions of high concentration to regions of low concentration. Water moves across lipid bilayers from regions of high concentration to regions of low concentration via osmosis—a special case of diffusion.

- In cells, membrane proteins are responsible for the passage of ions, polar molecules, and large molecules that can't cross the membrane on their own because they are not soluble in lipids.

The research discussed in previous chapters suggests that biological evolution began with an RNA molecule that could make a copy of itself. As the offspring of this molecule multiplied in the prebiotic soup, natural selection would have favored versions of the molecule that were particularly stable and efficient at catalysis. Another great milestone in the history of life occurred when a descendant of this replicator became enclosed within a membrane. This event created the first cell and thus the first organism.

The **cell membrane**, or **plasma membrane**, is a layer of molecules that surrounds the cell, separating it from the external environment and selectively regulating the passage of molecules and ions into or out of the cell. The evolution of the plasma membrane was a momentous development because it separated life from nonlife. Before plasma membranes existed, self-replicating molecules probably clung to clay-sized mineral particles, building copies of themselves as they randomly encountered the appropriate nucleotides in the prebiotic soup that washed over them. But the membrane made an internal environment possible—one that could have a chemical composition different from that of the external environment. This was important for two reasons. First, the chemical reactions necessary for life could occur much more efficiently in an enclosed area, because reactants could collide more frequently. Second, the membrane could serve as a selective barrier. That is, it could keep compounds out of the cell that might damage the replicator, but it might allow the entry of compounds required by the replicator. The membrane not only created the cell but also made it into an efficient and dynamic reaction vessel.

The goal of this chapter is to investigate how membranes behave, with an emphasis on how they differentiate the internal environment from the external environment. Let's begin by examining the structure and properties of the most abundant molecules in plasma membranes: the "oily" or "fatty" compounds called lipids. Then we can delve into analyzing the way lipids behave when they form membranes. Which ions and molecules can pass through a membrane that consists of lipids? Which cannot, and why? The chapter ends by exploring how proteins that become incorporated into a lipid membrane can control the flow of materials across the membrane.

6.1   Lipids

Most biochemists are convinced that the building blocks of membranes, called lipids, existed in the prebiotic soup. This conclusion is based on the observation that several types of lipids have been produced in experiments designed to mimic the chemical and energetic conditions that prevailed early in Earth's history. For example, the spark-discharge experiments reviewed in Chapter 3 succeeded in producing at least two types of lipid.

An observation made by A. D. Bangham illustrates why this result is interesting. In the late 1950s, Bangham performed experiments to determine how lipids behave when they are immersed in water. But until the electron microscope was invented, he had no idea what these lipid-water mixtures looked like. Once **transmission electron microscopes** became available, Bangham was able to produce high magnification, high resolution images of his lipid-water mixtures. (Transmission electron microscopy is introduced in **BioSkills 8**.) The images that resulted, called **micrographs**, were astonishing. As **Figure 6.1a** shows, the lipids had spontaneously formed enclosed compartments filled with water. Bangham called these membrane-bound structures vesicles and noted that they resembled cells (**Figure 6.1b**). Bangham had not done anything special to the lipid-water mixtures; he had merely shaken them by hand.

The experiment raises a series of questions: How could these structures have formed? Is it possible that vesicles like these existed in the prebiotic soup? If so, could they have surrounded a self-replicating molecule and become the first plasma membrane? Let's begin answering these questions by investigating what lipids are and how they behave.

What Is a Lipid?

Earlier chapters analyzed the structures of the organic molecules called amino acids, nucleotides, and monosaccharides and explored how these monomers polymerize to form macro-

(a) In solution, lipids form water-filled vesicles.

50 nm

(b) Red blood cells resemble vesicles.

50 µm

FIGURE 6.1 Lipids Can Form Cell-like Vesicles When in Water.
(a) Transmission electron micrograph showing a cross section through the tiny, bag-like compartments that formed when a researcher shook a mixture of lipids and water. **(b)** Scanning electron micrograph showing red blood cells from humans. Note the scale bars.

molecules. Here let's focus on another major type of mid-sized molecule found in living organisms: lipids.

Lipid is a catch-all term for carbon-containing compounds that are found in organisms and are largely nonpolar and hydrophobic—meaning that they do not dissolve readily in water. (Recall from Chapter 2 that water is a polar solvent.) Lipids do dissolve, however, in liquids consisting of nonpolar organic compounds.

To understand why lipids do not dissolve in water, examine the five-carbon compound called isoprene illustrated in **Figure 6.2a**; notice that it consists of a group of carbon atoms bonded to hydrogen atoms. Molecules that contain only carbon and hydrogen, such as isoprene or octane (see Chapter 2) are known as **hydrocarbons**. Hydrocarbons are nonpolar, because electrons are shared equally in carbon-hydrogen bonds. This property makes hydrocarbons hydrophobic. Thus, the reason lipids do not dissolve in water is that they have a significant hydrocarbon component. **Figure 6.2b** is a type of compound called a **fatty acid**, which consists of a hydrocarbon chain bonded to a carboxyl (COOH) functional group. Isoprene and fatty acids are key building blocks of the lipids found in organisms.

(a) Isoprene

(b) Fatty acid

Carboxyl group

Hydrocarbon chain

FIGURE 6.2 Hydrocarbon Groups Make Lipids Hydrophobic.
(a) Isoprenes are hydrocarbons. Isoprene subunits can be linked end to end to form long hydrocarbon chains. **(b)** Fatty acids typically contain a total of 14–20 carbon atoms, most found in their long hydrocarbon tails.

● **EXERCISE** Circle the hydrophobic portion of a fatty acid.

A Look at Three Types of Lipids Found in Cells

Unlike amino acids, nucleotides, and carbohydrates, lipids are defined by a physical property—their solubility—instead of their chemical structure. As a result, the structure of lipids varies widely. To drive this point home, consider the structures of the most important types of lipids found in cells: fats, steroids, and phospholipids.

- **Fats** are composed of three fatty acids that are linked to a three-carbon molecule called **glycerol**. Because of this structure, fats are also called triacylglycerols or triglycerides. As **Figure 6.3a** shows, fats form when a dehydration reaction occurs between a hydroxyl group of glycerol and the carboxyl group of a fatty acid. The glycerol and fatty-acid molecules become joined by an **ester linkage**, which is analogous to the peptide bonds, phosphodiester bonds, and glycosidic linkages in proteins, nucleic acids, and carbohydrates, respectively. Fats are not polymers, however, and fatty acids are not monomers. As **Figure 6.3b** shows, fatty acids are not linked together to form a macromolecule in the way that amino acids, nucleotides, and monosaccharides are. ● After studying the structure in Figure 6.3b, you should be able to explain why fats store a great deal of chemical energy, and why they are hydrophobic.

(a) Fats form via dehydration reactions.

Glycerol

Dehydration reaction

Fatty acid

(b) Fats consist of glycerol linked by ester linkages to three fatty acids.

Ester linkages

FIGURE 6.3 Fats Are One Type of Lipid Found in Cells.   (a) When glycerol and a fatty acid react, a water molecule leaves. (b) The covalent bond that results from this reaction is termed an ester linkage. The fat shown here as a structural formula and a space-filling model is tristearin, the most common type of fat in beef.

(a) A steroid

(b) A phospholipid

FIGURE 6.4 Amphipathic Lipids Contain Hydrophilic and Hydrophobic Elements. (a) All steroids have a distinctive four-ring structure. **(b)** All phospholipids consist of a glycerol that is linked to a phosphate group and to either two chains of isoprene or two fatty acids.

● **QUESTION** What makes cholesterol—the steroid shown in part (a)—different from other steroids?

● **QUESTION** If these molecules were in solution, where would water molecules interact with them?

- **Steroids** are a family of lipids distinguished by the four-ring structure shown in solid orange in **Figure 6.4a**. The various steroids differ from one another by the functional groups or side groups attached to those rings. The molecule pictured in Figure 6.4a is cholesterol, which is distinguished by a hydrocarbon "tail" formed of isoprene subunits. Cholesterol is an important component of plasma membranes in many organisms. In mammals, it is also used as the starting point for the synthesis of several of the signaling molecules called hormones. Estrogen, progesterone, and testosterone are examples of hormones derived from cholesterol. These molecules are responsible for regulating sexual development and activity in humans.

- **Phospholipids** consist of a glycerol that is linked to a phosphate group (PO_4^{2-}) and to either two chains of isoprene or two fatty acids. In some cases, the phosphate group is bonded to another small organic molecule, such as the choline shown on the phospholipid in **Figure 6.4b**. Phospholipids with isoprene tails are found in the domain Archaea introduced in Chapter 1; phospholipids composed of fatty acids are found in the domains Bacteria and Eukarya. In all three domains of life, phospholipids are critically important components of the plasma membrane.

To summarize, the lipids found in organisms have a wide array of structures and functions. In addition to storing chemical energy and serving as signals between cells, lipids act as pigments that capture or respond to sunlight, form waterproof coatings on leaves and skin, and act as vitamins used in an array of cellular processes. The most important lipid function, however, is their role in the plasma membrane. Let's take a closer look at the specific types of lipids found in membranes.

The Structures of Membrane Lipids

Not all lipids can form the artificial membranes that Bangham and his colleagues observed. In fact, just two types of lipids are usually found in plasma membranes. ◯ Membrane-forming lipids have a polar, hydrophilic region in addition to the nonpolar, hydrophobic region found in all lipids. To better understand this structure, take another look at the phospholipid illustrated in Figure 6.4b. Notice that the molecule has a "head" region containing highly polar covalent bonds as well as positive and negative charges. The charges and polar bonds in the head region interact with water molecules when a phospholipid is placed in solution. In contrast, the long isoprene or fatty-acid tails of a phospholipid are nonpolar. Water molecules cannot form hydrogen bonds with the hydrocarbon tail, so they do not interact with this part of the molecule.

Compounds that contain both hydrophilic and hydrophobic elements are **amphipathic** ("dual-sympathy"). Phospholipids

are amphipathic. As Figure 6.4a shows, cholesterol is also amphipathic. It has both hydrophilic and hydrophobic regions.

The amphipathic nature of phospholipids is far and away their most important feature biologically. It is responsible for their presence in plasma membranes.

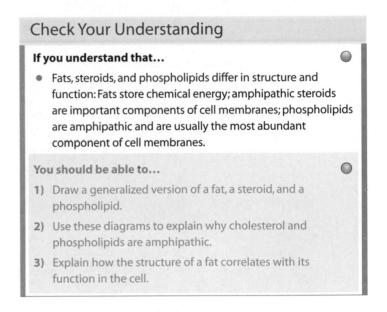

Check Your Understanding

If you understand that... ◯

- Fats, steroids, and phospholipids differ in structure and function: Fats store chemical energy; amphipathic steroids are important components of cell membranes; phospholipids are amphipathic and are usually the most abundant component of cell membranes.

You should be able to... ◯

1) Draw a generalized version of a fat, a steroid, and a phospholipid.

2) Use these diagrams to explain why cholesterol and phospholipids are amphipathic.

3) Explain how the structure of a fat correlates with its function in the cell.

6.2   Phospholipid Bilayers

Phospholipids do not dissolve when they are placed in water. Water molecules interact with the hydrophilic heads of the phospholipids, but not with their hydrophobic tails. Instead of dissolving in water, then, phospholipids may form one of two types of structures: micelles or lipid bilayers.

Micelles (**Figure 6.5a**) are tiny droplets created when the hydrophilic heads of phospholipids face the water and the hydrophobic tails are forced together, away from the water. Phospholipid bilayers, or simply, **lipid bilayers**, are created when two sheets of phospholipid molecules align. As **Figure 6.5b** shows, the hydrophilic heads in each layer face a surrounding solution while the hydrophobic tails face one another inside the bilayer. In this way, the hydrophilic heads interact with water while the hydrophobic tails interact with each other. Micelles tend to form from phospholipids with relatively slender tails; bilayers tend to form from phospholipids with bulkier tails.

Once you understand the structure of micelles and phospholipid bilayers, the most important point to recognize about them is that they form spontaneously. No input of energy is required. This concept can be difficult to grasp, because the formation of these structures clearly decreases entropy. Micelles and lipid bilayers are much more highly organized than phospholipids floating free in the solution. The key is to recognize that micelles and lipid bilayers are much more stable energetically than are independent molecules in solution.

(a) Lipid micelles   **(b)** Lipid bilayers

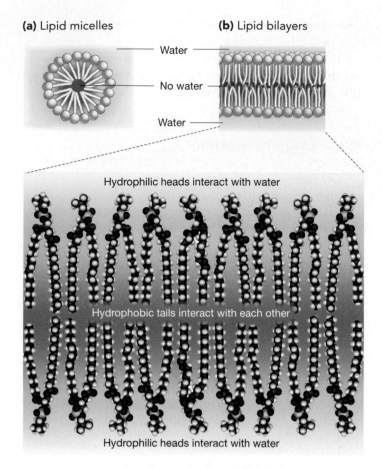

FIGURE 6.5 **Phospholipids Form Bilayers in Solution.** In **(a)** a micelle or **(b)** a lipid bilayer, the hydrophilic heads of lipids face out, toward water; the hydrophobic tails face in, away from water. Plasma membranes consist in part of lipid bilayers.

Hydrocarbon surrounded by water molecules

FIGURE 6.6 **Hydrocarbons Disrupt Hydrogen Bonds between Water Molecules.** Hydrocarbons are unstable in water because they disrupt hydrogen bonding between water molecules.

● **EXERCISE** Label the area where no hydrogen bonding is occurring between water molecules.

● **QUESTION** Hydrogen bonds pull water molecules closer together. Which way are the water molecules in this figure being pulled, relative to the hydrocarbon?

Stated another way, micelles and lipid bilayers have much lower potential energy than do independent phospholipids in solution. Independent phospholipids are unstable in water because their hydrophobic tails disrupt hydrogen bonds that otherwise would form between water molecules (**Figure 6.6**; see also Figure 2.13b). As a result, amphipathic molecules are much more stable in aqueous solution when their hydrophobic tails avoid water and instead participate in the hydrophobic (van der Waals) interactions introduced in Chapter 3. In this case, the loss of potential energy outweighs the decrease in entropy. Overall, the free energy of the system decreases. Lipid bilayer formation is exergonic and spontaneous.

● If you understand this reasoning, you should be able to add water molecules that are hydrogen-bonded to each hydrophilic head in Figure 6.5, and explain the logic behind your drawing.

Artificial Membranes as an Experimental System

When lipid bilayers are agitated by shaking, the layers break and re-form as small, spherical structures. This is what happened in

Bangham's experiment. The resulting vesicles had water on the inside as well as the outside because the hydrophilic heads of the lipids faced outward on each side of the bilayer.

Researchers have produced these types of vesicles by using dozens of different types of phospholipids. Artificial membrane-bound vesicles like these are called liposomes. The ability to create them supports an important conclusion: If phospholipid molecules accumulated during chemical evolution early in Earth's history, they almost certainly formed water-filled vesicles.

To better understand the properties of vesicles and plasma membranes, researchers began creating and experimenting with liposomes and other types of artificial bilayers. Some of the first questions they posed concerned the permeability of lipid bilayers. The **permeability** of a structure is its tendency to allow a given substance to pass across it. Once a membrane forms a water-filled vesicle, can other molecules or ions pass in or out? If so, is this permeability selective in any way? The permeability of membranes is a critical issue, because if certain molecules or ions pass through a lipid bilayer more readily than others, the internal environment of a vesicle can become different from the outside. This difference between exterior and interior environments is a key characteristic of cells.

Figure 6.7 shows the two types of artificial membranes that are used to study the permeability of lipid bilayers. Figure 6.7a shows liposomes, roughly spherical vesicles. Figure 6.7b illustrates planar bilayers, which are lipid bilayers constructed across a hole in a glass or plastic wall separating two aqueous (watery) solutions.

(a) Liposomes: Artificial membrane-bound vesicles

(b) Planar bilayers: Artificial membranes

(c) Artificial-membrane experiments

How rapidly can different solutes cross the membrane (if at all) when …

1. Different types of phospholipids are used to make the membrane?

2. Proteins or other molecules are added to the membrane?

FIGURE 6.7 Liposomes and Planar Bilayers Are Important Experimental Systems. (a) Electron micrograph of liposomes in cross section (left) and a cross-sectional diagram of the lipid bilayer in a liposome. **(b)** The construction of planar bilayers across a hole in a glass wall separating two water-filled compartments (left), and a close-up sketch of the bilayer. **(c)** A wide variety of experiments are possible with liposomes and planar bilayers; a few are suggested here.

Using liposomes and planar bilayers, researchers can study what happens when a known ion or molecule is added to one side of a lipid bilayer (Figure 6.7c). Does the ion or molecule cross the membrane and show up on the other side? If so, how rapidly does the movement take place? What happens when a different type of phospholipid is used to make the artificial membrane? Does the membrane's permeability change when proteins or other types of molecules become part of it?

Biologists describe such an experimental system as elegant and powerful because it gives them precise control over which factor changes from one experimental treatment to the next. Control, in turn, is why experiments are such an effective way to explore scientific questions. You might recall from Chapter 1 that good experimental design allows researchers to alter one factor at a time and determine what effect, if any, each has on the process being studied.

Equally important for experimental purposes, liposomes and planar bilayers provide a clear way to determine whether a given change in conditions has an effect. By sampling the solutions on both sides of the membrane before and after the treatment and then analyzing the concentration of ions and molecules in the samples, researchers have an effective way to determine whether the treatment had any consequences.

Using such systems, what have biologists learned about membrane permeability?

Selective Permeability of Lipid Bilayers

When researchers put molecules or ions on one side of a liposome or planar bilayer and measure the rate at which the molecules arrive on the other side, a clear pattern emerges: Lipid bilayers are *highly* selective. **Selective permeability** means that some substances cross a membrane more easily than other substances can. Small, nonpolar molecules move across bilayers quickly. In contrast, large molecules and charged substances cross the membrane slowly, if at all. According to the data in **Figure 6.8**, small, nonpolar molecules such as oxygen (O_2) move across selectively permeable membranes more than a billion times faster than do chloride ions (Cl^-). Very small and uncharged molecules such as water (H_2O) can also cross membranes relatively rapidly, even if they are polar. Small, polar molecules such as glycerol and urea have intermediate permeability.

The leading hypothesis to explain this pattern is that charged compounds and large, polar molecules can't pass through the nonpolar, hydrophobic tails of a lipid bilayer. Because of their electrical charge, ions are more stable in solution where they form hydrogen bonds with water than they are in the interior of membranes, which is electrically neutral. ● If you understand this hypothesis, you should be able to predict whether amino acids and nucleotides will cross a membrane readily. To test the hypothesis, researchers have manipulated the size and structure of the tails in liposomes or planar bilayers.

(a) Permeability scale (cm/sec)

High permeability

10^0

10^{-2}

10^{-4}

10^{-6}

10^{-8}

10^{-10}

10^{-12}

Low permeability

O_2, CO_2

H_2O

Glycerol, urea

Glucose

Cl^-

K^+

Na^+

(b) Size and charge affect the rate of diffusion across a membrane.

Phospholipid bilayer

Small, nonpolar molecules

Small, uncharged polar molecules

Large, uncharged polar molecules

Ions

O_2, CO_2, N_2

H_2O, urea, glycerol

Glucose, sucrose

Cl^-, K^+, Na^+

FIGURE 6.8 Selective Permeability of Lipid Bilayers. (a) The numbers represent "permeability coefficients," or the rate (cm/sec) at which an ion or molecule crosses a lipid bilayer. **(b)** The relative permeabilities of various molecules and ions, based on data like those presented in part (a).

● **QUESTION** About how fast does water cross the lipid bilayer?

Does the Type of Lipid in a Membrane Affect Its Permeability?

Theoretically, two aspects of a hydrocarbon chain could affect the way the chain behaves in a lipid bilayer: (1) the number of double bonds it contains and (2) its length. Recall from Chapter 2 that when carbon atoms form a double bond, the attached atoms are found in a plane instead of a (three-dimensional) tetrahedron. The carbon atoms involved are also locked into place. They cannot rotate freely, as they do in carbon-carbon single bonds. As a result, a double bond between carbon atoms produces a "kink" in an otherwise straight hydrocarbon chain (**Figure 6.9**).

When a double bond exists between two carbon atoms in a hydrocarbon chain, the chain is said to be **unsaturated**. Conversely, hydrocarbon chains without double bonds are said to be **saturated**. This choice of terms is logical, because if a hydrocarbon chain does not contain a double bond, it is saturated with the maximum number of hydrogen atoms that can attach to the carbon skeleton. If it is unsaturated, then fewer than the maximum number of hydrogen atoms are attached. Because they contain more C–H bonds, which have much more free energy than C=C bonds, saturated fats have much more chemical energy than unsaturated fats do. People who are dieting are

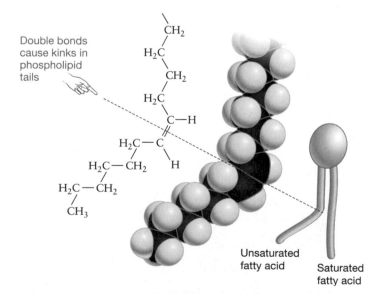

Double bonds cause kinks in phospholipid tails

CH_2

H_2C

CH_2

H_2C

$C-H$

H_2C-C

H_2C-CH_2 H

H_2C-CH_2

CH_3

Unsaturated fatty acid

Saturated fatty acid

FIGURE 6.9 Unsaturated Hydrocarbons Contain Carbon-Carbon Double Bonds. A double bond in a hydrocarbon chain produces a "kink." The icon on the right indicates that one of the hydrocarbon tails in a phospholipid is unsaturated and therefore kinked.

● **EXERCISE** Draw the structural formula and a schematic diagram for an unsaturated fatty acid containing two double bonds.

Lipid bilayer with **no** unsaturated fatty acids

Lower permeability

Lipid bilayer with **many** unsaturated fatty acids

Higher permeability

FIGURE 6.10 Fatty-Acid Structure Changes the Permeability of Membranes. Lipid bilayers containing many unsaturated fatty acids have more gaps and should be more permeable than are bilayers with few unsaturated fatty acids.

often encouraged to eat fewer saturated fats. Foods that contain lipids with many double bonds are said to be polyunsaturated and are advertised as healthier than foods with more-saturated fats.

Why do double bonds affect the permeability of membranes? When hydrophobic tails are packed into a lipid bilayer, the kinks created by double bonds produce spaces among the tightly packed tails. These spaces reduce the strength of hydrophobic interactions among the tails. Because the interior of the membrane is "glued together" less tightly, the structure should become more fluid and more permeable (**Figure 6.10**).

Hydrophobic interactions also become stronger as saturated hydrocarbon tails increase in length. Membranes dominated by phospholipids with long, saturated hydrocarbon tails should be

stiffer and less permeable because the interactions among the tails are stronger.

A biologist would predict, then, that bilayers made of lipids with long, straight, saturated fatty-acid tails should be much less permeable than membranes made of lipids with short, kinked, unsaturated fatty-acid tails. Experiments on liposomes have shown exactly this pattern. Phospholipids with long, saturated tails form membranes that are much less permeable than membranes consisting of phospholipids with shorter, unsaturated tails.

The central point here is that the degree of hydrophobic interactions dictates the behavior of these molecules. This is another example in which the structure of a molecule—specifically, the number of double bonds in the hydrocarbon chain and its overall length—correlates with its properties and function.

These data are also consistent with the basic observation that highly saturated fats are solid at room temperature (**Figure 6.11a**). Lipids that have extremely long hydrocarbon tails, as **waxes** do, form stiff solids at room temperature due to the extensive hydrophobic interactions that occur (**Figure 6.11b**). Birds, sea otters, and many other organisms synthesize waxes and spread them on their exterior surface as a waterproofing; plant cells secrete a waxy layer that covers the surface of leaves and stems and keeps water from evaporating. In contrast, highly unsaturated fats are liquid at room temperature (**Figure 6.11c**). Liquid triacylglycerides are called **oils**.

Besides exploring the role of hydrocarbon chain length and degree of saturation on membrane permeability, biologists have investigated the effect of adding cholesterol molecules. Because the steroid rings in cholesterol are bulky, adding cholesterol to

(a) Saturated lipids

(b) Saturated lipids with long hydrocarbon tails

(c) Unsaturated lipids

Butter

Beeswax

Safflower oil

FIGURE 6.11 The Fluidity of Lipids Depends on the Characteristics of Their Hydrocarbon Chains. The fluidity of a lipid depends on the length and saturation of its hydrocarbon chain. **(a)** Butter consists primarily of saturated lipids. **(b)** Waxes are lipids with extremely long hydrocarbon chains. **(c)** Oils are dominated by "polyunsaturates"—lipids with hydrocarbon chains that contain multiple double bonds.

● **QUESTION** Why are waxes so effective for waterproofing floors?

a membrane should increase the density of the hydrophobic section. As predicted, researchers found that adding cholesterol molecules to liposomes dramatically reduced the permeability of the liposomes. The data behind this claim are presented in **Figure 6.12**. The graph in this figure makes another important point, however: Temperature has a strong influence on the behavior of lipid bilayers.

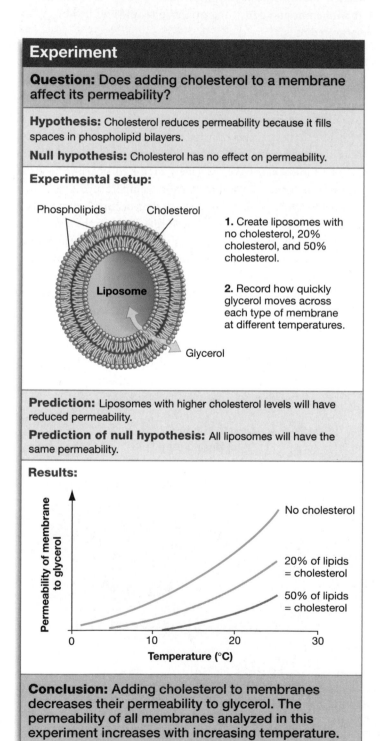

Experiment

Question: Does adding cholesterol to a membrane affect its permeability?

Hypothesis: Cholesterol reduces permeability because it fills spaces in phospholipid bilayers.

Null hypothesis: Cholesterol has no effect on permeability.

Experimental setup:

Phospholipids   Cholesterol

Liposome

Glycerol

1. Create liposomes with no cholesterol, 20% cholesterol, and 50% cholesterol.

2. Record how quickly glycerol moves across each type of membrane at different temperatures.

Prediction: Liposomes with higher cholesterol levels will have reduced permeability.

Prediction of null hypothesis: All liposomes will have the same permeability.

Results:

(Graph: Permeability of membrane to glycerol vs. Temperature (°C), showing three curves)

No cholesterol

20% of lipids = cholesterol

50% of lipids = cholesterol

Temperature (°C): 0, 10, 20, 30

Conclusion: Adding cholesterol to membranes decreases their permeability to glycerol. The permeability of all membranes analyzed in this experiment increases with increasing temperature.

FIGURE 6.12 The Permeability of a Membrane Depends on Its Composition.

Why Does Temperature Affect the Fluidity and Permeability of Membranes?

At about 25°C—or "room temperature"—the phospholipids found in plasma membranes are liquid, and bilayers have the consistency of olive oil. This fluidity, as well as the membrane's permeability, decreases as temperature decreases. As temperatures drop, individual molecules in the bilayer move more slowly. As a result, the hydrophobic tails in the interior of membranes pack together more tightly. At very low temperatures, lipid bilayers begin to solidify. As the graph in Figure 6.12 indicates, low temperatures can make membranes impervious to molecules that would normally cross them readily.

The fluid nature of membranes also allows individual lipid molecules to move laterally within each layer, a little like a person moving about in a dense crowd (**Figure 6.13**). By tagging individual phospholipids and following their movement, researchers have clocked average speeds of 2 micrometers (μm)/second at room temperature. At these speeds, phospholipids could travel the length of a small bacterial cell in a second.

These experiments on lipid and ion movement demonstrate that membranes are dynamic. Phospholipid molecules whiz around each layer while water and small, nonpolar molecules shoot in and out of the membrane. How quickly molecules move within and across membranes is a function of temperature and the structure of the hydrocarbon tails in the bilayer.

Given these insights into the permeability and fluidity of lipid bilayers, an important question remains: *Why* do certain molecules move across membranes spontaneously?

Phospholipids are in constant lateral motion, but rarely flip to the other side of the bilayer

FIGURE 6.13 Phospholipids Move within Membranes. Membranes are dynamic—in part because phospholipid molecules move within each layer in the structure.

Check Your Understanding

If you understand that...

- In solution, phospholipids form bilayers that are selectively permeable—meaning that some substances cross them much more readily than others do.

- Permeability is a function of temperature, the amount of cholesterol in the membrane, and the length and degree of saturation of the hydrocarbon tails in membrane phospholipids.

You should be able to...

Fill in a chart with rows called "Temperature," "Cholesterol," "Length of hydrocarbon tails," and "Saturation of hydrocarbon tails" and columns named "Factor," "Effect on permeability," and "Reason."

6.3 Why Molecules Move across Lipid Bilayers: Diffusion and Osmosis

A thought experiment can help explain why molecules and ions are able to move across membranes spontaneously. Suppose you rack up a set of blue billiard balls on a pool table containing many white balls and then begin to vibrate the table. Because of the vibration, the balls will move about randomly. They will also bump into one another. After these collisions, some blue balls will move outward—away from their original position. In fact, the overall (or net) movement of blue balls will be outward. This occurs because the random motion of the blue balls disrupts their original, nonrandom position—as they move at random, they are more likely to move away from each other than to stay together. Eventually, the blue billiard balls will be distributed randomly across the table. The entropy of the blue billiard balls has increased. Recall from Chapter 2 that entropy is a measure of the randomness or disorder in a system. The second law of thermodynamics states that in a closed system, entropy always increases.

This hypothetical example illustrates why molecules or ions located on one side of a lipid bilayer move to the other side spontaneously. The dissolved molecules and ions, or **solutes**, have thermal energy and are in constant, random motion. Movement of molecules and ions that results from their kinetic energy is known as **diffusion**. Because solutes change position randomly due to diffusion, they tend to move from a region of high concentration to a region of low concentration. A difference in solute concentrations creates a **concentration gradient**. ◉ Molecules and ions still move randomly in all directions when a concentration gradient exists, but there is a net movement from regions of high concentration to regions of low concentration. Diffusion along a concentration gradient is a spontaneous process because it results in an increase in entropy.

Once the molecules or ions are randomly distributed throughout a solution, equilibrium is established. For example, consider two aqueous solutions separated by a lipid bilayer. **Figure 6.14** shows how molecules that pass through the bilayer diffuse to the other side. At equilibrium, molecules continue to move back and forth across the membrane, but at equal rates—simply because each molecule or ion is equally likely to move in any direction. This means that there is no longer a net movement of molecules across the membrane.

What about water itself? As the data in Figure 6.8 showed, water moves across lipid bilayers relatively quickly. ◉ Like

DIFFUSION ACROSS A LIPID BILAYER

1. Start with different solutes on opposite sides of a lipid bilayer. Both molecules diffuse freely across bilayer.

Lipid bilayer

2. Solutes diffuse across the membrane—each undergoes a net movement along its own concentration gradient.

3. Equilibrium is established. Solutes continue to move back and forth across the membrane but at equal rates.

FIGURE 6.14 Diffusion across a Selectively Permeable Membrane.

◉ **EXERCISE** If a solute's rate of diffusion increases linearly with its concentration difference across the membrane, write an equation for the rate of diffusion across a membrane.

OSMOSIS

1. Start with more solute on one side of the lipid bilayer than the other, using molecules that cannot cross the selectively permeable membrane.

2. Water undergoes a net movement from the region of low concentration of solute (high concentration of water) to the region of high concentration of solute (low concentration of water).

FIGURE 6.15 Osmosis.

◉ **QUESTION** Suppose you doubled the number of molecules on the right side of the membrane (at the start). At equilibrium, would the water level on the right side be higher or lower than what is shown here?

other substances that diffuse, water moves along its concentration gradient—from higher to lower concentration. The movement of water is a special case of diffusion that is given its own name: **osmosis**. Osmosis occurs only when solutions are separated by a membrane that is permeable to some molecules but not others—that is, a selectively permeable membrane.

The best way to think about water moving in response to a concentration gradient is to focus on the concentration of solutes in the solution. Let's suppose the concentration of a particular solute is higher on one side of a selectively permeable membrane than it is on the other side (**Figure 6.15**, step 1). Further, suppose that this solute cannot diffuse through the membrane to establish equilibrium. What happens? Water will move from the side with a lower concentration of solute to the side with a higher concentration of solute (step 2). It dilutes the higher concentration and equalizes the concentrations on both sides. This movement is spontaneous. It is driven by the increase in entropy achieved when solute concentrations are equal on both sides of the membrane.

Another way to think about osmosis is to realize that water is at higher concentration on the left side of the beaker in Figure 6.15 than it is on the right side of the beaker. As water diffuses, then, there will be net movement of water molecules from the left side to the right side: from a region of high concentration to a region of low concentration.

The movement of water by osmosis is important because it can swell or shrink a membrane-bound vesicle. Consider the liposomes illustrated in **Figure 6.16**. If the solution outside the

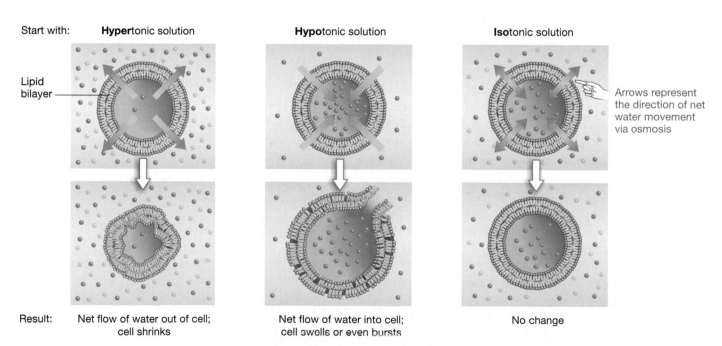

Start with: **Hyper**tonic solution   **Hypo**tonic solution   **Iso**tonic solution

Lipid bilayer

Arrows represent the direction of net water movement via osmosis

Result: Net flow of water out of cell;   Net flow of water into cell;   No change
cell shrinks   cell swells or even bursts

FIGURE 6.16 Osmosis Can Shrink or Burst Membrane-Bound Vesicles.

◉ **QUESTION** Some species of bacteria can live in extremely salty environments, such as saltwater-evaporation ponds. Is this habitat likely to be hypertonic, hypotonic, or isotonic relative to the interior of the cells?

membrane has a higher concentration of solutes than the interior has, and the solutes are not able to pass through the lipid bilayer, then water will move out of the vesicle into the solution outside. As a result, the vesicle will shrink and the membrane shrivel. Such a solution is said to be **hypertonic** ("excess-tone") relative to the inside of the vesicle. The word root *hyper* refers to the outside solution containing more solutes than the solution on the other side of the membrane. Conversely, if the solution outside the membrane has a lower concentration of solutes than the interior has, water will move into the vesicle via osmosis. The incoming water will cause the vesicle to swell or even burst. Such a solution is termed **hypotonic** ("lower-tone") relative to the inside of the vesicle. Here the word root *hypo* refers to the outside solution containing fewer solutes than the inside solution has. If solute concentrations are equal on either side of the membrane, the liposome will maintain its size. When the outside solution does not affect the membrane's shape, that solution is called **isotonic** ("equal-tone").

Note that the terms hypertonic, hypotonic, and isotonic are relative—they can be used only to express the relationship between a given solution and another solution. ● If you understand this concept, you should be able to draw liposomes in Figure 6.16 that change the relative "tonicity" of the surrounding solution. Specifically, draw (1) a liposome on the left such that the surrounding solution is hypotonic relative to the solution inside the liposome, and (2) a liposome in the center where the surrounding solution is hypertonic relative to the solution inside the liposome.

 Web Animation at www.masteringbio.com
Diffusion and Osmosis

To summarize, diffusion and osmosis move solutes and water across lipid bilayers. What does all this have to do with the first membranes floating in the prebiotic soup? Osmosis and diffusion tend to *reduce* differences in chemical composition between the inside and outside of membrane-bound structures. If liposome-like structures were present in the prebiotic soup, it's unlikely that their interiors offered a radically different environment from the surrounding solution. In all likelihood, the primary importance of the first lipid bilayers was simply to provide a container for self-replicating molecules. Experiments have shown that ribonucleotides can diffuse across lipid bilayers. Further, it is clear that cell-like vesicles grow as additional lipids are added and then divide if sheared by shaking, bubbling, or wave action. Based on these observations, it is reasonable to hypothesize that once a self-replicating ribozyme had become surrounded by a lipid bilayer, this simple lifeform and its descendants would continue to occupy cell-like structures that grew and divided.

Now let's investigate the next great event in the evolution of life: the formation of a true cell. How can lipid bilayers become

a barrier capable of creating and maintaining a specialized internal environment that is conducive to life? How could an effective plasma membrane—one that admits ions and molecules needed by the replicator while excluding ions and molecules that might damage it—evolve in the first cell?

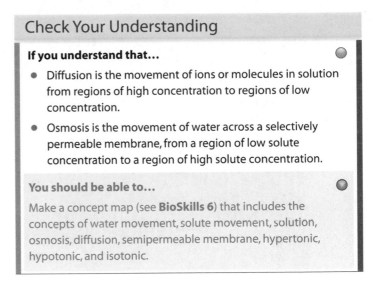

Check Your Understanding

If you understand that...

- Diffusion is the movement of ions or molecules in solution from regions of high concentration to regions of low concentration.

- Osmosis is the movement of water across a selectively permeable membrane, from a region of low solute concentration to a region of high solute concentration.

You should be able to...

Make a concept map (see **BioSkills 6**) that includes the concepts of water movement, solute movement, solution, osmosis, diffusion, semipermeable membrane, hypertonic, hypotonic, and isotonic.

6.4   Membrane Proteins

What sort of molecule could become incorporated into a lipid bilayer and affect the bilayer's permeability? The title of this section gives the answer away. Proteins that are amphipathic can be inserted into lipid bilayers.

Proteins can be amphipathic because they are made up of amino acids and because amino acids have side chains, or R-groups, that range from highly nonpolar to highly polar. (Some are even charged; see Figure 3.3 and Table 3.2.) It's conceivable, then, that a protein could have a series of nonpolar amino acids in the middle of its primary structure, but polar or charged amino acids on both ends of its primary structure, as illustrated in **Figure 6.17a**. The nonpolar amino acids would be stable in the interior of a lipid bilayer, while the polar or charged amino acids would be stable alongside the polar heads and surrounding water (**Figure 6.17b**). Further, because the secondary and tertiary structures of proteins are almost limitless in their variety and complexity, it is possible for proteins to form tubes and thus function as some sort of channel or pore across a lipid bilayer.

Based on these considerations, it is not surprising that when researchers began analyzing the chemical composition of plasma membranes in eukaryotes they found that proteins were just as common, in terms of mass, as phospholipids. How were these two types of molecules arranged? In 1935 Hugh Davson and James Danielli proposed that plasma membranes were struc-

(a) Proteins can be amphipathic.

The polar and charged amino acids are hydrophilic

The nonpolar amino acids are hydrophobic

(b) Amphipathic proteins can integrate into lipid bilayers.

Outside cell

Inside cell

FIGURE 6.17 Proteins Can Be Amphipathic.

● **QUESTION** Researchers can analyze the primary structure of a membrane protein and predict which portions are embedded in the membrane and which are exposed to the cell's interior or exterior. How is this possible?

● **QUESTION** What type of secondary structure is shown in part (b)?

(a) Sandwich model

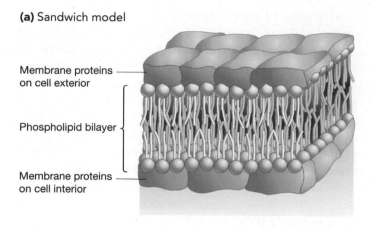

Membrane proteins on cell exterior

Phospholipid bilayer

Membrane proteins on cell interior

(b) Fluid-mosaic model

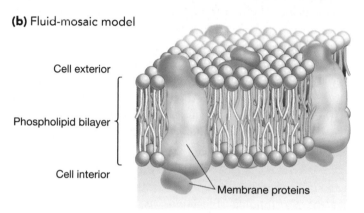

Cell exterior

Phospholipid bilayer

Cell interior

Membrane proteins

FIGURE 6.18 Past and Current Models of Membrane Structure. (a) The protein-lipid-lipid-protein sandwich model was the first hypothesis for the arrangement of lipids and proteins in plasma membranes. **(b)** The fluid-mosaic model was a radical departure from the sandwich hypothesis.

tured like a sandwich, with hydrophilic proteins coating both sides of a pure lipid bilayer (**Figure 6.18a**). Early electron micrographs of plasma membranes seemed to be consistent with the sandwich model, and for decades it was widely accepted.

The realization that membrane proteins could be amphipathic led S. Jon Singer and Garth Nicolson to suggest an alternative hypothesis, however. In 1972, they proposed that at least some proteins span the membrane instead of being found only outside the lipid bilayer. Their hypothesis was called the **fluid-mosaic model**. As **Figure 6.18b** shows, Singer and Nicolson suggested that membranes are a mosaic of phospholipids and different types of proteins. The overall structure was proposed to be dynamic and fluid.

The controversy over the nature of the plasma membrane was resolved in the early 1970s with the development of an innovative technique for visualizing the surface of plasma membranes. The method is called freeze-fracture electron microscopy, because the steps involve freezing and fracturing

the membrane before examining it with a **scanning electron microscope,** which produces images of an object's surface (see **BioSkills 8**). As **Figure 6.19** shows, the technique allows researchers to split plasma membranes and view the middle of the structure. The scanning electron micrographs that result show pits and mounds studding the inner surfaces of the lipid bilayer. Researchers interpret these structures as the locations of membrane proteins. As step 4 in Figure 6.19 shows, the pits and mounds are hypothesized to represent proteins that span the lipid bilayer.

These observations conflicted with the sandwich model but were consistent with the fluid-mosaic model. Based on these and subsequent observations, the fluid-mosaic model is now widely accepted.

Figure 6.20 summarizes the current hypothesis for where proteins and lipids are found in a plasma membrane. Note that some proteins span the membrane and have segments facing both the interior and exterior surfaces. Proteins such as these are called

VISUALIZING MEMBRANE PROTEINS

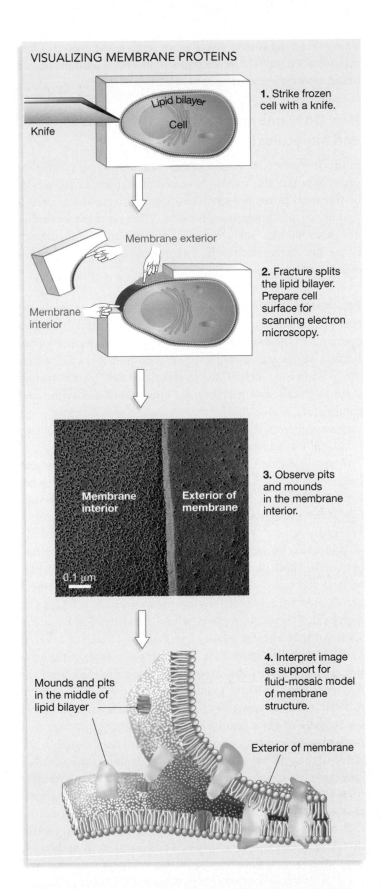

1. Strike frozen cell with a knife.

2. Fracture splits the lipid bilayer. Prepare cell surface for scanning electron microscopy.

3. Observe pits and mounds in the membrane interior.

4. Interpret image as support for fluid-mosaic model of membrane structure.

Knife

Lipid bilayer

Cell

Membrane exterior

Membrane interior

Membrane interior

Exterior of membrane

0.1 μm

Mounds and pits in the middle of lipid bilayer

Exterior of membrane

FIGURE 6.19 Freeze-Fracture Preparations Allow Biologists to View Membrane Proteins.

● **QUESTION** What would the micrograph in step 3 look like if the sandwich model of membrane structure were correct?

Outside cell

Peripheral membrane protein

Integral membrane protein

Peripheral membrane protein

Inside cell

FIGURE 6.20 Integral and Peripheral Membrane Proteins. Integral membrane proteins are also called transmembrane proteins because they span the membrane. Peripheral membrane proteins are often attached to integral membrane proteins.

● **QUESTION** Are the external and internal faces of a plasma membrane the same or different? Explain.

integral membrane proteins, or **transmembrane proteins**. Other proteins, called **peripheral membrane proteins**, are found only on one side of the membrane. Often, peripheral membrane proteins are attached to an integral membrane protein. In most cases, specific peripheral proteins are found only in the inside of the plasma membrane and thus inside the cell, while others are found only on the outside of the plasma membrane and thus facing the surrounding environment. The location of peripheral proteins is one of several reasons that the exterior surface of the plasma membrane is very different from the interior surface. It's also important to realize that the position of these proteins is not static. Like the phospholipids in the bilayer, membrane proteins are in constant motion, diffusing through the oily film.

What do all these proteins do? Later chapters will explore how certain membrane proteins act as enzymes or are involved in cell-to-cell signaling or making physical connections between cells. Here, let's focus on how integral membrane proteins are involved in the transport of selected ions and molecules across the plasma membrane.

Systems for Studying Membrane Proteins

The discovery of integral membrane proteins was consistent with the hypothesis that proteins affect membrane permeability. The evidence was not considered conclusive enough, though, because it was also plausible to claim that integral membrane proteins were structural components that influenced membrane strength or flexibility. To test whether proteins actually do affect

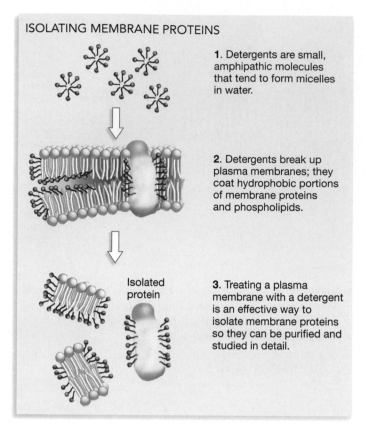

ISOLATING MEMBRANE PROTEINS

1. Detergents are small, amphipathic molecules that tend to form micelles in water.

2. Detergents break up plasma membranes; they coat hydrophobic portions of membrane proteins and phospholipids.

Isolated protein

3. Treating a plasma membrane with a detergent is an effective way to isolate membrane proteins so they can be purified and studied in detail.

FIGURE 6.21 Detergents Can Be Used to Get Membrane Proteins into Solution.

membrane permeability, researchers needed some way to isolate and purify membrane proteins.

Figure 6.21 outlines one method that researchers developed to separate proteins from membranes. The key to the technique is the use of detergents. A detergent is a small, amphipathic molecule. As step 1 of Figure 6.21 shows, the hydrophobic tails of detergents clump in solution, forming micelles. When detergents are added to the solution surrounding a lipid bilayer, the hydrophobic tails of the detergent molecule interact with the hydrophobic tails of the lipids. In doing so, the detergent tends to disrupt the bilayer and break it apart (step 2). If the membrane contains proteins, the hydrophobic tails of the detergent molecules also interact with the hydrophobic parts of the membrane proteins. The detergent molecules displace the membrane phospholipids and end up forming water-soluble, detergent-protein complexes (step 3).

To isolate and purify these membrane proteins once they are in solution, researchers use the technique called gel electrophoresis, introduced in **BioSkills 6**. When detergent-protein complexes are loaded into a gel and a voltage is applied, the larger protein complexes migrate more slowly than smaller proteins. As a result, the various proteins isolated from a plasma membrane separate from each other. To obtain a pure sample of a particular protein, the appropriate band is cut out of the gel.

The gel material is then dissolved to retrieve the protein. Once this protein is inserted into a planar bilayer or liposome, dozens of different experiments are possible.

How Do Membrane Proteins Affect Ions and Molecules?

In the 55 years since intensive experimentation on membrane proteins began, researchers have identified three broad classes of **transport proteins**—channels, transporters, and pumps—that affect membrane permeability. What do these molecules do? Can plasma membranes that contain these proteins create an internal environment more conducive to life than the external environment is?

Facilitated Diffusion via Channel Proteins One of the first membrane peptides to be investigated in detail is called gramicidin. Gramicidin is produced by a bacterium called *Bacillus brevis* and is used as a weapon: *B. brevis* cells release the protein just before a resistant coating forms around their cell wall and membrane. The gramicidin wipes out competitors, giving *B. brevis* cells more room to grow when they emerge from the resistant phase. Gramicidin is also used medicinally in humans as an antibiotic.

After observing that experimental cells treated with gramicidin seemed to lose large numbers of ions, researchers became interested in understanding how the molecule works. Could this protein alter the flow of ions across plasma membranes?

Biologists answered this question by inserting purified gramicidin into planar bilayers. The experiment they performed was based on an important fact about ion movement across membranes: Not only do ions move from regions of high concentration to regions of low concentration via diffusion, but they also flow from areas of like charge to areas of unlike charge. In **Figure 6.22**, for example, a large concentration gradient favors the movement of sodium ions from the outside of the cell to the inside. But in addition, the inside of this cell has a net negative charge while the outside has a net positive charge. As a result, there is also a charge gradient that favors the movement of sodium ions from the outside to the inside of the cell. Based on this example, it should be clear that ions move in response to a combined concentration and electrical gradient, or what biologists call an **electrochemical gradient**. ● If you understand this concept, you should be able to add an arrow to Figure 6.22, indicating the electrochemical gradient for chloride ions assuming that chloride concentrations are equal on both sides of the membrane.

To determine whether gramicidin affected the membrane's permeability to ions, the researchers measured the flow of electric current across the membrane. Because ions carry a charge, the movement of ions produces an electric current. This prop-

FIGURE 6.22 Electrochemical Gradients. When ions build up on one side of a membrane, they establish a combined concentration and electrical gradient.

● **EXERCISE** By adding sodium ions to this figure, illustrate a situation where there is no electrochemical gradient favoring the movement of either Na⁺ or Cl⁻.

erty provides an elegant and accurate test for assessing the bilayer's permeability to ions—one that is simpler and more sensitive than taking samples from either side of the membrane and determining the concentrations of solutes present. If gramicidin facilitates ion movement, then an investigator should be able to detect an electric current across planar bilayers that contain gramicidin.

The result? The graph in **Figure 6.23** shows that when gramicidin was absent, no electric current passed through the membrane. But when gramicidin was inserted into the membrane, current began to flow. Based on this observation, biologists proposed that gramicidin is an **ion channel**. An ion channel is a peptide or protein that makes lipid bilayers permeable to ions. (Recall from Chapter 3 that peptides are proteins containing fewer than 50 amino acids.) Follow-up work corroborated that gramicidin is selective. It allows only positively charged ions, or cations, to pass. Gramicidin does not allow negatively charged ions, or anions, to pass through the membrane. It was also established that gramicidin is most permeable to hydrogen ions (or protons, (H⁺) and somewhat less permeable to other cations, such as potassium (K⁺) and sodium (Na⁺).

Researchers gained additional insight into the way gramicidin works when they determined its amino acid sequence (that is, primary structure) and tertiary structure. **Figure 6.24** provides a view from the outside of a cell to the inside through gramicidin. The key observation is that the molecule forms a hole. The portions of amino acids that line this hole are hydrophilic, while regions on the exterior (in contact with the membrane phospholipids) are hydrophobic. The molecule's structure correlates with its function.

Experiment

Question: Does gramicidin affect the flow of ions across a membrane?

Hypothesis: Gramicidin increases the flow of cations across a membrane.

Null hypothesis: Gramicidin has no effect on membrane permeability.

Experimental setup:

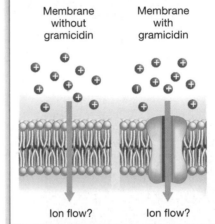

Prediction: Ion flow will be higher in membrane with gramicidin.

Prediction of null hypothesis: Ion flow will be the same in both membranes.

Results:

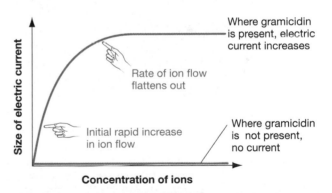

Conclusion: Gramicidin facilitates diffusion of cations along an electrochemical gradient. Gramicidin is an ion channel.

FIGURE 6.23 Measuring Ion Flow through the Channel Gramicidin. Experiment for testing the hypothesis that gramicidin is an ion channel.

● **QUESTION** Why does the curve in the Results section flatten out?

Subsequent research has shown that cells have many different types of **channel proteins** in their membranes, each featuring a structure that allows it to admit a particular type of ion or small molecule. For example, Peter Agre and co-workers recently discovered channels called **aquaporins** ("water-pores") that allow

(a) Top view of gramicidin

Hydrophobic exterior

Hydrophilic interior

(b) Side view of gramicidin

FIGURE 6.24 The Structure of a Channel Protein. Gramicidin is an α-helix consisting of only 15 amino acids. **(a)** In top view, the molecule forms a hole or pore. **(b)** In side view, a green helix traces the peptide-bonded backbone of the polypeptide. R-groups hang off the backbone to the outside. The interior of the channel is hydrophilic; the exterior is hydrophobic.

● **EXERCISE** In (a) and (b), add symbols indicating the locations of phospholipids relative to gramicidin in a plasma membrane.

water to cross the plasma membrane over 10 times faster than it does in the absence of aquaporins. **Figure 6.25a** shows a cutaway view from the side of an aquaporin, indicating how it fits in a plasma membrane. Like gramicidin, the channel has a pore that is lined with polar regions of amino acids—in this case, functional groups that interact with water. Hydrophobic groups form the outside of the structure and interact with the lipid bilayer. Unlike gramicidin, aquaporins are extremely selective. They admit water but not other small molecules or ions.

Selectivity turns out to be a prominent characteristic of most channel proteins. The vast majority of these proteins admit only a single type of ion. In many cases, researchers are now

(a) Water pores allow only water to pass through.

Hydrophilic interior

Outside cell

H_2O

Hydrophobic exterior

Inside cell

(b) Potassium channels allow only potassium ions to pass through.

Potassium ions can enter the channel, but cannot pass into the cell

Outside cell

Inside cell

Closed

When a change in electrical charge occurs outside the membrane, the protein changes shape and allows the ions to pass through

+ + + + + + +   + + + + + + +

Open

FIGURE 6.25 Most Membrane Channels Are Highly Selective and Highly Regulated (a) A cutaway view looking at the side of an aquaporin—a membrane channel that admits only water. Water moves through its pore via osmosis over 10 times faster than it can move through the lipid bilayer. **(b)** A model of a K^+ channel in the open and closed configurations.

able to identify exactly which amino acids are responsible for making the pore selective.

Recent research has also shown that the aquaporins and ion channels are **gated channels**—meaning that they open or close in response to the binding of a particular molecule or to a change in the electrical charge on the outside of the membrane. For example, **Figure 6.25b** shows a potassium channel in the open and closed configuration. When the electrical charge on the membrane becomes positive on the outside relative to the inside, the protein's structure changes in a way that opens the channel and allows potassium ions to cross. The important point here is that in almost all cases, the flow of ions and small molecules through membrane channels is carefully controlled.

In all cases, the movement of substances through channels is passive—meaning it does not require an expenditure of energy. **Passive transport** is powered by diffusion along an electrochemical gradient. Channel proteins enable ions or polar molecules to move across lipid bilayers efficiently. ● If you understand the nature of membrane channels, you should be able to (1) draw the structure of a channel that admits calcium ions (Ca^{2+}) when a signaling molecule binds to it, (2) label hydrophilic and hydrophobic portions of the channel, (3) add ions to the outside and inside of a membrane containing the channel to explain why an electrochemical gradient favors entry of Ca^{2+}, (4) sketch the channel in the open versus closed configuration, and (5) suggest a hypothesis to explain why it might be important for the channel to be selective.

To summarize, membrane proteins such as gramicidin, aquaporins, and potassium channels circumvent the lipid bilayer's impermeability to small, charged compounds. They are responsible for **facilitated diffusion**: the passive transport of substances that otherwise would not cross a membrane readily. The presence of channels *reduces* differences between the interior and exterior. Water molecules and ions are not the only substances that move across membranes through membrane proteins, however. Larger molecules can, too.

Facilitated Diffusion via Carrier Proteins
Even though facilitated diffusion does not require an expenditure of energy,

it is facilitated—aided—by the presence of a specialized membrane protein. Facilitated diffusion can occur through channels or through **carrier proteins**, also called **transporters**, that change shape during the process. Perhaps the best-studied transporter is specialized for moving glucose into cells.

Next to ribose, the six-carbon sugar glucose is the most important sugar found in organisms. Virtually all cells alive today use glucose as a building block for important macromolecules and as a source of stored chemical energy. But as Figure 6.8 showed, lipid bilayers are only moderately permeable to glucose. It is reasonable to expect, then, that plasma membranes have some mechanism for increasing their permeability to this sugar.

This prediction was supported when researchers compared the permeability of glucose across planar bilayers with its permeability across membranes from cells. The plasma membrane in this study came from human red blood cells, which are among the simplest cells known. A mature red blood cell consists of a membrane, about 300 million hemoglobin molecules, and not much else (**Figure 6.26**, step 1). When these cells are placed in a hypotonic solution (step 2), water rushes into them by osmosis. As water flows inward, the cells swell. Eventually they burst, releasing the hemoglobin molecules and other cell contents. This leaves researchers with pure preparations of plasma membranes called red blood cell "ghosts" (step 3). Experiments have shown that these membranes are much more permeable to glucose than are pure lipid bilayers. Why?

After isolating and analyzing many proteins from red blood cell ghosts, researchers found one protein that specifically increases membrane permeability to glucose. When this purified protein was added to liposomes, the artificial membrane transported glucose at the same rate as a membrane from a living cell. This experiment convinced biologists that a membrane protein was indeed responsible for transporting glucose across plasma membranes. Follow-up work showed that this glucose-

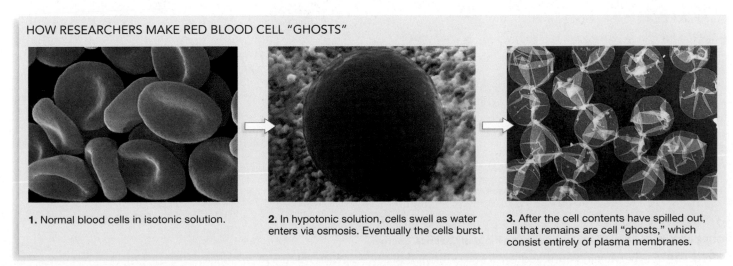

HOW RESEARCHERS MAKE RED BLOOD CELL "GHOSTS"

1. Normal blood cells in isotonic solution.

2. In hypotonic solution, cells swell as water enters via osmosis. Eventually the cells burst.

3. After the cell contents have spilled out, all that remains are cell "ghosts," which consist entirely of plasma membranes.

FIGURE 6.26 Red Blood Cell "Ghosts." Red blood cell ghosts are simple membranes that can be purified and studied in detail.

A HYPOTHESIS FOR HOW GLUT-1 FACILITATES GLUCOSE DIFFUSION

Outside cell

Glucose

GLUT-1

Inside cell

1. GLUT-1 is a transmembrane transport protein, shown with its binding site facing outside the cell.

2. Glucose binds to GLUT-1 from outside the cell.

3. A conformational change results, transporting glucose to the interior.

4. Glucose is released inside the cell.

FIGURE 6.27 A Hypothesis to Explain How Membrane Transport Proteins Work. This model suggests that the GLUT-1 transporter acts like an enzyme. It binds a substrate (in this case, a glucose molecule), undergoes a conformation change, and releases the substrate.

QUESTION GLUT's binding site has the same affinity for glucose in both of its conformations. Explain how this trait allows glucose to diffuse along its concentration gradient.

transporting protein, a carrier that is now called GLUT-1, facilitates transport of the "right-handed" optical isomer of glucose but not the left-handed form. Cells use only the right-handed form of glucose, and GLUT-1's binding site is specific for the right-handed form. To make sense of these observations, biologists hypothesize that GLUT proteins with binding sites that interact with the right-handed form of glucose were favored by natural selection. Stated another way, cells with proteins like GLUT-1 thrived much better than cells without a glucose-transport protein or with proteins that transported the left-handed form.

Exactly how GLUT-1 works is a focus of ongoing research. Biologists who are working on the problem have noted that because glucose transport by GLUT-1 is so specific, it is logical to predict that the mechanism resembles the action of enzymes. One hypothesis is illustrated in **Figure 6.27**. The idea is that glucose binds to GLUT-1 on the exterior of the membrane and that this binding induces a conformational change in the protein which transports glucose to the interior of the cell. Recall from Chapter 3 that enzymes frequently change shape when they bind substrates and that such conformational changes are often a critical step in the catalysis of chemical reactions.

HOW THE SODIUM-POTASSIUM PUMP (Na$^+$/K$^+$-ATPase) WORKS

Outside cell

Inside cell

ATP

Phosphate group

ADP

1. Three binding sites within the protein have a high affinity for sodium ions.

2. Three sodium ions from the inside of the cell bind to these sites.

3. A phosphate group from ATP binds to the protein. In response, the protein changes shape.

4. The sodium ions leave the protein and diffuse to the exterior of the cell.

FIGURE 6.28 Active Transport Depends on an Input of Chemical Energy.

EXERCISE Circle the two steps where addition or deletion of a phosphate group causes the protein to change conformation. Label each "Shape change."

Importing molecules into cells via carrier proteins is still powered by diffusion, however. When glucose enters a cell via GLUT-1, it does so because it is following its concentration gradient. If the concentration of glucose is the same on both sides of the plasma membrane, then no net movement of glucose occurs even if the membrane contains GLUT-1. A large array of molecules moves across plasma membranes via facilitated diffusion through specific carrier proteins.

Active Transport by Pumps   Whether diffusion is facilitated by channel proteins or by carrier proteins, it is a passive process that makes the cell interior and exterior more similar. But it is also possible for today's cells to import molecules or ions against their electrochemical gradient. Accomplishing this task requires energy, however, because the cell must counteract the entropy loss that occurs when molecules or ions are concentrated. It makes sense, then, that transport against an electrochemical gradient is called **active transport**.

In cells, the energy required to move substances against their electrochemical gradient is usually provided by a phosphate group (HPO_4^{2-}) from adenosine triphosphate, or ATP. ATP contains three phosphate groups. When one of these phosphate groups leaves ATP and binds to a protein, two negative charges are added to the protein. These charges repel other charges on the protein's amino acids. The protein's potential energy increases in response, and its conformation (shape) usually changes. As Chapter 9 will detail, proteins usually move when a phosphate group binds to them or when a phosphate group drops off. When a phosphate group leaves ATP, the resulting molecule is adenosine diphosphate (ADP), which has two phosphate groups.

Figure 6.28 shows how ions or molecules can move against an electrochemical gradient when membrane proteins called **pumps** change shape. The figure highlights the first pump that was discovered and characterized: a protein called the **sodium-potassium pump**, or more formally, Na^+/K^+-ATPase. The Na^+/K^+ part of this expression refers to the ions that are transported; ATP indicates that adenosine triphosphate is used; and –ase implies that the molecule acts like an enzyme. Unlike the situation with GLUT-1, the mechanism of action in Na^+/K^+-ATPase is now well known. When the protein is in the conformation shown in step 1 of Figure 6.28, binding sites with a high affinity for sodium ions are available. As step 2 shows, three sodium ions from the inside of the cell bind to these sites. A phosphate group from ATP then binds to the pump (step 3). When the phosphate group attaches, the pump's shape changes in a way that reduces its affinity for sodium ions. As a result, the ions leave the protein and diffuse to the exterior of the cell (step 4). In this conformation, though, the protein has binding sites with a high affinity for potassium ions (step 5). As step 6 shows, two potassium ions from outside the cell bind to the pump. When they do, the phosphate group drops off the protein and its shape changes in response—back to the original shape (step 7). In this conformation, the pump has low affinity for potassium ions. As step 8 shows, the potassium ions leave the protein and diffuse to the interior of the cell. The cycle then repeats.

This movement of ions can occur even if an electrochemical gradient exists that favors the outflow of potassium and the inflow of sodium. By exchanging three sodium ions for every two potassium ions, the outside of the membrane becomes positively charged relative to the inside. In this way, the

5. In this conformation, the protein has binding sites with a high affinity for potassium ions.

6. Two potassium ions bind to the pump.

7. The phosphate group drops off the protein. In response, the protein changes back to its original shape.

8. The potassium ions leave the protein and diffuse to the interior of the cell. These 8 steps repeat.

sodium-potassium pump sets up an electrical gradient as well as a chemical gradient across the membrane.

Similar pumps are specialized for moving protons (H^+), calcium ions (Ca^{2+}), or other ions or molecules. In this way, cells are capable of concentrating certain substances or setting up electrochemical gradients. It is difficult to overemphasize the importance of these gradients. For example, the electrochemical gradients produced by proton pumps allow plants to take up nutrients from the soil; the gradients established by the Na^+/K^+-ATPase and calcium pumps allow your nerve cells to transmit electrical signals throughout your body. You will encounter active transport, membrane pumps, and electrochemical gradients throughout this text.

(MB) **Web Animation** at www.masteringbio.com

Membrane Transport Proteins

Taken together, the lipid bilayer and the proteins involved in passive transport and active transport enable cells to create an internal environment that is much different from the external one. Membrane proteins allow ions and molecules to cross the plasma membrane, even though they are not lipid soluble. (**Figure 6.29**). When membrane proteins first evolved, then, the early cells acquired the ability to create an internal environment that was conducive to life—meaning that such an environment contained the substances required for manufac-

turing ATP and copying ribozymes. Cells with particularly efficient and selective membrane proteins would be favored by natural selection and would come to dominate the population. Cellular life had begun.

Some 3.5 billion years later, cells continue to evolve. What do today's cells look like, and how do they produce and store the chemical energy that makes life possible? How do they use ATP to move pumps and channels and other molecules and machines where they're needed? Answering these and related questions is the focus of Unit 2.

Check Your Understanding

If you understand that...

- Membrane proteins allow ions and molecules that ordinarily do not readily cross lipid bilayers to enter or exit cells.

- Substances may move along an electrochemical gradient via facilitated diffusion through channel proteins or transport proteins, or they may move against an electrochemical gradient in response to work done by pumps.

You should be able to...

1) Sketch a phospholipid bilayer.

2) Indicate how ions and large molecules cross it via each major type of membrane transport protein.

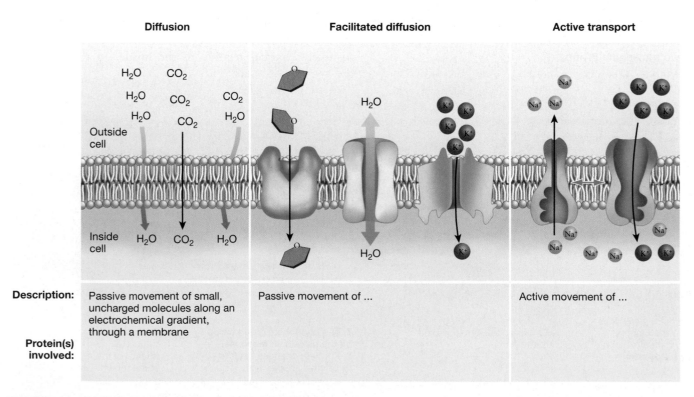

FIGURE 6.29 Mechanisms of Membrane Transport: A Summary.

EXERCISE Complete the chart.

Chapter Review

SUMMARY OF KEY CONCEPTS

⦿ **Phospholipids are amphipathic molecules—they have a hydrophilic region and a hydrophobic region. In solution, they spontaneously form bilayers that are selectively permeable—meaning that only certain substances cross them readily.**

The plasma membrane is a structure that forms a physical barrier between life and nonlife. The basic structure of plasma membranes is created by a phospholipid bilayer. Phospholipids have a polar head and a nonpolar tail. The nonpolar tail consists of a lipid—usually a fatty acid or an isoprene. Lipids do not dissolve in water.

Small, nonpolar molecules tend to move across membranes readily; ions and other charged compounds cross rarely, if at all. The permeability and fluidity of lipid bilayers depend on temperature and on the types of phospholipids present. For example, because phospholipids that contain long, saturated fatty acids form a dense and highly hydrophobic membrane interior, they tend to be less permeable than phospholipids containing shorter, unsaturated fatty acids.

You should be able to draw phospholipid bilayers that are highly permeable and fluid versus highly impermeable and lacking in fluidity. ⦿

⦿ **Ions and molecules diffuse spontaneously from regions of high concentration to regions of low concentration. Water moves across lipid bilayers from regions of high concentration to regions of low concentration via osmosis—a special case of diffusion.**

Diffusion is movement of ions or molecules due to their kinetic energy. Solutes move via diffusion from a region of high concentration to a region of low concentration. This is a spontaneous process driven by an increase in entropy. Water also moves across membranes spontaneously if a molecule or an ion that cannot cross the membrane is found in different concentrations on the two sides. In osmosis, water moves from the region with a lower concentration of solutes to the region of higher solute concentration. Osmosis is a passive process driven by an increase in entropy.

You should be able to draw a beaker with solutions on either side separated by a plasma membrane, and then predict what will happen after addition of a solute to one side if the solute (1) crosses the membrane readily, versus (2) is incapable of crossing the membrane. ⦿

(MB) **Web Animation** at www.masteringbio.com
Diffusion and Osmosis

⦿ **In cells, membrane proteins are responsible for the passage of ions, polar molecules, and large molecules that can't cross the membrane on their own because they are not soluble in lipids.**

The permeability of lipid bilayers can be altered significantly by membrane transport proteins, which are scattered throughout the plasma membrane. Channel proteins, for example, are molecules that provide holes in the membrane and facilitate the diffusion of specific ions into or out of the cell. Transport proteins are enzyme-like proteins that allow specific molecules to diffuse into the cell. In addition to these forms of facilitated diffusion, membrane proteins that act as energy-demanding pumps actively move ions or molecules against their electrochemical gradient. In combination, the selective permeability of phospholipid bilayers and the specificity of transport proteins make it possible to create an environment inside a cell that is radically different from the exterior.

You should be able to draw and label the membrane of a cell that pumps hydrogen ions to the exterior, has channels that admit calcium ions along an electrochemical gradient, and has carriers that admit lactose (a sugar) molecules along a concentration gradient. Your drawing should include arrows and labels indicating the direction of solute movement and the direction of the appropriate electrochemical gradients. ⦿

(MB) **Web Animation** at www.masteringbio.com
Membrane Transport Proteins

QUESTIONS

⦿ Test Your Knowledge

1. What does the term *hydrophilic* mean when it is translated literally?
   a. "oil loving"
   b. "water loving"
   c. "oil fearing"
   d. "water fearing"

2. If a solution surrounding a cell is hypotonic relative to the inside of the cell, how will water move?
   a. It will move into the cell via osmosis.
   b. It will move out of the cell via osmosis.
   c. It will not move, because equilibrium exists.
   d. It will evaporate from the cell surface more rapidly.

3. If a solution surrounding a cell is hypertonic relative to the inside of the cell, how will water move?
   a. It will move into the cell via osmosis.
   b. It will move out of the cell via osmosis.
   c. It will not move, because equilibrium exists.
   d. It will evaporate from the cell surface more rapidly.

4. When does a concentration gradient exist?
   a. when membranes rupture
   b. when solute concentrations are high
   c. when solute concentrations are low
   d. when solute concentrations differ on the two sides of a membrane

5. Which of the following must be true for osmosis to occur?
   a. Water must be at room temperature or above.
   b. Solutions with the same concentration of solutes must be separated by a selectively permeable membrane.
   c. Solutions with different concentrations of solutes must be separated by a selectively permeable membrane.
   d. Water must be under pressure.

6. Why are the lipid bilayers in cells called "selectively permeable"?
   a. They are not all that permeable.
   b. Their permeability changes with their molecular composition.
   c. Their permeability is temperature dependent.
   d. They are permeable to some substances but not others.

Test Your Knowledge answers: 1. b; 2. a; 3. b; 4. d; 5. c; 6. d

● Test Your Understanding

Answers are available at www.masteringbio.com

1. Cooking oil is composed of lipids that consist of long hydrocarbon chains. Would you expect these lipids to form membranes spontaneously? Why or why not? Describe, on a molecular level, how you would expect these lipids to interact with water.

2. Explain why phospholipids form a bilayer in solution.

3. Ethanol, the active ingredient in alcoholic beverages, is a small, polar, uncharged molecule. Would you predict that this molecule crosses plasma membranes quickly or slowly? Explain your reasoning.

4. Why can osmosis occur only if solutions are separated by a selectively permeable membrane? What happens in solutions that are *not* separated by a selectively permeable membrane?

5. The text claims that the portion of membrane proteins that spans the hydrophobic tails of phospholipids is itself hydrophobic (see Figure 6.17b). Why is this logical? Look back at Figure 3.3 and Table 3.2, and make a list of amino acids you might expect to find in these regions of transmembrane proteins.

6. Examine the membrane in the accompanying figure. Label the molecules and ions that will pass through the membrane as a result of osmosis, diffusion, and facilitated diffusion. Draw arrows to indicate where each of the molecules and ions will travel.

● Applying Concepts to New Situations

Answers are available at www.masteringbio.com

1. When phospholipids are arranged in a bilayer, it is theoretically possible for individual molecules in the bilayer to flip-flop. That is, a phospholipid could turn 180° and become part of the membrane's other surface. Sketch this process. From what you know about the behavior of polar heads and nonpolar tails, predict whether flip-flops are frequent or rare. Then design an experiment, using a planar bilayer made up partly of fatty acids that contain a dye molecule on their hydrophilic head, to test your prediction.

2. Unicellular organisms that live in extremely cold habitats have an unusually high proportion of unsaturated fatty acids in their plasma membranes. Some of these membranes even contain polyunsaturated fatty acids, which have more than one double bond in each hydrocarbon chain. Draw a picture of this type of membrane and comment on the hypothesis that membranes with unsaturated fatty-acid tails function better at cold temperatures than do membranes with saturated fatty-acid tails. Make a prediction about the structure of fatty acids found in organisms that live in extremely hot environments.

3. When biomedical researchers design drugs that must enter cells to be effective, they sometimes add methyl (CH_3) groups to make the drug molecules more likely to pass through plasma membranes. Conversely, when researchers design drugs that act on the exterior of plasma membranes, they sometimes add a charged group to decrease the likelihood that the drugs will pass through membranes and enter cells. Explain why these strategies are effective.

4. Advertisements frequently claim that laundry and dishwashing detergents "cut grease." What the ad writers mean is that the detergents surround oil droplets on clothing or dishes, making the droplets water soluble. When this happens, the oil droplets can be washed away. Explain how this happens on a molecular level.

www.masteringbio.com is also your resource for • Answers to text, table, and figure caption questions and exercises • Answers to *Check Your Understanding* boxes • Online study guides and quizzes • Additional study tools including the *E-Book for Biological Science* 3rd ed., textbook art, animations, and videos.

Inside the Cell

7

This cell has been treated with fluorescing molecules that bind to its fibrous skeleton. Microtubules (large protein fibers) are green; microfilaments (smaller fibers) are red. The cell's nucleus has been stained blue.

KEY CONCEPTS

- The structure of cell components is closely related to their function.

- Inside cells, materials are transported to their destinations with the help of molecular "zip codes."

- Cells are dynamic. Thousands of chemical reactions occur each second within cells; molecules constantly enter and exit across the plasma membrane; cell products are shipped along protein fibers; and elements of the cell's internal skeleton grow and shrink.

I n Chapter 1 you were introduced to the cell theory, which states that all organisms consist of cells and that all cells are derived from preexisting cells. Since this theory was initially developed and tested in the 1850s and 1860s, an enormous body of research has confirmed that the cell is the fundamental structural and functional unit of life. Life on Earth is cellular.

In a very real sense, then, understanding how an organism works is a matter of understanding how cells are structured and how they function. To drive this point home, recall from Chapter 1 that many eukaryotic organisms and virtually all bacteria and archaea are unicellular. In terms of sheer numbers, unicellular organisms dominate life on Earth. For researchers who study these unicellular species, understanding the cell is synonymous with understanding the organism as a whole. Even in

plants, animals, and other multicellular eukaryotes, complex behavior originates at the level of the cell. For example, your ability to read this page begins with changes in light-sensitive molecules located in cells at the back of your eyes. When these molecules change shape, they trigger changes in the membranes of nerve cells that connect your eyes to your brain. To understand complex processes such as vision, then, researchers often begin by studying the structure and function of the individual cells involved—the parts that make up the whole.

Chapter 6 introduced an essential part of the cell: the **plasma membrane**. Thanks to the selective permeability of phospholipid bilayers and the activity of membrane transport proteins, this structure creates an internal environment that differs from conditions outside the cell. Our task now is to

● Key Concept | Important Information | ● Practice It

explore the structures that are found inside the membrane and analyze what they do. The chapter focuses on several particularly dynamic structures and processes and introduces some of the experimental approaches that biologists use to understand them. Let's begin by surveying the structures that make up a cell, and then go on to analyze three of the most important cell systems: the nucleus, the protein manufacturing and shipping complex called the endomembrane system, and the network of protein fibers called the cytoskeleton.

7.1 What's Inside the Cell?

In Chapter 1 you read about the two fundamental types of cells observed in nature—eukaryotes and prokaryotes. Eukaryotic cells have a membrane-bound compartment called a nucleus, while prokaryotic cells do not. According to **morphology** ("form-science"), then, species fall into the two broad categories: (1) prokaryotes and (2) eukaryotes. But according to **phylogeny** ("tribe-source"), or evolutionary history, organisms fall into the three broad groups called (1) Bacteria, (2) Archaea, and (3) Eukarya. Members of the Bacteria and Archaea are prokaryotic; members of the Eukarya—including algae, fungi, plants, and animals—are eukaryotic.

In the late seventeenth century, biologists began using microscopes to study the structure of cells. Over time, improvements in optics and cell preparation techniques allowed researchers to catalogue the structures reviewed in this section. When electron microscopes became widely available in the 1950s, investigators described the internal anatomy of these structures in more detail. More recent advances in microscopy have allowed investigators to videotape certain types of cell processes in living cells (see **BioSkills 8**).

What have anatomical studies based on microscopy revealed? Let's look first at the general anatomy of prokaryotic cells and eukaryotic cells, and then consider how the structures that have been identified help cells function.

Prokaryotic Cells

Figure 7.1 shows the general structure of a prokaryotic cell. For most bacterial and archaeal species, the plasma membrane encloses a single compartment—meaning that the cell has few or no substructures separated from the rest of the cell by internal membranes. Closer examination reveals a series of important internal structures, however. Let's take a look at a typical prokaryotic cell, starting at the outside and working in.

As Chapter 6 pointed out, the cell membrane, or plasma membrane, consists of a phospholipid bilayer and proteins that either span the bilayer or attach to one side. Inside the membrane, the contents of a cell are collectively termed the **cytoplasm** ("cell-formed"). Because the cytoplasm contains a high concentration of solutes, in most habitats it is hypertonic relative to the surrounding environment. When this is the case,

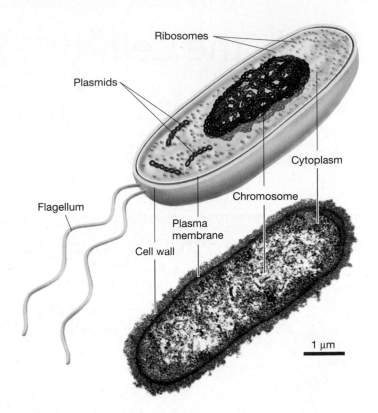

FIGURE 7.1 A Prokaryotic Cell. Prokaryotic cells are identified by a negative trait—the absence of a membrane-bound nucleus. Although there is wide variation in the size and shape of bacterial and archaeal cells, they all contain a plasma membrane, a chromosome and protein-synthesizing ribosomes; almost all have a stiff cell wall. Some prokaryotes have flagella that power swimming and/or inner membranes where photosynthesis takes place.

● **EXERCISE** Label the nucleoid in the drawing.

water enters the cell via osmosis and makes the cell's volume expand. In virtually all bacteria and archaea, this pressure is resisted by a stiff **cell wall**. Bacterial and archaeal cell walls are a tough, fibrous layer that surrounds the plasma membrane (**Figure 7.2**). The pressure of the plasma membrane against the cell wall is about the same as the pressure in an automobile tire. The cell wall protects the organism and gives it shape and rigidity. In addition, many bacteria have another protective layer outside the cell wall that consists of lipids with polysaccharides attached. Lipids that contain carbohydrate groups are termed **glycolipids**.

In the cytoplasm of a prokaryotic cell, the most prominent structure is the chromosome. Most prokaryotic species have a single, circular **chromosome** that consists of a large DNA molecule associated with a small number of proteins. The DNA molecule contains information, while the proteins provide structural support for the DNA. You might recall from Chapter 4 that the information in DNA is encoded in the form of its sequence of nitrogenous bases. More specifically, the primary structure of DNA contains the instructions for making the proteins, RNAs, and other molecules needed by the cell. A segment of DNA that contains the information for building an RNA

FIGURE 7.2 The Bacterial Cell Wall. In bacteria and archaea, the cell wall consists of peptidoglycan or similar polymers that are cross-linked into tough sheets. The inside of the cell wall contacts the plasma membrane, which pushes up against the wall. The outside of the cell wall makes direct contact with the outside environment, which is almost always filled with competitors and predators.

molecule or a polypeptide is called a **gene**. To summarize, chromosomes contain DNA, which contains genes.

Prokaryotic chromosomes are found in a localized area of the cell called the **nucleoid**. The nucleoid is usually found in the center of the cell and typically represents about 20 percent of the cell's total volume. It's important to note, though, that the genetic material in the nucleoid is not separated from the rest of the cytoplasm by a membrane. In the well-studied bacterium *Escherichia coli*, the circular chromosome would be over 1 mm long if it were linear—500 times longer than the cell itself. This situation is typical in prokaryotes. To fit into the cell, the DNA double helix coils on itself with the aid of enzymes to form the highly compact, "supercoiled" structure shown in **Figure 7.3**.

Supercoiled regions of DNA resemble a string that has been held at either end and then twisted until it coils back upon itself.

Depending on the species and population being considered, prokaryotic cells may also contain one to about a hundred small, usually circular, supercoiled DNA molecules called **plasmids**. Plasmids contain genes but are physically independent of the main, cellular chromosome. In most cases the genes carried by plasmids are not required under normal conditions; instead they help cells adapt to unusual circumstances, such as the sudden presence of a poison in the environment. As a result, plasmids can be considered auxiliary genetic elements.

Two other prominent cell structures found in prokaryotes are **ribosomes**, which manufacture proteins, and **flagella** (singular: **flagellum**), which rotate to power swimming in aquatic species. Ribosomes are observed in all prokaryotic cells and are found throughout the cytoplasm. Bacterial ribosomes are complex structures consisting of a total of three distinct RNA molecules and over 50 different proteins. These molecular components are organized into two major structural elements, called the large subunit and small subunit (**Figure 7.4**). It is not unusual for a single cell to contain 10,000 ribosomes. Both ribosomes and prokaryotic flagella lack a membrane. Not all bacterial species have flagella, however. When present they are usually few in number and are located on the surface of the cell. Over 40 different proteins are involved in building and controlling bacterial flagella. At top speed, flagellar movement can drive a bacterial cell through water at 60 cell lengths per

FIGURE 7.3 Bacterial DNA Is Supercoiled. The circular chromosomes of bacteria and archaea must be coiled extensively, into "supercoils," to fit in the cell.

FIGURE 7.4 The Bacterial Ribosome. Bacterial ribosomes are made of RNA and protein molecules that are organized into large and small subunits.

0.5 μm

FIGURE 7.5 Photosynthetic Membranes in Bacteria. The green stripes in this photosynthetic bacterium are infoldings of the plasma membrane. They are green because they contain the pigments and enzymes required for photosynthesis.

second. In contrast, the cheetah qualifies as the fastest land animal but can sprint at a mere 25 body lengths per second.

Prokaryotes lack a nucleus, but it is not correct to say that they lack membrane-bound internal structures. Many species contain membrane-bound storage containers, and extensive internal membranes occur in bacteria and archaea that perform photosynthesis. Photosynthesis is the suite of chemical reactions responsible for converting the energy in sunlight into chemical energy stored in sugars. The photosynthetic membranes observed in prokaryotes contain the enzymes and pigment molecules required for these reactions to occur and develop as infoldings of the plasma membrane. In some cases, vesicles pinch off as the plasma membrane folds in. In other cases, flattened stacks of photosynthetic membrane, like those shown in **Figure 7.5**, form from the infolded sections of the plasma membrane.

In addition, recent research indicates that several bacterial species have internal compartments that qualify as organelles ("little organs"). An **organelle** is a membrane-bound compartment in the cytoplasm that contains enzymes or structures specialized for a particular function. One type of bacterial

organelle is specialized for storing calcium ions; another contains crystals of the mineral magnetite. The magnetite crystals function like a compass needle to help cells orient themselves in a magnetic field and swim in a directed way.

Recent research has also shown that bacteria and archaea contain long, thin fibers that serve a structural role inside the cell. All bacterial species, for example, contain protein fibers. These filaments are essential for cell division to take place. Some species also have protein filaments that help maintain cell shape. Protein filaments such as these form the basis of the **cytoskeleton** ("cell skeleton").

Even though internal membranes and some cytoskeletal components are found in all prokaryotic cells studied to date, their extent pales in comparison with that in eukaryotes. When typical prokaryotic and eukaryotic cells are compared side by side, four outstanding differences jump out: (1) Eukaryotic chromosomes are found inside a membrane-bound compartment called the nucleus; (2) eukaryotic cells are often much larger; (3) they contain extensive amounts of internal membrane; and (4) they feature a diverse and dynamic cyto-skeleton (**Table 7.1**).

Eukaryotic Cells

The lineage called Eukarya includes forms ranging from unicellular species to 100-meter-tall redwoods. Brown algae, red algae, fungi, amoebae, and slime molds are all eukaryotic, as are green plants and animals. The first thing that strikes biologists about eukaryotic cells is how much larger they are on average than bacteria and archaea (**Figure 7.6**). Most eukaryotic cells range from about 5 to 100 μm in diameter, while most prokaryotic cells vary between 1 and 10 μm in diameter. A micrograph of an average eukaryotic cell, at the same scale as the bacterial cell in Figure 7.5, would fill this page. This difference in size inspired the hypothesis that when eukaryotes first evolved, they made their living by ingesting bacterial and archaeal cells whole. Stated another way, the evolution of large cell size is thought to have made it possible for eukaryotic cells to act as predators—organisms that kill and consume other organisms. Thousands of unicellular eukaryotic species alive today still survive by surrounding and taking in whole bacterial and archaeal cells.

SUMMARY TABLE **7.1 How Do the Structures of Prokaryotic and Eukaryotic Cells Differ?**

	Location of DNA	Internal Membranes and Organelles	Cytoskeleton	Overall Size
Bacteria and Archaea	In nucleoid (not membrane bound); plasmids also common	Extensive internal membranes only in photosynthetic species; limited types and numbers of organelles	Limited in extent, relative to eukaryotes	Usually small relative to eukaryotes
Eukaryotes	Inside nucleus (membrane bound); plasmids extremely rare	Large numbers of organelles; many types of organelles	Extensive—usually found throughout volume of cell	Most are larger than prokaryotes

(a) Generalized animal cell

(b) Generalized plant cell

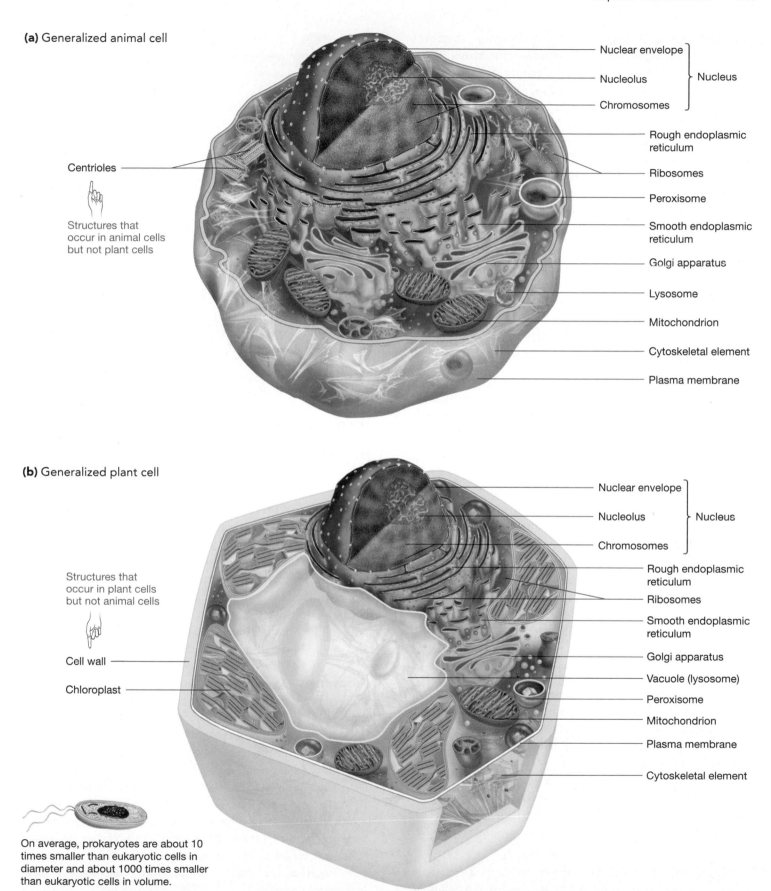

Centrioles

☞

Structures that
occur in animal cells
but not plant cells

Nuclear envelope ⎤
Nucleolus ⎬ Nucleus
Chromosomes ⎦

Rough endoplasmic
reticulum

Ribosomes

Peroxisome

Smooth endoplasmic
reticulum

Golgi apparatus

Lysosome

Mitochondrion

Cytoskeletal element

Plasma membrane

Structures that
occur in plant cells
but not animal cells

☞

Cell wall

Chloroplast

Nuclear envelope ⎤
Nucleolus ⎬ Nucleus
Chromosomes ⎦

Rough endoplasmic
reticulum

Ribosomes

Smooth endoplasmic
reticulum

Golgi apparatus

Vacuole (lysosome)

Peroxisome

Mitochondrion

Plasma membrane

Cytoskeletal element

On average, prokaryotes are about 10
times smaller than eukaryotic cells in
diameter and about 1000 times smaller
than eukaryotic cells in volume.

FIGURE 7.6 Animal and Plant Cells. Generalized, or "typical," **(a)** animal and **(b)** plant cells. (Compare with the prokaryotic
cell, shown at true relative size at bottom left.)

The evolution of large cells has a downside, however. Ions and small molecules such as adenosine triphosphate (ATP), amino acids, and nucleotides cannot diffuse across a large volume quickly. If the ATP that supplies chemical energy in cells is used up on one side of a large cell, ATP from the other side of the cell would take a long time to diffuse to that location. Prokaryotic cells are small enough that ions and small molecules arrive where they are needed via diffusion. In fact, the size of prokaryotic cells is probably limited by the distance that molecules must diffuse or be transported inside the cell.

How do eukaryotic cells solve the problems that size can engender? The answer lies in the numerous organelles observed in eukaryotic cells. In effect, the huge volume inside a eukaryotic cell is compartmentalized into a large number of bacterium-sized parts. Because eukaryotic cells are subdivided, the molecules required for specific chemical reactions are often located within a given compartment or organelle. Compartmentalization offers two key advantages:

- Incompatible chemical reactions can be separated. For example, new fatty acids can be synthesized in one organelle while excess or damaged fatty acids are degraded and recycled in a different organelle.

- The efficiency of chemical reactions is increased. First, the substrates required for particular reactions can be localized and maintained at high concentration within organelles. Second, if substrates are used up in a particular part of the organelle, they can be replaced by substrates that have only a short distance to diffuse. Third, groups of enzymes that work together can be clustered on internal membranes instead of floating free in the cytoplasm. When the product of a reaction is the substrate for a reaction catalyzed by another enzyme, clustering the enzymes increases the speed and efficiency of a reaction sequence.

If bacteria and archaea can be compared to small machine shops, then eukaryotic cells resemble sprawling industrial complexes. The organelles and other structures found in eukaryotes are analogous to highly specialized buildings that act as factories, power stations, warehouses, transportation corridors, and administrative centers. Figure 7.6 shows how organelles are arranged in a typical animal cell and plant cell. What are these structures, and what do they do? As you read about each of these cell components in the pages that follow, focus on identifying how their structure correlates with their function. Then use the summary table on page 132 as a study guide (**Table 7.2**).

The Nucleus   The **nucleus** is among the largest organelles and is highly organized (**Figure 7.7**). It is enclosed by a unique structure—a complex double membrane called the **nuclear envelope**. As section 7.2 will detail, the nuclear envelope is studded with pore-like openings, and its inside surface is

FIGURE 7.7 The Nucleus Is the Eukaryotic Cell's Information Storage and Retrieval Center. The genetic, or hereditary, information is encoded in DNA, which is a component of the chromosomes inside the nucleus.

associated with fibrous proteins that form a lattice-like sheet called the **nuclear lamina**. The nuclear lamina stiffens the structure and maintains its shape.

Chromosomes do not float free inside the nucleus—instead, each chromosome occupies a distinct area and is attached to the nuclear lamina in at least one location. The nucleus also includes a distinctive region called the **nucleolus**, where the RNA molecules found in ribosomes are manufactured and the large and small ribosomal subunits are assembled.

Ribosomes   In eukaryotes, the cytoplasm consists of everything inside the plasma membrane excluding the nucleus; the fluid portion of the cytoplasm is called the **cytosol**. Many of the cell's millions of ribosomes are scattered throughout the cytosol (**Figure 7.8**). Like the bacterial ribosome pictured in Figure 7.4, the ribosomes shown in Figure 7.8 comprise two subunits, one large and one small. Each subunit is composed of proteins and RNA. In eukaryotes the large subunit contains three RNA molecules while the small subunit contains one. (In prokaryotes the large subunit has just two RNA molecules.) Neither ribosomal subunit is enclosed by a membrane. When the two subunits come together, they form a complex molecular machine that synthesizes proteins.

Rough Endoplasmic Reticulum   In addition to the ribosomes found free in the cytosol, many ribosomes are

Ribosomes

Ribosome

100 nm

FIGURE 7.8 Ribosomes Are the Site of Protein Synthesis.
Eukaryotic ribosomes are similar in structure to bacterial and archaeal ribosomes—though not identical. They are composed of large and small subunits, each of which contains both RNA molecules and proteins.

Rough endoplasmic reticulum

Lumen of rough ER

Ribosomes on outside of rough ER

Free ribosomes in cytoplasm

200 nm

FIGURE 7.9 Rough ER Is a Protein Synthesis and Processing Complex.
Rough ER is a system of membrane-bound sacs and tubules with ribosomes attached. It is continuous with the nuclear envelope and with smooth ER.

associated with membranes. More specifically, hundreds of thousands of ribosomes are attached to a network of membrane-bound sacs and tubules called the **rough endoplasmic reticulum**, or **rough ER**. Translated literally, *endoplasmic reticulum* means "inside-formed network." Look back at Figure 7.6 and notice that the ER is continuous with the outer membrane of the nuclear envelope. From the nuclear envelope, the layers of sacs that make up the ER extend into the cytoplasm.

The ribosomes associated with the rough ER are responsible for synthesizing proteins that will be inserted into the plasma membrane, secreted to the cell exterior, or shipped to an organelle called the lysosome. As they are being manufactured by ribosomes, these proteins move to the interior of the sac-like component of the rough ER (**Figure 7.9**). The interior of the rough ER, like the interior of any sac-like structure in a cell or body, is called the **lumen**. In the lumen of the rough ER, newly manufactured proteins undergo folding and other types of processing.

The proteins produced in the rough ER have a variety of functions. Some carry messages to other cells; some act as membrane transport proteins or pumps; others are enzymes. The common theme is that rough ER products are destined for transport to various distant destinations—often to the surface of the cell or beyond.

Golgi Apparatus   In many cases, the products of the rough ER pass through the Golgi apparatus before they reach their final destination. The **Golgi apparatus** consists of flattened, membranous sacs called **cisternae** (singular: **cisterna**), which are stacked on top of one another (**Figure 7.10**). The organelle also has a distinct polarity, or sidedness. The *cis* ("this side") surface is closest to the rough ER and nucleus, and the *trans* ("across") surface is oriented toward the plasma membrane. The *cis* side of a Golgi apparatus receives products from the rough ER, and the *trans* side ships them out toward the cell surface. In between, within the cisternae, the rough ER's products are processed and packaged for delivery. Micrographs often show "bubbles" on either side of the Golgi stack. These are membrane-bound vesicles that carry proteins or other products to and from the organelle. Section 7.3 analyzes the intracellular movement of molecules from the rough ER to the Golgi apparatus and beyond in more detail.

Smooth Endoplasmic Reticulum   Although the ER forms one continuous structure, not all of it is associated with transport of material to the Golgi apparatus and not all of it has ribosomes attached. In electron micrographs, parts of the ER that contain ribosomes look dotted and rough while portions that are free of ribosomes appear smooth and even. Appropriately enough, these parts of the ER are called **smooth endoplasmic**

The *cis* face of the Golgi apparatus is oriented towards the rough ER

The *trans* face of the Golgi apparatus is oriented towards the plasma membrane

Golgi apparatus

cis face

Vesicle

Lumen of Golgi apparatus

Cisternae

Vesicles

trans face

100 nm

FIGURE 7.10 The Golgi Apparatus Is a Site of Protein Processing, Sorting, and Shipping. The Golgi apparatus is a collection of flattened vesicles called cisternae.

Smooth endoplasmic reticulum

Lumen of smooth ER

200 nm

FIGURE 7.11 Smooth ER Is a Lipid-Handling Center and a Storage Facility. Smooth ER is a system of membrane-bound sacs and tubules that lacks ribosomes.

reticulum, or **smooth ER** (**Figure 7.11**). The smooth ER membrane contains enzymes that catalyze reactions involving lipids. Depending on the type of cell, these enzymes may be involved in synthesizing lipids needed by the organism, or in breaking down lipids that are poisonous. Smooth ER is also the manufacturing site for phospholipids used in cell membranes. In addition to processing lipids, smooth ER functions as a reservoir for calcium ions (Ca^{2+}) that act as a signal triggering a wide array of activities inside the cell.

The structure of endoplasmic reticulum correlates closely with its function. Rough ER has ribosomes and functions primarily as a protein-manufacturing center; smooth ER lacks ribosomes and functions primarily as a lipid-processing center. Together with the Golgi apparatus and lysosomes, the endoplasmic reticulum forms the **endomembrane system**. The endomembrane ("inner-membrane") system is the primary center for protein and lipid synthesis in eukaryotic cells.

Peroxisomes Virtually all eukaryotic cells contain globular organelles called **peroxisomes** (**Figure 7.12**). These organelles have a single membrane and grow and divide independently of other organelles. Although different types of cells from the same individual may have distinct types of peroxisomes, these organelles all share a common function: Peroxisomes are centers for oxidation reactions. As Chapter 9 will explain in detail, oxidation reactions remove electrons from atoms and molecules. In many cases the products of these reactions include hy-

drogen peroxide (H_2O_2), which is highly corrosive. If hydrogen peroxide escaped from the peroxisome, the H_2O_2 would quickly damage organelle membranes and the plasma membrane. This is rare, however. Inside the peroxisome, the enzyme catalase quickly "detoxifies" hydrogen peroxide by converting it to water and oxygen.

Different types of peroxisomes contain different suites of oxidative enzymes. As a result, each is specialized for oxidizing particular compounds. For example, the peroxisomes in your liver cells contain enzymes that oxidize an array of toxins, including the ethanol in alcoholic beverages. The products of these oxidation reactions are usually harmless and are either excreted from the body or used in other reactions. Other peroxisomes contain enzymes that catalyze the oxidation of fatty acids. These reactions produce molecules that include acetyl CoA, which is used for the synthesis of important molecules elsewhere in the cell. In plant leaves, specialized peroxisomes called **glyoxisomes** are packed with enzymes that convert one of the products of photosynthesis into a sugar that can be used to store energy for the cell. In contrast, seeds do not perform photosynthesis and lack this type of peroxisome. Instead, they have peroxisomes with enzymes that oxidize fatty acids to yield glucose. The glucose is then used by the young plant as it begins to grow. In each case, there is a clear connection between structure and function: The enzymes found inside the peroxisome make a specialized set of oxidation reactions possible.

monomers. The monomers are then excreted or recycled into new polymers. The digestive enzymes are collectively called acid hydrolases because they catalyze hydrolysis reactions that break monomers from macromolecules most efficiently at pH 5.0. In the cytosol, where the pH is about 7.2, these enzymes are less active.

Figure 7.14 illustrates two ways that materials are delivered to lysosomes in animal cells. When **phagocytosis** (literally, "eat-cell-act") occurs, the plasma membrane of a cell surrounds a

Peroxisomes

Peroxisome membrane

Peroxisome lumen

100 nm

FIGURE 7.12 Peroxisomes Are the Site of Oxidation Reactions. Peroxisomes are globular organelles with a single membrane.

Lysosomes The major structures involved in solid-waste processing and materials storage in the cell are called **lysosomes.** The size and shape of these organelles vary widely, and in the cells of plants, fungi, and certain other groups they are referred to as vacuoles. In animal cells, lysosomes function as digestive centers (**Figure 7.13**). The organelle's interior, or lumen, is acidic because proton pumps in the lysosome membrane import enough hydrogen ions to maintain a pH of 5.0. This organelle also contains about 40 different enzymes. Each of these proteins is specialized for breaking up a different type of macromolecule—proteins, nucleic acids, lipids, or carbohydrates—into its component

Lysosomes

Material being digested within lysosomes

500 nm

FIGURE 7.13 Lysosomes Are Recycling Centers. Lysosomes are usually oval or globular and have a single membrane.

(a) PHAGOCYTOSIS

Phagosome

Lysosome

1. The plasma membrane detects a smaller cell or food particle and begins to engulf it.

2. The resulting membrane-bound particle is a phagosome, or food particle.

3. The phagosome is delivered to a lysosome, which takes it in and begins digesting it.

4. Small molecules from the digested food particles are released into the cytosol.

(b) AUTOPHAGY

Damaged organelle

Lysosome

1. A damaged organelle is surrounded by a membrane (of unknown origin).

2. The membrane-bound organelle is delivered to a lysosome, which takes it in and begins digesting it.

3. Small molecules from the digested organelle are recycled into the cytosol.

FIGURE 7.14 Two Ways to Deliver Materials to Lysosomes. Materials can be transported to lysosomes after phagocytosis or via autophagy.

smaller cell or a food particle and engulfs it, forming a structure called a phagosome or food vacuole. This structure is delivered to a lysosome, where it is taken in and digested. During **autophagy** ("same-eating"), damaged organelles are surrounded by a membrane and delivered to a lysosome. There the components are digested and recycled.

Materials can also be processed by lysosomes as a result of **receptor-mediated endocytosis**. As **Figure 7.15** shows, this sequence of events begins when macromolecules outside the cell bind to membrane proteins that act as receptors. More than 25 distinct receptors have now been characterized, each specialized for responding to a different macromolecule. Once binding occurs, the plasma membrane folds in and pinches off to form a membrane-bound vesicle called an **early endosome** ("inside-body"). Early endosomes undergo a series of processing steps that include the receipt of digestive enzymes from the Golgi apparatus and the activation of proton pumps that gradually lower their pH. In this way, early endosomes undergo a gradual maturation process that may lead to the formation of a **late endosome** and eventually a fully functioning lysosome.

Regardless of whether the materials in lysosomes originate via phagocytosis, autophagy, or receptor-mediated endocytosis, the result is similar: Molecules are hydrolyzed. The amino acids, nucleotides, sugars, and other molecules that result from acid hydrolysis leave the lysosome via transport proteins in the organelle's membrane. Once in the cytoplasm, they can be reused.

It is important to note, however, that not all of the materials that are surrounded by membrane and taken into a cell end up in lysosomes. **Endocytosis** ("inside-cell-act") refers to any pinching off of the plasma membrane that results in the uptake of material from outside the cell. Endocytosis can occur in three ways: (1) phagocytosis, (2) receptor-mediated endocytosis, and (3) **pinocytosis** ("drink-cell-act"). Pinocytosis brings fluid into the cytoplasm via tiny vesicles that form from infoldings of the plasma membrane. The fluid inside these vesicles is not transported to lysosomes, but is used elsewhere in the cell. In addition, most of the macromolecules that collect in early endosomes are selectively removed and used long before the structure becomes a lysosome.

Compared with the lysosomes of animal cells, the **vacuoles** of plant and fungal cells are large—sometimes taking up as much as 80 percent of a plant cell's volume (**Figure 7.16**). Although

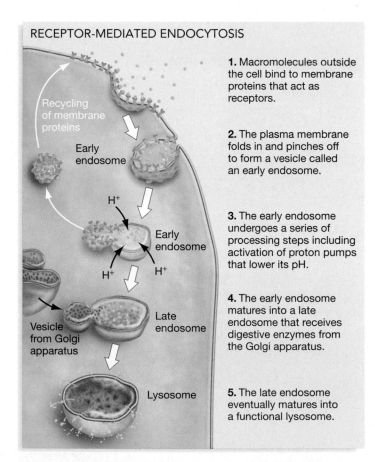

RECEPTOR-MEDIATED ENDOCYTOSIS

Recycling of membrane proteins

Early endosome

Early endosome

H⁺   H⁺   H⁺

Vesicle from Golgi apparatus

Late endosome

Lysosome

1. Macromolecules outside the cell bind to membrane proteins that act as receptors.

2. The plasma membrane folds in and pinches off to form a vesicle called an early endosome.

3. The early endosome undergoes a series of processing steps including activation of proton pumps that lower its pH.

4. The early endosome matures into a late endosome that receives digestive enzymes from the Golgi apparatus.

5. The late endosome eventually matures into a functional lysosome.

FIGURE 7.15 Receptor-Mediated Endocytosis May Lead to Lysosome Formation. Endosomes created by receptor-mediated endocytosis may mature into lysosomes.

● **QUESTION** Why is it significant that vesicles from the Golgi apparatus fuse with early endosomes?

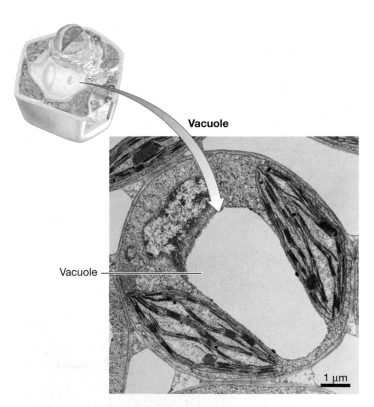

Vacuole

Vacuole

1 µm

FIGURE 7.16 Vacuoles Are Storage Centers. Vacuoles are variable in size and function. Some contain digestive enzymes and serve as recycling centers; most are large storage containers.

● **QUESTION** Why are toxins like nicotine, cocaine, and caffeine stored in vacuoles instead of the cytosol?

● **QUESTION** About what percentage of this cell's volume is taken up by the vacuole?

some vacuoles contain enzymes that are specialized for digestion, most of the vacuoles observed in plant and fungal cells act as storage depots. In many cases, the stored material is water, which maintains the cell's normal volume, or ions such as potassium (K^+) and chloride (Cl^-). Inside seeds, cells may contain a large vacuole filled with proteins. When the embryonic plant inside the seed begins to grow, enzymes begin digesting these proteins to provide amino acids for the growing individual. In cells that make up flower petals or fruits, vacuoles are filled with colorful pigments. Elsewhere in the plant, vacuoles may be packed with noxious compounds that protect leaves and stems from being eaten by predators. The type of chemical involved varies by species, ranging from bitter-tasting tannins to toxins such as nicotine, morphine, caffeine, or cocaine.

Mitochondria   The chemical energy required to build all of these organelles and do other types of work comes from adenosine triphosphate (ATP), most of which is produced in the cell's **mitochondria** (singular: mitochondrion). As **Figure 7.17** shows, each mitochondrion has two membranes. The outer membrane defines the organelle's surface, while the inner membrane is connected to a series of sac-like **cristae**. The solution inside the inner membrane is called the **mitochondrial matrix**. In eukaryotes, most of the enzymes and molecular machines responsible for synthesizing ATP are embedded in the membranes of the cristae or suspended in the matrix. Depending on the type of cell, from 50 to more than a million mitochondria may be present.

Each mitochondrion possesses a small chromosome that contains genes, independent of the main chromosomes in the nucleus. This mitochondrial DNA is a component of a circular and supercoiled chromosome that is similar in structure to bacterial chromosomes. Mitochondria also manufacture their own ribosomes. Like most organelles, mitochondria can grow and divide independently of nuclear division and cell division.

Chloroplasts   Most algal and plant cells possess an organelle called the **chloroplast**, in which sunlight is converted to chemical energy during photosynthesis. The chloroplast has a double membrane around its exterior, analogous to the structure of a mitochondrion (**Figure 7.18**). Instead of featuring sac-like cristae that connect to the inner membrane, though, the interior of the chloroplast is dominated by hundreds of membrane-bound, flattened vesicles called **thylakoids**, which are independent of the inner membrane. Thylakoids are stacked like pancakes into piles called **grana** (singular: *granum*). Many of the pigments, enzymes, and molecular machines responsible for converting light energy into carbohydrates are embedded in the thylakoid membranes. Certain critical enzymes and substrates, however, are found outside the thylakoids in the region called the **stroma**.

The number of chloroplasts per cell varies from none to several dozen. Like mitochondria, each chloroplast contains a

Mitochondrion

Outer and inner membranes

Matrix

Cristae

0.1 μm

FIGURE 7.17 Mitochondria Are Power-Generating Stations.
Mitochondria are variable in size and shape, but all have a double membrane with sac-like cristae inside.

Chloroplast

Stroma

Thylakoids

Granum

Outer and inner membranes

1 μm

FIGURE 7.18 Chloroplasts Are Sugar-Manufacturing Centers. Many of the enzymes and other molecules required for photosynthesis are located in membranes inside the chloroplast. These membranes are folded into thylakoids and stacked into grana.

circular chromosome. Chloroplast DNA is independent of the main genetic material inside the nucleus. Chloroplasts also grow and divide independently of nuclear division and cell division.

Cytoskeleton   The final major structural feature that is common to all eukaryotic cells is an extensive system of protein fibers called the cytoskeleton. The cytoskeleton contains several distinct types of proteins and fibers and has an array of functions which are analyzed in detail in Section 7.4. In addition to giving the cell its shape and structural stability, cytoskeletal proteins are involved in moving the cell itself and in moving materials within the cell.

The Cell Wall   In fungi, algae, and plants, cells possess an outer cell wall in addition to their plasma membrane (**Figure 7.19**). The cells of animals, amoebae, and other groups lack this feature. Although the composition of the cell wall varies among species and even between types of cells in the same individual, the general plan is similar: Rods or fibers composed of a carbohydrate run through a stiff matrix made of other polysaccharides and proteins (see Chapter 8 for details on this). In addition, some plant

FIGURE 7.19 Cell Walls Protect Plants and Fungi. Plants have cell walls that contain cellulose; in fungi the major structural component of the cell wall is chitin.

QUESTION Is the cell wall inside or outside of the plasma membrane?

cells produce a secondary cell wall that features a particularly tough molecule called lignin. Lignin forms a branching, cagelike network that is almost impossible for enzymes to attack. The combination of cellulose fibers and lignin in secondary cell walls makes up most of the material we call wood.

(MB) Web Animation   *BioFlix*   at www.masteringbio.com
Tour of an Animal Cell; Tour of a Plant Cell

How Does Cell Structure Correlate with Function?

The preceding discussion emphasized how the structure of each organelle fits with its role in the cell. As **Table 7.2** (page 132) indicates, an organelle's membrane and its complement of enzymes correlate closely with its function. If you understand this concept, you should be able to draw and label a generalized eukaryotic cell and indicate which components serve as the administrative/information hub, protein manufacturing and shipping system, recycling center, support beams, roads, storage depot, power station, food manufacturing facility, fatty-acid processing center, and perimeter fencing with secured gates.

The same connection between structure and function occurs at the level of the entire cell. Inside an individual plant or animal, cells are specialized for certain tasks and have a structure that correlates with those tasks. For example, the muscle cells in your upper leg are extremely long, tube-shaped structures. They are filled with protein fibers that slide past one another as the entire muscle flexes or extends. It is this sliding motion that allows your muscles to contract or extend as you run. Muscle cells are also packed with mitochondria, which produce the ATP required for the sliding motion to occur. In contrast, nearby fat cells are rounded, globular structures that store fatty acids. They consist of little more than a plasma membrane, a nucleus, and a fat droplet. Neither cell bears a close resemblance to the generalized animal cell pictured in Figure 7.6a

To drive home the correlation between the overall structure and function of a cell, examine the transmission electron micrographs in **Figure 7.20**. The animal cell in Figure 7.20a is from the pancreas and is specialized for the manufacture and export of digestive enzymes. It is packed with rough ER and Golgi, which make this function possible. The animal cell in Figure 7.20b is from the testis and synthesizes the steroid hormone called testosterone. This cell is dominated by smooth ER, where processing of steroids and other lipids takes place. The plant cell in Figure 7.20c is from the leaf of a potato and is specialized for absorbing light and manufacturing sugar; the cell in Figure 7.20d is from a potato tuber (part of an underground stem) and functions as a starch storage container. The leaf cell contains hundreds of chloroplasts, while the tuber cell has a prominent storage vacuole filled with carbohydrate. In each case, the type of organelles in each cell and their size and number correlate with the cell's specialized function.

(a) Animal pancreatic cell: Exports digestive enzymes.

(b) Animal testis cell: Exports lipid-soluble signals.

(c) Plant leaf cell: Manufactures ATP and sugar.

(d) Plant root cell: Stores starch.

FIGURE 7.20 Cell Structure Correlates with Function.

● **EXERCISE** In part (a), label the rough ER and secretory vesicles. (They are dark and round.) In (b), label the smooth ER. In (c), label the chloroplasts, vacuole, and nucleus. In (d), label the starch granules (they are white in this micrograph).

The Dynamic Cell

Biologists study the structure and function of organelles and cells with a combination of tools and approaches. Light microscopes and transmission electron microscopes have allowed researchers to see cells at increasingly high magnification and increasingly better resolution. These types of microscopy allowed biologists to characterize the basic size and shape of organelles and where they occurred in the cell; a technique called **differential centrifugation** made it possible to isolate particular cell components and analyze their chemical composition. As **Box 7.1** explains, differential centrifugation is based on breaking cells apart to create a complex mixture and then separating components in a centrifuge. The individual parts of the cell can then be purified and studied in detail, in isolation from other parts of the cell.

Although these techniques have led to an increasingly sophisticated understanding of how cells work, they have a limitation. Transmission electron microscopy is based on a fixed "snapshot" of the cell that is to be observed, and differential centrifugation is based on splitting cells into parts that are analyzed independently. Neither technique allows investigators to directly explore how things move from place to place in the cell or how parts interact. The information gleaned from these techniques can make cells seem static. In reality, cells are dynamic.

The amount of chemical activity and the speed of molecular movement inside cells is nothing short of fantastic. Bacterial ribosomes add up to 20 amino acids per second to a growing polypeptide, and eukaryotic ribosomes typically add two per second. Given that there are about 15,000 ribosomes in each bacterium and possibly a million in an average eukaryotic cell, hundreds or even thousands of new protein molecules can be finished each second in every cell. In the same amount of time, a typical cell in your body uses an average of 10 million ATP molecules and synthesizes just as many. It's not unusual for a cellular enzyme to catalyze 25,000 or more reactions per second; most cells contain hundreds or thousands of enzymes. A minute is more than enough time for each membrane phospholipid in your body to travel the breadth of the organelle or cell where it resides. The hundreds of trillions of mitochondria inside you are completely replaced about every 10 days, for as long as you live. The plasma membrane is fluid, and its composition is constantly changing.

Because humans are such large organisms, it's impossible for us to imagine what life is really like inside a cell. At the scale of a ribosome or an organelle or a cell, gravity is inconsequential. Instead, the dominant forces are the charge- or polarity-based electrostatic attractions between molecules and the kinetic energy of motion. At this level, events take nanoseconds, and speeds are measured in micrometers per second. Contemporary methods for studying cells, including some of the imaging techniques featured in **BioSkills 8**, capture

SUMMARY TABLE **7.2 Eukaryotic Cell Components**

		Structure		Function
		Membrane	**Components**	
	Nucleus	Double ("envelope"); openings called nuclear pores	Chromosomes Nucleolus Nuclear lamina	Genetic information Assembly of ribosome subunits Structural support
	Ribosomes	None	Large/small subunits—complex of RNA and proteins	Protein synthesis
	Endomembrane system			
	Rough ER	Single; contains receptors for entry of selected proteins	Network of branching sacs Ribosomes associated	Protein synthesis and processing
	Golgi apparatus	Single; contains receptors for products of rough ER	Stack of flattened cisternae	Protein processing (e.g., glycosylation)
	Smooth ER	Single; contains enzymes for synthesizing phospholipids	Network of branching sacs Enzymes for synthesizing lipids	Lipid synthesis
	Lysosomes	Single; contains proton pumps	Acid hydrolases (catalyze hydrolysis reactions)	Digestion and recycling
	Peroxisomes	Single; contains transporters for selected macromolecules	Enzymes that catalyze oxidation reactions Catalase (processes peroxide)	Processing of fatty acids
	Vacuoles	Single; contains transporters for selected molecules	Varies—pigments, oils, carbohydrates, water, or toxins	Varies—coloration, storage of oils, carbohydrates, water, or toxins
	Mitochondria	Double; outer contains enzymes for processing pyruvate; inner contains enzymes for ATP production	Enzymes that catalyze oxidation-reduction reactions, ATP synthesis	ATP production
	Chloroplasts	Double; plus membrane-bound sacs in interior	Pigments Enzymes that catalyze oxidation-reduction reactions	Production of ATP and sugars via photosynthesis
	Cytoskeleton	None	Actin filaments Intermediate filaments Microtubules	Structural support; movement of materials; in some species, movement of whole cell
	Plasma membrane	Single; contains transport and receptor proteins	Phospholipid bilayer with transport and receptor proteins	Selective permeability—maintains intracellular environment
	Cell wall	None	Carbohydrate fibers running through carbohydrate or protein matrix	Protection, structural support

this dynamism by tracking how organelles and molecules move and interact over time.

The rest of this chapter focuses on this theme of cellular dynamism and movement. To begin, let's look at how molecules move into and out of the cell's control center—the nucleus.

Then we'll consider how proteins move from ribosomes into the lumen of the rough ER and then to the Golgi apparatus and beyond. The chapter closes by analyzing how cytoskeletal elements help transport cargo inside the cell or move the cell itself.

BOX 7.1 How Does a Centrifuge Work?

For decades, the centrifuge was among the most common tools used by biologists who study life at the level of molecules and cells. It was vital to early studies of organelles and other cell structures because it can separate cell components efficiently. A centrifuge accomplishes this by spinning samples in a solution that allows molecules and other cell components to separate according to their density or size and shape.

The first step in preparing a cell sample for centrifugation is to release the organelles and cell components by breaking the cells apart. This can be done by putting them in a hypotonic solution, by exposing them to high frequency vibration, by treating cells with a detergent, or by grinding them up. Each of these methods breaks apart plasma membranes and releases the contents of the cells.

The pieces of plasma membrane broken up by these techniques quickly reseal to form small vesicles, often trapping cell components inside. The solution that results from the homogenization step is a mixture of these vesicles, free-floating macromolecules released from the cells, and organelles. A solution such as this is called a cell extract or cell homogenate.

When a cell homogenate is placed in a centrifuge tube and spun at high speed, the components that are in solution tend to move outward, along the red arrow in **Figure 7.21a**. The effect is similar to a merry-go-round, which seems to push you outward in a straight line away from the spinning platform. In response to this outward-directed force, the solution containing the cell homogenate exerts a centripetal ("center-seeking") force that pushes the homogenate away from the bottom of the tube. Larger, denser molecules or particles resist this inward force more readily than do smaller, less dense ones and so reach the bottom of the centrifuge tube faster.

To separate the components of a cell extract, researchers often perform a series

(a) How a centrifuge works

When the centrifuge spins, the macromolecules tend to move toward the bottom of the centrifuge tube (red arrow)

The solution in the tube exerts a centripetal force, which resists movement of the molecule to the bottom of the tube (blue arrow)

Motor

Very large or dense molecules overcome the centripetal force more readily than smaller, less dense ones. As a result, larger, denser molecules move toward the bottom of the tube faster.

(b) DIFFERENTIAL CENTRIFUGATION

Low-speed centrifugation → Medium-speed centrifugation → High-speed centrifugation

Supernatant

Pellet

1. Start with uniform cell homogenate in centrifuge tube.

2. After low-speed spin, pellet contains large components. Transfer supernatant to new tube.

3. After medium-speed spin, pellet contains medium components. Transfer supernatant to new tube.

4. After high-speed spin, pellet contains small components.

(c) DENSITY GRADIENT CENTRIFUGATION

Lower-density solution

Sample

Higher-density solution

1. Add sample to tube of variable-density solution.

2. Run centrifuge. Cell components separate by density into distinct bands.

3. To extract specific cell components for analysis, poke tube with needle and withdraw a specific band.

FIGURE 7.21 Cell Components Can Be Separated by Centrifugation. (a) The forces inside a centrifuge tube allow cell components to be separated. **(b)** Through a series of centrifuge runs made at increasingly higher speeds, an investigator can separate fractions of a cell homogenate by size via differential centrifugation. **(c)** A high-speed centrifuge run can achieve extremely fine separation among cell components by density gradient centrifugation.

of centrifuge runs. Steps 1 and 2 of **Figure 7.21b** illustrate how an initial treatment at low speed causes larger, heavier parts of the homogenate to move below smaller, lighter parts. The material that collects at the bottom of the tube is called the **pellet**, and the solution and solutes left behind form the **supernatant** ("above

(Continued on next page)

(continued)

swimming"). The supernatant is placed in a fresh tube and centrifuged at increasingly higher speeds and longer durations. Each centrifuge run continues to separate cell components based on their size and density.

To accomplish even finer separation of macromolecules or organelles, researchers frequently follow up with centrifugation at extremely high speeds. One strategy is based on filling the centrifuge tube with a series of sucrose solutions of increasing density. The density gradient allows cell components to separate on the basis of small differences in size, shape, and density. When the centrifuge run is complete, each cell component occupies a distinct band of material in the tube, based on how quickly each moves through the increasingly dense gradient of sucrose solution during the centrifuge run. A researcher can then collect the material in each band for further study.

Check Your Understanding

If you understand that...

- Each structure found in prokaryotic and eukaryotic cells performs a function vital to the cell's survival.

You should be able to...

1) Summarize the differences between the structure of a prokaryotic cell and a eukaryotic cell.

2) Describe the structure and function of the eukaryotic nucleus, endomembrane system, mitochondria, chloroplast, and cytoskeleton.

3) Explain how the structure of peroxisomes and lysosomes correlate with their function.

7.2 The Nuclear Envelope: Transport into and out of the Nucleus

The nucleus is the information center of eukaryotic cells. It is a corporate headquarters, design center, and library all rolled into one. Appropriately enough, its interior is highly organized. The organelle's overall shape and structure are defined by the mesh-like nuclear lamina, which also helps anchor each chromosome. The remainder of each chromosome occupies a well-defined region in the nucleus, and specific centers exist where the genetic information in DNA is decoded and processed. At these locations, large suites of enzymes interact to produce RNA messages from specific genes at specific times. Meanwhile, the nucleolus functions as the site of ribosome synthesis.

Consistent with its role as information repository and processing center, the nucleus is separated from the rest of the cell by the nuclear envelope. Biologists began to understand exactly how the nuclear envelope is structured when transmission electron microscopy became available in the 1950s. Micrographs of cross sections though the envelope showed that the structure is supported by the fibrous nuclear lamina and is bounded by two membranes, each consisting of a lipid bilayer. As the diagram in **Figure 7.22** indicates, the inner membrane and the outer membrane are separated by a space that is continuous with the lumen of the endoplasmic reticulum.

Later, scanning electron microscopes allowed biologists to view the surface of the envelope. As the micrograph in **Figure 7.23** shows, the envelope contains thousands of openings called **nuclear pores**. Because these pores extend through both inner

FIGURE 7.22 The Structure of the Nuclear Envelope. The nuclear envelope has a double membrane and is continuous with the endoplasmic reticulum.

FIGURE 7.23 The Nuclear Envelope Is Studded with Pores. Scanning electron micrograph of the surface of the nuclear envelope.

and outer nuclear membranes, they connect the inside of the nucleus with the cytoplasm. Follow-up research showed that each pore consists of over 50 different proteins. These protein molecules form an elaborate structure called the **nuclear pore complex** (**Figure 7.24a**).

A series of experiments in the early 1960s showed that molecules travel into and out of the nucleus through the nuclear pore complexes. The initial studies were based on injecting tiny gold particles into cells and then preparing the cells for electron microscopy. In electron micrographs, gold particles show up as black dots. One or two minutes after injection, the micrographs showed that most of the gold particles were in the cytoplasm. A few, however, were closely associated with nuclear pores. Ten minutes after injection, particles were inside the nucleus as well

(a) Cross-sectional view of nuclear pore

(b) Molecules move into and out of the nucleus through the nuclear pore

FIGURE 7.24 Structure and Function of the Nuclear Pore.
(a) The drawing is based on electron micrographs of the nuclear pore.
(b) Messenger RNAs are synthesized in the nucleus and must be exported to the cytoplasm. Proteins needed in the nucleus are synthesized in the cytoplasm and have to be imported to the nucleus.

as in the cytoplasm. These data supported the hypothesis that the pores function as the doors to the nucleus. Follow-up work confirmed that the nuclear pore complex is the only gate between the cytoplasm and the nucleus and that only certain molecules go in and out. Passage through the nuclear pore is selective.

What substances traverse nuclear pores? DNA clearly docs not—it never leaves the nucleus. But information coded in DNA is used to synthesize RNA inside the nucleus. Several types of RNA molecules are produced, each distinguished by size and function. For example, most ribosomal RNAs are manufactured in the nucleolus, where they bind to proteins to form completed ribosomal subunits. The functional ribosomes are then exported to the cytoplasm. Similarly, molecules called messenger RNAs carry the information required to manufacture proteins out to the cytoplasm, where protein synthesis takes place. Traffic in the other direction is also impressive. Nucleoside triphosphates that act as building blocks for DNA and RNA enter the nucleus, as do the proteins responsible for copying DNA, synthesizing RNAs, extending the nuclear lamina, assembling ribosomes, or building chromosomes (**Figure 7.24b**).

To summarize, ribosomal subunits and various types of RNAs exit the nucleus; proteins that are needed inside enter it. In a typical cell, over 500 molecules pass through each of the 3000–4000 nuclear pores every second. The traffic is intense. How is it regulated and directed?

How Are Molecules Imported into the Nucleus?

The first experiments on how molecules move through the nuclear pore focused on proteins that are produced by viruses. **Viruses** are parasites that use the cell's machinery to make copies of themselves. When a virus infects a cell, certain of its proteins enter the nucleus. Investigators noticed that if a particular amino acid in one of these proteins happens to be altered, the viral protein was no longer able to pass through the nuclear pore. This simple-sounding observation led to a key hypothesis: Proteins that are synthesized by ribosomes in the cytosol but are headed for the nucleus contain a "zip code"—a molecular address tag that marks them for transport through the nuclear pore complex. The idea was that viral proteins were able to enter the nucleus only if they have the same address tag as normal cellular proteins have. Thus, the proteins with the altered amino acid were thwarted. This zip code came to be called the **nuclear localization signal (NLS)**.

A series of experiments on a protein called nucleoplasmin helped researchers better understand the nature of this signal. Nucleoplasmin plays an important role in the assembly of chromosomes and happens to have a distinctive structure: It consists of a globular protein core surrounded by a series of extended protein "tails." When researchers labeled nucleoplasmin with a radioactive atom and injected it into the cytoplasm of living cells, they found that the radioactive signal quickly ended up in the nucleus.

Experiment

Question: Where is the "Send to nucleus" zip code in the nucleoplasmin protein?

Hypothesis: The "Send to nucleus" zip code is in either the tail region or the core region of the nucleoplasmin protein.

Null hypothesis: The zip code is not on the nucleoplasmin protein itself, or there is no zip code.

Experimental setup:

"Tails"   Core

1. Use protease to cleave tails off of nucleoplasmin protein core.

2. Attach radioactive label.

3. Inject labeled protein fragments into cytoplasm of cell.

4. Wait, then locate labeled fragments ....

Prediction:

Prediction of null hypothesis:

Results:

Tail fragments located in nucleus

Core fragments still located in cytoplasm

Conclusion:

FIGURE 7.25 Where Is the "Send to Nucleus" Zip Code in the Nucleoplasmin Protein?

● **EXERCISE** Without looking at the text, fill in the predictions and conclusion(s) in this experiment.

Figure 7.25 outlines how the nuclear localization signal in nucleoplasmin was found. Researchers began by using enzymes called proteases to separate the core sections of nucleoplasmin from the tails. After separating the two components, the researchers labeled each part with radioactive atoms and injected

them into the cytoplasm of different cells. When they examined the experimental cells with the electron microscope, they found that tail fragments were transported to the nucleus. Core fragments, in contrast, remained in the cytoplasm. These data suggested that the zip code must be somewhere in the tail part of the protein.

By analyzing different stretches of the tail section, the biologists eventually found a 17-amino-acid-long section that had to be present to direct proteins to the nucleus. The biologists therefore concluded that the NLS zip code consisted of 17 specific amino acids in the tail.

Follow-up work confirmed that other proteins bound for the nucleus have similar localization signals. More recent research has shown that the movement of proteins and other large molecules into and out of the nucleus is an energy-demanding process that involves transport proteins called importins and exportins. Importins and exportins function like trucks that haul cargo in and out of the nucleus, through the nuclear pore complex.

Currently, biologists are trying to unravel how all this traffic in and out of the nucleus is regulated to avoid back-ups and head-on collisions. The goal is to understand the precise mechanisms responsible for moving cargo into and out of the cell's headquarters.

Web Animation at www.masteringbio.com

Transport into the Nucleus

7.3 The Endomembrane System: Manufacturing and Shipping Proteins

The nuclear membrane is not the only place in cells where cargo moves in a regulated and energy-demanding fashion. For example, Chapter 6 highlighted how specific ions and molecules are pumped into and out of cells or transported across the plasma membrane by specialized membrane proteins. In addition, proteins that are synthesized by ribosomes in the cytosol for use inside mitochondria or chloroplasts contain special signal sequences, like the nuclear localization signal, that target the proteins for transport to these organelles. Ions, ATP, amino acids, and other small molecules diffuse randomly throughout the cell, but the movement of proteins and other large molecules is energy demanding and tightly regulated.

If you pause to think about it for a moment, the need to sort proteins and ship them to specific destinations should be clear. Proteins are produced by ribosomes in the cytosol or ribosomes on the ER. Each protein that is synthesized needs to be transported to one of the many compartments inside the eukaryotic cell. Acid hydrolases need to end up in lysosomes, catalase needs to shipped to peroxisomes, and ribosomal proteins need to be transported to the nucleolus. To get to the right location, each protein has to have an address tag and a transport and delivery system.

To get a better understanding of how protein sorting and transport occur in eukaryotic cells, let's consider perhaps the most intricate of all manufacturing and shipping complexes: the endomembrane system. In this system, proteins that are synthesized in the rough ER move to the Golgi apparatus for processing, and from there travel to the cell surface or other destinations. The idea that materials might move through the endomembrane system in an orderly way was inspired by a simple observation. According to electron micrographs, cells that secrete digestive enzymes, hormones, or other types of products have particularly large amounts of rough ER and Golgi. This correlation led to the idea that these cells have a "secretory pathway" that starts in the rough ER and ends with products leaving the cell (**Figure 7.26**). How does this hypothesized pathway work?

George Palade and colleagues did pioneering research on the secretory pathway with an experimental approach known as a **pulse-chase experiment**. The strategy is based on providing experimental cells with a large concentration of a radioactively labeled molecule for a short time. For example, if a cell receives a large amount of labeled amino acid for a short time, virtually all of the proteins synthesized during that interval will be labeled.

This "pulse" of labeled molecule is followed by a chase—large amounts of an unlabeled version of the same molecule, provided for a long time. If the chase consists of unlabeled amino acid, then the proteins synthesized during the chase period will *not* be labeled. The general idea is to mark a population of molecules at a particular interval and then follow their fate over time. This approach is analogous to adding a small amount of dye to a stream and then following the movement of the dye molecules.

In testing the secretory pathway hypothesis, Palade's team focused on pancreatic cells that were growing in culture, or in vitro.[1] These cells are specialized for secreting digestive enzymes into the small intestine and are packed with rough ER and Golgi. The basic experimental approach was to supply the cells with a 3-minute pulse of the amino acid leucine, labeled with a radioactive atom, followed by a long chase with nonradioactive leucine. Because the radioactive leucine was incorporated into all proteins being produced during the pulse, those proteins were labeled. Then the researchers prepared a sample of the cells for electron microscopy and autoradiography (see **BioSkills 7 and 8**). **Figure 7.27a** shows part of a single cell that was examined immediately after the pulse. A close look at this micrograph should convince you that most of the newly synthesized

---

[1]The term *in vitro* is Latin for "in glass." Experiments that are performed outside living organisms are done in vitro. The term *in vivo*, in contrast, is Latin for "in life." Experiments performed with living organisms are done in vivo.

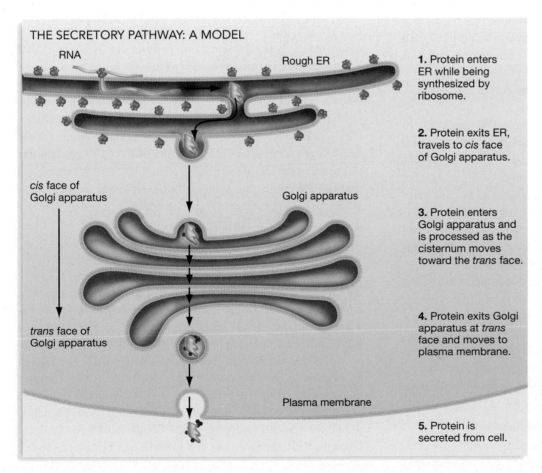

THE SECRETORY PATHWAY: A MODEL

RNA

Rough ER

cis face of Golgi apparatus

Golgi apparatus

trans face of Golgi apparatus

Plasma membrane

1. Protein enters ER while being synthesized by ribosome.

2. Protein exits ER, travels to *cis* face of Golgi apparatus.

3. Protein enters Golgi apparatus and is processed as the cisternum moves toward the *trans* face.

4. Protein exits Golgi apparatus at *trans* face and moves to plasma membrane.

5. Protein is secreted from cell.

FIGURE 7.26 The Secretory Pathway Hypothesis. The secretory pathway hypothesis proposes that proteins intended for secretion from the cell are synthesized and processed in a highly prescribed set of steps.

(a) Immediately after labeling.

(b) 7 minutes after end of labeling.

(c) 80 minutes after end of labeling.

Labeled proteins in rough ER

Labeled proteins in secretory vesicles

Labeled proteins in secretory vesicles

Labeled proteins in duct (where they will get carried away)

FIGURE 7.27 Results of a Pulse-Chase Experiment.

proteins are inside this cell's rough ER. **Figure 7.27b**, in contrast, shows part of a single cell 7 minutes later, after the pulse ended. Now the situation has changed. Few of the labeled proteins were in the rough ER. Instead, most of the labeled proteins are inside structures called secretory vesicles on the *trans* side of a Golgi apparatus (some are found inside the Golgi apparatus itself). The micrograph in **Figure 7.27c** is at lower magnification and shows parts of five cells. The structure in the middle is a duct that carries digestive enzymes from pancreatic cells toward their destination in the small intestine. This micrograph was taken 80 minutes after the pulse. If you look at the figure carefully, you'll note that most labeled proteins are in secretory vesicles or actually outside the cell, in the duct.

Because the labeled proteins move from the rough ER to Golgi apparatus to secretory vesicles to the cell exterior over time, the results support the hypotheses that a secretory pathway exists and that the rough ER and Golgi apparatus function as an integrated endomembrane system. The data suggest that proteins produced in the rough ER don't float around the cytoplasm aimlessly or drift randomly from organelle to organelle. Instead, traffic through the endomembrane system appears to be highly organized and directed.

Now let's break the system down and examine four of the steps in more detail. Recall that the ribosomes in rough ER are bound to the outside of the membrane. How do the proteins that they manufacture get into the lumen of the ER? How do they move from the ER to the Golgi apparatus? Once they're inside the Golgi, what happens to them? And finally, how do the finished proteins get to their destination? Let's consider each question in turn.

(MB) **Web Animation** at www.masteringbio.com

A Pulse Chase Experiment

Entering the Endomembrane System: The Signal Hypothesis

How do proteins enter the endomembrane system? The signal hypothesis predicted that proteins bound for the endomembrane system have a zip code analogous to the nuclear localization signal. The idea, proposed by Günter Blobel and colleagues, was that these proteins are synthesized by ribosomes that are attached to the outside of the ER, and that the first few amino acids in the growing polypeptide act as a signal that brings the protein into the lumen of the ER.

This hypothesis received important support when researchers made a puzzling observation: When proteins that are normally synthesized in the rough ER are instead manufactured by isolated ribosomes in vitro—with *no* ER present—they are 20 amino acids longer than usual. Blobel seized on these data. He claimed that the 20 amino acids are the "send-to-ER" signal and that the 20-amino-acid signal is removed inside the organelle. When the same protein is synthesized outside of the ER, the signal is not removed. His group went on to produce convincing data that supported the hypothesis: They identified the exact series of amino acids in the **ER signal sequence**.

More recent work has documented the mechanisms responsible for receiving the send-to-ER signal and inserting the protein into the rough ER (**Figure 7.28**). The action begins when a ribosome synthesizes the ER signal sequence, which then binds to a **signal recognition particle (SRP)** in the cytosol. An SRP is a complex of RNA and protein that acts as a receptor for the ER signal sequence. The ribosome + signal sequence + SRP complex then attaches to an SRP receptor in the ER membrane itself. You can think of the SRP as a key that is activated by an ER signal sequence. The SRP receptor in the ER membrane is the lock. Once the lock (the receptor) and key (the SRP) connect, the rest

FIGURE 7.28 The Signal Hypothesis Explains How Proteins Destined for Secretion Enter the Endomembrane System.
According to the signal hypothesis, proteins destined for secretion contain a short stretch of amino acids that interact with a signal recognition particle (SRP) in the cytoplasm. This interaction allows the protein to enter the ER.

of the protein is synthesized, and then the signal sequence is removed. If the finished polypeptide will eventually be shipped to an organelle or secreted from the cell, then it enters the lumen of the rough ER. If the finished polypeptide is a membrane protein, it remains in the rough ER membrane while it is being processed.

Once proteins are inside the rough ER or inserted into its membrane, they fold into their three-dimensional shape with the help of chaperone proteins introduced in Chapter 3. In addition, proteins that enter the lumen interact with enzymes that catalyze the addition of carbohydrate side chains. Because carbohydrates are polymers of sugar monomers, the addition of one or more carbohydrate groups is called **glycosylation** ("sugar-together"). The resulting molecule is called a **glycoprotein** ("sugar-protein"; see Chapter 5). As **Figure 7.29** shows, proteins that enter the ER often gain a specific carbohydrate that consists of 14 sugar subunits. Thus, proteins that are synthesized in the rough ER are also folded and modified by glycosylation there. The completed glycoproteins are ready for shipment to the Golgi apparatus.

Getting from the ER to the Golgi

How do proteins travel from the ER to the Golgi apparatus? Palade's group thought they had the answer, based on data from the pulse-chase experiments that first confirmed the existence of the endomembrane system. When labeled proteins appeared in a region between the rough ER and the Golgi apparatus, they appeared to be inside small membrane-bound structures. Based on these observations, the biologists suggested that proteins are transported between the two organelles in vesicles. The idea was that vesicles bud off of the ER, move away, fuse with the membrane on the *cis* face of the Golgi apparatus, and dump their contents inside.

This hypothesis was supported when other researchers used differential centrifugation to isolate and characterize the vesicles that contained labeled proteins. Using this approach, investigators have established that a distinctive type of vesicle carries proteins from the rough ER to the Golgi apparatus.

Taken together, these results support a general conclusion about the endomembrane system: It is a sophisticated, highly organized complex. Proteins move through it in a directed way

FIGURE 7.29 Glycosylation Adds Carbohydrate Groups to Proteins.
When proteins enter the ER, most acquire the 14 sugar residues shown here. Some of these sugars may be removed or others added as proteins pass through the Golgi apparatus.

and undergo a series of manufacturing, transport, and processing steps.

What Happens inside the Golgi Apparatus?

Recall from Section 7.1 that the Golgi apparatus consists of a stack of flattened vesicles called cisternae, and that cargo enters one side of the structure and exits the other. Recent research has shown that the composition of the Golgi apparatus is dynamic. New cisternae are constantly created at the *cis* face, while old cisternae break apart at the *trans* face. Once a cisternum forms, it gradually moves toward the *trans* face. As it does, it changes in composition and activity.

By separating individual cisternae and analyzing their contents, researchers have found that cisternae at various stages of maturation contain different suites of enzymes. These enzymes catalyze glycosylation reactions. As a result, proteins undergo further modification as a cisterna matures. While cisternae are still near the *cis* face, some of the proteins inside have sugar-phosphate groups added. Later, the carbohydrate group that was added in the rough ER is removed. Near the *trans* face, various types of carbohydrate chains are attached that may protect the protein or help it attach to surfaces.

If the rough ER is like a foundry and stamping plant where rough parts are manufactured, then the Golgi is like a finishing area where products are polished and painted and readied for shipping.

How Are Products Shipped from the Golgi?

The rough ER and Golgi apparatus function as an impressive assembly line. Some of the proteins they produce stay in the endomembrane system itself, replacing worn-out molecules. But if proteins are processed to the end of the line, they will be sent to one of several destinations, including lysosomes, the plasma membrane, or the outside of the cell. How are these finished products put into the right shipping containers, and how are the different containers addressed?

Studies on enzymes that are shipped to lysosomes have provided some answers to both questions. A key finding was that lysosome-bound proteins have a phosphate group attached to a specific sugar subunit on their surface, forming the compound mannose-6-phosphate. If mannose-6-phosphate is removed from these proteins, they are not transported to a lysosome. This is strong evidence that the phosphorylated sugar serves as a zip code, analogous to the nuclear localization and rough ER signals discussed earlier. More specifically, data indicate that mannose-6-phosphate binds to a protein in the membranes of certain vesicles. These vesicles, in turn, have proteins on their surface that interact specifically with proteins in the lysosomal membranes. In this way, the presence of mannose-6-phosphate targets proteins for vesicles that deliver their contents to lysosomes.

Figure 7.30 pulls these observations together into a comprehensive model explaining how the products of the endomem-

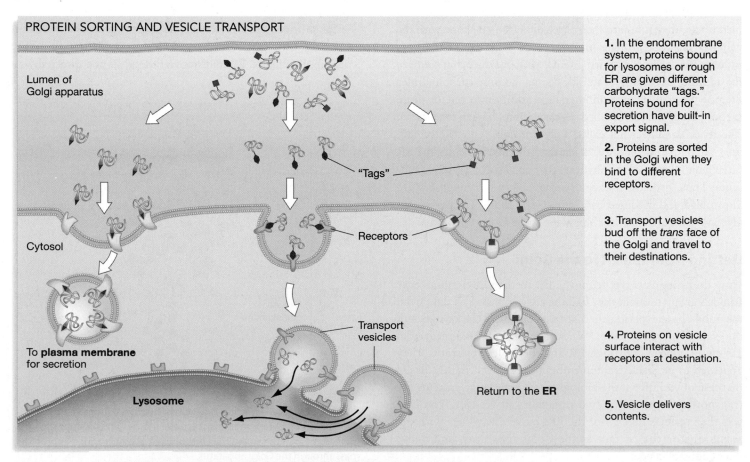

PROTEIN SORTING AND VESICLE TRANSPORT

Lumen of Golgi apparatus

"Tags"

Cytosol

Receptors

To **plasma membrane** for secretion

Transport vesicles

Lysosome

Return to the **ER**

1. In the endomembrane system, proteins bound for lysosomes or rough ER are given different carbohydrate "tags." Proteins bound for secretion have built-in export signal.

2. Proteins are sorted in the Golgi when they bind to different receptors.

3. Transport vesicles bud off the *trans* face of the Golgi and travel to their destinations.

4. Proteins on vesicle surface interact with receptors at destination.

5. Vesicle delivers contents.

FIGURE 7.30 In the Golgi Apparatus, Proteins Are Sorted into Vesicles That Are Targeted to a Destination.

brane system are loaded into specific vesicles and shipped to their correct destination. The key observation is that each protein that comes out of the Golgi apparatus has a molecular tag that places it in a particular type of transport vesicle. Each type of transport vesicle, in turn, has a tag that allows it to be transported to the correct destination—the plasma membrane, a lysosome, or the ER.

In particular, notice that the transport vesicle shown on the left of Figure 7.30 is bound for the plasma membrane, where it will secrete its contents to the outside. This process is called **exocytosis** ("outside-cell-act"). When exocytosis occurs, the vesicle membrane and plasma membrane make contact and fuse, analogous to the way that the vesicle membrane and lysosome membrane are shown fusing in the center, bottom of the figure. At the plasma membrane, the two sets of lipid bilayers rearrange in a way that exposes the interior of the vesicle to the outside of the cell. The vesicle's contents then diffuse away from the cell into the space outside the cell. In this way, cells in your pancreas deliver digestive enzymes to the duct that leads to the small intestine—where digestion actually takes place.

● Proteins that are synthesized in the cytoplasm also have zip codes directing them to mitochondria, chloroplasts, or other destinations. To summarize, then, the array of proteins produced in a cell are sorted via distinctive zip codes. These molecular addresses allow proteins to be shipped to the compartment where they function.

If vesicles function like shipping containers for products that move between organelles, do they travel along some sort of road or track? What molecule or molecules function as the delivery truck, and does ATP supply the gas? In general, what physical mechanisms are responsible for moving vesicles to their destination? Let's delve into these questions in Section 7.4.

Check Your Understanding

If you understand that...

- In cells, the transport of proteins and other large molecules is energy demanding and tightly regulated.
- Proteins must have the appropriate molecular zip code to enter or leave the nucleus, enter the lumen of the rough ER, or become incorporated into vesicles destined for lysosomes or the plasma membrane.
- In many cases, proteins and other types of cargo are shipped in vesicles that contain molecular zip codes on their surface.

You should be able to...

1) Predict what happens to proteins that lack an ER signal sequence.
2) Predict the outcome of an experiment where secreted proteins are placed inside vesicles with a zip code associated with shipment to lysosomes.

7.4 The Dynamic Cytoskeleton

Based on early observations with light microscopes, biologists concluded that the cytoplasm of eukaryotic cells was a fluid-filled space devoid of structure. As microscopy improved, however, researchers realized that cell cytoplasm contains an extremely dense and complex network of fibers. This cytoskeleton helps maintain cell shape by providing structural support. It's important to recognize, though, that the cytoskeleton is not a static structure like the scaffolding used at construction sites. The fibrous proteins that make up the cytoskeleton move and change to alter the cell's shape, to move materials from place to place, and to move the entire structure. Like the rest of the cell, the cytoskeleton is dynamic.

As **Summary Table 7.3** shows, there are several distinct types of cytoskeletal elements: actin filaments (also known as microfilaments), intermediate filaments, and microtubules. Each of these elements has a distinct size, structure, and function. Let's look at each one in turn.

Actin Filaments

Actin filaments are sometimes referred to as **microfilaments** because they are the cytoskeletal element with the smallest diameter. As Summary Table 7.3 indicates, **actin filaments** are fibrous structures made of a globular protein called actin. In animal cells, actin is often the most abundant of all proteins—typically it represents 5–10 percent of the total protein in the cell. Each of your liver cells contains about half a billion of these molecules.

Actin filaments form when individual actin molecules polymerize. The completed structure resembles two long strands that coil around each other. Because each actin subunit in the strand is asymmetrical, the structure as a whole has a distinct polarity. The two ends of an actin filament are different and are referred to as plus and minus ends. Each filament grows and shrinks as actin subunits are added or subtracted from each end of the structure. This phenomenon is called treadmilling, because the dynamics of the fibers resemble those of a treadmill. Actin filaments tend to grow at the plus end, however, because polymerization occurs fastest there.

In animal cells, actin filaments are particularly abundant just under the plasma membrane. **Figure 7.31a** shows a fluorescence micrograph of the actin filaments in a mammalian cell. Notice that groups of actin filaments are organized into long bundles or dense networks. Whether they are arranged in parallel as part of bundles or crisscrossed in networks, individual actin filaments are linked to one another by other proteins. In combination, bundles and networks of actin filaments help stiffen the cell and define its shape.

In addition to providing structural support, actin filaments are involved in movement when they interact with the specialized protein **myosin**. Myosin is a **motor protein**: a protein that

SUMMARY TABLE **7.3** **Cytoskeletal Filaments**

	Actin Filaments (Microfilaments)	Intermediate Filaments	Microtubules
	The three types of filaments found in the cytoskeleton are distinguished by their size and structure, and by the protein subunit of which they are made.		
Protein subunits	Actin	Keratin, vimentin, lamin, others	α-tubulin and β-tubulin dimers
Structure	Strands in double helix − end + end 7 nm ● Actin subunit	Fibers wound into thicker cables 10 nm Keratin subunits	Hollow tube − end + end 25 nm α-tubulin and β-tubulin dimer
Functions	• maintain cell shape by resisting tension (pull) • motility via muscle contraction or cell crawling • cell division in animals • movement of organelles and cytoplasm in plants, fungi, and animals	• maintain cell shape by resisting tension (pull) • anchor nucleus and some other organelles	• maintain cell shape by resisting compression (push) • motility via flagella or cilia • move chromosomes during cell division • formation of cell plate during plant cell division • move organelles • growth of plant cell walls

converts the chemical energy in ATP into the mechanical work of movement, just as a car's motor converts the chemical energy in gasoline into movement. The interaction between actin and myosin that produces movement is discussed in detail in Chapter 46. For now, it's enough to recognize that when ATP binds to myosin and is then hydrolyzed to ADP, the "head" region of the myosin molecule binds to actin and moves. The movement of this protein causes the actin filament to slide (**Figure 7.32a**). This type of movement is analogous to a line of people who are passing along a long log or pole. The people are myosin molecules; the log or pole is actin.

As **Figure 7.32b** shows, the (ATP-powered) interaction between actin and myosin is the basis for an array of cell movements:

• **Cell crawling** occurs in a wide array of organisms and cell types, including amoebae, slime molds, and certain types of human cells. Cell crawling is based on three processes: a directional extension of actin filaments that pushes the plasma membrane into bulges called **pseudopodia** ("false-feet"), adherence to a solid substrate, and a myosin-driven contraction of actin filaments at the cell's other end. In combination, the three events result in directed movement by whole cells.

(a) Actin filaments in mammalian cell

(b) Intermediate filaments in mammalian cell

(c) Microtubules in mammalian cell

  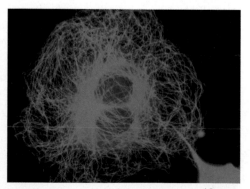

10 μm

FIGURE 7.31 How Are Cytoskeletal Elements Distributed in the Cell? To make these fluorescence micrographs, researchers attached a fluorescent compound to **(a)** actin, the protein subunit of actin filaments, to **(b)** a protein found in intermediate filaments, and to **(c)** tubulin dimers.

(a) Actin and myosin interact to cause movement.

(b) Actin-myosin interactions produce several types of movement.

Cell crawling

Cell division in animals

Cytoplasmic streaming in plants

FIGURE 7.32 Many Cellular Movements Are Based on Actin-Myosin Interactions. (a) When the "head" region of the myosin protein interacts with ATP, myosin attaches to actin and changes shape. The movement causes the actin filament to slide. **(b)** Actin-myosin interactions can move cells, divide cells, and move organelles and cytoplasm.

- **Cytokinesis** ("cell-moving") is the process of cell division in animals. For these cells to divide in two, actin filaments that are arranged in a ring under the plasma membrane must slide past one another. Because they are connected to the plasma membrane, the movement of the actin fibers pinches the cell in two (see Chapter 11).

- **Cytoplasmic streaming** is the directed flow of cytosol and organelles around plant and fungal cells. The movement occurs along actin filaments and is powered by myosin. It is especially common in large cells, where the circulation of cytoplasm facilitates material transport.

Extension of actin filaments is also responsible for the expansion of long, thin fungal cells into soil or rotting wood, toward fresh supplies of nutrients. The same mechanism causes structures called pollen tubes to grow toward the egg cells of plants so sperm can be delivered prior to fertilization. You'll see actin and myosin moving cells and organisms around throughout this text.

Intermediate Filaments

Unlike actin filaments and microtubules, **intermediate filaments** (**Figure 7.31b**) are defined by size rather than composition. Many types of intermediate filaments exist, each consisting of a different protein. In many cases, different types of cells in the same organism contain different types of intermediate filaments. This is in stark contrast to actin filaments and mi-

crotubules, which are made from the same protein subunits in all eukaryotic cells. In addition, intermediate filaments are not polar; instead, each end of these filaments is identical. As a result, intermediate filaments do not treadmill, and they are not involved in directed movement driven by myosin or related proteins. Intermediate filaments serve a purely structural role in eukaryotic cells.

The intermediate filaments that you are most familiar with belong to a family of molecules called the keratins. The cells that make up your skin and that line surfaces inside your body contain about 20 types of keratin. The presence of these intermediate filaments provides the mechanical strength required for these cells to resist pressure and abrasion. Skin cells manufacture another 10 distinct forms of keratin. Depending on the location of the skin cell and keratins involved, the secreted filaments form fingernails, toenails, or hair.

Nuclear lamins, which make up the nuclear lamina layer introduced in Section 7.1, also qualify as intermediate filaments. These fibers form a dense mesh under the nuclear envelope. Recall that in addition to giving the nucleus its shape, they anchor the chromosomes. They are also involved in the breakup and reassembly of the nuclear envelope when cells divide. Some intermediate filaments project from the nucleus through the cytoplasm to the plasma membrane, where they are linked to intermediate filaments that run parallel to the cell surface. In this way, intermediate filaments form a flexible skeleton that helps shape the cell surface and hold the nucleus in place.

Microtubules

Microtubules are composed of the proteins α-tubulin and β-tubulin and are the largest cytoskeletal components in terms of diameter (**Figure 7.31c**). Molecules of α-tubulin and β-tubulin bind to form **dimers** ("two-parts"), compounds formed by the joining of two monomers. Tubulin dimers then polymerize to form the large, hollow tube called a microtubule. Because each end of a tubulin dimer is different, each end of a microtubule has a distinct polarity. Like actin filaments, microtubules are dynamic and usually grow at their plus end. Microtubules grow and shrink in length as tubulin dimers are added or subtracted.

Microtubules originate from a structure called the **microtubule organizing center** and grow outward, radiating throughout the cell. Although plant cells typically have hundreds of these organizing centers, most animal and fungal cells have just one. In animals, the microtubule organizing center has a distinctive structure and is called a **centrosome**. As **Figure 7.33** shows, animal centrosomes contain two bundles of microtubules called **centrioles**. The centrioles are not required for microtubules to form and begin growing from the centrosome, however. Although biologists have been aware for decades that they exist, their role in the cell is still unknown.

In function, microtubules are similar to actin filaments: They provide stability and are involved in movement. The microtubules that radiate out from an organizing center function like steel girders in a skyscraper. They stiffen the cell by resisting compression forces. Microtubules may also provide a structural framework for organelles. If microtubules are prevented from forming, the ER no longer assembles in its normal network-like configuration.

During cell division, microtubules from the organizing center are responsible for moving chromosomes from the original cell to each of the two resulting cells (see Chapters 11 and 12). But microtubules are involved in many other types of cellular movement as well. The remainder of this chapter focuses on how microtubules function in moving materials inside cells and in moving the entire cell.

Studying Vesicle Transport   Materials are transported to a wide array of destinations inside cells via vesicles. To study how this movement happens, Ronald Vale and colleagues focused on a cell in squid called the giant axon. The giant axon is an extremely large nerve cell that runs the length of a squid's body. If the animal is disturbed, the cell signals muscles to contract so the individual can jet away to safety.

The researchers decided to study this particular cell for three reasons. First, the giant axon is so large that it is relatively easy to see and manipulate. Second, large numbers of molecules are synthesized in the cell's ER and then transported in vesicles down the length of the cell, where they are released. As a result, a large amount of cargo moves a long distance. Third, the researchers found that if they gently squeeze the cytoplasm out of the cell, vesicle transport still occurs in the cytoplasmic material. This allowed them to do experiments on vesicle transport without the plasma membrane being in the way.

In short, the squid giant axon provided a system that could be observed and manipulated efficiently in the lab. What did the biologists find out?

Microtubules Act as "Railroad Tracks"   To watch vesicle transport in action, the researchers mounted a video camera to a microscope. As **Figure 7.34** shows, this technique allowed them to document that vesicle transport was occurring along a filamentous track. A simple experiment convinced the group that this movement is an energy-dependent process. If they depleted the amount of ATP in the cytoplasm, vesicle transport stopped.

(a) Electron micrograph

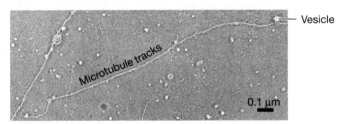

Vesicle

Microtubule tracks

0.1 μm

(b) Video image

Vesicle

Microtubule tracks

0.1 μm

FIGURE 7.34 Transport Vesicles Move along Microtubule Tracks. The images are of extruded cytoplasm from a squid giant axon. **(a)** An electron micrograph that allowed researchers to measure the diameter of the filaments and confirm that they are microtubules. In the upper right of this image, you can see a vesicle on a "track." **(b)** A slightly fuzzy but higher-magnification videomicroscope image, in which researchers actually watched vesicles move.

Centrosome

200 μm

Centrioles (oriented at 90° to each other)

FIGURE 7.33 Centrosomes Are a Type of Microtubule Organizing Center. Microtubules emanate from microtubule organizing centers, which in animals are called centrosomes. The centrioles inside a centrosome are made of microtubules.

To identify the type of filament involved, the biologists measured the diameter of the tracks and analyzed their chemical composition. Both types of data indicated that the tracks consist of microtubules. Microtubules also appear to be required for movement of materials elsewhere in the cell. If experimental cells are treated with a drug that disrupts microtubules, the movement of vesicles from the rough ER to the Golgi apparatus is impaired.

The general message of these experiments is that transport vesicles move through the cell along microtubules. How? Do the tracks themselves move, like a conveyer belt, or are vesicles carried along on some sort of molecular truck?

A Motor Protein Generates Motile Forces To study the way vesicles move along microtubules, Vale's group set out to take the squid axon's transport system apart and then put it back together. To begin, they assembled microtubule fibers from purified α-tubulin and β-tubulin. Then they used differential centrifugation to isolate transport vesicles. But when they mixed purified microtubules and vesicles with ATP, no transport occurred.

Something had been left out—but what?

To find the missing element or elements, the researchers purified one subcellular part after another, using differential centrifugation, and added it to the microtubule + vesicle + ATP system. Through trial and error, they found something that triggered movement. After further purification steps, the researchers finally succeeded in isolating a protein that generated vesicle movement. They named the molecule **kinesin**, from the Greek word *kinein* ("to move"). Like myosin, kinesin is a **motor protein**. Kinesin converts chemical energy in ATP into mechanical energy in the form of movement. More specifically, when ATP is added to kinesin or drops off, the protein moves.

Biologists began to understand how kinesin works when X-ray diffraction studies—similar to those that revealed the helical nature of DNA—revealed the three-dimensional structure of kinesin. As **Figure 7.35a** shows, the protein consists of two large, intertwined polypeptide chains associated with two small polypeptides. It has three major regions: a head section with two globular pieces, a tail associated with the small polypeptides, and a stalk that connects the head and tail. Follow-up studies confirmed that the two globular components of the head bind to the microtubule. The tail region binds to the transport vesicle. A kinesin molecule is like a delivery truck that carries transport vesicles along microtubule tracks. Cells contain several different kinesin proteins, each specialized for carrying a different type of vesicle.

How does kinesin move? More detailed studies of this protein's structure indicated that each of the globular components of the molecule's head has a site for binding ATP as well as a site that binds to the microtubule. To pull these observations together, biologists propose that kinesin transports vesicles by "walking" along a microtubule. The idea is that each part of the head region changes shape when it binds ATP. As Chapter 3 showed, these types of shape changes often alter the activity of a protein. As **Figure 7.35b** shows, the ATP-dependent conformational change in kinesin results in a step forward. As each head alternately binds and hydrolyzes ATP, the protein and its cargo move down the microtubule track.

In short, kinesins move molecular cargo to destinations throughout the cell. They are not the only type of motor protein active inside cells, however. Recall that myosin causes actin filaments to slide, resulting in the movement of cells or cytoplasm. Myosin is also involved in the movement of organelles along tracks made of actin. And a third motor protein, called dynein, powers the transport of certain organelles as well as swimming movements that move the entire cell. Let's take a closer look at how cells swim.

(a) Structure of kinesin

(b) Kinesin "walks" along a microtubule track.

FIGURE 7.35 A Motor Protein Moves Vesicles along Microtubules. (a) Kinesin has three major segments. **(b)** The current model depicting how kinesin "walks" along a microtubule track to transport vesicles. The two head segments act like feet that alternately attach and release in response to the gain or loss of a phosphate group.

Cilia and Flagella: Moving the Entire Cell

Flagella are long, hairlike projections from the cell surface that function in movement. Flagella are found in many bacteria and eukaryotes. The structure of flagella is completely different in the two groups, however. Bacterial flagella are made of a protein called flagellin; eukaryotic flagella are constructed from microtubules (tubulin). Bacterial flagella move the cell by rotating like a ship's propeller; eukaryotic flagella move the cell by undulating—they whip back and forth. Eukaryotic flagella are surrounded by plasma membrane; bacterial flagella are not. Based on these observations, biologists conclude that the two structures evolved independently, even though their function is similar.

To understand how cells move, let's focus on eukaryotic flagella. Eukaryotic flagella are closely related to structures called **cilia** (singular: cilium), which are short, filament-like projections that are also found in some eukaryotic cells. Unicellular eukaryotes may have either flagella or cilia, while some multicellular organisms have both. In humans, for example, the cells that line the respiratory tract have cilia; sperm cells have flagella.

Flagella are generally longer than cilia, and a cell will typically have just one or two flagella but many cilia (**Figure 7.36**). But when researchers examined the two structures with the electron microscope, they found that their underlying organization is identical.

How Are Cilia and Flagella Constructed?

In the 1950s, anatomical studies established that both cilia and flagella have a characteristic "9 + 2" arrangement of microtubules. As **Figure 7.37a** shows, nine microtubule pairs, or doublets, surround two central microtubules. The doublets consist of one complete and one incomplete microtubule and are arranged around the periphery of the structure. The entire 9 + 2 structure is called the **axoneme** ("axle-thread"). The axoneme attaches to the cell at a structure called the **basal body**. The basal body is identical in structure to a centriole. But unlike centrioles, the function of the basal body is well understood: It plays a central role in the growth of the axoneme.

As electron microscopy improved, biologists gained a more detailed view of the structure. As the sketch in **Figure 7.37b** illustrates, spoke-like structures connect each doublet to the central pair of microtubules. In addition, molecular links connect the nine doublets to one another. Finally, each of the doublets has a set of arms that project toward an adjacent doublet.

Microtubules are complex. How do their components interact to generate motion?

A Motor Protein in the Axoneme In the 1960s Ian Gibbons began studying the cilia of a common unicellular eukaryote called *Tetrahymena*, which lives in pond water. Gibbons found that by using a detergent to remove the plasma membrane that surrounds cilia and then subjecting the resulting solution to differential centrifugation, he could isolate axonemes. Further, the isolated structures would beat if Gibbons supplied them with ATP. These results confirmed that the beating of cilia is an energy-demanding process. They also provided Gibbons with a cell-free system for exploring the molecular mechanism of movement. Cell-free systems are elegant to study because they are isolated—they are relatively easy to manipulate, and no other cell components are present that might confuse experimental results.

(a) Transmission electron micrograph of axoneme

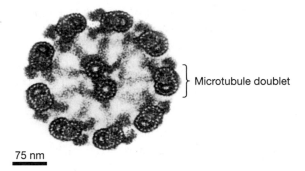

Microtubule doublet

75 nm

(b) Diagram of axoneme

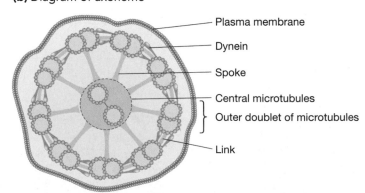

Plasma membrane
Dynein
Spoke
Central microtubules
Outer doublet of microtubules
Link

FIGURE 7.37 The Structure of Cilia and Flagella. (a) Transmission electron micrograph of a cross section through an axoneme. **(b)** The major structural elements in cilia and flagella. The microtubules are connected by links and spokes, and the entire structure is surrounded by the plasma membrane.

● **EXERCISE** In part (a), write the numbers 1–9 next to the microtubule doublets around the outside of the axoneme.

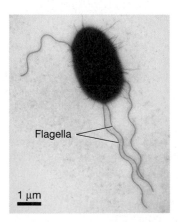

Cilia

Flagella

50 μm

1 μm

FIGURE 7.36 Cilia and Flagella Differ in Length and Number. Cilia are relatively short and large in number; flagella are relatively long and few in number.

In an early experiment with isolated axonemes, Gibbons treated the structures with a molecule that affects the ability of proteins to bind to one another. The axonemes that resulted from this treatment could not bend or use ATP. When Gibbons examined them in the electron microscope, he found that the arms had fallen off. This observation led to the hypothesis that the arms are required for movement. Follow-up work showed that the arms are made of a large protein that Gibbons named **dynein** (from the Greek word *dyne,* meaning "force").

Like myosin and kinesin, dynein is a motor protein. Structural and chemical studies have shown that dynein changes shape when a phosphate group from ATP attaches to it. More specifically, the end of a dynein molecule changes shape when it is phosphorylated. This shape change moves the molecule along the nearby microtubule. When the dynein molecule reattaches, it has succeeded in walking up the microtubule. This walking motion allows the microtubule doublets to slide past one another.

The outcome of dynein walking is very different from the outcome of a myosin-actin interaction, however. To understand why, remember that each of the nine doublets in the axoneme is connected to the central pair of microtubules by a spoke, and that all of the doublets are connected to each other by molecular links. As a result, the sliding motion that results from dynein walking is constrained—if one doublet slides, it transmits force to the rest of the axoneme via the links and spokes (**Figures 7.38**).

● If you understand this concept, you should be able to study the top drawing in Figure 7.38 and describe what happens when movement by dynein arms causes the doublet at the bottom of the axoneme to move toward the bottom right-hand corner of the page. If the dynein arms on just one side of the axoneme walk and cause some of the doublets to slide while those on the other side are at rest, the result of the constrained, localized movement is bending (see Figure 7.38, bottom). The result of the bending of cilia or flagella is a swimming motion.

Scaled for size, flagellar-powered swimming can be rapid. In terms of the number of body or cell lengths traveled per second, a sperm cell from a bull moves faster than a human world-record-holder does when swimming freestyle. At the level of the cell, life is fast paced.

● Taken together, the data reviewed in this chapter can be summed up in three words: Cells are dynamic. Chemical reactions take place at mind-boggling speeds. Microfilaments and microtubules grow and shrink. The endomembrane system synthesizes, sorts, and ships an array of products in a highly regulated manner. How does all this activity inside the cell relate to what is going on outside? This is the issue we'll take up in Chapter 8.

Dynein arms are at rest: Flagellum is straight

5 μm

Dynein arms on this side of flagellum walk to the left, pushing their microtubules to the right. As a result, the flagellum bends

FIGURE 7.38 How Do Flagella Bend? Researchers attached a pair of gold beads to a flagellum and photographed its movement over a short time sequence. When dynein arms walk along the microtubule doublets on one side of a flagellum, the structure bends.

Check Your Understanding

If you understand that... ●

- Each component of the cytoskeleton has a unique structure and set of functions. In addition to providing structural support, actin filaments and microtubules work in conjunction with motor proteins to move the cell or materials inside the cell. Intermediate filaments provide structural support.

- Most elements of the cytoskeleton are dynamic—they grow and shrink over time.

You should be able to... ●

Predict what will happen when experimental cells are treated with drugs that inhibit formation of actin filaments or microtubules.

Chapter Review

SUMMARY OF KEY CONCEPTS

⬤ **The structure of cell components is closely correlated with their function.**

Because all organisms consist of cells, many questions in biology can be answered by understanding the structure and function of cells and cell components. There are two basic cellular designs: prokaryotic and eukaryotic. Eukaryotic cells are usually much larger and more structurally complex than prokaryotic cells. Most prokaryotic cells consist of a single membrane-bound compartment in which nearly all cellular functions occur. Eukaryotic cells contain numerous membrane-bound compartments called organelles. Organelles allow eukaryotic cells to compartmentalize functions and grow to a large size.

Eukaryotic organelles are specialized for carrying out different functions, and their structure is often correlated closely with their function. The defining organelle of eukaryotic cells is the nucleus, which contains the cell's chromosomes and serves as its control center. Mitochondria and chloroplasts have extensive internal membrane systems where the enzymes responsible for ATP generation and photosynthesis reside. Rough ER is named for the ribosomes that attach to it. Ribosomes are protein-making machines, and rough ER is a site for protein synthesis and processing. Smooth ER lacks ribosomes because it is a center for lipid synthesis and processing.

You should be able to predict what would happen to cells that are exposed to a drug that poisons mitochondria, an enzyme that degrades the cell wall, or a drug that prevents ribosomes from functioning. ⬤

(MB) Web Animation *BioFlix* **at www.masteringbio.com**
Tour of an Animal Cell; Tour of a Plant Cell

⬤ **Inside cells, materials are transported to their destinations with the help of molecular "zip codes."**

Cells have sophisticated systems for making sure that proteins and other products end up in the right place. For example, traffic across the nuclear envelope occurs through nuclear pores, which contain a multiprotein nuclear pore complex that serves as gatekeeper. Both passive and active transports of materials occur through these nuclear pore complexes. Movement of proteins and RNAs is highly regulated, allowing proteins into the nucleus only if they contain a specific molecular signal.

The endomembrane system is an extensive, interconnected system of membranes and membrane-bound compartments that can extend from the nucleus to the plasma membrane. Two principal organelles in the endomembrane system are the endoplasmic reticulum (ER) and the Golgi apparatus. The ER is the site of synthesis for a wide array of proteins and lipids. Most ER products are shipped to the Golgi apparatus, which serves as a

processing and dispatching station. In many proteins the major processing step is glycosylation, or the addition of carbohydrate groups.

The movement of materials through the endomembrane system is highly organized and takes place inside membrane-bound transport organelles called vesicles. Before products leave the endomembrane system, they are sorted with molecular zip codes that direct them to vesicles headed for their final destination. The vesicles contain proteins that interact with receptor proteins on the surface of a target organelle or the plasma membrane so that the contents can be delivered.

You should be able to explain, based on information in Chapter 3, why proteins are the molecules responsible for "reading" the array of molecular zip codes in cells. ⬤

(MB) Web Animation **at www.masteringbio.com**
Transport into the Nucleus; A Pulse-Chase Experiment

⬤ **Cells are dynamic. Thousands of chemical reactions occur each second within cells; molecules constantly enter and exit across the plasma membrane; cell products are shipped along protein fibers; and elements of the cell's internal skeleton grow and shrink.**

The cell is a membrane-bound structure with a highly organized, dynamic interior. Inside the cell, thousands of different chemical reactions take place at incredible speeds. The products of these chemical reactions allow the cell to acquire resources from the environment, synthesize additional molecules, dispose of wastes, and reproduce.

The cytoskeleton is an extensive system of fibers that serves as a structural support for eukaryotic cells. Elements of the cytoskeleton also provide paths for moving vesicles inside cells and the machinery for moving the cell as a whole through the beating of flagella or cilia, or cell crawling. Both the movement of vesicles inside cells and cell motility depend on motor proteins, which can convert chemical energy stored in ATP into movement. For example, movement of transport vesicles occurs as the motor protein kinesin "walks" along microtubule tracks. Cilia and flagella bend as the motor protein dynein "walks" along microtubule doublets. The bending motion allows these structures to beat back and forth, enabling cells to swim or generate water currents.

You should be able to compare and contrast the structure and function of actin filaments, intermediate filaments, and microtubules. You should also be able to explain why actin filaments and microtubules have directionality and change length, and why it is misleading to refer to the cytoskeleton as "scaffolding." ⬤

QUESTIONS

Test Your Knowledge

1. Which of the following best describes the nuclear envelope?
   a. It is continuous with the endomembrane system.
   b. It is continuous with the nucleolus.
   c. It is continuous with the plasma membrane.
   d. It contains a single membrane and nuclear pores.

2. What is a nuclear localization signal?
   a. a stretch of amino acids that directs proteins from the nucleus to the ER
   b. a molecule that is attached to nuclear proteins so that they are retained inside the nucleus
   c. a signal built into a protein that directs it to the nucleus
   d. a component of the multiprotein nuclear pore complex

3. Which of the following is *not* true of secreted proteins?
   a. They are synthesized in ribosomes.
   b. They are transported through the endomembrane system in membrane-bound transport organelles.
   c. They are transported from the Golgi apparatus to the ER.
   d. They contain a signal sequence that directs them into the ER.

4. To find the nuclear localization signal in the protein nucleoplasmin, researchers separated the molecule's core and tail segments, labeled both with a radioactive atom, and injected them into the cyto-plasm. Why did the researchers conclude that the signal is in the tail region of the protein?
   a. The protein reassembled and folded into its normal shape spontaneously.
   b. Only the tail segments appeared in the nucleus.
   c. With a confocal microscope, tail segments were clearly visible in the nucleus.
   d. The tail and head segments both appeared in the nucleus.

5. Molecular zip codes direct molecules to particular destinations in the cell. How are these signals read?
   a. They bind to receptor proteins.
   b. They enter transport vesicles.
   c. They bind to motor proteins.
   d. They are glycosylated by enzymes in the Golgi apparatus.

6. The number and size of organelles in a cell correlates with that cell's function. Propose a function for cells that contain extensive rough ER.
   a. rapid cell division in growing bones or muscle tissues
   b. production and processing of fatty acids and other lipids
   c. movement via cell crawling
   d. acquiring ions and other nutrients via specialized membrane proteins

Test Your Knowledge answers: 1. a; 2. c; 3. c; 4. b; 5. a; 6. d

Test Your Understanding

Answers are available at www.masteringbio.com

1. Compare and contrast the structure of a generalized plant cell, animal cell, and prokaryotic cell. Which features are common to all cells? Which are specific to just prokaryotes, or just plants, or just animals?

2. Make a concept map that traces the movement of a secreted protein from its site of synthesis to the outside of a eukaryotic cell (see **BioSkills 6**). Identify all of the organelles that the protein passes through. Your linking verbs should indicate what is happening to the protein at each step.

3. Describe how a motor protein such as kinesin can move a transport vesicle down a microtubule track. Include all necessary steps and components.

4. Describe the logic of a pulse-chase experiment. How was this approach used to document the pattern of protein transport through the endomembrane system?

5. Briefly describe how researchers use centrifugation to isolate particular cell components for further study.

6. Structurally, what is the difference between microtubules involved in structural support, vesicle movement, chromosome movement during cell division, and flagella?

Applying Concepts to New Situations

Answers are available at www.masteringbio.com

1. In addition to delivering cellular products to specific organelles, eukaryotic cells can take up material from the outside and route it to specific organelles. For example, specialized cells of the human immune system ingest bacteria and viruses and then deliver them to lysosomes for degradation. Suggest a hypothesis for how this material is tagged and directed to lysosomes. How would you test this hypothesis?

2. The leading hypothesis to explain the origin of the nuclear envelope is that a deep infolding of the plasma membrane occurred in an ancient prokaryote. Draw a diagram that illustrates this infolding hypothesis. Does your model explain the existence of the structure's inner and outer membranes? Explain.

3. Propose a function for cells that contain (a) a large number of lysosomes, (b) a particularly extensive cell wall, and (c) many peroxisomes.

4. Suggest a hypothesis or a series of hypotheses to explain why bacteria, archaea, algae, and plants have cell walls. Suppose that mutant individuals from each group lacked a cell wall. How could you use these individuals to test your idea(s)?

www.masteringbio.com is also your resource for • Answers to text, table, and figure caption questions and exercises • Answers to *Check Your Understanding* boxes • Online study guides and quizzes • Additional study tools including the *E-Book for Biological Science* 3rd ed., textbook art, animations, and videos.

8

Cell-Cell Interactions

KEY CONCEPTS

- Extracellular material strengthens cells and helps bind them together.

- Cell-cell connections help adjacent cells adhere. Cell-cell gaps allow adjacent cells to communicate.

- Intercellular signals are responsible for creating an integrated whole from many thousands of independent parts.

Transmission electron micrograph highlighting the cell walls (colored yellow) of several plant cells. The cell surface is where adjacent cells join and where signals from distant cells arrive.

Chapter 6 introduced the structure and function of the plasma membrane, which is the defining feature of the cell. Chapter 7 surveyed the organelles, molecular machines, and cytoskeletal elements that fill the space inside that membrane and explored how cargo moves from sources to destinations within the cell. Both chapters highlighted the breathtaking speed and diversity of events that take place at the cellular level.

The cell is clearly a bustling enterprise. But it would be a mistake to think that cells are self-contained—that they are worlds in and of themselves. Instead, cells constantly interact with other cells and the surrounding environment. They continuously adjust their activities in response to stimulation from other cells and to changes in environmental conditions.

To understand the life of a cell thoroughly, it is critical to analyze how the cell interacts with the world outside its membrane. How do cells obtain information about the outside world and respond to that information? In particular, how do cells interact with nearby cells?

For most unicellular species, the outside environment is made up of soil or water that is teeming with other organisms, either of the same or different species. In your mouth and gut and on your skin, hundreds of billions of bacterial cells are jostling for space and resources. Similar numbers of single-celled organisms are found in every tablespoon of good-quality soil. In addition to interacting with these individuals, each unicellular organism must contend with constantly shifting aspects of the physical environment, such

● Key Concept   Important Information   ● Practice It

as heat, light, ion concentrations, and food supplies. If unicellular organisms are unable to sense these conditions and respond appropriately, they die.

In multicellular species, the environment outside the cell is made up of other cells, both neighboring and distant. The cells that make up a redwood tree, an *Amanita* mushroom, or your body are intensely social. Although biologists often study cells in isolation, an individual tree, fungus, or person is actually an interdependent community of cells. If those cells do not communicate and cooperate, the whole will break into dysfunctional parts and die.

To introduce the ways that cells interact with each other, this chapter focuses on how cells in a multicellular organism communicate. Let's begin with the cell surface—with the molecules that separate the cell from its environment.

8.1 The Cell Surface

Chapter 6 introduced the currently accepted model for the structure of the plasma membrane. The fluid-mosaic hypothesis contends that the plasma membrane is a phospholipid bilayer with interspersed proteins. Many of these phospholipids and proteins have carbohydrate groups attached, forming glycoproteins and glycolipids (**Figure 8.1**). Data from electron micrographs and chemical treatments suggest that membrane proteins can be either integral, meaning that they are embedded in the bilayer, or peripheral, meaning that they are attached to one surface. Recall that if an integral protein spans the entire membrane, it is termed a transmembrane protein. Membrane proteins may float independently in the lipid bilayer or be anchored by connections to the cytoskeleton or materials outside the cell.

Chapter 6 also analyzed the primary function of the plasma membrane: to create an environment inside the cell that is different from conditions outside. Ions and molecules move across plasma membranes by direct diffusion through the phospholipid bilayer or via several types of membrane proteins. Transport of materials across the membrane can be energy demanding or passive, but it is always selective.

This picture—of a dynamic, complex plasma membrane that selectively admits or blocks passage of specific substances—is accurate but not complete. The plasma membrane does not exist in isolation. Cytoskeletal elements introduced in Chapter 7 attach to the interior face of the bilayer, and a complex array of extracellular structures exists outside. Let's consider the nature of the material outside the cell first and then analyze how the cell interacts with other cells.

The Structure and Function of an Extracellular Layer

In species from across the tree of life, it is extremely rare for cells to be bounded simply by a plasma membrane. Most cells possess a layer or wall that forms just beyond the membrane. The extracellular material helps define the cell's shape and either attaches it to another cell or acts as a first line of defense against the outside world. This observation holds for both unicellular organisms and the various types of cells in multicellular species.

Virtually all types of extracellular structures, in turn—from the cell walls of bacteria, algae, fungi, and plants to the extracellular matrix that surrounds most animal cells—follow the same fundamental design principle. Like reinforced concrete and fiberglass, they are "fiber composites": They consist of a cross-linked network of long filaments embedded in a stiff surrounding material, or ground substance (**Figure 8.2**). The molecules that make up the rods and the encasing material vary from group to group, but the engineering principle is the same.

Extracellular compartment

Carbohydrate groups of glycoproteins

Carbohydrate groups of glycolipids

Cholesterol

Integral protein

Phospholipids

Cytoskeletal elements

Peripheral proteins

Cytoplasm

FIGURE 8.1 The Plasma Membrane Is a Mosaic of Phospholipids and Proteins. According to the fluid-mosaic model, the plasma membrane consists of integral and peripheral proteins scattered throughout a phospholipid bilayer.

Steel rods (the "fibers") resist tension

Concrete (the "ground substance") resists compression

FIGURE 8.2 Fiber Composites Resist Tension and Compression. Fiber composites such as reinforced concrete consist of a massive ground substance that fills spaces between cross-linked rods.

Fiber composites are a successful design because rods and filaments are extremely effective at withstanding stretching and straining forces, or tension. The steel rods in reinforced concrete and the cellulose fibers in a plant cell wall are unlikely to break as a result of being pulled or pushed lengthwise. Scaled for size, in fact, steel and cellulose are *equally* unlikely to break. Cellulose fibers have the same tensile strength—or resistance to tearing—as steel. In addition, the stiff surrounding substance is effective at withstanding the pressing forces called compression. Concrete performs this function in highways, and a semisolid, gelatinous mixture of polysaccharides achieves the same end in plant cell walls.

Thanks to the combination of tension- and compression-resisting elements, fiber composites are particularly rugged. And in many living cells, the fiber and composite elements are flexible as well as sturdy. When this is the case, the extracellular material is both supple and strong.

What molecules make up the rods and the ground substance found on the surface of plant and animal cells? How are these extracellular layers synthesized, and what do they do?

The Plant Cell Wall

Most plant cells are surrounded by a layer of extracellular material called a cell wall. When new cells first form, they secrete an initial fiber composite designated a **primary cell wall**. As Chapter 5 indicated, the fibrous component of the cell wall consists of long strands of the polysaccharide cellulose, which is cross-linked by polysaccharide filaments and bundled into stout, cable-like structures termed microfibrils. Microfibrils form a crisscrossed network that becomes filled with gelatinous polysaccharides (**Figure 8.3**). Chief among these gelatin-forming carbohydrates are **pectins**—the molecules that are used to thicken jams and jellies. Pectins are not as strong as cellulose, because the polysaccharides that make up their structure do not form tightly packed filaments strengthened with cross-linkages. Instead, the polysaccharides in pectin are hydrophilic. They attract and hold large amounts of water to help keep the cell wall moist.

Pectins and other gelatinous polysaccharides that form the ground substance of the cell wall are synthesized in the rough endoplasmic reticulum (ER) and Golgi apparatus and secreted to the extracellular space. The cellulose microfibrils, in contrast, are synthesized by a complex of enzymes in the plasma membrane itself.

The primary cell wall defines the shape of a plant cell. Under normal conditions, the nucleus and cytoplasm fill the entire volume of the cell and push the plasma membrane up against the wall. Because the concentration of solutes is higher inside the cell than outside, water tends to enter the cell via osmosis. The incoming water inflates the plasma membrane, exerting a force against the wall that is known as **turgor pressure**. Although plant cells experience turgor pressure throughout their lives, it is particularly important in young cells that are actively growing. Young plant cells secrete enzymes named expansins into their cell-wall matrix. Expansins catalyze reactions that allow the cellulose microfibrils in the matrix to slide past one another. Turgor pressure then forces the wall to elongate and expand. The result is cell growth.

As certain plant cells mature and stop growing, they secrete a layer of material—a **secondary cell wall**—inside the primary cell wall. The structure of the secondary cell wall varies from cell to cell in the plant and correlates with that cell's function. For example, cells on the surface of a leaf have cell walls that are impregnated with waxes that form a waterproof coating. Cells that furnish structural support for the plant's stem have secondary

Side view   Top view

Primary cell wall
Plasma membrane

Cellulose microfibrils
Cross-links
Pectin

50 nm

1 μm

FIGURE 8.3 Primary Cell Walls of Plants Are Fiber Composites. In the primary plant cell wall, cellulose microfibrils are cross-linked by polysaccharide chains. The spaces between the microfibrils are filled with pectin molecules, which form a gelatinous solid.

cell walls that contain a great deal of cellulose. In some species, the secondary cell walls of support cells include a complex and extremely tough substance named **lignin**. Lignin forms a rigid network that is exceptionally stiff and strong. Cells that form wood secrete a secondary cell wall that contains large amounts of lignin. Cells that have thick cell walls of cellulose and lignin help plants withstand the forces of gravity and wind.

Plant cell walls are also dynamic. If they are damaged by insect attack, they may release signaling molecules that trigger the reinforcement of walls in nearby cells. When a seed germinates (sprouts), the walls of cells that store oils or starch are actively broken down, releasing these nutrients to the growing plant. Cell walls are also degraded in a controlled way as fruits ripen, making the fruits softer and more digestible for the animals that disperse the seeds inside.

Although animal cells do not make a cell wall, they do form a fiber composite outside their plasma membrane. What is this substance, and what does it do?

The Extracellular Matrix in Animals

Virtually all animal cells secrete a fiber composite called the **extracellular matrix (ECM)**. The design of this substance follows the same principles observed in the cell walls of bacteria, archaea, algae, fungi, and plants. The fibrous component of the animal ECM consists of protein fibers instead of polysaccharide filaments, however. **Figure 8.4a** illustrates the most abundant protein found in the ECM—a fibrous, cable-like molecule termed **collagen**. As in cell walls, the matrix that surrounds collagen and the other fibrous components consists of gel-forming polysaccharides—most of which are attached to a protein core. But because collagen and the other common ECM proteins are much more elastic and bendable than cellulose or lignin, the structure as a whole is relatively pliable. Most ECM components are synthesized in the rough ER, processed in the Golgi apparatus, and secreted from the cell via exocytosis. Some of the protein-bound polysaccharides that form the composite material are synthesized by membrane proteins, however.

The amount of ECM varies among different types of cells in the same organism. In bone and cartilage, for example, the number of cells is relatively small and the amount of ECM very large (**Figure 8.4b**). Skin cells, in contrast, are packed together with a minimal amount of ECM between them. The composition of the ECM also varies among cell types. For example, the ECM of lung cells contains large amounts of a rubber-like protein called elastin, which allows the ECM to expand and contract during breathing movements. The structure of a cell's ECM correlates with that cell's function.

⬤ Wherever it is found in the body, one of the ECM's most important functions is structural support—a common theme in the extracellular materials found in other organisms. ECM provides support because the combination of protein fil-

(a) Collagen molecules are made of three chains that wind around each other.

3 chains

1.5 nm

Collagen molecule

(b) Collagen fibrils in the extracellular matrix

Cell in connective tissue

1 μm

Collagen fibrils running lengthwise

Collagen fibrils in cross section

Each collagen fibril is composed of many collagen molecules

FIGURE 8.4 The Extracellular Matrix Is a Fiber Composite. (a) Although several types of fibrous proteins are found in the ECM of animal cells, the most abundant is collagen. Groups of collagen molecules coalesce to form collagen microfibrils, and bundles of microfibrils link up to form collagen fibers. **(b)** A cross section of ECM from monkey cartilage tissue, showing a cell surrounded by abundant ECM. The spaces between the collagen fibers are filled with gelatinous polysaccharides.

⬤ **QUESTION** In part (b), collagen fibrils appear lengthwise and in cross section. How does this observation relate to the diagram in part (a) of Figure 8.3?

aments and surrounding polysaccharide gel creates a strong external layer, and because transmembrane proteins "grab" proteins in the ECM. As **Figure 8.5** shows, actin filaments in the cytoskeleton are connected to transmembrane proteins called **integrins**. The integrins bind to nearby proteins in the ECM named **fibronectins**, which in turn bind to collagen fibers. This direct linkage between the cytoskeleton and ECM is critical. In addition to keeping individual cells in place, it helps adjacent cells adhere to each other via their common connection to the ECM.

The cytoskeleton-ECM linkage is also important in the development of cancer. When the cell-ECM connection is lost—through mechanisms that are not well understood—cells that are growing in an uncontrolled fashion can leave the site

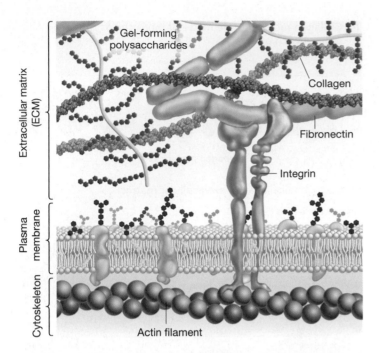

FIGURE 8.5 The Extracellular Matrix Connects to the Cytoskeleton.

🔵 **EXERCISE** Circle the direct physical connection among actin filaments, integrins, fibronectin, and collagen fibers.

of a tumor and begin migrating throughout the body. Eventually these cells settle in a new region of the body and seed the formation of new tumors. This increases the severity of disease, because cancerous cells begin growing at many sites instead of just one. **Box 8.1** details other types of maladies that result from breakdown of the ECM.

Check Your Understanding

If you understand that...

- Most cells secrete a layer of structural material that stiffens the cell and helps define its shape. The extracellular material is usually a fiber composite—a combination of cross-linked filaments surrounded by a ground substance.

You should be able to...

1) Diagram the general structure of a fiber composite.
2) Compare and contrast the molecular composition of a plant cell wall and the ECM of animal cells.

8.2 How Do Adjacent Cells Connect and Communicate?

To understand a cell fully, it is important to appreciate it as a social entity. Cells continually interact with other cells. This is true even for unicellular organisms, which usually live in habitats packed with other unicellular species. For example, the 500 species of bacteria and archaea that inhabit your

FIGURE 8.6 Groups of Unicellular Organisms May Secrete Biofilms. The yellow band consists of *Pseudomonas aeruginosa* cells—a common soil bacterium that also causes infections in cystic fibrosis patients. The red layer is a biofilm secreted by the cells that helps them stick to the substrate.

mouth compete with each other for space and nutrients. Like many other prokaryotes that are adapted for life on a solid surface, certain oral bacteria secrete a hard, polysaccharide-rich substance that accumulates outside their cell walls. This substance creates a biofilm that encases the cells and attaches them to surfaces. Biofilms that form in the mouth are termed dental plaque, and they contain an array of species. In some cases, biofilm formation is actually a product of cooperation among members of the same species. When cells of the bacterium *Pseudomonas aeruginosa* settle on a lung cell, for instance, they secrete a signaling molecule that recruits other *P. aeruginosa* cells to the site. The cells then collectively secrete a biofilm that attaches them to the surface of the lung and protects them from attack by your immune system cells or antibiotics (**Figure 8.6**).

Although unicellular species may live in close proximity and even communicate with each other, they do not make physical connections regularly. Physical connections between cells are the basis of **multicellularity** (**Figure 8.7**). The billions of cells that make up a rose bush, a mushroom-forming fungus, or a mountain gorilla must be held together somehow. And because cells in multicellular species are specialized for particular tasks, the cells must be able to communicate with each other to function as an integrated whole. It is logical, then, that rose and fungus and gorilla cells are organized into functional units, called tissues. **Tissues** consist of groups of similar cells that perform a similar function. In your body, individual muscle cells are grouped into muscle tissue that contracts and relaxes to make movement possible. Several tissues, in turn, combine to make up the integrated structures called **organs**. Organs perform specialized functions such as reproduction, digestion, and support. Flowers, roots, and stems are organs; your gonads, small intestine, and bones also are organs.

A look at the phylogeny in **Figure 8.8** should convince you that multicellularity has evolved numerous times, in an array of independent groups. (For help interpreting this tree, see

Plasma membranes of adjacent cells are connected

This cell contains vesicles filled with mucus

5 μm

FIGURE 8.7 In Multicellular Organisms, Cells Are Connected.
A cross section through cells that line the small intestine. Notice that cells are physically connected and that even adjacent cells can have different structures and functions. The cell in the middle secretes mucus that neutralizes stomach acid. Secretions from the cells to the left and right help in the digestive process.

BioSkills 2.) Based on this observation, it is logical to predict that brown algae, land plants, animals, and other multicellular groups do not share the same types of intercellular (literally, "between-cell") connections and communication mechanisms. Instead of inheriting the same structures and processes for cell-cell interactions from a common ancestor that was multicellular, each multicellular group independently evolved ways of organizing their tissues and organs.

Many of the structures responsible for cell-cell adhesion and communication in these groups were first described in the 1950s, when transmission electron microscopy made it possible to examine tissues at high magnification. In the decades that followed, researchers made exciting progress in understanding the molecular nature of cell-cell attachments and openings. Let's look first at the structures that physically attach adjacent cells to each other, then at the openings that allow nearby cells to exchange materials and information.

Cell-Cell Attachments

The structures that hold cells together vary among multicellular organisms. To illustrate this diversity, consider the intercellular connections observed in the best-studied groups of organisms: plants and animals. Transmission electron microscopy showed that a distinctive layer of material joins plant cells together. (For an introduction to electron microscopy, see **BioSkills 8.**) The same tool allowed scientists to examine the structure of cell-cell attachment points in animal cells.

FIGURE 8.8 Multicellularity Has Evolved Numerous Times. A phylogeny of eukaryotes, indicating lineages where multicellular forms exist. Based on these data, biologists conclude that multicellularity evolved many times independently.

● **EXERCISE** Suppose that multicellularity arose just once. If this hypothesis is correct, put a mark on the tree to indicate where multicellularity evolved.

● **QUESTION** Why is it more logical to infer that multicellularity arose many times, instead of just once?

The extracellular space between adjacent plant cells comprises three layers (**Figure 8.9**). The primary cell walls of adjacent cells sandwich a central layer designated the middle lamella, which consists primarily of gelatinous pectins. Because this gel layer is continuous with the primary cell walls of the adjacent cells, it serves to glue them together. The two cell walls are like slices of bread; the middle lamella is like a layer of peanut butter. If enzymes degrade the middle lamella, as they do when flower petals and leaves detach and fall, the surrounding cells separate.

In many animal tissues, integrins connect the cytoskeleton of each cell to the extracellular matrix (see Section 8.1). A middle-lamella-like layer of gelatinous polysaccharides runs between adjacent animal cells, so cytoskeleton-ECM connections help hold individual animal cells together. In addition, in certain animal tissues the polysaccharide glue is reinforced by cable-like proteins that span the ECM to connect adjacent cells. These structures are particularly important in **epithelia** (singular: **epithelium**)—tissues that form external and internal surfaces. Epithelial cells form layers that separate organs and other structures. These epithelial layers must be sealed to prevent mixing of solutions from adjacent organs or structures. A variety of cell-cell attachment structures exist in epithelia and other tissues. Here, however, we'll analyze just two: tight junctions and desmosomes.

Tight Junctions   The electron micrograph in **Figure 8.10a** shows a cross section through epithelial cells that are linked by tight junctions. A **tight junction** is a cell-cell attachment composed of specialized proteins in the plasma membranes of adjacent animal cells. As the drawing in **Figure 8.10b** indicates, these proteins line up and bind to one another. The resulting structure resembles quilting, with the proteins acting as stitches.

Because tight junctions form a watertight seal, this type of junction is common in cells that form a barrier, such as the cells lining your stomach and intestines. There, they prevent the ions and molecules in your gut contents from leaking between stomach cells or intestinal cells and from diffusing into your body. Instead, only selected nutrients enter the cells. These ions and molecules are admitted via specialized transport proteins and channels in the plasma membrane.

Although tight junctions are tight, they are also variable and dynamic. For example, the tight junctions in the cells lining your bladder are much tighter than tight junctions in the cells lining your small intestine, because they consist of different proteins. As a result, small ions can pass through the surface of the small intestine much more easily than they can

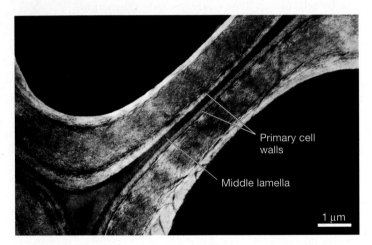

FIGURE 8.9 The Middle Lamella Connects Adjacent Plant Cells. The middle lamella contains gelatinous polysaccharides called pectins.

BOX 8.1   What Happens When the Extracellular Matrix Is Defective?

Defects in ECM have serious consequences. Consider, for example, the effects of collagen breakdown. One of collagen's most common constituents is the amino acid proline. When collagen fibers first assemble in the ECM, many of the proline residues participate in a reaction that forms hydroxyproline. This reaction depends on ascorbic acid, also called vitamin C. If vitamin C is lacking in the diet, hydroxyproline formation slows and collagen fibers begin to weaken or disintegrate. The ECM surrounding capillaries—the smallest blood vessels in the body—is particularly sensitive to this problem. When collagen fibers in the ECM of capillary cells weaken, the structure as a whole begins to lose its integrity. Blood cells begin leaking out, forming bruises. If vitamin C deficiency continues, capillaries weaken to the point where extensive internal bleeding occurs. The potentially fatal disease called scurvy results. Scurvy is readily cured by eating citrus fruits or other foods rich in vitamin C.

Skin is also sensitive to collagen damage. Skin cells are layered over a thick, collagen-rich ECM called the **basal lamina**. In response to intense sunlight, skin cells produce and secrete the enzyme metalloproteinase-1, which causes collagen fibers to break down. As a result, people who work for long periods in bright sunlight or who cultivate a tanned skin are almost certain to experience photoaging—changes in skin caused by the loss of collagen.

Defects in other ECM components can also cause problems. Marfan syndrome develops when a genetic defect prevents people from manufacturing the normal form of the protein fibrillin. Along with elastin, fibrillin is a key component of the elastic fibers in the ECM. People with Marfan syndrome tend to have exceptionally long fingers and be extremely tall and thin. They are also susceptible to severe heart problems. Biomedical researchers have hypothesized that Abraham Lincoln may have had Marfan syndrome.

(a) Electron micrograph of a tight junction

A tight junction forms a watertight seal between epithelial cells

Tight junction

0.1 μm

(b) Three-dimensional view of a tight junction

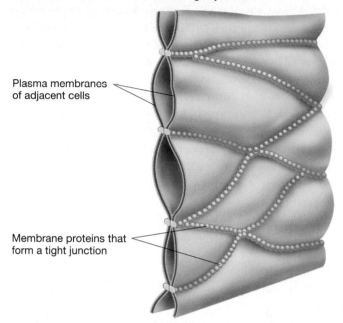

Plasma membranes of adjacent cells

Membrane proteins that form a tight junction

FIGURE 8.10 In Animals, Tight Junctions May Form a Seal between Adjacent Cells. (a) A cross section through adjacent cells that are attached by a tight junction. **(b)** Proteins in the membranes of each cell are arranged in a quilted pattern and bind to one another to form the junction.

○ **EXERCISE** Add lines connecting the structures in part (a) to the labels in part (b).

pass through the surface of the bladder—helping you absorb ions in your food and eliminate them in your waste. The structures are dynamic because they can be loosened in some cases to permit more transport between epithelial cells—for example, in the small intestine after a meal—and then "retightened" later.

Desmosomes   Figure 8.11a illustrates another type of cell-cell connection found in animals: desmosomes. **Desmosomes** are particularly common cell-cell attachments in epithelial cells and in certain types of muscle cells. The structure and function of a desmosome are analogous to the rivets that hold pieces of sheet metal together. As **Figure 8.11b** indicates, though, desmosomes

(a) Micrograph of desmosome

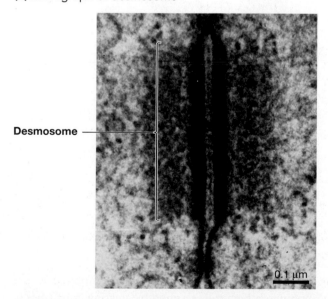

Desmosome

0.1 μm

(b) Three-dimensional view of desmosome

Plasma membranes of adjacent cells

Anchoring proteins inside cell

Membrane proteins that link cells

Intermediate filaments

FIGURE 8.11   Adjacent Animal Cells Are Linked by Desmosomes, Which Bind Cytoskeletons Together.
(a) A cross section through a desmosome, showing the connections between adjacent cells and the intermediate filaments that link the desmosome to the cytoskeleton of each cell. **(b)** The major components of a desmosome.

○ **EXERCISE** Add lines connecting the structures in part (a) to the labels in part (b).

○ **EXERCISE** Label the desmosome in Figure 8.10a, left micrograph.

are extremely sophisticated connections. At their heart are proteins that form a physical link between the cytoskeletons of the adjacent cells. In addition to binding to each other, these proteins bind to larger proteins that anchor intermediate filaments in the cytoskeletons of the two cells. In this way, desmosomes bind together the cytoskeletons of the adjacent cells.

The proteins that form desmosomes may be involved in attaching animal cells long before a desmosome is complete, however. Early in the development of an animal, an embryo's first tissues form when selected cells begin to adhere to each other. Mature desmosomes and adult tissues and organs form much later. Which proteins are involved in these most basic types of cell-cell adhesion, early in development?

Selective Adhesion   Long before electron micrographs revealed the presence of desmosomes, biologists realized that some sort of molecule must bind animal cells to each other. This insight grew out of experiments conducted on the cells of sponges in the early 1900s. Sponges are aquatic animals, and the sponge species used in this study consists of just two basic types of cells. When a biologist treated adult sponges with chemicals that made the cells dissociate from each other and separate, the result was a jumbled mass of individual and unconnected cells. But when normal chemical conditions were restored, the cells gradually began to move and stick to other cells. With time, cells of each type began to combine and aggregate, because they adhered to cells of the same tissue type. This phenomenon came to be called **selective adhesion**. In this experiment, the cells eventually re-formed complete and functional adult sponges.

In an even more dramatic experiment, the researcher dissociated the cells of adult sponges from two differently pigmented sponge species and randomly mixed them together in a culture dish. As **Figure 8.12** shows, the cells eventually sorted themselves out into distinct aggregates containing cells from only one species and from only one cell type. These results implied that the cells had some way of physically linking to each other and that the linkage was specific to the species and cell type involved. Based on these observations, researchers hypothesized that there must be some sort of molecule on the surface of animal cells that attaches only to cells of the same species and cell or tissue type.

The Discovery of Cadherins   What is the molecular nature of selective adhesion? The initial hypothesis, proposed in the 1970s, was that certain membrane proteins might be involved. The idea was that if different types of cells produce different types of adhesion proteins in their membranes, the molecules might interact in a way that anchors cells of the same type to each other.

This hypothesis was tested through experiments that relied on molecules called antibodies. An **antibody** is a protein that binds specifically to a section of another protein. Because they stick to certain proteins so specifically, antibodies can be used to block certain portions of proteins or mark proteins so they can be seen. To test the hypothesis that cell-cell adhesion takes

Experiment

Question: Do animal cells adhere selectively?

Hypothesis: Cells of the same type and from the same species have a mechanism for selectively adhering to each other.

Null hypothesis: Cells do not adhere or they adhere to each other randomly, not selectively.

Experimental setup:

1. Start with two adult sponges of different species.

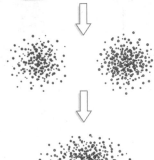

2. Use a chemical treatment to dissociate the cells.

3. Mix cells from the two species.

Prediction: Cells of the same cell type will adhere to one another, and cells from the same species will adhere to one another.

Prediction of null hypothesis:

Results:

Cells spontaneously reaggregate by cell type and species.

Conclusion: Cells of the same cell type and species have specific adhesion molecules or mechanisms.

FIGURE 8.12 Do Animal Cells Adhere Selectively?

● **EXERCISE** In the space provided, fill in the prediction made by the null hypothesis.

place via interactions between membrane proteins, researchers pursued the following strategy (**Figure 8.13**):

1. Isolate the membrane proteins from a certain cell type. Create pure preparations of each protein.

2. Inject one of the membrane proteins into a rabbit. The rabbit's immune system cells respond by creating antibodies to the membrane protein. Purify antibodies that bind tightly to the injected membrane protein. Repeat this procedure for the other membrane proteins that were isolated. In this way,

Experiment

Question: Do animal cells have adhesion proteins on their surfaces?

Hypothesis: Selective adhesion is due to specific membrane proteins.

Null hypothesis: Selective adhesion is not due to specific membrane proteins.

Experimental setup:

Isolated membrane proteins

1. Isolate the membrane proteins from a certain cell type.

2. Create antibodies that bind to specific membrane proteins. Purify the antibodies.

Experiment 1: Experiment 2: Experiment 3:

3. Treat cells with an antibody, one type at a time. Wait; then observe whether cells adhere normally.

Prediction:

Prediction of null hypothesis:

Results:

Experiment 1:
Cells adhere normally.

Experiment 2:
Cells do not adhere.

Experiment 3:
Cells adhere normally.

Conclusion: The protein that was blocked in Experiment 2 (called a cadherin) is involved in cell-cell adhesion.

FIGURE 8.13 Do Animal Cells Have Adhesion Proteins on Their Surfaces?

● **EXERCISE** Fill in the prediction made by each hypothesis.

obtain a large collection of antibodies—each of which binds specifically to one (and only one) type of membrane protein.

3. Add one antibody type to the mixture of dissociated cells. Observe whether the cells are able to reaggregate normally. Repeat this experiment with the other antibody types, one type at a time. (Three different antibodies are shown in Figure 8.13, so three experimental treatments are required.)

4. If treatment with a particular antibody prevents the cells from attaching to each other, the antibody is probably bound to an adhesion protein. The hypothesis is that the interaction between antibody and adhesion protein blocks the adhesion protein and keeps it from functioning properly. In effect, the antibody "shakes hands" with the adhesion protein, which prevents the adhesion protein from "shaking hands" with other adhesion proteins to attach cells.

Based on studies such as this, researchers have identified several major classes of cell adhesion proteins. The molecules in desmosomes belong to a group that came to be named **cadherins**. Further research has shown that each major type of cell in the body has a different type of cadherin in its plasma membrane, and each type of cadherin can bind only to cadherins of the same type. In this way, cells of the same tissue type attach specifically to one another.

To summarize: Animal cells attach to each other in a selective manner because different types of cell adhesion proteins can bind and rivet certain cells together. Cadherins provide the physical basis for selective adhesion in many cells and are

(a) Plasmodesmata create gaps that connect plant cells.

Tubule of endoplasmic reticulum passing through **plasmodesmata**

Cell walls

0.1 µm

Cell wall of cell 1 | Cell wall of cell 2

Membrane of cell 1

Membrane of cell 2

Smooth endoplasmic reticulum

(b) Gap junctions create gaps that connect animal cells.

Gap junctions

20 µm

Membrane proteins from adjacent cells line up to form a channel

FIGURE 8.14 Adjacent Plant Cells and Adjacent Animal Cells Communicate Directly. (a) Plasmodesmata are holes in plasma membranes and cell walls of adjacent plant cells. A tubule of smooth ER runs through the hole. **(b)** Gap junctions are collections of protein-lined holes in the membranes of adjacent animal cells.

a critical component of the desmosomes that join mature cells. ● If you understand cell-cell attachments, you should be able to predict (1) whether molecules can pass between adjacent cells through middle lamellae and tight junctions, and (2) what would happen if you treated cells in a developing frog embryo with a molecule that blocked a cadherin observed in muscle tissue.

Cell-Cell Gaps

● The presence of a middle lamella, continuous ECM, tight junctions, desmosomes, and cadherins bind adjacent cells to each other. How do they communicate? In both plants and animals, direct connections between cells in the same tissue help the cells work in a coordinated fashion. The first clue emerged from careful analyses of electron micrographs.

In plants, gaps in cell walls create direct connections between the cytoplasm of adjacent cells. At these connections, named **plasmodesmata** (singular: **plasmodesma**), the plasma membrane and the cytoplasm of the two cells are continuous. Smooth endoplasmic reticulum (ER) runs through these holes (**Figure 8.14a**). Growing evidence suggests that plasmodesmata also contain proteins that regulate the passage of specific proteins, making the connections similar in function to the nuclear pore complex introduced in Chapter 7. At least some of the proteins that are transported through plasmodesmata are involved in coordinating the activity of adjacent cells. Plasmodesmata are communication portals.

In most animal tissues, adjacent cells are connected by structures called **gap junctions**. The key feature of gap junctions is the specialized proteins that create channels between cells (**Figure 8.14b**). These channels allow water, ions, and small molecules such as amino acids, sugars, and nucleotides to move between adjacent cells. The flow of small molecules can help adjacent cells communicate and coordinate their activities by allowing the rapid passage of regulatory ions or molecules. In the muscle cells of your heart, for example, a flow of ions through gap junctions acts as a signal that coordinates contractions. Without this cell-cell communication, a normal heartbeat would be impossible.

In effect, adjacent plant cells and adjacent animal cells share most or all of the ions and small molecules in their cytoplasm but retain their own organelles, proteins, and nucleic acids. The existence of cell-cell gaps makes it possible for cells that are immediate neighbors to communicate with each other efficiently. A tissue can act as an integrated whole if its component cells are connected by gaps in their extracellular material and in their plasma membranes (**Figure 8.15**).

How do more distant cells in a multicellular organism communicate? For example, suppose that leaf cells in a maple tree are being attacked by caterpillars or that the muscle cells in your arm are exercising so hard that they begin to run low

FIGURE 8.15 An Array of Structures Is Involved in Cell-Cell Adhesion and Communication.

● **EXERCISE** In the blanks on the right, write the name of each type of cell-cell attachment.

___ seal cells together.

___ connect the cytoskeletons of cells.

___ act as channels between cells.

Space between cells

on sugar. How do these cells signal tissues or organs elsewhere in the body to release materials that might be needed to fend off caterpillars or exhaustion? Distant cell communication is the subject of Section 8.3.

Check Your Understanding

If you understand that...

- In plants and animals, adjacent cells are physically connected and communicate with each other through openings in their plasma membranes.

You should be able to...

1) Compare and contrast the structure and function of the middle lamella of plants and the tight junctions and desmosomes of animals.

2) Describe the structure and function of plasmodesmata and gap junctions.

3) Explain the molecular mechanism of selective adhesion.

8.3 How Do Distant Cells Communicate?

Humans, sequoia trees, and other large multicellular organisms typically contain trillions of cells, hundreds of different organs, and dozens of different tissue types. For all of these parts to work together as a unified whole, information has to be exchanged. As an example, suppose that cells in your brain sense that you are becoming dehydrated. Brain cells can't do much about the water you lose during urination, but kidney cells can. In response to dehydration, certain brain cells release a signaling molecule that travels to the kidneys. The arrival of the signal changes the activity of cells that control the amount of water being excreted in urine. The response to the signal leads to less water being lost and more water being retained—an important aspect of fighting dehydration.

Thanks to long-distance signals, the activities of cells, tissues, and organs in different parts of a multicellular body are coordinated. Even as you read this, in fact, molecules that carry information about conditions throughout your body are traveling through your bloodstream and arriving at cells from your head to your toes.

How do cells receive and respond to signals that arrive from distant cells? Let's take a step-by-step approach to answering this question, starting with a quick look at the nature of long-distance chemical signals.

Hormones Are Long-Distance Messengers

Intercellular (or "cell-cell") signals are chemical messengers. Several distinctive types of cell-cell signals exist, but here we'll focus on hormones—the molecules involved in long-distance signaling. A **hormone** is an information-carrying molecule that is secreted from a cell, circulates in the body, and acts on target cells far from the original cell that sent the signal. Hormones are usually small molecules and are typically present in minute concentrations. Even so, they have a large impact on the activity of target cells and the condition of the body as a whole. Hormones are like a fleeting scent or whispered phrase from someone you are attracted to—a tiny signal, but one that makes your cheeks flush and your heart pound.

As **Table 8.1** indicates, hormones have a wide array of effects. In humans they are responsible for regulating most aspects of body chemistry and triggering sexual maturation and sexual activity. In plants they direct most aspects of growth and development and mobilize defenses against attacks by insects and other threats.

Table 8.1 should also convince you that hormones vary widely in chemical structure. The important point about a hormone's structure, though, is not whether it is a gas or peptide or glycoprotein, but whether it is lipid soluble. Steroids and other types of lipid-soluble hormones diffuse across the plasma membrane and enter the cytoplasm of their target cells. Hydrophilic hormones, in contrast, do not cross the plasma membrane. This is important because a hormone has to be recognized to have an effect on a target cell. Where does this recognition and reception step occur—inside the cell or outside?

TABLE 8.1 Hormones Have Diverse Structures and Functions

Hormone Name	Chemical Structure	Function of Signal
Ethylene	C_2H_4 (a gas)	Stimulates fruit ripening, regulates aging
Insulin	Protein, 51 amino acids	Stimulates glucose uptake in animal bloodstream
Systemin	Peptide, 18 amino acids	Stimulates plant defenses against herbivores
Estrogen	Steroid	Stimulates development of female characteristics in animals
Brassinosteroids	Steroid	Stimulates plant cell elongation
Prostaglandins	Modified fatty acid	Perform a variety of functions in animal cells
Thyroxine (T4)	Modified amino acid	Regulates metabolism in animals
FSH	Glycoprotein	Stimulates egg maturation, sperm production in animals
Auxin	Small organic compound	Signals changes in long axis of plant body

Signal Reception

Hormones and other types of cell-cell signals deliver their message by binding to receptor molecules. When the molecule that carries the message: "We're getting dehydrated—conserve water" reaches your kidneys, only certain cells respond because only certain cells have the appropriate receptor. The presence of an appropriate receptor dictates which cells will be able to respond to a particular hormone. Bone and muscle cells don't respond to the "conserve water" message, because they don't have a receptor for it.

The association between signal and receptor brings up another important point about long-distance signaling between cells. In many cases, cells in a wide array of tissues and organs respond to the same signal. This is possible because they all have the appropriate receptor. For example, if you are startled by a loud noise, cells in your adrenal glands secrete a hormone that carries the message: "Get ready to fight or run." In response, your heart rate increases, your breathing rate increases, and cells in your liver release sugars that your muscles can use to power rapid movement. This is the basis of an "adrenalin rush." Heart, lung, and liver cells respond to adrenalin because they have an adrenalin receptor. ● The general message here is that long-distance signaling coordinates the activities of cells throughout a multicellular organism.

If receptors are the key to a cell's response, how do they work? **Signal receptors** are proteins that change conformation—meaning, overall shape—or activity after binding to a signaling molecule. The signal receptors that respond to steroid hormones are located inside the cell, because steroid hormones are lipid soluble and thus readily diffuse through the plasma membrane. But most signal receptors are located in the plasma membrane because most hormones are not lipid soluble. No matter where they are located, it's important to recognize two additional points about signal receptors:

- Receptors are dynamic. The number of receptors in a particular cell may decline if hormonal stimulation occurs at high levels over a long period of time. The ability of a receptor molecule to bind tightly to a hormone may also decline in response to intensive stimulation. As a result, the sensitivity of a cell to a particular hormone may change over time.

- Receptors can be blocked. The drugs called beta-blockers, for example, bind to certain receptors for the hormone adrenalin (adrenalin is also called epinephrine). When adrenalin binds to receptors in heart cells, the cells are stimulated to contract. So if a physician wants to reduce heart cell contraction as a way to lower heart rate and thus blood pressure, she is likely to prescribe a beta-blocker.

The most important and most general characteristic of signal receptors, though, is that their physical conformation changes when a hormone binds to them. This is a critical event in cell-cell signaling. The change in receptor structure means that the signal has been received. It's like throwing an "on" switch. What happens next?

Signal Processing

Once a cell receives a signal, something has to happen to initiate the cell's response. In some cases the "something" occurs directly. For example, steroid hormones such as testosterone and estrogen diffuse through the plasma membrane and enter the cytoplasm, where they bind to a receptor protein. The hormone-receptor complex is then transported to the nucleus, where it triggers changes in the genes being expressed in the cell (**Figure 8.16**). Early in human development, the arrival of testosterone in cells leads to the expression of genes that trigger the development of male reproductive organs. (In the absence of testosterone, female reproductive organs develop.) Later in life, a surge of testosterone triggers the changes associated with puberty in boys, while the arrival of estrogen molecules spurs the development of adult female characteristics in girls.

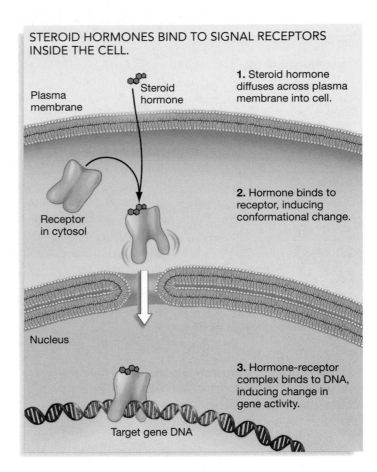

STEROID HORMONES BIND TO SIGNAL RECEPTORS INSIDE THE CELL.

Plasma membrane

Steroid hormone

1. Steroid hormone diffuses across plasma membrane into cell.

Receptor in cytosol

2. Hormone binds to receptor, inducing conformational change.

Nucleus

3. Hormone-receptor complex binds to DNA, inducing change in gene activity.

Target gene DNA

FIGURE 8.16 Some Cell-Cell Signals Enter the Cell and Bind to Receptors in the Cytoplasm. Because they are lipids, steroid hormones can diffuse across the plasma membrane and bind to signal receptors inside the cell. The hormone-receptor complex is transported to the nucleus and binds to genes, changing their activity.

The general point here is that when lipid-soluble hormones bind to receptors located inside the cell, they usually trigger a change in the cell's activity directly. The change in activity might involve certain types of genes being activated or shut down, or certain pumps in the plasma membrane being stimulated. In each case, the hormone-receptor complex directly triggers the change by binding to the genes or to the pumps.

The effects of hormones that *cannot* diffuse across the plasma membrane and enter the cytoplasm are not nearly so direct. It is not possible for hormones that bind to receptors on the plasma membrane to directly change the activity of pumps or genes or other molecules and structures inside the cell. Instead, the signal that arrives at the surface of the cell has to be changed to an intracellular signal. The knock at the door has to be answered from inside. Thus, a hormone's arrival at the cell surface triggers **signal transduction**—the conversion of the signal from one form to another. A long and often complex series of events ensues, collectively called a signal transduction pathway.

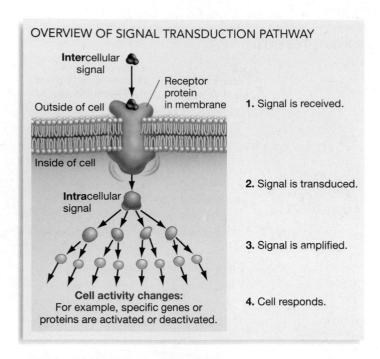

FIGURE 8.17 Signal Transduction Converts an Intercellular Signal to an Intracellular Signal. Signal transduction is a multistep process.

● **QUESTION** What do the word roots inter– and intra– mean?

Signal transduction is a common occurrence in everyday life. For example, the e-mail messages you receive are transmitted from one computer to another over cables or wireless transmissions. These types of electronic or electromagnetic signals can be transmitted efficiently over long distances. But to you, they are meaningless. Software in your computer has to transduce the signals to letters and words on the screen—into a form that you can understand and respond to. If you own a portable music player, songs that are encoded as 1s and 0s in a digital file are transduced to sound waves that you listen to. Reading the 1s and 0s directly would be meaningless to you.

Signal transduction pathways work the same way. In a cell, signal transduction converts an extracellular signal to an intracellular signal (**Figure 8.17**). As in an e-mail transmission, a signal that is easy to transmit is converted to a signal that is easily understood and that triggers a response. There is another important aspect of signal transduction, though, in addition to this change in form. When a hormone arrives at the cell surface, the message it transmits is amplified as it changes form. Recall that hormones are present in miniscule concentrations but trigger a large response from cells. Signal amplification is one reason this is possible. In everyday life, the amplifier in a portable music player performs an analogous function. Once it is amplified, a tiny sound signal can get a whole roomful of people dancing.

In cells, signal transduction occurs at the plasma membrane; amplification occurs inside. One major type of signal transduction system consists of membrane channels that open to allow a flow of ions into the cell, changing the electrical properties of the membrane. This type of signal transduction is analyzed in detail in Chapter 45. Here let's focus on the other major types of signal transduction systems. Each involves a distinctive type of membrane protein: (1) G proteins and (2) enzyme-linked receptors. G proteins work by triggering the production of an intracellular or "second" messenger. Enzyme-linked receptors work by triggering the activation of a series of proteins inside the cell, through the addition of phosphate groups. There are many types of G proteins and enzyme-linked receptors, but each of these protein families works in the same general way. Let's look at these two signal transduction systems in turn.

G Proteins   Many signal receptors span the plasma membrane and are closely associated with peripheral membrane proteins inside the cell called **G proteins**. The G is appropriate because these proteins bind guanosine triphosphate (GTP) and guanosine diphosphate (GDP).

GTP is a nucleoside triphosphate that is similar in structure to adenosine triphosphate (ATP). As Chapter 4 noted, nucleoside triphosphates have high potential energy because their three phosphate groups have four negative charges close together. The cluster of charges results in electrons repulsing each other and moving farther from nearby nuclei. And as Chapter 6 noted when discussing the sodium-potassium pump, ATP's high potential energy allows it to make things happen in cells—such as pumping ions against an electrochemical gradient. When ATP binds to a protein, or when a single phosphate group from ATP binds to a protein, the addition of the negative charges makes the protein change shape. Changes in shape produce changes in activity. GTP works the same way. When GTP binds to a protein, the protein changes its structure and function. G proteins are turned on or activated when they bind GTP; they are off or inactivated when a phosphate group drops off to form GDP. An activated G protein triggers the production of a second messenger inside the cell.

To understand how G proteins work, follow the events diagrammed in **Figure 8.18**. As step 1 in the figure shows, the sequence begins when a hormone arrives and binds to a receptor in the plasma membrane. Notice that the receptor is a transmembrane protein, and that it is coupled to a peripheral G protein on the membrane's inner surface. In response to hormone binding, the receptor changes shape. The shape change is a switch that activates its G protein. More specifically, the G protein releases the GDP molecule that was keeping it in an inactive state and binds GTP instead. Compared to GDP, GTP has an additional phosphate group with two negative charges. When GTP is attached, the G protein responds by undergoing

HOW DO G PROTEINS WORK?

Signal

Receptor   Enzyme

1. G protein binds GDP. Signal arrives and binds to receptor.

GDP

G protein in "off" conformation

2. Signal-receptor complex changes conformation. G protein binds GTP and splits into two parts.

GTP

G protein in "on" conformation

Enzyme

3. In response to binding of activated G protein, enzyme catalyzes a reaction that produces a second messenger.

GTP

Substrate

Second messenger

Triggers response

FIGURE 8.18 G Proteins Trigger the Production of a Second Messenger. If a receptor that is coupled to a G protein is activated by a signal, the G protein binds guanosine triphosphate (GTP) and splits into two parts. The change in the G protein activates a nearby enzyme and results in the production of a second messenger.

a radical shape change: It splits into two parts (see step 2). One of these parts activates a nearby enzyme that is embedded in the plasma membrane.

The enzymes that are activated by G proteins catalyze the production of small molecules called second messengers. A "first messenger" is a hormone that arrives at the cell surface;

second messengers are nonprotein signaling molecules that increase in concentration inside a cell and elicit a response to the first messenger. They are intracellular signals that spread the message carried by the extracellular signal—the hormone.

Second messengers are effective because they are small and because they diffuse rapidly to spread the signal throughout the cell. In addition, they can be produced in large quantities in a short time. This characteristic is important. Because the arrival of a single hormone molecule can stimulate the production of many second messenger molecules, the signal transduction event amplifies the original signal.

Several types of small molecules act as second messengers in cells. **Table 8.2** lists some of the best-studied second messengers and provides an example of how specific types of cells respond to each second messenger. Note that several second messengers activate **protein kinases**—enzymes that activate or inactivate other proteins by adding a phosphate group to them. If you read the table closely, you'll see evidence for two important conclusions: (1) Second messengers aren't restricted to a single role or single cell type—the same second messenger can initiate dramatically different events in different cell types; and (2) it is common for more than one second messenger to be involved in triggering a cell's response to the same extracellular signal.

To make sure that you understand how G proteins and second messengers work, imagine the following movie scene: A spy arrives at a castle gate. The guard receives a note from the spy, but he cannot read the coded message. Instead, the guard turns to the queen. She reads the note and summons a nearby teacher, telling him to send children from the school throughout the castle to warn everyone of approaching danger. ● You should be able to identify which characters in the scene correspond to the second messenger, G protein, hormone, receptor, and enzyme activated by the G protein.

It's also important to appreciate what happens when a signal transduction system like this doesn't work. For example,

TABLE **8.2 Examples of Second Messengers**

Name	Type of response
Cyclic guanosine monophosphate (cGMP)	Opens ion channels; activates certain protein kinases
Diacylglycerol (DAG)	Activates certain protein kinases
Inositol triphosphate (IP$_3$)	Opens calcium channels—mobilizes stored calcium ions
Cyclic adenosine monophosphate (cAMP)	Activates certain protein kinases
Calcium ions (Ca^{2+})	Binds to a receptor called calmodulin; Ca^{2+}/calmodulin complex then activates proteins

suppose the guard in the movie scene refuses to acknowledge the arrival of the spy. Would the inhabitants of the castle respond to the advancing army? An analogous defect can cause the condition in humans known as type II diabetes mellitus—a disease that Chapter 43 discusses in detail. In many cases, Type II diabetes develops when the cell-surface receptors for the hormone insulin do not function normally. As a result, cells do not receive the message carried by insulin and do not initiate their normal response—importing glucose from the bloodstream. Thus, glucose concentrations in the blood remain high, causing tissue damage and disease.

Enzyme-Linked Receptors

Another major class of signal transduction systems is based on what biologists call enzyme-linked receptors. Instead of activating a nearby G protein, enzyme-linked receptors transduce the signal from a hormone by directly catalyzing a reaction inside the cell. To see how this works and understand the consequences, follow the events illustrated in **Figure 8.19**. The figure focuses on what is perhaps the best-studied group of enzyme-linked receptors: the **receptor tyrosine kinases (RTKs)**.

When a hormone binds to an RTK, the protein responds by forming a dimer (step 1). In this conformation, the receptor has a binding site for a phosphate group from ATP inside the cell (see step 2). Once it is phosphorylated, an RTK becomes an active enzyme. Proteins inside the cell then form a bridge between the activated RTK and a peripheral membrane protein called **Ras,** which functions like a G protein (step 3). The formation of the RTK bridge activates Ras. Specifically, GTP binds to Ras. Once Ras binds GTP, it triggers the phosphorylation of another protein, activating it (step 4). The phosphorylated protein then catalyzes the phosphorylation of still other proteins, which phosphorylate yet another population of proteins (step 5).

This sequence of events is termed a **phosphorylation cascade.** Because each enzyme in the cascade catalyzes the phosphorylation of numerous "downstream" enzymes, the original signal is amplified many times over. One activated receptor tyrosine kinase can trigger the activation of thousands of proteins at the end of the cascade. In most cases, the proteins that take part in a phosphorylation cascade are held in close physical proximity by scaffolding proteins. The presence of this structure increases the speed and efficiency of the reaction sequence.

To make sure that you understand how RTKs and phosphorylation cascades work, imagine that you had one red domino, one black domino, and a large supply of green, blue, yellow, pink, and orange dominos. The red domino represents an RTK, and the black domino represents Ras. Each of the other colors represents a type of protein in a phosphorylation cascade. ● You should be able to explain (1) how you would set up the dominos to simulate a phosphorylation cascade and (2) why tipping the red domino simulates what

HOW DO ENZYME-LINKED RECEPTORS WORK?

1. Signal arrives and binds to receptor.

2. Signal-receptor complex changes conformation and is phosphorylated.

3. Proteins form a bridge to Ras. Ras exchanges its GDP for a GTP.

4. Activated Ras triggers phosphorylation of an intracellular protein, activating it.

5. Phosphorylation cascade results, each protein phosphorylating another until a response is triggered in the cell.

FIGURE 8.19 Enzyme-Linked Receptors Trigger a Phosphorylation Cascade.

happens when RTK becomes phosphorylated in response to hormone binding.

Although this discussion linked G proteins with the production of second messengers and enzyme-linked receptors with phosphorylation cascades, it's important to recognize that some G proteins trigger phosphorylation cascades and that some phosphorylation cascades result in the production of second messengers. In addition, it's common for many signaling pathways to exist in the same cell, and for intracellular signals that are part of one pathway to participate in other pathways as well. As a result, signal transduction pathways intersect and connect. Instead of being strictly linear like the pathways illustrated in Figures 8.17 through 8.19, in reality signal transduction pathways form more of a network. This complexity is important: It allows cells to respond to an array of extracellular signals in an integrated way.

To summarize: Many of the key signal transduction events observed in cells occur via G proteins or enzyme-linked receptors. The signal transduction event has two results: (1) It converts an easily transmitted extracellular message into an intracellular message, and (2) it amplifies the original message many times over. Although the hormone at the surface of the cell eventually disengages from its receptor and diffuses away or disintegrates, a "louder" intracellular signal—whether it is a large number of second messenger molecules or a large population of phosphorylated proteins—carries the information throughout the cell and triggers a response.

Signal Response

What is the ultimate response to the messages carried by hormones? The answer varies from signal to signal and from cell to cell. Recall that steroid hormones bind to intracellular receptors and directly change which genes are active in the cell. As a result, the cell's activity changes. Similar types of changes in gene expression can occur in response to second messengers or a cascade of protein phosphorylation events.

In wheat seeds, for example, embryos secrete a hormone called GA_1 when they start to grow. The hormone binds to receptors in cells near the starch-storing part of the seed. Once hormone-receptor binding occurs, the concentration of a second messenger called cGMP rises inside the cells. The second messenger triggers the production of a protein that activates the gene for a starch-digesting enzyme. Once the gene is turned on, large quantities of the starch-digesting enzyme α-amylase begin to enter the cell's endomembrane system. But the G protein activated by GA_1 also triggers a rise in intracellular Ca^{2+}. High Ca^{2+} concentrations allow vesicles that are packed with α-amylase to fuse with the plasma membrane and release their contents into the starch storage area. Thus, target cells begin producing *and* secreting an enzyme that digests stored starch. When the enzymes begin to work on the starch, sugars are released. By releasing a hormone, the ger-

minating embryo has signaled for the release of the nutrients it needs to grow.

Second messengers and phosphorylation cascades result in the activation or deactivation of a particular target protein that already exists in the cell—an enzyme, a membrane channel, or a protein that activates certain genes. Whatever the mechanism, the activity of the target cell changes dramatically in response to the arrival of the signal.

We've analyzed long-distance cell-cell communication as a three-step system: signal reception, signal transduction and amplification, and response. Now the question is, how is the signal turned off? Cells near the starch storage area of wheat seeds do not need to secrete unlimited quantities of α-amylase. Abnormalities would result if the morphological changes induced by testosterone and estrogen continued past your early teenage years into adulthood. What keeps signals from being amplified indefinitely?

Signal Deactivation

Cells have built-in systems for turning off intracellular signals. For example, activated G proteins convert GTP to GDP in a reaction that releases a phosphate group. The G protein's conformation changes as a result. Activation of its associated enzyme stops, and production of the second messenger ceases. The messengers in the cytosol are also short lived. For example, pumps in the membrane of the smooth ER return calcium ions to storage, and enzymes called phosphodiesterases convert active cAMP (see Table 8.2) and cGMP to inactive AMP and GMP.

Phosphorylation cascades wind down in a similar way. Enzymes called phosphatases are always present in cells, where they catalyze reactions that remove phosphate groups from proteins. If hormone stimulation of a receptor tyrosine kinase ends, phosphatases are able to dephosphorylate enough components of the phosphorylation cascade that the response begins to slow. Eventually it stops.

Although an array of specific mechanisms are involved, here is the general observation: Signal transduction systems trigger a rapid response and can be shut down quickly. As a result, they are exquisitely sensitive to small changes in the concentration of hormones or in the number and activity of signal receptors.

It is critical, though, to appreciate what happens when a signal transduction system does not shut down properly. For example, recall that Ras activates a phosphorylation cascade when it binds GTP but is deactivated when it binds GDP. With surprising frequency, human cells begin producing Ras proteins that are defective and no longer convert GTP to GDP once they are activated. As a result, GTP stays bound and the defective Ras proteins stay in the "on" position. They continue stimulating a phosphorylation cascade even when no growth hormones are present. Cells with defective Ras are likely to keep dividing as a result, which may lead to the development of cancer. It is estimated that 25–30 percent of all human cancers involve cells with this type of defective Ras protein. The family of diseases called cancer is explored in much more detail in Chapter 11 and Chapter 18.

The four steps of cell-cell signaling—reception, processing, response, and deactivation—allow hormone-secreting cells to elicit a specific response from cells in nearby or distant tissues. Despite this diversity in structure and function, all hormones have the same general role in the organism: They coordinate the activities of cells in response to information from outside or inside the body. In multicellular organisms ranging from wheat to humans, cell-cell signaling helps millions of individual cells function as an integrated whole. As a result, multicellular organisms can respond to changing conditions in an appropriate way.

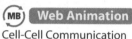 **Web Animation** at www.masteringbio.com
Cell-Cell Communication

Check Your Understanding

If you understand that...

- Intercellular signals coordinate the activities of cells throughout the body in response to changes in internal or external conditions.
- If intercellular signals do not enter the cell, they bind to a receptor on the plasma membrane. In response, the intercellular signal is transduced to an intracellular signal that the cell responds to.

You should be able to...

1) Explain what happens during each of the four steps involved in signal transduction (reception, processing, response, and deactivation).

2) Diagram a phosphorylation cascade versus production of a second messenger.

3) Explain why only certain cells respond to particular signals.

4) Explain how signals are amplified.

Chapter Review

SUMMARY OF KEY CONCEPTS

● **Extracellular material strengthens cells and helps bind them together.**

The vast majority of cells secrete an extracellular layer. In bacteria, archaea, algae, and plants, the extracellular material is stiff and is called a cell wall. In animals, the secreted layer is flexible and is called the extracellular matrix (ECM). Although the types of molecules present in the external coating vary widely, the basic structure and function of the extracellular layer are the same: It is a fiber composite that defines the cell's shape and helps protect it from damage. Fiber composites consist of cross-linked filaments that provide tensile strength and a ground substance that fills space and resists compression. In plants the extracellular filaments are cellulose microfibrils; in animals the most abundant filaments are made of the protein collagen. In both plants and animals, the ground substance is composed of gel-forming polysaccharides.

You should be able to predict what happens to animal cells when (1) they are treated with an enzyme that cuts integrin molecules, or (2) collagen fibers degrade. ●

● **Cell-cell connections help adjacent cells adhere. Cell-cell gaps allow adjacent cells to communicate.**

In both unicellular and multicellular organisms, cell-cell (intercellular) interactions are mediated by molecules in the extracellular layer and plasma membrane. Most interactions among unicellular species are competitive in nature, but the cells of multicellular organisms are intensely social and cooperate with each other. Many cells of multicellular organisms are physically bound to one another via the glue-like middle lamella that forms between plant cells or the tight junctions and desmosomes observed between animal cells. The cytoplasm of adjacent cells

is in direct communication, via openings called plasmodesmata in plants and gap junctions in animals.

You should be able to explain why the hypothesis that multicellularity evolved independently in plants and animals is supported by the observation that they have different structures serving similar functions in cell-cell adhesion and communication. ●

● **Intercellular signals are responsible for creating an integrated whole from many thousands of independent parts.**

Distant cells in multicellular organisms communicate through signaling molecules that bind to receptors found on or in specific target cells. Once these signals are received, they are often transduced to a new type of intracellular signal, which is amplified. This internal signal triggers a sequence of events that leads to the activation of certain enzymes, the release or uptake of specific ions or molecules, or a change in the activity of target genes. Because enzymes inside the cell quickly deactivate the signal, the cell's response is tightly regulated. A continued response usually depends on continued stimulation by the signaling molecule. Thus, cell-cell communication occurs by a four-step process: (1) signal reception, (2) signal processing, (3) signal response, and (4) signal deactivation. As a result, cells and tissues throughout the body can alter their activity in response to changing conditions, and do so in a coordinated way.

You should be able to explain why the hormone adrenalin can stimulate cells in both the heart and liver, yet trigger different responses (increasing heart rate versus release of glucose). ●

MB **Web Animation** at www.masteringbio.com
Cell-Cell Communication

QUESTIONS

● Test Your Knowledge

1. Which of the following statements represents a fundamental difference between the fibers found in the extracellular layers of plants and those of animals?
   a. Plant fibers are thicker; they are also stronger because they have more cross-linkages.
   b. Animal fibers consist of proteins; plant fibers consist of polysaccharides instead.
   c. Plant extracellular fibers never move; animal fibers can slide past one another.
   d. Cellulose microfibrils run parallel to each other; collagen filaments crisscross.

2. In animals, where are most components of the extracellular material synthesized?
   a. smooth ER
   b. the rough ER and Golgi apparatus
   c. in the extracellular layer itself
   d. in the plasma membrane

3. Treating dissociated cells with certain antibodies makes the cells unable to reaggregate. Why?
   a. The antibodies bind to cell adhesion proteins called cadherins.
   b. The antibodies bind to the fiber component of the extracellular matrix.
   c. The antibodies bind to receptors on the cell surface.
   d. The antibodies act as enzymes that break down desmosomes.

4. What does it mean to say that a signal is transduced?
   a. The signal enters the cell directly and binds to a receptor inside.
   b. The physical form of the signal changes between the outside of the cell and the inside.
   c. The signal is amplified, such that even a single molecule evokes a large response.
   d. The signal triggers a sequence of phosphorylation events inside the cell.

5. Why are tight junctions found in only certain types of tissues, while desmosomes are found in a wide array of cells?
   a. Tight junctions are required only in cells where communication between adjacent cells is particularly important.
   b. Tight junctions are not as strong as desmosomes.
   c. Tight junctions have different structures but the same functions.
   d. Tight junctions are found only in epithelial cells that must be watertight.

6. What physical event represents the receipt of an intercellular signal?
   a. the passage of ions through a desmosome
   b. the activation of the first protein in a phosphorylation cascade
   c. the binding of a hormone to a signal receptor, which changes conformation in response
   d. the activation of a G protein associated with a signal receptor

Test Your Knowledge answers: 1. b; 2. b; 3. a; 4. b; 5. d; 6. c

● Test Your Understanding

Answers are available www.masteringbio.com

1. Why is it difficult to damage a fiber composite?

2. Why does a phosphorylation cascade amplify an intercellular signal?

3. Compare and contrast a tight junction and a desmosome. How are they similar, and how are they different? Then compare and contrast desmosomes and plasmodesmata.

4. Animal cells adhere to each other selectively. Summarize experimental evidence that supports this statement. Explain the molecular basis of selective adhesion.

5. The cell-cell signal called insulin acts via the receptor tyrosine kinase. Make a concept map showing the reception, processing, and response steps in the signal transduction pathway for insulin. Your diagram should also include the concepts of amplification, phosphorylation cascade, activated transport proteins, and importing glucose from the bloodstream.

6. Why do researchers use the term *pathway* in reference to signal transduction? What is the significance of the observation that many signal transduction pathways intersect or overlap, creating a network?

● Applying Concepts to New Situations

Answers are available at www.masteringbio.com

1. Suppose that an animal species and a plant species each lacked the ability to secrete an extracellular matrix. What would these organisms look like, and how would they live?

2. Suppose that a signal that is released from one cell triggered a response in another cell in the *absence* of signal transduction. Diagram the sequence of events that would occur between the arrival of the signal and the cell's response. Compare and contrast the response to a signal that does involve signal transduction.

3. In most species of fungi, chitin is a major polysaccharide found in cell walls. Review the structure of chitin as described in Chapter 5, and then make a sketch predicting the structure of the fungal cell wall.

4. Suppose you created an antibody that bound to the receptor illustrated in Figure 8.18. How would the signal transduction pathway be affected? How would the signal transduction pathway be affected by a drug that bound permanently to the receptor?

9

Cellular Respiration and Fermentation

KEY CONCEPTS

● In cells, the endergonic reactions required for life occur in conjunction with an exergonic reaction involving ATP.

● Cellular respiration produces ATP from molecules with high potential energy— often glucose.

● Glucose processing has four components: (1) glycolysis, (2) pyruvate processing, (3) the Krebs cycle, and (4) electron transport coupled with oxidative phosphorylation.

● Fermentation pathways allow glycolysis to continue when the lack of an electron acceptor shuts down electron transport chains.

● Respiration and fermentation are carefully regulated.

When table sugar is heated, it undergoes the uncontrolled oxidation reaction known as burning. Burning gives off heat. In cells, the simple sugar called glucose is oxidized through a long series of carefully regulated reactions. Instead of being given off as heat, some of the energy produced by these reactions is used to synthesize ATP.

Cells are dynamic. Vesicles move cargo from the Golgi apparatus to the plasma membrane and other destinations, enzymes synthesize a complex array of macromolecules, and millions of membrane proteins pump ions and molecules to create an environment conducive to life. These cell activities change almost constantly in response to signals from other cells or the environment. What fuels all this action? The answer lies in the molecule **adenosine triphosphate (ATP)**. ATP has high potential energy and acts as the major energy currency in cells.

Earlier chapters explored the role that ATP plays in moving materials through cells, in moving cells themselves through the environment, and in responding to signals. Recall that phosphate groups contain a negative charge. When a phosphate group from ATP is transferred to a protein, the addition of the negative charge causes the newly phosphorylated protein to change shape. When a protein pump is phosphorylated, its shape changes in a way that transports ions or molecules across the plasma membrane against an electrochemical gradient (see Chapter 6). In some cases, an entire ATP molecule— which contains four tightly clustered negative charges—is added to a protein. When ATP binds to a motor protein such as kinesin or dynein or myosin, the protein moves in a way that transports materials or changes the cell's shape (Chapter 7). When ATP or a phosphate group from ATP binds to a specific site on an enzyme, the enzyme's shape and activity can change in a way that triggers synthesis of a second messenger or contributes to a phosphorylation cascade (Chapter 8).

● Key Concept   Important Information   ● Practice It

In short, ATP allows cells to do work. Because staying alive takes work, there is no life without ATP.

The goal of this chapter is to investigate how cells make adenosine triphosphate, starting from sugars and other compounds that have high potential energy. The energy that is released as sugar is processed in cells and is used to transfer a phosphate group to **adenosine diphosphate (ADP)**, generating ATP. Section 9.1 introduces the events involved by reviewing some fundamental points about free energy, analyzing how most ATP is used in cells, and exploring the principles of reduction-oxidation reactions, which are responsible for producing ATP. Section 9.2 follows up with an overview of the reactions involved in processing glucose, the most common fuel used by organisms. Sections 9.3 through 9.6 delve into the glucose-processing reactions in detail. Section 9.7 introduces an alternative route for ATP production—called fermentation—that occurs in many bacteria, archaea, and eukaryotes. Section 9.8, which closes the chapter, examines how cells divert certain carbon-containing compounds away from ATP production and into the synthesis of DNA, RNA, amino acids, and other molecules. This chapter is your introduction to **metabolism**—the chemical reactions that occur in living cells.

9.1 The Nature of Chemical Energy and Redox Reactions

Recall from Chapter 2 that chemical energy is a form of potential energy. Potential energy is energy that is associated with position or configuration. In cells, electrons are the most important source of chemical energy. The amount of potential energy that an electron has is based on its position relative to

other electrons and to the protons in the nuclei of nearby atoms. If an electron is close to negative charges on other electrons and far from the positive charges in nuclei, it has high potential energy. In general, the potential energy of a molecule is a function of the way its electrons are configured or positioned.

ATP makes things happen in cells because it has a great deal of potential energy. To understand why this is so, study the structural formula in **Figure 9.1a**. Notice that the three phosphate groups in ATP contain a total of four negative charges, and that these charges are confined to a small area. In part because these negative charges repel each other, the potential energy of the electrons in the phosphate groups is extraordinarily high.

When ATP reacts with water during a hydrolysis reaction, the bond between ATP's outermost phosphate group and its neighbor is broken, resulting in the formation of ADP and inorganic phosphate, P_i, which has the formula $H_2PO_4^-$ (**Figure 9.1b**). This reaction is highly exergonic. Recall that **exergonic** reactions release energy, while **endergonic** reactions require an input of energy. Under standard conditions of temperature and pressure in the laboratory, a total of 7.3 kilocalories (kcal) of energy, per mole of ATP, is released during the reaction.

You might remember from Chapter 2 that changes in free energy dictate whether a reaction is exergonic or endergonic. Changes in free energy, in turn, depend on the relationship between reactants and products in terms of potential energy and entropy (disorder). The hydrolysis of ATP is exergonic because the entropy of the product molecules is much higher than that of the reactants, and because there is a large drop in potential energy when ADP and P_i are formed from ATP. The change in

(a) ATP consists of three phosphate groups, ribose, and adenine.

(b) Energy is released when ATP is hydrolyzed.

FIGURE 9.1 Adenosine Triphosphate (ATP) Has High Potential Energy. **(a)** ATP has high potential energy, in part because four negative charges are clustered in its three phosphate groups. The negative charges repel each other, raising the potential energy of the electrons. **(b)** When ATP is hydrolyzed to ADP and inorganic phosphate, a large free-energy change occurs.

potential energy occurs in part because the electrons from ATP's phosphate groups are now spread between two molecules instead of being clustered on one molecule, meaning that there is less electrical repulsion. In addition, the negative charges on ADP and P_i are stabilized much more efficiently by interactions with the partial positive charges on surrounding water molecules than are the charges on ATP.

If the reaction diagrammed in Figure 9.1b occurred in a test tube, the energy released would be lost as heat. But cells don't lose that 7.3 kcal/mole as heat. Instead, they use it to make things happen. More specifically, things start to happen when ATP is hydrolyzed and a phosphate group is transferred to a protein. The addition of a phosphate group to a substrate is called **phosphorylation**. Phosphorylation of proteins is exergonic because the electrons in ADP and the phosphate group have much less potential energy than they did in ATP.

When phosphorylation adds a negative charge to a protein, the electrons in the protein change configuration in response. The molecule's overall shape, or conformation, is usually altered as well (**Figure 9.2**). Part of the protein moves. Protein movement—either in response to phosphorylation or to binding of an entire ATP molecule—is what transports materials inside cells, powers flagella or cilia, and pumps ions across membranes. It is also responsible for driving the endergonic reactions required for life.

Phosphorylation leads to shape change (even a small change can affect protein activity)

Active form

Inactive form

FIGURE 9.2 Phosphorylation Changes the Shape and Activity of Proteins. When proteins are phosphorylated or an ATP molecule binds to them, they often change shape in a way that alters their activity. The shape changes might seem subtle to you, but at the atomic scale, where reactions occur, they are huge. The figure shows an enzyme called MAP kinase. Note that two phosphate groups (yellow) activate this molecule.

How Does ATP Drive Endergonic Reactions?

In the time it takes to read this sentence, millions of endergonic reactions will have occurred in your cells. This chemical activity is possible because entire ATP molecules or phosphate groups from ATP are being added to reactant molecules or enzymes. To see how this process works, consider an endergonic reaction between a compound A and compound B that results in a product AB needed by your cells. This reaction can happen only when ATP reacts with the substrate to produce a phosphorylated molecule and ADP. If the reactant that is phosphorylated is compound B, it is referred to as an activated substrate. Activated substrates contain a phosphate group and have high free energy—high enough that the reaction between compound A and the activated form of compound B is exergonic. The two compounds then go on to react and form the product molecule AB.

In some cases, the enzyme that catalyzes the reaction is phosphorylated instead of a reactant. ● When either a substrate or an enzyme is phosphorylated, the exergonic phosphorylation reaction is said to be coupled to an endergonic reaction. In cells, endergonic reactions become exergonic when the substrates or enzymes involved are phosphorylated.

Figure 9.3 graphs how **energetic coupling** between an exergonic and endergonic reaction works. To begin reading this graph, note that the reaction between A and B is endergonic—ΔG is positive. But when the exergonic reaction occurs that results in a phosphate group leaving ATP and binding to B, the free energy of the reactants A and B is now high enough to make the reaction that forms AB exergonic. This is what coupling between phosphorylation and endergonic reactions does.

● If you understand the principles of energetic coupling, consider the following endergonic reaction: ribulose + $CO_2 \rightarrow$ 2 glycerate. You should be able to:

(1) Draw an enzyme and add ribulose and CO_2 nearby. (Don't worry about what ribulose actually looks like.) Explain why the reaction can't occur.

(2) Show 2 ATPs adding 2 phosphate groups to the ribulose. Explain why the resulting molecule, called ribulose bisphosphate, is called an activated substrate. Draw the activated substrate and CO_2 binding to the enzyme.

(3) Then draw the outcome of the reaction—two molecules of glycerate, each of which has a phosphate group attached, leaving the enzyme. Summarize how the coupling of an exergonic reaction and endergonic reaction occurred.

It is hard to overstate the importance of energetic coupling. Without it, life is impossible. If you run out of ATP, then enzymes and reactants can no longer be phosphorylated. If this happens, you will die within minutes. The energetic coupling going on in your cells right now is keeping you alive.

Now the issue is, if a great deal of energy is released when ATP loses a phosphate group and a protein is phosphorylated,

FIGURE 9.3 Exergonic Phosphorylation Reactions Are Coupled to Endergonic Reactions. In cells, endergonic reactions occur if one reactant (or an enzyme) undergoes phosphorylation—the addition of a phosphate group from ATP. The phosphorylated reactant molecule (or enzyme) has high enough free energy that the subsequent reaction is exergonic.

then a great deal of energy must be required to synthesize ATP from ADP by adding P_i. Where does this energy come from? The answer is redox reactions.

What Is a Redox Reaction?

Reduction-oxidation reactions, or **redox reactions**, are a class of chemical reactions that involve the loss or gain of one or more electrons. Many reactions, like the acid-base reactions introduced in Chapter 2, do not involve the gain or loss of an electron. Redox reactions are central in biology, though, because they drive the formation of ATP.

In a redox reaction, the atom that loses one or more electrons is said to be oxidized, and the atom that gains one or more electrons is said to be reduced. To help keep these terms straight, chemists use the mnemonic "LEO the lion goes GER"—Loss of Electrons is **Oxidation**; Gain of Electrons is **Reduction**. (An alternative is OIL RIG—Oxidation Is Loss; Reduction Is Gain.) Oxidation events are always coupled with a reduction; if one atom loses an electron, another has to gain it. Stated another way, a reactant that acts as an **electron donor** is always paired with a reactant that acts as an **electron acceptor**.

The gain or loss of an electron can be relative, however. During a redox reaction, electrons can be transferred completely from one atom to another, or the electrons can simply shift their positions in covalent bonds. For example, consider the overall reaction for photosynthesis diagrammed in **Figure 9.4**. Plants take in carbon dioxide (CO_2) and water (H_2O); and

FIGURE 9.4 Redox Reactions Involve the Gain or Loss of One or More Electrons. This diagram shows how the position of electrons changes in the overall reaction of photosynthesis. During photosynthesis, carbon atoms in CO_2 are reduced to form glucose and other sugars. The process is endergonic and requires an input of energy. Glucose has much higher potential energy than carbon dioxide does.

● **EXERCISE** Draw the rest of the structure of glucose (the four CH_2O groups, with the last carbon also bonded to an H), and add the positions of the electrons. Circle and label an atom that is reduced (has "gained" electrons); circle and label an atom that is oxidized (has "lost" electrons).

with the aid of sunlight, they synthesize carbohydrate—in this example, the sugar glucose ($C_6H_{12}O_6$)—and release molecular oxygen (O_2) and water. The orange dots in the illustration represent the positions of the electrons involved in covalent bonds. Now compare the position of the electrons in the first reactant, carbon dioxide, with their position in the first product, glucose. Notice that many of the electrons have moved closer to the carbon nucleus in glucose. This means that carbon has been reduced. It has "gained" electrons. The change occurred because the carbon and oxygen atoms in CO_2 do not share electrons equally, while the carbon and hydrogen atoms in glucose do. In CO_2, the high electronegativity of the oxygen atoms pulled electrons away from the carbon atom. Now compare the position of the electrons in the reactant water molecules with their position in the O_2 molecules that are produced. In O_2, the electrons have moved farther from the oxygen nuclei than they were in the water molecules, meaning that the oxygen atoms have been oxidized. Oxygen has "lost" electrons. In the overall reaction for photosynthesis, carbon atoms are reduced while oxygen atoms are oxidized.

These shifts in electron position change the amount of chemical energy in the reactants and products. When photosynthesis occurs, electrons are held much more loosely in the product molecules than in the reactant molecules. This means their potential energy has increased. The entropy of the products is also much lower than the reactants. As a result, the reaction is endergonic. It can take place only with an input of energy from sunlight.

Some Other Approaches to Understanding Redox

Redox reactions are absolutely fundamental to life, so it's critical to gain a solid understanding of redox chemistry. To give you a deeper understanding of how these reactions work, here are some other descriptions of redox dynamics:

- The electrons and bonds involved in redox reactions can be likened to a seesaw. In this analogy, an electron pair that is being shared equally by two atoms is like a seesaw balanced on its fulcrum.

Because this state is inherently unstable, it has high potential energy. The electrons involved in a covalent bond between carbon and hydrogen correspond to this situation—they are shared equally, because carbon and hydrogen are equally electronegative. Now suppose that a redox reaction occurs, and electrons are transferred toward a more electronegative atom such as oxygen. In this case, the bond is like a seesaw with one end lowered to the ground.

In this state, the seesaw is stable—meaning that it has much lower potential energy. This is what happens in redox reactions when electrons that were formerly involved in

covalent bonds between carbon and hydrogen instead participate in bonds between carbon and oxygen. In a reaction like this, the difference in potential energy between the two states is released as heat or transferred to another atom.

- During the redox reactions that occur in cells, electrons (e^-) are often transferred from an atom in one molecule to an atom in a different molecule. When this occurs, the electron is usually accompanied by H^+. The proton transfer is not a reduction reaction, but the upshot is that the molecule that contains the reduced atom gains a hydrogen (H) atom. Such molecules tend to have high potential energy, because the electrons in C–H bonds are relatively far from the positive charges in a nucleus. This observation should sound familiar, from the introduction to carbohydrates in Chapter 5. Molecules that have a large number of C–H bonds, such as carbohydrates and fats, store a great deal of potential energy.

Conversely, molecules that are oxidized in cells often lose a proton along with an electron. Instead of having many C–H bonds, oxidized molecules in cells tend to have many C–O bonds. Such molecules also have lower potential energy. To understand why, remember that oxygen atoms have extremely high electronegativity. Because oxygen atoms hold electrons so tightly, the electrons involved in bonds with oxygen atoms have low potential energy.

In many redox reactions in biology, understanding where oxidation and reduction have occurred becomes a matter of following hydrogen atoms. In many biological systems, reduction means "adding H's" and oxidation means "removing H's."

What Happens When Glucose Is Oxidized? To test your understanding of redox reactions, consider what happens when glucose undergoes the uncontrolled oxidation reaction called burning:

$$C_6H_{12}O_6 + 6\ O_2 \rightarrow 6\ CO_2 + 6\ H_2O + energy$$

The photograph at the start of the chapter shows this reaction occurring, and **Figure 9.5** provides a diagram. If you understand the fundamental principles of reduction-oxidation, you should be able to fill in the electron positions in each bond (see Figure 9.5 exercise) and answer the following questions:

1. Are the carbon atoms in glucose oxidized or reduced?

2. Are the oxygen atoms in the oxygen molecule (O_2) oxidized or reduced?

3. Glucose is the molecule that acts as an electron donor in this reaction. Which molecule acts as the electron acceptor?

4. Which has higher potential energy: the reactants or the products? Based on your answer, add "Energy" to the appropriate side of the equation; then add a label indicating "Input of energy" or "Release of energy."

COH

H — C — OH + O═══O $\longrightarrow$ O═══C═══O + (water H-O-H structure)

(CH₂O)₄H

Potential
energy decreases

Glucose   6 O₂   6 CO₂   6 H₂O
        (oxygen)   (carbon dioxide)   (water)

FIGURE 9.5 Tracking Electron Transfer during the Oxidation of Glucose.

● **EXERCISE** Fill in the electron positions in each bond shown; then use these data to explain why the reaction is exergonic. (Check your work, using the electron positions diagrammed in Figure 9.4).

When glucose burns, the change in potential energy is converted to kinetic energy in the form of heat. More specifically, a total of 686 kcal of heat is released when one mole of this sugar is oxidized.

Glucose does not burn in cells, however. Instead, the glucose in cells is oxidized through a long series of carefully controlled redox reactions. These reactions are occurring, millions of times per minute, in your cells right now. Instead of being given off as heat, much of the energy released by the drop in potential energy is being used to make the ATP you need to read, think, move, and stay alive. In cells, the change in free energy that occurs during the oxidation of glucose is used to synthesize ATP from ADP and P_i.

9.2 An Overview of Cellular Respiration

The first thing to appreciate about glucose processing and ATP production in cells is just how dynamic it is. In general, a cell contains only enough ATP to last from 30 seconds to a few minutes. Like many other cellular processes, the production and use of ATP is fast.

A second point to consider is where glucose comes from. Most glucose is produced by plants and other photosynthetic species, which use the energy in sunlight to reduce carbon dioxide (CO_2) to glucose and other carbohydrates. When photosynthetic species decompose or are eaten, they provide glucose to animals, fungi, and many bacteria and archaea. All organisms use glucose as a building block in the synthesis of fats, carbohydrates such as starch and glycogen, and other energy-storage compounds. The chemical energy stored in fats and carbohydrates acts like a savings account. ATP, in contrast, is like cash. To make ATP and get cash, carbohydrates must first be converted to glucose (**Figure 9.6**). When a cell needs energy, glucose is used to produce ATP through processes called cellular respiration and fermentation.

Chapter 10 focuses on how plants and other photosynthetic organisms produce glucose from carbon dioxide, water, and sunlight. This chapter focuses on cellular respiration and fermentation. Some of the chemistry involved in processing glucose actually predates the evolution of photosynthesis.

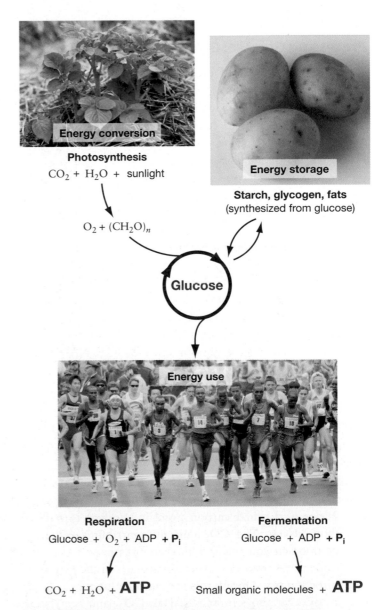

Energy conversion

Photosynthesis

$CO_2 + H_2O +$ sunlight

$O_2 + (CH_2O)_n$

Energy storage

Starch, glycogen, fats
(synthesized from glucose)

Glucose

Energy use

Respiration

Glucose + O₂ + ADP + **Pᵢ**

$CO_2 + H_2O +$ **ATP**

Fermentation

Glucose + ADP + **Pᵢ**

Small organic molecules + **ATP**

FIGURE 9.6 Glucose Is the Hub of Energy Processing in Cells.
Glucose is the end-product of photosynthesis. Both plants and animals store glucose and oxidize it to provide chemical energy in the form of ATP.

● **EXERCISE** The CO₂ produced by cellular respiration is used by plants during photosynthesis. The O₂ produced by photosynthesis is used during cellular respiration. Add arrows to this diagram to indicate these relationships.

Indeed, the reactions analyzed early in this chapter were probably occurring as the first cell sprang to life. The ancestry of some of the enzymes introduced in this chapter dates back about 3.5 billion years.

How does the oxidation of glucose take place in a way that supports the production of ATP? ● In many organisms, cellular respiration is a four-step process: (1) Glucose is broken down to a three-carbon compound called pyruvate; (2) pyruvate is processed to form a compound called acetyl-CoA; (3) acetyl-CoA is oxidized to CO_2; and (4) compounds that were reduced in steps 1–3 are oxidized in reactions that lead to the production of ATP. Let's get an overview of the way these steps work, and then pursue a more detailed analysis in Sections 9.3 through 9.6.

Processing Glucose: Glycolysis

An enormous diversity of organisms use the monosaccharide **glucose** as their primary fuel. In virtually all species examined to date, the first step in the oxidation of glucose is a sequence of 10 chemical reactions that are collectively called **glycolysis** (literally, "sugar breakdown"). These reactions occur in the cytosol of cells. During glycolysis, one molecule of the six-carbon sugar glucose is broken into two molecules of the three-carbon compound pyruvate. Some of the potential energy released by this sequence of reactions is used to phosphorylate ADP molecules, forming ATP. In addition, one of the reactions in the sequence reduces a molecule called **nicotinamide adenine dinucleotide**, symbolized **NAD+**. As **Figure 9.7** shows, the addition of two electrons and a proton to NAD+ produces **NADH**. We'll soon see that NADH readily donates electrons to other molecules. As a result, it is called an **electron carrier** and is said to have "reducing power."

Figure 9.8a summarizes the results of the reactions of glycolysis. Notice that glucose goes in and that ATP, NADH, and pyruvate come out.

Pyruvate Processing

What happens to the pyruvate produced by glycolysis? If an electron acceptor such as oxygen is present in the cell, pyruvate undergoes a sequence of reactions that results in the product molecule acetyl-CoA. These redox reactions have two results: (1) another molecule of NADH is synthesized, and (2) one of the three carbon atoms in pyruvate is oxidized to carbon dioxide. This CO_2 molecule contributes to the carbon dioxide that you exhale with every breath.

In eukaryotes, pyruvate processing occurs in the matrix inside mitochondria. During this step in the glucose oxidation system, pyruvate goes in and NADH, CO_2, and acetyl-CoA come out (**Figure 9.8b**).

The Krebs Cycle

The acetyl CoA produced by pyruvate processing enters a key sequence of reactions called the **Krebs cycle**. In eukaryotic cells,

the enzymes involved in the Krebs cycle are located inside mitochondria. Some of these enzymes are found in the matrix inside the organelle, while others are associated with the surface of the mitochondrion's inner membrane.

During the Krebs cycle, each acetyl-CoA molecule is oxidized to two molecules of carbon dioxide. Some of the potential energy released by these reactions is used to (1) reduce NAD+ to NADH; (2) reduce another electron carrier, called **flavin adenine dinucleotide** (**FAD**), to **FADH$_2$**; and (3) phosphorylate ADP, yielding ATP (**Figure 9.8c**). Acetyl-CoA enters this sequence of reactions; carbon dioxide, NADH, FADH$_2$, and ATP come out of the sequence.

With the completion of the Krebs cycle, glucose has been completely oxidized to CO_2. This observation raises several questions, though. According to the overall reaction for the oxidation of glucose ($C_6H_{12}O_6 + 6 O_2 \rightarrow 6 CO_2 + 6 H_2O +$ energy), molecular oxygen (O_2) is a reactant. Where does oxygen come into play? And what happens to all of the electrons that have been transferred from glucose to NADH and FADH$_2$?

Electron Transport

In cells, the high potential energy of electrons carried by NADH and FADH$_2$ is gradually "stepped down" by molecules that participate in a series of redox reactions. Stated another way, the potential energy in the electrons is gradually decreased via a series of step-by-step reactions. The proteins involved in the redox reactions make up what is known as an **electron transport chain (ETC)**. As electrons are passed along the electron transport chain, they gradually fall from a higher to a lower potential energy. In eukaryotes, the ETC components are located in the inner membrane of mitochondria; in bacteria and archaea, they are found in the plasma membrane.

The transfer of an electron to each molecule in the electron transport chain changes the molecule's shape and activity. In several cases, the molecule that receives the electron responds by changing conformation in a way that transfers a proton across the mitochondrion's inner membrane. This is crucial. In effect, the movement of electrons through the chain results in the pumping of protons to the outside of the membrane. It's a little bit like using electricity to pump water up into a storage tank above a water wheel. The subsequent buildup of protons results in a strong electrochemical gradient across the mitochondrial membrane. The gradient drives the protons through a membrane protein called ATP synthase. In eukaryotes, ATP synthase is located in the inner membrane of mitochondria. The force generated by the flow of protons through ATP synthase makes part of the protein spin. This change in conformation drives the phosphorylation of ADP. Indirectly, then, the sequence of redox reactions in the electron transport chain results in the production of ATP (**Figure 9.8d**).

To summarize: A series of reduction-oxidation reactions allows the proteins in the electron transport chain to pump

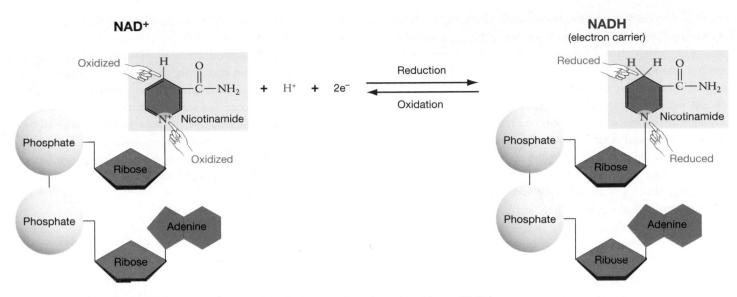

FIGURE 9.7 NAD$^+$ and NADH and Oxidized and Reduced Forms. NADH is the reduced form of NAD$^+$.

● **QUESTION** Compare the structure of NAD$^+$ with the structure of ATP in Figure 9.1. Which portions of the two molecules are identical?

FIGURE 9.8 An Overview of Glucose Oxidation. Cells produce ATP from glucose via a series of processes: **(a)** glycolysis, **(b)** pyruvate processing, **(c)** the Krebs cycle, and **(d)** electron transport combined with oxidative phosphorylation. Each component produces at least some ATP or NADH. Because the four components are connected, glucose oxidation is an integrated metabolic pathway. Glycolysis, pyruvate processing, and the Krebs cycle complete the oxidation of glucose. The NADH and FADH$_2$. they produce then feed the electron transport chain.

● **EXERCISE** Fill in the chart along the bottom.

protons across the inner membrane of the mitochondria. When the protons flow back across the membrane through ATP synthase, ATP is synthesized from ADP. ADP diffuses into mitochondria along a concentration gradient; ATP diffuses out along a concentration gradient.

Most people (me included!) initially have trouble understanding how the electron transport chain works. To make sure that you understand what is happening, consider this analogy: The proteins in the electron transport chain are like an electric pump—they use a flow of electrons to pump water up to a storage tank above a water wheel (**Figure 9.9**). When the water falls down through the water wheel, it turns and drives a millstone that grinds flour. ● You should be able to identify which parts of the water-pumping analogy correspond to the electron transport chain, ATP production, the pumping of protons, ATP synthase, and the proton gradient.

Once the electrons donated by NADH and FADH₂ have passed through the electron transport chain, they are transferred to a final electron acceptor, which in many organisms is oxygen. A transfer of these electrons to oxygen, along with protons, results in the formation of water as an end-product. The four molecules in the overall reaction for the oxidation of glucose—glucose, oxygen, carbon dioxide, and water—are now accounted for. Chemical energy has been transferred from

glucose to ATP, via NADH and FADH₂. These reactions are keeping you alive right now.

Most of the ATP production in cells results from the reactions in the electron transport chain. Hence, glycolysis and the Krebs cycle can be thought of as mechanisms for stripping electrons from glucose and feeding them to the electron transport chain. ● Biologists use the term **cellular respiration** for any process of ATP production that involves each of the following: a compound that acts as an electron donor, an electron transport chain, and an electron acceptor.

Methods of Producing ATP

The phosphorylation events that are catalyzed by ATP synthase are powered by a stream of protons. The gradient that drives these protons is set up by an electron transport chain that uses a molecule with low free energy—usually oxygen—as the final electron acceptor. Because this mode of ATP production links the phosphorylation of ADP with the oxidation of NADH and FADH₂, it is called **oxidative phosphorylation**.

During glycolysis and the Krebs cycle, however, ATP is produced by enzyme-catalyzed phosphorylation reactions. These reactions are called **substrate-level phosphorylation**. In substrate-level phosphorylation, an enzyme catalyzes the transfer of a phosphate group from a phosphorylated substrate to ADP, resulting in the production of ATP (**Figure 9.10**). The key distinction here is that the energy to produce the ATP comes from the phosphorylated substrate—not from a proton gradient, as it does in oxidative phosphorylation.

In many organisms, cellular respiration begins with the oxidation of glucose and ends with a flow of electrons through an electron transport chain to a final electron acceptor. The process starts with glycolysis and the Krebs cycle, which oxidize glucose to CO_2. As **Figure 9.11** shows, these reactions produce dramatic free-energy changes. Notice that relative to the other compounds in glycoysis and the Krebs cycle, glucose has a great deal of potential energy and low entropy. During glycolysis and the Krebs cycle, the free energy of each subsequent compound involves drops—at times in small steps and at other times in large steps.

FIGURE 9.9 An Analogy for Oxidative Phosphorylation.

● **QUESTION** Which parts of the diagram represent endergonic reactions? Which parts represent exergonic reactions?

● **EXERCISE** Once you understand which components of this drawing correspond to the ETC, protons, proton gradient, and ATP synthase, sketch in the inner mitochondrial membrane where you think it exists.

FIGURE 9.10 Substrate-Level Phosphorylation Involves an Enzyme and a Phosphorylated Substrate. Substrate-level phosphorylation occurs when an enzyme catalyzes the transfer of a phosphate group from a phosphorylated substrate to ADP, forming ATP.

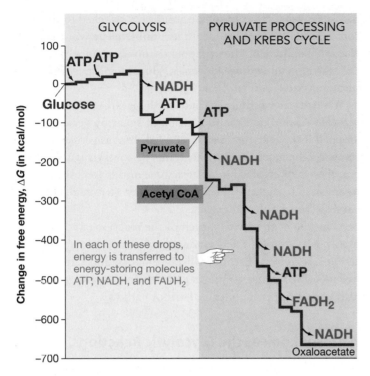

FIGURE 9.11 Free Energy Changes as Glucose Is Oxidized.
Graph of free-energy changes during glycolysis and the Krebs cycle.
Drops in free energy are associated with the production of ATP, NADH,
and $FADH_2$. Oxaloacetate is the endpoint of the Krebs cycle.

QUESTION Which is associated with larger changes in free energy,
production of ATP or production of NADH and $FADH_2$? In which parts of
the cell do glycolysis, pyruvate processing, and the Krebs cycle take place?

The small drops in free energy are coupled with the production
of ATP; the large drops are associated with the synthesis of
NADH and $FADH_2$. In effect, then, the free energy present in
glucose is transferred to ATP, NADH, and $FADH_2$. In the third
(and final) stage of cellular respiration, NADH and $FADH_2$
carry these electrons to the electron transport chain.

If the molecule that functions as the final electron acceptor
is unavailable, however, then cellular respiration cannot occur.
In this case, alternative pathways called fermentation reactions
take over (**Figure 9.12**). In eukaryotes, for example, fermenta-

tion occurs when oxygen is lacking. As we'll see in Section 9.5,
fermentation consists of reactions that allow glycolysis to continue even if pyruvate is not drawn off to be processed into
acetyl-CoA. By-products such as ethanol (the active ingredient
in alcoholic beverages) and lactic acid result from the fermentation pathways present in various species of eukaryotes. In
bacteria and archaea, other fermentation pathways and products are possible.

This overview of cellular respiration is intended simply to
introduce the processes responsible for most ATP production
in cells. Now we need to explore glycolysis, pyruvate processing, the Krebs cycle, and electron transport chains in more
detail. Let's take a closer look at each process in turn.

Check Your Understanding

If you understand that...

- Glucose oxidation starts with glycolysis and continues with
  pyruvate processing and the Krebs cycle. The products of
  these processes include CO_2, ATP, and the electron carriers
  NADH and $FADH_2$.

- NADH and $FADH_2$ donate electrons to an electron transport
  chain, which pumps protons to create a proton gradient.
  Oxygen or another compound with low free energy serves
  as the final electron acceptor.

- Protons flowing down the gradient established by the
  electron transport chain allow ATP synthase to make ATP.

You should be able to...

Explain how oxidative phosphorylation and substrate-level
phosphorylation differ in terms of

1) the nature of the enzyme or physical mechanism involved,

2) where they occur during the processing of glucose, and

3) the source of the energy to drive the endergonic reaction
   forming ATP.

(MB) **Web Animation**   *BioFlix* at www.masteringbio.com
Cellular Respiration

FIGURE 9.12 Cellular Respiration and Fermentation Are Alternative Pathways for Producing Energy.
When oxygen or another electron acceptor used by the ETC is present in a cell, the pyruvate produced by glycolysis enters
the Krebs cycle and the electron transport system is active. But if no electron acceptor is available to keep the ETC running,
the pyruvate undergoes reactions known as fermentation.

9.3 Glycolysis

Glycolysis may be the most fundamental of all metabolic pathways, but it was discovered by accident. In the late 1890s Hans and Edward Buchner were working out techniques for manufacturing extracts of baker's yeast for commercial or medicinal use. (Yeast extracts are still added to some foods as a flavor enhancer or nutritional supplement.) In one set of experiments the Buchners added sucrose, or table sugar, to their extracts. Sucrose is a disaccharide consisting of glucose linked to fructose, another six-carbon sugar. At the time, sucrose was commonly used as a preservative—a substance used to preserve food from decay. But the Buchners found that instead of preserving the yeast extracts, the sucrose was quickly broken down and fermented, with alcohol appearing as a by-product. This was a key finding because it showed that fermentation and other types of cellular metabolism could be studied in vitro—outside the organism. Until then, researchers thought that metabolism could take place only in intact organisms.

The Buchners and other researchers followed up on this observation by trying to determine how the sugar was being processed. An important early observation was that the reactions could go on much longer than normal if inorganic phosphate was added to the mixture. This finding implied that some of the compounds involved were being phosphorylated. Soon after, a molecule called fructose bisphosphate was isolated. (The prefix *bis*– means that two phosphate groups are attached to the fructose molecule at distinct locations.) Subsequent work showed that all but two of the compounds involved in glycolysis—the starting and ending molecules, glucose and pyruvate—are phosphorylated.

A third major finding came to light in 1905, when researchers found that the processing of sugar by yeast extracts stopped if the reaction mix was boiled. Because enzymes were known to be inactivated by heat, this discovery suggested that enzymes were involved in at least some of the processing steps. Years later, investigators realized that each step in glycolysis is catalyzed by a different enzyme.

Over the next 35 years, each of the reactions and enzymes involved in glycolysis was gradually worked out by several different researchers. Because Gustav Embden and Otto Meyerhof identified many of the steps in glycolysis, this process is also known as the Embden–Meyerhof pathway.

A Closer Look at the Glycolytic Reactions

By breaking open cells, separating the cell components via differential centrifugation (introduced in Chapter 7), and testing which components could sustain glycolysis, biologists discovered something interesting: All 10 reactions of glycolysis occur in the cytosol. **Figure 9.13** details the 10 reactions involved in glycolysis. There are three key points to note about the reaction sequence:

FIGURE 9.13 Glycolysis Pathway. Glucose is oxidized to pyruvate through this sequence of 10 reactions. Each reaction is catalyzed by a different enzyme. The products are two net ATP (four ATP are produced, but two are invested), two molecules of NADH, and two molecules of pyruvate.

● **EXERCISE** Enzymes called kinases catalyze the addition of a phosphate group to a substrate. Circle the reactions that are catalyzed by a kinase.

All 10 reactions of glycolysis occur in cytosol

GLYCOLYSIS

What goes in:

ATP

ATP

HOCH₂
Glucose

P OCH₂
Glucose-6-phosphate

P OCH₂   H₂COH
Fructose-6-phosphate

P OCH₂   H₂CO P
Fructose-1,6-bisphosphate

P OCH₂
C=O
H₂COH

H
C=O
HCOH
H₂CO P

What comes out: ADP   ADP

Glycolysis begins with an energy-investment phase of 2 ATP

1. Contrary to what most researchers expected, glycolysis starts by *using* ATP, not producing it. In the initial step, glucose is phosphorylated to form glucose-6-phosphate. After an enzyme rearranges this molecule to fructose-6-phosphate in the second step, the third step in the reaction sequence adds a second phosphate group, forming the fructose-1,6-bisphosphate observed by early researchers. Thus, two ATP molecules are used up before any ATP is produced by glycolysis.

2. Once this energy-investment phase of glycolysis is complete, the subsequent reactions represent an energy payoff phase. The sixth reaction in the sequence results in the reduction of two molecules of NAD$^+$; the seventh produces two molecules of ATP and erases the energy "debt" of two molecules of ATP invested early in glycolysis. The final reaction in the sequence produces another two ATPs. For each molecule of glucose processed, the net yield is two molecules of NADH, two of ATP, and two of pyruvate.

3. The production of ATP during glycolysis occurs by substrate-level phosphorylation. In reactions 7 and 10 of Figure 9.13, an enzyme catalyzes the transfer of a phosphate group from a phosphorylated intermediate in glycolysis directly to ADP.

The discovery and elucidation of the glycolytic pathway ranks as one of the great achievements in the history of biochemistry. Because the enzymes involved have been observed in nearly every bacterium, archaean, and eukaryote examined, it is logical to infer that the ancestor of all organisms living today made ATP by glycolysis. The reactions outlined in Figure 9.13 are among the most ancient and fundamental of life processes.

How Is Glycolysis Regulated?

Once the glycolytic pathway was worked out, researchers focused on the structures and functions of the enzymes involved and the way the sequence is regulated. It seemed logical to predict that glycolysis does not occur at the maximum rate at all times. Instead, researchers predicted that glycolysis would proceed only when cells needed fresh supplies of ATP. An important advance occurred when biologists observed that high levels of ATP inhibit a key glycolytic enzyme called phosphofructokinase. **Phosphofructokinase** catalyzes step 3 in Figure 9.13—the synthesis of fructose-1,6-bisphosphate from fructose-6-phosphate.

The discovery that ATP inhibits phosphofructokinase was important because this enzyme catalyzes a key step in the reaction sequence. After steps 1 and 2 occur, an array of enzymes that aren't involved in glycolysis can convert the products of the reactions to reactants used in other metabolic pathways. Before step 3, then, the sequence can be interrupted and the intermediates used efficiently elsewhere in the cell. This is not true of later steps in glycolysis. Once fructose-1,6-bisphosphate is synthesized, then, there is no point in stopping the process. Based on these observations, it makes sense that the pathway is turned on or off at step 3.

The observation that ATP inhibits phosphofructokinase raises an important question, however: Why would ATP—a substrate that is required for the reaction in step 3—also inhibit the reaction? In the vast majority of cases, the addition of a substrate *speeds* the rate of a chemical reaction instead of slowing it.

To explain this unusual situation, biologists hypothesized that high levels of ATP are a signal that the cell does not need to produce more ATP. When an enzyme in a pathway is

The "2" indicates that glucose has been split into two 3-carbon sugars

During the energy payoff phase, 4 ATP are produced for a net gain of 2 ATP

Feedback inhibition:
Presence of product inhibits enzyme 1

FIGURE 9.14 Metabolic Pathways May Be Regulated by Feedback Inhibition. Feedback inhibition occurs when the product of a metabolic pathway inhibits an enzyme that is active early in the pathway.

● **QUESTION** Why would regulation be less effective if the product inhibited enzyme 3 instead of enzyme 1?

inhibited by the product of the reaction sequence, **feedback inhibition** is said to occur (**Figure 9.14**). The product molecule "feeds back" to stop the reaction sequence when the product is abundant. In glycolysis, cells that are able to stop the glycolytic reactions when ATP is abundant can conserve their stores of glucose for times when ATP is scarce. As a result, natural selection should favor individuals who have phosphofructokinase molecules that are inhibited by high concentrations of ATP.

This hypothesis provided a satisfying explanation for why ATP inhibits the key step of glycolysis, but it did not explain *how* high levels of the substrate inhibit the enzyme. The answer came later, when researchers were able to determine the three-dimensional structure of phosphofructokinase. As **Figure 9.15** shows, phosphofructokinase has two distinct binding sites for ATP. ATP can bind at the enzyme's active site or at a site that changes the enzyme's activity—a **regulatory site**. At the ac-

When ATP binds here, the reaction rate slows dramatically

Fructose-1,6-bisphosphate at active site

ADP at active site

FIGURE 9.15 Phosphofructokinase Has Two Binding Sites for ATP. A model of one of the four identical subunits of phosphofructokinase. Notice the active site, where a phosphate group is transferred from ATP to form fructose-1,6-bisphosphate, and the regulatory site, where ATP binds.

● **QUESTION** The active site has much higher affinity for ATP than the regulatory site does. What would be the consequences if the regulatory site had higher affinity for ATP than the active site did?

tive site, ATP is converted to ADP and the phosphate group is transferred to fructose-6-phosphate. This reaction results in the synthesis of fructose-1,6-bisphosphate. ● If ATP concentrations are high, however, the molecule also binds at a regulatory site on phosphofructokinase. When ATP binds at this second location, the enzyme's conformation changes in a way that makes the reaction rate at the active site drop dramatically. In this case, ATP acts as an allosteric regulator (see Chapter 3).

Thanks to advances such as these, glycolysis is among the best understood of all metabolic pathways. Now the question is, what happens to its product? How is pyruvate processed?

9.4 Pyruvate Processing

In eukaryotes, the pyruvate produced by glycolysis is transported from the cytosol to mitochondria. As Chapter 7 noted, mitochondria are organelles found in virtually all eukaryotes. **Figure 9.16** shows a diagram and an image of this organelle generated with a technique called cryo-electron tomography.[1] Notice that a mitochondrion has two membranes and that the interior is filled with layers of sac-like structures called **cristae**. Cristae are connected to the main part of the inner membrane by short tubes. The region inside the inner membrane but outside the cristae is called the **mitochondrial matrix**.

The pyruvate produced by glycolysis moves across the mitochondrion's outer membrane and into the intermembrane space through small pores. Entry into the matrix occurs via active transport, through a membrane protein called the pyruvate carrier, located in the inner membrane. Because the transport of pyruvate into the mitochondrion is an active process, it represents an energy-consuming step in glucose oxidation.

What happens to pyruvate once it enters the mitochondrion? This issue remained unanswered until it was discovered that an organic compound called **coenzyme A (CoA)** serves as a cofactor in a wide variety of reactions catalyzed by cellular enzymes. You might recall from Chapter 3 that many enzymes require some type of cofactor to function. Cofactors are often a metal ion or a relatively small organic molecule called a coenzyme. Coenzymes often bind to the enzyme's active site and stabilize the reaction's transition state. Some coenzymes, for example, donate or accept electrons during a redox reaction; others transfer an acetyl ($-COCH_3$) group or carbon chains. CoA acts as a coenzyme by accepting and then transferring an acetyl group to a substrate. In fact, the "A" in "CoA" stands for acetylation. When an acetyl group bonds to a sulfur atom on one end of the CoA molecule, **acetyl CoA** forms. When an acetyl group is bound to CoA, the acetyl group becomes activated and can be easily transferred to an acceptor molecule.

---

[1] Cryo-electron tomography has provided a much more accurate picture of mitochondrial morphology compared to inferences based on transmission electron microscopy. For a recent review, see C. Mannella, *Biochemica et Biophysica Acta* 1762 (2006):140–147.

FIGURE 9.16 The Structure of the Mitochondrion. These images are based on recent research using an imaging technique called cryo-electron tomography. Notice that the mitochondria have outer and inner membranes, and that the inner membrane is connected by short tubes to sac-like structures called cristae. Pyruvate processing occurs within the mitochondrial matrix.

Soon after CoA's role as a coenzyme was revealed, other investigators found that pyruvate reacts with CoA to produce **acetyl CoA**. Follow-up work showed that the conversion of pyruvate to acetyl CoA occurs in a series of steps inside an enormous and intricate enzyme complex called **pyruvate dehydrogenase.** In eukaryotes, pyruvate dehydrogenase is located in the mitochondrial matrix. In bacteria and archaea, pyruvate dehydrogenase is located in the cytosol.

As pyruvate is being processed, NAD^+ is reduced to NADH and one of the carbons in the pyruvate is oxidized to CO_2. The remaining two-carbon acetyl unit is transferred to CoA (**Figure 9.17**). Acetyl CoA is the final product of the pyruvate processing step in glucose oxidation.

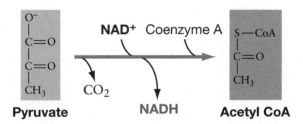

FIGURE 9.17 Pyruvate Is Oxidized to Acetyl CoA. The reaction shown here is catalyzed by pyruvate dehydrogenase.

⬤ **EXERCISE** Above the reaction arrow, list 3 molecules whose presence speeds up the reaction. Label them "Positive control." Below the reaction arrow, list 3 molecules whose presence slows down the reaction. Label them "Feedback inhibition."

⬤ **EXERCISE** Circle the acetyl group ($-C_2H_3O$) on acetyl CoA. Label the sulfur atom on CoA where the acetyl group bonds.

When supplies of ATP are abundant, however, the events just described fail to occur. Pyruvate processing stops when the pyruvate dehydrogenase complex becomes phosphorylated. In this case, phosphorylation changes the shape of pyruvate dehydrogenase in a way that inhibits catalytic activity. High concentrations of other products—specifically acetyl CoA and NADH—also increase the rate of phosphorylation of the enzyme complex. Like the control of phosphofructokinase activity in glycolysis, these are examples of feedback inhibition. Reaction products feed back to stop or slow down the pathway.

On the contrary, the presence of high concentrations of NAD^+, CoA, or adenosine monophosphate (AMP)—which indicates low ATP supplies—speeds up the reactions catalyzed by the pyruvate dehydrogenase complex. ⬤ To capture this point, biologists say that pyruvate processing is under both positive and negative control. Large supplies of products inhibit the enzyme complex; large supplies of reactants and low supplies of products stimulate it. Pyruvate processing is a key regulatory point in glucose oxidation.

If ATP supplies are low in a cell, though, the complex actively churns out acetyl CoA. This compound participates in a reaction that kicks off the next major step in glucose oxidation: the Krebs cycle.

9.5 The Krebs Cycle

While Embden, Meyerhof, and others were working out the sequence of reactions in glycolysis, biologists in several laboratories were focusing on a different set of redox reactions that take place in actively respiring cells. These reactions involve small organic acids such as citrate, malate, and succinate. Because they have the form R-COOH, these molecules are called carboxylic acids. In some cases, the redox reactions that produce carboxylic acids also result in the production of carbon dioxide. Recall from Section 9.1 that carbon dioxide is a highly oxidized form of carbon and that it is the endpoint of glucose metabolism. Thus, it was logical for researchers to propose that the oxidation of small carboxylic acids could be an important component of glucose oxidation.

Early researchers made several key observations about these reactions. First, a total of eight small carboxylic acids are oxidized rapidly enough to imply that they are involved in glucose metabolism—the most rapid set of oxidation reactions predicted to occur in cells. Second, when one of these carboxylic acids is added to cells, the rate of glucose oxidation increases. The added molecules did not appear to be used up, however. Instead, virtually all of the carboxylic acid added seemed to be recovered later. Why this should be so was a puzzle. Even while this observation remained unexplained, however, the biochemists working on the reactions were able to determine the order in which the eight acids were oxidized. The result is the reaction pathway shown in **Figure 9.18a**.

The mystery of why intermediates in the pathway were not used up was resolved when Hans Krebs realized that the reaction sequence might occur in a cyclical fashion instead of a linear pathway. Krebs had another crucial insight when he suggested that the reaction sequence was directly tied to the processing of pyruvate produced by glycolysis. To test these hypotheses, Krebs and a colleague set out to determine whether pyruvate—the endpoint of the glycolytic pathway—could react with oxaloacetate—the most odidixed carboxylic

acid. Because oxaloacetate was the endpoint of the pathway in Figure 9.18a, the cycle hypothesis predicted that a reaction of oxaloacetate with pyruvate should produce the starting point of the pathway—the six-carbon carboxylic acid called citrate (**Figure 9.18b**).

The result? As predicted, citrate formed when the biologists added oxaloacetate and pyruvate to cells. Based on this result, Krebs proposed that pyruvate is oxidized to carbon dioxide through a cycle of reactions. In honor of this insight, the pathway became known as the Krebs cycle.[2]

When radioactive isotopes of carbon became available in the early 1940s, researchers in several labs used them to confirm the cyclical nature of the Krebs cycle. For example, by adding radioactively labeled citrate or pyruvate to cells and analyzing the radioactive compounds that resulted, it was possible to show that carbon atoms cycle through the sequence of reactions just as Krebs had proposed. Later, the discovery of the pyruvate dehydrogenase complex showed that acetyl CoA is the key compound that feeds the Krebs cycle, not pyruvate.

To summarize Sections 9.4 and 9.5, the reactions catalyzed by pyruvate dehydrogenase act as a preparatory step for the Krebs cycle. The acetyl CoA that is produced by the pyruvate processing goes on to react with oxaloacetate, forming citrate. When processing of pyruvate in the mitochondrion is complete, all three carbons in pyruvate are oxidized to carbon dioxide. One

---

[2]The Krebs cycle is also known as the *citric acid cycle* because it starts with citrate, which becomes citric acid when protonated. Because citric acid has three carboxyl groups, the reaction sequence is also called the *tricarboxylic acid (TCA) cycle*.

(a) The carboxylic acids that are oxidized during cellular respiration

(b) The Krebs cycle hypothesis

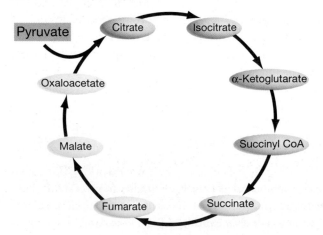

FIGURE 9.18 A Series of Carboxylic Acids Is Oxidized in the Krebs Cycle. **(a)** The eight molecules that are oxidized in actively respiring cells are pictured in order of their tendency to donate or accept electrons in redox reactions. **(b)** Hans Krebs proposed that if the pyruvate from glycolysis reacted with oxaloacetate to form citrate, then the eight reactions in part (a) would form a cycle.

● **EXERCISE** Krebs turned out to be wrong about pyruvate starting the cycle. In part (b), add the correct molecule that links pyruvate to the cycle.

THE KREBS CYCLE

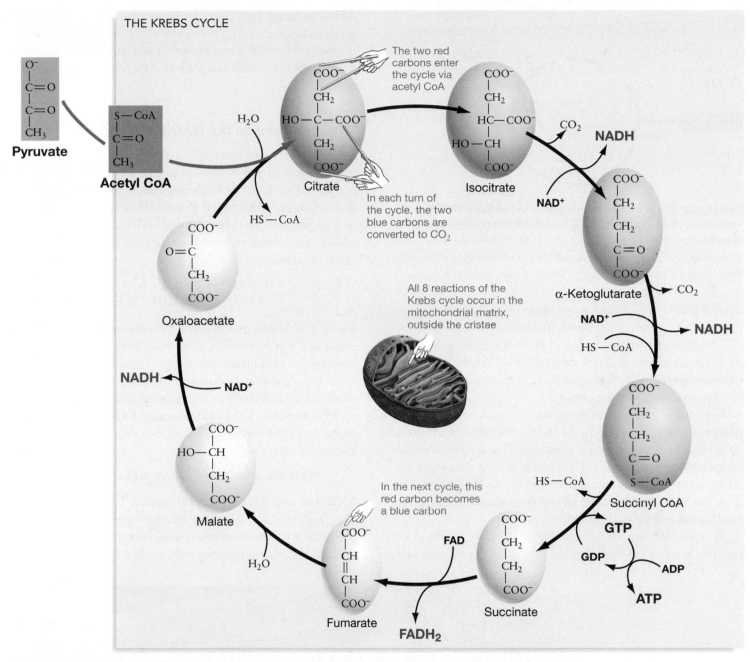

FIGURE 9.19 The Krebs Cycle Completes the Oxidation of Glucose. Acetyl CoA goes into the Krebs cycle, and carbon dioxide, NADH, FADH$_2$, and GTP come out. The GTP is produced by substrate-level phosphorylation and then converted to ATP. If you follow individual carbon atoms around the cycle several times, you'll come to an important conclusion: All the carbons in the cycle are eventually positioned as "blue carbons" that are released as CO$_2$. This occurs because the "red carbons" in fumarate—which is symmetric—can flip position when malate forms.

of the carbons is oxidized in the pyruvate dehydrogenase complex, and two are oxidized in the Krebs cycle itself. As **Figure 9.19** indicates, the energy released by the oxidation of one molecule of acetyl CoA is used to produce three molecules of NADH, one of FADH$_2$, and one of **guanosine triphosphate (GTP)** through substrate-level phosphorylation. GTP is then converted to ATP. Pyruvate and CoA enter the sequence of reactions; CoA, carbon dioxide, ATP, NADH, and FADH$_2$ come out. In bacteria and archaea, the enzymes responsible for pyruvate processing

and the Krebs cycle are located in the cytosol. In eukaryotes, most of the enzymes responsible for the Krebs cycle are located in the mitochondrial matrix.

How Is the Krebs Cycle Regulated?

Like glycolysis and pyruvate processing, the Krebs cycle is carefully regulated. Reaction rates are high when ATP is scarce; reaction rates are low when ATP is abundant. The major

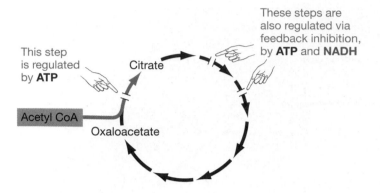

This step is regulated by **ATP**

These steps are also regulated via feedback inhibition, by **ATP** and **NADH**

Citrate

Acetyl CoA

Oxaloacetate

FIGURE 9.20 The Krebs Cycle Is Regulated by Feedback Inhibition. The Krebs cycle slows down when ATP and NADH are plentiful. ATP acts as an allosteric regulator, while NADH acts as a competitive inhibitor.

● **QUESTION** How do allosteric regulation and competitive inhibition differ? (Your answer should include a sketch of how they work.)

control points are highlighted in **Figure 9.20**. Notice that the enzyme that converts acetyl CoA to citrate is shut down when ATP binds to it. Like the control of phosphofructokinase activity in glycolysis and pyruvate dehydrogenase activity, this is an example of feedback inhibition. Reaction products feed back to stop or slow down the pathway.

As Figure 9.20 indicates, feedback inhibition also occurs at two points later in the cycle. At the first of these two points, NADH binds to the enzyme's active site. This is an example of competitive inhibition (see Chapter 3). At the second point, ATP binds to an allosteric regulatory site. ● Thus, the Krebs cycle can be turned off at multiple points, via several different mechanisms of feedback inhibition.

By speeding up the oxidation of glucose when ATP supplies are low and by slowing down the reactions when

ATP is plentiful, cells carefully match the rate of cellular respiration to their energy requirements. Natural selection has favored the evolution of enzymes that allow cells to conserve glucose, pyruvate, and acetyl CoA. ATP is produced on an as-needed basis.

What Happens to the NADH and FADH$_2$?

Figure 9.21 reviews the relationships of glycolysis, pyruvate processing, and the Krebs cycle and identifies where each process takes place in eukaryotic cells. For each molecule of glucose that is fully oxidized to 6 carbon dioxide molecules, the cell produces 10 molecules of NADH, 2 of FADH$_2$, and 4 of ATP. The overall reaction for glycolysis and the Krebs cycle can be written as

$$C_6H_{12}O_6 + 10\ NAD^+ + 2\ FAD + 4\ ADP + 4\ P_i \rightarrow$$
$$6\ CO_2 + 10\ NADH + 2\ FADH_2 + 4\ ATP$$

The ATP molecules are produced by substrate-level phosphorylation and can be used to drive endergonic reactions, power movement, or run membrane pumps. The carbon dioxide molecules are a gas that is disposed of as waste—you exhale it; plants release it or use it as a reactant in photosynthesis.

What happens to the NADH and FADH$_2$ produced by glycolysis and the Krebs cycle? Recall that the overall reaction for glucose oxidation is

$$C_6H_{12}O_6 + 6\ O_2 \rightarrow 6\ CO_2 + 6\ H_2O + energy$$

Glycolysis and the Krebs cycle account for the glucose, the CO_2, and—because ATP is produced—some of the chemical energy that results from the overall reaction. But the O_2 and the H_2O that appear in the overall reaction for the oxidation of

SUMMARY OF GLUCOSE OXIDATION

2 NADH

2 NADH

6 NADH
2 FADH$_2$

Glucose ----→ 2 Pyruvate ──→ 2 Acetyl CoA ── KREBS CYCLE ──→ 4 CO$_2$

2 CO$_2$

2 ATP

2 ATP

Cytoplasm   Mitochondrion

FIGURE 9.21 A Summary of Glucose Oxidation. Glucose is completely oxidized to carbon dioxide via glycolysis, the subsequent oxidation of pyruvate, and then the Krebs cycle. In eukaryotes, glycolysis occurs in the cytosol; pyruvate oxidation and the Krebs cycle take place in the mitochondrial matrix.

● **EXERCISE** Label the outer membrane, the inner membrane, and the sac-like cristae of the mitochondrion.

glucose are still unaccounted for. As it turns out, so is much of the chemical energy. The reaction that has yet to occur is

$$NADH + FADH_2 + O_2 + ADP + P_i \rightarrow$$
$$NAD^+ + FAD + H_2O + ATP$$

In this reaction, oxygen is reduced to form water. The electrons that drive the redox reaction come from NADH and $FADH_2$. These molecules are oxidized to $NAD^+ + FAD$.

In effect, glycolysis, pyruvate processing, and the Krebs cycle transfer electrons from glucose to NAD^+ and FAD, creating NADH and $FADH_2$. These molecules then carry the electrons to oxygen, which serves as the final electron acceptor in eukaryotic cells. When oxygen accepts electrons, water is produced. All the components of the overall reaction for glucose oxidation are accounted for.

How does this final part of the process occur? Specifically, how is ATP generated as electrons are transferred from NADH or $FADH_2$ to O_2? In the 1960s—decades after the details of glycolysis and the Krebs cycle had been worked out—a startling and highly controversial answer to these questions emerged.

Check Your Understanding

If you understand that...   ◯

- During glycolysis, pyruvate processing, and the Krebs cycle, one molecule of glucose ($C_6H_{12}O_6$) is completely oxidized to six molecules of carbon dioxide (CO_2).

You should be able to...   ◯

Model the following components of cellular respiration by pretending that a large piece of paper is a cell. Draw a large mitochondrion inside it. Cut out small squares of paper and label them as glucose, glycolytic reactions, Krebs cycle reactions, pyruvate dehydrogenase complex, pyruvate, acetyl CoA, CO_2, ADP → ATP, NAD → NADH, and FAD → $FADH_2$.

1) Put each of the squares in the appropriate location in the cell.

2) Draw arrows to connect the appropriate molecules and reactions.

3) Using dimes for electrons, demonstrate how glucose is oxidized to CO_2.

4) Label points where regulation occurs.

Once your model is working, you'll be ready to consider what happens to the NADH and $FADH_2$ you've produced.

9.6 Electron Transport and Chemiosmosis

The answer to one fundamental question about the oxidation of NADH and $FADH_2$ turned out to be relatively straightforward. To determine where the oxidation reactions take place in eukaryotes, researchers isolated mitochondria by using differential centrifugation techniques (introduced in Chapter 7). Then they broke open the organelles and separated the inner and outer membranes from the mitochondrial matrix and the solution in the intermembrane space. They found that NADH could be oxidized in the isolated membranes, but not in the matrix or in the intermembrane fluid. Follow-up work showed that the oxidation process takes place on the inner membrane of the mitochondria and the membranes of cristae. In prokaryotes, the oxidation of NADH occurs in the plasma membrane.

Biologists who analyzed the components of these membranes made a key discovery when they isolated molecules that switch between a reduced and an oxidized state during respiration. The molecules were hypothesized to be the key to processing NADH and $FADH_2$. What are these molecules, and how do they work?

Components of the Electron Transport Chain

Collectively, the molecules responsible for the oxidation of NADH and $FADH_2$ are designated the electron transport chain (ETC). As electrons are passed from one protein to another in the chain, the energy released by the redox reactions is used to pump protons across the inner membrane of mitochondria. Once this proton gradient is established, a stream of protons through the enzyme ATP synthase drives the production of ATP from ADP and Pi. Once the electrons at the bottom of the chain are accepted by oxygen to form water, the oxidation of glucose is complete.

Several points are fundamental to understanding how the ETC works:

- Most of these molecules are proteins that contain distinctive chemical groups where the redox events take place. The active groups include ring-containing structures called flavins or iron-sulfur complexes or iron-containing heme groups. Each of these groups is readily reduced or oxidized.

- The inner membrane of the mitochondrion also contains a molecule called **ubiquinone**, which is not a protein. Ubiquinone got its name because it belongs to a family of compounds called quinones and because it is nearly ubiquitous in organisms. It is also called **coenzyme Q** or simply **Q**. Ubiquinone consists of a carbon-containing ring attached to a long tail made up of isoprene subunits. The structure of Q determines the molecule's function. The long, isoprene-rich tail is hydrophobic. As a result, Q is lipid soluble and can move throughout the mitochondrial membrane efficiently. In contrast, all but one of the proteins in the ETC are embedded in the membrane.

- The molecules involved in processing NADH and $FADH_2$ differ in electronegativity, or their tendency to hold electrons.

Because Q and the ETC proteins can cycle between a reduced state and an oxidized state, and because they differ in electronegativity, investigators realized that it should be possible to arrange these molecules into a logical sequence. The idea

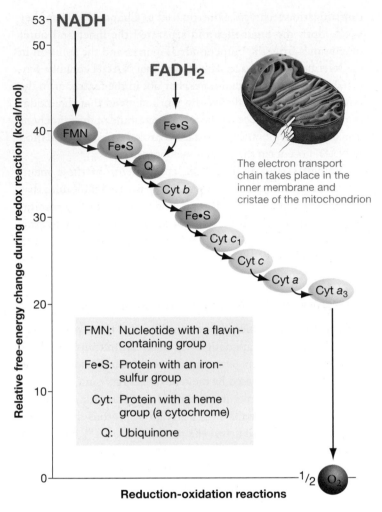

FMN: Nucleotide with a flavin-containing group

Fe•S: Protein with an iron-sulfur group

Cyt: Protein with a heme group (a cytochrome)

Q: Ubiquinone

The electron transport chain takes place in the inner membrane and cristae of the mitochondrion

FIGURE 9.22 A Series of Reduction-Oxidation Reactions Occurs in an Electron Transport Chain. Electrons step down in potential energy from the electron carriers NADH and FADH$_2$ through an electron transport chain to a final electron acceptor. When oxygen is the final electron acceptor, water is formed. The overall free-energy change of 53 kcal/mol (from NADH to oxygen) is broken into small steps.

● **EXERCISE** Label the most and least electronegative molecules in the figure.

was that electrons would pass from a molecule with lower electronegativity to one with higher electronegativity, via a redox reaction. As electrons moved through the chain, they would be held in covalent bonds more and more tightly. A small amount of energy would be released in each reaction, and the potential energy in each successive bond would lessen.

Figure 9.22 shows how electrons step down in potential energy from the electron carriers NADH and FADH$_2$ to O$_2$. Researchers worked out the sequence of compounds in the chain by experimenting with poisons that inhibit particular proteins in the inner membrane. For example, when an electron transport chain is treated with the drug antimycin A, cytochrome *b* and Q are reduced; but all of the other elements in the chain remain oxidized. This pattern makes sense only if electrons flow from NADH and FADH$_2$ to cytochrome *b* and Q before being passed on to other components.

Experiments with other poisons showed that NADH donates an electron to a flavin-containing protein at the top of the chain, while FADH$_2$ donates electrons to an iron-sulfur-containing protein that then passes them directly to Q. After passing through each of the remaining components in the chain, the electrons are finally accepted by oxygen. Under standard conditions of temperature and pressure in the laboratory, the total potential energy difference from NADH to oxygen is a whopping 53 kilocalories/mole (abbreviated kcal/mol).

To summarize, electrons from NADH and FADH$_2$ pass through an electron transport chain consisting of a series of proteins and Q. Each successive molecule in the chain has slightly higher electronegativity than the previous one. As the electrons move from one link in the chain to the next, they are held tighter and tighter and their potential energy lessens. A molecule with particularly high electronegativity—oxygen, in plants, animals, fungi, and many bacteria and archaea—acts as the final electron acceptor. Electron transport takes place within the inner mitochondrial membrane.

Once the nature of the electron transport chain became clear, biologists understood the fate of the electrons carried by NADH and FADH$_2$ and how oxygen acts as the final electron acceptor. All of the electrons that were originally present in glucose were now accounted for. This is satisfying, except for one crucial question: How do these redox reactions generate ATP? Does substrate-level phosphorylation take place in the electron transport chain, just as it does in glycolysis and the Krebs cycle?

The Chemiosmotic Hypothesis

Throughout the 1950s most biologists working on cellular respiration assumed that electron transport chains include enzymes that catalyze substrate-level phosphorylation. Despite intense efforts, however, no one was able to find a component of the ETC that phosphorylated ADP to produce ATP.

In 1961 Peter Mitchell made a radical break with prevailing ideas by proposing that the connection between electron transport and ATP production is indirect. His hypothesis: The real job of the electron transport chain is to pump protons from the matrix of the mitochondrion through the inner membrane and out to the intermembrane space or the interior of cristae. According to Mitchell, the pumping activity of the electron transport chain would lead to a buildup of protons in these areas. In this way, the intermembrane space or inside of cristae would become positively charged relative to the matrix and would have a much higher concentration of protons. The result would be a strong electrochemical gradient favoring the movement of protons back into the matrix. This **proton-motive force**, he hypothesized, is used by an enzyme in the inner membrane to synthesize ATP.

Mitchell called the production of ATP via a proton gradient **chemiosmosis**, and his proposal of an indirect link between electron transport and ATP production is the chemiosmotic hypothesis. Although proponents of a direct link between

Experiment

Question: How are the electron transport chain and ATP production linked?

Chemiosmotic hypothesis: The linkage is indirect. The ETC creates a proton-motive force that drives ATP synthesis by a mitochondrial protein.

Alternative hypothesis: The linkage is direct. The ETC is associated with enzymes that perform substrate-level phosphorylation.

Experimental setup:

Vesicle

ATP-synthesizing enzyme

1. Create vesicles from artificial membranes; add ATP-synthesizing enzyme found in mitochondria.

Bacterio-rhodopsin

2. Add bacteriorhodopsin, a protein that acts as a light-activated proton pump.

Light   Light

H^+

3. Illuminate vesicle so that bacteriorhodopsin pumps protons out of vesicle, creating a proton gradient.

Prediction of chemiosmotic hypothesis: ATP will be produced within the vesicle.

Prediction of alternative hypothesis: No ATP will be produced.

Results:

ATP is produced within the vesicle, in the absence of the electron transport chain.

Conclusion: The linkage between electron transport and ATP synthesis is indirect; the movement of protons drives the synthesis of ATP.

FIGURE 9.23 Evidence for the Chemiosmotic Hypothesis.

● **QUESTION** Do you regard this as a convincing test of the chemiosmotic hypothesis? Why or why not?

electron transport and substrate-level phosphorylation objected vigorously to Mitchell's idea, several key experiments supported it.

One of these experiments demonstrated the existence of a key element in Mitchell's scheme (**Figure 9.23**): A mitochondrial enzyme can use a proton gradient to synthesize ATP. The biologists who did this experiment began by making vesicles from artificial membranes that contained an ATP-synthesizing enzyme found in mitochondria. Along with this enzyme, they inserted bacteriorhodopsin. Bacteriorhodopsin is a well-studied membrane protein that acts as a light-activated proton pump. When light strikes this protein, the protein absorbs some of the light energy and changes conformation in a way that results in the pumping of protons from the interior of a cell or membrane-bound vesicle to the outside. As a result, the experimental vesicles established a strong electrochemical gradient favoring proton movement to the interior. When the vesicles were illuminated to initiate proton pumping, ATP began to be produced from ADP inside the vesicles. This result provided strong support for the chemiosmosis hypothesis. Mitchell's prediction was correct: In this situation, ATP production depended solely on the existence of a proton-motive force. It could occur in the *absence* of an electron transport chain.

Based on this experiment and others, most biologists accepted the chemiosmotic hypothesis as valid. Instead of being produced by enzymes inside the mitochondrion, ATP is produced by a flow of protons. In Section 9.1 you considered an analogy between the ETC system and a waterwheel. You can also liken chemiosmosis to a hydroelectric dam. The electron transport chain is analogous to a series of gigantic pumps that force water up and behind a dam. The inner mitochondrial membrane functions as the dam, and ATP synthase is like the turbines inside the dam. At a hydroelectric dam, the movement of water makes turbines spin and generate electricity. In a mitochondrion, protons are pumped instead of water. When protons move through ATP synthase, the protein spins and generates ATP. ● If you understand chemiosmosis, you should be able to explain why ATP production during cellular respiration is characterized as indirect. More specifically, you should be able to explain the relationship between glucose oxidation, the proton gradient, and ATP synthase.

Electron transport chains and ATP synthase occur in organisms throughout the tree of life. They are humming away in your cells now. Let's look in more detail at how they function.

How Is the Electron Transport Chain Organized?

Once the predictions of the chemiosmotic hypothesis were verified, researchers focused on understanding the three-dimensional structure of the components of the electron transport chain and on determining how electron transport is coupled to proton pumping. This work has shown that the

The electron transport chain occurs in the inner membrane of the mitochondrion (membranes of cristae)

FIGURE 9.24 How Is the Electron Transport Chain Organized? The individual components of the electron transport chain diagrammed in Figure 9.22 are grouped into large multiprotein complexes. Electrons are carried from one complex to another by Q and by cytochrome *c*; Q also shuttles protons across the membrane. The orange arrow indicates Q moving back and forth. Complexes I and IV use the potential energy released by the redox reactions to pump protons from the mitochondrial matrix to the intermembrane space.

● **EXERCISE** Add an arrow across the membrane and label it "Proton gradient." In the boxes at the bottom, list "What goes in" and "What comes out" for each complex.

ETC components are organized into four large complexes of proteins and cofactors (**Figure 9.24**). Protons are pumped by two of the complexes. Q and the protein **cytochrome *c*** act as shuttles that transfer electrons between complexes. Q also carries a proton across the membrane along with an electron.

The structural studies completed to date confirm that in complexes I and IV, protons actually pass directly through a sequence of electron carriers. The exact route taken by the protons is still being worked out. It is also not clear how the redox reactions taking place inside each complex—as electrons step down in potential energy—make proton movement possible.

The best-understood interaction between electron transport and proton pumping takes place in complex III. Research has shown that when Q accepts electrons from complex I or complex II, it also gains protons. The reduced form of Q then diffuses to the outer side of the inner membrane, where its electrons are used to reduce a component of complex III near the intermembrane space. The protons held by Q are released to the intermembrane space. In this way, Q shuttles electrons and protons from one side of the membrane to the other. The electrons proceed down the transport chain, and the protons released to the intermembrane space contribute to the proton-motive force.

Now, how does this proton gradient make the production of ATP possible?

The Discovery of ATP Synthase

In 1960 Efraim Racker made several key observations about how ATP is synthesized in mitochondrial membranes. When he used mitochondrial membranes to make vesicles, Racker noticed that some happened to form with their membrane inside out. Electron microscopy revealed that the inside-out membranes had numerous large proteins studded along their surfaces. Each of these proteins appeared to have a base in the membrane with a lollipop-shaped stalk and a knob that projected out from the membrane (**Figure 9.25a**). If the solution was vibrated or treated with a compound called urea, the stalks and knobs fell off.

Racker seized on this technique to isolate the stalks and knobs and do experiments with them. For example, he found that isolated stalks and knobs could hydrolyze ATP, forming

(a) A vesicle formed from an "inside-out" mitochondrial membrane

Membrane proteins with a lollipop-shaped stalk and knob

(b) The F_o unit is the base; the F_1 unit is the knob.

THE STRUCTURE OF ATP SYNTHASE

Intermembrane space

F_o unit

Stalk

Mitochondrial matrix

H^+

F_1 unit

$ADP + P_i$ → **ATP**

FIGURE 9.25 The Structure of ATP Synthase. (a) When patches of mitochondrial membrane turn inside out and form vesicles, proteins that have a lollipop-shaped stalk-and-knob structure face outward. Normally, the stalk and knob face inward, toward the mitochondrial matrix. **(b)** ATP synthase has two major components, designated F_o and F_1, connected by a stalk.

ADP and inorganic phosphate. The vesicles that contained just the base component, without the stalks and knobs, could not process ATP. The base components were, however, capable of transporting protons across the membrane.

Based on these observations, Racker proposed that the stalk-and-knob component of the protein was an enzyme that both hydrolyzes and synthesizes ATP. To test this idea, he added the stalk-and-knob components back to vesicles that had been stripped of them and confirmed that the vesicles were then capable of synthesizing ATP. Follow-up work also confirmed his hypothesis that the membrane-bound base component is a proton channel.

As **Figure 9.25b** shows, the structure of this protein complex is now reasonably well understood. The ATPase "knob"

component is called the F_1 unit; the membrane-bound, proton-transporting base component is the F_o unit. The F_1 and F_o units are connected by the stalk. The entire complex is known as **ATP synthase**. According to the current model, here is how this enzyme functions: A flow of protons through the F_o unit causes the stalk connecting the two subunits to spin. By attaching long actin filaments to the stalk and examining them with a videomicroscope, researchers have been able to see the rotation, which can reach speeds of 350 revolutions per second. As the F_1 unit rotates along with the stalk, its subunits are thought to change conformation in a way that catalyzes the phosphorylation of ADP to ATP. Understanding exactly how this reaction occurs is currently the focus of intense research. ATP synthase makes most of the ATP that keeps you alive.

Oxidative Phosphorylation

As discussed earlier, formation of ATP through the combination of proton pumping by electron transport chains and the action of ATP synthase is called oxidative phosphorylation. The term is appropriate because the phosphorylation of ADP is based on the oxidation of NADH and $FADH_2$.

Figure 9.26 summarizes glucose oxidation and cellular respiration by tracing the fate of the carbon atoms and electrons in glucose. Notice that electrons from glucose are transferred to NADH and $FADH_2$, passed through the electron transport chain, and accepted by oxygen. The pumping of protons during electron transport creates the proton-motive force that drives ATP synthesis. The diagram also indicates the approximate yield of ATP from each component of the process. Recent research has shown that about 30 ATP molecules are produced from each molecule of glucose.[3] Of these, 26 ATP molecules are produced by ATP synthase. The fundamental message here? The vast majority of the "payoff" from the oxidation of glucose occurs via oxidative phosphorylation.

It is important to recognize, though, that cellular respiration can occur without oxygen. Oxygen is the electron acceptor used by all eukaryotes and a wide diversity of bacteria and archaea. Species that depend on oxygen as an electron acceptor for the ETC are said to use **aerobic** respiration and are called aerobic organisms. (The Latin root *aero–* means "air.") But many thousands of bacterial and archaeal species rely on electron acceptors other than oxygen and electron donors other than glucose. As Chapter 28 will show, some bacteria and archaea use H_2, H_2S, CH_4, or other inorganic compounds as electron donors. For species in oxygen-poor environments, nitrate (NO_3^{2-}) and sulfate (SO_4^{2-}) are particularly common electron acceptors. Cells that depend on

---

[3]Traditionally, biologists thought that 36 ATP would be synthesized for every mole of glucose oxidized. More recent work has shown that actual yield is about 30 ATP (see P.C. Hinkle et al. *Biochemistry* 30 (1991): 3576–3582). Also, it's important to note that yield varies with conditions in the cell.

SUMMARY OF CELLULAR RESPIRATION

FIGURE 9.26 A Summary of Cellular Respiration

● **EXERCISE** Across the top of the figure, write the overall reaction for the oxidation of glucose. Explain what happens to each reactant. Identify the source of each product.

electron acceptors other than oxygen are said to use **anaerobic** ("no air") respiration. Even though the starting and ending points of cellular respiration are different, these cells are still able to use electron transport chains to create a proton-motive force that drives the synthesis of ATP. In bacteria and archaea, the ETC and ATP synthase are located in the plasma membrane.

Oxygen is the most effective of all electron acceptors, however, because of its high electronegativity. Because oxygen holds electrons so tightly, the potential energy of electrons in a bond between an oxygen atom and a non-oxygen atom is low. As a result, there is a large difference between the potential energy of electrons in NADH and the potential energy of electrons bonded to an oxygen atom (see Figure 9.22). The large differential in potential energy means that the electron transport chain can generate a large proton-motive force. Cells that do not use oxygen as an electron acceptor cannot generate such a large potential energy difference. As a result, they cannot make as much ATP as do cells that use aerobic respiration. This is important: It means that anaerobic organisms tend to grow much more slowly than aerobic organisms. If cells that use anaerobic respiration compete with cells using aerobic respiration, the cells that use oxygen as an electron acceptor almost always grow faster and reproduce more. Because of this, species that rely on anaerobic respiration usually live in environments where oxygen is not present. In oxygen-free habitats, they do not have to compete with aerobic species.

What happens when oxygen or other electron acceptors are temporarily used up and are unavailable? Without oxygen or another electron acceptor in place, the electrons carried by

NADH have no place to go. When this happens, the electron transport chain stops. All of the NAD$^+$ in the cell quickly becomes NADH. This situation is life threatening. When there is no longer any NAD$^+$ to supply the reactions of glycolysis, no ATP can be produced. If NAD$^+$ cannot be regenerated somehow, the cell will die. How do cells cope?

Check Your Understanding

If you understand that...

- As electrons from NADH and FADH$_2$ move through the electron transport chain, protons are pumped into the intermembrane space of mitochondria.

- The electrochemical gradient across the inner mitochondrial membrane drives protons through ATP synthase, resulting in the production of ATP from ADP.

You should be able to...

Add paper cutouts labeled ETC and ATP synthase to the model you made in Section 9.3, and then explain the steps in electron transport and chemiosmosis, using dimes to represent electrons and pennies to represent protons.

9.7   Fermentation

● **Fermentation** is a metabolic pathway that regenerates NAD+ from stockpiles of NADH and allows glycolysis to continue producing ATP in the absence of the electron acceptor required by the ETC. It occurs when pyruvate or a molecule derived from pyruvate—*not* oxygen—accepts electrons from NADH. When NADH gets rid of electrons in this way, NAD$^+$ is produced. With NAD$^+$ present, glycolysis can continue to produce ATP via substrate-level phosphorylation and the cell can stay alive, even when electron transport chains are shut down for lack of an electron acceptor. In many cases, the molecule that is formed by the addition of an electron to pyruvate (or another electron acceptor) cannot be used by the cell. In some cases, this by-product is toxic and is excreted from the cell as waste (**Figure 9.27a**).

In organisms that usually use oxygen as an electron acceptor, fermentation is an alternative mode of energy production when oxygen supplies temporarily run out. For example, if you sprint a long distance, your muscles begin metabolizing glucose so fast that your lungs and circulatory system cannot supply oxygen rapidly enough to keep electron transport chains active. When oxygen is absent, the electron transport chains shut down and NADH cannot donate its electrons there. The pyruvate produced by glycolysis then begins to accept electrons from NADH, and fermentation takes place (**Figure 9.27b**). The result of this process, called **lactic acid fermentation**, is the formation of a product molecule called lactate and regeneration of NAD$^+$.

Figure 9.27c illustrates a different type of fermentation pathway. These reactions, called **alcohol fermentation**, occur in the fungus *Saccharomyces cerevisiae*—baker's and brewer's yeast. When these fungal cells are placed in an environment such as bread dough or a bottle of grape juice and begin growing there, they quickly use up all the available oxygen. They continue to use glycolysis to metabolize sugar, however, by enzymatically converting pyruvate to the two-carbon compound acetaldehyde. This reaction gives off carbon dioxide, which causes bread to rise and creates the bubbles in champagne and beer. Acetaldehyde then accepts electrons from NADH, forming the NAD$^+$ required to keep glycolysis going. The addition of electrons to acetaldehyde results in the formation of ethanol as a waste product. Ethanol is excreted from yeast cells as waste but functions as the active ingredient in alcoholic beverages.

Bacteria and archaea that exist exclusively through fermentation are present in phenomenal numbers in the oxygen-free environment of your small intestine and in the rumen (first stomach) of cows. The rumen is a specialized digestive organ that contains over 10^{10} (10 billion) bacterial and archaeal cells per *milliliter* of fluid. The fermentations that occur in these cells result in the production of an array of fatty acids. Cows use these fermentation by-products as a source of energy. Cattle don't actually live off grass directly—they only eat it to feed the bacteria and archaea that they do live off. Cells that employ other types of fermentations are used commercially in the production of soy sauce, tofu, yogurt, cheese, vinegar, and other products.

Even though fermentation is a widespread and commercially important type of metabolism, it is extremely inefficient compared with cellular respiration. Fermentation produces just

(a) Fermentation pathways allow cells to regenerate NAD$^+$ for glycolysis.

(b) Lactic acid fermentation occurs in humans.

(c) Alcohol fermentation occurs in yeast.

FIGURE 9.27 Fermentation Regenerates NAD$^+$ So that Glycolysis Can Continue. Among the bacteria, archaea, and eukaryotes, many types of fermentation occur.

2 molecules of ATP for each molecule of glucose metabolized, while cellular respiration produces about 30—approximately 15 times more energy per glucose molecule than fermentation. The reason for this disparity is that oxygen has much higher electronegativity than electron acceptors such as pyruvate and

acetaldehyde. As a result, the potential energy drop between the start and end of fermentation is a tiny fraction of the potential energy change that occurs during cellular respiration.

Based on these observations, it is not surprising that organisms capable of both processes never use fermentation when an appropriate electron acceptor is available for cellular respiration. Organisms that can switch between fermentation and cellular respiration that uses oxygen as an electron acceptor are called **facultative aerobes**. The term aerobe refers to using oxygen, while the adjective facultative reflects the ability to use cellular respiration when oxygen is present and fermentation when it is absent.

Check Your Understanding

If you understand that...

- Fermentation occurs in the absence of an ETC. It consists of reactions that oxidize NADH and regenerate the NAD$^+$ required for glycolysis.

You should be able to...

1) Explain why organisms that have an ETC as well as fermentation pathways never ferment pyruvate if an electron acceptor is available.

2) Predict whether organisms exist that lack an ETC entirely and produce ATP solely through fermentation.

(MB) **Web Animation** at www.masteringbio.com
Glucose Metabolism

9.8 How Does Cellular Respiration Interact with Other Metabolic Pathways?

The enzymes, products, and intermediates involved in cellular respiration and fermentation do not exist in isolation. Instead, they are part of a huge and dynamic inventory of chemicals inside the cell. Because metabolism includes thousands of different chemical reactions, the amounts and identities of molecules inside cells are constantly in flux (**Figure 9.28**). Fermentation pathways, electron transport, and other aspects of carbohydrate metabolism may be crucial to the life of a cell, but they also have to be seen as parts of a whole.

To make sense of the chemical inventory inside cells and the full scope of metabolism, it is critical to recognize that cells have two fundamental requirements to stay alive, grow, and reproduce: energy and carbon. More formally, cells need a source of high-energy electrons for generating chemical energy in the form of ATP, and a source of carbon-containing molecules that can be used to synthesize DNA, RNA, proteins, fatty acids, and other molecules. Reactions that result in the breakdown of molecules and the production of ATP are called **catabolic pathways**;

FIGURE 9.28 Metabolic Pathways Interact. A representation of a few of the thousands of chemical reactions that occur in cells. The dots represent molecules, and the lines represent enzyme-catalyzed reactions.

● **EXERCISE** Label the red dot, and circle the 10 reactions of glycolysis. Draw a box around the Krebs cycle.

reactions that result in the synthesis of larger molecules from smaller components are called **anabolic pathways**.

How do glycolysis and the Krebs cycle interact with other catabolic pathways and with anabolic pathways? Let's consider how molecules other than carbohydrates are used as fuel in eukaryotes, and then examine how molecules involved in glycolysis and the Krebs cycle are sometimes used as building blocks in the synthesis of cell components.

Processing Proteins and Fats as Fuel

Most organisms ingest, synthesize, or absorb a wide variety of carbohydrates. These molecules range from sucrose, maltose, and other simple sugars to large polymers such as glycogen and starch. As Chapter 5 noted, glycogen is the major form of stored carbohydrate in animals, while starch is the major form of stored carbohydrate in plants. Recall that both glycogen and starch are polymers of glucose, but they differ in the way their long chains of glucose branch. Using enzyme-catalyzed reactions, cells can produce glucose from glycogen, starch, and most simple sugars. Glucose and fructose can then be processed by the enzymes of the glycolytic pathway.

Carbohydrates are not the only important source of carbon compounds used in the catabolic pathways, however. As Chapter 6 pointed out, fats are highly reduced macromolecules consisting of glycerol bonded to chains of fatty acids. In cells, fats are routinely broken down by enzymes to form glycerol and acetyl CoA. Glycerol enters the glycolytic pathway once it has been oxidized and phosphorylated to form glyceraldehyde-3-phosphate—one of the intermediates in the 10-reaction sequence. Acetyl CoA enters the Krebs cycle.

Proteins can also be catabolized, meaning that they can be broken down and used to produce ATP. Once they are broken

FIGURE 9.29 Proteins, Carbohydrates, and Fats Can All Furnish Substrates for Cellular Respiration. A variety of carbohydrates can be converted to glucose and processed by glycolysis. If carbohydrates are scarce, cells can obtain high-energy compounds from fats or even proteins for ATP production. These are catabolic reactions.

down to their constituent amino acids, the amino (−NH₂) groups are removed in enzyme-catalyzed reactions. These amino groups are excreted in urine as waste. The carbon compounds that remain after this catabolic step are converted to pyruvate, acetyl CoA, and other intermediates in glycolysis and the Krebs cycle.

Figure 9.29 summarizes the catabolic pathways of carbohydrates, fats, and proteins and shows how their breakdown products feed an array of steps in glucose oxidation and cellular respiration. When all three types of molecules are available in the cell to generate ATP, carbohydrates are used up first, then fats, and finally proteins.

Anabolic Pathways Synthesize Key Molecules

Where do cells get the precursor molecules required to synthesize amino acids, RNA, DNA, phospholipids, and other cell components? Not surprisingly, the answer often involves intermediates in carbohydrate metabolism. For example,

- In humans, about half of the 21 required amino acids can be synthesized from molecules siphoned off from the Krebs cycle.

- Acetyl CoA is the starting point for anabolic pathways that result in the synthesis of fatty acids.

- The molecule that is produced by the first reaction in glycolysis can be oxidized to start the synthesis of ribose-5-phosphate—a key intermediate in the production of ribonucleotides and deoxyribonucleotides. These nucleotides, in turn, are required for manufacturing RNA and DNA.

- If ATP is abundant, pyruvate and lactate (from fermentation) can be used as a substrate in the synthesis of glucose. Excess glucose is converted to glycogen and stored.

Figure 9.30 summarizes how intermediates in carbohydrate metabolism are drawn off to synthesize macromolecules. The take-home message is that the same molecule can serve many different functions in the cell. As a result, catabolic and anabolic pathways are closely intertwined.

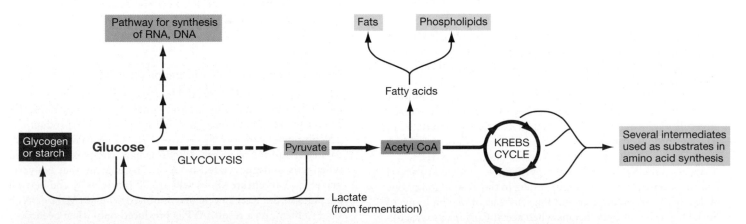

FIGURE 9.30 Intermediates in Carbohydrate Metabolism Can Be Drawn Off to Synthesize Cell Components. Several of the intermediates in carbohydrate metabolism act as precursor molecules in anabolic reactions leading to the synthesis of RNA, DNA, glycogen or starch, amino acids, and fatty acids.

Chapter Review

SUMMARY OF KEY CONCEPTS

● **In cells, the endergonic reactions required for life occur in conjunction with an exergonic reaction involving ATP.**

ATP is the currency that cells use to pump ions, drive endergonic reactions, move cargo, and perform other types of work. When proteins bind ATP or are phosphorylated, they respond by changing conformation, or shape, in a way that allows them to act as enzymes, pumps, or motors.

Almost all of the reactions that occur inside cells are endergonic. When ATP or a phosphate group from ATP is added to a substrate or enzyme that participates in an endergonic reaction, the potential energy of the substrate or enzyme is raised enough to make the reaction exergonic and thus spontaneous. In this way, ATP drives reactions that otherwise would not occur.

You should be able to explain why the addition of phosphate groups raises the potential energy of proteins, and why phosphorylation often causes proteins to change shape. ●

● **Cellular respiration produces ATP from molecules with high potential energy—often glucose.**

Cells produce ATP from sugars or other compounds with high free energy by using one of two general pathways: (1) cellular respiration or (2) fermentation. Cellular respiration is based on redox reactions that transfer electrons from a compound with high free energy, such as glucose, to a molecule with lower free energy, such as oxygen, through an electron transport chain. Fermentation involves the transfer of electrons from one organic compound to another without participation by an electron transport chain.

You should be able to explain how cellular respiration occurs in organisms that use electron donors other than glucose and in organisms that use electron acceptors other than oxygen. Which of the four steps in glucose oxidation would also occur in these organisms? Could fermentation also occur? ●

● **Glucose oxidation has four components: (1) glycolysis, (2) pyruvate processing, (3) the Krebs cycle, and (4) electron transport coupled with oxidative phosphorylation.**

In eukaryotes, glycolysis takes place in the cytosol. The Krebs cycle occurs in the mitochondrial matrix, and electron transport and oxidative phosphorylation proceed in the inner membranes of mitochondria.

Glycolysis is a 10-step reaction sequence in which glucose is broken down into two molecules of pyruvate. For each molecule of glucose processed during glycolysis, two molecules of ATP are produced by substrate-level phosphorylation and two molecules of NAD^+ are reduced to two molecules of NADH.

Prior to the Krebs cycle, each pyruvate is oxidized to acetyl CoA by a series of reactions that results in the production of one molecule of carbon dioxide and the synthesis of one molecule of NADH. The Krebs cycle begins when acetyl CoA reacts with oxaloacetate to form citrate. The ensuing series of reactions results in the regeneration of oxaloacetate, the production of 2 molecules of carbon dioxide, the synthesis of 1 molecule of ATP by substrate-level phosphorylation, the reduction of 1 molecule of FAD to $FADH_2$, and the reduction of 3 molecules of NAD^+ to 3 of NADH.

In essence, glycolysis, pyruvate processing, and the Krebs cycle strip electrons from glucose and use them to reduce NAD^+ and FAD. The resulting electron carriers—NADH and $FADH_2$—then donate electrons to an electron transport chain, which gradually steps the electrons down in potential energy until they are finally accepted by oxygen or by another final electron acceptor, resulting in the formation of water in aerobic organisms. The components of the electron transport chain use the energy released by the oxidation of NADH and $FADH_2$ to pump protons across the inner mitochondrial membrane. The pumping activity creates an electrochemical gradient, or proton-motive force, that ATP synthase uses to produce ATP. Oxidative phosphorylation is the production of ATP by electron transport and the generation of a proton-motive force.

You should be able to describe what would happen to NADH levels in a cell in the first few seconds after a drug has poisoned phosphofructokinase, the enzyme that converts acetyl CoA to citrate, cytochrome c, or ATP synthase. ●

(MB) Web Animation at www.masteringbio.com
Cellular Respiration

● **Fermentation pathways allow glycolysis to continue when the lack of an electron acceptor shuts down electron transport chains.**

If no electron acceptor such as oxygen is available, cellular respiration stops, because electron transport chains cannot continue. As a result, all NAD^+ is converted to NADH and glycolysis must stop. Fermentation pathways regenerate NAD^+ when an organic molecule such as pyruvate accepts electrons from NADH. Depending on the molecule that acts as an electron acceptor, fermentation pathways produce lactate, ethanol, or other reduced organic compounds as a by-product.

You should be able to explain why organisms that are producing ATP only via fermentation grow much more slowly than organisms that are producing ATP via cellular respiration. ●

(MB) Web Animation at www.masteringbio.com
Glucose Metabolism

● **Cellular respiration and fermentation are carefully regulated.**

When supplies of ATP, NADH, and $FADH_2$ are high in the cell, feedback regulation occurs—that is, product molecules bind to and inhibit enzymes involved in ATP production. The glycolytic pathway slows when ATP binds to phosphofructokinase, and the pyruvate dehydrogenase complex is inhibited when it is phosphorylated by ATP. The enzyme that converts acetyl CoA to citrate slows when ATP binds to it, and certain enzymes in the Krebs cycle are inhibited when NADH or ATP bind to them. As a result, ATP is produced only when needed.

You should be able to draw a graph predicting how the rate of ATP production changes as a function of ATP concentration. (Write "ATP concentration" on the x-axis and "ATP production" on the y-axis.) ●

QUESTIONS

● Test Your Knowledge

1. When does feedback inhibition occur?
   a. when lack of an appropriate electron acceptor makes an electron transport chain stop
   b. when an enzyme that is active early in a metabolic pathway is inhibited by a product of the pathway
   c. when ATP synthase reverses and begins pumping protons out of the mitochondrial matrix
   d. when cellular respiration is inhibited and fermentation begins

2. Where does the Krebs cycle occur in eukaryotes?
   a. in the cytosol
   b. in the matrix of mitochondria
   c. in the inner membrane of mitochondria
   d. in the intermembrane space of mitochondria

3. What does the chemiosmotic hypothesis claim?
   a. Substrate-level phosphorylation occurs in the electron transport chain.
   b. Substrate-level phosophorylation occurs in glycolysis and the Krebs cycle.
   c. The electron transport chain is located in the inner membrane of mitochondria.
   d. Electron transport chains generate ATP indirectly, by the creation of a proton-motive force.

4. What is the function of the reactions in a fermentation pathway?
   a. to generate NADH from NAD^+, so electrons can be donated to the electron transport chain
   b. to synthesize pyruvate from lactate
   c. to generate NAD^+ from NADH, so glycolysis can continue
   d. to synthesize electron acceptors, so that cellular respiration can continue

5. When do cells switch from cellular respiration to fermentation?
   a. when electron acceptors required by the ETC are not available
   b. when the proton-motive force runs down
   c. when NADH and $FADH_2$ supplies are low
   d. when pyruvate is not available

6. Why are NADH and $FADH_2$ said to have "reducing power"?
   a. They are the reduced forms of NAD^+ and FAD.
   b. They donate electrons to components of the ETC, reducing those components.
   c. They travel between the cytosol and the mitochondrion.
   d. They have the power to reduce carbon dioxide to glucose.

Test Your Knowledge answers: 1. b; 2. b; 3. d; 4. c; 5. a; 6. b

● Test Your Understanding

Answers are available at www.masteringbio.com

1. Explain why NADH and $FADH_2$ are called electron carriers. Where do these molecules get electrons, and where do they deliver them? In eukaryotes, what molecule do these electrons reduce?

2. Compare and contrast substrate-level phosphorylation and oxidative phosphorylation.

3. What is the relationship between cellular respiration and fermentation? Why does cellular respiration produce so much more ATP than fermentation does? Why does aerobic respiration produce much more ATP than anaerobic respiration?

4. Diagram the relationship among the four components of cellular respiration: glycolysis, pyruvate processing, the Krebs cycle, and electron transport. What molecules connect these three processes? Where does each process occur in a eukaryotic cell?

5. Explain the relationship between electron transport and oxidative phosphorylation. What does ATP synthase look like, and how does it work?

6. Describe the relationship among carbohydrate metabolism, the catabolism of proteins and fats, and anabolic pathways.

● Applying Concepts to New Situations

Answers are available at www.masteringbio.com

1. Cyanide ($C \equiv N^-$) blocks complex IV of the electron transport chain. Suggest a hypothesis for what happens to the ETC when complex IV stops working. Your hypothesis should explain why cyanide poisoning is fatal.

2. The presence of many sac-like cristae results in a large amount of membrane inside mitochondria. Suppose that some mitochondria had few cristae. How would their output of ATP compare with that of mitochondria with many cristae? Explain your answer.

3. When yeast cells are placed into low-oxygen environments, the mitochondria in the cells become reduced in size and number. Suggest an explanation for this observation.

4. Most agricultural societies have come up with ways to ferment the sugars in barley, wheat, rice, corn, or grapes to produce alcoholic beverages. Historians argue that this was an effective way for farmers to preserve the chemical energy in grains and fruits in a form that would not be eaten by rats or spoiled by bacteria or fungi. Why does a great deal of chemical energy remain in the products of fermentation pathways?

www.masteringbio.com is also your resource for • Answers to text, table, and figure caption questions and exercises • Answers to *Check Your Understanding* boxes • Online study guides and quizzes • Additional study tools including the *E-Book for Biological Science* 3rd ed., textbook art, animations, and videos.

10 | Photosynthesis

KEY CONCEPTS

- Photosynthesis consists of two linked sets of reactions. In reactions driven directly by light, ATP and the electron carrier NADPH are produced. In subsequent reactions called the Calvin cycle, the ATP and NADPH are used to reduce carbon dioxide (CO_2) to carbohydrate $(CH_2O)_n$. In eukaryotic cells, both processes take place in chloroplasts.

- The light-dependent reactions transform the energy in sunlight to chemical energy in the form of electrons with high potential energy. Excited electrons either are used to produce NADPH or are donated to an electron transport chain, which results in the production of ATP via chemiosmosis.

- The Calvin cycle starts with the enzyme rubisco, which catalyzes the addition of CO_2 to a five-carbon molecule. The resulting compound undergoes a series of reactions that use ATP and NADPH and lead to the production of sugar.

Plants and other photosynthetic organisms convert the energy in sunlight to chemical energy in the bonds of sugar. The sugar produced by photosynthetic organisms fuels cellular respiration and growth. Photosynthetic organisms, in turn, are consumed by animals, fungi, and a host of other organisms. Directly or indirectly, most organisms on Earth get their energy from photosynthesis.

About three billion years ago, a novel combination of light-absorbing molecules and enzymes gave a bacterial cell the capacity to convert light energy into chemical energy in the carbon-carbon and carbon-hydrogen bonds of sugar. When sunlight is used to manufacture carbohydrate, **photosynthesis** is said to occur. The origin of photosynthesis ranks as one of the great events in the history of life. Since this process evolved, photosynthetic organisms have dominated the Earth in terms of abundance and mass.

The vast majority of organisms alive today rely on photosynthesis, either directly or indirectly, to stay alive. Photosynthetic organisms such as trees and mosses and ferns are termed **autotrophs** (literally, "self-feeders"), because they make all of their own food from ions and simple molecules. Non-photosynthetic organisms such as humans and fungi and bacteria like *Escherichia coli* are called **heterotrophs** ("different-feeders")

because they have to obtain the sugars and many of the other macromolecules they need from other organisms. Because there could be no heterotrophs without autotrophs, photosynthesis is fundamental to almost all life. Glycolysis may qualify as the most ancient set of energy-related chemical reactions from an evolutionary viewpoint; but ecologically—meaning, in terms of how organisms interact with each other—photosynthesis is easily the most important.

All organisms perform respiration or fermentation, but only selected groups are also capable of photosynthesis. This chapter presents a step-by-step analysis of how photosynthetic species manufacture sugar from sunlight, carbon dioxide, and water. After studying how algae, plants, and other photosynthetic organisms do this remarkable chemistry, you should have a deeper appreciation for the food you eat and the oxygen you breathe.

- Key Concept   Important Information   ● Practice It

10.1   An Overview of Photosynthesis

Research on photosynthesis began early in the history of biological science. Starting in the 1770s, a series of experiments showed that photosynthesis takes place only in the green parts of plants; that sunlight, carbon dioxide (CO_2), and water (H_2O) are required; and that oxygen (O_2) is produced as a by-product. By the early 1840s enough was known about this process for biologists to propose that photosynthesis allows plants to convert electromagnetic energy in the form of sunlight into chemical energy in the bonds of carbohydrates. Eventually the overall reaction was understood to be

$$CO_2 + 2\,H_2O + \text{light energy} \rightarrow (CH_2O)_n + H_2O + O_2$$

The $(CH_2O)_n$ is a generic carbohydrate. (The subscript "n" indicates that different carbohydrates have different multiples of CH_2O.) In essence, energy from light is transformed to chemical energy in the C–H and C–C bonds of carbohydrates. When glucose is the carbohydrate that is eventually produced, the reaction can be written as

$$6\,CO_2 + 12\,H_2O + \text{light energy} \rightarrow$$

$$C_6H_{12}O_6 + 6\,O_2 + 6\,H_2O$$

Take a moment to look at this summary reaction carefully, and make note of a key point: It is the reverse of cellular respiration. Photosynthesis is an endergonic suite of reactions that reduces carbon dioxide to glucose or other sugars. Cellular respiration is an exergonic suite of reactions that oxidizes glucose to carbon dioxide and results in the production of ATP.

How does photosynthesis happen? Based on the overall reaction, early investigators assumed that CO_2 and H_2O react directly to form CH_2O, and that the oxygen atoms in carbon dioxide are given off as oxygen gas (O_2). Both hypotheses proved to be incorrect. Let's see why.

Photosynthesis: Two Linked Sets of Reactions

During the 1930s two independent lines of research on photosynthesis converged, leading to a major advance. The first research program, led by Cornelius van Niel, focused on how photosynthesis occurs in purple sulfur bacteria. Van Niel and his group found that these cells can grow in the laboratory on a food source that lacks sugars. Based on this observation, he concluded they must be autotrophs that manufacture their own carbohydrates. But to grow, the cells had to be exposed to sunlight and hydrogen sulfide (H_2S). Van Niel also showed that these cells did not produce oxygen as a by-product of photosynthesis. Instead, elemental sulfur (S) accumulated in their medium. In these organisms, the overall reaction for photosynthesis was

$$CO_2 + 2\,H_2S + \text{light energy} \rightarrow (CH_2O)_n + H_2O + 2\,S$$

Van Niel's work was crucial for two reasons. First, it showed that CO_2 and H_2O do *not* combine directly during photosynthesis. Instead of acting as a reactant in photosynthesis in these species, H_2O is a product of the process. Second, van Niel's data showed that the oxygen atoms in CO_2 are not released as oxygen gas (O_2). This conclusion was logical because no oxygen was produced by the purple sulfur bacteria, even though carbon dioxide participated in the reaction—just as it did in plants.

Based on these findings, biologists hypothesized that the oxygen atoms released during plant photosynthesis must come from H_2O. This proposal was supported by experiments with isolated chloroplasts, which produced oxygen in the presence of sunlight even if no CO_2 was present. The hypothesis was confirmed when heavy isotopes of oxygen—^{18}O compared with the normal isotope, ^{16}O—became available to researchers. Biologists then exposed algae or plants to H_2O that contained ^{18}O, collected the oxygen gas that was given off as a by-product of photosynthesis, and confirmed that the released oxygen gas contained the heavy isotope. As predicted, the reaction that produced this oxygen occurred only in the presence of sunlight. The light-dependent reactions of photosynthesis result in the production of oxygen from water.

A second major line of research helped support these discoveries. Between 1945 and 1955, a team led by Melvin Calvin began feeding labeled carbon dioxide ($^{14}CO_2$) to algae and identifying the molecules that subsequently became labeled with the radioisotope. These experiments allowed researchers to identify the exact sequence of reactions involved in reducing CO_2 to sugars. Because Calvin played an important role in this research, the reactions that reduce carbon dioxide and result in the production of sugar came to be known as the **Calvin cycle**. Later research showed that the Calvin cycle can function only if the light-dependent reactions are occurring.

To summarize: ● Early research on photosynthesis showed that the process consists of two linked sets of reactions. One set is triggered by light; the other set—the Calvin cycle—requires the products of the light-dependent reactions. The light-dependent reactions result in the production of oxygen from water; the Calvin cycle results in the production of sugar from carbon dioxide.

How are the two sets of reactions linked? The short answer is by electrons. More specifically, electrons are released when water is split to form oxygen gas. During the light-dependent reactions, these electrons are promoted to a high energy state by light and then transferred to a phosphorylated version of NAD^+, called $NADP^+$, that is abundant in photosynthesizing cells. This reaction forms NADPH. Like NADH, NADPH is an electron carrier. ATP is also produced in the light-dependent reactions (see the middle portion of **Figure 10.1**). During the Calvin cycle, the electrons in NADPH and the potential energy in ATP are used to reduce

FIGURE 10.1 Photosynthesis Has Two Linked Components. In the light-dependent reactions of photosynthesis, light energy is transformed to chemical energy in the form of ATP and NADPH. During the Calvin cycle, the ATP and NADPH produced in the light-dependent reactions are used to reduce carbon dioxide to carbohydrate.

CO_2 to carbohydrate (the right-hand side of Figure 10.1). The resulting sugars are used in cellular respiration to produce ATP for the cell. Plants oxidize sugars in their mitochondria and consume O_2 in the process, just like animals and other eukaryotes do.

Where does all this activity take place?

The Structure of the Chloroplast

Experiments with various plant tissues established that photosynthesis takes place only in the green portions of plants.

Follow-up work with light microscopes suggested that the reactions occur inside the bright green organelles called **chloroplasts** ("green-formed"). Leaf cells typically contain from 40 to 50 chloroplasts, and a square millimeter of leaf averages about 500,000 chloroplasts (**Figure 10.2a**). When membranes derived from chloroplasts were found to release oxygen after exposure to sunlight, the hypothesis that chloroplasts are the site of photosynthesis became widely accepted.

When electron microscopy became available in the 1950s, researchers observed that chloroplasts are extremely membrane rich. As **Figure 10.2b** shows, the organelle is enclosed by an

(a) Leaves contain millions of chloroplasts.

(b) Chloroplasts are highly structured, membrane-rich organelles.

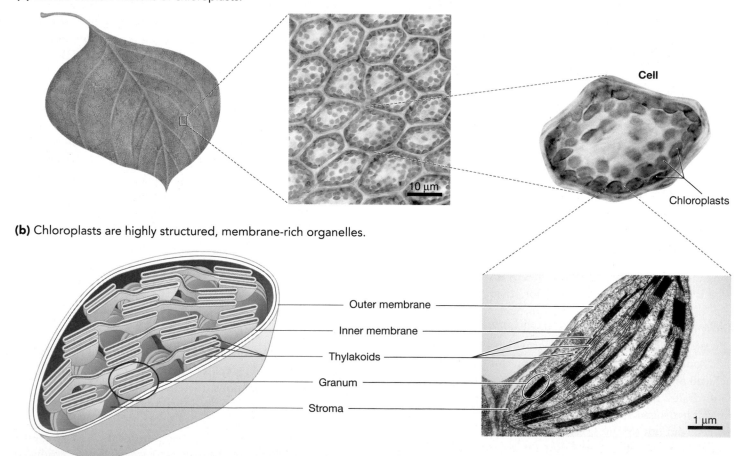

FIGURE 10.2 Photosynthesis Takes Place in Chloroplasts. (a) In plants, photosynthesis takes place in organelles called chloroplasts. **(b)** The internal membranes of chloroplasts form flattened, vesicle-like structures called thylakoids, some of which form stacks called grana.

outer membrane and an inner membrane. The interior is dominated by vesicle-like structures called **thylakoids**, which often occur in interconnected stacks called **grana** (singular: **granum**). The space inside a thylakoid is its **lumen**. (Recall that *lumen* is a general term for the interior of any sac-like structure. Your stomach and intestines have a lumen.) The fluid-filled space between the thylakoids and the inner membrane is the **stroma**.

When researchers analyzed the chemical composition of thylakoid membranes, they found huge quantities of pigments. **Pigments** are molecules that absorb only certain wavelengths of light—other wavelengths are either transmitted or reflected. Pigments have colors because we see the wavelengths that pass through or bounce off them. The most abundant pigment found in the thylakoid membranes turned out to be chlorophyll. Chlorophyll ("green-leaf") reflects or transmits green light. As a result, it is responsible for the green color of plants, some algae, and many photosynthetic bacteria.

Developmental studies showed that chloroplasts are derived from colorless organelles called proplastids ("before-plastids"). Proplastids are found in the cells of embryonic plants and in the rapidly dividing tissues of mature plants. As cells mature, proplastids develop into chloroplasts or other types of plastids specialized for that cell's particular task (see **Box 10.1**).

Before plunging into the details of how photosynthesis occurs inside a chloroplast, take a moment to consider just how astonishing the process is. Chemists have synthesized an amazing diversity of compounds from relatively simple starting materials, but their achievements pale in comparison to cells that synthesize sugar from just carbon dioxide, water, and sunlight. If it is not *the* most sophisticated chemistry on Earth, photosynthesis is certainly a contender. What's more, photosynthetic cells accomplish this feat in environments ranging from mountaintop snowfields to the open ocean, tropical rain forests, and polar ice caps. Photosynthesis can occur in virtually any habitat where light is available.

10.2 How Does Chlorophyll Capture Light Energy?

Photosynthesis begins with the light-dependent reactions. The light-dependent reactions, in turn, begin with the simple act of sunlight striking chlorophyll. To understand the consequences of this event, it's helpful to review the nature of light. Light is a type of electromagnetic radiation, which is a form of energy. The essence of photosynthesis is converting electromagnetic

---

BOX 10.1 Types of Plastids

Plastids are a family of double-membrane-bound organelles found in plants. As **Figure 10.3** shows, they develop from small, unspecialized organelles called proplastids. As a cell matures and takes on a specialized function in the plant, its proplastids also develop a distinctive form and function. If a developing plant cell in a stem or leaf is exposed to light, for example, its proplastids are usually stimulated to develop into chloroplasts.

There are three major types of plastids: (1) chloroplasts, (2) leucoplasts, and (3) chromoplasts. Figure 10.2 detailed the structure of chloroplasts. Leucoplasts ("white-formed") often function as energy storehouses. More specifically, leucoplasts may store chemical energy in the bonds of molecules with high potential energy. Some leucoplasts store oils; others synthesize and sequester starch; still others store proteins. Chromoplasts ("color-formed") are brightly colored

because they synthesize and hoard large amounts of orange, yellow, or red pigments in their vacuoles. High concentrations of chromoplasts are responsible for many of the bright colors of fruits and some flowers.

Proplastid

Has DNA, but poorly developed internal membranes

Storage
Leucoplasts are found in storage organs

Photosynthesis
Chloroplasts are found in leaves

Color
Chromoplasts are found in fruit and some flowers

FIGURE 10.3 Chloroplasts Are a Specialized Type of Plastid. Each type of plastid has a distinctive structure and function, but all plastids develop from organelles called proplastids.

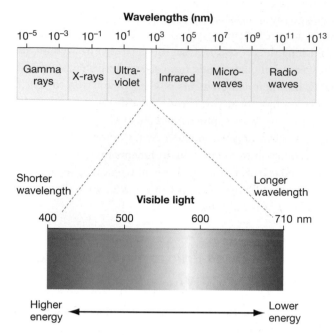

Wavelengths (nm)

Visible light

400   500   600   710 nm

Shorter wavelength   Longer wavelength

Higher energy   Lower energy

FIGURE 10.4 The Electromagnetic Spectrum. Electromagnetic energy radiates through space in the form of waves. Humans can see radiation at wavelengths between about 400 nm and about 710 nm. The shorter the wavelength of electromagnetic radiation, the higher its energy.

energy in the form of sunlight into chemical energy in the C–C and C–H bonds of sugar.

Physicists describe light's behavior as both wavelike and particle-like. As is true of all waves, including waves of water or air, electromagnetic radiation is characterized by its **wavelength**—the distance between two successive wave crests (or wave troughs). The wavelength determines the type of electromagnetic radiation.

Figure 10.4 illustrates the **electromagnetic spectrum**—the range of wavelengths of electromagnetic radiation. Humans

cannot see all these wavelengths, however. Electromagnetic radiation that humans can see is called **visible light**; it ranges in wavelength from about 400 to about 710 nanometers (nm, or 10^{-9} m). Shorter wavelengths of electromagnetic radiation contain more energy than longer wavelengths do, so blue light and ultraviolet light contain much more energy than red light and infrared light do.

To emphasize the particle-like nature of light, physicists point out that it exists in discrete packets called **photons**. In understanding photosynthesis, the important point to consider is that each photon and each wavelength of light have a characteristic amount of energy. Pigment molecules absorb this energy. How?

Photosynthetic Pigments Absorb Light

When a photon strikes an object, the photon may be absorbed, transmitted, or reflected. A pigment molecule absorbs particular wavelengths of light. Sunlight is white light, which consists of all wavelengths in the visible portion of the electromagnetic spectrum at once. If a pigment absorbs all of the visible wavelengths, no visible wavelength of light is reflected back to your eye, and the pigment appears black. If a pigment absorbs many or most of the wavelengths in the blue and green parts of the spectrum but transmits or reflects red wavelengths, it appears red.

What wavelengths do various plant pigments absorb? In one approach to answering this question, researchers grind up leaves and add a liquid that acts as a solvent for lipids. The solvent extracts pigment molecules from the leaf mixture. As **Figure 10.5a** shows, the pigments in the extract can then be separated from each other using a technique called thin layer chromatography. To begin, spots of a raw leaf extract are placed

(a) ISOLATING PIGMENTS VIA THIN LAYER CHROMATOGRAPHY

1. Grind leaves, add solvent. Pigment molecules move from leaves into solvent.

2. Spot pigments on a thin layer of porous material that coats a solid support.

3. Separate pigments in solvent.

(b) A finished chromatograph

Carotene

Pheophytin
Chlorophyll *a*
Chlorophyll *b*

Carotenoids

FIGURE 10.5 Pigments Can Be Isolated via Chromatography. (a) Thin layer chromatography is an effective way to isolate the various pigments in photosynthetic tissue. **(b)** Photosynthetic tissues, including those from the grass leaves used to make this chromatograph, typically contain several pigments. Different species of photosynthetic organisms may contain different types and quantities of pigments.

near the bottom of a stiff support that is coated with a thin layer of silica gel, cellulose, or similar porous material. The coated support is then placed in a solvent solution. As the solvent wicks upward through the coating, the pigment molecules in the mixture are carried along. Because the pigment molecules in the extract vary in size, solubility, or both, they are carried along with the solvent at different rates.

Figure 10.5b shows a chromatograph from a grass-leaf extract. Notice that this leaf contains an array of pigments. To find out which wavelengths are absorbed by each of these molecules, researchers cut out a single region (color band) of the filter paper, extract the pigment, and use an instrument called a spectrophotometer to record the wavelengths absorbed (**Box 10.2**).

Using this approach, biologists have produced data like those shown in **Figure 10.6a**. This graph is a plot of light absorbed versus wavelength and is called an **absorption spectrum**. Research based on these techniques has confirmed that there are two major classes of pigment in plant leaves: chlorophylls and carotenoids. The **chlorophylls**, designated chlorophyll *a* and chlorophyll *b*, absorb strongly in the blue and red regions of the visible spectrum. The presence of chlorophylls makes plants look green because they reflect and transmit green light, which they do not absorb. Beta-carotene (*β*-carotene)and other **carotenoids** constitute a different family of pigments, which absorb in the blue and green parts of the visible spectrum. Thus, carotenoids appear yellow, orange, or red.

Which of these wavelengths drive photosynthesis? T. W. Engelmann answered this question by laying a filamentous alga across a glass slide that was illuminated with a spectrum of colors. The idea was that the alga would begin performing photosynthesis in response to the various wavelengths of light and produce oxygen as a by-product. To determine exactly where oxygen was being produced, Engelmann added bacterial cells from a species that is attracted to oxygen. As **Figure 10.6b** shows, most of the bacteria congregated in the blue and red regions of the slide. Because wavelengths in these parts of the spectrum were associated with high oxygen concentrations, Engelmann concluded that they defined the **action spectrum** for photosynthesis. The data suggested that blue and red photons are the most effective at driving photosynthesis. Because the chlorophylls absorb these wavelengths, the data also suggested that chlorophylls are the main photosynthetic pigments.

Before analyzing what happens during the absorption event itself, let's look quickly at the structure and function of the chlorophylls and carotenoids.

What Is the Role of Carotenoids and Other Accessory Pigments?

Carotenoids are called accessory pigments because they absorb light and pass the energy on to chlorophyll. The carotenoids found in plants belong to two classes, called carotenes and xanthophylls. **Figure 10.8a** shows the structure of *β*-carotene, which gives carrots their orange color. A

(a) Different pigments absorb different wavelengths of light.

(b) Pigments that absorb blue and red photons are the most effective at triggering photosynthesis.

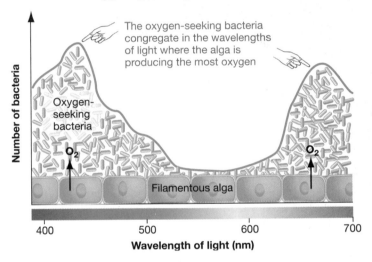

FIGURE 10.6 There Is a Strong Correlation between the Absorption Spectrum of Pigments and the Action Spectrum for Photosynthesis.

● **EXERCISE** Some photosynthetic species have versions of chlorophyll pigments that absorb most strongly at 390 nm and 880 nm. Draw what part (b) would look like if, instead of the alga shown here, these organisms were placed along the bottom.

xanthophyll called zeaxanthin, which gives corn kernels their bright yellow color, is nearly identical to *β*-carotene, except that the ring structures on either end of the molecule contain a hydroxyl (–OH) group.

Both xanthophylls and carotenes are found in chloroplasts. In autumn, when the leaves of deciduous trees begin to die, their chlorophyll degrades first. The wavelengths scattered by the carotenoids that remain turn northern forests into spectacular displays of yellow, orange, and red.

Carotenoids absorb wavelengths of light that are not absorbed by chlorophyll. As a result, they extend the range of wavelengths that can drive photosynthesis. But researchers

BOX 10.2 How Do Researchers Measure Absorption Spectra?

Researchers use an instrument called a **spectrophotometer** to measure the wavelengths of light absorbed by a particular pigment. As **Figure 10.7** shows, a solution of purified pigment molecules is exposed to a specific wavelength of light. If the pigment absorbs a great deal of this wavelength, then little light will be transmitted through the sample. But if the pigment absorbs the wavelength poorly, then most of the incoming light will be transmitted. The light that passes through the sample strikes a photoelectric tube, which converts it to an electric current. The amount of the electric current that is generated, as measured by a meter, is proportional to the intensity of the incoming light. High current readings indicate low absorption at a particular wavelength, while low current readings signal high absorption. By testing one wavelength after another, investigators can measure the full absorption spectrum of a pigment.

HOW DOES A SPECTROPHOTOMETER WORK?

White light

Prism

Screen with a slit

Pigment

Photoelectric tube

Record low absorption of green light

WAVELENGTH DATA TRANSMITTANCE ABSORBANCE CONCENTRATION FACTOR

1. Insert tube of purified pigment into spectrophotometer.

2. Instrument refracts a narrow beam of white light with a prism.

3. Slitted screen is moved to select a wavelength of light, such as 525 nm (green), to shine through the sample.

4. Pigment sample absorbs some fraction of the light; remainder is transmitted.

5. Photoelectric tube within the instrument converts light to electric current, displayed by a meter.

FIGURE 10.7 Spectrophotometers Can Measure the Light Wavelengths Absorbed by Pigments

● **QUESTION** Suppose that you were analyzing a sample of the green pigment shown here and that you changed the wavelength selector in the spectrophotometer to 480 nm (blue light). How would the amount of absorption change?

(a) β-carotene

(b) Chlorophylls *a* and *b*

CH₃ in chlorophyll *a*
CHO in chlorophyll *b*

Ring structure in "head"
(absorbs light)

discovered an even more important function by analyzing what happens to leaves when carotenoids are destroyed. Many herbicides, for example, work by inhibiting enzymes that are involved in carotenoid synthesis. Plants lacking carotenoids rapidly lose their chlorophyll, turn white, and die. Based on these results, researchers have concluded that carotenoids also serve a protective function.

To understand the molecular basis of carotenoid function, recall from Chapter 2 that energy from photons—especially the high-energy, short-wavelength photons in the ultraviolet part of the electromagnetic spectrum—contain enough energy to knock electrons out of atoms and create free radicals. Free radicals, in turn, trigger reactions that degrade molecules. Fortunately, carotenoids can "quench" free radicals by accepting or stabilizing unpaired electrons. As a result, they protect

FIGURE 10.8 Photosynthetic Pigments Contain Ring Structures. (a) β-Carotene is an orange pigment found in carrot roots and other plant tissues. **(b)** Although chlorophylls *a* and *b* are very similar structurally, they have the distinct absorption spectra shown in Figure 10.6a.

chlorophyll molecules from harm. When carotenoids are absent, chlorophyll molecules are destroyed. As a result, photosynthesis stops. Starvation and death follow.

Carotenoids are not the only molecules that protect plants from the damaging effects of sunlight, however. Researchers have recently analyzed individuals of the mustard plant *Arabidopsis thaliana* that are unable to synthesize pigments called flavonoids, which are normally stored in the vacuoles of leaf cells. Because flavonoids absorb ultraviolet radiation, individuals that lack flavonoids are subject to damage from the free radicals resulting from exposure to UV light. In effect, these pigments function as a sunscreen for leaves and stems. Without them, chlorophyll molecules would be broken apart by high-energy radiation.

The general message here is that the energy in sunlight is a double-edged sword. It makes photosynthesis possible, but it can also lead to the formation of free radicals that damage cells. The role of carotenoids and flavonoids as protective pigments is crucial.

The Structure of Chlorophyll

As **Figure 10.8b** shows, chlorophyll *a* and chlorophyll *b* are very similar in structure as well as in absorption spectra. Both have two fundamental parts: a long tail made up of isoprene subunits (introduced in Chapter 6) and a "head" consisting of a large ring structure with a magnesium atom in the middle. The tail keeps the molecule embedded in the thylakoid membrane; the head is where light is absorbed.

But just what is "absorption"? Stated another way, what happens when a photon of a particular wavelength—say, red light with a wavelength of 680 nm—strikes a chlorophyll molecule?

When Light Is Absorbed, Electrons Enter an Excited State

When a photon strikes a chlorophyll molecule, the photon's energy can be transferred to an electron in the chlorophyll molecule's head region. In response, the electron is "excited," or raised to a higher electron shell—one with greater potential energy. As **Figure 10.9** shows, the excited electron states that are possible in a particular pigment are discrete—meaning, incremental rather than continuous—and can be represented as lines on an energy scale. In the figure, the ground state, or unexcited state, is shown as 0 and the higher energy states are designated 1 and 2. If the difference between the possible energy states is the same as the energy in the photon, then the photon can be absorbed and an electron is excited to that energy state.

In chlorophyll, for example, the energy difference between the ground state and state 1 is equal to the energy in a red photon, while the energy difference between state 0 and state 2 is equal to the energy in a blue photon. Thus, chlorophyll can readily absorb red photons and blue photons. Chlorophyll does not absorb green light well, because there is no discrete

FIGURE 10.9 Electrons Are Promoted to High-Energy States When Photons Strike Chlorophyll. When a photon strikes chlorophyll, an electron can be promoted to a higher energy state, depending on the energy in the photon.

● **QUESTION** Suppose a pigment had a discrete energy state that corresponded to the energy in green light. Where would you draw this energy state on this diagram?

step—no difference in possible energy states for its electrons—that corresponds to the amount of energy in a green photon.

Wavelengths in the ultraviolet part of the spectrum have so much energy that they may actually eject electrons from a pigment molecule. In contrast, wavelengths in the infrared regions have so little energy that in most cases they merely increase the movement of atoms in the pigment, generating heat rather than exciting electrons.

If an electron is excited by a photon, it means that energy in the form of electromagnetic radiation is transferred to an electron, which now has high potential energy. What happens to the potential energy in this excited electron? If it simply falls right back to its ground state, some of the absorbed energy is released as heat—meaning molecular movement—while the rest is released as electromagnetic radiation. This is the phenomenon known as **fluorescence** (**Figure 10.10**). Because some

FIGURE 10.10 Fluorescence Occurs When Excited Electrons Fall Back to the Ground State. A pure solution of chlorophyll exposed to ultraviolet light. Electrons are excited to a high-energy state but immediately fall back to a low-energy state, emitting red photons and heat.

of the energy in the original photon is transformed to heat, the electromagnetic radiation that is given off during fluorescence has lower energy and a longer wavelength than the original photon does. When chlorophyll is in a chloroplast, though, only about 2 percent of the red and blue photons that it absorbs normally produce fluorescence. What happens to the other 98 percent of the excited electrons?

An answer to this question began to emerge when experiments showed that chlorophyll molecules work in groups—not individually. In the thylakoid membrane, 200–300 chlorophyll molecules and accessory pigments such as carotenoids are grouped together in an array of proteins, forming a complex called a **photosystem**. Each photosystem, in turn, has two major elements: an antenna complex and a reaction center.

The Antenna Complex   When a red or blue photon strikes a pigment molecule in the antenna complex, the energy is absorbed and an electron is excited in response. This energy—but not the electron itself—is passed along to a nearby chlorophyll molecule, where another electron is excited in response. This phenomenon is known as resonance. Chlorophyll molecules in the antenna complex transmit resonance energy.

Once the energy is transmitted, the original excited electron falls back to its ground state. In this way, energy is transmitted from one chlorophyll molecule to the next inside the antenna complex. The system acts like a radio antenna that receives a specific wavelength of electromagnetic radiation and transmits it to a receiver. In a photosystem, the receiver is called the reaction center.

The Reaction Center   When energy from the antenna complex reaches the reaction center of a photosystem, an all-important energy transformation event occurs. At the **reaction center**, excited electrons are transferred to a molecule that acts as an electron acceptor. When this molecule becomes reduced, the energy transformation event that started with the absorption of light becomes permanent: Electromagnetic energy is transformed to chemical energy. It cannot be reemitted as fluorescence. The redox reaction that occurs in the reaction center results in the production of chemical energy from sunlight.

The key to understanding the reaction center is that, in the absence of light, chlorophyll cannot reduce the electron acceptor because the reactions are endergonic. But when light excites electrons in chlorophyll to a high-energy state, the reactions become exergonic. In this way, a light-induced, exergonic reaction is coupled to an endergonic reaction. The chemistry of the reaction center is analogous to the way that exergonic phosphorylation reactions drive other endergonic reactions in cells (see Chapter 9).

Figure 10.11 summarizes how chlorophyll interacts with light energy by illustrating the three possible fates of electrons that are excited by photons in photosynthetic pigments. They can (1) drop back down to a low energy level and cause fluorescence, (2) excite an electron in a nearby pigment and induce resonance, or (3) be transferred to an electron acceptor in a redox reaction. Fluorescence is typical of isolated pigments, resonance occurs in antenna complex pigments, and redox occurs in reaction center pigments.

Now the question is, what happens to the high-energy electrons that are transferred to the reaction center? Specifically, how are they used to manufacture sugar?

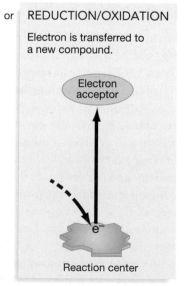

FIGURE 10.11 Three Fates for Excited Electrons in Photosynthetic Pigments. When sunlight promotes electrons in pigments to a high-energy state, three things can happen: They can fluoresce, pass energy to a nearby pigment via resonance, or transfer the electron to an electron acceptor.

Check Your Understanding

If you understand that...

- Pigments absorb certain wavelengths of light.
- When a chlorophyll molecule in the antenna complex of a chloroplast membrane absorbs red or blue light, one of its electrons is promoted to a high-energy state.
- In the antenna complex, high-energy electrons transmit their energy among chlorophyll molecules. When energy is transferred to a chlorophyll molecule in the reaction center, the electron that is excited in response reduces an electron acceptor. In this way, light energy is transformed to chemical energy.

You should be able to...

Explain which of these processes is analogous to what happens when you strike a tuning fork and then touch the vibrating tuning fork against another tuning fork.

10.3   The Discovery of Photosystems I and II

During the 1950s the fate of the high-energy electrons in photosystems was the central issue facing biologists interested in photosynthesis. Ironically, a central insight into this issue came from a simple experiment on how green algae responded to various wavelengths of light.

The experimental setup was based on the observation that photosynthetic cells respond to specific wavelengths of light. In particular, the algal cells being studied responded strongly to wavelengths of 700 nm and 680 nm, which are in the far-red and red portions of the visible spectrum, respectively. In a key experiment, Robert Emerson found that if cells were illuminated with either far-red light or red light, the photosynthetic response was moderate (**Figure 10.12**). But if cells were exposed to a combination of far-red and red light, the rate of photosynthesis increased dramatically. When both wavelengths were present, the photosynthetic rate was much more than the sum of the rates produced by each wavelength independently. This phenomenon was called the enhancement effect. Why it occurred was a complete mystery at the time.

The puzzle posed by the enhancement effect was eventually solved by Robin Hill and Faye Bendall. These biologists synthesized data emerging from an array of labs and proposed that green algae and plants have two distinct types of reaction centers rather than just one. Hill and Bendall proposed that one reaction center, which came to be called **photosystem II**, interacts with a different reaction center, now referred to as **photosystem I** (because it was discovered first). According to the two-photosystem hypothesis, the enhancement effect occurs because photosynthesis is much more efficient when both photosystems are operating together.

Experiment

Question: Light at both the red and far-red wavelengths stimulates photosynthesis. How does a combination of these wavelengths affect the rate of photosynthesis?

Hypothesis: When red light and far-red light are combined, the rate of photosynthesis will double.

Null Hypothesis: When red light and far-red light are combined, the rate of photosynthesis will not double.

Experimental setup:

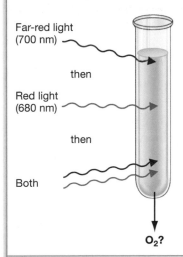

Far-red light (700 nm)

then

Red light (680 nm)

then

Both

O_2?

1. Expose algal cells to an intensity of far-red (700 nm) or red (680 nm) light that maximizes rate of photosynthesis at each wavelength. Then expose same algal cells to same intensities of each wavelength at same time.

2. Record the rate of photosynthesis as the amount of oxygen produced.

Prediction: When the wavelengths are combined, the rate of photosynthesis will be double the maximum rate observed for each wavelength independently.

Prediction of null hypothesis: When the wavelengths are combined, the rate of photosynthesis will not be double the maximum rate observed for each wavelength independently.

Results:

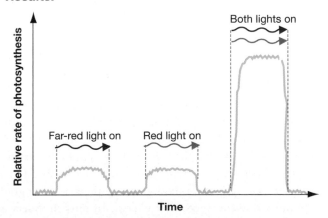

Conclusion: There is an enhancement effect for red and far-red light. The combination of 700 nm and 680 nm wavelengths more than doubles the rate of photosynthesis.

FIGURE 10.12 Discovery of the "Enhancement Effect" of Red and Far-Red Light.

Subsequent work has shown that the two-photosystem hypothesis is correct. In green algae and land plants, thylakoid membranes contain photosystems that differ in structure and function but complement each other.

To figure out how the two photosystems work, investigators chose not to study them together. Instead, they focused on species of photosynthetic bacteria that have photosystems similar to either photosystem I or II, but not both. Once each type of photosystem was understood in isolation, they turned to understanding how the two photosystems work in combination in green algae and land plants. Let's do the same—we'll analyze photosystem II, then photosystem I, and then how the two interact.

How Does Photosystem II Work?

To analyze photosystem II, researchers focused on studying species from the purple nonsulfur bacteria and the purple sulfur bacteria. These cells have a single photosystem that has many of the same components observed in photosystem II of cyanobacteria ("blue-green bacteria"), algae, and plants.

In photosystem II, the action begins when the antenna complex transmits energy to the reaction center and the molecule pheophytin comes into play. Structurally, **pheophytin** is identical to chlorophyll except that pheophytin lacks a magnesium atom in its head region. Functionally, though, the two molecules are different. Instead of acting as a pigment that promotes an electron when it absorbs a photon, pheophytin acts as an electron acceptor. When an electron in the reaction center chlorophyll is excited energetically, the electron binds to pheophytin and the reaction center chlorophyll is oxidized. When pheophytin is reduced in this way, the energy transformation step that started with the absorption of light is completed.

Electrons from Pheophytin Enter an Electron Transport Chain Electrons that reach pheophytin are passed to an electron transport chain in the thylakoid membrane. In both structure and function, this group of molecules is similar to the electron transport chain in the inner membrane of mitochondria (Chapter 9). For example, the electron transport chain associated with photosystem II contains several quinones and cytochromes. Electrons in both chains participate in a series of reduction-oxidation reactions and are gradually stepped down in potential energy. In mitochondria as well as chloroplasts, the redox reactions result in protons being pumped from one side of an internal membrane to the other. In both organelles, the resulting proton gradient drives ATP production via ATP synthase. ● Photosystem II triggers chemiosmosis and ATP synthesis in the chloroplast.

Figure 10.13a details the sequence of events in the electron transport chain of thylakoids. One of the key molecules involved is a quinone called **plastoquinone**, symbolized PQ.

Photosystem II and the cytochrome complex are located in the thylakoid membranes

(a) In photosystem II, excited electrons feed an electron transport chain.

(b) Plastoquinone carries protons to the inside of thylakoids, creating a proton-motive force.

FIGURE 10.13 Photosystem II Feeds an Electron Transport Chain That Pumps Protons. (a) When an excited electron leaves the chlorophyll molecule in the reaction center of photosystem II, the electron is accepted by pheophytin, transferred to plastoquinone (PQ), and then stepped down in energy along an electron transport chain. **(b)** PQ carries electrons from photosystem II along with protons from the stroma. The electrons are passed to the cytochrome complex, and the protons are released in the thylakoid lumen.

● **EXERCISE** Add an arrow to part (b), indicating the direction of the proton-motive force.

Recall from Chapter 9 that quinones are small hydrophobic molecules. Because plastoquinone is lipid soluble and not anchored to a protein, it is free to move from one side of the thylakoid membrane to the other. When it receives electrons from pheophytin, plastoquinone carries them to the other side of the membrane and delivers them to more electronegative molecules in the chain. These electron acceptors are found in a complex that contains a cytochrome similar to the types of cytochromes present in mitochondrial membranes.

In this way, plastoquinone shuttles electrons from pheophytin to the cytochrome complex. The electrons are then passed through a series of iron- and copper-containing proteins in the cytochrome complex. The potential energy released by these reactions allows protons to be added to other plastoquinone molecules, which carry them to the lumen side of the thylakoid membrane. As **Figure 10.13b** shows, the protons transported by plastoquinone result in a large concentration of protons in the thylakoid lumen. When photosystem II is active, the pH of the thylakoid interior reaches 5 while the pH of the stroma hovers around 8. Because the pH scale is logarithmic, the difference of 3 units means that the concentration of H^+ is $10 \times 10 \times 10 = 1000$ times higher in the lumen than in the stroma. In addition, the stroma becomes negatively charged relative to the thylakoid lumen. The net effect of electron transport, then, is to set up a large proton gradient that will drive H^+ out of the thylakoid lumen and into the stroma. Based on your reading of Chapter 9, it should come as no surprise that this proton-motive force drives the production of ATP.

ATP Synthase Uses the Proton-Motive Force to Phosphorylate ADP In mitochondria, NADH and $FADH_2$ donate electrons to an electron transport chain, and the redox reactions that occur in the chain result in protons being pumped out of the matrix and into the intermembrane space. In chloroplasts, pheophytin molecules that are part of photosystem II donate electrons to an electron transport chain, and the redox reactions that occur result in protons being pumped out of the stroma and into the thylakoid lumen.

In both mitochondria and chloroplasts, protons diffuse down the resulting electrochemical gradient. This is an exergonic process that drives the endergonic synthesis of ATP. More specifically, the flow of protons through the enzyme ATP synthase causes conformational changes that drive the phosphorylation of ADP. In plants, chemiosmosis occurs in chloroplasts as well as mitochondria.

(MB) Web Animation at www.masteringbio.com
Chemiosmosis

In photosystem II, then, the light energy captured by chlorophyll is transformed to chemical energy stored in ATP.

This process is called **photophosphorylation**. Substrate-level phosphorylation, oxidative phosphorylation, and photophosphorylation all result in the production of ATP.

To summarize, photosystem II starts with an electron being promoted to a high-energy state and ends with the production of ATP via chemisosmosis. ● If you understand these aspects of photosystem II, you should be able to compare and contrast photosystem II to the electron transfer chain (ETC) of mitochondria. You also should be able to explain where high-energy electrons come from in each system and what the eventual product is. The photosystem II story is not yet complete, however, because we haven't accounted for the electrons that flow through the system. The electron that was transferred from chlorophyll to pheophytin in the photosystem II reaction center needs to be replaced. In addition, the electron transport chain of photosystem II needs to donate its electrons to some final electron acceptor.

Where do the electrons required by photosystem II come from, and where do they go? The parallel between the photosystem in purple sulfur bacteria and photosystem II of plants ends here. In purple sulfur bacteria, cytochrome donates an electron back to the reaction center, so the same electron can again be promoted to a high-energy state when a photon is absorbed. In this way, electrons cycle through the system. But in photosystem II, electrons come from a different source.

Photosystem II Obtains Electrons by Oxidizing Water To understand where the electrons that feed photosystem II come from, think back to the overall reaction for photosynthesis: $CO_2 + 2 H_2O$ + light energy $\rightarrow (CH_2O)_n + H_2O + O_2$. In the presence of sunlight, carbon dioxide and water are used to produce carbohydrate, water, and oxygen gas. Recall that experiments with radioisotopes of oxygen showed that the oxygen atoms in O_2 come from water, not from carbon dioxide. As it turns out, the electrons that enter photosystem II also come from water. The oxygen-generating reaction can be written as

$$2 H_2O \rightarrow 4 H^+ + 4 e^- + O_2$$

Because electrons are removed from water, the molecule becomes oxidized. This reaction is referred to as "splitting" water. It supplies a steady stream of electrons for photosystem II and is catalyzed by enzymes that are physically integrated into the photosystem II complex. When excited electrons leave photosystem II and enter the ETC, the photosystem becomes so electronegative that enzymes can strip electrons away from water, leaving protons and oxygen. The oxygen-generating reaction is highly endergonic. It is possible only because the energy in sunlight drives it by removing electrons from photosystem II.

Among all life-forms, photosystem II is the only protein complex that can catalyze the splitting of water molecules.

Organisms such as cyanobacteria, algae, and plants that have this type of photosystem are said to perform **oxygenic** ("oxygen-producing") photosynthesis, because they generate oxygen as a by-product of the process. The purple sulfur and purple nonsulfur bacteria cannot oxidize water. They perform **anoxygenic** ("no oxygen-producing") photosynthesis.

It is difficult to overstate the importance of this unique reaction. The oxygen that is keeping you alive right now was produced by it. O_2 was, in fact, almost nonexistent on Earth before enzymes evolved that could catalyze the oxidation of water. According to the fossil record, oxygen levels in the atmosphere and oceans began to rise only about 2 billion years ago, as organisms that perform oxygenic photosynthesis increased in abundance. This change was a disaster for anaerobic organisms, because oxygen is toxic to them. But as oxygen became even more abundant, certain bacterial cells evolved the ability to use it as an electron acceptor during cellular respiration. This was a momentous development. O_2 is so electronegative that it creates a huge potential energy drop for the electron transport chains involved in cellular respiration. As a result, organisms that use O_2 as an electron acceptor in cellular respiration can produce much more ATP than can organisms that use other electron acceptors (see Chapter 9). Aerobic organisms grow so efficiently that they have long dominated our planet. Biologists rank the evolution of the oxygen-rich atmosphere as one of the most important events in the history of life.

Despite its fundamental importance, though, the mechanism responsible for the oxygen-generating reaction is not yet understood. Determining exactly how photosystem II splits water may be the greatest challenge currently facing researchers interested in photosynthesis. This issue has important practical applications as well, because if human chemists could replicate the reaction in an industrial setting, it might be possible to produce huge volumes of O_2 and hydrogen gas (H_2) from water. If this could be accomplished inexpensively, the H_2 produced could be used as a clean fuel for cars and trucks.

● If you understand the essence of photosystem II, you should be able to make a concept map summarizing how it works. Your diagram should include a reaction center, pheophytin, an ETC, the proton gradient, ATP synthase, and the enzyme complex that splits water.

How Does Photosystem I Work?

Recall that researchers dissected photosystem II by studying similar, but simpler, photosystems in purple nonsulfur and purple sulfur bacteria. To understand the structure and function of photosystem I, they turned to heliobacteria ("sun-bacteria").

Like purple nonsulfur and purple sulfur bacteria, heliobacteria use the energy in sunlight to promote electrons to a high-energy state. But instead of being passed to an electron transport chain that pumps protons across a membrane, the high-energy electrons in heliobacteria are used to reduce NAD^+. When NAD^+ gains two electrons and a proton, NADH is produced. In the cyanobacteria and algae and land plants, a similar set of reactions reduces a phosphorylated version of NAD^+, symbolized $NADP^+$, yielding NADPH. Both NADH and NADPH function as electron carriers.

Figure 10.14 shows how photosystem I works. When pigments in the antenna complex absorb photons and pass the energy on to the reaction center in photosystem I, excited electrons are passed through an electron transport chain. Specifically, they pass along a series of iron- and sulfur-containing proteins inside the photosystem, and then to a molecule called **ferredoxin**. The electrons then move from ferredoxin to the enzyme ferredoxin/$NADP^+$ oxidoreductase—also called $NADP^+$ reductase—which transfers two electrons and a proton to $NADP^+$. This reaction forms NADPH. The photosystem itself and $NADP^+$ reductase are anchored in the thylakoid membrane; ferredoxin is closely associated with the bilayer.

To summarize: ● Photosystem I results in the production of NADPH. **NADPH** is similar in function to the NADH and $FADH_2$ produced by the Krebs cycle. It is an electron carrier that can donate electrons to other compounds and thus reduce them. Photosystem II, in contrast, results in the production of a proton gradient that drives the synthesis of ATP.

In combination, then, photosystems I and II produce chemical energy stored in ATP as well as reducing power in the form of NADPH. Although several groups of bacteria have just one of the two photosystems, the cyanobacteria, algae, and plants have both. In these organisms, how do the two photosystems interact?

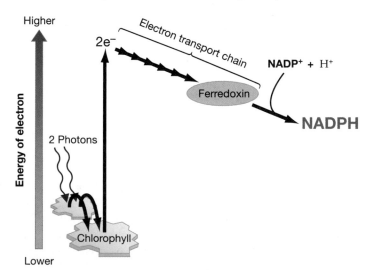

FIGURE 10.14 Photosystem I Produces NADPH. When excited electrons leave the chlorophyll molecule in the reaction center of photosystem I, they pass through a series of iron- and sulfur-containing proteins until they are accepted by ferredoxin. In an enzyme-catalyzed reaction, the reduced form of ferredoxin reacts with $NADP^+$ to produce NADPH.

● **QUESTION** Is $NADP^+$ reduced or oxidized in photosystem I?

The Z Scheme: Photosystems I and II Work Together

When they realized that photosystems I and II have distinct but complementary functions, Robin Hill and Faye Bendall proposed that these systems interact as shown in **Figure 10.15**. The diagram illustrates a model, known as the **Z scheme**, that furnished a breakthrough in research on photosynthesis. The name was inspired by the shape of the proposed path of electrons through the two photosystems, when that path was plotted on a vertical axis representing the changes occurring in their potential energy.

Following the path of electrons through the Z scheme—by tracing the route of electrons through Figure 10.15 with your finger—will help drive home how photosynthesis works. The process starts when photons excite electrons in the chlorophyll molecules of photosystem II's antenna complex. When the energy in the excited electron is transmitted to the reaction center, a special pair of chlorophyll molecules named P680 passes excited electrons to pheophytin. From there the electron is gradually stepped down in potential energy through redox reactions among a series of quinones and cytochromes, which act as an electron transport chain. Using the energy released by the redox reactions, plastoquinone (PQ) carries protons across the thylakoid membrane. ATP synthase uses the resulting proton-motive force to phosphorylate ADP, creating ATP.

When electrons reach the end of photosystem II's electron transport chain, they are passed to a small diffusible protein called **plastocyanin** (symbolized PC in Figure 10.15). Plastocyanin picks up an electron from the cytochrome complex, diffuses through the lumen of the thylakoid, and donates the electron to photosystem I. A single plastocyanin molecule can shuttle over 1000 electrons per second between photosystems. In this way, plastocyanin forms a physical link between photosystem II and photosystem I.

The flow of electrons between photosystems, by means of plastocyanin, is important because it replaces electrons that are carried away from a chlorophyll molecule called P700 in the photosystem I reaction center. The electrons flowing from photosystem II to P700 are eventually transferred to the protein ferredoxin, which then passes electrons to an enzyme that catalyzes the reduction of NADP$^+$ to NADPH. The electrons that initially left photosystem II are replaced by electrons that are stripped away from water, producing oxygen gas as a by-product.

To summarize, the Z scheme traces the path of electrons from water through photosystem II to photosystem I to NADPH. ● You should be able to explain (1) the role of plastocyanin in linking the two photosystems, (2) where the electrons that flow through the system have their highest potential energy, and (3) why the Z scheme is sometimes referred to as **noncyclic electron flow.**

The Z scheme helps explain the enhancement effect in photosynthesis documented in Figure 10.12. When algal cells are illuminated with wavelengths at 680 nm, in the red portion of the spectrum, only photosystem II can run at a maximum rate. The overall rate of electron flow through the Z scheme is moderate because photosystem I's efficiency is reduced. Similarly, when cells receive only wavelengths at 700 nm, in the far red, only photosystem I is capable of peak efficiency; photosystem II is working at a below-maximum rate, so the overall rate of electron flow is reduced. But when both wavelengths are available

FIGURE 10.15 The Z Scheme Links Photosystems I and II. The Z scheme proposes that electrons from photosystem II enter photosystem I, where they are promoted to an energy state high enough to make the reduction of NADP$^+$ possible.

● **QUESTION** How do the electron transport chains in photosystems I and II compare and contrast?

FIGURE 10.16 Cyclic Photophosphorylation Produces ATP. Cyclic electron transport is an alternative to the Z scheme. Instead of being donated to $NADP^+$, electrons cycle through the system, and this results in the production of additional ATP via photophosphorylation.

at the same time, both photosystem II and photosystem I are activated by light and work at a maximum rate, leading to enhanced efficiency.

Recent evidence indicates that a different electron path also occurs in green algae and plants. This pathway, called **cyclic photophosphorylation**, is illustrated in **Figure 10.16**. During cyclic photophosphorylation, photosystem I transfers electrons back to the electron transport chain in photosystem II, to augment ATP generation through photophosphorylation. This "extra" ATP is required for the chemical reactions that reduce carbon dioxide (CO_2) and produce sugars. In this way, cyclic photophosphorylation coexists with the Z scheme and produces additional ATP.

Although the Z-scheme model has held up well under experimental tests, a major mystery remains. As **Figure 10.17** shows, photosystem II is much more abundant in the interior, stacked membranes of grana, while photosystem I and ATP synthase are much more common in the exterior, unstacked membranes. Notice that the proton gradient established by photosystem II drives protons into the stroma, meaning that the stroma is the site of ATP production. The physical separation between the photosystems is perplexing, though, considering that their functions are so tightly integrated according to the Z scheme. Why they are found in different parts of the thylakoid is currently the focus of intense research and debate.

In contrast to the Z scheme, the fate of the ATP and NADPH produced by photosystems I and II is well documented. Chloroplasts use ATP and NADPH to reduce carbon dioxide to sugar. Your life, and the life of most other organisms, depends on this process. How does it happen?

Web Animation *BioFlix* at www.masteringbio.com
Photosynthesis

	Photosystem II	Found in thylakoid membrane facing inside of grana
	Photosystem I	Found in thylakoid membrane facing stroma
	ATP synthase	
	Cytochrome complex	Equally common in both types of membrane

FIGURE 10.17 Photosystems I and II Occur in Separate Regions of Thylakoid Membranes within Grana.

● **EXERCISE** On the figure, draw the path of an electron that follows the Z scheme from photosystem II to photosystem I. Then draw the path of an electron that participates in cyclic photophosphorylation.

Check Your Understanding

If you understand that...

- Photosystem II contributes high-energy electrons to an electron transport chain that pumps protons, creating a proton-motive force that drives ATP synthase.
- Photosystem I makes NADPH.

You should be able to...

1) Make a model of the Z scheme using paper cutouts representing the following elements: the antenna systems of photosystems II and I, the ETCs in photosystems I and II, pheophytin, plastoquinone, plastocyanin, cytochrome complex, ferredoxin, the reaction that splits water, and the reduction of $NADP^+$.

2) Using dimes to represent electrons, explain how they flow through the photosystems.

10.4 How Is Carbon Dioxide Reduced to Produce Glucose?

The reactions analyzed in Section 10.3 are triggered by light. This is logical, because their entire function is focused on transforming energy—converting electromagnetic energy in the form of sunlight to chemical energy in the phosphate bonds of ATP and the electrons of NADPH. The reactions that lead to the production of sugar from carbon dioxide, in contrast, are not triggered directly by light. Instead, they depend on the ATP and NADPH produced by the light-dependent reactions of photosynthesis.

The realization that the energy transformation and carbon dioxide reduction components of photosynthesis are two separate but linked processes was a fundamental insight. Research on the CO_2-reduction reactions gained momentum just after World War II, when radioactive isotopes of carbon became available for research purposes.

The Calvin Cycle

To unravel the reaction sequence that reduces carbon dioxide, Melvin Calvin's group used the pulse-chase strategy introduced in Chapter 7. Recall that pulse-chase experiments introduce a pulse of labeled compound followed by a chase of unlabeled compound. The fate of the labeled compound is then followed through time. In this case, the researchers fed green algae a pulse of $^{14}CO_2$ followed by a large amount of unlabeled CO_2 (**Figure 10.18**). After waiting a specified amount of time, they ground up the cells to form a crude extract, separated individual molecules in the extract via chromatography, and laid X-ray film over the chromatography surface. If radioactively labeled molecules were present, the energy they emitted would expose the film and create a dark spot. The labeled compounds could then be isolated and identified.

By varying the amount of time between starting the pulse of labeled $^{14}CO_2$ and analyzing the cells, Calvin and co-workers began to piece together the sequence in which various intermediates formed. For example, when the team analyzed cells almost immediately after starting the $^{14}CO_2$ pulse, they found that the three-carbon compound 3-phosphoglycerate predominated. This result suggested that 3-phosphoglycerate was the initial product of carbon reduction. Stated another way, it appeared that carbon dioxide reacted with some unknown molecule to produce 3-phosphoglycerate.

This was an interesting result, because 3-phosphoglycerate is one of the ten intermediates in glycolysis. The finding that glycolysis and carbon reduction share intermediates was intriguing because of the relationship between the two pathways. The ATP and NADPH reactions lead to the manufacture of carbohydrate; glycolysis breaks it down. Because the two processes are related in this way, it was logical that at least some intermediates in glycolysis and CO_2 reduction are the same.

Experiment

Question: What intermediates are produced as carbon dioxide is reduced to sugar?

Hypothesis: No specific hypothesis.

Experimental setup:

$^{14}CO_2$
CO_2

1. Feed algae a pulse of $^{14}CO_2$ followed by CO_2.

2. Wait 5–60 seconds; then homogenize cells by immersing in hot alcohol.

3. Separate molecules via chromatography.

4. Lay X-ray film on chromatograph to locate radioactive label.

Prediction: No specific prediction.

Results:

3-Phosphoglycerate

Compounds produced after 5 seconds

Compounds produced after 60 seconds

Conclusion: 3-Phosphoglycerate is the first intermediate product. Other intermediates appear later.

FIGURE 10.18 Experiments Revealed the Reaction Pathway Leading to Reduction of CO_2.

● **QUESTION** Why wasn't this experiment based on a specific hypothesis and set of predictions?

As Calvin's group pieced together the sequence of events in carbon dioxide reduction, an important question remained unanswered: What compound reacts with CO_2 to produce 3-phosphoglycerate? This was the key, initial step. The group searched in vain for a two-carbon compound that might serve as the initial carbon dioxide acceptor and yield 3-phosphoglycerate. Then, while Calvin was running errands one day, it occurred to him that the molecule reacting with carbon dioxide might contain five carbons, not two. The idea was that adding CO_2 to a five-carbon molecule would produce a six-carbon compound, which could then split in half to form 2 three-carbon molecules.

Experiments to test this hypothesis confirmed that the five-carbon compound **ribulose bisphosphate (RuBP)** is the initial reactant. Eventually three phases of CO_2 reduction were worked out and became known as the Calvin cycle (**Figure 10.19**):

1. *Fixation phase.* The events begin when CO_2 reacts with RuBP. This phase "fixes" carbon dioxide by attaching it to a more complex molecule. It also leads to the production of two molecules of 3-phosphoglycerate. **Carbon fixation** is the addition of carbon dioxide to an organic compound, putting CO_2 into a biologically useful form.

2. *Reduction phase.* Next, 3-phosphoglycerate is phosphorylated by ATP and then reduced by electrons from NADPH. The product is the phosphorylated sugar **glyceraldehyde-3-phosphate (G3P)**. Some of the resulting G3P is drawn off to manufacture glucose and fructose, which are linked to form the disaccharide sucrose.

3. *Regeneration phase.* The rest of the G3P keeps the cycle going by serving as the substrate for the third phase in the cycle: reactions that result in the regeneration of RuBP.

All three phases take place in the stroma of chloroplasts. One turn of the Calvin cycle fixes one molecule of CO_2; three turns of the cycle are required to produce one molecule of G3P. ● If you understand the essence of the Calvin cycle, you should be able to explain the relationship between CO_2, G3P, RuBP, and 3-phosphoglycerate.

● The discovery of the Calvin cycle clarified how the ATP and NADPH produced by light-dependent reactions allow cells to reduce CO_2 to carbohydrate $(CH_2O)_n$. Because sugars store a great deal of potential energy, producing them takes a great deal of chemical energy. During photosynthesis, the energy required to reduce CO2 to sugar is provided by ATP and NADPH synthesized in the light-dependent reactions.

Once the reaction sequence in the Calvin cycle was confirmed, attention focused on the initial phase—the reaction between RuBP and CO_2. The reaction between RuBP and CO_2 is one of only two reactions that are unique to the Calvin cycle. Most reactions involved in reducing CO_2 also occur during glycolysis or other metabolic pathways.

The reaction between CO_2 and RuBP starts the transformation of carbon dioxide gas from the atmosphere into sugars. Plants use sugars to fuel cellular respiration and build leaves, roots, flowers, seeds, tree trunks, and other structures. Millions of non-photosynthesizer organisms—including fish, insects, fungi, and mammals—also depend on this reaction to provide the sugars they need for cellular respiration. Ecologically, the

(a) The Calvin cycle has three phases.

All three phases of the Calvin cycle take place in the stroma of chloroplasts

Fixation: 3 RuBP + 3 CO_2 ⟶ 6 3-phosphoglycerate

Reduction: 6 3-phosphoglycerate + 6 **ATP** + 6 **NADPH** ⟶ 6 G3P

Regeneration: 5 G3P + 3 **ATP** ⟶ 3 RuBP

(b) The reaction occurs in a cycle.

3 **CO₂**

Carbons are symbolized as red balls to help you follow them through the cycle

3 P ●●●●● P
RuBP

6 ●●● P
3-phosphoglycerate

Fixation of carbon dioxide

3 ADP + 3 P_i

3 **ATP**

Regeneration of RuBP from G3P

Reduction of 3-phospho-glycerate to G3P

6 **ATP**

6 ADP + 6 P_i

6 **NADPH**

6 NADP⁺ + 6 H⁺

6 ●●● P
G3P

5 G3P

1 G3P ⟹ Glucose, fructose

FIGURE 10.19 Carbon Dioxide Is Reduced in the Calvin Cycle. The reactions of the Calvin cycle depend on the products of the light-dependent reactions.

addition of CO_2 to RuBP may be the most important chemical reaction on Earth. The enzyme that catalyzes it is fundamental to all life. How does this molecule work?

The Discovery of Rubisco

To find the enzyme that fixes CO_2, Arthur Weissbach and colleagues ground up spinach leaves, purified a large series of proteins from the resulting cell extracts, and then tested each protein to see if it could catalyze the incorporation of $^{14}CO_2$ into RuBP to form 3-phosphoglycerate. Eventually they were able to isolate an enzyme that catalyzes the reaction. The enzyme turned out to be extremely abundant in leaf tissue. The researchers' data suggested that the enzyme constituted at least 10 percent of the total protein found in spinach leaves.

The CO_2-fixing enzyme was eventually purified and analyzed. Ribulose-1,5-bisphosphate carboxylase/oxygenase is its full name, but it is commonly referred to as **rubisco**. Rubisco is found in all photosynthetic organisms that use the Calvin cycle to fix carbon, and it is thought to be the most abundant enzyme on Earth. Its three-dimensional structure has been determined as well. The molecule is shaped like a cube and has a total of eight active sites where CO_2 is fixed.

Even though it has a large number of active sites, rubisco is a slow enzyme. Each active site catalyzes just three reactions per second; other enzymes typically catalyze thousands of reactions per second. Plants synthesize huge amounts of rubisco, possibly as an adaptation compensating for its lack of speed.

Besides being slow, rubisco is extremely inefficient. Rubisco is inefficient because it catalyzes the addition of O_2 to RuBP as well as the addition of CO_2 to RuBP. This is a key point. Oxygen and carbon dioxide compete at the enzyme's active sites, which slows the rate of CO_2 reduction. Why would an active site of rubisco accept both molecules? Given rubisco's importance in producing food for photosynthetic species, the observation is puzzling. It appears to be **maladaptive**—a trait that reduces the fitness of individuals. One hypothesis to explain the dual nature of the active site is based on the observation that rubisco was present in photosynthetic organisms long before the evolution of oxygenic photosynthesis. As a result, O_2 was extremely rare in the atmosphere when rubisco evolved. According to this hypothesis, rubisco's inefficiency is a historical artifact. The idea is that rubisco is adapted to an atmosphere that no longer exists—one that was extremely rich in CO_2 and poor in O_2.

Unfortunately, the reaction of O_2 with RuBP does more than simply compete with the reaction of CO_2 at the same active site. One of the molecules that results from the addition of oxygen to RuBP is processed in reactions that consume ATP and release CO_2. Part of this pathway occurs in chloroplasts, and part in peroxisomes and mitochondria. The reaction sequence resembles respiration, because it consumes

Reaction with carbon dioxide during photosynthesis:

RuBP + CO_2 ⟶ two 3-phosphoglycerate
↓
used in Calvin cycle

Reaction with oxygen during "photorespiration":

RuBP + O_2 ⟶ one 3-phosphoglycerate + one 2-phosphoglycolate
↓ ↓
used in Calvin cycle when processed, CO_2 is released and **ATP** is used

FIGURE 10.20 Photorespiration Competes with Photosynthesis. Rubisco catalyzes competing reactions with very different outcomes.

QUESTION After studying these reactions, explain why biologists say that photorespiration "undoes" photosynthesis.

oxygen and produces carbon dioxide. As a result, it is called **photorespiration**. Because photorespiration consumes energy and undoes carbon fixation, it can be considered a reverse photosynthesis (**Figure 10.20**). When photorespiration occurs, the overall rate of photosynthesis declines.

The oxygenation reaction that triggers photorespiration is favored when oxygen concentrations are high and CO_2 concentrations are low. But as long as carbon dioxide concentrations in leaves are high, the CO_2-fixation reaction is favored and photorespiration is relatively rare.

How Is Carbon Dioxide Delivered to Rubisco?

Carbon dioxide is present in the atmosphere and is continuously used as a reactant in photosynthesizing cells. It would seem straightforward, then, for CO_2 to diffuse directly into plants along a concentration gradient. But the situation is not this simple, because plants are covered with a waxy coating called a cuticle. This lipid layer prevents water from evaporating out of tissues, but it also prevents CO_2 from entering them.

How does CO_2 get into photosynthesizing tissues? A close look at a leaf surface, such as the one in **Figure 10.21a**, provides the answer. The leaf surface is dotted with openings bordered by two distinctively shaped cells. The paired cells are called **guard cells**, the opening is called a pore, and the entire structure is called a **stoma** (plural: **stomata**). If CO_2 concentrations inside the leaf are low as photosynthesis gets under way, chemical signals activate proton pumps in the membranes of guard cells. These pumps establish a charge gradient across the membrane. In response, potassium ions (K^+) move into the guard cells. When water follows along the newly created osmotic gradient, the cells swell and create a pore. As **Figure 10.21b** shows, an open stoma allows CO_2 from the atmosphere to diffuse into

(a) Leaf surfaces contain stomata.

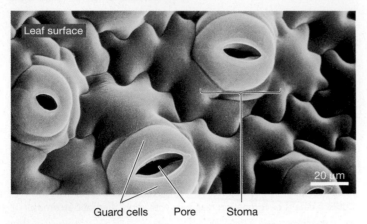

Leaf surface

20 μm

Guard cells   Pore   Stoma

(b) Carbon dioxide diffuses into leaves through stomata.

Interior of leaf

O_2
H_2O

Leaf surface

Photosynthetic   Extracellular CO_2 Stoma
cells   space

FIGURE 10.21 Leaf Cells Obtain Carbon Dioxide through Stomata.
(a) Stomata consist of two guard cells and a pore. **(b)** When a stoma is open, CO_2 diffuses into the leaf along a concentration gradient.

filled spaces inside the leaf, and from there into the extracellular fluid surrounding photosynthesizing cells. Eventually the CO_2 diffuses along a concentration gradient into the chloroplasts of the cells. A strong concentration gradient favoring entry of CO_2 is maintained by the Calvin cycle, which constantly uses up the CO_2 in chloroplasts.

Stomata are normally open during the day, when photosynthesis is occurring, and closed at night. But if the daytime is extremely hot and dry, leaf cells may begin losing a great deal of water to evaporation through their stomata. When this occurs, they must either close the openings and halt photosynthesis or risk death from dehydration. When conditions are hot and dry, then, photosynthesis and growth stop. How do plants that live in hot, dry environments cope? An answer emerged as biologists struggled to understand a surprising experimental result.

C_4 Photosynthesis After the Calvin cycle had been worked out in algae, researchers in a variety of labs used the same pulse-

C_3 plants:
RuBP + CO_2 $\xrightarrow{\text{Rubisco}}$ two 3-phosphoglycerate (3-carbon sugar)

C_4 plants:
3-carbon compound + CO_2 $\xrightarrow{\text{PEP carboxylase}}$ 4-carbon organic acids

FIGURE 10.22 Initial Carbon Fixation in C_4 Plants Is Different from That in C_3 Plants.

chase approach to investigate how carbon fixation occurs in other species. Just as Calvin had done, Hugo Kortschak and colleagues and Y. S. Karpilov and associates exposed leaves of sugarcane and maize (corn) to radioactive carbon dioxide ($^{14}CO_2$) and sunlight, and then isolated and identified the product molecules. Both research teams expected to find the first of the radioactive carbon atoms in 3-phosphoglycerate—the normal product of carbon fixation by rubisco. Instead, they found that in their species the radioactive carbon atom ended up in four-carbon compounds such as malate and aspartate—not in three-carbon sugars.

These experiments revealed a twist on the usual pathway for carbon fixation. Instead of creating a three-carbon sugar, it appeared that in their species CO_2 fixation produced four-carbon sugars. The two pathways became known as **C_3 photosynthesis** and **C_4 photosynthesis,** respectively (**Figure 10.22**).

Researchers who followed up on the initial reports found that, in some plant species, carbon dioxide can be added to RuBP by rubisco *or* to three-carbon compounds by an enzyme called **PEP carboxylase**. They also showed that the two enzymes are found in distinct cell types within the same leaf. PEP carboxylase is common in **mesophyll cells** near the surface of leaves, while rubisco is found in **bundle-sheath cells** that surround the vascular tissue in the interior of the leaf (**Figure 10.23a**). **Vascular tissue** conducts water and nutrients in plants.

Based on these observations, Hal Hatch and Roger Slack proposed a three-step model to explain how CO_2 that is fixed to a four-carbon sugar feeds the Calvin cycle (**Figure 10.23b**):

1. PEP carboxylase fixes CO_2 in mesophyll cells.

2. The four-carbon organic acids that result travel to bundle-sheath cells.

3. The four-carbon organic acids release a CO_2 molecule that rubisco uses as a substrate to form 3-phosphoglycerate. This step initiates the Calvin cycle.

In effect, then, the C_4 pathway acts as a CO_2 pump. The reactions that take place in mesophyll cells require energy in the form of ATP, but they increase CO_2 concentrations in cells where rubisco is active. Because it increases the ratio of carbon

(a) C₄ plant

FIGURE 10.23 In C₄ plants, Carbon Fixation Occurs Independently of the Calvin Cycle. (a) The carbon-fixing enzyme PEP carboxylase is located in mesophyll cells, while rubisco is in bundle-sheath cells. **(b)** CO_2 is fixed to the three-carbon compound PEP by PEP carboxylase, forming a four-carbon organic acid. A CO_2 molecule from the four-carbon sugar then feeds the Calvin cycle.

dioxide to oxygen in photosynthesizing cells, less O_2 binds to rubisco's active sites. Stated another way, CO_2 fixation is favored over O_2 fixation when carbon dioxide concentrations in leaves are high. As a result, the C₄ pathway limits the damaging effects of photorespiration. The pathway is an adaptation that keeps CO_2 concentrations in leaves high. Later experiments supported the Hatch and Slack model in almost every detail.

Follow-up work showed that the C₄ pathway is found almost exclusively in plants that thrive in hot, dry habitats. To understand why, it is critical to realize that plants have to close their stomata on hot, dry days, to avoid catastrophic water loss. When stomata close, oxygen levels build up in leaves, CO_2 concentration declines, and photorespiration is favored. PEP carboxylase is not affected by high oxygen levels, however, and has a higher affinity for CO_2 than rubisco does. Thus, PEP carboxylase can continue to deliver CO_2 to rubisco and minimize

the impact of photorespiration even when oxygen is plentiful and CO_2 is rare inside leaves. As a result, C₄ plants can continue to produce sugars on hot, dry days much more efficiently than C₃ plants can.

Sugarcane, maize (corn), and crabgrass are some familiar C₄ plants, but the pathway is actually found in several thousand species in 19 distinct lineages of flowering plants. These observations suggest that the C₄ pathway has evolved independently several times. It is not the only mechanism that plants use to continue growth under hot, dry conditions, however.

CAM Plants   Some years after the discovery of C₄ photosynthesis, researchers studying a group of flowering plants called the Crassulaceae came across a second mechanism for limiting the effects of photorespiration. This photosynthetic pathway became known as **crassulacean acid metabolism**, or **CAM**. Like the C₄ pathway, CAM is a CO_2 pump that acts as an additional, preparatory step to the Calvin cycle. It has the same effect: It increases the concentration of CO_2 inside photosynthesizing cells. It also involves an organic acid with four carbons. But unlike the C₄ pathway, CAM occurs at a different time than the Calvin cycle does—not in a different place.

CAM occurs in cacti and other species living in such hot, dry environments that individuals routinely keep their stomata closed all day. When night falls and conditions become cooler and moister, CAM plants open their stomata and take in huge quantities of CO_2. These molecules are temporarily fixed to organic acids and stored in the central vacuoles of photosynthesizing cells. During the day, the molecules are processed in reactions that release the CO_2 and feed the Calvin cycle.

Figure 10.24 summarizes the similarities and differences between C₄ photosynthesis and CAM. Both C₄ photosynthesis and CAM function as CO_2 pumps to minimize the amount of photorespiration that occurs when stomata are closed and CO_2 cannot diffuse in directly from the atmosphere. Both are found in flowering plant species that live in hot, dry environments. But while C₄ plants stockpile CO_2 in cells where rubisco is not active, CAM plants store CO_2 when rubisco is inactive. In C₄ plants, the reactions catalyzed by PEP-carboxylase and rubisco are separated in space; in CAM plants, the reactions are separated in time.

Obtaining and reducing CO_2 is fundamental to photosynthesis. In a larger sense, photosynthesis is fundamental to the millions of species that depend on plants, algae, and cyanobacteria for food. What do photosynthetic organisms do with the sugar they synthesize? More specifically, what happens to the G3P that is drawn off from the Calvin cycle?

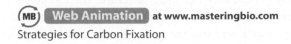 **Web Animation** at www.masteringbio.com
Strategies for Carbon Fixation

(a) C₄ plants sequester CO₂ in certain cells.

(b) CAM plants sequester CO₂ at night.

CO₂ stored in one cell ...

CO₂ stored at night ...

... and used in another.

... and used during the day.

FIGURE 10.24 C₄ Photosynthesis and CAM Accomplish the Same Task in Different Ways. (a) In C₄ plants, CO₂ is fixed to organic acids in some cells and then released to other cells where the Calvin cycle enzymes are located. **(b)** CAM plants open their stomata at night and fix CO₂ to organic acids.

What Happens to the Sugar That Is Produced by Photosynthesis?

The G3P molecules that exit the Calvin cycle enter one of several reaction pathways. The most important of these pathways results in the production of the monosaccharides glucose and fructose, which in turn combine to form the disaccharide **sucrose** (**Figure 10.25**). The reaction sequence starts with G3P from the Calvin cycle, involves a series of other phosphorylated three-carbon sugars, includes the synthesis of the familiar six-carbon sugar glucose, and ends with the production of sucrose. An alternative pathway results in the production of glucose molecules that polymerize to form **starch.** Starch production

occurs inside the chloroplast; sucrose synthesis takes place in the cytosol.

All of the intermediates involved in the production of glucose, as well as of G3P, also occur in glycolysis. The general observation here is that many of the enzymes and intermediates involved in glucose processing are common to both respiration and photosynthesis.

When photosynthesis is taking place slowly, almost all the glucose that is produced is used to make sucrose. As Chapter 5 noted, sucrose is a disaccharide ("two-sugar") consisting of a glucose molecule bonded to a fructose molecule. Sucrose is water soluble and is readily transported to other parts of the plant. If the sucrose is delivered to rapidly growing parts of the plant, it is broken down to fuel cellular respiration and growth. If it is transported to storage cells in roots, it is converted to starch and stored for later use.

When photosynthesis is proceeding rapidly and sucrose is abundant, glucose is used to synthesize starch in the chloroplasts of photosynthetic cells. Recall from Chapter 5 that starch is a polymer of glucose. In photosynthesizing cells, starch acts as a temporary sugar-storage product. Starch is not water soluble, so it cannot be transported from photosynthetic cells to other areas of the plant. At night, the starch that is temporarily stored in leaf cells is broken down and used to manufacture sucrose molecules. The sucrose is then used by the photosynthetic cell in respiration or transported to other parts of the plant. In this way, chloroplasts provide sugars for cells throughout the plant by day and by night.

If a mouse eats the starch that is stored in a chloroplast or root cell, however, the chemical energy in the C–C and C–H bonds of the starch is used to fuel the mouse's growth and reproduction. If the mouse is then eaten by an owl, the chemical energy in the mouse's tissues fuels the predator's growth and reproduction. In this way, virtually all cell growth and reproduction can be traced back to the chemical energy that was originally captured by photosynthesis. Photosynthesis is the staff of life.

Check Your Understanding

If you understand that...

- The Calvin cycle is a three-phase process: CO₂ fixation (synthesis of 3-phosphoglycerate), production of carbohydrate (synthesis of G3P), and regeneration of RUBP.

You should be able to...

1) Describe three ways that CO₂ is delivered to rubisco: (a) via organic acids in mesophyll cells, (b) via organic acids synthesized at night and stored in vacuoles, and (c) directly.

2) Explain how the solubility of starch and sucrose affect their use by plants.

FIGURE 10.25 Sucrose and Starch Are the Main Photosynthetic Products. In plants, sugars are transported in the form of sucrose and stored in the form of starch.

● **EXERCISE** Sucrose can be converted to starch, and starch to sucrose. Add an element to the diagram to indicate this.

Chapter Review

SUMMARY OF KEY CONCEPTS

Photosynthesis is the conversion of light energy to chemical energy, stored in the bonds of carbohydrates. The sucrose generated by photosynthesis fuels cellular respiration and is a substrate for the synthesis of complex carbohydrates, amino acids, fatty acids, and other cell components. As the primary food source for a diverse array of heterotrophs, photosynthetic organisms provide the energy that sustains most life on Earth.

● **Photosynthesis consists of two linked sets of reactions. In reactions driven directly by light, ATP and the electron carrier NADPH are produced. In subsequent reactions called the Calvin cycle, the ATP and NADPH are used to reduce carbon dioxide (CO_2) to carbohydrate [$(CH_2O)_n$]. In eukaryotic cells, both processes take place in chloroplasts.**

The light-dependent reactions occur in internal membranes of the chloroplast that are organized into structures called thylakoids in stacks known as grana. The Calvin cycle takes place in a fluid portion of the chloroplast called the stroma.

You should be able to explain why biologists have stopped using the phrase "light-independent reactions" to describe the Calvin cycle. ●

● **The light-dependent reactions transform the energy in sunlight to chemical energy in the form of electrons with high potential energy. Excited electrons either are used to produce NADPH or are donated to an electron transport chain, which results in the production of ATP via chemiosmosis.**

The energy transformation step of photosynthesis begins when a pigment molecule in an antenna complex absorbs a photon in the blue or red part of the visible spectrum. When absorption occurs, the energy in the photon is transferred to an electron in the pigment molecule. The energy in the excited electron is eventually transferred to chlorophyll molecules that act as reaction centers. There a high-energy electron is transferred to an electron acceptor, which becomes reduced. In this way, light energy is transformed to chemical energy.

Plants and algae have two types of reaction centers, which are part of larger complexes called photosystem I and photosystem II. Each photosystem consists of an antenna complex with 200–300 chlorophyll and carotenoid molecules, a reaction center, and an electron acceptor that completes energy transformation.

In photosystem II, high-energy electrons are accepted by the electron acceptor pheophytin. Electrons are then passed along an electron transport chain, where they are gradually stepped down in potential energy. The energy released by these redox reactions is used to pump protons across the thylakoid membrane. The resulting proton gradient drives the synthesis of ATP by ATP synthase. This method of producing ATP is called photophosphorylation. Electrons donated to the electron transport chain by photosystem II are replaced by electrons taken from water, resulting in the production of oxygen as a by-product.

In photosystem I, high-energy electrons are accepted by iron- and sulfur-containing proteins and passed to ferredoxin. In an enzyme-catalyzed reaction, the reduced form of ferredoxin passes electrons to $NADP^+$ to form NADPH.

The Z scheme describes how photosystems I and II are thought to interact. The scheme begins with the movement of an electron from photosystem II to the electron transport chain. At the end of the chain, the protein plastocyanin carries electrons to photosystem I. There the electrons are promoted to a high energy state in response to the absorption of a photon, and subsequently used to reduce $NADP^+$. Electrons from photosystem I may occasionally be passed to the electron transport chain instead of being used to reduce $NADP^+$, resulting in a cyclic flow of electrons between the two photosystems to produce the additional ATP needed to reduce carbon dioxide.

 at www.masteringbio.com

Chemiosmosis; Photosynthesis

You should be able to predict what happens to NADP levels in the first few seconds after a drug poisons photosystem I or photosystem II. You should also be able to explain why the rate of photosynthesis can be estimated by measuring the rate of oxygen production in chloroplasts. ●

● **The Calvin cycle starts with the enzyme rubisco, which catalyzes the addition of CO_2 to a five-carbon molecule. The compound that results undergoes a series of reactions that use ATP and NADPH and lead to the production of sugar.**

The CO_2-reduction reactions of photosynthesis depend on the products of the light-dependent reactions. The process begins when CO_2 is attached to a five-carbon compound called ribulose bisphosphate (RuBP). This reaction is catalyzed by the enzyme rubisco. The six-carbon compound that results immediately splits in half to form two molecules of 3-phosphoglycerate. Subsequently, 3-phosphoglycerate is reduced to a sugar called glyceraldehyde-3-phosphate (G3P). Some G3P is used to synthesize glucose and fructose, which combine to form sucrose; the rest participates in reactions that regenerate RuBP so the cycle can continue.

Rubisco catalyzes the addition of oxygen as well as carbon dioxide to RuBP. The reaction with oxygen leads to a loss of fixed CO_2 and ATP and is called photorespiration. Photorespiration is particularly problematic in hot, dry conditions, when stomata close to prevent excessive water loss. Because the closure of stomata reduces CO_2 levels in photosynthesizing cells, the reaction of O_2 with RuBP is favored. C_4 and CAM plants have distinct but functionally similar mechanisms for augmenting CO_2 concentrations in photosynthesizing cells and thus for limiting photorespiration.

 at www.masteringbio.com

Strategies for Carbon Fixation

You should be able to predict what would happen in a chloroplast containing an unusual form of rubisco—either a form that did not bind oxygen or a form that worked 10 times as fast as most forms of the enzyme. ●

QUESTIONS

● Test Your Knowledge

1. What is the stroma of a chloroplast?
   a. the inner membrane
   b. the pieces of membrane that connect grana
   c. the interior of a thylakoid
   d. the fluid inside the chloroplast but outside the thylakoids

2. Why is chlorophyll green?
   a. It absorbs all wavelengths in the visible spectrum, transmitting ultraviolet and infrared light.
   b. It absorbs wavelengths only in the red and far-red portions of the spectrum (680 nm, 700 nm).
   c. It absorbs wavelengths in the blue and red parts of the visible spectrum.
   d. It absorbs wavelengths only in the blue part of the visible spectrum and transmits all other wavelengths.

3. What does it mean to say that CO_2 becomes fixed?
   a. It becomes bonded to an organic compound.
   b. It is released during cellular respiration.
   c. It acts as an electron acceptor.
   d. It acts as an electron donor.

4. What do the light-dependent reactions of photosynthesis produce?
   a. G3P
   b. RuBP
   c. ATP and NADPH
   d. sucrose or starch

5. Why do the absorption spectrum for chlorophyll and the action spectrum for photosynthesis coincide?
   a. Photosystems I and II are activated by different wavelengths of light.
   b. Wavelengths of light that are absorbed by chlorophyll trigger the light-dependent reactions.
   c. Energy from wavelengths absorbed by carotenoids is passed on to chlorophyll.
   d. The rate of photosynthesis depends on the amount of light received.

6. What happens when an excited electron is passed to an electron acceptor in a photosystem?
   a. It drops back down to its ground state, resulting in the phenomenon known as fluorescence.
   b. The chemical energy in the excited electron is released as heat.
   c. The electron acceptor is oxidized.
   d. Energy in sunlight is transformed to chemical energy.

Test Your Knowledge answers: 1. d; 2. c; 3. a; 4. c; 5. b; 6. d

Test Your Understanding

Answers are available at www.masteringbio.com

1. Explain how the energy transformation step of photosynthesis occurs. How is light energy converted to chemical energy in the form of ATP and NADPH?

2. Explain how the carbon reduction step of photosynthesis occurs. How is carbon dioxide fixed? Why are both ATP and NADPH required to produce sugar?

3. Sketch the Z scheme. Explain how photosystem I and photosystem II interact by tracing the path of an electron through the Z scheme. What molecule connects the two photosystems?

4. When does photorespiration occur? What are its consequences for the plant?

5. Make a sketch showing how C_4 photosynthesis and CAM separate CO_2 acquisition from the Calvin cycle in space and time, respectively.

6. Why do plants need both chloroplasts and mitochondria?

Applying Concepts to New Situations

Answers are available at www.masteringbio.com

1. Compare and contrast mitochondria and chloroplasts. In what ways are their structures similar and different? What molecules or systems function in both types of organelles? Which enzymes or processes are unique to each organelle?

2. Do you agree with the hypothesis that photorespiration is an evolutionary "holdover?" Why or why not?

3. In addition to providing their protective function, carotenoids absorb certain wavelengths of light and pass the energy on to the reaction centers of photosystem I and II. Based on their function, predict exactly where carotenoids are located in the chloroplast. Explain your rationale. How would you test your hypothesis?

4. Consider plants that occupy the top, middle, or ground layer of a forest, and algae that live near the surface of the ocean or in deeper water. Would you expect the same photosynthetic pigments to be found in species that live in these different habitats? Why or why not? How would you test your hypothesis?

11 | The Cell Cycle

KEY CONCEPTS

- After chromosomes are copied, mitosis distributes one chromosome copy to each of two daughter cells. Mitosis and cytokinesis produce two cells that are identical to the parent cell.

- Over their life span, eukaryotic cells go through a cycle that consists of four carefully controlled phases.

- In multicellular organisms, uncontrolled cell growth leads to cancer. Different types of cancer result from different types of defects in control over the cell cycle.

This cell, from a hyacinth, is undergoing a type of cell division called mitosis. Understanding how mitosis occurs is a major focus of this chapter.

Chapter 1 introduced the cell theory, which maintains that all organisms are made of cells and that all cells arise from preexisting cells. Although the cell theory was widely accepted among biologists by the 1860s, a great deal of confusion remained about how cells reproduced. Most proponents of the cell theory believed that new cells arose within preexisting cells by a process that resembled the growth of mineral crystals. But Rudolf Virchow proposed that new cells arise through the division of preexisting cells—that is, **cell division**.

In the late 1800s, careful microscopic observations of newly developing individuals, or **embryos**, confirmed Virchow's hypothesis. Research documented that multicellular eukaryotes start life as single-celled embryos and grow through a series of cell divisions. As better microscopes became available, biologists were able to observe and describe the division process in more detail. In particular, researchers focused their attention on the nuclei of dividing cells and the fate of the chromosomes. Chromosomes are the carriers of hereditary material—the instructions for building and operating the cell. These early studies revealed two fundamentally different ways that nuclei divide prior to cell division. In animals, one type of nuclear division leads to the production of sperm and eggs, and the other type of nuclear division leads to the production of all other cell types. Sperm and eggs are male and female reproductive cells, termed **gametes**; all other cell types are referred to as **somatic** (literally, "body-belonging") **cells**. In both kinds of cell division, a so-called parent cell is said to give rise to daughter cells.

Key Concept   Important Information   Pratice It

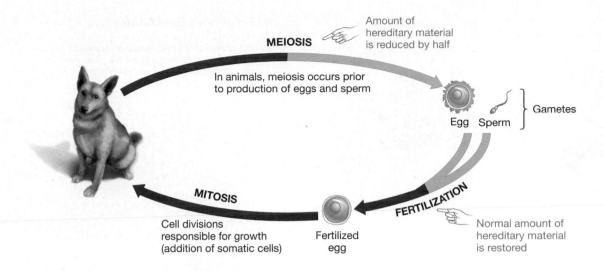

FIGURE 11.1 Two Types of Nuclear Division Occur in Many Species. In animals, meiosis leads to the production of eggs and sperm. Mitosis is responsible for producing somatic cells.

During the type of nuclear division that leads to the production of sperm and eggs, the amount of hereditary material found in the parent cell nucleus is reduced by half. As a result, the daughter cells that become sperm or eggs do not contain the same genetic material as the parent cell. This type of nuclear division is called **meiosis**. In animals, meiosis occurs only prior to the formation of gametes (**Figure 11.1**). In general, meiosis produces reproductive cells and is the basis of sexual reproduction. Chapter 12 explores the mechanism and consequences of meiosis in detail.

When new somatic cells form in eukaryotes, the amount of hereditary material in the original cell and the daughter cells remains constant. **Mitosis** is a division of the genetic material that produces daughter cells that are genetically identical to their parent cell. Mitosis is usually accompanied by **cytokinesis** ("cell movement")—the division of the cytoplasm into the two daughter cells. Together, mitosis and cytokinesis are the processes responsible for three key events in multicellular eukaryotes.

1. *Growth* The ancestry of the trillions of genetically identical cells that make up your body can be traced back through a series of mitotic divisions to a single fertilized egg—the result of the union of a sperm and an egg from your parents.

2. *Wound repair* When you suffer a scrape or cut, the cells that repair your skin and heal the wound are generated via mitosis and cytokinesis.

3. *Reproduction* When yeast cells greatly increase in number in a pile of bread dough or in a vat of beer, they are reproducing by mitosis and cytokinesis. In both unicellular and multicellular species, mitosis followed by cytokinesis is the basis of asexual reproduction. **Asexual reproduction** results in the production of offspring that are genetically identical to the parent.

This chapter's goals are to explore how mitosis occurs and how cell division is regulated. The first section introduces the relationship between mitosis and other major stages in a cell's life cycle. The next two sections provide an in-depth look at each event in mitosis and explore how mitosis and other events in the cell's life cycle are regulated. The chapter concludes by examining why uncontrolled cell division and cancer can result when the systems that regulate cell division break down.

11.1 Mitosis and the Cell Cycle

In the course of studying cell division, nineteenth-century biologists found that certain chemical dyes made threadlike structures visible in the nuclei of dividing eukaryotic cells. In 1879 Walther Flemming followed up on this discovery by documenting how the threadlike structures in salamander embryos change as the cells divide. As **Figure 11.2a** shows, the threads that Flemming observed were paired when they first appeared, just before cell division. During cell division, each pair of threads split to produce single, unpaired threads in the daughter cells. Flemming introduced the term *mitosis*, from the Greek *mitos* ("thread"), to describe this division process.

Soon after Flemming made that discovery, similar observations were reported for the roundworm *Ascaris*. In addition to confirming that each pair of threads split in this species, investigators reported that the total number of threads in a cell remained constant during subsequent divisions. Thus, all of the cells in a roundworm's body had the same number and types of threads.

In 1888 Wilhelm Waldeyer coined the term **chromosome** ("colored-body") to refer to the threadlike structures observed in dividing cells (**Figure 11.2b**). Chromosomes are

(a) **(b)**

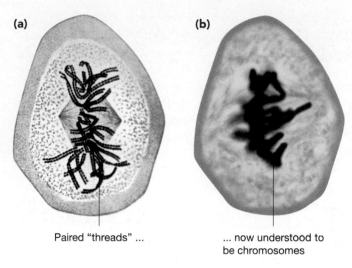

Paired "threads" ...

... now understood to
be chromosomes

FIGURE 11.2 Chromosomes Move during Mitosis. (a) Walther Flemming's 1879 drawing of mitosis in the salamander embryo. The black threads are chromosomes. **(b)** Chromosomes can be stained with dyes and observed in the light microscope.

made up in part of deoxyribonucleic acid (DNA). More specifically, a chromosome consists of a single, long DNA double helix that is wrapped around proteins in a highly organized manner. DNA encodes the cell's hereditary information, or genetic material. A gene is a length of DNA that codes for a particular protein or ribonucleic acid (RNA) found in the cell. By observing how chromosomes moved during mitosis, biologists realized that the purpose of mitosis was to distribute the parent cell's genetic material to daughter cells during cell division. Prior to mitosis, each chromosome is copied. As mitosis starts, the chromosomes also condense from long, thin, fibrous forms into compact structures that can be moved around the cell efficiently. In the remainder of mitosis, one of the chromosome copies is distributed to each of two daughter cells. **Figure 11.3** illustrates unreplicated chromosomes, replicated chromosomes before they have condensed prior to mitosis, and replicated chromosomes that have condensed at the start of mitosis.

Before delving into the details of how mitosis occurs, though, let's examine how it fits into the other events in the life of a cell.

M Phase and Interphase

As early researchers studied the fate of chromosomes during cell division, they realized that even rapidly growing plant and animal cells do not divide continuously. Instead, growing cells cycle between a dividing phase called the **mitotic** (or **M**) **phase** and a nondividing phase called **interphase** ("between-phase"). Chromosomes can be stained and observed with a light microscope only during M phase, when they are condensed into compact structures. Cells actually spend most of their time in interphase, however. No dramatic changes are observed in the nucleus during interphase, when chromosomes are uncoiled into extremely long, thin structures. Even when stains are used,

(a) Unreplicated chromosome

The unreplicated chromosome consists of a single, long DNA double helix wrapped around proteins (proteins not shown).

Gene 1   Gene 2

1 μm

The DNA replicates, resulting in two copies of the same chromosome.

(b) Replicated chromosome

Gene 1   Gene 2

Copies of same chromosome

Gene 1   Gene 2

The DNA condenses around its associated proteins, resulting in a compact chromosome that is 10,000 times shorter than its original length.

(c) Condensed replicated chromosome

Copies of same chromosome, condensed

Centromere   1μm

FIGURE 11.3 Chromosome Morphology Changes before and during Mitosis. (a) Chromosomes consist of long DNA molecules associated with proteins. Genes are segments of DNA. **(b)** Chromosomes are replicated prior to mitosis, forming two identical copies that stay attached to each other. **(c)** Early in mitosis, replicated chromosomes condense into highly compact structures, attached at a region called the centromere.

individual interphase chromosomes are not normally visible in the light microscope.

Nineteenth-century biologists knew that chromosomes moved to daughter cells during mitosis, and that each daughter cell ended up with the same number of chromosomes as the parent cell. Thus, it was logical to infer that the chromosomes had to be duplicated in the parent cell at some point in the cell cycle. When did this replication step occur?

The Discovery of the Cell Cycle

The question of when chromosomes were copied was not resolved until the early 1950s, when radioactive isotopes became available. To test hypotheses about the timing of chromosome replication, researchers exposed growing cells to a radioactive isotope that would be incorporated into DNA as it was being synthesized. The specific compounds used were radioactive phosphorus or radioactive thymidine. Phosphorus is a component of deoxyribonucleotides such as thymidine, and deoxyribonucleotides are components of DNA (see Chapter 4). The idea was to label DNA as chromosomes were being copied, then wash away any radioactive isotope that hadn't been incorporated. The labeled DNA could then be visualized by exposing the treated cells to X-ray film. Emissions from radioactive phosphorus or thymidine create a black dot in the film. This is the technique called autoradiography (see **BioSkills 7**).

When Alma Howard and Stephen Pelc did the first experiments using this approach, they simply looked to see if black dots—indicating active DNA synthesis—were present in the nuclei of M-phase cells or interphase cells. When they looked at cells soon after the exposure to a radioactive isotope ended, they found that black dots appeared only in interphase cells, but never in M-phase cells. This was strong evidence that DNA replication occurs during interphase.

Howard and Pelc had identified a new stage in the life of a cell; they called it **synthesis** (or **S**) **phase**, for DNA synthesis. S phase is part of interphase. Their data showed that duplication of the genetic material occurs independently of mitosis—the process that distributes chromosome copies to daughter cells. To describe the regular alternation between M phase and interphase, Howard and Pelc began referring to the cell cycle. The **cell cycle** is the orderly sequence of events that occurs starting from the formation of a eukaryotic cell through the duplication of its chromosomes to the time it undergoes division itself. During the cycle, two key events are (1) the replication, or copying, of the hereditary material in chromosomes and (2) the partitioning of copied chromosomes to the two daughter cells during M phase. The hereditary material is duplicated, with one copy going to each daughter cell during mitosis. As a result, daughter cells contain genetic information identical to that of the parent cell.

The Discovery of the Gap Phases

Howard and Pelc, along with researchers in other labs, followed up on these early results by asking how long the S phase lasted. The researchers used the following experimental strategy to answer this question: They exposed growing cells to radioactive phosphorus or thymidine and then waited various lengths of time before looking at the cells. In one experiment, researchers worked with cells that were growing in culture and stopped the labeling of DNA after 30 minutes of exposure to radioactive thymidine. Cultured cells are powerful experimental tools because they can be manipulated much more easily than cells in an intact organism can (see **Box 11.1**). To stop DNA from being labeled with radioactive thymidine, they flooded the solution surrounding the cultured cells with nonradioactive

BOX 11.1 Cell-Culture Methods

For researchers, there are important advantages to growing plant and animal cells outside the organism itself. Cell cultures provide large populations of a single type of cell and the opportunity to control experimental conditions precisely.

The first successful attempt to culture animal cells occurred in 1907, when a researcher cultivated amphibian nerve cells in a drop of fluid from the spinal cord. But it was not until the 1950s and 1960s that biologists could routinely culture plant and animal cells in the laboratory. The long lag time was due to the difficulty of recreating conditions that exist in the intact organism.

To grow in culture, animal cells must be provided with a liquid mixture containing the nutrients, vitamins, and hormones that stimulate growth. Initially, this mixture was provided through the use of serum, which is the liquid portion of blood; now serum-free media are available for certain cell types. Serum-free media are preferred because they are much more precisely defined chemically than serum. In addition, many types of animal cells will not grow in culture unless they are provided with a solid surface that mimics the types of surfaces to which cells in the intact organisms adhere. As a result, cells are typically cultured in flasks like the one shown in Figure 11.4.

Even under optimal conditions, though, normal cells display a finite life span in culture. In contrast, many cultured cancerous cells grow indefinitely. In culture, cancerous cells also do not adhere tightly to the surface of the culture flask. These characteristics correlate with two features of cancerous cells in organisms: they grow in a continuous and uncontrolled fashion and can break away from the site where they originated and infiltrate new tissues and organs.

Because of their immortality and relative ease of growth, cultured cancer cells are commonly used in research on basic aspects of cell structure and function. For example, the first human cell type to be grown in culture was isolated in 1951 from a malignant tumor of the uterine cervix. These cells are called HeLa cells in honor of their source, Henrietta Lacks, who died soon thereafter from the cervical cancer. HeLa cells continue to grow in laboratories around the world. They have been used in numerous studies of human cell function, including the experiments that documented chromosome replication occurring independently of mitosis.

thymidine. Stated another way, they used the pulse-chase approach introduced in Chapter 7 to label a small population of cells. Then they analyzed the labeled cells 2 hours after the end of labeling, then 4 hours after labeling, 6 hours after labeling, and so on. When they graphed the percentage of marked cells that were undergoing mitosis versus the time after exposure, they got the result indicated in **Figure 11.4**.

The cells that these researchers were studying take about 24 hours to complete one cell cycle. The cells do not divide synchronously, though—at any time, the different cells in a sample are in all different parts of interphase and M phase. The key to interpreting Figure 11.4, then, is to realize that when the pulse ended, some cells should have been just completing S phase. If they went into M phase immediately, some labeled M-phase cells should have been present just after the pulse ended. But instead, no labeled dividing cells appeared for about 4 hours. The roughly 4-hour time lag between the end of the pulse and the appearance of the first labeled mitotic nuclei corresponds to a time lag that occurs between the end of S phase and the beginning of M phase. Stated another way, there is a gap in the cycle. The gap represents the period when chromosome replication is complete but mitosis has not yet begun.

This "lag" in the cell cycle came to be called **G₂ phase**, for second gap. It was considered the second gap because data indicate that another gap exists. To follow the logic, note in Figure 11.4 that labeled nuclei undergoing mitosis are observed over a period of about 6 to 8 hours. Because all of these cells had to be somewhere in S phase—ranging from just beginning DNA synthesis to almost ending—when radioactive thymidine was available, it is logical to conclude that S phase lasts 6 to 8 hours. When the times to get through the S, G₂, and M phases are added up and compared with the 24 hours it takes these cells to complete one cell cycle, though, there is a discrepancy of 7 to 9 hours. This discrepancy represents the gap called **G₁ phase**, for first gap. As **Figure 11.5** shows, G₁ phase occurs after M phase but before S phase. In these cells, the G₁ phase is about twice as long as G₂

Why do the gap phases exist? Besides copying their chromosomes during S phase, dividing cells also must replicate organelles and manufacture additional cytoplasm. Before mitosis can take place, the parent cell must grow large enough and synthesize enough organelles that its daughter cells will be normal in size and function. The two gap phases provide the time required to accomplish these tasks. They allow the cell to complete all the requirements for cell division other than chromosome replication.

Given this overview of the major events in the life of a cell, let's turn now to M phase and delve into the process of mitosis. Once the genetic material has been copied, how do cells divide it up between daughter cells? Recall that a gene is a segment of DNA that contains the information for synthesizing a particular protein or RNA molecule. How do cells ensure that each daughter cell receives an identical complement of chromosomes

Experiment

Question: How long does S phase last?

Hypothesis: No explicit hypothesis—this experiment was exploratory in nature.

Experimental setup:

1. Feed radioactive thymidine (T*) to cells growing in culture. Only the cells synthesizing DNA (in S phase) will incorporate the T* label.

2. After 30 minutes, wash unincorporated T* out of cell culture.

3. Spread out cells and expose to photographic emulsion 2, 4, 6,...16 hours after end of labeling.

Prediction: No explicit predictions.

Results:

Because all of these M phase cells had to be somewhere in S phase when they were labeled, S phase must last about 8 hours

Conclusion: S phase is about 8 hours long. More important, there is a gap between S phase and M phase.

FIGURE 11.4 A Gap Occurs between Chromosome Replication and Mitosis. Because thymidine is incorporated into newly synthesized DNA, a short pulse of radioactive thymidine labels a population of cells that are in S phase. The graph shows the percentage of cells undergoing mitosis that are labeled versus the time since the pulse ended.

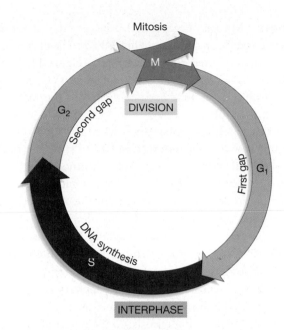

FIGURE 11.5 The Cell Cycle Has Four Phases. A representative cell cycle. The time required for the G₁ and G₂ phases varies dramatically among cells and organisms.

and thus an identical complement of genes? (The discussion that follows describes mitosis in plant cells and in animal cells; **Box 11.2** on page 230 describes cell division in bacteria.)

11.2 How Does Mitosis Take Place?

Early observations of cell division focused on the fate of the parent cell's chromosomes. Because chromosomes that are undergoing mitosis are visible under the light microscope when

they are stained, investigators could watch mitosis occur. As a result, the major events in mitosis were well understood long before the cell cycle was fully described.

Mitosis results in the division of replicated chromosomes and the formation of two daughter nuclei with identical chromosomes and genes. Mitosis is usually accompanied by cytokinesis—cytoplasmic division and the formation of two daughter cells. Let's take a closer look at the events in mitosis, beginning with observations about the nature of a eukaryotic cell's chromosomes.

Events in Mitosis

The first drawing in **Figure 11.6** shows the chromosomes found in a hypothetical plant cell or animal cell. The number of chromosomes in each cell varies widely among species. Both humans and potato plants have a total of 46 chromosomes in each cell; a maize (corn) plant has 20, dogs have 66, and fruit flies have 8. In Figure 11.6 there are a total of four chromosomes in the cell shown. (Although this cell is in interphase, the chromosomes are shown partially condensed simply to make them visible.) Eukaryotic chromosomes normally exist as extremely long, threadlike strands consisting of DNA associated with globular proteins called **histones**. In eukaryotes the DNA-protein material is called **chromatin**.

The second drawing in Figure 11.6 shows chromosomes that have been copied prior to mitosis. Each of the DNA copies in a replicated chromosome is called a **chromatid**. The two chromatids are joined together along their entire length as well as at a specialized region of the chromosome called the **centromere**. Chromatids from the same chromosome are referred to as **sister chromatids**. Sister chromatids represent

FIGURE 11.6 An Overview of Mitosis. Chromosomes are replicated prior to mitosis. During mitosis, the replicated chromosomes are partitioned to the two daughter nuclei. In most cases, mitosis is followed by cytokinesis.

● **QUESTION** In the daughter cells of mitosis, are chromosomes replicated or unreplicated?

exact copies of the same genetic information. Each chromatid contains one long DNA double helix. At the start of M phase, then, each chromosome consists of two sister chromatids that are attached to one another at the centromere. ● You should be able to explain the relationship between chromosomes and (1) genes, (2) chromatin, and (3) sister chromatids.

The third drawing in Figure 11.6 shows that as mitosis begins, chromatin condenses to form a much more compact structure. Replicated, condensed chromosomes correspond to the paired threads observed in salamander cells by early biologists (see Figure 11.2a).

● The final drawing in Figure 11.6 shows that during mitosis, the two sister chromatids separate to form independent chromosomes, and one copy of each chromosome goes to each of the two daughter cells. As a result, each daughter cell receives a copy of the genetic information that is contained in each chromosome. Every daughter cell ends up with exactly the same complement of chromosomes as the parent cell had prior to replication, and thus every daughter cell receives the same genetic information.

Although mitosis is a continuous process, biologists routinely identify several subphases within M phase on the basis of distinctive events that occur. These subphases of mitosis are designated prophase, prometaphase, metaphase, anaphase, and telophase. Some students use the mnemonic device IPPMAT to remind themselves that interphase is followed by the mitotic subphases prophase, prometaphase, metaphase, anaphase, and telophase.

To understand how mitosis proceeds, let's look at each subphase in turn.

Prophase   Mitosis begins with the events of **prophase** ("before-phase"), as shown in **Figure 11.7**. The chromosomes have already replicated during interphase (Figure 11.7, step 1); during prophase (step 2), they condense into compact structures. Chromosomes first become visible in the light microscope during prophase.

In the cytoplasm, prophase is marked by the formation of the mitotic spindle. The **mitotic spindle** is a structure that produces mechanical forces that pull chromosomes into the daughter cells during mitosis. The mitotic spindle consists of an array of microtubules—components of the cytoskeleton that were introduced in Chapter 7. Groups of microtubules attach to the chromosomes and are called **spindle fibers**. In all eukaryotes, spindle fibers originate from a microtubule organizing center. Although the nature of this organizing region varies among plants, animals, fungi, and other eukaryotic groups, the function of the spindle itself is the same. Figure 11.7 illustrates an animal cell undergoing mitosis, so the microtubule organizing center is a **centrosome**—a structure that contains a pair of **centrioles** (see Chapter 7). During prophase in all eukaryotes, the mitotic spindles either begin moving to opposite sides of the cell or form on opposite sides.

Prometaphase   Once chromosomes have condensed, the nucleolus disappears and the nuclear envelope fragments. After the nuclear envelope has disintegrated, spindle fibers from each mitotic spindle attach to one of the two sister chromatids of each chromosome. These events occur during **prometaphase** ("before middle-phase"); see Figure 11.7, step 3.

PRIOR TO MITOSIS

Chromosomes replicate.

Centrosomes   Chromosomes

Centrioles

1. Interphase: After chromosome replication, each chromosome is composed of two sister chromatids. Centrosomes have replicated.

MITOSIS

Sister chromatids separate; one chromosome copy goes to each daughter nucleus.

Early mitotic spindle   Kinetochore

Spindle fibers

2. Prophase: Chromosomes condense, and mitotic spindle begins to form.

3. Prometaphase: Nuclear envelope breaks down. Spindle fibers contact chromosomes at kinetochore.

4. Metaphase: Chromosomes complete migration to middle of cell.

FIGURE 11.7 Mitosis and Cytokinesis.

The attachment between the spindle fibers and each chromatid is made at a structure called the **kinetochore**. Kinetochores are located at the centromere region of the chromosome, where sister chromatids are attached to each other. Each chromosome has two kinetochores where spindle fibers attach—one on each side.

During prometaphase in animals, the centrosomes continue their movement to opposite poles of the cell. In all groups, the microtubules that are attached to the kinetochores begin moving the chromosomes to the middle of the cell.

Metaphase During **metaphase** ("middle-phase"), animal centrosomes complete their migration to the opposite poles of the cell (Figure 11.7, step 4). In all eukaryotes, the kinetochore microtubules finish moving the chromosomes to the middle of the cell. When metaphase is over, the chromosomes are lined up along an imaginary plane called the **metaphase plate**. At this point, the formation of the mitotic spindle is complete. Each chromatid is attached to spindle fibers that run from its kinetochore to one of the two poles of the cell. Each chromosome is held by kinetochore spindle fibers reaching to opposite poles and exerting the same amount of tension or pull. A tug of war is occurring, with kinetochore spindle fibers pulling each chromosome in opposite directions.

Anaphase At the start of **anaphase** ("against-phase"), the centromeres that are holding sister chromatids together split (Figure 11.7, step 5). Because they are under tension, sister chromatids are pulled apart equally—with the same amount of force—to create independent chromosomes. The kinetochore spindle fibers then begin to shorten. As they do, motor proteins pull the chromosomes to opposite poles of the cell. The two poles of the cell are also pushed away from each other by motor proteins associated with microtubules that are not attached to chromosomes.

During anaphase, replicated chromosomes split into two identical sets of unreplicated chromosomes. The separation of sister chromatids to opposite poles is a critical step in mitosis, because it ensures that each daughter cell receives the same complement of chromosomes. When anaphase is complete, each pole of the cell has an equivalent and complete collection of chromosomes that are identical to those present in the parent cell prior to chromosome replication.

Telophase During **telophase** ("end-phase"), a nuclear envelope begins to form around each set of chromosomes (Figure 11.7, step 6). The mitotic spindle disintegrates, and the chromosomes begin to de-condense. Once two independent nuclei have formed, mitosis is complete.

Cytokinesis

Prior to the onset of M phase, mitochondria, lysosomes, chloroplasts, and other organelles have replicated, and the rest of the cell contents have grown. During cytokinesis (Figure 11.7, step 7), the cytoplasm divides to form two daughter cells, each with its own nucleus and complete set of organelles. Cytokinesis normally occurs immediately following mitosis.

5. Anaphase: Sister chromatids separate. Chromosomes are pulled to opposite poles of the cell.

6. Telophase: The nuclear envelope re-forms, and the spindle apparatus disintegrates.

CYTOKINESIS

Cytoplasm is divided.

7. Cell division begins: Actin-myosin ring causes the plasma membrane to begin pinching in.

8. Cell division is complete: Two daughter cells form.

In animals, fungi, and slime molds, cytokinesis begins with the formation of a **cleavage furrow** (**Figure 11.9a**). The furrow appears because a ring of actin filaments forms just inside the plasma membrane, in a plane that bisects the cell. A motor protein called myosin binds to these actin filaments. When myosin binds to ATP or ADP, part of the protein moves in a way that causes actin filaments to slide (see Chapter 46). As myosin moves the ring of actin filaments on the inside of the plasma membrane, the ring shrinks in size and tightens. Because the ring is attached to the plasma membrane, the shrinking ring pulls the membrane with it. As a result, the plasma membrane pinches inward. The actin and myosin filaments continue to slide past each other, tightening the ring further, until the original membrane is pinched in two and cell division is complete.

In plants, the mechanism of cytokinesis is different. A series of microtubules and other proteins define and organize the region where the new plasma membranes and cell walls will form. Vesicles from the Golgi apparatus are then transported to the middle of the dividing cell, where they form a structure called the **cell plate** (**Figure 11.9b**). The vesicles carry components of the cell wall and plasma membrane that gradually build up, completing the cell plate and dividing the two daughter cells.

To help you review the major events in cell division, **Table 11.1** summarizes the key structures involved, and **Figure 11.10** shows photographs of cells in interphase and undergoing mitosis and cytokinesis. ● After you've studied this table and figure and reviewed Figure 11.7, you should be able to make a table with rows titled (1) microtubule organizing center and spindle fibers, (2) nuclear envelope, and (3) chromosomes, and columns titled with the five phases of mitosis. Fill in the table by summarizing what happens to the three structures during each phase of mitosis.

Once mitosis and cytokinesis had been described in detail, biologists turned their attention to understanding the molecular mechanisms that are responsible for them. In particular, researchers wanted to know two things: (1) How do sister chromatids separate to become independent chromosomes, and (2) how do those chromosomes move to daughter cells? The exact and equal partitioning of genetic material to the two daughter cells is the most fundamental aspect of mitosis. How does this process occur?

(MB) **Web Animation** *BioFlix* at www.masteringbio.com

The Phases of Mitosis; Mitosis

BOX 11.2 How Do Bacteria Divide?

To reproduce, bacterial cells divide into two genetically identical daughters. This process is called **binary fission**. **Figure 11.8** sketches the major events in involved. Recall that most bacteria contain a single, circular chromosome composed of DNA and that this chromosome is coiled upon itself. The bacterial chromosome is located in a distinctive region of the cytosol called the nucleoid, but a nucleus is lacking. Recent research has shown that after the chromosome is replicated, the two daughter chromosomes stay attached to each other for a period of time and then separate to opposite sides, or poles, of the cell. Understanding how this movement happens is currently an active area of research. As the chromosomes part, a contractile ring composed of FtsZ fibers forms between the two chromosomes. FtsZ fibers are a major component of the bacterial cytoskeleton and are similar in structure to microtubules. As the FtsZ ring closes, the bacterial cytoplasm is divided in two, completing cell division.

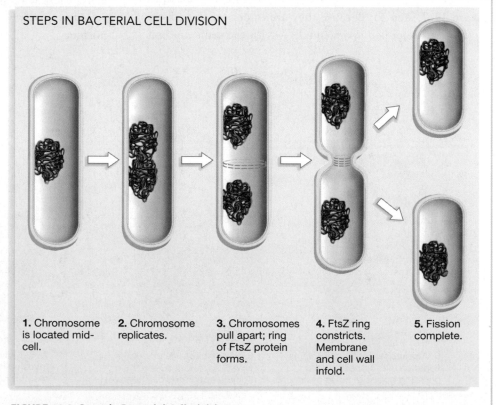

STEPS IN BACTERIAL CELL DIVISION

1. Chromosome is located mid-cell.

2. Chromosome replicates.

3. Chromosomes pull apart; ring of FtsZ protein forms.

4. FtsZ ring constricts. Membrane and cell wall infold.

5. Fission complete.

FIGURE 11.8 Steps in Bacterial Cell Division.

● **QUESTION** Are the daughter cells of bacterial cell division identical to each other in chromosomal makeup, or are they different?

(a) Cytokinesis in animals

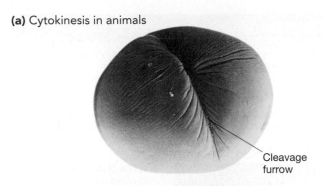

Cleavage furrow

(b) Cytokinesis in plants

Cell plate

FIGURE 11.9 The Mechanism of Cytokinesis Varies among Eukaryotes.
(a) In animals, the cytoplasm is divided by a cleavage furrow that pinches
the parent cell in two. **(b)** In plants, the cytoplasm is divided by a cell plate
that forms in the middle of the parent cell.

SUMMARY TABLE **11.1 Structures Involved in Mitosis**

Structure	Definition
Chromosome	A structure composed of a DNA molecule and associated proteins.
Chromatin	The material that makes up eukaryotic chromosomes. Consists of a DNA molecule complexed with histone proteins.
Chromatid	One strand of a replicated chromosome, with its associated proteins.
Sister chromatids	The two strands of a replicated chromosome. When chromosomes are replicated, they consist of two sister chromatids. The genetic material in sister chromatids is identical. When sister chromatids separate during mitosis, they become independent chromosomes.
Centromere	The structure that joins sister chromatids.
Kinetochore	The structure on sister chromatids where spindle fibers attach.
Microtubule organizing center	Any structure that organizes microtubules.
Centrosome	The microtubule organizing center in animals.
Centriole	Cylindrical structures that comprise micro-tubules, located inside animal centrosomes.

Interphase   Prophase   Prometaphase

Metaphase   Anaphase   Telophase   Cytokinesis

FIGURE 11.10 Mitosis and Cytokinesis in Action. Micrographs showing newt cells in interphase and undergoing mitosis
and cytokinesis. Chromosomes are stained blue, microtubules are green, and actin filaments are red.

● **EXERCISE** Next to each image, describe what is happening in less than five words. In at least two of the photographs,
label the chromosomes, centrosomes, and mitotic spindle. In the metaphase cell, label the metaphase plate.

How Do Chromosomes Move during Mitosis?

To understand how sister chromatids separate and move to daughter cells, biologists have focused on understanding how the mitotic spindle functions. Do spindle microtubules act as railroad tracks, the way they do in vesicle transport? Is some sort of motor protein involved? And what is the nature of the kinetochore, where the chromosome and microtubules are joined?

Mitotic Spindle Forces   Spindle fibers are composed of microtubules. Recall from Chapter 7 that microtubules are composed of α-tubulin and β-tubulin dimers, that the length of a microtubule is determined by the number of tubulin dimers it contains, that microtubules are asymmetric—meaning they have a plus end and a minus end—and that microtubules grow at their plus end. During mitosis, spindle fibers grow from the microtubule organizing center until their plus end attaches to the kinetochore.

These observations suggest two possible mechanisms for the movement of chromosomes during anaphase. Is the spindle microtubule shortening due to a loss of tubulin dimers from one end? Alternatively, it is possible that microtubules slide past each other, much as actin and myosin filaments do during cytokinesis in animal cells, and that this sliding action pulls chromosomes toward the poles of the cell.

To test these contrasting hypotheses, biologists introduced fluorescently labeled tubulin subunits into prophase or metaphase cells. This treatment made the entire mitotic spindle visible. (See step 1 in the "Experimental setup" section of **Figure 11.11**.) Then, once anaphase had begun, the researchers marked a region of the spindle with a bar-shaped beam of laser light. The laser quenched the fluorescence in the exposed region, making that section of the spindle dark, or "photobleached," although it was still functional (see step 2 of the "Experimental setup"). The "Results" section of Figure 11.11 shows that as anaphase progressed, two things happened: (1) the photobleached region remained stationary, and (2) The spindle fibers got shorter between the photobleached region and the kinetochore.

To explain this result, the biologists concluded that the kinetochore microtubules must remain stationary during anaphase. The hypothesis suggested that, instead of sliding past each other as actin fibers do, microtubules shorten at the kinetochore because tubulin subunits are lost from the plus ends. As kinetochore microtubules shorten at the kinetochore, the chromosomes are pulled along. How does this happen?

A Kinetochore Motor   **Figure 11.12** shows the current model for the structure and function of the kinetochore during chromosome movement. Research is continuing, however, and it is likely that the model presented in the figure will be modified as additional data become available.

Experiment

Question: How do microtubules shorten to pull sister chromatids apart at anaphase?

Hypothesis: Microtubules shorten at one end.

Alternate hypothesis: Microtubules slide past each other like actin filaments.

Experimental setup:

1. Use fluorescent labels to make the metaphase chromosomes fluoresce blue and the microtubules fluoresce yellow.

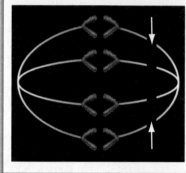

2. At the start of anaphase, photobleach a section of microtubules to mark them without changing their function.

Prediction: The photobleached section will still be visible as chromosomes begin to move.

Prediction of alternative hypothesis: The photobleached section will disappear when chromosomes begin to move.

Results:

The photobleached section remained visible, but the distance between chromosomes and photobleached section lessened.

Conclusion: Microtubules shorten at one end—at the kinetochore.

FIGURE 11.11 During Anaphase, Microtubules Shorten at the Kinetochore.

● **EXERCISE** Add drawings to show the outcomes you would expect if microtubules shorten at the end opposite the chromosome or if microtubules slide past each other.

FIGURE 11.12 How Do Microtubules Move Chromosomes during Mitosis? The kinetochore consists of an inner plate, an outer plate, and associated fibers that contain motor proteins. The fibers are thought to act as "legs" that walk the chromosomes down the length of the kinetochore microtubules as the microtubules shorten, due to loss of tubulin dimers at the kinetochore.

Note that the kinetochore is thought to have a base that attaches to the centromere region of the chromosome and a "crown" of fibrous proteins projecting outward. Recent research has shown that the kinetochore contains dyneins and other motor proteins. These motor proteins are thought to be attached to the kinetochore's fibrous crown and to be capable of "walking" down microtubules—from their plus ends near the kinetochore toward their minus ends at the spindle. This observation suggests that the mechanism of chromosome movement is reminiscent of the way that kinesin walks down microtubules during vesicle transport or that dynein walks down the microtubule doublets in flagella. As Chapter 7 explained, motor proteins such as kinesin and dynein not only attach or detach from microtubules when they bind to ATP versus ADP, they also change shape. The combination of shape changes and attachment-detachment cycles causes motor proteins to walk down the length of a microtubule or other fiber.

To put all of these ideas together, biologists hypothesize that as anaphase gets under way, proteins in the kinetochore catalyze the loss of tubulin subunits at the plus end of the spindle fiber, while dyneins and other kinetochore motor proteins walk toward the minus end. As the microtubule shortens and the detach-move-reattach cycle of the motor proteins repeats, the chromosome is pulled to one end of the mitotic spindle.

Efforts to understand spindle structure and movement bring us to the frontier of research on mitosis. Having explored how the process occurs, let's focus on how it is controlled. When does a cell divide, and when does it stop dividing? How is cell division regulated? These questions are fundamental. When cell division occurs in an uncontrolled manner, cancerous tumors can form.

11.3 Control of the Cell Cycle

Although the events of mitosis are virtually identical in all eukaryotes, other aspects of the cell cycle are variable. For example, the length of the cell cycle can vary enormously among different cell types, even in the same individual. In humans, intestinal cells routinely divide more than twice a day to renew tissue that is lost during digestion; mature human nerve and muscle cells do not divide at all. Most of these differences are due to variation in the length of the G_1 phase. In rapidly dividing cells, G_1 is essentially eliminated. Most nondividing cells, in contrast, are permanently stuck in G_1. Researchers refer to this arrested stage as the G_0 state, or simply "G zero." Cells that are in G_0 have effectively exited the cell cycle and are sometimes referred to as post-mitotic. Nerve cells, muscle cells, and many other cell types enter G_0 once they have matured.

A cell's division rate can also vary in response to changes in conditions. For example, human liver cells normally divide about once per year. But if part of the liver is damaged or lost, the remaining cells divide every one or two days until repair is accomplished. Cells of unicellular organisms such as yeasts, bacteria, or archaea divide rapidly only if the environment is rich in nutrients; otherwise, they enter a quiescent (inactive) state.

To explain the existence of so much variability, biologists hypothesized that the cell cycle must be regulated in some way and that regulation varies among cells and organisms. Understanding how the cell cycle is controlled is now the most prominent issue in research on cell division—partly because defects in control can lead to uncontrolled, cancerous growth.

The Discovery of Cell-Cycle Regulatory Molecules

The first solid evidence for cell-cycle control molecules came to light in 1970, when researchers published the results of experiments on fusing pairs of mammalian cells that were growing in cell culture. In the presence of certain chemicals, viruses, or an electric shock, the membranes of two cells that are growing in culture can be made to fuse, creating a single cell with two nuclei.

How did cell fusion experiments point to the existence of cell-cycle control molecules? When investigators fused cells that were in different stages of the cell cycle, certain nuclei changed phases. For example, when a cell in M phase was fused with one in interphase, the nucleus of the interphase cell initiated M phase (**Figure 11.13a**). The biologists hypothesized that the cytoplasm of M-phase cells contains a regulatory molecule that induces interphase cells to enter M phase.

This hypothesis was supported by experiments on the South African claw-toed frog, *Xenopus laevis*. As an egg of these frogs matures, it changes from a cell called an **oocyte**, which is arrested in a phase similar to G_2, to a mature egg that has entered M phase. The eggs are attractive to study, partly because they are more than 1 mm in diameter. Their large size makes it relatively easy to purify large amounts of cytoplasm and to use syringes with extremely fine needles to inject the eggs with cytoplasm from eggs in different stages of development. When biologists purified cytoplasm from M-phase frog eggs and injected it into the cytoplasm of frog oocytes arrested in the G_2 phase, the immature oocytes entered M phase (**Figure 11.13b**). But when cytoplasm from interphase cells was injected into G_2 oocytes, the cells remained in the G_2 phase. The researchers concluded that the cytoplasm of M-phase cells—but not the cytoplasm of interphase cells—contains a factor that drives immature oocytes into M phase to complete their maturation.

Experiment

Question: Do regulatory molecules control entry into specific phases of the cell cycle?

Hypothesis: A regulatory molecule triggers mitosis (M phase).

Null hypothesis: No regulatory molecule triggers mitosis.

Experimental setup:

(a) Cell fusion experimental setup:

M-phase cell   Interphase cell

Fuse M-phase cells with G_1, S, or G_2 phase cells.

(b) Microinjection experimental setup:

M-phase cytoplasm   Interphase cytoplasm

Inject cytoplasm from M-phase cell into one frog oocyte and cytoplasm from interphase cell into another frog oocyte.

Prediction: Interphase cells will begin M phase.

Prediction of null hypothesis: Interphase cells will not begin M phase.

Prediction: One or both of the frog oocytes will begin M phase.

Prediction of null hypothesis: Neither of the frog oocytes will begin M phase.

Results:

Chromosomes in interphase cell condense, signaling start of M phase.

Results:

Oocyte is driven into M phase (early mitotic spindle appears).

Oocyte remains in G_2 phase.

Conclusion: M-phase cytoplasm contains a regulatory molecule that induces M phase in interphase cells.

FIGURE 11.13 Experimental Evidence for Cell-Cycle Control Molecules. (a) When M-phase cells are fused with cells in G_1 S, or G_2 phase, the interphase chromosomes condense and begin M phase. **(b)** Microinjection experiments supported the hypothesis that a regulatory molecule induces M phase.

This factor was eventually purified and is now called **mitosis-promoting factor**, or **MPF**. Subsequent experiments showed that MPF induces mitosis in all eukaryotes. For example, injecting M-phase cytoplasm from mammalian cells into immature frog eggs results in egg maturation. Human MPF can trigger mitosis in yeast cells. The molecule appears to be a general signal that says "Start mitosis." How does it work?

MPF Contains a Protein Kinase and a Cyclin Once MPF had been isolated and purified, researchers found that it is made up of two distinct polypeptide subunits. One of the subunits is a **protein kinase**—an enzyme that catalyzes the transfer of a phosphate group from ATP to a target protein. Recall from Chapter 9 that proteins can be activated or inactivated by phosphorylation. Because the addition of a phosphate group changes the target protein's shape and activity, protein kinases frequently act as regulatory elements in the cell.

These observations suggested that MPF acts by phosphorylating a protein that triggers the onset of mitosis. But research showed that the concentration of MPF protein kinase is more or less constant throughout the cell cycle. How can MPF trigger mitosis if the protein kinase subunit is always present?

The answer to this question lies in the second MPF subunit, which belongs to a family of proteins called the **cyclins**. Cyclins got their name because their concentrations fluctuate throughout the cell cycle. As **Figure 11.14a** shows, the cyclin associated with MPF builds up in concentration during interphase and peaks during M phase. This increase in concentration is important because the protein kinase subunit in MPF can be active only when it is bound to the cyclin subunit. As a result, the protein kinase subunit of MPF is called a **cyclin-dependent kinase**, or **Cdk**. MPF is a dimer consisting of a cyclin subunit and a cyclin-dependent kinase subunit. The cyclin subunit functions as a regulatory protein; the kinase subunit is the part that catalyzes the phosphorylation of other proteins to start mitosis.

According to Figure 11.14a, the number of complete MPF dimers builds up steadily during interphase. Why doesn't this increasing concentration of MPF trigger the onset of M phase? The answer is that MPF's Cdk subunit becomes phosphorylated at two sites after it binds to cyclin. When Cdk is phosphorylated, its conformation changes in a way that renders the protein inactive. Late in G_2 phase, however, enzymes cause one of the phosphate groups on the Cdk subunit to drop off. This dephosphorylation reaction changes MPF's shape in a way that activates it.

Once MPF is activated, it binds to the proteins listed in **Figure 11.14b** and catalyzes their phosphorylation. Chromosomal proteins activated by MPF cause chromosomes to condense into the threads visible during M phase. In addition, microtubule-associated proteins phosphorylated by MPF

(a) Cyclin concentration regulates MPF concentration.

(b) Activated MPF has an array of effects.

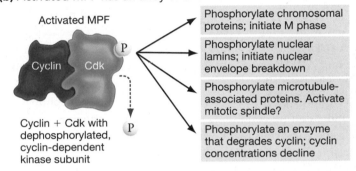

FIGURE 11.14 M-Phase Promoting Factor Is Created When a Cyclin Binds to a Protein Kinase. (a) Cyclin concentrations cycle in dividing cells, reaching a peak in M phase. **(b)** Cyclin binds to a cyclin-dependent kinase, creating MPF. When MPF is activated, it activates genes that initiate M phase. It also activates proteins that degrade cyclin.

● **QUESTION** Why is it important that MPF activates proteins that degrade cyclin?

may be involved in assembling the mitotic spindle apparatus. In this way, MPF triggers the onset of M phase.

MPF also activates an enzyme complex that promotes the degradation of MPF's own cyclin subunit, however. By activating this enzyme complex, MPF triggers its own destruction. This is an example of **negative feedback**—similar to the mechanisms of feedback inhibition introduced in Chapter 9. In response to MPF activity, then, the concentration of cyclin declines rapidly. Slowly, it builds up again during interphase. In this way, an oscillation in cyclin concentration is set up. ● If you understand this aspect of cell-cycle regulation, you should be able to describe what MPF does. You should also be able to explain the relationship between MPF and (1) cyclin, (2) Cdk, and (3) the enzymes that phosphorylate MPF, dephosphorylate MPF, and degrade MPF.

The dramatic oscillation in cyclin concentration and activation acts as a clock that drives the ordered events of the cell cycle. These events are happening millions of times each day in locations throughout your body. Over a 24-hour period, you swallow millions of cheek cells. Millions of additional cells are lost from your intestinal lining each day and leave your body as waste. To replace them, cells in your cheek and intestinal tissue are constantly making and degrading cyclin and pushing themselves through the cell cycle.

Cell-Cycle Checkpoints

MPF is only one of many protein complexes involved in regulating the cell cycle. For example, a different cyclin and protein kinase are involved in triggering the passage from G_1 phase into S phase, and several regulatory proteins are involved in maintaining the G_0 state of quiescent cells. A complex array of regulatory molecules is involved in either holding cells in particular stages or in stimulating passage to the next phase.

To make sense of these observations, Leland Hartwell and Ted Weinert introduced the concept of a cell-cycle checkpoint. A **cell-cycle checkpoint** is a critical point in the cell cycle that is regulated. Hartwell and Weinert identified checkpoints by analyzing yeast cells with defects in the cell cycle. The defective cells lacked a specific checkpoint and, as a result, kept dividing under culture conditions when normal cells stopped growing. In the body, cells that keep growing in this way form a mass of cells called a **tumor**.

As **Figure 11.15** indicates, there are three distinct checkpoints during the four phases of the cell cycle. In effect, interactions among regulatory molecules at each checkpoint allow a cell to "decide" whether to proceed with division. If these regulatory molecules are defective, the checkpoint may fail. As a result, cells may start growing in an uncontrolled fashion.

The first cell-cycle checkpoint occurs late in G_1. For most cells, this checkpoint is the most important in establishing whether the cell will continue through the cycle and divide or exit the cycle and enter G_0. What determines whether a cell passes the G_1 checkpoint?

- Because a cell must reach a certain size before its daughter cells will be large enough to function normally, biologists hypothesize that some mechanism exists to arrest the cell cycle if the cell is too small.

- Unicellular organisms arrest at the G_1 checkpoint if nutrient conditions are poor.

- Cells in multicellular organisms pass (or do not pass) through the G_1 checkpoint in response to signaling molecules from other cells, or what are termed social signals.

- If DNA is physically damaged, the protein **p53** activates genes that either stop the cell cycle until the damage can be repaired or lead to the cell's programmed, controlled destruction—a phenomenon known as **apoptosis**. In this way, p53 acts as a brake on the cell cycle. Cancer can develop if "brake" molecules such as p53 are defective, because the damaged DNA remains unrepaired. If the damage occurs

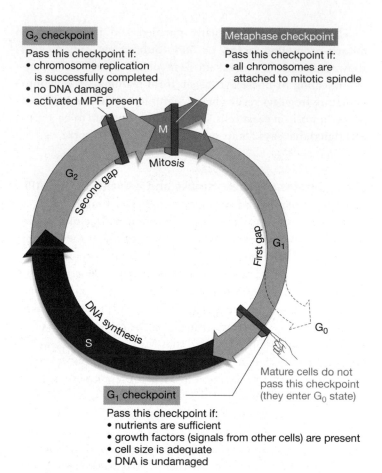

G₂ checkpoint

Pass this checkpoint if:
- chromosome replication is successfully completed
- no DNA damage
- activated MPF present

Metaphase checkpoint

Pass this checkpoint if:
- all chromosomes are attached to mitotic spindle

Mature cells do not pass this checkpoint (they enter G_0 state)

G₁ checkpoint

Pass this checkpoint if:
- nutrients are sufficient
- growth factors (signals from other cells) are present
- cell size is adequate
- DNA is undamaged

FIGURE 11.15 The Three Cell-Cycle Checkpoints

QUESTION Why is it advantageous for a cell to arrest at the M-phase checkpoint if not all chromosomes are attached to the mitotic spindle?

in certain genes, cells can begin growing in an uncontrolled fashion and spread to other locations in the body. Consequently, regulatory proteins like p53 are called **tumor suppressors**.

The general message here is that the components of the G_1 checkpoint have the same function: ensuring that the cell is healthy and should replicate its DNA and divide.

The second checkpoint occurs after S phase, at the boundary between the G_2 and M phases. Specifically, cells appear to arrest at the G_2 checkpoint if chromosome replication has not been completed properly or if DNA is damaged. Because MPF is the key signal triggering the onset of M phase, investigators were not surprised to find that it is involved in the G_2 checkpoint. Although much remains to be learned, data suggest that if DNA is damaged or if chromosomes are not replicated correctly, the dephosphorylation and activation of MPF are blocked. When MPF is not activated, cells remain in G_2 phase. Some data indicate that cells at this checkpoint may also respond to signals from other cells and to internal signals relating to their size.

The final checkpoint occurs during mitosis. If not all chromosomes are properly attached to the mitotic spindle, M phase arrests at metaphase. Specifically, anaphase is delayed until all kinetochores are properly attached to mitotic spindle fibers. If this checkpoint did not exist, some chromosomes might not separate properly, and daughter cells would receive an incorrect number of chromosomes during anaphase. Because they would receive too much or too little genetic material, the effect on the daughter cells could be disastrous.

To summarize, the three cell-cycle checkpoints have the same purpose: They prevent the division of cells that are damaged or that have other problems. The G_1 checkpoint also prevents the growth of mature cells that are in the G_0 state and should not grow any more. If one of the checkpoints fails, the affected cells may begin growing in an uncontrolled fashion. For the organism as a whole, the consequences of uncontrolled cell division are dire: cancer.

Check Your Understanding

If you understand that...

- The cell cycle consists of four carefully controlled phases.

You should be able to...

Diagram the cell cycle and indicate

1) the locations of cell-cycle checkpoints,
2) the level of MPF Cdk at each stage,
3) the level of MPF cyclin at each stage, and
4) the levels of MPF activity during each stage.

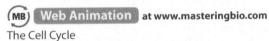
Web Animation at www.masteringbio.com

The Cell Cycle

11.4 Cancer: Out-of-Control Cell Division

Few diseases inspire more fear than cancer does. The fear springs from the difficulty of treating many forms of cancer, the potentially fatal nature of many cancers, and their frequency of occurrence. Most of us know someone who has had some form of cancer, and most of us have known someone who died from the disease. According to the American Cancer Society, 50 percent of American men and 33 percent of American women will develop cancer during their lifetime. In the United States, one in four of all deaths are from cancer. It is the second leading cause of death, exceeded only by heart disease.

Cancer is a general term for disease caused by cells that are growing in an uncontrolled fashion, that invade nearby tissues, and that spread to other sites in the body. Cancerous cells cause disease because they use nutrients and space needed by normal cells and disrupt the function of normal tissues. Humans suffer from at least 200 types of cancer. Stated another way, cancer is

(a) Cancer death rates in **males** have changed over time.

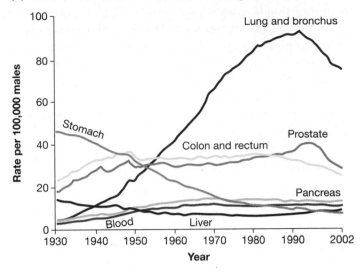

(b) Cancer death rates in **females** have changed over time.

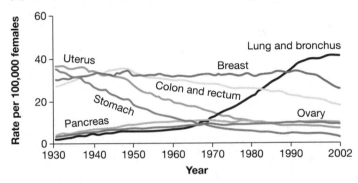

FIGURE 11.16 Changes in Cancer Death Rates over Time. Changes in the death rates from various types of cancer in **(a)** men and **(b)** women in the United States.

● **QUESTION** In this population, how do changes over time in deaths due to lung cancer compare in males and females? Suggest a hypothesis to explain the similarities and differences.

not a single illness but a complex family of diseases that affect an array of organs, including the breast, colon, brain, lung, and skin. In addition, several types of cancer can affect the same organ. Skin cancers, for example, come in multiple forms. Some are relatively easy to treat; others are often fatal. **Figure 11.16** provides data on how the number of deaths due to different types of cancer has changed through time in the United States.

Although cancers vary in time of onset, growth rate, seriousness, and cause, they have a unifying feature: ● Cancers arise from cells in which cell-cycle checkpoints have failed. More specifically, cancerous cells have two types of defects: (1) defects that make the proteins required for cell growth active at all times, and (2) defects that prevent tumor suppressor genes from shutting down the cell cycle. The protein called Ras, for example, was introduced in Chapter 8 as a key component in signal transduction systems—including phosphorylation cascades that trigger cell growth. Many cancerous cells have defective forms of Ras that do not become inactivated when they should. Instead, the defective Ras constantly send signals that trigger mitosis and cell division. Likewise, a large percentage of cancerous cells have defective forms of the tumor suppressor p53. Instead of being arrested or destroyed, cells that have damaged DNA are allowed to continue growth.

To get an understanding of why cancer occurs, let's review general characteristics of these diseases and then delve into the details of why regulatory mechanisms become defective.

Properties of Cancer Cells

When even a single cell in a multicellular organism begins to divide in an uncontrolled fashion, a mass of cells called a tumor results. For example, most cells in the adult human brain do not divide. But if a single abnormal brain cell begins unrestrained division, the growing tumor that results may disrupt the brain's function. What can be done? If the tumor can be removed without damaging the affected organ, a cure might be achieved. This is why surgical removal of the tumor is usually the first step in the treatment of a cancer. Often, though, surgery does not cure cancer. Why?

In addition to growing quickly, cancer cells are invasive—meaning that they are able to spread throughout the body via lymphatic vessels (introduced in Chapter 49), which collect excess fluid from tissues and return it to the bloodstream, or via the bloodstream itself. Invasiveness is a defining feature of a **malignant tumor**—one that is cancerous. Masses of noninvasive cells are noncancerous and form **benign tumors**. Some noninvasive tumors, like warts, grow slowly and are largely harmless. Others grow quickly and can cause problems if they are located in the brain or other sensitive parts of the body.

Cells become malignant and cancerous if they gain the ability to detach from the original tumor and invade other tissues. By spreading from the primary tumor site where uncontrolled growth originated, cancer cells can establish secondary tumors elsewhere in the body (**Figure 11.17**). This process is called **metastasis**. If metastasis has occurred by the time the original tumor is detected, then secondary tumors have begun to form and surgical removal of the primary tumor will not lead to a cure. As a result, the disease can be very difficult to treat. This is why early detection is the key to treating cancer most effectively.

Cancer Involves Loss of Cell-Cycle Control

If cancer is caused by uncontrolled cell growth, what is the molecular nature of the disease? Recall that when many cells mature, they enter the G_0 phase—meaning their cell cycle is arrested at the G_1 checkpoint. In contrast, cells that do pass through the G_1 checkpoint are irreversibly committed to replicating their DNA and entering G_2. Based on this observation, biologists hypothesized that many or even most types of cancer involve defects in the G_1 checkpoint. To understand the molecular nature of the disease, then, researchers focused

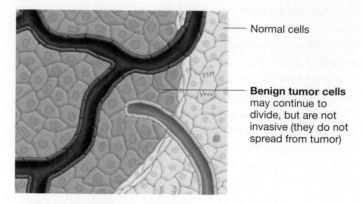

(a) Benign tumor

- Normal cells
- **Benign tumor cells** may continue to divide, but are not invasive (they do not spread from tumor)

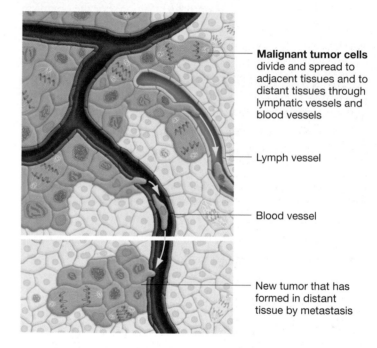

(b) Malignant tumor

- **Malignant tumor cells** divide and spread to adjacent tissues and to distant tissues through lymphatic vessels and blood vessels
- Lymph vessel
- Blood vessel
- New tumor that has formed in distant tissue by metastasis

FIGURE 11.17 Cancers Spread to New Locations in the Body.
(a) Benign tumors grow in a single location. **(b)** Malignant tumors are metastatic—meaning that their cells can spread to distant parts of the body and initiate new tumors. Malignant tumors cause cancer.

on understanding the normal mechanisms that operate at that checkpoint. In this way, cancer research and research on the normal cell cycle have become two sides of the same coin.

Social Control In unicellular organisms, passage through the G_1 checkpoint is thought to depend primarily on cell size and the availability of nutrients. If nutrients are plentiful, cells pass through the checkpoint and grow rapidly. In multicellular organisms, however, cells receive a constant supply of adequate nutrients via the bloodstream. Because nutrients generally are not limiting, most cells in multicellular organisms divide in response to some other type of signal. Because these signals arrive from other cells, biologists refer to *social control* over cell division. The general idea is that individual cells should be allowed to divide only when their growth is in the best interests of the organism as a whole.

Social control of the cell cycle is based on the types of cell-cell signals introduced in Chapter 8—more specifically, signals called growth factors. **Growth factors** are polypeptides or small proteins responsible for stimulating cell division. Many growth factors were discovered in the course of working out techniques for growing cells in culture. When researchers isolated mammalian cells in culture and provided them with adequate nutrients, the cells arrested in G_1 phase. They began to grow only when biologists added **serum**—the liquid that remains after blood clots and the blood cells have been removed. Some component of serum allowed cells to pass through the G_1 checkpoint. What was it?

In 1974 biologists succeeded in identifying one of these serum components. The component was a protein called platelet-derived growth factor (PDGF). As its name implies, PDGF is released by blood components called **platelets**, which promote blood clotting at wound sites. PDGF binds to receptor tyrosine kinases in the plasma membranes of cells in the area. When these cells receive the growth signal, they are stimulated to divide. The increased cell numbers facilitate wound healing.

Researchers subsequently found that PDGF is produced by an array of cell types. Investigators also succeeded in isolating and identifying a diverse array of other growth factors. For different types of cells to grow in culture, different combinations of growth factors must be supplied. Based on this result, biologists infer that different types of cells in an intact multicellular organism are controlled by different combinations of growth factors. Cancer cells are another story, however. Cancerous cells can often be cultured successfully without externally supplied growth factors. This observation suggests that the normal social controls on the G_1 checkpoint have broken down in cancer cells.

Social Controls and Cell-Cycle Checkpoints   Why don't the social controls of the G_1 checkpoint work in cancer cells? Let's start by understanding the relationship between growth factors and the G_1 checkpoint in normal cells.

Recall from Section 11.3 that for an interphase cell to begin mitosis, a specific cyclin-Cdk complex must be activated. In cells that are in the G_0 resting state, the arrival of growth factors stimulates the production of cyclins. The arrival of growth factors also stimulates the production of a key regulatory protein called E2F. E2F is analogous to MPF, which activates genes required for M phase. When E2F is activated, it triggers the expression of genes required for S phase.

When E2F is first produced, however, it binds to a tumor suppressor protein called Rb. **Rb protein** is one of the key molecules that enforces the G_1 checkpoint. When E2F is bound to Rb, it is in the "off" position—it can't activate the genes required for S phase. In effect, a cell's commitment to dividing depends on whether Rb stays bound to the E2F protein. As long as Rb binds to E2F, the cell remains in G_0.

This situation changes dramatically if growth factors continue to arrive. With continued social stimulation, production of cyclins and thus cyclin-Cdk complexes continues. The cyclins involved are different from the cyclins involved in initiating M phase, however. They are specific to the G_1 checkpoint. (There are also cyclins specific to the G_2 checkpoint.) When G_1 cyclin-Cdk complexes are activated, they begin catalyzing the phosphorylation of Rb (**Figure 11.18**). When Rb is phosphorylated, it changes shape and no longer binds to E2F. The unbound E2F is free to activate its target genes, and

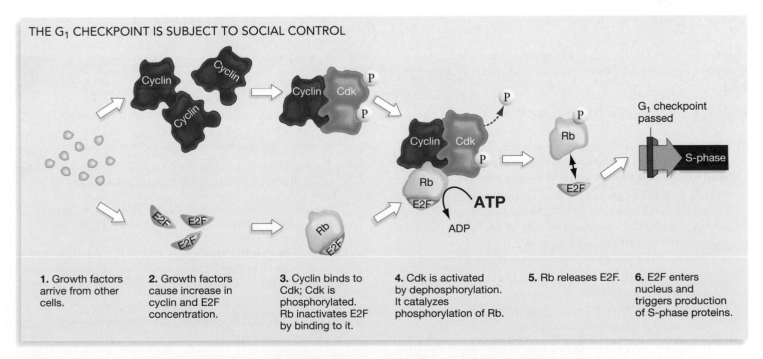

THE G_1 CHECKPOINT IS SUBJECT TO SOCIAL CONTROL

1. Growth factors arrive from other cells.

2. Growth factors cause increase in cyclin and E2F concentration.

3. Cyclin binds to Cdk; Cdk is phosphorylated. Rb inactivates E2F by binding to it.

4. Cdk is activated by dephosphorylation. It catalyzes phosphorylation of Rb.

5. Rb releases E2F.

6. E2F enters nucleus and triggers production of S-phase proteins.

FIGURE 11.18 The G_1 Checkpoint Is Subject to Social Control. Cells that are in G_0 do not exit and begin dividing again unless they receive signals from other cells, in the form of growth factors. If enough signals arrive, the proteins that respond override the inhibitory effects of the Rb protein and begin S phase.

S phase gets under way. The growth factors thus function as a social signal that says, "It's OK to override Rb. Go ahead and pass the G_1 checkpoint and divide."

Cells can become cancerous when social controls fail—meaning, when cells begin dividing in the *absence* of a go-ahead signal from growth factors. One of two things go wrong.

- In some cancers the G_1 cyclin is overproduced. When cyclins stay at high concentrations, the Cdk that binds to cyclin phosphorylates Rb continuously, activating E2F and sending the cell into S phase. Cyclin overproduction results from the presence of excessive amounts of growth factors or from cyclin production in the absence of growth signals—usually due to a defect in a signaling pathway. The specific pathway involved starts with a receptor tyrosine kinase that responds to growth factors and includes the Ras protein highlighted in Chapter 8. This is why it is common to find defective Ras proteins in cancerous cells.

- The second possibility is that Rb is defective. If the Rb protein is missing or if it does not bind to E2F normally, any E2F that is present pushes the cell through the G_1 checkpoint and into S phase. In this way, defects in Rb lead to uncontrolled cell division.

The link between defective Rb protein and cancer was originally discovered in children with retinoblastoma, a cancer that occurs in about 1 in 20,000 children. The disease is characterized by the appearance of malignant tumors in the light-sensing tissue, or retina, of the eye. But follow-up work showed that most human cancer cells have defects in the Rb protein, not just cells that lead to retinoblastoma.

To summarize, overproduction of cyclins or defects in Rb itself can lead to the inactivation of Rb as a tumor suppressor. When Rb not longer works properly, the result can be failure of social control over the G_1 checkpoint, rampant cell growth, and tumor formation. ◉ If you understand the relationship between the failure of social controls and cancer, you should be able to make diagrams like Figure 11.18 showing the consequences of (1) constant G_1 cyclin production and (2) defective Rb proteins.

Cancer Is a Family of Diseases   One of the broadest messages to come out of research on p53, Rb, and other molecules involved in arresting cell growth is that a wide variety of defects can lead to the failure of the G_1 checkpoint and the onset of cancer. Because many proteins are essential to the G_1 checkpoint, many different defects can cause the checkpoint to fail. In addition, researchers now realize that cancer is seldom due to a single defect. Most cancers develop only after several genes have been damaged. The combined damage is then enough to break cell-cycle control and induce uncontrolled growth and metastasis. Each type of cancer is due to a unique combination of errors. Stated another way, cancer can be caused by hundreds if not thousands of different defects.

Because cancer is actually a family of diseases with a complex and highly variable molecular basis, there will be no "magic bullet," or single therapy that cures all forms of the illness. Still, recent progress in understanding the cell cycle and the molecular basis of cancer has been dramatic, and cancer prevention and early detection programs are increasingly effective. The prognosis for many cancer patients is remarkably better now than it was even a few years ago.

Chapter Review

SUMMARY OF KEY CONCEPTS

◯ **After chromosomes are copied, mitosis distributes one chromosome copy to each of two daughter cells. Mitosis and cytokinesis produce two cells that are identical to the parent cell.**

Most unicellular eukaryotes and some multicellular eukaryotes reproduce by mitosis followed by cytokinesis. Mitotic divisions are also responsible for building the bodies of multicellular organisms from a single-celled embryo.

Mitosis can be described as a sequence of five phases:

Prophase Chromosomes condense, and the mitotic spindle apparatus begins to form.

Prometaphase The nuclear envelope disintegrates, and spindle fibers make contact with chromosomes.

Metaphase Spindle fibers move chromosomes to the metaphase plate, an imaginary plane in the middle of the cell.

Anaphase Spindle fibers pull sister chromatids apart.

Telophase Chromosomes are pulled to opposite poles of the cell, and a nuclear envelope forms around each set.

In most cells, mitosis is followed immediately by division of all cell contents, or cytokinesis.

You should be able to explain why the nuclear envelope has to disintegrate prior to spindle fibers contacting chromosomes, and what would happen if the sister chromatids of one chromosome failed to separate at anaphase. ◉

(MB) **Web Animation**   *BioFlix* at www.masteringbio.com
The Phases of Mitosis; Mitosis

● **Over their life span, eukaryotic cells go through a cycle that consists of four carefully controlled phases.**

Dividing cells alternate between the dividing phase, called M phase, and a nondividing phase known as interphase. Chromosome synthesis and replication, or S phase, occurs during interphase. S phase and M phase are separated by gap phases called G_1 and G_2. Cell growth and replication of nonnuclear cell components occur during the gap phases. The order of cell-cycle phases is $G_1 \rightarrow S \rightarrow G_2 \rightarrow M \rightarrow G_1$ and so forth.

Progression through the cell cycle is regulated at three checkpoints. Passage through the G_1 checkpoint represents a commitment to divide and is contingent on cell size, nutrient availability, lack of DNA damage, and/or growth signals from other cells. Mature cells do not pass this checkpoint. At the G_2 checkpoint, progression through the cycle is delayed until chromosome replication has been successfully completed. At the metaphase checkpoint during M phase, anaphase is delayed until all chromosomes are correctly attached to the mitotic spindle.

An array of different molecules regulates progression through the cell-cycle checkpoints. The most important of these regulatory molecules are cyclin-dependent kinases (Cdks) and cyclins, which are found in all eukaryotic cells. Active Cdks are enzymes that trigger progress through a checkpoint by phosphorylating important target proteins. Cyclins are proteins whose concentrations oscillate during the cell cycle, regulating the activity of Cdks. In multicellular organisms, cyclin concentrations are partially controlled by growth factors from other cells. As a result, the G_1 checkpoint is said to be under social control.

You should be able to predict what happens if an influx of growth factors leads to an increase in the production of cyclins. ●

(MB) Web Animation at www.masteringbio.com

The Cell Cycle

● **In multicellular organisms, uncontrolled cell growth leads to cancer. Different types of cancer result from different types of defects in control over the cell cycle.**

Cancer is a common disease that is characterized by loss of control at the G_1 checkpoint, resulting in cells that divide in an uncontrolled fashion. Cancer cells also have the ability to spread throughout the body, which makes treatment difficult. In normal cells, growth-factor signals are required to produce G_1 cyclin-Cdk complexes and trigger division. An important target of activated G_1 cyclin-Cdk complexes is the retinoblastoma (Rb) protein, which regulates a protein responsible for initiating S phase. Defects in G_1 cyclin and Rb protein are common in human cancer cells.

You should be able to explain why cancer is considered a family of diseases, and why there is unlikely to be a single cure for cancer. ●

QUESTIONS

● **Test Your Knowledge**

1. Which statement about the daughter cells of mitosis is correct?
   a. They differ genetically from one another and from the parent cell.
   b. They are genetically identical to one another and to the parent cell.
   c. They are genetically identical to one another but are different from the parent cell.
   d. Only one of the two daughter cells is genetically identical to the parent cell.

2. Progression through the cell cycle is regulated by oscillations in the concentration of which type of molecule?
   a. p53, Rb, and other tumor suppresssors
   b. receptor tyrosine kinases
   c. cyclin-dependent kinases
   d. cyclins

3. According to the data in Figure 11.4, the first labeled mitotic cells appear about 4 hours after the labeling period ends. From these data, researchers concluded that G_2 lasted about 4 hours. Why?
   a. The lengths of the total cell cycle and the G_1, S, and M phases were known, so the length of G_2 could be found by subtraction.
   b. It confirmed other data indicating that G_2 lasts about 4 hours in cultured cells of this type.
   c. Cyclins were labeled, so their concentration had to build up for 4 hours.
   d. Labeled cells are in S phase, so 4 hours passed between the end of S phase and the onset of M phase.

4. What major events occur during anaphase of mitosis?
   a. Chromosomes replicate, so each chromosome consists of two identical sister chromatids.
   b. Chromosomes condense and the nuclear envelope disappears.
   c. The chromosomes end up at opposite ends of the cell and two nuclear envelopes form.
   d. Sister chromatids separate, forming independent chromosomes.

5. What evidence suggests that during anaphase, spindle fibers shorten at the kinetochore and not at the base of the mitotic spindle?
   a. Motor proteins are located at the kinetochore.
   b. Motor proteins are located at the kinetochore *and* at the base of the mitotic spindle.
   c. When fluorescing microtubules are bleached in the middle, the bleached segment stays stationary as the fibers shorten near the kinetochore.
   d. When fluorescing microtubules are bleached in the middle, the bleached segment moves toward the base of the mitotic spindle as the fibers shorten near the microtubule organizing center.

6. The normal function of the Rb protein is to bind to E2F, inactivating it. What are the consequences of defects in Rb?
   a. It binds more tightly to E2F, so E2F cannot stimulate the production of molecules needed in S phase. The cell cannot progress through the cell cycle.
   b. It no longer responds to phosphorylation—it cannot be regulated.
   c. It does not bind to E2F properly, so E2F continuously stimulates the production of molecules needed in S phase.
   d. Rb acts as a tumor suppressor, similar in function to p53.

Test Your Knowledge answers: 1. b; 2. d; 3. d; 4. d; 5. c; 6. c

Test Your Understanding

Answers are available at www.masteringbio.com

1. Sketch the phases of mitosis in a cell with four chromosomes listing the major events of each phase. Identify at least two events that must be completed successfully for daughter cells to share an identical complement of chromosomes.

2. Make a concept map illustrating normal and abnormal events at the G_1 checkpoint. Your diagram should include p53, Rb, E2F, social signals, G_1 Cdk, G_1 cyclin, S-phase proteins, uncontrolled growth, phosphorylated (inactivated) cyclin-Cdk, dephosphorylated (activated) cyclin-Cdk, phosphorylated (inactivated) Rb.

3. Explain how cell fusion and microinjection experiments supported the hypothesis that specific molecules are involved in the transition from interphase to M phase.

4. Why are most protein kinases considered regulatory proteins?

5. Why are cyclins called cyclins? Explain their relationship to cyclin-dependent kinases and to growth factors.

6. Early detection is the key to surviving most cancers. Why?

Applying Concepts to New Situations

Answers are available at www.masteringbio.com

1. In multicellular organisms, nondividing cells stay in G_1 phase. For the cell, why is it better to be held in G_1 rather than S, G_2, or M phase?

2. When fruit fly embryos first begin to develop, mitosis occurs without cytokinesis. What is the result?

3. Predict the outcome of an experiment involving the fusion of a cell in G_1 phase with a cell in G_2 phase. What would happen to the G_1–phase nucleus? To the G_2–phase nucleus? Why?

4. Cancer is primarily a disease of older people. Further, a group of individuals may share a genetic predisposition to developing certain types of cancer, yet vary a great deal in time of onset—or not get the disease at all. Discuss these observations in light of the claim that several defects usually have to occur for cancer to develop.

www.masteringbio.com is also your resource for • Answers to text, table, and figure caption questions and exercises • Answers to *Check Your Understanding boxes* • Online study guides and quizzes • Additional study tools including the *E-Book for Biological Science* 3rd ed., textbook art, animations, and videos.

Meiosis

12

Scanning electron micrograph showing human sperm attempting to enter a cell that will become an egg. This chapter introduces the type of nuclear division called meiosis, which in animals occurs prior to the formation of sperm and eggs.

KEY CONCEPTS

○ Meiosis is a type of nuclear division. It results in cells that have half as many chromosomes as the parent cell, and in animals it is involved in the formation of eggs and sperm. When an egg and a sperm cell combine to form an offspring, the original number of chromosomes is restored.

○ Each cell produced by meiosis receives a different combination of chromosomes. Because genes are located on chromosomes, each cell produced by meiosis receives a different complement of genes. Meiosis leads to offspring that are genetically distinct from each other and from their parents.

○ The leading hypothesis to explain meiosis is that genetically variable offspring are more likely to thrive in changing environments.

○ If mistakes occur during meiosis, the resulting egg and sperm cells may contain the wrong number of chromosomes. It is rare for offspring with an incorrect number of chromosomes to develop normally.

Why sex?

Simple questions—such as why sexual reproduction exists—are sometimes the best. This chapter asks what sexual reproduction is and why some organisms employ it. Understanding sex has been one of the great mysteries of classical and contemporary biology.

Before plunging into the question of why sex exists, it is important to realize that there are two basic levels of explanation in biology: ultimate explanation and proximate explanation. An **ultimate explanation** is evolutionary in nature. It explains *why* something happens. In terms of sex, biologists want to know why it evolved. This is an important question because most

organisms never reproduce sexually, or they do so only occasionally. It's also a difficult question, because asexual reproduction is much more efficient than sexual reproduction. Organisms that reproduce asexually don't have to flash colorful ornaments, exude exotic scents, or dance or sing to woo a mate and reproduce sexually. Instead, they just clone themselves. In eukaryotes that reproduce asexually, the cloning is done via mitosis.

In searching for an ultimate explanation for sex, biologists focus on a simple observation: Asexually produced offspring are genetically identical to their parents. Sexually produced offspring, in contrast, are genetically unlike their parents. Presumably, sexual reproduction is worth all the time and trouble involved because genetic variation is a good thing for offspring in certain types of environments. Why is sex

○ Key Concept   Important Information   ○ Practice It

advantageous for some species? This question is the focus of Sections 12.2 and 12.3.

A **proximate explanation** is mechanistic in nature. It explains *how* something happens. For example, biologists want to know how the process of sexual reproduction occurs in terms of the molecular and cellular structures and events that are involved. It has long been understood that during sexual reproduction, a male reproductive cell—a **sperm**—and a female reproductive cell—an **egg**—unite to form a new individual. The process of uniting sperm and egg is called **fertilization**. The first biologists to observe fertilization studied the large, translucent eggs of sea urchins. Due to the semi-transparency of the sea urchin egg cell, researchers were able to see the nuclei of a sperm and an egg fuse.

When these results were published in 1876, they raised an important question at the proximate level, concerning the number of chromosomes found in parents, sperm, eggs, and offspring. Cell biologists had already established that the number of chromosomes is constant from cell to cell within a multicellular organism. It was also accepted that chromosome number is the same in the parent and daughter cells of mitosis. Biologists confirmed that all of the cells in a newly growing offspring, or **embryo**, are the products of mitotic divisions, and that all cells in the body are the direct descendants of the nucleus that formed at fertilization. The question is, How can the chromosomes from a sperm cell and an egg cell combine, but form an offspring that has the same chromosome number as its mother and its father?

A hint at the answer came in 1883, when a researcher noted that cells in the body of roundworms of the genus *Ascaris* have four chromosomes, while their sperm and egg nuclei have only two chromosomes apiece. Four years later, August Weismann formally proposed a hypothesis to explain the riddle: During the formation of **gametes**—reproductive cells such as sperm and eggs—there must be a distinctive type of cell division that leads to a reduction in chromosome number. Specifically, if the sperm and egg contribute an equal number of chromosomes to the fertilized egg, Weismann reasoned, they must each contain half of the usual number of chromosomes. Then, when sperm and egg combine, the resulting cell has the same chromosome number as its mother's cells and its father's cells have.

In the decades that followed, biologists confirmed this hypothesis by observing gamete formation in a wide variety of plant and animal species. Eventually this form of cell division came to be called meiosis (literally, "lessening-act"). **Meiosis** is nuclear division that leads to a halving of chromosome number. It precedes the formation of eggs and sperm in animals, and it provides a satisfying proximate explanation for how sexual reproduction occurs. To a biologist, asking "Why sex?" is equivalent to asking "Why meiosis?" Let's delve in by looking at how meiosis occurs.

12.1 How Does Meiosis Occur?

When cell biologists began to study the cell divisions that lead to gamete formation, they made an important observation: Each organism has a characteristic number of chromosomes. Consider the drawing in **Figure 12.1**, based on a paper published by Walter Sutton in 1902. It shows the chromosomes of the lubber grasshopper during the cell divisions leading up to the formation of a sperm. In total, there are 23 chromosomes in the cell. Sutton realized, however, that there are just 12 distinct types of chromosomes based on size and shape. In this species, there were two chromosomes of each type.

Sutton designated 11 of the chromosomes by the letters *a* through *k* and the twelfth by the letter *X*. Some years later, Nettie Stevens established that the X chromosome is associated with the sex of the individual and is called a **sex chromosome**. Non-sex chromosomes, such as *a–k* in Sutton's grasshopper cell, are known as **autosomes**. It turns out that lubber grasshoppers have just one type of sex chromosome. In this species, females have two sex chromosomes and are designated XX; males have just one sex chromosome and are designated XO, where the O refers to the "missing" chromosome. Two types of sex chromosomes, known as X and Y, exist in humans and other mammals. Human females have two X chromosomes while males have one X and one Y chromosome.

Sutton also introduced an important term for the pairs of sex chromosomes or autosomes. He referred to the two chromosomes of each type as **homologous** ("same proportion") **chromosomes**. (They can also be called **homologs**.) The two chromosomes labeled *c*, for example, have the same size and shape and are homologous.

FIGURE 12.1 Cells Contain Different Types of Chromosomes, and Chromosomes Come in Pairs. Letters designate each of the 12 distinct types of chromosomes found in lubber grasshopper cells. There are two of each type of chromosome (except the X, as this is a male). The two members of a chromosome pair are called homologs.

Later work showed that homologous chromosomes are similar not only in size and shape but also in content. Homologous chromosomes carry the same genes. A **gene** is a section of DNA that influences one or more hereditary traits in an individual. A trait is a characteristic. For example, each copy of chromosome *c* found in lubber grasshoppers might carry genes that influence eye formation, body size, singing behavior, or jumping ability. The versions of a gene found on homologous chromosomes may differ, however. Biologists use the term **allele** to denote different versions of the same gene. For example, the alleles on each copy of chromosome *c* in a lubber grasshopper might contribute to rounder eyes versus narrower eyes, larger body size versus smaller body size, faster songs versus slower songs, and so on. Homologous chromosomes carry the same genes, but each homolog may contain different alleles.

At this point in his study, Sutton had succeeded in determining the lubber grasshopper's **karyotype**—meaning the number and types of chromosomes present (see **Box 12.1** on page 252). As karyotyping studies expanded, cell biologists realized that, like lubber grasshoppers, the vast majority of plants and animals have more than one of each type of chromosome. These investigators designated terms to identify the number of chromosome copies they observed. Organisms such as lubber grasshoppers, humans, and cedar trees are called **diploid** ("double-form"), because they have two versions of each type of chromosome. Diploid organisms have two alleles of each gene—one on each of the homologous pairs of chromosomes. Organisms such as bacteria, archaea, and many algae are called **haploid** ("single-form"), because their cells contain just one of each type of chromosome. Haploid organisms do not contain homologous chromosomes. They have just one allele of each gene.

Researchers also invented a compact notation to indicate the number of chromosomes and chromosome sets in a particular organism or type of cell. By convention, the letter *n* stands for the number of distinct types of chromosomes in a given cell and is called the **haploid number**. If sex chromosomes are present, they are counted as a single type in the haploid number. Diploid cells have a haploid number, which indicates the number of different types of chromosomes present. To indicate the number of complete chromosome sets observed, a number is placed before the *n*. Thus, a cell can be *n*, or 2*n*, or 3*n*, and so on. The combination of a number and *n* is termed the cell's **ploidy**. Diploid cells or species are designated 2*n*, because two chromosomes of each type are present—one from each parent. Haploid cells or species are labeled simply *n*, because they have just one set of chromosomes—no homologs are observed. In haploid cells, the number 1 in front of *n* is implied and is not written out.

Later work revealed that it is common for species in some lineages—particularly land plants such as ferns—to contain more than two of each type of chromosome. Instead of having

TABLE 12.1 The Number of Chromosomes Found in Some Familiar Organisms

Organism	Number of Different Types of Chromosomes (haploid number *n*)	Diploid Chromosome Number (2*n*)
Humans	23	46
Domestic dog	36	72
Fruit fly	4	8
Chimpanzee	24	48
Bulldog ant	1	2
Garden pea	7	14
Corn (maize)	10	20

two homologous chromosomes per cell, as many organisms do, **polyploid** ("many-form") species may have three or more of each type of chromosome in each cell. Depending on the number of homologs present, such species are called triploid (3*n*), tetraploid (4*n*), hexaploid (6*n*), octoploid (8*n*), and so on. Why some species are haploid versus diploid or tetraploid is currently the subject of debate and research.

To summarize, the haploid number *n* indicates the number of distinct types of chromosomes present. Different species have different haploid numbers (see **Table 12.1**). Human cells contain 23 distinct types of chromosomes, so *n* = 23. In the grasshopper cells that Sutton studied, *n* = 12. In contrast, a cell's ploidy (*n*, 2*n*, 3*n*, etc.) indicates the number of each type of chromosome present. Stating a cell's ploidy is the same as stating the number of haploid chromosome sets present. Because most human and grasshopper cells contain two of each type of chromosome, they are diploid. In humans 2*n* = 46; in Sutton's grasshoppers 2*n* = 24. Ploidy refers to the number of each type of chromosome present; haploid number identifies how many different types of chromosomes occur.

Table 12.2 summarizes the vocabulary that biologists use to describe the number and types of chromosomes found in a cell.

Sutton and the other early cell biologists did more than describe the karyotypes observed in their study organisms, however. Through careful examination, they were able to track how chromosome numbers change during meiosis. These studies confirmed Weismann's hypothesis that a special type of cell division occurs during gamete formation. This result was a major advance in understanding sex at the proximate level.

An Overview of Meiosis

Recall that cells replicate each of their chromosomes before undergoing meiosis. At the start of the process, then, chromo-

SUMMARY TABLE **12.2 Vocabulary Used to Indicate the Chromosomal Makeup of a Cell**

Term	Definition	Example or Comment
Chromosome	Structure made up of DNA and proteins; carries the cell's hereditary information (genes)	Eukaryotes have linear, threadlike chromosomes; most bacteria and archaea have just one circular chromosome
Sex chromosome	Chromosome associated with an individual's sex	X and Y chromosomes of humans: Males are XY, females XX. Also, Z and W chromosomes of birds and buttterflies: Males are ZZ, females ZW.
Autosome	A non-sex chromosome	Chromosomes 1–22 in humans.
Unreplicated chromosome	A chromosome that consists of a single copy (in eukaryotes, a single "thread")	
Replicated chromosome	A chromosome that has been copied; consists of two linear structures joined at the centromere	Centromere
Sister chromatids	The chromosome copies in a replicated chromosome	Sister chromatids
Homologous chromosomes (homologs)	In a diploid organism, chromosomes that are similar in size, shape, and gene content	You have a chromosome 22 from your mother (red) and a chromosome 22 from your father (blue). Homologous chromosomes
Non-sister chromatids	The chromosome copies in homologous chromosomes	Non-sister chromatids
Tetrad	Homologous chromosomes that are joined together	Tetrad
Haploid number	The number of different types of chromosomes in a cell; symbolized n	Humans have 23 different types of chromosomes ($n = 23$)
Diploid number	The number of chromosomes present in a diploid cell; symbolized $2n$	In humans all cells except gametes are diploid and contain 46 chromosomes ($2n = 46$)
Ploidy	The number of each type of chromosome present; equivalent to the number of haploid chromosome sets present	
Haploid	Having one of each type of chromosome (n)	Bacteria and archaea are haploid, as are many algae; plant and animal gametes are haploid
Diploid	Having two of each type of chromosome ($2n$)	Most familiar plants and animals are diploid
Polyploid	Having more than two of each type of chromosome; cells may be triploid ($3n$), tetraploid ($4n$), hexaploid ($6n$), and so on	Seedless bananas are triploid; many ferns are tetraploid; bread wheat is hexaploid

somes are in the same state they are in prior to mitosis. When chromosome replication is complete, each chromosome consists of two identical **sister chromatids**. Sister chromatids contain the same genetic information and are physically joined at a portion of the chromosome called the **centromere** as well as along their entire length (**Figure 12.2a**).

To understand meiosis, it is critical to understand the relationship between chromosomes and sister chromatids. An unreplicated chromosome consists of a single DNA molecule with its associated proteins, while a replicated chromosome consists of two sister chromatids. The trick is to recognize that unreplicated and replicated chromosomes are each considered *single*

(a) Each chromosome replicates prior to undergoing meiosis.

(b) During meiosis, chromosome number in each cell is reduced.

Four daughter cells contain one chromosome each. In animals, these cells become gametes.

(c) A full complement of chromosomes is restored during fertilization.

FIGURE 12.2 The Major Events in Meiosis. Meiosis reduces chromosome number by half. In diploid organisms, the products of meiosis are haploid.

● **EXERCISE** In parts (b) and (c), write *n* or 2*n* next to each cell to indicate its ploidy.

chromosomes—even though the replicated chromosome comprises *two* sister chromatids. A chromosome is still just one chromosome whether it is unreplicated (consisting of a single strand) or replicated (consisting of two identical strands). ● If you understand this concept, you should be able to draw the same chromosome in the replicated and unreplicated state, explain why both structures represent a single chromosome, and then label the sister chromatids in the replicated chromosome.

Meiosis consists of two cell divisions, called **meiosis I** and **meiosis II**. As **Figure 12.2b** shows, the two divisions occur consecutively but differ sharply. During meiosis I, the homologs in each chromosome pair separate from each other. One homolog goes to one daughter cell; the other homolog goes to the other daughter cell. The homolog that came from the individual's mother is colored red in Figure 12.2; the homolog that came from the father is colored blue. It is a matter of chance which daughter cell receives which homolog. The end result is that the daughter cells from meiosis I have one of each type of chromosome instead of two, and thus half as many chromosomes as the parent cell has. During meiosis I, the diploid (2*n*) parent cell produces two haploid (*n*) daughter cells. Each chromosome still consists of two sister chromatids, however.

During meiosis II, sister chromatids from each chromosome separate. One sister chromatid goes to one daughter cell; the other sister chromatid goes to the other daughter cell. The cell that starts meiosis II has one of each type of chromosome, but each chromosome has been replicated (meaning it still consists of two sister chromatids). The cells produced by meiosis II also have one of each type of chromosome, but now the chromosomes are unreplicated.

To reiterate, sister chromatids separate during meiosis II, just as they do during mitosis. Meiosis II is actually equivalent to mitosis occurring in a haploid cell. As in mitosis, chromosome movements during meiosis I and II are caused by spindle fibers that attach at the centromere of each chromosome and pull sister chromatids apart.

Sutton and a host of other early cell biologists worked out this sequence of events through careful observation of cells with the light microscope. Based on these studies, they came to a key realization: The outcome of meiosis is a reduction in chromosome number. For this reason, meiosis is known as a reduction division. In most plants and animals, the original cell is diploid and the four daughter cells are haploid. These four haploid daughter cells, each containing one of each homologous chromosome, eventually go on to form egg cells or sperm cells via a process called **gametogenesis** ("gamete-origin"), which is described in Chapter 22, Chapter 40, and Chapter 48. ● When two gametes fuse during fertilization, a full complement of chromosomes is restored (**Figure 12.2c**). The cell that results from fertilization is diploid and is called a **zygote**. In this way, each diploid individual receives both a haploid chromosome set from its mother and a haploid set from its father. Homologous chromosomes are therefore referred to as either maternal or paternal

FIGURE 12.3 Ploidy Changes during the Life Cycle of a Dog. This type of life cycle is called diploid dominant, because the organism is diploid throughout most of its life cycle.

● **EXERCISE** In many algae, the only diploid phase is the zygote. Zygotes undergo meiosis to form haploid cells, which then grow into haploid adults by the process of mitosis. Adults produce haploid sperm and eggs. Sketch this "haploid-dominant" life cycle.

in origin. **Maternal chromosomes** come from the mother; **paternal chromosomes** come from the father.

Figure 12.3 puts these events into the context of an animal's **life cycle**—the sequence of events that occurs over the life span of an individual, from fertilization to the production of offspring. As you study the figure, note how ploidy changes as the result of meiosis and fertilization. In the case of the dog illustrated here, meiosis in a diploid adult results in the formation of haploid gametes, which combine to form a diploid zygote.

It's also important to realize that meiosis and gametogenesis occur in specialized organs—the testes or ovaries of a dog, for example—and that meiosis represents a tiny fraction of all cell division events that occur during the life cycle.

Once Sutton and others had published their work on the nature of meiosis and changes in ploidy, the mystery of fertilization was finally solved. To appreciate the consequences of meiosis fully, though, let's now analyze the events in more detail.

PRIOR TO MEIOSIS

Chromosomes replicate, forming sister chromatids.

MEIOSIS I

Homologous chromosomes separate.

1. Interphase: Chromosomes replicate in parent cell, in uncondensed state.

2. Early Prophase I: Chromosomes condense, nuclear envelope breaks up, spindle apparatus forms. Synapsis of homologous chromosomes.

3. Late Prophase I: Crossing over of non-sister chromatids (often multiple cross-overs between the same chromatids).

4. Metaphase I: Tetrads migrate to metaphase plate.

5. Anaphase I: Homologs separate and begin moving to opposite sides of cell.

FIGURE 12.4 The Phases of Meiosis.

● **EXERCISE** In step 3, circle all of the other chiasmata.

The Phases of Meiosis I

Meiosis begins after chromosomes have been replicated during S phase. Prior to the start of meiosis, chromosomes are extremely long structures, just as they are during interphase of the normal cell cycle. The major steps that occur once meiosis begins are shown in **Figure 12.4**, using a diploid species with a haploid number of 2 as an example ($n = 2$; $2n = 4$). As in Figure 12.3, maternal chromosomes are red and paternal chromosomes are blue.

During early prophase I the chromosomes condense, the spindle apparatus forms, and the nuclear envelope begins to disappear. The next event illustrated, still during early prophase of meiosis I, is crucial: Homologous chromosome pairs come together. This pairing process is called **synapsis** and is illustrated in step 2 of Figure 12.4. Synapsis is possible because regions of homologous chromosomes that are similar at the molecular level attract one another, via mechanisms that are currently the subject of intense research. The structure that results from synapsis is called a **tetrad** (*tetra* means four in Greek). A tetrad consists of two homologous chromosomes, with each homolog consisting of two sister chromatids. The chromatids from the homologs are referred to as **non-sister chromatids**. In the figure, the red-colored chromatids are non-sister chromatids with respect to the blue-colored chromatids.

During late prophase I, the non-sister chromatids begin to separate at many points along their length. They stay joined at certain locations, however, and look as if they cross over one another. Each crossover forms an X-shaped structure called a **chiasma** (plural: *chiasmata*). (In the Greek alphabet, the letter *X* is "chi.") Normally, at least one chiasma forms in every pair of homologous chromosomes; usually there are several chiasmata. As step 3 of Figure 12.4 shows, the chromatids involved in chiasma formation are homologous but not sisters. Consistent with this observation, Thomas Hunt Morgan proposed that a physical exchange of paternal and maternal chromosomes occurs at chiasmata. According to this hypothesis, paternal and maternal chromatids break and rejoin at each chiasma, producing chromatids that have both paternal and maternal segments. Morgan called this process of chromosome exchange **crossing over**. In step 4 of Figure 12.4, the result of crossing over is illustrated by chromosomes with a combination of red and blue segments. When crossing over occurs, the chromosomes that result have a mixture of maternal and paternal alleles.

The next major stage in meiosis I occurs during metaphase I, when pairs of homologous chromosomes (tetrads) are moved to a region called the **metaphase plate** by spindle fibers (step 4). Two points are key here: Each tetrad moves to the metaphase plate independently of the other tetrads, and the alignment of maternal and paternal homologs from each chromosome is random. During anaphase I, the homologous chromosomes in each tetrad separate and begin moving to opposite sides of the cell (step 5). Meiosis I concludes with telophase I, when the homologs finish moving to opposite sides of the cell (step 6). When meiosis I is complete, **cytokinesis** (division of cytoplasm) occurs and two haploid daughter cells form.

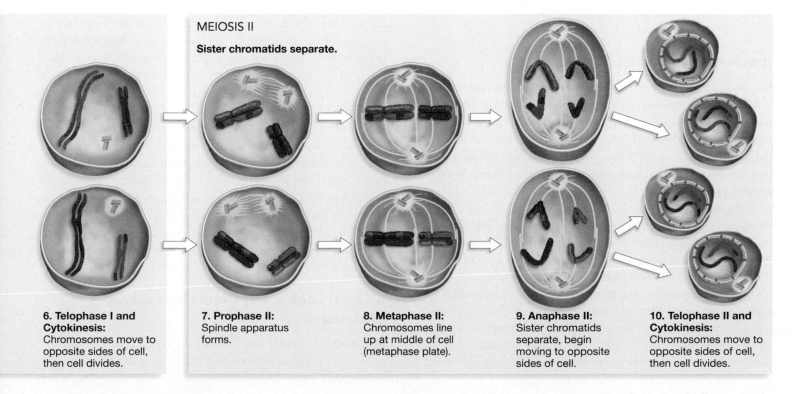

MEIOSIS II

Sister chromatids separate.

6. Telophase I and Cytokinesis: Chromosomes move to opposite sides of cell, then cell divides.

7. Prophase II: Spindle apparatus forms.

8. Metaphase II: Chromosomes line up at middle of cell (metaphase plate).

9. Anaphase II: Sister chromatids separate, begin moving to opposite sides of cell.

10. Telophase II and Cytokinesis: Chromosomes move to opposite sides of cell, then cell divides.

The end result of meiosis I is that one chromosome of each homologous pair is distributed to a different daughter cell. A reduction division has occurred: The daughter cells of meiosis I are haploid. The sister chromatids remain attached in each chromosome, however, meaning that the haploid daughter cells produced by meiosis I still contain replicated chromosomes. The chromosomes in each cell are a random assortment of maternal and paternal chromosomes as a result of crossing over and the random distribution of maternal and paternal homologs during metaphase.

Although meiosis I is a continuous process, biologists summarize the events by identifying these distinct phases:

- Early Prophase I—Replicated chromosomes condense, the spindle apparatus forms, and the nuclear envelope disappears. Synapsis of homologs forms pairs of homologous chromosomes (tetrads). Spindle fibers attach to the kinetochores at the centromeres of chromosomes.

- Late Prophase I—Crossing over results in a mixing of chromosome segments from maternal and paternal chromosomes.

- Metaphase I—Pairs of homologous chromosomes (tetrads) migrate to the metaphase plate and line up.

- Anaphase I—Homologs separate and begin moving to opposite ends of the cell.

- Telophase I—Homologs finish moving to opposite sides of the cell. In some species, a nuclear envelope re-forms around each set of chromosomes.

When meiosis I is complete, the cell divides.

The Phases of Meiosis II

Recall that chromosome replication occurred prior to meiosis I. Throughout meiosis I, sister chromatids remain attached. Because no further chromosome replication occurs between meiosis I and meiosis II, each chromosome consists of two sister chromatids at the start of meiosis II. And because only one member of each homologous pair of chromosomes is present, the cell is haploid.

Next, during prophase II, a spindle apparatus forms in both daughter cells. Spindle fibers attach to each side of the chromosomes—one spindle fiber to each sister chromatid—and begin moving the chromosomes toward the middle of each cell (step 7 of Figure 12.4). In metaphase II, the chromosomes are lined up at the metaphase plate (step 8). The sister chromatids of each chromosome separate during anaphase II (step 9) and move to different daughter cells during telophase II (step 10). Once they are separated, each chromatid is considered an independent chromosome. Meiosis II results in four haploid cells, each with one chromosome of each type.

As in meiosis I, biologists routinely designate distinct phases in meiosis II:

- Prophase II—The spindle apparatus forms. If a nuclear envelope formed at the end of meiosis I, it breaks apart.

- Metaphase II—Replicated chromosomes, consisting of two sister chromatids, are lined up at the metaphase plate.

- Anaphase II—Sister chromatids separate. The unreplicated chromosomes that result begin moving to opposite sides of the cell.

- Telophase II—Chromosomes finish moving to opposite sides of the cell. A nuclear envelope forms around each haploid set of chromosomes.

When meiosis II is complete, each cell divides to form two daughter cells. Because meiosis II occurs in both daughter cells of meiosis I, the process results in a total of four daughter cells from each original, parent cell. In a nutshell, one diploid cell with replicated chromosomes gives rise to four haploid cells with unreplicated chromosomes. It should make sense, after you examine the right side of Figure 12.4 in detail, that the movement of chromosomes during meiosis II is virtually identical to what happens in a mitotic division in a haploid cell.

In male animals the cells produced by meiosis go on to form sperm through a series of events that are detailed in later chapters. In females of at least some animal species, though, meiosis begins in cells called oocytes and then stops and restarts at several points as oocytes mature into eggs. In mammals, for example, meiosis is not completed until after fertilization takes place. And in plants, the products of meiosis do not form gametes—instead, the haploid cells that result form reproductive cells called spores (see Chapter 40). For the purposes of this chapter, however, you can think of meiosis as a process involved in gamete formation.

Figure 12.5 and **Table 12.3** provide a detailed comparison of mitosis and meiosis. The key difference between the two processes is that homologous chromosomes pair early in meiosis, but not during mitosis. Because homologs pair in prophase of meiosis I, they can migrate to the metaphase plate together and then separate during anaphase of meiosis I, resulting in a reduction division. ● If you understand this key distinction between meiosis and mitosis, you should be able to describe the consequences for meiosis if homologs do not pair. Now let's delve into the details of this critical event.

A Closer Look at Key Events in Prophase of Meiosis I

Figure 12.7 provides more detail on how several important events in prophase of meiosis I occur. Step 1 of Figure 12.7 shows that after chromosome replication is complete, sister

FIGURE 12.5 A Comparison of Mitosis and Meiosis. Mitosis produces two daughter cells with chromosomal complements identical to the parent cell. Meiosis produces four haploid cells with chromosomal complements unlike each other and unlike the diploid parent cell.

SUMMARY TABLE **12.3 Key Differences between Mitosis and Meiosis**

Feature	Mitosis	Meiosis
Number of cell divisions	One	Two
Number of chromosomes in daughter cells, compared with parent cell	Same	Half
Synapsis of homologs	No	Yes
Number of crossing-over events	None	One or more per pair of homologous chromosomes
Makeup of chromosomes in daughter cells	Identical	Different—only one of each chromosome type present, paternal and maternal segments mixed within chromosomes
Role in life cycle	Asexual reproduction in eukaryotes; cell division for growth of multicellular organisms	Precedes production of gametes in sexually reproducing animals

BOX 12.1   Karyotyping Techniques

Although chromosomes maintain their individuality and physical integrity throughout the cell cycle, they are readily visible only during mitosis or meiosis, when they are condensed. To describe an individual's karyotype, then, biologists must study cells undergoing cell division.

The first step in generating a karyotype is to obtain a sample of cells from the individual being studied. Cancer researchers might collect cells from a tumor; physicians who are concerned about the possibility of birth defects might obtain a few cells from the developing embryo inside the mother. The next step is to grow the cells in culture, using techniques introduced in Chapter 11. When the cultured cells are dividing rapidly, they are treated with a compound called colchicine. Colchicine stops mitosis at metaphase by disrupting the formation of the mitotic spindle. At this stage the chromosomes are relatively easy to study, because they are condensed and consist of sister chromatids.

The chromosomes of colchicine-treated cells are then stained and examined with the light microscope.

Researchers can distinguish condensed chromosomes by size; by the position of the centromere, which holds sister chromatids together; and by striping or banding patterns that appear in response to some stains. Subtler differences among chromosomes are apparent when a higher-resolution technique for karyotyping called **spectral karyotyping (SKY)**, or **chromosome painting**, is used. The "painting" is done with fluorescent dyes that are attached to short DNA molecules. The dyed pieces of DNA bind to particular regions of particular chromosomes. By using a combination of dyes, technicians can give each pair of homologous chromosomes a distinctive suite of colors (**Figure 12.6a**). The high-resolution image produced by this technique allows clinicians to diagnose an array of chromosomal abnormalities. Consider some examples:

- **Figure 12.6b** shows chromosomal changes associated with a cancer called chronic myelogenous leukemia. The defect is a **translocation**—a swapping of chromosome segments. In this case, a piece of chromosome 9 and a piece of chromosome 22 have changed places. The translocation causes a genetic change that leads to uncontrolled cell growth.

- **Figure 12.6c** shows the sex chromosomes of an individual with **Klinefelter syndrome**, which develops in people who have two X chromosomes and a Y chromosome instead of a single X and a single Y. People with this syndrome have male sex organs but are sterile. They may also develop some female-like traits, such as enlarged breasts.

Karyotyping is an important diagnostic tool. As techniques for generating karyotypes improve, our ability to detect and interpret chromosomal defects also improves.

(a) Normal human karyotype

(b) Translocations are abnomalities

Pieces of chromosomes have been swapped

(c) Sex chromosomes in Klinefelter syndrome

Two X chromosomes and one Y chromosome

FIGURE 12.6 Human Chromosomes. (a) Unlike the orderly arrangement seen here, condensed chromosomes that are undergoing mitosis are arranged randomly when first observed with the microscope. To determine a karyotype, a technician uses a computer to separate the image of each condensed chromosome, place homologous pairs side by side, and arrange the homologs by number. Note that even though these are replicated chromosomes, the fluorescent labeling is so strong that individual chromatids can't be distinguished. **(b)** In this individual, a piece of chromosome 9 (white) and a piece of chromosome 22 (purple) have been swapped via translocation. **(c)** Individuals with Klinefelter syndrome have two X chromosomes and a Y chromosome.

● **QUESTION** Is the karyotype in part (a) of a male or female?

A CLOSER LOOK AT THREE KEY EVENTS IN MEIOSIS

Sister chromatids

Chromosomes

Centromere

1. Replication, during interphase. Sister chromatids are held together by proteins along the chromosome "arms" and at the centromere. Shown: early prophase of meiosis I, when chromosomes have condensed.

One homolog

Synaptonemal complex

Second homolog

2. Synapsis, during prophase I. Homologous chromosomes are held together by proteins in the synaptonemal complex.

Non-sister chromatids

Protein complex

3. Crossing over, during prophase I. Complex of proteins forms where crossing over will occur. Chromosome segments are swapped between non-sister chromatids.

Crossing over usually occurs at least once in each non-sister chromatid, but is only shown on 1 pair here

FIGURE 12.7 A Closer Look at Chromosome Replication, Synapsis, and Crossing Over.

chromatids stay very tightly joined along their entire length. When homologs synapse, two pairs of non-sister chromatids are brought close together and are held there by a network of proteins called the **synaptonemal complex** (step 2). Crossing over can occur at many locations along the length of this paired structure. In most species crossing over typically occurs at least once in each pair of homologs; in some species it occurs two or three times. During the process, a complex of proteins cuts the chromosomes and then reattaches segments from homologs. The key to understanding crossing over is to recognize that at each point where it occurs—as in the circles labeled "protein complex" in step 3 of Figure 12.7—the non-sister chromatids from each homolog get physically broken at the same point and *attached to each other*. As a result, segments of maternal and paternal chromosomes are swapped, as shown at the bottom of Figure 12.7. ○ If you understand what is happening during crossing over, you should be able to make two chains of paper clips to simulate non-sister chromatids and then simulate a crossing over event between them.

This more detailed look at replication, synapsis, and crossing over should help you interpret the micrographs in **Figure 12.8**, which show grasshopper cells undergoing meiosis. Once you have a solid understanding of how meiosis occurs, you will be ready to consider ultimate explanations for why

meiosis and sexual reproduction exist. Meiosis is an intricate, tightly regulated process that involves dozens if not hundreds of different proteins. Given this complexity, it is logical to hypothesize that this type of nuclear division is extremely important. Why?

(MB) **Web Animation** *BioFlix* at www.masteringbio.com
Meiosis

Check Your Understanding

If you understand that...

- Meiosis is called a reduction division because the total number of chromosomes present is cut in half.
- During meiosis, a single diploid parent cell with replicated chromosomes gives rise to four haploid daughter cells.

You should be able to...

1) Demonstrate the phases of meiosis illustrated in Figure 12.4 by using pipe cleaners or pieces of cooked spaghetti.

2) Identify the event that makes meiosis a reduction division, unlike mitosis, and explain why it is responsible for reduction division.

(a) Prophase of meiosis I

(b) Anaphase of meiosis I

(c) Metaphase of meiosis II

(d) Anaphase of meiosis II

FIGURE 12.8 Meiotic Chromosomes in Grasshoppers.

● **EXERCISE** (a) Label several chiasmata, sister chromatids, tetrads, and non-sister chromatids; (b) add the spindle apparatus; (c) label sister chromatids and the metaphase plate, and add a spindle apparatus; (d) label the haploid number and ploidy of each nucleus.

12.2   The Consequences of Meiosis

The cell biologists who worked out the details of meiosis in the late 1800s and early 1900s realized that the process solved the riddle of fertilization. Weissman's hypothesis—that a reduction division precedes gamete formation in animals—was confirmed. This was an important advance in providing a proximate explanation for sexual reproduction. But researchers also understood that meiosis has another important outcome: Thanks to the independent shuffling of maternal and paternal chromosomes and crossing over during meiosis I, the chromosomes in gametes are different from the chromosomes in parental cells. Subsequently, fertilization brings haploid sets of chromosomes from a mother and father together to form a diploid offspring. The chromosome complement of this offspring is unlike that of either parent. It is a random combination of genetic material from each parent.

This change in chromosomal complement is crucial. The critical observation is that changes in chromosome configuration occur only during sexual reproduction—*not* during asexual reproduction. **Asexual reproduction** refers to any mechanism of producing offspring that does not involve the fusion of gametes. Asexual reproduction in eukaryotes is usually based on mitosis, and the chromosomes in the daughter cells of mitosis are identical to the chromosomes in the parental cell (see Chapter 11). In contrast, **sexual reproduction** refers to the production of offspring through the fusion of gametes. Sexual reproduction results in offspring that have chromosome complements unlike each other and their parents. Why is this difference important?

Chromosomes and Heredity

The changes in chromosomes produced by meiosis and fertilization are significant because chromosomes contain the cell's hereditary material. Stated another way, chromosomes

contain the instructions for specifying what a particular trait might be in an individual. These inherited traits range from eye color and height in humans to the number or shape of the bristles on a fruit fly's leg to the color or shape of the seeds found in pea plants.

In the early 1900s biologists began using the term gene to refer to the inherited instructions for a particular trait. Chapter 13 explores the experiments that confirmed that each chromosome is composed of a series of genes encoding information for different traits. In humans, for example, a single chromosome might contain genes that influence height, hair color, the spacing of teeth, and the tendency to develop colon cancer; another chromosome might include genes that affect eye color, susceptibility to allergies, and predisposition to schizophrenia. In most cases there are hundreds or thousands of genes on each chromosome. Recall from Section 12.1 that the term *allele* refers to a particular version of a gene and that homologous chromosomes may carry different alleles. You might have a chromosome containing alleles that tend to produce medium height, black hair color, and a low predisposition to colon cancer, as well as a homologous chromosome containing alleles that contribute to short height, blond hair color, and a predisposition to colon cancer. Other individuals you know might have alleles associated with extreme height, red hair color, and a moderate predisposition to colon cancer.

Chromosomes are composed of genes, and identical copies of chromosomes are distributed to daughter cells during mitosis. Thus, cells that are produced by mitosis are genetically identical to the parent cell, and offspring produced during asexual reproduction are genetically identical to one another as well as to their parent. The offspring of asexual reproduction are **clones**—or exact copies—of their parent. But the offspring produced by sexual reproduction are genetically different from one another and unlike either their mother or their father. Let's analyze three aspects of meiosis that create variation among chromosomes—and hence the genetic makeup—of sexually produced offspring: (1) separation and distribution of homologous chromosomes, (2) crossing over, and (3) fertilization.

How Does the Separation and Distribution of Homologous Chromosomes Produce Genetic Variation?

Each cell in your body contains 23 homologous pairs of chromosomes and 46 chromosomes in total. Half of these chromosomes came from your mother, and half came from your father. Each chromosome is composed of genes that influence particular traits. For example, one gene that affects your eye color might be located on one chromosome, while one of the genes that affects your hair color might be located on a different chromosome (**Figure 12.9a**).

Suppose that the chromosomes you inherited from your mother contain alleles that tend to produce brown eyes and black hair, but the chromosomes you inherited from your father include alleles that tend to specify green eyes and red hair. (This is a simplification for the purpose of explanation. In reality, several genes with various alleles interact in complex ways to produce human eye color and hair color.) Will any particular gamete you produce contain the genetic instructions inherited from your mother or the instructions inherited from your father?

To answer this question, study the diagram of meiosis in **Figure 12.9b**. It shows that when pairs of homologous chromosomes line up during meiosis I and the homologs separate, a variety of combinations of maternal and paternal chromosomes can result. Each daughter cell gets a random assortment of maternal and paternal chromosomes. As Chapter 13 will explain in detail, this phenomenon is known as the principle of independent assortment. In the example given here, meiosis results in gametes with alleles for brown eyes and black hair, like your mother, and green eyes and red hair, like your father. But two additional combinations also occur: brown eyes and red hair, or green eyes and black hair. Four different combinations of paternal and maternal chromosomes are possible when two chromosomes are distributed to daughter cells during meiosis I. ● If you understand how independent assortment produces genetic variation in the daughter cells of meiosis, you should be able to explain how genetic variation would be affected if maternal chromosomes always lined up together on one side of the metaphase plate during meiosis I and paternal chromosomes always lined up on the other side.

How many different combinations of maternal and paternal homologs are possible when more chromosomes are involved? In an organism with three chromosomes per haploid set ($n = 3$) eight types of gametes can be generated by randomly grouping maternal and paternal chromosomes. In general, a diploid organism can produce 2^n combinations of maternal and paternal chromosomes, where n is the haploid chromosome number. This means a human ($n = 23$) can produce 2^{23}, or about 8.4 million, gametes that differ in their combination of maternal and

(a) Example: individual who is heterozygous at two genes

(b) During meiosis I, tetrads can line up two different ways before the homologs separate.

OR

Brown eyes Black hair   Green eyes Red hair   Brown eyes Red hair   Green eyes Black hair

FIGURE 12.9 Separation of Homologous Chromosomes Results in Variable Combinations of Genes. (a) A hypothetical example: Genes that influence eye color and hair color in humans are on different chromosomes. **(b)** The cells along the bottom are products of meiosis. Notice that each cell has a different combination of genes, due to the separation of homologous chromosomes during meiosis I.

paternal chromosome sets. Clearly, the random assortment of whole chromosomes generates an impressive amount of genetic variation among gametes.

The Role of Crossing Over

Recall from Section 12.1 that segments of paternal and maternal chromatids exchange at each chiasma that forms during meiosis I. Thus, crossing over produces new combinations of alleles on the same chromosome—combinations that did not exist in either parent. This phenomenon is known as recombination. **Genetic recombination** is any change in the combination of alleles on a given chromosome. In species that reproduce sexually, recombination occurs via crossing over during meiosis. But genetic recombination also occurs in haploid organisms such as bacteria, which cannot undergo meiosis (see **Box 12.2**).

Crossing over and recombination are important because they dramatically increase the genetic variability of gametes

BOX 12.2 How Does Recombination Occur in Bacteria?

In many eukaryotes, sexual reproduction occurs when haploid cells that are formed by meiosis combine to form a new diploid individual. Sexual reproduction does not exist in bacteria, however. Bacteria are haploid throughout their lives and do not undergo meiosis. Genes can move from one individual to another, however, via a mechanism called conjugation.

Conjugation takes place when a physical connection, called a conjugation tube, forms between two bacterial cells. In many cases conjugation is followed by an event called plasmid transfer: As **Figure**

12.10a shows, a small loop of DNA called a plasmid is copied in one cell and transferred to the other cell. Plasmids were introduced briefly in Chapter 7 and are analyzed in more detail in Chapter 19.

As **Figure 12.10b** shows, however, conjugation can also result in genetic recombination. This occurs when a plasmid that has become integrated into the main bacterial chromosome is copied and transferred, along with genes from the main bacterial chromosome, through the conjugation tube to a recipient cell. In some cases the transferred portion of

chromosome replaces the homologous portion in the recipient cell's original chromosome, resulting in a new combination of alleles on that chromosome—meaning, recombination.

The figure highlights two key points about bacterial "sex": (1) It is a one-way transfer of genetic material instead of an exchange between individuals, and (2) instead of involving all of the genes present, the transfer is limited to a plasmid or a small portion of the genes in the main chromosome.

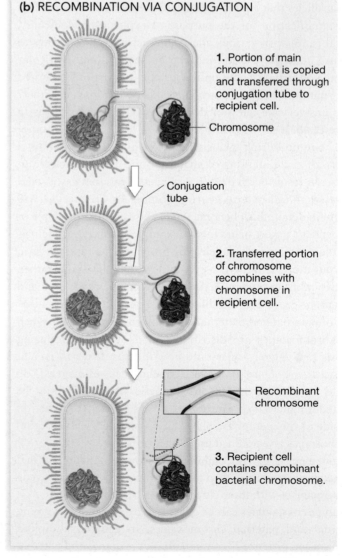

(a) PLASMID TRANSFER VIA CONJUGATION

Plasmid

Chromosome

1. Two bacterial cells come into contact. One cell contains a plasmid.

Conjugation tube

2. Copy of plasmid is transferred from donor cell to recipient cell through a conjugation tube.

3. Recipient cell contains plasmid.

(b) RECOMBINATION VIA CONJUGATION

1. Portion of main chromosome is copied and transferred through conjugation tube to recipient cell.

Chromosome

Conjugation tube

2. Transferred portion of chromosome recombines with chromosome in recipient cell.

Recombinant chromosome

3. Recipient cell contains recombinant bacterial chromosome.

FIGURE 12.10 Bacterial "Sex" Is One-Way and Involves Only a Few Genes.

produced by meiosis. Recall that the separation and distribution of homologous chromosomes during meiosis varies the combination of chromosomes present. But in addition, crossing over varies the combinations of alleles within each chromosome. Recent data on humans, for example, indicate that an average of about 50 chiasmata occur in each cell undergoing meiosis I. As a result, the number of genetically different gametes (PE) that an individual can produce is much more than the 8.4 million produced by the separation and distribution of homologs. When crossing over occurs a total of 50 or more times throughout the entire suite of tetrads, the number of genetically distinct gametes that you can produce is virtually limitless.

Biologists say that meiosis "shuffles" alleles, because each daughter cell gets a random assortment of maternal and paternal chromosomes and because most chromosomes contain both maternal and paternal alleles. As Section 12.3 will show, the genetic variation produced by meiosis has profound consequences for the ability of offspring to survive and reproduce.

How Does Fertilization Affect Genetic Variation?

● Crossing over and the random mixing of maternal and paternal chromosomes ensure that each gamete is genetically unique. Even if two gametes produced by the same individual fuse to form a diploid offspring—meaning **self-fertilization**, or "selfing," takes place—the offspring are very likely to be genetically different from the parent (**Figure 12.11**). Selfing is common in some plant species. It also occurs in the many animal species in which single individuals contain both male and female sex organs.

Self-fertilization is rare or nonexistent in many sexually reproducing species, however. Instead, gametes from different individuals combine to form offspring. This is called **outcrossing**. Outcrossing increases the genetic diversity of offspring because it combines chromosomes from different individuals, which are likely to contain different alleles.

How many genetically distinct offspring can be produced when outcrossing occurs? Let's answer this question using humans as an example. Recall that a single human can produce about 8.4 million different gametes—even in the absence of crossing over. When a person mates with a member of the opposite sex, the number of different genetic combinations that can result is equal to the product of the number of different gametes produced by each parent. In humans this means that potentially 8.4 million $\times$ 8.4 million $= 70.6 \times 10^{12}$ genetically distinct offspring can result from any one mating. This number

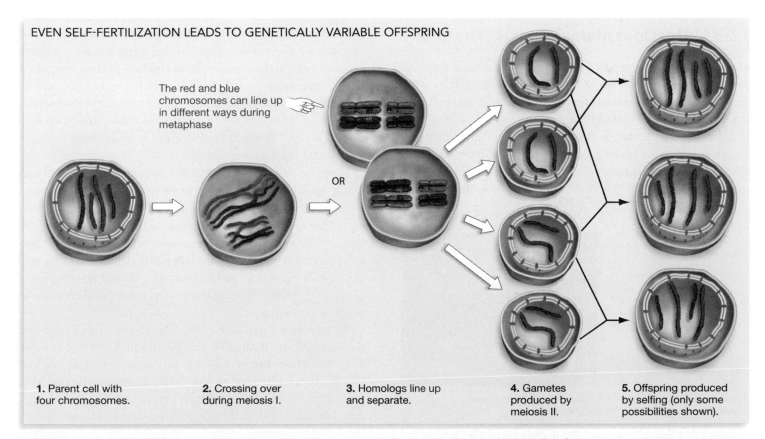

EVEN SELF-FERTILIZATION LEADS TO GENETICALLY VARIABLE OFFSPRING

The red and blue chromosomes can line up in different ways during metaphase

OR

1. Parent cell with four chromosomes.

2. Crossing over during meiosis I.

3. Homologs line up and separate.

4. Gametes produced by meiosis II.

5. Offspring produced by selfing (only some possibilities shown).

FIGURE 12.11 Even If Self-Fertilization Takes Place, Offspring Are Genetically Variable. Some possible results of self-fertilization in an organism with four chromosomes (2n = 4).

● **EXERCISE** In step 5, only a few of the many types of offspring that could be produced are shown. Sketch two additional types that are different from those shown.

is far greater than the total number of people who have ever lived—and the calculation does not even take into account variation generated by crossing over, which occurs at least once along each chromosome. Sexual reproduction results in genetically diverse offspring.

Check Your Understanding

If you understand that...

- The daughter cells produced by meiosis are genetically different from the parent cell because maternal and paternal homologs align randomly at metaphase of meiosis I and because crossing over leads to recombination within chromosomes.

You should be able to...

1) Draw a diploid parent cell with $n = 3$ (three types of chromosomes), and then sketch six of the many genetically distinct types of daughter cells that may result when this parent cell undergoes meiosis.

2) Compare and contrast the degree of genetic variation that results from asexual reproduction, selfing, and outcrossing.

12.3   Why Does Meiosis Exist? Why Sex?

Meiosis and sexual reproduction occur in only a small fraction of the lineages on the tree of life. Bacteria and archaea normally undergo only asexual reproduction; most algae, fungi, and some animals and land plants reproduce asexually as well as sexually. Recall that asexual reproduction in eukaryotes occurs via mitosis. Quaking aspen trees, for example, can produce new individuals by sending up shoots from underground stems (**Figure 12.12**). Asexual reproduction is found even

New shoots sprout from underground stems.

A genetically identical clone of trees results.

FIGURE 12.12 Asexual Reproduction in Quaking Aspen. Quaking aspen trees send out underground stems, from which new individuals sprout. Both the underground stems and the new shoots are produced by mitosis. As this process continues, a large group of genetically identical individuals forms. The photograph shows such a group in autumn, when quaking aspen leaves turn bright yellow.

among the vertebrates. For example, several species of guppy in the genus *Poeciliopsis* reproduce exclusively via mitosis.

Sexual reproduction is common among multicellular ("many-celled") organisms, however. It is the major mode of reproduction in the insects, which number over 43 million species, as well as in species-rich groups like the molluscs (clams, snails, squid) and vertebrates.

Even though sex plays an important role in the life of many organisms, until recently it was not clear why it occurs. On the basis of theory, biologists had good reason to think that sexual reproduction should not exist. Let's take a closer look.

The Paradox of Sex

In 1978 John Maynard Smith pointed out that the existence of sexual reproduction presents a paradox. Maynard Smith developed a mathematical model showing that because asexually reproducing individuals do not have to produce male offspring, their progeny can produce twice as many grand-offspring as can individuals that reproduce sexually. **Figure 12.13** diagrams this result by showing the number of females (♀) and males (♂) produced over several generations by asexual versus sexual reproduction. In this example, each individual produces four offspring over the course of his or her lifetime. In the asexual population, each individual is a female that produces four offspring. But in the sexual population, it takes two individuals—one male and one female—to produce four offspring. Thus, two out of every four children that each female produces sexually—the males—cannot have children of their own. As a result, generation 2 of the sexual population has just half as many child-producing offspring as generation 2 in the asexual population. Maynard Smith referred to this as the "two-fold cost of males." Asexual reproduction is much more efficient than sexual reproduction because no males are produced.

Based on this analysis, what will happen when asexual and sexual individuals exist in the same population and compete with one another? If all other things are equal, individuals that reproduce asexually should increase in frequency in the population while individuals that reproduce sexually should decline in frequency. In fact, Maynard Smith's model predicts that sexual reproduction is so inefficient that it should be completely eliminated. At the ultimate level of explanation, the existence of sex is a paradox.

To resolve the paradox of sex, biologists began examining the assumption "If all other things are equal." Stated another way, biologists began looking for ways that meiosis and outcrossing could lead to the production of offspring that reproduce more than asexually produced individuals do. After decades of debate and analysis, two solid answers are beginning to emerge.

The Purifying Selection Hypothesis

The first response to the paradox of sex hinges on a simple observation: If a gene is damaged or changed in a way that

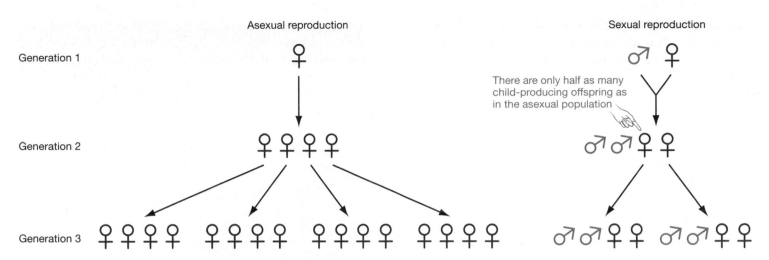

FIGURE 12.13 Asexual Reproduction Confers a Large Numerical Advantage. Each female symbol (♀) and male symbol (♂) represents an individual. In this hypothetical example, every individual produces four offspring over the course of a lifetime, sexually reproducing individuals produce half males and half females, and all offspring survive to breed.

● **QUESTION** How many asexually produced offspring would be present in generation 4? How many sexually produced offspring?

causes it to function poorly, it will be inherited by *all* of that individual's offspring when asexual reproduction occurs. Suppose the damaged gene arose in generation 1 of Figure 12.13. If the damaged gene is important enough, it might cause the four asexual females present in generation 2 to produce fewer than four offspring apiece—perhaps because they die young. If so, then generation 3 will not have twice as many individuals in the asexual lineage compared to the sexual lineage. An allele that functions poorly and lowers the fitness of an individual is said to be deleterious. Asexual individuals are doomed to transmitting all of their deleterious alleles to all of their offspring.

Suppose, however, that the same deleterious allele arose in the sexually reproducing female in generation 1 of Figure 12.13. If the female also has a normal copy of the gene, and if she mates with a male that has normal copies of the gene, then on average half of her offspring will lack the deleterious allele. Sexual individuals are likely to have offspring that lack deleterious alleles present in the parent.

Natural selection against deleterious alleles is called purifying selection. Over time, purifying selection should steadily reduce the numerical advantage of asexual reproduction.

To test this hypothesis, researchers recently compared the same genes in closely related species of *Daphnia*—a common inhabitant of ponds and lakes—that reproduce asexually versus sexually. As predicted, they found that individuals in the asexual species contained many more deleterious alleles than individuals in the sexual species. Results like these have convinced biologists that purifying selection is an important factor limiting the success of asexual reproduction.

The Changing Environment Hypothesis

The second hypothesis to explain sexual reproduction focuses on the benefits of producing genetically diverse offspring.

Here's the key idea: If the environment changes from one generation to the next in a way that makes parents poorly adapted, then offspring that are genetically different from their parents and from each other are more likely to survive and produce offspring of their own. Conversely, offspring that are genetic clones of their parents are less likely to thrive if the environment changes.

What type of environmental change would favor genetically diverse offspring? The possibilities include changes in temperature, moisture, predators, competitors, and food sources. But recently researchers have focused on one particular component of environmental change—the emergence of new strains of disease-causing agents.

The genetic characteristics of disease-causing organisms and viruses tend to change very quickly over time. In your own lifetime, for example, several new disease-causing agents have emerged that afflict humans; they include the SARS virus, new strains of HIV, the parasite that causes malaria, and the tuberculosis bacterium. In addition to employing drugs to combat these types of agents, humans have hundreds of genes that are involved in defense. At many of these genes, certain alleles help individuals fight off particular strains of bacteria, parasitic eukaryotes, or viruses. As you might predict, the presence of certain disease-fighting alleles is crucial for plants and animals that cannot rely on drug therapies for help.

What happens if all of the offspring produced by an individual are genetically identical? ● If a new strain of disease-causing agent evolves, then all of the asexually produced offspring are likely to be susceptible to that new strain. But if the offspring are genetically variable, then it is likely that at least some offspring will have combinations of alleles that enable them to fight off the new disease and produce offspring of their own.

The logic of the hypothesis is sound. Do any data support it?

Testing the Changing-Environment Hypothesis Curtis Lively and colleagues tested the changing-environment hypothesis recently by studying a species of snail that is native to New Zealand. This type of snail lives in ponds and other freshwater habitats and is susceptible to infection by over a dozen species of parasitic trematode worms. Snails that become infected cannot reproduce—the worms eat their reproductive organs. The parasites are rare in some habitats and common in others.

The biologists were interested in working on this snail species because some individuals reproduce only sexually while others reproduce only asexually. If the changing-environment hypothesis for the advantage of sex is correct, then the frequency of sexually reproducing individuals should be much higher in habitats where parasites are common than it is in habitats where parasites are rare (**Figure 12.14**). The logic here is that asexually reproducing individuals should have high fitness in environments where parasites are rare. Because they produce so many offspring in parasite-free environment, alleles that are associated with asexual reproduction should increase in frequency. In contrast, sexually reproducing individuals should have high fitness in habitats where parasites are common. In these environments, alleles that are associated with sexual reproduction should be frequent.

To test these predictions, the researchers collected a large number of individuals from different habitats. Lively and co-workers examined snails in habitats where parasites were more or less common and calculated the frequency of individuals that reproduce sexually versus those that reproduce asexually. The results are plotted in Figure 12.14. The data show that habitats where parasite infection rates are high have a relatively large number of sexually reproducing individuals compared with habitats that have low parasite incidence.

This result and a variety of other studies support the changing-environment hypothesis. Although the paradox of sex remains an active area of research, more biologists are becoming convinced that sexual reproduction is an adaptation that reduces the impact of purifying selection and increases the fitness of individuals in environments where disease-causing organisms are common. It's a way to rid your lineage of bad alleles and make sure your offspring have good alleles.

12.4   Mistakes in Meiosis

When homologous chromosomes separate during meiosis I, a complete set of chromosomes is transmitted to each daughter cell. But what happens if there is a mistake and the chromosomes are not properly distributed? What are the consequences for offspring if gametes contain an abnormal set of chromosomes?

In 1866 Langdon Down described a distinctive suite of co-occurring conditions observed in some humans. The syndrome was characterized by mental retardation, a high risk for heart problems and leukemia, and a degenerative brain disorder

Experiment

Question: Why does sexual reproduction occur?

Hypothesis: In habitats where parasitism is common, sexually produced offspring have higher fitness than do asexually produced individuals.

Null hypothesis: There is no relationship between the presence of parasites and reproduction method.

Experimental setup:

1. Collect snails from a wide array of habitats.

2. Document percentage of males in each population, as an index of frequency of sexual reproduction. More males means that more sexual reproduction is occurring.

3. Note two types of populations: In one, males are common; in the other, males are almost nonexistent. Infer that sexual reproduction is either common or almost nonexistent.

4. Document percentage of individuals infected with parasites in sexually versus asexually reproducing populations.

Prediction: In populations where sexual reproduction is common, parasitism rates are high. In populations with only asexual reproduction, infection rates are low.

Prediction of null hypothesis: No difference in parasitism rate between populations that reproduce sexually versus asexually.

Results:

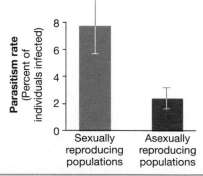

Conclusion: Sexual reproduction is common in habitats where parasitism is common. Asexual reproduction is common in habitats where parasitism is rare.

FIGURE 12.14 Is Sexual Reproduction Favored when Disease or Parasitism Rates Are High?

similar to Alzheimer's disease. **Down syndrome**, as the disorder came to be called, is observed in about 0.15 percent of live births (1 infant in every 666). For over 80 years the cause of the syndrome was unknown. Then in the late 1950s a researcher published observations on the chromosome sets of nine Down syndrome children. The data suggested that the condition is associated with the presence of an extra copy of chromosome 21. This situation is called a **trisomy** ("three-bodies")—in this case, trisomy-21—because each cell has three copies of the chromosome. To explain why trisomy occurred, the biologist proposed that the extra chromosome resulted from a mistake during meiosis in one of the parents.

How Do Mistakes Occur?

For a gamete to get one complete set of chromosomes, two steps in meiosis must be perfectly executed.

1. Each pair of homologous chromosomes must separate from each other during the first meiotic division, so that only one homolog ends up in each daughter cell. If both homologs move to the same pole of the parent cell, the products of meiosis will be abnormal. This sort of meiotic error, illustrated in **Figure 12.15**, is referred to as **nondisjunction**, because the homologs do not separate or disjoin. Notice that two daughter cells have two copies of the same chromosome (blue in Figure 12.15), while the other two lack that chromosome entirely. Gametes that contain an extra chromosome are symbolized as $n + 1$; gametes that lack one chromosome are symbolized as $n - 1$. If an $n + 1$ gam-

ete (PE) is fertilized by a normal n gamete, the resulting zygote will be $2n + 1$. This situation is a trisomy. If the $n - 1$ gamete is fertilized by a normal n gamete, the resulting zygote will be $2n - 1$. This situation is called **monosomy**. Cells that have too many or too few chromosomes are said to be **aneuploid** ("without-form").

2. Sister chromatids must separate from each other and move to opposite poles of the dividing cell during meiosis II. If this step fails, then the resulting daughter cells will be $n + 1$ and $n - 1$. It is relatively rare, though, for abnormal $n + 1$ and $n - 1$ gametes to be produced during nondisjunction in the second meiotic division.

Taken together, these two types of meiotic mistakes occur at a relatively high frequency. In humans, for example, researchers estimate that nondisjunction events occur in as many as 10 percent of meiotic divisions. The types of mistakes vary, but the consequences are almost always severe when defective gametes participate in fertilization. In a recent study of human pregnancies that ended in early embryonic or fetal death, 38 percent of the 119 cases involved atypical chromosome complements that resulted from mistakes in meiosis. Trisomy accounted for 36 percent of the abnormal karyotypes found. It was also common to find the incorrect number of complete chromosome sets, called triploidy ($3n$). Less common were abnormally sized or shaped chromosomes and monosomy ($2n - 1$). ● Based on these data and a large number of other studies, it is clear that mistakes in meiosis are a major cause of spontaneous abortion in humans.

NONDISJUNCTION

$2n = 4$
$n = 2$

$n + 1$

$n + 1$

$n - 1$

$n - 1$

1. Meiosis I starts normally. Tetrads line up in middle of cell.

2. Then one set of homologs does *not* separate (= nondisjunction).

3. Meiosis II occurs normally.

4. All gametes have an abnormal number of chromosomes—either one too many or one too few.

FIGURE 12.15 Nondisjunction Leads to Gametes with Abnormal Chromosome Numbers. If homologous chromosomes fail to separate during meiosis I, the gametes that result will have an extra chromosome or will lack a chromosome.

● **EXERCISE** Nondisjunction also results when meiosis I occurs normally but two of the sister chromatids illustrated in step 3 fail to separate (nondisjunction at meiosis II). Redraw steps 2–4 to show how this mechanism of nondisjunction occurs.

Why Do Mistakes Occur?

The leading hypothesis to explain the incidence of trisomy and other meiotic mistakes is that they are accidents—random errors that occur during meiosis. Consistent with this proposal, there does not seem to be any genetic or inherited predisposition to trisomy or other types of dysfunction. Most cases of Down syndrome, for example, occur in families with no history of the condition.

Even though meiotic errors may be random, there are still strong patterns in their occurrence:

- With the exception of trisomy-21, most of the trisomies and monosomies observed in humans involve the sex chromosomes. Klinefelter syndrome, which develops in XXY individuals—as described in Box 12.1—occurs in about 1 in 1000 live male births. Trisomy X (karyotype XXX) occurs in about 1 in 1000 live births and results in females who may or may not have symptoms such as impaired mental function and sterility. **Turner syndrome** develops in XO individuals, where the "O" stands for lack of a second X, and occurs in about 1 in 5000 live births. Individuals with this syndrome are female but are sterile.

- **Table 12.4** shows data collected on trisomies in the autosomes of human fetuses and infants. Three observations deserve mention: (1) Trisomy is much more common in the smaller chromosomes (numbers 13–22) than it is in the larger chromosomes (numbers 1–12); (2) trisomy-21 is far and away the most common type of trisomy observed; and

TABLE **12.4** **The Incidence of Trisomy in Humans: Effects of Chromosome Number and Paternal versus Maternal Origin**

Trisomy (chromosome number)	Total Number of Cases	Due to Error in Sperm	Due to Error in Egg	Maternal Errors (%)
2–12	16	3	13	81
13	7	2	5	71
14	8	2	6	75
15	11	3	8	73
16	62	0	62	100
18	73	3	70	96
21	436	29	407	93
22	11	0	11	100

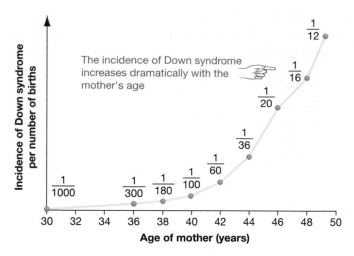

FIGURE 12.16 The Frequency of Down Syndrome Increases as a Function of a Mother's Age.

● **QUESTION** Suppose that you are an obstetrician. Based on these data, at what age would you recommend that pregnant mothers undergo procedures to check the karyotype of the embryos they carry?

(3) maternal errors account for most incidences of trisomy. For example, over 90 percent of cases of Down syndrome are due to chromosomal defects in eggs.

- Maternal age is an important factor in the occurrence of trisomy. As **Figure 12.16** shows, the incidence of Down syndrome increases dramatically in mothers over 35 years old.

Why do these patterns occur? Biologists still do not have a good explanation for why most cases of aneuploidy in humans are due to problems with meiosis in females. In addition, it is still a mystery why there is such a strong correlation between maternal age and frequency of trisomy-21. But to explain why most instances of aneuploidy involve chromosome 21 or the sex chromosomes, biologists offer two hypotheses: (1) Individuals with other types of aneuploidy do not develop normally and are spontaneously aborted, long before birth; and (2) the sex chromosomes and chromosome 21 are more susceptible to aneuploidy than are other types of chromosomes. These hypotheses are not mutually exclusive, so both may be correct.

Studies on the mechanism of aneuploidy continue. In the meantime, one overall message is clear: Undergoing meiosis correctly is critical to the health and welfare of offspring.

(MB) **Web Animation** at www.masteringbio.com
Mistakes in Meiosis

Chapter Review

SUMMARY OF KEY CONCEPTS

● Meiosis is a type of nuclear division. It results in cells that have half as many chromosomes as the parent cell, and in animals it is involved in the formation of eggs and sperm. When an egg and a sperm cell combine to form an offspring, the original number of chromosomes is restored.

When biologists confirmed that sperm and egg nuclei fuse during fertilization, it led to the hypothesis that a special type of cell division must precede gamete formation. Specifically, the proposal was that a sperm and egg must each have half of the normal number of chromosomes found in other cells. This hypothesis was confirmed when researchers observed meiosis and established that it results in gametes with half the normal chromosome number.

As the details of meiosis were being worked out, biologists realized that chromosomes exist in sets. In diploid organisms, individuals have two versions of each type of chromosome. One of the versions is inherited from the mother, and one from the father. The similar, paired chromosomes are called homologs. Haploid organisms, in contrast, have just one of each type of chromosome.

Each chromosome is replicated before meiosis begins. At the start of meiosis I, each chromosome consists of a pair of sister chromatids joined along their length and at a centromere. Homologous pairs of chromosomes synapse early in meiosis I, forming a tetrad—a group of two homologous chromosomes. After non-sister chromatids from the homologous chromosomes undergo crossing over, the pair of homologous chromosomes migrates to the metaphase plate. At the end of meiosis I, the homologous chromosomes separate and are distributed to two daughter cells. During meiosis II, sister chromatids separate and are distributed to two daughter cells.

You should be able to explain why meiosis does not occur in bacteria—most of which have a singular, circular chromosome. ●

(MB) Web Animation *BioFlix* at www.masteringbio.com
Meiosis

● Each cell produced by meiosis receives a different combination of chromosomes. Because genes are located on chromosomes, each cell produced by meiosis receives a different complement of genes. Meiosis leads to offspring that are genetically distinct from each other and from their parents.

When meiosis and outcrossing occur, the chromosome complements of offspring differ from one another and from their parents, for three reasons: (1) Maternal and paternal homologs are distributed randomly when chromosomes separate at the end of meiosis I; (2) maternal and paternal homologs exchange segments during crossing over; and (3) outcrossing results in a combination of chromosome sets from different individuals. The consequences of these differences became clear when biologists realized that chromosomes contain the hereditary material. Meiosis leads to genetic differences among offspring and between parents and offspring.

You should be able to explain why monozygotic twins, which develop from the same fertilized egg, are much more similar than dizygotic twins, which develop from two different eggs that were fertilized by two different sperm. ●

● The leading hypothesis to explain meiosis is that genetically variable offspring are more likely to thrive in changing environments.

Asexual reproduction is much more efficient than sexual reproduction because males do not have to be produced and no time or energy has to be spent in courtship. Sexual reproduction is favored in many groups, however, because it allows parents to produce offspring that lack deleterious alleles and because genetically diverse offspring are better able to resist parasites than are genetically uniform offspring. Some species alternate between sexual and asexual reproduction over the course of the year.

You should be able to predict whether sexual reproduction occurs during seasons when conditions are stable or seasons when conditions change rapidly. ●

● If mistakes occur during meiosis, the resulting egg and sperm cells may contain the wrong number of chromosomes. It is rare for offspring with an incorrect number of chromosomes to develop normally.

Mistakes during meiosis lead to gametes and offspring with an unbalanced set of chromosomes. Children with Down syndrome, for example, have an extra copy of chromosome 21. The leading hypothesis to explain these mistakes is that they are random accidents resulting in a failure of homologous chromosomes or sister chromatids to separate properly during meiosis.

You should be able to explain what happens when none of the homologous chromosomes present separate at anaphase of meiosis, but their sister chromatids separate normally at meiosis II. ●

(MB) Web Animation at www.masteringbio.com
Mistakes in Meiosis

QUESTIONS

● Test Your Knowledge

1. In the roundworm *Ascaris*, eggs and sperm have two chromosomes, but all other cells have four. Observations such as this inspired which important hypothesis?
   a. Before gamete formation, a special type of cell division leads to a quartering of chromosome number.
   b. Before gamete formation, a special type of cell division leads to a halving of chromosome number.
   c. After gamete formation, half of the chromosomes are destroyed.
   d. After gamete formation, either the maternal or the paternal set of chromosomes disintegrates.

2. What are homologous chromosomes?
   a. chromosomes that are similar in their size, shape, and gene content
   b. similar chromosomes that are found in different individuals of the same species
   c. the two "threads" in a replicated chromosome (they are identical copies)
   d. the products of crossing over, which contain a combination of segments from maternal chromosomes and segments from paternal chromosomes

3. What is a tetrad?
   a. the "X" that forms when chromatids from homologous chromosomes cross over
   b. a group of four chromatids produced when homologs synapse
   c. the four points where homologous chromosomes touch as they synapse
   d. the group of four genetically identical daughter cells produced by mitosis

4. What is genetic recombination?
   a. the synapsing of homologs during prophase of meiosis I

b. the new combination of maternal and paternal chromosome segments that results when homologs cross over
c. the new combinations of chromosome segments that result when outcrossing occurs
d. the combination of a haploid phase *and* a diploid phase in a life cycle

5. What is meant by a paternal chromosome?
   a. the largest chromosome in a set
   b. a chromosome that does not separate correctly during meiosis I
   c. the member of a homologous pair that was inherited from the mother
   d. the member of a homologous pair that was inherited from the father

6. Meiosis II is similar to which process?
   a. mitosis in haploid cells
   b. nondisjunction
   c. outcrossing
   d. meiosis I

Test Your Knowledge answers: 1. b; 2. a; 3. b; 4. b; 5. d; 6. a

---

● Test Your Understanding

Answers are available at www.masteringbio.com

1. Triploid ($3n$) watermelons are produced by crossing a tetraploid ($4n$) strain with a diploid ($2n$) plant. Briefly explain why this mating produces a triploid individual. Why can mitosis proceed normally in triploid cells, but meiosis cannot?

2. Meiosis is called a reduction division, but all of the reduction occurs during meiosis I—no reduction occurs during meiosis II. Explain why meiosis I is a reduction division but meiosis II is not.

3. Some plant breeders are concerned about the resistance of asexually cultivated plants, such as seedless bananas, to new strains of disease-causing bacteria, viruses, or fungi. Briefly explain their concern by discussing the differences in the genetic "outcomes" of asexual and sexual reproduction.

4. Explain why nondisjunction leads to trisomy and other types of abnormal chromosome complements. Draw an example of how an abnormal chromosome complement is produced during meiosis. In what sense are these chromosome complements "unbalanced?"

5. Explain the relationship between homologous chromosomes and the relationship between sister chromatids.

6. Lay four pens and four pencils on a tabletop, and imagine that they represent replicated chromosomes in a diploid cell with $n = 2$. Explain the phases of meiosis by moving the pens and pencils around. (If you don't have enough pens and pencils, use strips of paper or fabric.)

● Applying Concepts to New Situations

Answers are available at www.masteringbio.com

1. The gibbon has 44 chromosomes per diploid set, and the siamang has 50 chromosomes per diploid set. In the 1970s a chance mating between a male gibbon and a female siamang produced an off-spring. Predict how many chromosomes were observed in the somatic cells of the offspring. Do you predict that this individual would be able to form viable gametes? Why or why not?

2. Meiosis results in a reassortment of maternal and paternal chromosomes. If $n = 3$ for a given organism, there are eight different combinations of paternal and maternal chromosomes. If no crossing over occurs, what is the probability that a gamete will receive *only* paternal chromosomes?

3. Some researchers predict that spontaneous abortion should be rare in older females, because they are less likely than young females to be able to have offspring in the future. How does this claim relate to Figure 12.16, which graphs mother's age versus the incidence of Down syndrome?

4. The data on snail populations that were used to test the changing-environment hypothesis have been criticized because they are

observational and not experimental in nature. As a result, they do not control for factors other than parasites that might affect the frequency of sexually reproducing individuals.
   a. Design an experimental study that would provide stronger evidence that the frequency of parasite infection causes differences in the frequency of sexually versus asexually reproduced individuals in this species of snail.
   b. In defense of the existing data, comment on the value of observing patterns like this in nature, versus under controlled conditions in the laboratory.

---

www.masteringbio.com is also your resource for • Answers to text, table, and figure caption questions and exercises • Answers to *Check Your Understanding* boxes • Online study guides and quizzes • Additional study tools including the *E-Book for Biological Science* 3rd ed., textbook art, animations, and videos.

Mendel and the Gene

13

KEY CONCEPTS

- Mendel discovered that in garden peas, individuals have two alleles, or versions, of each gene. Prior to the formation of eggs and sperm, the two alleles of each gene separate so that one allele is transmitted to each egg or sperm cell.

- Genes are located on chromosomes. The separation of homologous chromosomes during anaphase of meiosis I explains why alleles of the same gene segregate to different gametes.

- If genes are located on different chromosomes, then the alleles of each gene are transmitted to egg cells and sperm cells independently of each other. This occurs because chromosomes line up randomly in metaphase of meiosis I.

- There are important exceptions and extensions to the basic patterns of inheritance that Mendel discovered.

Experiments on garden peas and sweet peas (shown here) helped launch the science of genetics.

The science of biology is built on a series of great ideas. Two of these—the cell theory and the theory of evolution—were introduced in Chapter 1. The cell theory describes the basic structure of organisms; the theory of evolution by natural selection clarifies why species change through time. These theories explain fundamental features of the natural world and answer some of our most profound questions about the nature of life: What are organisms made of? Where did species come from?

A third great idea in biology addresses an equally important question: Why do offspring resemble their parents? An Austrian monk named Gregor Mendel provided part of the answer in 1865 when he announced that he had worked out the rules of inheritance through a series of experiments on garden peas. The other part of the answer was provided during the final decades of the nineteenth century by the biologists who de-

scribed the details of meiosis. The chromosome theory of inheritance, formulated in 1903 by Walter Sutton and Theodor Boveri, linked these two results. This theory contends that meiosis, introduced in Chapter 12, causes the patterns of inheritance that Mendel observed. It also asserts that the hereditary factors called genes are located on chromosomes.

This chapter focuses on the evidence for the chromosome theory of inheritance. Let's begin with a basic question: What are the rules of inheritance that Mendel discovered?

13.1 Mendel's Experiments with a Single Trait

Gregor Mendel was a monk who lived and worked in the city of Brünn, located 70 miles north of Vienna. (Brünn was then part of Austria. Today the city is called Brno and is part of the

○ Key Concept   Important Information   ● Practice It

Czech Republic.) Mendel was educated in the natural sciences at the University of Vienna and also studied physics and mathematics under Christian Doppler, who discovered the Doppler effect for sound and light waves.

In Mendel's day, questions about **heredity**—meaning inheritance, or the transmission of traits from parents to offspring— were primarily the concern of animal breeders and horticulturists. A **trait** is any characteristic of an individual, ranging from overall height to the primary structure of a particular membrane protein. In Brünn, for example, there was a particular interest in how selective breeding could result in hardier and more productive varieties of sheep, fruit trees, and vines. To that end, an Agricultural Society had been formed. Its members emphasized the importance of research that would help breeding programs become more efficient. Mendel was an active member of this society; the monastery he belonged to was also devoted to scientific teaching and research.

What Questions Was Mendel Trying to Answer?

Mendel set out to address the most fundamental of all issues concerning heredity: What are the basic patterns in the transmission of traits from parents to offspring? At the time, two hypotheses had been formulated to answer this question. The first, called blending inheritance, claimed that the traits observed in a mother and father blend together to form the traits observed in their offspring. As a result, an offspring's traits are intermediate between the mother's and father's traits. For example, blending inheritance contended that black sheep have hereditary determinants for black wool and that white sheep have hereditary determinants for white wool. When these individuals mate, their hereditary determinants blend to form a new hereditary determinant for gray wool—their offspring should be gray.

The second hypothesis was called the inheritance of acquired characters, which claimed that traits present in parents are modified, through use, and passed on to their offspring in the modified form. The classical prediction of this hypothesis is that adult giraffes acquire longer necks by straining to reach leaves high in the tops of trees and that they subsequently produce longernecked offspring. The idea here is that the genetic determinants present in an individual are modified through use, and passed on to offspring in the modified form. Jean-Baptiste Lamarck was a prominent proponent of this hypothesis in the eighteenth century, and it had many adherents until the late 1800s.

These hypotheses were being promoted by the greatest scientists of Mendel's time. But are they correct?

Garden Peas Serve as the First Model Organism in Genetics

Mendel was certainly not the first scientist interested in studying the basic mechanisms of heredity. Why was he successful where others failed? Several factors came into play. One of the most important was that Mendel chose an appropriate model organism to study. A **model organism** for genetics studies is a species where individuals are small, short-lived, inexpensive to care for, able to produce large numbers of offspring, and easy to manipulate experimentally. Such species are called models because the conclusions drawn from studying them turn out to apply to many other species as well.

After investigating and discarding several candidates, Mendel chose to study the pea plant *Pisum sativum*. His reasons were practical: Peas are inexpensive, are easy to grow from seed, have a relatively short reproductive cycle, and produce reasonably large numbers of seeds. These features made it possible for Mendel to continue experiments over several generations and to collect data from a large number of individuals. Because of his choice, garden peas became the first model organism in genetics. **Genetics** is the branch of biology that focuses on the inheritance of traits.

Two additional features of the pea made it possible for Mendel to design his experiments: He could control which parents were involved in a mating, and he could arrange matings between individuals that differed in easily recognizable traits such as flower color or seed shape. Why was this important?

How Did Mendel Arrange Matings?   **Figure 13.1a** shows a garden pea flower, including its male and female reproductive organs. Sperm cells are produced in **pollen grains**, which are small sacs that mature in the male reproductive structure of the plant. Eggs are produced in the female reproductive structure. Fertilization begins when pollen grains are deposited on a tubelike section of the female reproductive organs. Sperm cells travel down this tube to the egg cells, where fertilization takes place.

Under normal conditions, garden peas pollinate themselves rather than requiring pollen from other pea plants for fertilization to occur. **Self-fertilization** (or selfing) takes place when pollen from one flower falls on the female reproductive organ of that same flower. Selfing is common because pollen from other plants rarely reaches the flower—its petals form a compartment that encloses the male and female reproductive organs and tends to exclude bees and other types of pollinating insects.

As **Figure 13.1b** shows, however, Mendel could circumvent this arrangement by removing the male reproductive organs from a flower before any pollen formed. Later he could transfer pollen from another flower to that flower's female reproductive organ with a brush. This type of mating is referred to as a **cross-pollination**, or simply a cross. Using this technique, Mendel could control the matings of his model organism.

What Traits Did Mendel Study?   Mendel conducted his experiments on varieties of peas that differed in seven traits: seed shape, seed color, pod shape, pod color, flower color, flower and pod position, and stem length. As **Figure 13.2** shows, each trait

(a) Self-pollination

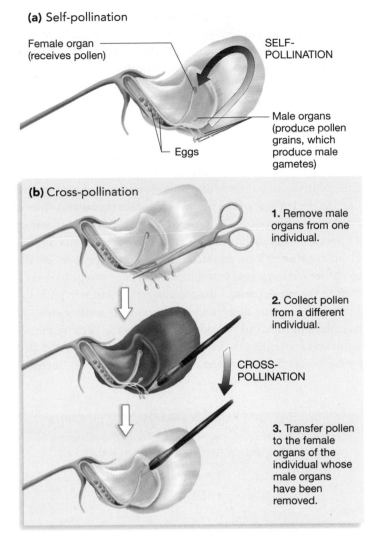

Female organ (receives pollen)

SELF-POLLINATION

Male organs (produce pollen grains, which produce male gametes)

Eggs

(b) Cross-pollination

1. Remove male organs from one individual.

2. Collect pollen from a different individual.

CROSS-POLLINATION

3. Transfer pollen to the female organs of the individual whose male organs have been removed.

FIGURE 13.1 Peas Can Be Self-Pollinated or Cross-Pollinated.
(a) The petals of a pea form an enclosed compartment. As a result, most fertilization takes place when pollen grains from the male reproductive organ of one flower fall on the female reproductive organ of the same flower. **(b)** Mendel arranged matings between individuals by removing the male organs from one flower and then using a paintbrush to dust its female organ with pollen collected from a different flower.

Trait	Phenotypes	
Seed shape	Round	Wrinkled
Seed color	Yellow	Green
Pod shape	Inflated	Constricted
Pod color	Green	Yellow
Flower color	Purple	White
Flower and pod position	Axial (on stem)	Terminal (at tip)
Stem length	Tall	Dwarf

FIGURE 13.2 Mendel Studied Seven Traits That Were Variable in Garden Peas. Two distinct phenotypes existed for each of the seven traits that Mendel studied in garden peas.

exhibited one of two forms. Biologists refer to the observable features of an individual, such as the shape of a pea seed or the eye color of a human, as its **phenotype** (literally, "show-type"). In the pea populations that Mendel studied, two distinct phenotypes existed for all seven traits.

Mendel began his work by obtaining individuals from what breeders called pure lines or true-breeding lines. A **pure line** consists of individuals that produce offspring identical to themselves when they are self-pollinated or crossed to another member of the pure-line population. For example, earlier breeders had developed pure lines for wrinkled seeds and round seeds. During two years of trial experiments, Mendel confirmed that individuals that germinated from his wrinkled seeds produced only wrinkled-seeded offspring when they were mated to them-

selves or to another pure-line individual that germinated from a wrinkled seed; individuals from his round seeds produced only round-seeded offspring when they were mated to themselves or to another pure-line individual from a round seed.

Why is this result important? Remember that Mendel wanted to find out how traits are transmitted from parents to offspring. Once he had confirmed that he was working with pure lines, he could predict how matings within each line would turn out; in other words, he knew what the offspring from these matings would look like. He could then compare these results with the outcomes of crosses between individuals from different pure lines. For example, suppose he arranged matings between a pure-line individual with round seeds and a pure-line individual with wrinkled seeds. He knew that one parent carried a hereditary determinant for round seeds, while the other carried a hereditary determinant for wrinkled seeds. But the offspring that resulted from this mating would have both hereditary determinants. They would be **hybrids**—offspring from matings between true-breeding parents that differ in one or more traits. Would they have wrinkled seeds, round seeds, or a blended combination of wrinkled and round? What would be the seed shape in subsequent generations when hybrid individuals self-pollinated or were crossed with members of the pure lines?

Inheritance of a Single Trait

Mendel's first set of experiments involved crossing pure lines that differed in just one trait, such as seed shape. Working with single traits was important because it made the results of the matings easier to interpret. Once he understood how a single trait was transmitted from parents to offspring, Mendel could then explore what happened when crosses were performed between individuals that differed in two traits.

Mendel began his single-trait crosses by crossing individuals from round- and wrinkled-seeded pure lines. The adults used in an initial experimental cross such as this represent the **parental generation**. Their progeny (that is, offspring) are called the **F₁ generation**. F₁ stands for "first filial"; the Latin roots *fili* and *filia* mean *son* and *daughter*. Subsequent generations from the F₁ parents are called the F₂ generation, F₃ generation, and so on.

Certain Traits "Recede"   In his first set of crosses, Mendel took pollen from round-seeded plants and placed it on the female reproductive organs of plants from the wrinkled-seeded line. As the left side of the Results section in **Figure 13.3** shows, all of the seeds produced by progeny from this cross were round. This was remarkable, for two reasons. First, the traits did not blend together to form an intermediate phenotype. Instead, the round-seeded form appeared intact. This result was in stark contrast to the predictions of the blending-inheritance hypothesis. Second, the genetic determinant for wrinkled seeds seemed to have disappeared. Did this disappearance occur because the determinant for wrinkled seeds was located in the egg (produced by the female flower) instead of in the sperm cells produced by

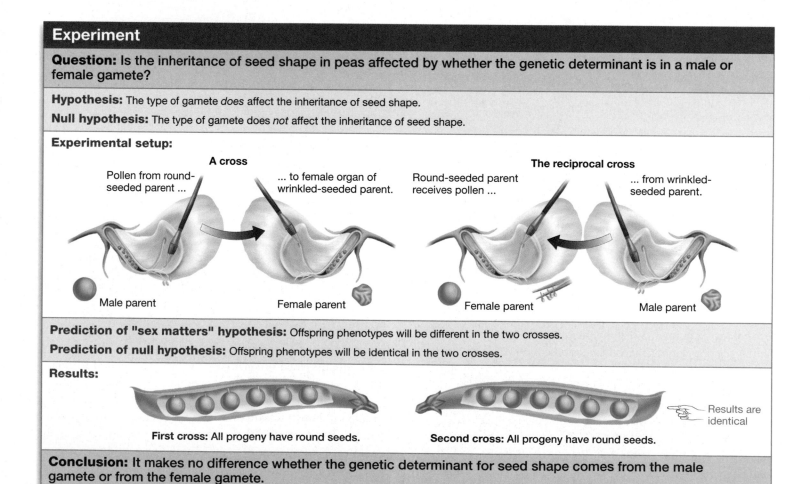

Experiment

Question: Is the inheritance of seed shape in peas affected by whether the genetic determinant is in a male or female gamete?

Hypothesis: The type of gamete *does* affect the inheritance of seed shape.

Null hypothesis: The type of gamete does *not* affect the inheritance of seed shape.

Experimental setup:

A cross

Pollen from round-seeded parent ...

... to female organ of wrinkled-seeded parent.

Male parent   Female parent

The reciprocal cross

Round-seeded parent receives pollen ...

... from wrinkled-seeded parent.

Female parent   Male parent

Prediction of "sex matters" hypothesis: Offspring phenotypes will be different in the two crosses.

Prediction of null hypothesis: Offspring phenotypes will be identical in the two crosses.

Results:

First cross: All progeny have round seeds.

Second cross: All progeny have round seeds.

Results are identical

Conclusion: It makes no difference whether the genetic determinant for seed shape comes from the male gamete or from the female gamete.

FIGURE 13.3 A Reciprocal Cross.

● **QUESTION** What is the purpose of a reciprocal cross?

pollen from the male flower? In general, did it matter which parent and gamete type had a particular genetic determinant?

To answer these questions, Mendel performed a second set of crosses—this time with pollen taken from an individual germinated from a wrinkled-seeded pea (see Figure 13.3). These crosses completed a **reciprocal cross**—a set of matings where the mother's phenotype in the first cross is the father's phenotype in the second cross, and the father's phenotype in the first cross is the mother's phenotype in the second cross.

In this case the results of the reciprocal crosses were identical: All of the F$_1$ progeny in the second cross had round seeds, just as in the first cross. This second cross established that it does not matter whether the genetic determinants for seed shape are located in the male or female parent. But a question still remained: What had happened to the genetic determinant for wrinkled seeds?

Dominant and Recessive Traits Mendel planted the F$_1$ seeds and allowed the individuals to self-pollinate when they matured. He collected the seeds that were produced by many plants in the resulting F$_2$ generation and observed that 5,474 were round and 1,850 were wrinkled. This observation was striking. The wrinkled seed shape reappeared in the F$_2$ generation after disappearing completely in the F$_1$ generation. This was astonishing. No one had observed the phenomenon before, simply because it had been customary for biologists to stop their breeding experiments with F$_1$ offspring.

Mendel invented some important terms to describe this result. He designated the genetic determinant for the wrinkled shape as **recessive**. This was an appropriate term because none of the F$_1$ individuals had wrinkled seeds—meaning the determinant for wrinkled seeds appeared to recede or temporarily become latent or hidden. In contrast, Mendel referred to the genetic determinant for round seeds as **dominant**. This term was apt because the round-seed determinant appeared to dominate over the wrinkled-seed determinant when both were present. It's important to note, though, that in genetics the term *dominant* has nothing to do with the everyday English usage as powerful or superior. Subsequent research has shown that individuals with the dominant phenotype do not necessarily have higher fitness than do individuals with the recessive phenotype. Nor are dominant genetic determinants necessarily more common than recessive ones. For example, a rare, dominant allele in humans causes a fatal illness—a type of brain degeneration called Huntington's disease. In genetics, the terms *dominance* and *recessiveness* identify *only* which phenotype is observed in individuals carrying two different genetic determinants. ● If you understand this concept, you should be able to explain why the allele for polydactyly, which causes extra digits to form in humans, is considered dominant while alleles associated with formation of five digits are considered recessive.

Mendel also noticed that the round and wrinkled seeds of the F$_2$ generation were present in a ratio of 2.96:1, or essentially 3:1. The 3:1 ratio means that for every four individuals, on average three had the dominant phenotype and one had the recessive phenotype. In other words, about 3/4 of the F$_2$ seeds were round and 1/4 were wrinkled.

Before trying to interpret this pattern, however, it was important for Mendel to establish that the results were not restricted to inheritance of seed shape. So he repeated the experiments with each of the six other traits of pea plants. In each case, he obtained similar results: The products of reciprocal crosses were the same; one form of the trait was always dominant regardless of the parent it came from; the F$_1$ progeny showed only the dominant trait and did not exhibit an intermediate phenotype; and in the F$_2$ generation, the ratio of individuals with dominant and recessive phenotypes was about 3 to 1.

How could these patterns be explained? Mendel answered this question with a series of propositions about the nature and behavior of the hereditary determinants. These hypotheses rank as some of the most brilliant insights in the history of biological science.

The Nature and Behavior of the Hereditary Determinants

Mendel's results were clearly inconsistent with the hypothesis of blending inheritance. To explain the patterns that he observed, Mendel proposed a competing hypothesis called particulate inheritance. He maintained that the hereditary determinants for traits do not blend together or acquire new or modified characteristics through use. In fact, hereditary determinants maintain their integrity from generation to generation. Instead of blending together, they act like discrete entities or particles.

Mendel's hypothesis was the only way to explain the observation that phenotypes disappeared in one generation and reappeared intact in the next. It also represented a fundamental break with ideas that had prevailed for hundreds of years.

What Are Genes, Alleles, and Genotypes? Today geneticists use the word **gene** to indicate the hereditary determinant for a trait. For example, the hereditary factor that determines the difference between round and wrinkled seeds in garden peas is referred to as the gene for seed shape.

Mendel's insights were even more penetrating, however. He also proposed that each individual has two versions of each gene. Today different versions of the same gene are called **alleles**. Different alleles are responsible for the variation in the traits that Mendel studied. In the case of the gene for seed shape, one allele of this gene is responsible for the round form of the seed while another allele is responsible for the wrinkled form. The alleles that are found in a particular individual are called its **genotype**. An individual's genotype has a profound effect on its phenotype—its physical traits.

The hypothesis that alleles exist in pairs was important because it gave Mendel a framework for explaining dominance and recessiveness. He proposed that some alleles are dominant and others are recessive. Recall that dominance and recessiveness identify which phenotype actually appears in an individual when both alleles are present. In garden peas, the allele for round seeds is dominant; the allele for wrinkled seeds is recessive. Therefore, as long as one allele for round seeds is present, seeds are round. When both alleles present are for wrinkled seeds (thus no allele for round seeds is present), seeds are wrinkled.

These hypotheses explain why the phenotype for wrinkled seeds disappeared in the F_1 generation and reappeared in the F_2 generation. But why did round- and wrinkled-seeded plants exist in a 3:1 ratio in the F_2 generation?

The Principle of Segregation ● To explain the 3:1 ratio of phenotypes in F_2 individuals, Mendel reasoned that the two alleles of each gene must segregate—that is, separate—into different gamete cells during the formation of eggs and sperm in the parents. As a result, each gamete contains one allele of each gene. This idea is called the **principle of segregation**.

To show how this principle works, Mendel used a letter to indicate the gene for a particular trait. For example, R represents the gene for seed shape. He used uppercase (R) to symbolize a dominant allele and lowercase (r) to symbolize a recessive allele. (Note that the symbols for genes are always italicized.)

Using this notation, Mendel could describe the genotype of the individuals in the pure line with round seeds (dominant) as RR. The genotype of the pure line with wrinkled seeds (recessive) is rr. Because RR and rr individuals have two copies of the same allele, they are said to be **homozygous** for the seed-shape gene (*homo* is the Greek root for "same," while *zygo* means "yoked together"). Pure-line individuals always produce offspring with the same phenotype because they are homozygous—no other allele is present.

Figure 13.4a diagrams what happened to these alleles when Mendel crossed the RR and rr pure lines. According to his analysis, RR parents produce eggs and sperm that carry the R allele, while rr parents produce gametes with the r allele. When two gametes—one from each parent—are fused together, they create offspring with the Rr genotype. Such individuals, with two different alleles for the same gene, are said to be **heterozygous** (*hetero* is the Greek root for "different"). Because the R allele is dominant, all of these F_1 offspring produced round seeds.

Why do the two phenotypes appear in a 3:1 ratio in the F_2 generation? A mating between parents that are both heterozygous at the gene in question is called a **monohybrid cross**. Mendel proposed that during gamete formation in the F_1 (heterozygous) individuals, the paired Rr alleles separate into different gamete cells. As a result, about half of the gametes

(a) A cross between two homozygotes

Offspring genotypes: All Rr (heterozygous)
Offspring phenotypes: All round seeds

(b) A cross between two heterozygotes

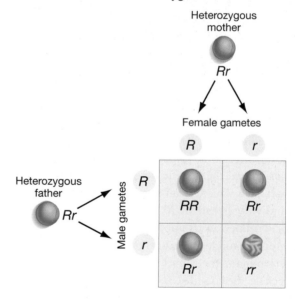

Offspring genotypes: 1/4 RR : 1/2 Rr : 1/4 rr
Offspring phenotypes: 3/4 round : 1/4 wrinkled

FIGURE 13.4 Mendel Analyzed the F_1 and F_2 Offspring of a Cross between Pure Lines. Notice that when you construct a Punnett square, you only need to list each unique type of gamete at the head of the rows and columns. For example, even though the RR alleles segregate in the male parent of part (a), you only have to list the R allele in the square—not R and R.

● **QUESTION** In constructing a Punnett square, does it matter whether the male or female gametes go on the left or across the top? Why or why not?

carry the R allele and half carry the r allele (**Figure 13.4b**). During self-fertilization, a given sperm has an equal chance of fertilizing either an R-bearing egg or an r-bearing egg.

Years after Mendel published his work, R. C. Punnett invented a straightforward technique for predicting the genotypes and phenotypes that should appear in the resulting offspring. A **Punnett square** is based on the following steps:

1. Write each unique gamete genotype produced by one parent to make a new row.

2. Write each unique gamete genotype produced by the other parent in a column.

3. Create boxes in the square or rectangle composed of the rows and columns.

4. Fill in the boxes with the offspring genotypes that result from fusion of the gamete genotypes in the appropriate row and column.

5. Predict the proportions of each offspring genotype and phenotype by tallying the offspring genotypes and phenotypes present in the boxes. **BioSkills 9** explains why this tallying process works.

⬤ If you understand these concepts, you should be able to state the purpose of a Punnett square and explain why it is not necessary to create a 2 × 2 square in Figure 13.4a, with the two r alleles that segregated from the rr mother as columns and the two R alleles that segregated from the RR father as rows.

As an example of the concluding step in analyzing a cross, the Punnett square in Figure 13.4b predicts that 1/4 of the F_2 offspring will be RR, 1/2 will be Rr, and 1/4 will be rr. Because the R allele is dominant to the r allele, 3/4 of the offspring should be round seeded and 1/4 should be wrinkled seeded. These results are *exactly* what Mendel found in his experiments with peas. In the simplest and most elegant fashion possible, his interpretation explains the 3:1 ratio of round to wrinkled seeds observed in the F_2 offspring and the mysterious reappearance of the wrinkled seeds.

The term **genetic model** refers to a set of hypotheses that explains how a particular trait is inherited. **Table 13.1** summarizes Mendel's model for explaining the basic patterns in the transmission of traits from parents to offspring. These hypotheses are sometimes referred to as Mendel's rules. His genetic model was a radical break from the hypotheses of blending inheritance and inheritance of acquired characters that previously dominated scientific thinking about heredity.

Testing the Model

Mendel's model explained his results in a logical way. But is it correct? To answer this question, Mendel conducted a series of experiments with the F_2 progeny described in Figure 13.4b. These experiments tested two important predictions:

1. Plants with the recessive phenotype—wrinkled seeds—are rr. Thus, they should produce only rr offspring when they are self-pollinated or crossed with another individual with wrinkled seeds.

2. Plants with the dominant phenotype (round seeds) may be either Rr or RR. These two genotypes should be present in the ratio 2:1. (That is, there should be twice as many heterozygotes as homozygotes among individuals with round seeds.) Individuals with the RR genotype should produce only RR offspring when they are self-pollinated. In contrast, Rr individuals should produce offspring with the same 3:1 ratio of round:wrinkled phenotypes observed in the cross diagrammed in Figure 13.4b.

Mendel planted the F_2 seeds and allowed the plants to self-pollinate when they matured. He then examined the phenotypes of the F_3 seeds. He confirmed the first prediction: F_2 plants with wrinkled seeds always produced offspring with wrinkled seeds. This result was consistent with the claim that these F_2 plants have an rr genotype.

What about the offspring of parents with the dominant phenotype? Mendel let 565 round-seeded plants self-pollinate. Of these, 193 plants produced only round-seeded offspring. Mendel inferred that these parents had the RR genotype. In contrast, 372 of the round-seeded parents produced seeds that

---

SUMMARY TABLE **13.1 Mendel's Model to Explain the Results of a Cross between Pure Lines**

Mendel's Claim	Notes
1. Peas have two versions, or alleles, of each gene.	This also turns out to be true for many other organisms.
2. Alleles do not blend together.	The hereditary determinants maintain their integrity from generation to generation.
3. Each gamete contains one allele of each gene.	This is due to the principle of segregation—the alleles of each gene segregate during the formation of gametes.
4. Males and females contribute equally to the genotype of their offspring.	When gametes fuse, offspring acquire a total of two alleles for each gene—one from each parent.
5. Some alleles are dominant to other alleles.	When a dominant allele and a recessive allele for the same gene are found in the same individual, that individual has the dominant phenotype.

⬤ **QUESTION** What is the difference between genes and alleles, a genotype and a phenotype, a homozygous individual and a heterozygous individual, and dominant alleles and recessive alleles?

were either round or wrinkled. Mendel inferred that in these individuals, the round-seeded parental genotype was *Rr*. In this experiment, the ratio of *Rr* to *RR* in the parents—based on the number of round and wrinkled seeds they produced—was 372:193 or 1.93:1. This is extremely close to the prediction of 2:1. Mendel observed the same patterns when he let F_2 individuals from the other six crosses (involving seed color and the other traits) self-fertilize. These results furnished overwhelming support for his model.

13.2 Mendel's Experiments with Two Traits

Working with one trait at a time allowed Mendel to establish that blending inheritance does not occur. It also allowed him to infer that each gene he was working with had two alleles and to recognize the principle of segregation. His next step was to extend these results. The most important question he addressed was whether the principle of segregation holds true if parental lines differ with respect to two traits.

To explore this issue, Mendel crossed a pure-line parent that produced round, yellow seeds with a pure-line parent that produced wrinkled, green seeds. The F_1 offspring of this cross should be heterozygous at both genes. A mating between parents that are both heterozygous for two traits is called a **dihybrid cross**.

Mendel's earlier experiments had established that the allele for yellow seeds was dominant to the allele for green seeds; these alleles were designated *Y* for yellow and *y* for green. As **Figure 13.5** indicates, two distinct possibilities existed for how the alleles of these two different genes—the gene for seed shape and the gene for seed color—would be transmitted to offspring. The first possibility was that the allele for seed shape and the allele for seed color present in each parent would separate from one another and be transmitted independently. This hypothesis is called independent assortment, because the two alleles would separate and sort themselves into gametes independently of each other (Figure 13.5a). The second possibility was that the allele for seed shape and the allele for seed color would be transmitted to gametes together. This hypothesis can be called dependent assortment, because the transmission of one allele would depend on the transmission of another (Figure 13.5b).

As Figure 13.5 shows, the F_1 offspring of Mendel's mating are expected to have the dominant round and yellow phenotypes whether the alleles were transmitted together or independently. When Mendel did the cross and observed the F_1 individuals, this is exactly what he found. All of the F_1 offspring had round, yellow seeds. All of these individuals were heterozygous at both genes.

In contrast to the situation in the F_1 generation, however, the two hypotheses make radically different predictions about

what Mendel should have observed when the F_1 individuals were allowed to self-fertilize and produce an F_2 generation. If the alleles assort independently and combine randomly to form gametes, then each heterozygous parent should produce four different gamete genotypes, as illustrated in Figure 13.5a. A 4-row-by-4-column Punnett square results, and it predicts that there should be 9 different offspring genotypes and 4 phenotypes. Further, the yellow-round, green-round, yellow-wrinkled, and green-wrinkled phenotypes should be present in the frequencies 9/16, 3/16, 3/16, and 1/16, respectively. This is equivalent to a ratio of 9:3:3:1. But if the alleles from each parent stay together, then a 2-row-by-2-column Punnett square would predict only three possible offspring genotypes and just two phenotypes, as Figure 13.5b shows. The hypothesis of dependent assortment predicts that F_2 offspring should be yellow-round or green-wrinkled, present in a ratio of 3:1. Note that the Punnett squares are making explicit predictions about the outcome of an experiment, based on a specific hypothesis about which alleles are present in each parent and how they are transmitted.

When Mendel examined the phenotypes of the F_2 offspring, he found that they conformed to the predictions of the hypothesis of independent assortment. Four phenotypes were present in frequencies that closely approximated the predicted frequencies of 9/16, 3/16, 3/16, and 1/16 and the predicted ratio of 9:3:3:1 (Figure 13.5c). Based on these data, he accepted the hypothesis that alleles of different genes are transmitted independently of one another. This result became known as the **principle of independent assortment**.

● If you understand the principle of independent assortment, it should make sense to you that an individual with the genotype *AaBb* produces gametes with the genotypes *AB*, *Ab*, *aB*, and *ab*. You should be able to predict the genotypes of the gametes produced by individuals with the genotypes *AABb*, *PpRr*, and *AaPpRr*.

Using a Testcross to Confirm Predictions

Mendel did experiments with combinations of traits other than seed shape and color and obtained results similar to those in Figure 13.5c. Each paired set of traits produced a 9:3:3:1 ratio of progeny phenotypes in the F_2 generation. He even did a limited set of crosses examining three traits at a time. Although all of these data were consistent with the principle of independent assortment, his most powerful support for the hypothesis came from a different type of experiment.

In designing this study, Mendel's goal was to test the prediction that an *RrYy* plant produces four different types of gametes in equal proportions. To accomplish this, Mendel invented a technique called a testcross. A **testcross** uses a parent that contributes only recessive alleles to its offspring to help determine the unknown genotype of the second parent. Testcrosses are useful because the genetic contribution of the homozygous

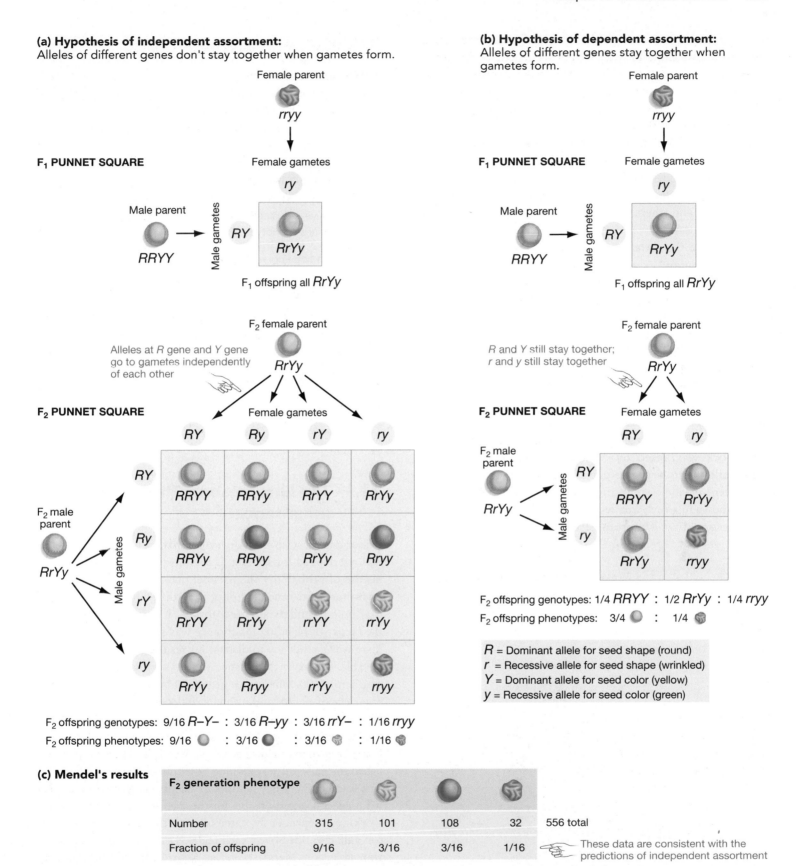

(a) Hypothesis of independent assortment:
Alleles of different genes don't stay together when gametes form.

(b) Hypothesis of dependent assortment:
Alleles of different genes stay together when gametes form.

F₂ offspring genotypes: 9/16 *R–Y–* : 3/16 *R–yy* : 3/16 *rrY–* : 1/16 *rryy*
F₂ offspring phenotypes: 9/16 : 3/16 : 3/16 : 1/16

F₂ offspring genotypes: 1/4 *RRYY* : 1/2 *RrYy* : 1/4 *rryy*
F₂ offspring phenotypes: 3/4 : 1/4

R = Dominant allele for seed shape (round)
r = Recessive allele for seed shape (wrinkled)
Y = Dominant allele for seed color (yellow)
y = Recessive allele for seed color (green)

(c) Mendel's results

F₂ generation phenotype				
Number	315	101	108	32	556 total
Fraction of offspring	9/16	3/16	3/16	1/16	These data are consistent with the predictions of independent assortment

FIGURE 13.5 Mendel Analyzed the F₁ and F₂ Offspring of a Cross between Pure Lines for Two Traits. Either of two events could occur when alleles of different genes are transmitted to offspring: Alleles could be sorted into gametes independently of each other, or alleles from the same parent could be transmitted together, generation after generation.

recessive parent is easy to predict and analyze. As a result, a testcross allows experimenters to test the genetic contribution of the other parent. If the other parent has the dominant phenotype but an unknown genotype, the results of the testcross allow researchers to infer whether that parent is homozygous or heterozygous for the dominant allele.

In this case, Mendel performed a testcross between parents that were *RrYy* and *rryy*. The types and proportions of offspring that should result can be predicted with the Punnett square shown in **Figure 13.6**. If the principle of independent assortment is valid, there should be four types of offspring in equal proportions.

What were the actual proportions observed? Mendel did this experiment and examined the seeds produced by the progeny. He found that 31 were round and yellow, 26 were round and green, 27 were wrinkled and yellow, and 26 were wrinkled and green. As predicted, these numbers are nearly identical to the 27.5 individuals expected with each genotype, given the total of 110 individuals. The predicted ratio of phenotypes was 1:1:1:1, which matched the observed ratio. The testcross had confirmed the principle of independent assortment.

Mendel's work provided a powerful conceptual framework for thinking about transmission genetics—the patterns that occur as alleles pass from one generation to the next. This framework was based on (1) the segregation of discrete, paired alleles into separate gametes and (2) the independent assortment of alleles that affect different traits. The experiments you've just reviewed were brilliant in design, execution, and interpretation. Unfortunately, they were ignored for 34 years.

(MB) **Web Animation** at www.masteringbio.com

Mendel's Experiments
The Principle of Independent Assortment

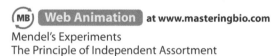

FIGURE 13.6 The Predictions Made by the Principle of Independent Assortment Can Be Evaluated in a Testcross. If the principle of independent assortment is correct and *RrYy* parents produce four types of gametes in equal proportions, then a mating between *RrYy* and *rryy* parents should produce four types of offspring in equal proportions, as this Punnett square shows.

Check Your Understanding

If you understand that...

- Mendel discovered that individuals have two alleles of each gene, and that each gamete receives one of the two alleles present in a parent. This is the principle of segregation.

- Mendel found that alleles from different genes are transmitted to gametes independently of each other. This is the principle of independent assortment.

- The alleles that Mendel analyzed were either dominant or recessive, meaning heterozygous individuals had the dominant phenotype.

You should be able to...

Use the genetic problems at the end this chapter to practice the following skills:

1) Starting with parents of known genotypes, create and analyze Punnett squares to predict the genotypes and phenotypes that will occur in their F_1 and F_2 offspring, and then calculate the expected frequency of each genotype and phenotype (Genetics Problems 12, 17, 19, 21).

2) Given the outcome of a cross, infer the genotypes and phenotypes of the parents (Genetics Problems 13, 15, 23, 24, 25).

13.3 The Chromosome Theory of Inheritance

Historians of science frequently debate why Mendel's work was overlooked for so long. It is almost undoubtedly true that his use of probability theory in explaining his results and his quantitative treatment of data were difficult for biologists of that time to understand and absorb. It may also be true that the theory of blending inheritance was so well entrenched that there was a tendency to dismiss his results as peculiar or unbelievable. Whatever the reason, Mendel's work was not appreciated until 1900, when three groups of biologists, working with a variety of plants and animals, independently "discovered" Mendel's main conclusions.

The rediscovery of Mendel's work, 16 years after his death, ignited the young field of genetics. Mendel's experiments established the basic rules governing how traits are passed from parents to offspring. They described the pattern of inheritance. But what process is responsible for these patterns? Two biologists, working independently, came up with the answer. Walter Sutton and Theodor Boveri each realized that meiosis could be responsible for Mendel's rules. When this hypothesis was published in 1903, research in genetics exploded.

Recall from Chapter 12 that meiosis is the type of cell division that precedes gamete formation. The details of the process were worked out in the final decades of the nineteenth century. What Sutton and Boveri grasped is that meiosis not only reduces

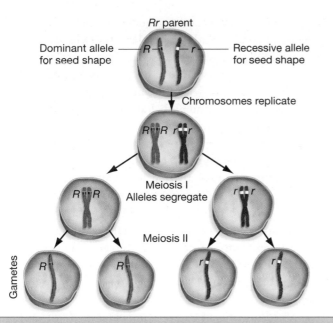

Principle of segregation: Each gamete carries only one allele for seed shape, because the alleles have segregated during meiosis.

FIGURE 13.7 Meiosis Is Responsible for the Principle of Segregation. A parent's two alleles segregate into different gametes, as Mendel hypothesized, because homologous chromosomes separate during meiosis I.

chromosome number by half but also explains the principle of segregation and the principle of independent assortment.

The cell nucleus at the top of **Figure 13.7** illustrates Sutton and Boveri's central insight—the hypothesis that chromosomes are composed of Mendel's hereditary determinants, or genes. In this example, the gene for seed shape is shown at a particular position along a chromosome. This location is known as a **locus** ("place"; plural: *loci*). A genetic locus is the physical location of a gene. The paternal and maternal chromosomes shown in Figure 13.7 happen to possess different alleles of the gene for seed shape: One allele specifies round seeds (*R*), while the other specifies wrinkled seeds (*r*).

The subsequent steps in Figure 13.7 show how these alleles segregate into different daughter cells during meiosis I, when homologous chromosomes separate. ● The physical separation of alleles during anaphase of meiosis I is responsible for Mendel's principle of segregation.

● **Figure 13.8** follows the fate of the alleles for two different genes—in this case, for seed shape and seed color—as meiosis proceeds. Because these genes are located on different nonhomologous chromosomes, they assort independently of one another at meiosis I. Four types of gametes, produced in equal proportions, result. This is the physical basis of Mendel's principle of independent assortment.

Sutton and Boveri formalized these observations in the **chromosome theory of inheritance**. Like other theories in

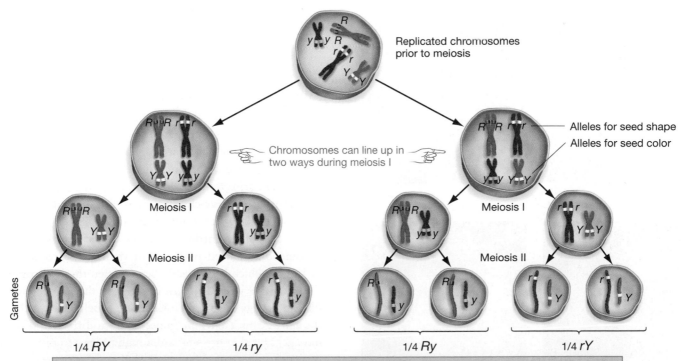

Principle of independent assortment: The genes for seed shape and seed color assort independently, because they are located on different chromosomes.

FIGURE 13.8 Meiosis Is Responsible for the Principle of Independent Assortment. The alleles for different traits assort independently, as Mendel hypothesized, because nonhomologous chromosomes assort independently during meiosis I. Maternal and paternal chromosomes are shown in different colors for clarity biology, the chromosome theory consists of a pattern—a set of observations about the natural world—and a process that explains the pattern. The chromosome theory states that Mendel's rules can be explained by the independent alignment and separation of homologous chromosomes at meiosis I.

When Sutton and Boveri published their findings, however, the hypothesis that chromosomes consist of genes was untested. What experiments confirmed that chromosomes contain genes?

13.4 Testing and Extending the Chromosome Theory

During the first decade of the twentieth century, an unassuming insect rose to prominence as a model organism for testing the chromosome theory of inheritance. This organism—the fruit fly *Drosophila melanogaster*—has been at the center of genetic studies ever since. *Drosophila melanogaster* has all the attributes of a useful model organism for experimental studies in genetics: small size, ease of rearing in the lab, a short reproductive cycle (about 10 days), and abundant offspring (up to a few hundred per mating). The elaborate external anatomy of this insect also makes it possible to identify interesting phenotypic variation among individuals (**Figure 13.9a**).

(a) The fruit fly *Drosophila melanogaster*

(b) Eye color is a variable trait.

FIGURE 13.9 The Fruit Fly *Drosophila melanogaster* Is an Important Model Organism in Genetics.

● **EXERCISE** In part (b), label the phenotype that is considered wild type. Label the phenotype that is a rare mutant.

Drosophila was adopted as a model organism by Thomas Hunt Morgan and his students. But because *Drosophila* is not a domesticated species like the garden pea, Morgan had no readily available phenotypic variants such as Mendel's round and wrinkled seeds. Consequently, an early goal of Morgan's research was simply to find and characterize individuals with different phenotypes.

Morgan's group referred to individuals with the most common phenotype as **wild type**. But while examining his cultures, Morgan discovered a male fly that had white eyes rather than the wild-type red eyes (**Figure 13.9b**). This individual had a discrete and easy-to-recognize phenotype different from the normal phenotype. Morgan inferred that the white-eyed phenotype resulted from a **mutation**—a change in a gene (in this case, a gene that affects eye color). Individuals with white eyes (or other traits attributable to mutation) are referred to as **mutants**.

To explore how the white-eye trait is inherited in fruit flies, Morgan mated a red-eyed female fly with the mutant white-eyed male fly. All of the F_1 progeny had red eyes. But when Morgan did the reciprocal cross, by mating white-eyed females to red-eyed males, he got a different result: All F_1 females had red eyes, but all F_1 males had white eyes. Recall that Mendel's reciprocal crosses had always given results that were similar to each other. But Morgan's reciprocal crosses did not. The experiment suggested a definite relationship between the sex of the progeny and the inheritance of eye color. When the female parent was white-eyed, the white-eyed phenotype appeared in F_1 males. How could these observations be reconciled with Mendel's rules of inheritance?

The Discovery of Sex Chromosomes

Nettie Stevens began studying the karyotypes of insects about the time that Morgan began his work with *Drosophila*. One of her outstanding observations was of a striking difference in the chromosome complements of males and females in the beetle *Tenebrio molitor*. In females of this species, diploid cells contain 20 large chromosomes. But diploid cells in males contain 19 large chromosomes and 1 small one. Stevens called the small chromosome the Y chromosome. This Y chromosome paired with one of the large chromosomes at meiosis I, which had already been named the X chromosome. The X and Y were different in size and shape, but they acted like homologs during meiosis. Later work showed that even though X and Y chromosomes contain different genes, they have regions that are similar enough to lead to proper pairing during prophase of meiosis I. The X and Y are now called sex chromosomes.

Stevens also noticed that female beetles had two X chromosomes, while male beetles had an X and a Y. Based on this observation, she proposed that an individual's genotype at the sex chromosomes (XX in females, XY in males) determines an individual's gender.

X-Linked Inheritance and the Chromosome Theory

To explain the results of his crosses with white-eyed flies, Morgan put his genetic data together with Stevens's observations on sex chromosomes. *Drosophila* females, like *Tenebrio* females, have two X chromosomes; male fruit flies carry an X and a Y, just as *Tenebrio* males do. As a result, half the gametes produced by a male fruit fly should carry an X chromosome; the other half, a Y chromosome (**Figure 13.10**).

Morgan realized that the transmission pattern of the X chromosome in males and females could account for the results of his reciprocal crosses. Specifically, he proposed that the gene for white eye color in fruit flies is located on the X chromosome and that the Y chromosome does not carry an allele of this gene. This hypothesis is called **X-linked inheritance** or **X-linkage**. (Proposing that a gene resides on the Y chromosome is called **Y-linked inheritance** or **Y-linkage**. Proposing that a gene is on either sex chromosome is termed **sex-linked inheritance** or **sex-linkage**.) The key observation is that even though the X and Y chromosomes synapse during prophase of meiosis I, they differ in size, shape, and gene content.

According to the hypothesis of X-linkage, a female has two copies of the gene that specifies eye color because she has two X chromosomes. One of these chromosomes came from her female parent, and the other from her male parent. A male, in contrast, has only one copy of the eye-color gene because he has only one X chromosome, which comes from his mother.

The Punnett squares in **Figure 13.11** show that Morgan's experimental results can be explained if the gene for eye color is located on the X chromosome, and if the allele for red color is

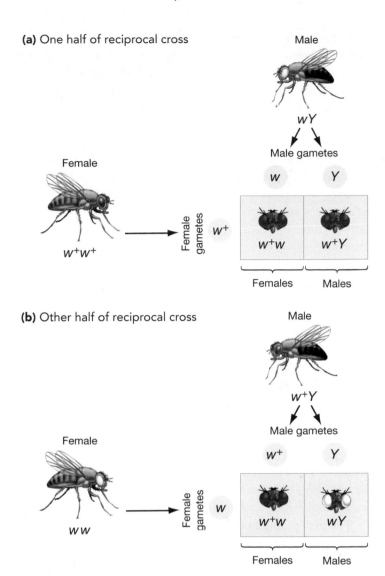

(a) One half of reciprocal cross

(b) Other half of reciprocal cross

FIGURE 13.11 Reciprocal Crosses Confirm that Eye Color in *Drosophila* Is an X-Linked Trait. When Morgan crossed red-eyed females with white-eyed males and then crossed white-eyed females with red-eyed males, he observed strikingly different results. This was consistent with his hypothesis that eye color is an X-linked trait in fruit flies.

● **EXERCISE** Morgan also crossed red-eyed w^+w females with red-eyed males, w^+Y. Create a Punnett square to predict the types and proportions of offspring genotypes and phenotypes that would result.

dominant to the allele for white color. In this figure, the allele for red eyes is denoted w^+, while the allele for white eyes is denoted w. (In fruit-fly genetics, the + symbol always indicates the wild-type trait and the gene symbol often refers to the mutant phenotype—for example, w refers to the mutant white-eye phenotype.) The Y chromosome present in males is simply designated as Y. Using this notation, the genotypes used in the experiment are written as $w^+ w^+$ for red-eyed females; wY for white-eyed males; ww for white-eyed females; and w^+Y for red-eyed males. If you study the offspring genotypes, you should be convinced that the results predicted by the hypothesis of X-linkage match the observed results.

When reciprocal crosses give different results, such as those illustrated in Figure 13.11, it is likely that the gene in question

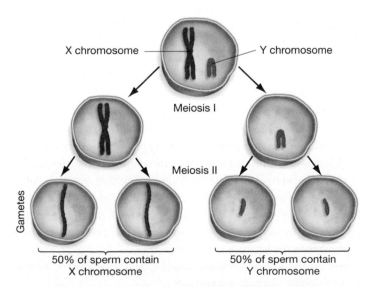

FIGURE 13.10 Sex Chromosomes Pair during Meiosis I, Then Segregate to Form X-Bearing and Y-Bearing Gametes. Sex chromosomes synapse at meiosis I in male fruit flies, even though the X and Y chromosomes differ in size and shape. (No crossing over occurs in male fruit flies.) Thus, half the sperm cells that result from meiosis bear an X chromosome; half have a Y chromosome.

is located on a sex chromosome. Recall from Chapter 12 that non-sex chromosomes are called **autosomes**. Genes on non-sex chromosomes are said to show **autosomal inheritance**.

Morgan's discovery of X-linked inheritance carried an even more important message, however. In *Drosophila*, the gene for white eye color is clearly correlated with inheritance of the X chromosome. This correlation was important evidence in support of the hypothesis that chromosomes contain genes. The discovery of X-linked inheritance convinced most biologists that the chromosome theory of inheritance was correct. ● If you understand this reasoning, you should be able to explain why sex-linked inheritance was consistent with predictions of the chromosome theory.

What Happens When Genes Are Located on the Same Chromosome?

When later experiments confirmed that genes are indeed the physical components of chromosomes, the result prompted Morgan and other geneticists to reevaluate Mendel's principle of independent assortment. The key issue was that genes could not undergo independent assortment if they were located on the same chromosome. The physical association of genes that are found on the same chromosome is called **linkage**. (Notice that the terms *linkage* and *sex-linkage* are different in meaning. If two or more genes are linked, it means that they are located on the same chromosome. If a single gene is sex-linked, it means that it is located on a sex chromosome.)

The first examples of linked genes involved the X chromosome of fruit flies. After Morgan established that the white-eye gene was located on *Drosophila*'s X chromosome, he and colleagues established that one of the several genes that affects body color is also located on the X. Red eyes and gray body are the wild-type phenotypes in this species; white eyes and a yellow body occur as rare mutant phenotypes. The alleles for red eyes (w^+) and gray body (y^+) are dominant to the alleles for white eyes (w) and yellow body (y). (Be sure not to confuse the notation for the Y chromosome in males, Y, with the yellow body allele, y.)

Thus, it seemed logical to predict that the linked genes would always be transmitted together during gamete formation. Stated another way, linked genes should violate the principle of independent assortment. Recall from Section 13.2 that independent assortment is observed when genes are on different chromosomes, because the alleles of unlinked genes segregate to gametes independently of one another during meiosis I. But when genes are on the same chromosome, their alleles are carried to gametes together. **Figure 13.12** shows that a female fruit fly with one X chromosome carrying the w and y^+ alleles and with a second X chromosome carrying the w^+ and y alleles should generate just two classes of gametes in equal numbers during meiosis, instead of the four classes that are predicted under the principle of independent assortment. Is this what actually occurs?

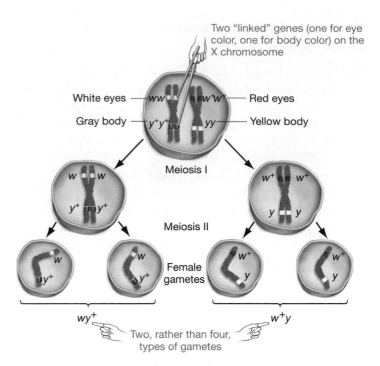

FIGURE 13.12 Linked Genes Are Inherited Together. If the *w* and *y* genes were found on different chromosomes, then this female would generate four different types of gametes instead of just two types as shown here.

● **EXERCISE** List the four genotypes that would be generated if the *w* and *y* genes were not linked.

The First Studies of Linked Genes To determine whether linked traits behave as predicted, Morgan performed crosses like the one described in the Experimental setup section of **Figure 13.13**. This figure introduces some new notation. When you express a genotype in writing, you are simply listing the relevant alleles that are present. When referring to linked genes, biologists use a slash symbol (/) to separate the alleles that are found on homologous chromosomes. For example, the female fruit flies in this experiment had wy^+ alleles on one X chromosome and w^+y alleles on the other X chromosome. This genotype is written as wy^+/w^+y.

The Results table in Figure 13.13 summarizes the phenotypes and genotypes observed in the male offspring of this experimental cross. Notice that most of these males carried an X chromosome with one of the two combinations of alleles found in their mothers: wy^+ or w^+y. Thus, *white* and *yellow* alleles do not segregate independently of each other most of the time. But a small percentage of males had novel phenotypes and genotypes: wy and w^+y^+. Morgan referred to these individuals as **recombinant** because the combination of alleles on their X chromosome was different from the combinations of alleles present in the parental generation.

To explain this result, Morgan proposed that gametes with new, recombinant genotypes were generated when crossing over occurred during prophase of meiosis I in the females. Recall from Chapter 12 that crossing over involves a physical exchange

Experiment

Question: Will genes undergo independent assortment if they are on the same chromosome?

Linkage hypothesis: Linked genes will violate the principle of independent assortment.

Null hypothesis: Linked genes will adhere to the principle of independent assortment.

Experimental setup:

Red-eyed gray-bodied female wy^+/w^+y

White-eyed gray-bodied male $wy^+/$ Y

Prediction: Because these two genes are X-linked, male offspring will have only one copy of each gene, from their mother; the two possible male offspring genotypes are wy^+/Y and w^+y/Y.

Prediction of null hypothesis: Four male genotypes are possible ($wy^+/Y : w^+y/Y : wy/Y : w^+y^+/Y$) and will occur in equal frequency.

Results:

	Male offspring		
	Phenotype	Genotype	Number
		$wy^+/$ Y	4292
		$w^+y/$ Y	4605
Recombinant genotypes		$wy/$ Y	86
		$w^+y^+/$ Y	44

Four male genotypes were observed (rather than two), but not in equal frequency as predicted by independent assortment

Conclusion: Neither hypothesis is fully supported. Independent assortment does not apply to linked genes—linked genes segregate together *except* when crossing over and genetic recombination have occurred.

FIGURE 13.13 Linked Genes Are Inherited Unless Recombination Occurs.

of segments from homologous chromosomes. Crossing over occurs at least once in every synapsed pair of homologous chromosomes, and usually multiple times. (Male fruit flies are an exception to this rule. For unknown reasons, no crossing over occurs in male fruit flies.) As **Figure 13.14** shows, a crossing-over event occurred somewhere between the *w* and *y* genes in the wy^+/w^+y females. The recombinant chromosomes that resulted would have the genotypes *wy* and w^+y^+ (see the middle of the bottom row of the figure). If these chromosomes ended up in a male offspring, they would produce individuals with yellow bodies and white eyes, along with individuals with gray bodies and red eyes. This is exactly what Morgan observed.

The take-home message of Morgan's experiments was that linked genes are inherited together unless crossing over occurs. When crossing over takes place, genetic recombination occurs. These results cemented the connection between the events of meiosis I and Mendel's laws. Linked alleles segregate together unless there is a physical crossover between homologous chromosomes.

Linkage Mapping   In an experiment like the one diagrammed in Figure 13.13, researchers calculate the recombination frequency as the number of offspring with recombinant phenotypes

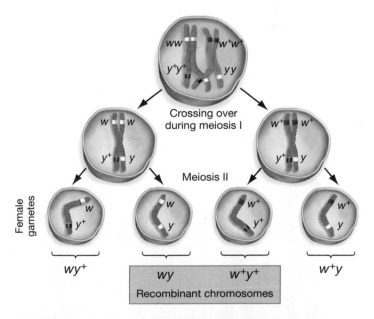

FIGURE 13.14 Genetic Recombination Results from Crossing Over. To explain the results in Figure 13.13, Morgan proposed that crossing over occurred between the *w* genes and the *y* genes in a small percentage of the F_1 females during meiosis I. The recombinant chromosomes that resulted would produce the recombinant phenotypes observed in F_2 males. (Note that the *w* and *y* genes are actually much closer than shown here.)

divided by the total number of offspring. With crosses involving white and yellow, about 1.4 percent of offspring have recombinant phenotypes, which indicates that they have genotypes. But when Morgan and his co-workers performed the same types of crosses with different pairs of X-linked traits, the team found that the fraction of recombinant gametes varied. For example, when crosses involved fruit flies with X-linked genes for white eyes and a mutant phenotype called singed bristles (*sn*), males with recombinant chromosomes for these two genes were produced about 19.6 percent of the time.

Morgan explained these observations by making a conceptual breakthrough: He proposed that genes are arranged in a linear array along a chromosome. According to Morgan's hypothesis, the physical distance between genes determines how frequently crossing over occurs between them. Crossing over occurs at random and can take place at locations all along the length of the chromosome. Based on these observations, it is logical to predict that the shorter the distance between any two genes on a chromosome, the lower the probability that crossing over will take place somewhere in between. The fundamental idea here is that greater physical distance between genes increases the chance that crossing over will take place between them (**Figure 13.15**).

A. H. Sturtevant, an undergraduate who was studying with Morgan, realized that variation in recombination frequency had an important implication: If genes are lined up along the chromosome and if the frequency of crossing over is a function of the physical distance between genes, then it should be possible to figure out where genes are in relation to each other based on the frequency of recombinants between various pairs. That is, it should be possible to create a **genetic map**. A genetic map is a diagram that shows the relative positions of genes along a particular chromosome.

Sturtevant proposed that in constructing a genetic map, the unit of distance along a chromosome should simply be the percentage of offspring that have recombinant phenotypes with respect to two genes. Sturtevant called this unit the centiMorgan (cM), in honor of his mentor. One map unit, or 1 centiMorgan, represents the physical distance that produces 1 percent recombinant offspring. For example, he proposed that the eye-color and bristle-shape genes of fruit flies are 19.6 map units apart on the X chromosome, because recombination between these genes results in 19.6 percent recombinant offspring, on average (**Figure 13.16a**). The yellow (*y*) gene for body color and the gene (*w*) for white eye color, in contrast, are just 1.4 map units apart.

Where is the *yellow* gene relative to the *singed-bristles* gene? Recombinants occurred in 21 percent of the gametes produced by females that are $y^+sn^+/y\ sn$, meaning that *y* and *sn* are 21 cM apart. Sturtevant inferred that the gene for white eyes must be located *between* the genes for yellow body and singed bristles, as shown in Figure 13.16a. Mapping genes relative to each other is like fitting puzzle pieces, and mapping *w* between *y* and *sn* is the only way to make the pairwise distances sum correctly. The key observation is that 21 cM—the distance between *y* and *sn*—is equal to 1.4 cM + 19.6 cM, or the sum of the distances between *y* and *w* and *w* and *sn*. ● If you understand this concept, you should be able to explain why *w* can't be mapped *above* *y* on Figure 3.16a.

Figure 13.16b provides a partial genetic map of the X chromosome in *Drosophila melanogaster* and the data on which the map positions are based. Using this logic and data like these, Sturtevant assembled the first genetic map. ● If you understand this concept, you should be able to add the gene called *crossveinless wings* to the chromosome in Figure 13.16b. It is 13.7 cM from *yellow*, 12.3 cM from *white*, 7.3 cM from *singed bristles*, and 22.4 cm from *miniature wings*.

Do *All* Linked Genes Violate the Principle of Independent Assortment?

When crossing over occurs, the new genotypes that result match the ones that would be produced if the genes involved were located on different chromosomes. Stated another way, crossing over breaks up linkage and makes it appear as if independent assortment is occurring.

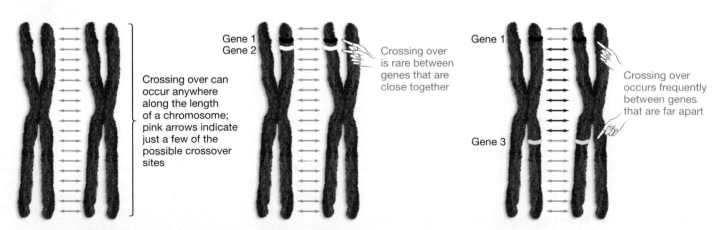

FIGURE 13.15 The Physical Distance between Genes Determines the Frequency of Crossing Over. Crossing over generally occurs at least once in every meiosis in each homologous pair of chromosomes, and often multiple times.

Crossing over can occur anywhere along the length of a chromosome; pink arrows indicate just a few of the possible crossover sites

Gene 1
Gene 2

Crossing over is rare between genes that are close together

Gene 1

Crossing over occurs frequently between genes that are far apart

Gene 3

(a) Frequency of recombinant offspring can be used to map genetic distance.

Yellow body
White eyes

Singed bristles

1.4

21

19.6

The frequency of recombinant offspring is directly correlated with the distance between the two genes, so 19.6% recombinant offspring translates to 19.6 map units (centimorgans, cM)

(b) Constructing a genetic map

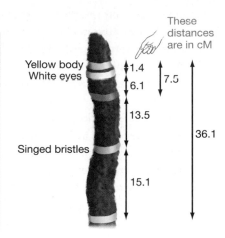

These distances are in cM

Yellow body
White eyes

Singed bristles

1.4
6.1
13.5
15.1

7.5

36.1

Frequency of crossing over (%) between some genes on the X chromosome of fruit flies		
	Miniature Wings	**Ruby Eyes**
Yellow body	36.1	7.5
White eyes	34.7	6.1
Singed bristles	15.1	13.5
Miniature wings	—	28.6

FIGURE 13.16 The Locations of Genes Can Be Mapped by Analyzing the Frequency of Recombination.
(a) Recombinants between the yellow-body locus and the white-eye locus occur 1.4 percent of the time. Therefore, these genes are 1.4 map units apart on the chromosome. Recombinants between the yellow-body locus and the singed-bristles locus of another X-linked gene occur 21 percent of the time. But recombinants between the white-eye locus and the singed-bristles locus occur just 19.6 percent of the time. Therefore, the genes must be arranged as shown. **(b)** A partial genetic map of the X chromosome in fruit flies.

● **EXERCISE** On the chromosome in part (b), label the orange and blue genes. (Which is *ruby* and which is *miniature wings?*)

To convince yourself that this is so, look back at the four gamete genotypes listed at the bottom of Figure 13.14. Recall that these were produced by a female with genotype wy^+/w^+y. The four gamete genotypes listed are w^+y^+, w^+y, wy^+, and wy. It should be clear that these are the same gamete genotypes that meiosis would produce if the female's genotype were w^+wy^+y, meaning the *white* and *yellow* genes were not linked.

If crossing over occurs frequently enough that all four of these gamete genotypes are produced in equal proportions, then the results of a cross involving linked genes would be indistinguishable from the predictions of independent assortment. When genes are 50 or more map units apart on the same chromosome, that is exactly what happens. Genes behave as if they assort independently when 50 percent of the gametes are recombinant with respect to the genes. As it turns out, this situation was occurring in Mendel's experiments. The genes for two of the traits he worked with are located on chromosome number 1 in peas, and at least two of the genes are located on chromosome 4. In each case, however, the genes he analyzed are located far apart. As a result, crossing over was frequent enough to produce the 9:3:3:1 ratios of F_2 phenotypes predicted by the hypothesis of independent assortment.

Before analyzing some extensions to Mendel's rules, let's step back and consider the events we've just surveyed—from the rediscovery of Mendel's work in 1900 to Sturtevant's first chromosome map, which was published in 1913. In 1899 geneticists did not understand the basic rules of heredity. But by 1913 they could map the locations of genes on chromosomes. A remarkable knowledge explosion had occurred.

Check Your Understanding

If you understand that... ○

- Meiosis is the process responsible for Mendel's principle of segregation. It occurs because alleles on homologous chromosomes separate at anaphase of meiosis I.

- Meiosis is the process responsible for Mendel's principle of independent assortment. Alleles of different genes go to gametes independently because pairs of homologous chromosomes line up randomly at metaphase of meiosis I.

- Because the frequency of crossing over between genes correlates with the physical distance between them, genes can be mapped on chromosomes, using the percentage of recombinant offspring observed as a unit.

You should be able to... ○

1) Draw the chromosomes involved in a cross between *Pp* and *Pp* peas, and use the diagram to explain why the principle of segregation occurs.

2) Draw the chromosomes involved in a cross between *YyRr* and *YyRr* peas, and use the diagram to explain why the principle of independent assortment occurs.

13.5 Extending Mendel's Rules

Biologists point out that Mendel analyzed the simplest possible genetic system. The traits that he was studying were not sex-linked and were influenced by just two alleles of a single gene, with alleles that were completely dominant or recessive.

Even so, Mendel was able to discover the most fundamental rules of inheritance. Investigating simple model systems is an extremely important research strategy in biological science. Researchers almost always choose to analyze the simplest situation possible before going on to explore more complicated systems. Mendel probably would have failed, as so many others had done before him, had he worked with more complex patterns of inheritance.

Once Mendel's work was rediscovered, though, researchers began to analyze traits and alleles whose inheritance was more complicated. If experimental crosses produced F_2 progeny that did not conform to the expected 3:1 or 9:3:3:1 ratios in phenotypes, researchers had a strong hint that something interesting was going on. In many cases it turned out that unraveling the cause of surprising results led to new insights into how genes and heredity worked. ● The discovery of linkage and sex-linkage were among several important exceptions and extensions to Mendel's work. How can traits that don't appear to follow Mendel's rules contribute to a more complete understanding of heredity?

Multiple Alleles and Polymorphic Traits

Mendel worked with a total of seven traits and just 14 alleles—two for each trait. In most populations, however, dozens of alleles can be identified at each gene. The existence of more than two alleles of the same gene is known as **multiple allelism**. As an example, consider the gene for the β-globin protein in humans. β-globin makes up part of the hemoglobin protein, which carries oxygen from the lungs to tissues. Over the past few decades, biologists have identified and named over 500 different alleles of the β-globin gene. Many of these alleles are associated with distinctive phenotypes. Some alleles produce polypeptides with normal oxygen-carrying capacity, while others lead to reduced oxygen-carrying capacity and various types of anemia. Still other β-globin alleles are associated with adaptation to living at high altitudes, decreased stability at high temperatures, or resistance to the parasites that infect red blood cells and cause malaria.

When more than two distinct phenotypes are present in a population due to multiple allelism, the trait is **polymorphic** ("many-formed"). Oxygen-carrying capacity in humans is a highly polymorphic trait, because many alleles exist at the β-globin gene.

Incomplete Dominance and Codominance

The terms *dominant* and *recessive* describe which phenotype is observed when two different alleles of a gene occur in the same individual. In all seven traits that Mendel studied, only the phenotype associated with one allele—the "dominant" one—appeared in heterozygous individuals. But consider the flowers called four-o'clocks, pictured in **Figure 13.17a**. In this species, biologists have developed a pure line that has purple flowers and a pure line that has white flowers. When individuals from these strains are mated, all of their offspring are lavender (**Figure 13.17b**). In Mendel's peas, crosses between purple- and

(a) Flower color is variable in four-o'clocks.

(b) Incomplete dominance in flower color

FIGURE 13.17 When Incomplete Dominance Occurs, Heterozygotes Have Intermediate Phenotypes. (a) Four-o'clocks got their name because their flowers open in the late afternoon. Flower color is highly variable. **(b)** The hypothesis illustrated is that flower color is controlled by a single gene with two alleles, symbolized R and r. These alleles exhibit incomplete dominance.

white-flowered parents produced all purple-flowered offspring. Why the difference?

Biologists answered this question by examining the phenotypes of F_2 offspring. These are the progeny of self-fertilization in lavender-flowered F_1 individuals. Of the F_2 plants, 1/4 have purple flowers, 1/2 have lavender flowers, and 1/4 have white flowers. This 1:2:1 ratio of *phenotypes* is unlike any we have seen to date, but it exactly matches the 1:2:1 ratio of *genotypes* that is produced when flower color is controlled by one gene with two alleles.

To convince yourself that this explanation is sound, study the genetic model shown in Figure 13.17b. According to the hypothesis shown in the diagram, the inheritance of flower color in four-o'clocks and peas is identical, except that the four-o'clock alleles show incomplete dominance rather than complete dominance. When **incomplete dominance** occurs, heterozygotes have an intermediate phenotype. In the case of four-o'clocks, neither purple nor white alleles dominate. Instead, the F_1 progeny—all heterozygous—show a phenotype intermediate between the two parental strains.

Incomplete dominance illustrates an important general point: Dominance is not necessarily an all-or-none phenomenon. In fact, many alleles show a relationship called **codominance**. When **codominance** occurs, heterozygotes have the phenotype associated with *both* alleles present. As an example, consider the *L* gene in humans. This gene codes for a plasma membrane glycoprotein that is found in red blood cells. Different alleles of the *L* gene lead to the appearance of membrane glycoproteins with different amino acid sequences. In most human populations, only two alleles are present. These alleles are designated L^M and L^N and code for glycoproteins called M and N. Thus, three genotypes are found: L^ML^M, L^ML^N, and L^NL^N. In L^ML^M individuals, all of the MN-type membrane glycoproteins present in red blood cells have the primary sequence associated with the L^M allele. In L^NL^N individuals, all of the MN-type proteins have the L^N allele's primary sequence. But in L^ML^N individuals, red blood cells have some membrane glycoproteins with the M phenotype and some with the N phenotype. Because both alleles are represented in the phenotype, the alleles are considered codominant.

Why Do Dominance Relationships Vary Among Alleles?

Once you understand what genes do on the molecular level, it is straightforward to see why some alleles are dominant, recessive, codominant, or intermediate in dominance. Genes contain the information required to manufacture the molecules that are found in cells. Some of these molecules are proteins that form structures in the cell or body, such as the MN proteins in the plasma membranes of red blood cells. Because the *M* and *N* alleles code for proteins with different amino acid sequences, both are present in the cells of heterozygous individuals. Both gene products are detected in the phenotype, so the alleles are considered codominant. Codominance is common in alleles that are associated with membrane proteins such as the MN protein.

Complete and incomplete dominance tend to occur in alleles that code for enzymes. The plant-height alleles that Mendel studied, for example, code for enzymes that are involved in the synthesis of a key growth hormone. A recessive allele has a defect that keeps the enzyme from functioning. Homozygous recessive individuals lack the growth hormone and are dwarfed. The dominant allele, in contrast, codes for an enzyme that functions normally. In heterozygotes, the one normal allele that is present produces enough enzyme to make a normal amount of growth hormone. Normal height results, and the allele is considered dominant.

The situation in four-o'clock flowers is different. The *R* gene analyzed in Figure 13.17b codes for an enzyme involved in producing a purple pigment in flower petals. The recessive allele codes for a dysfunctional enzyme and leads to a complete lack of pigment. Recessive homozygotes are white as a result. In heterozygotes, though, the one normal copy of the gene can produce enough enzyme to result in a small amount of purple pigment. The lavender phenotype results. Homozygous dominant individuals, in contrast, produce large amounts of normal enzyme and thus enough pigment to produce purple petals.

The message here is that alleles can interact in many different ways, depending on what their products do and how they work. In addition, recall that dominance relationships have nothing to do with fitness—the ability of an individual to produce offspring in a particular environment. Dominant alleles do not necessarily confer high fitness, and they are not necessarily more common than recessive alleles. Recall that the allele that causes Huntington's disease in humans is fatal, dominant, and rare.

Pleiotropy

As far as is known, the alleles that Mendel analyzed affect just a single trait. The gene for seed color in garden peas, for example, does not appear to affect other aspects of the individual's phenotype. In contrast, many cases have been documented in which a single allele affects a wide variety of traits. A gene that influences many traits, rather than just one trait, is said to be **pleiotropic** ("more-turning"). The gene responsible for **Marfan syndrome** in humans, called *FBN1,* is a good example. Although current research suggests that just a single gene is involved, individuals with Marfan syndrome exhibit a wide array of phenotypic effects: increased height, disproportionately long limbs and fingers, an abnormally shaped chest, and potentially severe heart problems. A large percentage of these individuals also suffer from problems with their backbone. The gene associated with Marfan syndrome is pleiotropic.

Genes Are Affected by the Physical Environment and Genetic Environment

When Mendel analyzed height in his experiments, he ensured that each plant received a similar amount of sunlight and grew in similar soil. This was important because individuals with alleles for tallness are stunted if they are deprived of nutrients, sunlight, or water—so much so that they look similar to individuals with alleles for dwarfing. For Mendel to analyze the hereditary determinants of height, he had to control the environmental determinants of height. Let's consider how two aspects of the environment affect phenotypes: (1) the individual's physical surroundings and (2) the alleles present at other genes.

The Physical Environment Has a Profound Effect on Phenotypes The phenotypes produced by most genes and alleles are strongly affected by the individual's physical environment. Consequently, an individual's phenotype is often as much a product of its physical environment as it is a product of its genotype. For example, people with the genetic disease called **phenylketonuria** (PKU) lack an enzyme that helps convert the amino acid phenylalanine to the amino acid tyrosine. As a result, phenylalanine and a related molecule, phenylpyruvic acid, accumulate in these people's bodies. The molecules interfere with the development of the nervous system and produce profound mental retardation. But if PKU individuals are identified at birth and placed on a low-phenylalanine diet, then they develop normally. In many countries, newborns are routinely tested for the defect. PKU is a genetic disease, but it is neither inevitable nor invariant. Through a simple change in their environment (their diet), individuals with a PKU genotype can have a normal phenotype.

Interactions with Other Genes Have a Profound Effect on Phenotypes

In Mendel's pea plants, a single locus influenced seed shape. Further, Mendel's data showed that the seed-shape phenotype does not appear to be affected by the action of genes for seed color, seed-pod color, seed-pod shape, or other traits. The pea seeds he analyzed were round or wrinkled regardless of the types of alleles present at other loci.

In many cases, however, genes are not as independent as the gene for seed shape in peas. Consider a classic experiment published in 1905 on comb morphology in chickens. The researchers, William Bateson and R. C. Punnett, crossed parents from pure-breeding lines with rose and pea combs and found that the F_1 offspring had a different phenotype, called walnut combs. When these individuals bred, their offspring were walnut, rose, pea, and a fourth phenotype—called single—in a 9:3:3:1 ratio (**Figure 13.18a**). The genetic model in **Figure 13.18b** explains the data. If comb morphology results from interactions between two genes, symbolized R and P, if a dominant and a recessive allele exist at each gene, and if the four comb phenotypes are associated with the genotypes indicated at the bottom of the figure, then a cross between $RRpp$ and $rrPP$ parents would give the results that Bateson and Punnett observed.

The key point is that the alleles of different genes affect each other in a way that influences the phenotype observed in these individuals. When these types of gene interactions occur, the phenotype produced by an allele depends on the action of alleles of other genes. If a chicken has an R allele, its phenotype depends on the allele present at the P gene. A special case of this phenomenon is known as **epistasis** ("stopping" or "diminishing"). Epistasis occurs when an allele of one gene masks the effect of an allele at another gene.

Quantitative Traits

Mendel worked with **discrete traits**—characteristics that are qualitatively different. In garden peas, seed color is either yellow or green—no intermediate phenotypes exist. But many traits in peas and other organisms don't fall into discrete categories.

(a) Crosses between chickens with different comb phenotypes give odd results.

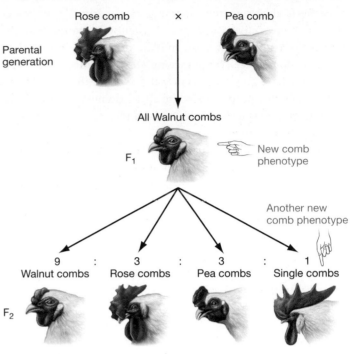

(b) A genetic model to explain the results

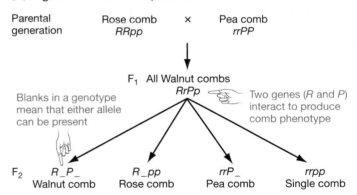

FIGURE 13.18 Different Genes Can Interact to Influence a Trait.
(a) This cross is unusual because new phenotypes show up in the F_1 and the F_2 generation. **(b)** To explain the results, researchers hypothesized that comb shape depends on two genes that interact. The phenotype associated with the alleles at each gene depends on the genetic environment—specifically, the alleles at the other gene.

In humans, for example, height, weight, and skin color fall anywhere on a continuous scale of measurement. People are not 160 cm tall or 180 cm tall, with no other heights possible. Height and many other characteristics exhibit quantitative variation—meaning that individuals differ by degree—and are called **quantitative traits**. Like discrete traits, quantitative traits are highly influenced by the physical environment. For example, the effects of nutrition on human height and intelligence have been well documented.

Quantitative traits share a common characteristic: When the frequencies of different values observed in a population are

(a) A phenotype distribution that forms a bell-shaped curve.

(b) Normal distribution—bell-shaped curve

FIGURE 13.19 Quantitative Traits Are Normally Distributed. (a) A histogram plotting the heights of male undergraduates at Connecticut Agricultural College in 1914. **(b)** In most cases, the distribution of a quantitative trait in a population approximates a bell-shaped curve, called a normal distribution.

● **QUESTION** The shortest student plotted on the histogram is 4'10" (147 cm), and the tallest is 6'2" (188 cm). How would a histogram of men in your class compare with the distribution in part (b)?

plotted on a histogram, or frequency distribution, they usually form a bell-shaped curve (**Figure 13.19a**). This distribution is observed so frequently that it is often called a normal distribution. In a normal distribution, high and low values occur at low frequency; intermediate values occur at high frequency (**Figure 13.19b**).

In 1909 Herman Nilsson-Ehle showed that if many genes each contribute a small amount to the value of a quantitative trait, then a continuous, bell-shaped (normal) distribution results. Nilsson-Ehle established this finding using strains of wheat that differed in kernel color. **Figure 13.20a** shows the results of a

cross he performed between pure lines of white wheat and dark-red wheat. Notice that the frequency of colors in F_2 progeny forms a bell-shaped curve. To explain these results, Nilsson-Ehle proposed the set of hypotheses illustrated in **Figure 13.20b**:

● The parental strains differ with respect to three genes that control kernel color: *AABBCC* produces dark-red kernels, and *aabbcc* produces white kernels.

● The three genes assort independently. When the *AaBbCc* F_1 individuals self-fertilize, white F_2 individuals would occur at a frequency of 1/4 (*aa*) × 1/4 (*bb*) × 1/4 (*cc*) = 1/64 *aabbcc*.

(a) Wheat kernel color is a quantitative trait.

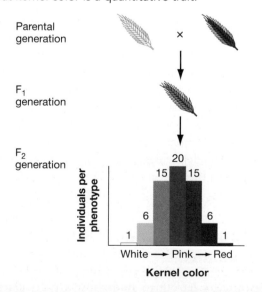

(b) Hypothesis to explain inheritance of kernel color

FIGURE 13.20 Quantitative Traits Result from the Action of Many Genes. (a) When wheat plants with white kernels were crossed with wheat plants with red kernels, the F_2 offspring showed a range of kernel colors. The frequency of these phenotypes approximates a normal distribution. **(b)** This model attempts to explain the results of part (a).

● **EXERCISE** Confirm that the distribution of genotypes shown in part (b) is correct by drawing a Punnett square based on the gametes produced by the F_1 parents and then filling in the F_2 offspring genotypes that result. (The square will be 8 × 8 with 64 boxes!)

SUMMARY TABLE **13.2 Some Extensions to Mendel's Rules**

Type of Inheritance	Definition	Consequences or Comments
Sex-linkage	Genes located on sex chromosomes.	Patterns of inheritance in males and females differ.
Linkage	Two genes found on same chromosome.	Linked genes violate principle of independent assortment.
Incomplete dominance	Heterozygotes have intermediate phenotype.	Polymorphism—heterozygotes have unique phenotype.
Codominance	Heterozygotes have phenotype of both alleles.	Polymorphism is possible—heterozygotes have unique phenotype.
Multiple allelism	In a population, more than two alleles present at a locus.	Polymorphism is possible.
Polymorphism	In a population, more than two phenotypes present.	Can result from actions of multiple alleles, incomplete dominance, and/or codominance.
Pleiotropy	A single allele affects many traits.	This is common.
Variation in the genetic environment	In discrete traits, the phenotype associated with an allele depends on which alleles are present at another gene.	One allele can be associated with different. phenotypes
Variation in the physical environment	Phenotype influenced by environment experienced by individual.	Same genotypes can be associated with different phenotypes.
Polygenic inheritance of quantitative traits	Many genes are involved in specifying traits that exhibit continuous variation.	Unlike alleles that determine discrete traits, each allele adds a small amount to phenotype.

- The *a*, *b*, and *c* alleles do not contribute to pigment production, but the *A*, *B*, and *C* alleles contribute to pigment production in an equal and additive way. As a result, the degree of red pigmentation is determined by the number of *A*, *B*, or *C* alleles present. Each uppercase (dominant) allele that is present makes a wheat kernel slightly darker red.

Check Your Understanding

If you understand that...

- Genes on the same chromosome violate the principle of independent assortment. They are not transmitted to gametes independently of each other unless crossing over occurs between them.

- Incomplete dominance, codominance, multiple allelism, pleiotropy, environmental effects, gene interactions, and polygenic inheritance are aspects of inheritance that Mendel did not study. When they occur, monohybrid and dihybrid crosses do not result in classical Mendelian ratios of offspring phenotypes.

You should be able to...

1) Define linkage, incomplete dominance, and the other phenomena listed above and give an example of each.

2) Explain why each of these phenomena makes it unlikely to observe the 3:1 and 9:3:3:1 phenotype ratios that Mendel observed in his monohybrid and dihyrid crosses.

Later work showed that Nilsson-Ehle's model hypotheses were correct in virtually every detail. Quantitative traits are produced by the independent actions of many genes, although it is now clear that some genes have much greater effects on the trait in question than other genes do. As a result, the transmission of quantitative traits is said to result from polygenic ("many-genes") inheritance. In **polygenic inheritance**, each gene adds a small amount to the value of the phenotype.

In the decades immediately after the rediscovery of Mendel's work, analyses of phenomena such as sex-linkage, multiple allelism, incomplete dominance, environmental effects, gene interactions, and polygenic inheritance provided a fairly comprehensive answer to the question of why offspring resemble their parents. **Table 13.2** summarizes some of the key exceptions and extensions to Mendel's rules and gives you a chance to compare and contrast their effects on patterns of inheritance.

13.6 Applying Mendel's Rules to Humans

When researchers set out to study how a particular gene is transmitted in wheat or fruit flies or garden peas, they begin by making a series of controlled experimental crosses. For obvious reasons, this research strategy is not possible with humans. But suppose you are concerned about an illness that runs in your family, and that you go to a genetic counselor to find out how likely your children are to have the disease. To advise you, the counselor needs to know how the trait is transmitted, including whether the gene involved is autosomal or sex-linked and what type of dominance is shown by the disease allele. To understand the transmission of human traits, investigators have to analyze human crosses that already exist. They do so by constructing a **pedigree**, or family tree, of affected individuals.

A pedigree records the genetic relationships among the individuals in a family along with each person's sex and phenotype with respect to the trait in question. If the trait is due to a single gene, then analyzing the pedigree may reveal whether the trait is due to a dominant or recessive allele and whether the gene responsible is located on a sex chromosome or on an autosome. Let's look at a series of specific case histories to see how this work is done.

Are Alleles Recessive or Dominant?

To analyze the inheritance of a trait that shows discrete variation, biologists begin by assuming that a single autosomal gene is involved and that the alleles present in the population have a simple dominant-recessive relationship. This is the simplest possible situation. If the pattern of inheritance fits this model, then the assumptions—of inheritance via a single gene and simple dominance—are supported. Let's first analyze the pattern of inheritance that is typical of autosomal recessive traits and then examine patterns that emerge in pedigrees for autosomal dominant traits.

Patterns of Inheritance: Autosomal Recessive Traits

In analyzing the inheritance of traits, it's helpful to distinguish conditions that *must* be met when a particular pattern of inheritance occurs versus conditions that are *likely* to be met. For example, if a phenotype is due to an autosomal recessive allele, then individuals with the trait must be homozygous. If the parents of an affected individual do not have the trait, then the parents must be heterozygous for the trait. Heterozygous individuals who carry a recessive allele for an inherited disease are referred to as **carriers** of the disease. These individuals carry the allele and transmit it even though they do not exhibit signs of the disease. When two carriers mate, they should produce offspring with the recessive phenotype about 25 percent of the time.

Figure 13.21 is the pedigree from a family in which an autosomal recessive trait, such as sickle cell disease, occurs. The key feature to notice in this pedigree is that some boys and girls exhibit the trait even though their parents do not. This is the pattern you would expect to observe under the model of an autosomal recessive trait if the parents of individuals with the trait are heterozygous. It is also logical to observe that when an affected individual has children, those children do not necessarily have the trait. This pattern is predicted if

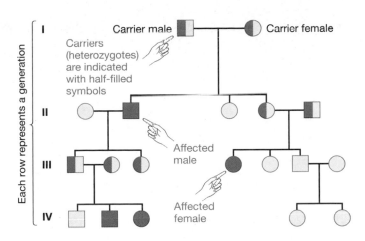

FIGURE 13.21 Pedigree of a Family with an Autosomal Recessive Disease. Diseases like sickle cell, which are inherited as autosomal recessives, appear in both males and females. For an individual to be affected, both parents must carry the allele responsible. If the recessive allele is rare, affected individuals may not appear in every generation.

● **QUESTION** In a pedigree where an autosomal recessive disease appears, an affected parent frequently has unaffected children. But unaffected parents can have affected offspring. Why?

affected people marry individuals who are homozygous for the wild-type allele, which is likely to be true if the recessive allele is rare in the population.

The recessive phenotype should show up in offspring only when both parents carry that recessive allele and pass it on to their offspring. By definition, a recessive allele produces a given phenotype only when the individual is homozygous for that allele.

Patterns of Inheritance: Autosomal Dominant Traits

By definition, individuals who are homozygous or heterozygous for an autosomal dominant trait will have the dominant phenotype. For autosomal dominant traits, individuals with a single copy of the allele must have the dominant phenotype. Even if one parent is heterozygous and the other is homozygous recessive, on average half of their children should show the dominant phenotype. And unless a new mutation has occurred, any child with the trait must have a parent with this trait. The latter observation is in strong contrast to the pattern seen in autosomal recessive traits.

Figure 13.22 shows the consequences of autosomal dominant inheritance, in the pedigree of a family affected by a degenerative brain disorder called **Huntington's disease**. The pedigree has two features that indicate this disease is passed to the next generation as an autosomal dominant allele. First, if a child shows the trait, then one of its parents shows the trait as well. Second, if families have a large number of children, the trait usually shows up in every generation—due to the high probability of heterozygous parents having affected children.

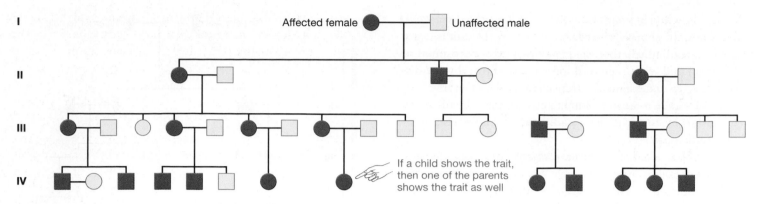

FIGURE 13.22 A Pedigree of a Family with an Autosomal Dominant Disease. Diseases like Huntington's, which are inherited as autosomal dominants, may appear in both males and females and tend to appear every generation.

Is the Trait Autosomal or Sex-Linked?

When it is not possible to arrange reciprocal crosses, can data in a pedigree indicate whether a trait is autosomal or sex-linked? The answer is based on a simple premise: If a trait appears about equally often in males and females, then it is likely to be autosomal. For example, the data in the Huntington's pedigree indicates that the disease appears in both males and females at about equal rates. But if males are much more likely to have the trait in question than females are, then the gene responsible may be found on the X chromosome. Because so few genes occur on the Y chromosome, Y-linked inheritance is rare.

To understand why a sex bias in phenotypes implicates sex-linked inheritance, recall from Section 13.4 that sex-linked genes are located on one of the sex chromosomes. Because human males have one X chromosome and one Y chromosome, they have just one copy of each X-linked gene. But because human females have two X chromosomes, they have two copies of each X-linked gene. These simple observations are critical. In humans—just as in fruit flies and in every other species that has sex chromosomes—the pattern of inheritance in sex-linked traits is different in males and females, because the complement of sex chromosomes differs in the two sexes.

What does the pedigree of an X-linked trait look like? To answer this question, let's consider the pedigree of a classic X-linked trait—the occurrence of hemophilia in the descendants of Queen Victoria and Prince Albert, who were monarchs of England in the mid-to-late 1800s. **Hemophilia** is caused by a defect in an important blood-clotting factor. Hemophiliacs are at a high risk of bleeding to death, because even minor injuries result in prolonged bleeding. The key point to notice in the pedigree of Queen Victoria's descendants is that only males developed hemophilia (**Figure 13.23**). Also note the affected male in generation II (the second square from the right in row II). His two sons were unaffected, but the trait reappeared in a grandson. Stated another way, the occurrence of hemophilia skipped a generation.

This pattern is logical because hemophilia is due to an X-linked *recessive* allele. Because males have only one X chromo-

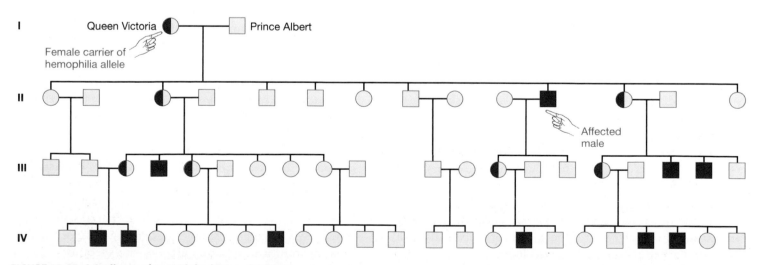

FIGURE 13.23 A Pedigree of an X-Linked Recessive Disease.

● **EXERCISE** Redraw this pedigree, showing the pattern of inheritance if the allele for hemophilia were X-linked dominant.

some, the phenotype associated with an X-linked recessive allele appears in every male that carries it. Further, the appearance of an X-linked recessive trait skips a generation in a pedigree because the affected male passes his only X chromosome on to his daughters. But because his daughters almost always received a wild-type allele from their mother, the daughters don't show the trait. They will pass the defective allele on to about half of their sons, however.

In contrast, X-linked traits that are dominant appear in every individual who has the defective allele. A good indicator of an X-linked dominant trait is a pedigree in which an affected male has all affected daughters but no affected sons. This pattern is logical because every female offspring of an affected fa-

ther gets an X chromosome from him and will herself be affected, while his sons get their only X chromosome from their unaffected mother. Besides the inherited form of a bone disease called rickets, however, very few diseases are known to be due to X-linked dominant alleles.

By analyzing pedigrees in this way, biomedical researchers have been able to characterize how most of the common genetic diseases in humans are inherited. By the 1940s, though, the burning question in genetics was no longer the nature of inheritance, but the nature of the gene itself. What are genes made of, and how are they copied so that parents pass their alleles on to their offspring? These are the questions we turn to in Chapter 14.

Chapter Review

SUMMARY OF KEY CONCEPTS

● **Mendel discovered that in garden peas, individuals have two alleles, or versions, of each gene. Prior to the formation of eggs and sperm, the two alleles of each gene separate so that one allele is transmitted to each egg or sperm cell.**

Gregor Mendel discovered that inheritance is particulate and proposed two basic principles of transmission genetics. The first was the principle of segregation, which claimed that each trait was specified by paired hereditary determinants (alleles of genes) that separate from each other during gamete formation.

You should be able to cite data that explain why observing intermediate phenotypes in hybrid offspring, due to incomplete dominance or polygenic inheritance, is consistent with the principle of segregation and does not support the hypothesis of blending inheritance. ●

● **Genes are located on chromosomes. The separation of homologous chromosomes during anaphase of meiosis I explains why alleles of the same gene segregate to different gametes.**

The chromosome theory of inheritance claimed that genes are located on chromosomes and that the movements of chromosomes during meiosis provide a physical basis for Mendel's observations. The chromosome theory provided a physical basis for the principle of segregation. Alleles of the same gene segregate from each other and end up in different gametes because homologous chromosomes segregate during meiosis I.

You should be able to draw the phases of meiosis in a diploid organism with $n = 2$, where the individual has the genotype $BbRr$, and identify where and why the B and b and the R and r each separate into different daughter cells and thus gametes. ●

● **If genes are located on different chromosomes, then the alleles of each gene are transmitted to egg cells and sperm cells independently of each other. This occurs because chromosomes line up randomly in metaphase of meiosis I.**

Mendel's second basic conclusion was the principle of independent assortment, which stated that the segregation of one

pair of genes—controlling a given trait—was not influenced by the segregation of other gene pairs. The chromosome theory provided a physical basis for the principle of independent assortment. Genes located on different chromosomes move to gametes independently of each other during meiosis. The principle of independent assortment holds because chromosomes line up independently of each other at the metaphase plate during meiosis I. Thus, gametes receive random combinations of chromosomes.

You should be able to add to your drawing of the phases of meiosis, indicating that the individual has the genotype $BP/bp\ Rr$, and explain why the B and R genes assort independently while the B and P genes do not. ●

(MB) Web Animations at www.masteringbio.com
Mendel's Experiments
The Principle of Independent Assortment

● **There are important exceptions and extensions to the basic patterns of inheritance that Mendel discovered.**

Thomas Hunt Morgan and colleagues extended Mendel's work by describing X-linked inheritance and by showing that genes located on the same chromosome do not exhibit independent assortment. Studies of X-linked traits helped confirm that genes are found on chromosomes, while studies of linked traits led to the first genetic maps showing the locations of genes on chromosomes. Later studies confirmed that many traits are influenced by the interaction of several genes and that phenotypes are influenced by the environment an individual has experienced, as well as by its genotype. Analysis of pedigrees allowed researchers to work out the basic modes of inheritance for most of the common genetic diseases in humans.

You should be able to explain why crossing over is said to break up linkage between alleles, why males do not pass their X-linked traits on to their sons, why highly homozygous populations are used to control for gene interactions in experiments, and why raising individuals in similar environments is important in genetics experiments. ●

GENETICS QUESTIONS

1. In studies of how traits are inherited, what makes certain species candidates for model organisms?
   a. They are the first organisms to be used in a particular type of experiment, so they are a historical "model" of what researchers expect to find.
   b. They are easy to study because a great deal is already known about them.
   c. They are the best or most fit of their type.
   d. They are easy to maintain, have a short life cycle, produce many offspring, and yield data that are relevant to many other organisms.

2. Why is the allele for wrinkled seed shape in garden peas considered recessive?
   a. It "recedes" in the F_2 generation when homozygous parents are crossed.
   b. The trait associated with the allele is not expressed in heterozygotes.
   c. Individuals with the allele have lower fitness than that of individuals with the dominant allele.
   d. The allele is less common than the dominant allele. (The wrinkled allele is a rare mutant.)

3. The alleles found in haploid organisms cannot be dominant or recessive. Why?
   a. Dominance and recessiveness describe interactions between two alleles of the same gene in the same individual.
   b. Because only one allele is present, alleles in haploid organisms are always dominant.
   c. Alleles in haploid individuals are transmitted like mitochondrial DNA or chloroplast DNA.
   d. Most haploid individuals are bacteria, and bacterial genetics is completely different from eukaryotic genetics.

4. Why can you infer that individuals that are "pure line" are homozygous for the gene in question?
   a. They are highly inbred.
   b. Only two alleles are present at each gene in the populations to which these individuals belong.
   c. In a pure line, phenotypes are not affected by environmental conditions or gene interactions.
   d. No other phenotype arises in a pure-line population because no other alleles are present.

5. The genes for the traits that Mendel worked with are either located on different chromosomes or so far apart on the same chromosome that crossing over almost always occurs between them. How did this circumstance help Mendel recognize the principle of independent assortment?
   a. Otherwise, his dihybrid crosses would not have produced a 9:3:3:1 ratio of F_2 phenotypes.
   b. The occurrence of individuals with unexpected phenotypes led him to the discovery of recombination.
   c. It led him to the realization that the behavior of chromosomes during meiosis explained his results.
   d. It meant that the alleles involved were either dominant or recessive, which gave 3:1 ratios in the F_1 generation.

6. The text claims that Mendel worked with the simplest possible genetic system. Is this claim legitimate?
   a. Yes—discrete traits, two alleles, simple dominance and recessiveness, no sex chromosomes, and unlinked genes are the simplest situation known.
   b. Yes—the ability to self-fertilize or cross-pollinate made it simple for Mendel to set up controlled crosses.

   c. No—Mendel was unaware of meiosis and the chromosome theory of inheritance, so it was not easy to reach the conclusions he did.
   d. No—Mendel's experimental designs and his rules of inheritance are actually quite complex and sophisticated.

7. Mendel's rules do not correctly predict patterns of inheritance for tightly linked genes or the inheritance of alleles that show incomplete dominance or epistasis. Does this mean that his hypotheses are incorrect?
   a. Yes, because they are relevant to only a small number of organisms and traits.
   b. Yes, because not all data support his hypotheses.
   c. No, because he was not aware of meiosis or the chromosome theory of inheritance.
   d. No, it just means that his hypotheses are limited to certain conditions.

8. The artificial sweetener NutraSweet consists of a phenylalanine molecule linked to aspartic acid. The labels of diet sodas that contain NutraSweet include a warning to people with PKU. Why?
   a. NutraSweet stimulates the same taste receptors that natural sugars do.
   b. People with PKU have to avoid phenylalanine in their diet.
   c. In people with PKU, phenylalanine reacts with aspartic acid to form a toxic compound.
   d. People with PKU cannot lead normal lives, even if their environment is carefully controlled.

9. When Sutton and Boveri published the chromosome theory of inheritance, research on meiosis had not yet established that paternal and maternal homologs assort independently of each other. Then, in 1913, Elinor Carothers published a paper about a grasshopper species with an unusual karyotype: One chromosome had no homolog (meaning no pairing partner at meiosis I); another chromosome had homologs that could be distinguished under the light microscope. If chromosomes assort independently, how often should Carothers have observed each of the four products of meiosis shown in the following figure?

Grasshopper chromosomes at meiosis I — No pairing partner — Maternal and paternal homologs look different

Four types of gametes possible
(each meiotic division can produce only two of the four)

   a. Only the gametes with one of each type of chromosome would occur.
   b. The four types of gametes should be observed to occur at equal frequencies.
   c. The chromosome with no pairing partner would disintegrate, so only gametes with one copy of the other chromosome would be observed.

d. Gametes with one of each type of chromosome would occur twice as often as gametes with just one chromosome.

10. Which of the following is the strongest evidence that a trait might be influenced by polygenic inheritance?

a. F_1 offspring of parents with different phenotypes have an intermediate phenotype.

b. F_1 offspring of parents with different phenotypes have the dominant phenotype.

c. The trait shows qualitative (discrete) variation.

d. The trait shows quantitative variation.

Genetics Questions answers: 1. d; 2. b; 3. a; 4. d; 5. a; 6. a; 7. d; 8. b; 9. b; 10. d

SOLVING GENETICS PROBLEMS

The best way to test and extend your knowledge of transmission genetics is to work problems. Most genetics problems are set up as follows: You are given some information about the genotypes or phenotypes of one or both parents, along with data on the phenotypes of F_1 or F_2 offspring. Your task is to generate a set of hypotheses—or what biologists call a genetic model—to explain the results. More specifically, you will need to generate a hypothesis to address each of the following questions:

- Is the trait involved discrete or quantitative?

- Is the phenotype a product of one gene or many genes?

- For each gene involved, how many alleles are present—one, two, or many?

- Do the alleles involved show complete dominance, incomplete dominance, or codominance?

- Are the genes involved sex-linked or autosomal?

- If more than one gene is involved, are they linked or unlinked? If they are linked, does crossing over occur frequently?

It's also helpful to ask yourself whether gene interactions or pleiotropy might be occurring and whether it is safe to assume that the experimental design carefully controlled for effects of variation in other genes or the environment.

In working the problem, be sure to start with the simplest possible explanation. For example, if you are dealing with a discrete trait, you might hypothesize that the cross involves a single autosomal gene with two alleles that show complete dominance. Your next step is to infer what the parental genotypes are (if they are not already given), and then do a Punnett square to predict the offspring phenotypes and their frequencies based on your hypotheses. Next, check whether these predictions match the observed results given in the problem. If the answer is yes, you have a valid solution. But if the answer is no, you need to go back and change one of your hypotheses, redo the Punnett square, and check to see if the predictions and observations match. Keep redoing these steps until you have a model that fits the data.

Example Problems

Example 1 *Plectritis congesta* plants produce fruits that either have or do not have prominent structures called wings. The alleles involved are W^+ = winged fruit; W^- = wingless fruit. Researchers collected an array of individuals from the field and performed a series of crosses. The results are given in the following table. Complete the table by writing down the genotype of the parent or parents involved in each cross.

Parental Phenotype(s)	Number of Offspring with Winged Fruits	Number of Offspring with Wingless Fruits	Parental Genotype(s)
Wingless (self-fertilized)	0	80	
Winged (self-fertilized)	90	30	
Winged × wingless	46	0	
Winged × winged	44	0	

Solution Here you're given offspring phenotypes and you're asked to infer parental genotypes. To do this you have to hypothesize that certain parental genotypes are correct, make a Punnett square to predict the offspring genotypes, and then see if the offspring phenotypes match the data. In this case, coming up with a hypothesis for the parental genotypes is relatively straightforward, because the problem states that the trait is due to one gene and two alleles. No information on sex is given, so assume the gene is autosomal. Now look at the second entry in the chart. It shows a 3:1 ratio of offspring from a winged individual that self-fertilizes. This result is consistent with the hypothesis that W^+ is dominant and W^- recessive and that this parent's genotype is W^+W^-.

Now let's look at the first cross in the chart. If W^+ is dominant, then a wingless parent must be W^-W^-. When you do the Punnett square to predict offspring genotypes from selfing, you find that all the offspring will produce wingless fruits, consistent with the data. In the third cross, all the offspring make winged fruits even though one of the parents produces wingless fruits and thus is W^-W^-. This would happen only if the winged parent is W^+W^+. (If this reasoning isn't immediately clear to you, work the Punnett square.) In the fourth cross, you could get offspring that all make winged fruits if the parents were W^+W^+ and W^+W^+, or if the parents were W^+W^+ and W^+W^-. Both answers are correct. Again, you can write out the Punnett squares to convince yourself that this is correct.

Example 2 Two black female mice are crossed with a brown male. In several litters, female I produced 9 blacks and 7 browns; female II produced 57 blacks. What deductions can you make concerning the inheritance of black and brown coat color in mice? What are the genotypes of the parents in this case?

Solution Here you are given parental and offspring phenotypes and are asked to infer the parental genotypes. As a starting point, assume that the coat colors are due to the simplest genetic system possible: one autosomal gene with two alleles, where one allele is dominant and the other recessive. Because female II produces only black offspring, it's logical to suppose that black is dominant to brown. Let's use B for black and b for brown. Then the male parent is bb. To produce offspring with a 1:1 ratio of black : brown coats, female I must be Bb. But to produce all black offspring, female II must be BB. This model explains the data, so you can accept it as correct.

GENETICS PROBLEMS

11. *Tay-Sachs disease* causes nerve cells to malfunction and results in death by age 4. Two healthy parents know from blood tests that each parent carries a recessive allele responsible for Tay-Sachs. If their first three children have the disease, what is the probability that their fourth child will not? Assuming they have not yet had a child, what is the probability that, if they have four children, all four will have the disease? If their first three children are male, what is the probability that their fourth child will be male?

12. Suppose that in garden peas the genes for seed color and seed-pod shape are linked, and that Mendel crossed *YYII* parents (which produce yellow seeds in inflated pods) with *yyii* parents (which produce green seeds in constricted pods).

    • Draw the F$_1$ Punnett square and predict the expected F$_1$ phenotype(s).

    • List the genotype(s) of gametes produced by F$_1$ individuals if no crossing over occurs.

    • Draw the F$_2$ Punnett square if no crossing over occurs. Based on this Punnett square, predict the expected phenotype(s) in the F$_2$ generation and the expected frequency of each phenotype.

    • If crossing over occurs during gamete formation in F$_1$ individuals, give the genotype of the recombinant gamete(s) that result.

    • Add the recombinant gametes to the F$_2$ Punnett square above. Will any additional phenotypes be observed at low frequency in the F$_2$ generation? If so, what are they?

13. In Jimson weed, the allele that results in violet flowers (*V*) is dominant to one that results in white flowers (*v*); at another locus, the allele that makes prickly seed capsules (*P*) is dominant to one that results in smooth capsules (*p*). A plant with white flowers and prickly capsules is crossed with one that has violet flowers and smooth capsules. The F$_1$ consisted of 47 plants with white flowers and prickly capsules, 45 plants with white flowers and smooth capsules, 50 plants with violet flowers and prickly capsules, and 46 plants with violet flowers and smooth capsules. What are the genotypes of the parents? Are these two genes linked?

14. In cats, the *Manx* allele (*M*) causes a short or absent tail, while a recessive allele *m* confers a normal, long tail. Cats of genotype *MM* die as embryos. If two Manx cats mate, what is the probability that each *surviving* kitten has a long tail?

15. A plant with orange-spotted flowers was grown in the greenhouse from a seed collected in the wild. The plant was self-pollinated and gave rise to the following progeny: 88 orange with spots, 34 yellow with spots, 32 orange with no spots, and 8 yellow with no spots. What can you conclude about the dominance relationships of the alleles responsible for the spotted and unspotted phenotypes? Of orange and yellow phenotypes? What can you conclude about the genotype of the original plant that had orange, spotted flowers?

16. As a genetic counselor, you routinely advise couples about the possibility of genetic disease in their offspring based on their family histories. This morning you met with an engaged couple, both of whom are phenotypically normal. The man, however, has a brother who died of Duchenne-type muscular dystrophy, an X-linked condition that results in death before the age of 20. The allele responsible for this disease is recessive. His prospective bride, whose family has no history of the disease, is worried that the couple's sons or daughters might be afflicted.

    • How would you advise this couple?

    • The sister of this man is planning to marry his fiancé's brother. How would you advise this second couple?

17. Suppose you are heterozygous for two genes that are located on different chromosomes. You carry alleles *A* and *a* for one gene and alleles *B* and *b* for the other. Draw a diagram illustrating what happens to these genes and alleles when meiosis occurs in your reproductive tissues. Label the stages of meiosis, the homologous chromosomes, sister chromatids, nonhomologous chromosomes, genes, and alleles. Be sure to list all of the genetically different gametes that could form and indicate how frequently each type should be observed. On the diagram, identify the events responsible for the principle of segregation and the principle of independent assortment.

18. In humans, the ABO blood type is a polymorphic trait due to alleles *i*, I^A, and I^B. $I^A i$ and $I^A I^A$ individuals have blood type A; $I^B i$ and $I^B I^B$ individuals have blood type B; $I^A I^B$ individuals have blood type AB; *ii* individuals have blood type O. Suppose a woman with blood type O married a man with blood type AB. What phenotypes and genotypes would you expect to observe in their offspring, and in what proportions? Answer the same question for a heterozygous mother with blood type A and a heterozygous father with blood type B.

19. Mr. Spock's father came from the planet Vulcan, and his mother was from planet Earth. A Vulcan has pointed ears and a heart located on the right side of the chest. Mr. Spock has both of these traits, which are known to be determined by two different genes, each with two alleles. Suppose that Mr. Spock married an Earth woman and that they had many children. About half of their children have pointed ears and a right-sided heart, like Spock, and about half have rounded ears and a left-sided heart, like their mother.

    • What would Mendel predict the progeny phenotypes and ratios to be? Explain your answer using formal genetic terminology.

    • How do you explain the actual results?

20. In Klingons, one gene determines hair texture; another gene determines whether the individual will have a sagittal crest (a protrusion on the forehead). The two genes are not linked.

    K = curly Klingon hair (dominant)

    k = silky earthling-like hair (recessive)

    S = large sagittal crest (dominant)

    s = smooth, flat, earthling-like forehead (recessive)

    Kayless has the genotype *KkSs*. He mates with an individual who is also heterozygous for both genes.

    • Set up a Punnett square for this dihybrid cross.

    • What are the four possible phenotypes that may result from this mating? Include a description of both hair and forehead for each phenotype.

    • What is the expected phenotypic ratio from the dihybrid cross?

    • What fraction of the progeny are expected to be heterozygous for both genes?

- What fraction are expected to be homozygous for both genes?

- Are Kayless and his mate more likely to see an actual ratio close to the predicted values if they have 16 children or 160? Why? Explain why sample size does or does not affect observed phenotypic ratios.

21. The blending-inheritance hypothesis proposed that the genetic material from parents is irreversibly mixed in the offspring. As a result, offspring should always appear intermediate in phenotype to the parents. Mendel, in contrast, proposed that genes are discrete and that their integrity is maintained in the offspring and in subsequent generations. Suppose the year is 1890. You are a horse breeder and have just read Mendel's paper. You don't believe his results, however, because you often work with cremello (very light-colored) and chestnut (reddish-brown) horses. You know that if you cross a cremello individual from a pure-breeding line with a chestnut individual from a pure-breeding line, the offspring will be palomino—meaning they have an intermediate (golden-yellow) body color. What additional crosses would you do to test whether Mendel's model is valid in the case of genes for horse color? List the crosses and the offspring genotypes and phenotypes you'd expect to obtain. Explain why these experimental crosses would provide a test of Mendel's model.

22. Two mothers give birth to sons at the same time in a busy hospital. The son of couple 1 is afflicted with hemophilia A, which is a recessive X-linked disease. Neither parent has the disease. Couple 2 has a normal son even though the father has hemophilia A. The two couples sue the hospital in court, claiming that a careless staff member swapped their babies at birth. You appear in court as an expert witness. What do you tell the jury? Make a diagram that you can submit to the jury.

23. You have crossed two *Drosophila melanogaster* individuals that have long wings and red eyes—the wild-type phenotype. In the progeny, the mutant phenotypes, called curved wings and lozenge eyes, appear as follows:

Females
600 long wings, red eyes
200 curved wings, red eyes

Males
300 long wings, red eyes
300 long wings, lozenge eyes
100 curved wings, red eyes
100 curved wings, lozenge eyes

- According to these data, is the curved-wing allele autosomal recessive, autosomal dominant, sex-linked recessive, or sex-linked dominant?

- Is the lozenge-eye allele autosomal recessive, autosomal dominant, sex-linked recessive, or sex-linked dominant?

- What is the genotype of the female parent?

- What is the genotype of the male parent?

24. In parakeets, two autosomal genes that are located on different chromosomes control the production of feather pigment. Gene *B* codes for an enzyme that is required for the synthesis of a blue pigment, and gene *Y* codes for an enzyme required for the synthesis of a yellow pigment. Recessive, loss-of-function mutations are known for both genes. Suppose that a bird breeder has two green parakeets and mates them. The offspring are green, blue, yellow, and albino (unpigmented).

- Based on this observation, what are the genotypes of the green parents? What is the genotype of each type of offspring? What fraction of the total progeny should exhibit each type of color?

- Suppose that the parents were the progeny of a cross between two true-breeding strains. What two types of crosses between true-breeding strains could have produced the green parents? Indicate the genotypes and phenotypes for each cross.

25. The pedigree shown below is for the human trait called osteopetrosis, which is characterized by bone fragility and dental abscesses. Is the gene that affects bone and tooth structure autosomal or sex-linked? Is the allele for osteopetrosis dominant or recessive?

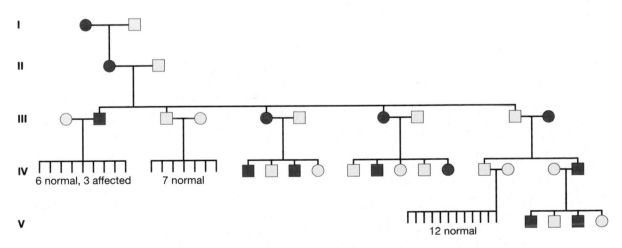

Pedigree of the trait osteopetrosis

Genetics Problems answers:

11. 3/4; 1/256 (see **BioSkills 9**); 1/2 (the probabilities of transmitting the alleles or having sons does not change over time). 12. Your answer should conform to the cross diagrammed in Figure 13.5b, except that different alleles and traits are being analyzed. 13. Parental genotypes are $vvPp \times Vvpp$. Based on these data, we can't tell if they are linked or not. Both hypotheses make the same prediction. 14. 1/3 (the question asks about the *surviving* offspring). 15. This is a dihybrid cross that yields progeny phenotypes in a 9:3:3:1 ratio. Let O stand for the allele for orange petals and o the allele for yellow petals; let S stand for the allele for spotted petals and s the allele for unspotted petals. Start with the hypothesis that O is dominant to o, that S is dominant to s, that the two genes are found on different chromosomes so they assort independently, and that the parent individual's genotype is $OoSs$. If you do a Punnett square for the $OoSs \times OoSs$ mating, you'll find that progeny phenotypes should be in the observed 9:3:3:1 proportions. 16. Let D stand for the normal allele and d stand for the allele responsible for Duchenne-type muscular dystrophy. The woman's family has no history of the disease, so her genotype is almost certainly DD. The man is not afflicted, so he must be DY. (The trait is X-linked, so he has only one allele; the "Y" stands for the Y chromosome.) Their children are not at risk. The man's sister could be a carrier, however—meaning she has the genotype Dd. If so, then half of the second couple's male children are likely to be affected. 17. Your stages of meiosis should look like Figure 12.4. The A and a alleles could be on the red and blue versions of the longer chromosome, and the B and b alleles could be on the red and blue versions of the smaller chromosomes. The places you draw them are the locations of the A and B genes, but each chromosome only has one allele. Each pair of red and blue chromosomes is a homologous pair. Sister chromatids bear the same allele (e.g., both sister chromatids of the long blue chromosomes might bear the a allele). Chromatids from the long and short chromosomes are not homologous. To identify the events that cause the principles of segregation and independent assortment, see Figures 13.7 and 13.8, and substitute A and B for R and Y. 18. Half of their offspring should have the genotype iI^A and the type A blood phenotype. Half of their offspring should have the genotype iI^B and the type B blood phenotype. Second case: the genotype and phenotype ratios would be 1:1:1:1 I^AI^B (type AB): I^Ai (type A): I^Bi (type B); ii (type O). 19. Presumably, Vulcans and earthlings are homozygous for both traits because they are "pure lines." Spock, then, should be heterozygous at both loci. Because he has the Vulcan phenotype, though, we can infer that pointed ears and a right-sided heart are dominant to rounded ears and a left-sided heart. Let P be the allele for pointed ears and p be the allele for rounded ears; let R be the allele for a right-sided heart and r the allele for a left-sided heart. Spock is $PpRr$; his earthling wife is $pprr$. Mendel would predict that their offspring would have the genotypes $PpRr$, $Pprr$, $ppRr$, and

$pprr$ in equal proportions. Thus a quarter of their children should have pointed ears and left-sided hearts or rounded ears and right-sided hearts. The only logical explanation for the actual results is that the two genes are closely linked. If so, then Spock's gametes are either PR or pr, and his wife's gametes are all pr. 20. Four phenotypes should be observed in the ratio 9:3:3:1, as follows: curly and large: curly and smooth: silky and large: silky and smooth. One-quarter of the progeny should be heterozygous for both traits. One-eighth should be homozygous for both traits. They are more likely to see the predicted ratios with a large family, because there would be fewer fluctuations due to chance. 21. According to Mendel's model, palomino individuals should be heterozygous at the locus for coat color. If you mated palomino individuals, you would expect to see a combination of chestnut, palomino, and cremello offspring. If blending inheritance occurred, however, all of the offspring should be palomino. 22. Because this is an X-linked trait, the father who has hemophilia could not have passed the trait on to his son. Thus, the mother in couple 1 must be a carrier and have passed the recessive allele on to her son, who is XY and affected. To educate a jury about the situation, you should draw what happens to the X and Y during meiosis, then make a drawing showing the chromosomes in couple 1 and couple 2, with a Punnett square showing how these chromosomes get passed to the affected and unaffected children. 23. The curved-wing allele is autosomal recessive; the lozenge-eye allele is sex-linked (specifically, X-linked) recessive. Let l^+ be the allele for long wings and l be the allele for curved wings; let r^+ be the allele for red eyes and r the allele for lozenge eyes. The female parent is l^+lr^+r; the male parent is l^+lr^+Y. 24. Albinism indicates the absence of pigment, so let b stand for an allele that gives the absence of blue and y for an allele that gives the absence of yellow pigment. If blue and yellow pigment blend to give green, then both green parents are $BbYy$. The green phenotype is found in $BBYY$, $BBYy$, $BbYY$, and $BbYy$ offspring. The blue phenotype is found in $BByy$ or $Bbyy$ offspring. The yellow phenotype is observed in $bbYY$ or $bbYy$ offspring. Albino offspring are $bbyy$. The phenotypes of the offspring should be in the ratio 9:3:3:1 as green:blue:yellow:albino. Two types of crosses yield $BbYy$ F_1 offspring: $BByy \times bbYY$ (blue × yellow) and $BBYY \times bbyy$ (green × albino). 25. Autosomal dominant.

DNA and the Gene: Synthesis and Repair

14

Electron micrograph showing DNA in the process of replication. The "Y" shape results from a structure called the replication fork, where DNA synthesis is taking place. The DNA double helix on the right is being replicated into two DNA double helices on the left.

KEY CONCEPTS

- Genes are made of DNA. When DNA is copied, each strand of a DNA double helix serves as the template for the synthesis of a complementary strand.

- DNA is synthesized only in the $5' \rightarrow 3'$ direction. When a DNA molecule is being copied, a large suite of specialized enzymes is involved in unwinding the double helix, continuously synthesizing the "leading strand" in the $5' \rightarrow 3'$ direction and synthesizing the "lagging strand" as a series of fragments that are then linked together.

- Specialized enzymes repair mistakes in DNA synthesis and damaged DNA. If these repair enzymes are defective, the mutation rate increases. Mutations in many types of genes can lead to cancer.

What are genes made of, and how are they copied so that they are faithfully passed on to offspring? These questions dominated biology during the middle of the twentieth century. Since Mendel's time, the predominant research strategy in genetics had been to conduct a series of experimental crosses, create a genetic model to explain the types and proportions of phenotypes that resulted, and then test the model's predictions through reciprocal crosses, testcrosses, or other techniques. This strategy was extremely productive. It led to virtually all of the discoveries analyzed in Chapter 13, including Mendel's rules, sex-linkage, linkage, genetic mapping, incomplete dominance and codominance, multiple allelism and polymorphic traits, pleiotropy, and quantitative inheritance.

But the molecular nature of Mendel's hereditary factors—what came to be called genes—remained a mystery. Further, biologists knew that genes and chromosomes were replicated during S phase of the cell cycle (see Chapter 11), with copies distributed to daughter cells during mitosis and meiosis. No one had the slightest clue about how the copying occurred, however.

The goal of this chapter is to explore how researchers solved these mysteries. Let's begin with studies that identified deoxyribonucleic acid (DNA) as the genetic material, explore how this molecule is copied during the synthesis phase of the cell cycle, and conclude by analyzing how damaged or incorrectly copied genes are repaired. Once the molecular nature of the gene was known, the nature of biological science changed forever.

○ Key Concept   Important Information   ● Practice It

14.1   DNA as the Hereditary Material

The first hint of what genes might be made of was published in 1928, when Frederick Griffith reported the discovery of a mysterious phenomenon involving hereditary traits. Griffith referred to this phenomenon as **transformation**. His transformation experiments appeared to isolate the hereditary material.

In the 1920s Griffith had been doing experiments in an attempt to develop a vaccine against the *Streptococcus pneumoniae* bacterium, which is a leading cause of pneumonia, earaches, sinusitis, and meningitis in humans. For his experiments, however, he worked with strains of *S. pneumoniae* that infect mice. A **strain** is a population of genetically identical individuals. As is the case with the strains that affect humans, the strains that affect mice vary in their **virulence**—their ability to cause disease and death. Virulent strains cause disease; avirulent (or benign) strains do not.

As **Figure 14.1** shows, the strains that Griffith happened to work with can be identified by eye when grown on a nutrient-containing medium in a petri dish. A **medium** is a liquid or solid that is suitable for growing cells. On a solid medium, cells from the nonvirulent strain form colonies that look rough; cells from the virulent strain form colonies that look smooth. Logically enough, Griffith called the nonvirulent strain R for rough and the virulent strain S for smooth.

FIGURE 14.1 There Are Two Strains of *Streptococcus pneumoniae*. Nonvirulent (R) and virulent (S) strains of *Streptococcus* differ in colony appearance and at the cellular level.

To understand how the strains interact, Griffith carried out the experiment illustrated in **Figure 14.2**. This figure shows four experimental treatments that he designed. In the first treatment, he injected mice with cells of the nonvirulent R strain. As he expected, these mice lived. In the second treatment, he injected

FIGURE 14.2 The Discovery of "Transformation." Mice died after being injected with a combination of live R cells and heat-killed S cells.

mice with cells of the virulent S strain. Not surprisingly, these mice died of pneumonia. So far, so good: These first two treatments were controls showing the effect that each strain of *S. pneumoniae* has on mice. In the third treatment, Griffith killed cells of the virulent S strain by heating them and then injected them into mice. These mice lived. This experimental treatment was interesting because it showed that dead S cells do not cause disease. In the final treatment, Griffith injected mice with heat-killed S cells *and* live nonvirulent R cells. Unexpectedly, these mice died. Autopsies confirmed pneumonia as the cause of death. When Griffith isolated and grew *Streptococcus* from these dead mice, he found virulent S cells, *not* nonvirulent R cells.

What was going on? Griffith proposed that something from the heat-killed S cells had "transformed" the nonvirulent R cells. Stated another way, something had changed the appearance and behavior of the R cells from R-like to S-like. Because this "something" appeared in the growing population of cells that Griffith isolated from the dead mice, it had been passed on to the offspring of the transformed cells. It was clearly some sort of hereditary, or genetic, factor. Next the question became, What is it?

Is DNA the Genetic Material?

The chromosome theory of inheritance, introduced in Chapter 13, proposed that chromosomes are composed of genes. It had been known since the late 1800s that chromosomes are a complex of DNA and proteins. Because the chromosome theory had been confirmed around 1920, it was clear that Griffith's transforming factor had to consist of either protein or DNA.

Initially, most biologists backed the hypothesis that genes are made of proteins. The arguments in favor of this hypothesis were compelling. Hundreds, if not thousands, of complex and highly regulated chemical reactions occur in even the simplest living cells. The amount of information required to specify and coordinate these reactions is almost mind-boggling. With their almost limitless variation in structure and function, proteins are complex enough to contain this much information.

DNA, in contrast, was known to be composed of just four types of deoxyribonucleotides—the monomers introduced in Chapter 4. It was also thought to be a simple molecule with some sort of repetitive and uninteresting structure. So when researchers published the first experimental evidence that DNA was the hereditary material, most biologists had the same reaction: They didn't believe it.

The Avery et al. Experiment   In the early 1940s Oswald Avery, Colin M. MacLeod, and Maclyn McCarty set out to understand the molecular basis of Griffith's result. They used an elegant experimental strategy to isolate the transforming factor. To determine whether protein, ribonucleic acid (RNA—see Chapter 4), or DNA was responsible for transformation, they grew quantities of virulent S cells in culture. A **culture** is a collection of cells that grows under controlled conditions—usually suspended in a liquid medium or on the surface of a solid

growth medium. Avery's group killed the cultured cells with heat, broke them open to create a cell extract, and then used chemical treatments to remove the lipids and carbohydrates from the extracts. These steps left a mixture containing protein, RNA, and DNA from the virulent S cells. The researchers divided the sample into three treatments and used different enzymes to destroy a specific macromolecule in each. One sample was treated with proteases, enzymes that destroy proteins. Another sample was treated with ribonuclease, an enzyme that breaks apart RNA. A third sample was treated with DNAase, an enzyme that cuts up DNA (**Figure 14.3**). When small quantities of the three resulting solutions were added to cultures containing nonvirulent R cells, virulent S cells appeared in all of the cultures that still contained S-cell DNA; no S cells appeared in the sample that lacked DNA. The biologists concluded that DNA—not protein or RNA—must be the transforming factor.

Experiment

Question: Does the transforming factor consist of protein, RNA, or DNA?

Hypothesis 1: Transforming factor is protein.
Hypothesis 2: Transforming factor is RNA.
Hypothesis 3: Transforming factor is DNA.

Experimental setup:

~~Lipid~~
~~Carbohydrate~~
Protein
RNA
DNA

1. Start with heat-killed *Streptococcus* cells, which contain different types of macromolecules. Remove lipids and carbohydrates.

~~Protein~~   Protein   Protein
RNA   ~~RNA~~   RNA
DNA   **DNA**   ~~DNA~~
+   +   +

2. Add enzyme to digest one type of macromolecule in each treatment. Add each treatment to a mixture of R cells.

Prediction: (No explicit prediction; all three alternatives are considered equally likely.)

Results:

Transformation requires presence of DNA

~~Protein~~   Protein   Protein
RNA   ~~RNA~~   RNA
DNA   **DNA**   ~~DNA~~

Transformation   Transformation   **No transformation**

Conclusion: DNA is the transforming factor.

FIGURE 14.3 Experimental Evidence that DNA Is the Transforming Factor.

● **QUESTION** If genes were made of protein, how would the results of this experiment change?

This experiment provided strong evidence that DNA is the hereditary material. But the result was not widely accepted. Researchers who advocated the protein hypothesis claimed that the enzymatic treatments were not sufficient to remove all of the proteins present and that enough protein could remain to transform R cells.

The Hershey-Chase Experiment Alfred Hershey and Martha Chase took up the question of whether genes were made of protein or DNA by studying how a virus called T2 infects the bacterium *Escherichia coli*. Hershey and Chase knew that T2 infections begin when the virus attaches to the cell wall of *E. coli* and injects its genes into the cell's interior (**Figure 14.4a**). These genes then direct the production of a new generation of virus particles inside the infected cell, which acts as a **host** for the parasitic virus. During the infection, the protein coat of the original, parent virus is left behind, still attached to the exterior of the host cell as a "ghost" (**Figure 14.4b**). Hershey and Chase also knew that T2 is made up almost exclusively of protein and DNA. But was it protein or DNA that entered the host cell and directed the production of new viruses?

(a) ONLY VIRAL GENES ENTER A CELL THAT IS BEING INFECTED.

Virus protein coat

Virus genes

Host cell

1. Start of infection. Virus genes enter host cell. Protein coat does not.

2. Virus genes direct the production of new virus particles.

3. End of infection. New generation of virus particles bursts from host cell.

(b) The virus's protein coat stays on the outside of the host cell.

Virus protein coat

New virus particles

0.2 μm

FIGURE 14.4 Viruses Inject Genes into Bacterial Cells and Leave a Protein Coat Behind. (a) Viruses that infect a bacterial cell start by injecting their genes into the cell. **(b)** The protein coat of the original virus is left behind, attached to the bacterial cell wall.

Hershey and Chase's strategy for determining which part of the virus enters the cell and acts as the hereditary material was based on two facts: (1) Proteins present in T2 contain sulfur but not phosphorus, and (2) DNA contains phosphorus but not sulfur. As **Figure 14.5** shows, the researchers began their work by growing viruses in the presence of either the radioactive isotope

Experiment

Question: Do viral genes consist of DNA or protein?

DNA hypothesis: Viral genes consist of DNA.

Protein hypothesis: Viral genes consist of protein.

Experimental setup:

★ Viral DNA is radioactive.

⁎ Viral protein is radioactive.

E. coli

E. coli

Protein coats outside

Genes inside

Protein coats in solution

Viral genes within bacterial cells in pellet

1. Grow some viruses in presence of ^{32}P (P is in DNA but not in viral protein) and some viruses in presence of ^{35}S (S is in protein but not in DNA).

2. Allow viruses with labeled DNA to infect one culture of *E. coli* cells and viruses with labeled protein to infect another.

3. Agitate cultures in kitchen blender to separate empty viral protein coats from bacterial cells in each culture.

4. Centrifuge solutions of bacterial cells from each culture to force cells into a pellet. Record location of radioactive labels.

Prediction of DNA hypothesis: Radioactive DNA will be located within pellet.

Prediction of protein hypothesis: Radioactive protein will be located within pellet.

Results:

Radioactive DNA is in pellet

Radioactive protein is in solution

DNA

Protein

Conclusion: Viral genes consist of DNA. Viral coats consist of protein.

FIGURE 14.5 More Experimental Evidence that DNA Is the Hereditary Material. The experiment shown here convinced most biologists that DNA is indeed the hereditary material.

of sulfur (^{35}S) or the radioactive isotope of phosphorus (^{32}P). Because these molecules were incorporated into newly synthesized proteins and DNA, this step produced a population of viruses with radioactive proteins and a population with radioactive DNA. Then Hershey and Chase allowed each set of radioactive viruses to infect *E. coli* cells. If genes consist of DNA, then the radioactive protein should be found in the empty, or "ghost," protein coats outside the infected host cells, while the radioactive DNA should be located inside the cells. But if genes consist of proteins, the opposite should be true.

To test these predictions, Hershey and Chase sheared the ghosts off the cells by agitating each of the cultures in kitchen blenders. When the researchers spun the samples in a centrifuge, the ghosts stayed in the solution while the cells formed a pellet at the bottom of the centrifuge tube. As predicted by the DNA hypothesis, the biologists found that virtually all of the radioactive protein was in the ghosts, while virtually all of the radioactive DNA was inside the host cells. Because the injected component of the virus directs the production of a new generation of virus particles, it is this component that represents the virus's genes.

After these results were published, proponents of the protein hypothesis had to admit that DNA, not protein, must be the hereditary material. In combination, the evidence from the bacterial transformation experiments and the virus-labeling experiments was convincing. The claim that a seemingly simple molecule contained all the information for life's complexity was finally accepted.

The realization that genes are made of DNA was one of the great advances of twentieth-century biology. The insight raised two crucial questions, however: (1) How did the simple primary structure and secondary structure of DNA hold the information required to make life possible? (2) How is DNA copied, so that genetic information is faithfully passed from one cell to another during growth and from parents to offspring during reproduction? Understanding how DNA contains information is the focus of Chapter 15. The remainder of this chapter concentrates on how genes are replicated prior to mitosis and meiosis, and how damaged genes are repaired.

14.2 Testing Early Hypotheses about DNA Synthesis

The DNA inside each cell is like an ancient text that has been painstakingly copied and handed down, generation after generation. But while the most ancient of all human texts contain messages that are thousands of years old, the DNA in living cells has been copied and passed down for billions of years. And instead of being copied by monks or clerks, DNA is replicated by molecular scribes. What molecules are responsible for copying DNA, and how do they work?

Chapter 4 introduced Watson and Crick's model for the secondary structure of DNA, which was proposed in 1953.

Recall that DNA is a long, linear polymer made up of monomers called deoxyribonucleotides, which consist of a deoxyribose molecule, a phosphate group, and a nitrogenous base (**Figure 14.6a**). Deoxyribonucleotides link together into a polymer when a phosphodiester bond forms between a hydroxyl group on the 3′ carbon of deoxyribose and the phosphate group attached to the 5′ carbon of deoxyribose. As **Figure 14.6b** shows, the primary structure of a DNA molecule has two major components: (1) a "backbone" made up of the sugar and phosphate groups of deoxyribonucleotides and (2) a series of nitrogen-containing bases that project from the backbone. A strand of DNA has a directionality or polarity: One end has an exposed hydroxyl group on the 3′ carbon of deoxyribose, while the other has an exposed phosphate group on a 5′ carbon. Thus, the molecule has a 3′ end and a 5′ end.

As they explored different models for the secondary structure of DNA, Watson and Crick hit on the idea of lining up two of these long strands in the opposite direction, or in what

(a) Structure of a deoxyribonucleotide

(b) Primary structure of DNA

FIGURE 14.6 DNA's Primary Structure. (a) Deoxyribonucleotides are monomers that polymerize to form DNA. **(b)** DNA's primary structure is made up of a sequence of deoxyribonucleotides. Notice that the structure has a sugar-phosphate "backbone" with nitrogen-containing bases attached.

● **EXERCISE** Write the base sequence of the DNA in part (b), in the 5′ → 3′ direction.

is called antiparallel fashion (**Figure 14.7a**). They realized that antiparallel strands will twist around each other into a spiral or helix because certain of the nitrogen-containing bases fit together in pairs inside the spiral and form hydrogen bonds. The double-stranded molecule that results is called a **double helix** (**Figure 14.7b**). This secondary structure is stabilized by hydrogen bonds that form between the nitrogenous bases adenine (A) and thymine (T) and between the bases guanine (G) and cytosine (C), along with hydrophobic interactions among the nitrogenous bases inside the helix. The specific pairing rules for hydrogen bonding of nitrogen-containing bases are called **complementary base pairing**.

Watson and Crick realized that the A-T and G-C pairing rules suggested a way for DNA to be copied when chromosomes are replicated during S phase of the cell cycle—prior to mitosis and meiosis. They suggested that the existing strands of DNA served as a template (pattern) for the production of new strands, with bases being added to the new strands according to complementary base pairing. For example, if the template strand contained a T, then an A would be added to the new strand to pair with that T. Similarly, a G on the template strand would dictate the addition of a C on the new strand.

Biologists had three alternative hypotheses about how the old and new strands might interact during replication, however:

1. If the old strands of DNA separated, each could then be used as a template for the synthesis of a new, daughter strand. This hypothesis is called semiconservative replication, because each new daughter DNA molecule would consist of one old strand and one new strand.

2. If the bases temporarily turned outward so that complementary strands no longer faced each other, they could serve as a template for the synthesis of an entirely new double helix all at once. This hypothesis is called conservative replication. Conservative replication results in the intact parental strands and a daughter DNA molecule consisting entirely of newly synthesized strands.

3. If the parent helix was cut and unwound in short sections before being copied and put back together, then new and old strands would intermingle—old sections of DNA would be interspersed with new DNA down the length of each daughter molecule. This possibility is called dispersive replication.

Which of these three hypotheses is correct?

The Meselson–Stahl Experiment

Matthew Meselson and Frank Stahl realized that if they could tag parental and daughter strands of DNA in a way that would make them distinguishable from each other, they could determine whether replication was conservative, semiconservative, or dispersive. Soon after their results were published in 1958, the Meselson-Stahl work became recognized as a classic experiment in biological science.

Before they could do any tagging or publishing, however, they needed to choose an organism to study. They decided to work with a common inhabitant of the human gastrointestinal tract, the bacterium *Escherichia coli*. Because *E. coli* is small and grows quickly and readily in the laboratory, it had become a favored model organism in studies of biochemistry and molecular genetics.

Like all organisms, bacterial cells copy their entire complement of DNA, or their **genome**, before every cell division. To distinguish parental strands of DNA from daughter strands when *E. coli* replicates, Meselson and Stahl grew the cells for many generations in the presence of one of two isotopes of nitrogen: either ^{15}N or ^{14}N. Because ^{15}N contains an extra neutron, it is heavier than the normal isotope, ^{14}N.

This difference in mass, which creates a difference in density of ^{14}N-containing and ^{15}N-containing DNA, was the key to the experiment. **Figure 14.8** summarizes Meselson and Stahl's experimental strategy. The biologists reasoned that if different nitrogen isotopes were available in the growth medium when

(a) Complementary base pairing   **(b)** The double helix

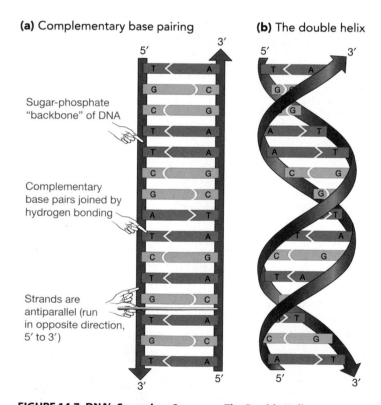

Sugar-phosphate "backbone" of DNA

Complementary base pairs joined by hydrogen bonding

Strands are antiparallel (run in opposite direction, 5′ to 3′)

FIGURE 14.7 DNA's Secondary Structure: The Double Helix.
(a) Deoxyribonucleic acid normally has a secondary structure consisting of two strands, each with a sugar-phosphate backbone. Nitrogen-containing bases project from each strand and form hydrogen bonds. Only A-T and G-C pairs fit together in a way that allows hydrogen bonding to occur. **(b)** That bonding between complementary bases twists the molecule into a double helix.

● **QUESTION** Why are some bases colored blue while others are green?

Experiment

Question: Is replication semiconservative, conservative, or dispersive?

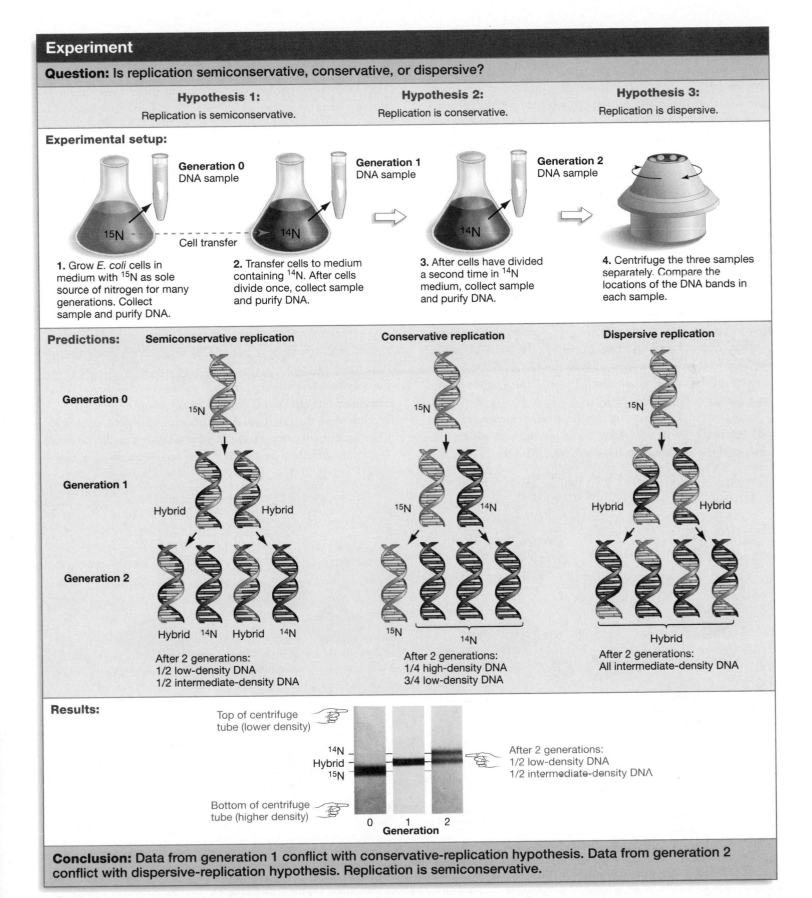

Hypothesis 1:
Replication is semiconservative.

Hypothesis 2:
Replication is conservative.

Hypothesis 3:
Replication is dispersive.

Experimental setup:

Generation 0
DNA sample

Generation 1
DNA sample

Generation 2
DNA sample

^{15}N — Cell transfer → ^{14}N ⇒ ^{14}N ⇒

1. Grow *E. coli* cells in medium with ^{15}N as sole source of nitrogen for many generations. Collect sample and purify DNA.

2. Transfer cells to medium containing ^{14}N. After cells divide once, collect sample and purify DNA.

3. After cells have divided a second time in ^{14}N medium, collect sample and purify DNA.

4. Centrifuge the three samples separately. Compare the locations of the DNA bands in each sample.

Predictions:

Semiconservative replication

Conservative replication

Dispersive replication

Generation 0
^{15}N

Generation 1
Hybrid   Hybrid

^{15}N ^{14}N

Hybrid   Hybrid

Generation 2
Hybrid ^{14}N Hybrid ^{14}N

^{15}N ^{14}N

Hybrid

After 2 generations:
1/2 low-density DNA
1/2 intermediate-density DNA

After 2 generations:
1/4 high-density DNA
3/4 low-density DNA

After 2 generations:
All intermediate-density DNA

Results:

Top of centrifuge tube (lower density)

^{14}N
Hybrid
^{15}N

After 2 generations:
1/2 low-density DNA
1/2 intermediate-density DNA

Bottom of centrifuge tube (higher density)

0   1   2
Generation

Conclusion: Data from generation 1 conflict with conservative-replication hypothesis. Data from generation 2 conflict with dispersive-replication hypothesis. Replication is semiconservative.

FIGURE 14.8 The Meselson–Stahl Experiment.

● **EXERCISE** Meselson and Stahl actually let their experiment run for 4 generations, with cultures growing in the presence of ^{14}N. Sketch what data from 3rd and 4th generation DNA should look like—that is, where the DNA band(s) should be.

parental and daughter strands of DNA were produced, then the two types of strands should behave differently during centrifugation. More specifically, when intact, double-stranded DNA molecules are added to a solution that forms a gradient from low to high density during centrifugation, DNA strands that contain ^{14}N should form a band in the lower-density part of the centrifuge tube. In contrast, DNA strands that contain ^{15}N should form a band in the higher-density part of the centrifuge tube. Because the highest-density solution is at the bottom of the tube, DNA that contains ^{15}N should be found lower in the tube than DNA containing ^{14}N. In this way, DNA strands containing ^{15}N or ^{14}N should form separate bands. How could this tagging system be manipulated to test whether replication is semiconservative, conservative, or dispersive?

Meselson and Stahl began by growing *E. coli* cells with nutrients that contained only ^{15}N. They purified DNA from a sample of these cells and transferred the rest of the culture to a growth medium containing only the ^{14}N isotope. After enough time had elapsed for these experimental cells to divide once, they removed a sample and isolated the DNA. After the remainder of the culture had divided again, they removed another sample and purified the DNA.

As Figure 14.8 shows, the conservative, semiconservative, and dispersive models make distinct predictions about the makeup of the DNA molecules after replication occurred in the first and second generation. For example, if replication is conservative, then the daughter cells should have double-stranded DNA with either ^{14}N or ^{15}N, but not both. As a result, two distinct DNA bands should form in the centrifuge tube—one high-density band and one low-density band. But if replication is semiconservative or dispersive, then all of the experimental DNA should contain an equal mix of ^{14}N or ^{15}N after one generation, and one intermediate-density band should form in the centrifuge tube. But after two generations, half of the daughter cells should contain only ^{14}N if replication is semiconservative—meaning a second, lower-density band should appear in the centrifuge tube. But the dispersive model predicts that there will be just one band at intermediate density.

The photograph at the bottom of Figure 14.8 shows the experiment's results. After one generation, the density of the DNA molecules was intermediate. These data suggested that the hypothesis of conservative replication was wrong. After two generations, a lower-density band appeared in addition to the intermediate-density band. This result offered strong support for the hypothesis that DNA replication is not dispersive but **semiconservative;** that is, each newly made DNA molecule comprises one old strand and one new strand.

How does the DNA synthesis reaction proceed? Does it require an input of energy in the form of ATP, or it is spontaneous? Is it catalyzed by an enzyme, or does it occur quickly on its own?

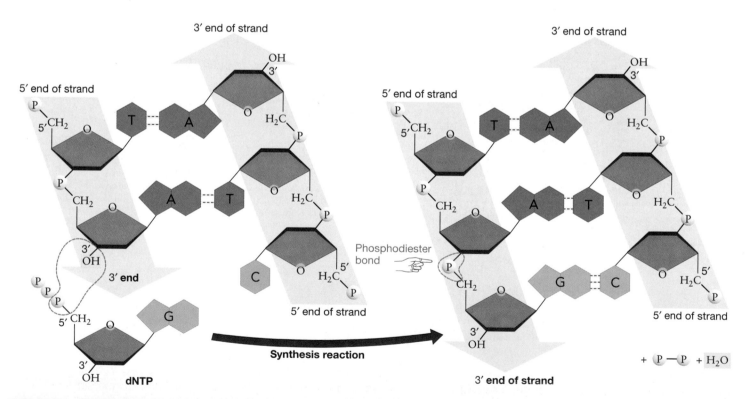

FIGURE 14.9 The DNA Synthesis Reaction. A dNTP (deoxyribonucleoside triphosphate) monomer is added to a DNA polymer when a phosphodiester bond forms between the 3′ carbon on the end of a DNA strand and the 5′ carbon on the dNTP in a condensation reaction.

QUESTION Does DNA synthesis proceed from the 5′ end toward the 3′ end, or in the 3′ to 5′ direction?

EXERCISE The "P–P" group reacts with H_2O, forming 2 phosphate ions (HPO_4^-) and releasing energy. Add this reaction to the figure.

14.3 A Comprehensive Model for DNA Synthesis

The initial breakthrough in research on DNA replication came with the discovery of an enzyme named **DNA polymerase**, because it polymerizes deoxyribonucleotides to DNA. This protein catalyzes DNA synthesis. Follow-up work showed that organisms contain several types of DNA polymerases. DNA polymerase III, for example, is the enzyme that is primarily responsible for copying *E. coli*'s chromosome prior to cell division.

Figure 14.9 illustrates a critical point about DNA polymerases: They can work in only one direction. More specifically, DNA polymerases can add deoxyribonucleotides to only the 3′ end of a growing DNA chain. As a result, DNA synthesis always proceeds in the 5′ → 3′ direction. ● If you understand this concept, you should be able to draw two lines indicating a DNA molecule, assign the 3′ to 5′ polarity of each strand, and then label the direction that DNA synthesis will proceed for each strand.

Figure 14.9 makes another important point about DNA synthesis. You might recall from earlier chapters that polymerization reactions are endergonic. But in cells, the reaction is exergonic because the monomers that are added to the growing strand are **deoxyribonucleoside triphosphates**, or **dNTPs**. The *N* in dNTP stands for any of the four bases found in DNA:

adenine, thymine, guanine, or cytosine. Because they have three phosphate groups close together, dNTPs have high potential energy (see Chapter 9)—high enough to make the formation of phosphodiester bonds in a growing DNA strand exergonic.

Once the basic synthesis reaction was understood, researchers could delve into questions about how DNA replication gets started and proceeds.

How Does Replication Get Started?

A second major insight into the mechanism of DNA synthesis emerged when electron micrographs caught chromosome replication in action. As **Figure 14.10a** shows, a "bubble" forms in a chromosome when DNA is actively being synthesized. Bacterial chromosomes have a single location where the replication process begins, and thus a single bubble forms. Initially, the replication bubble forms at a specific sequence of bases called the **origin of replication** (**Figure 14.10b**). Replication bubbles grow as DNA replication proceeds, because synthesis is bidirectional—that is, it occurs in both directions (though always 5′ → 3′ because the strands are antiparallel) at the same time. Eukaryotes also have bidirectional replication; but they have multiple sites along each chromosome where DNA synthesis begins, and thus multiple replication bubbles (**Figure 14.10c**). ● If you understand this concept, you should be able to draw a replication bubble, assign the 3′ and 5′ polarity of each

(a) A chromosome being replicated

(b) Bacterial chromosomes have a single point of origin.

Old DNA
New DNA
Replication proceeds in both directions
Origin of replication

(c) Eukaryotic chromosomes have multiple points of origin.

Replication fork
Replication bubble
Old DNA
New DNA
Replication proceeds in both directions from each starting point

FIGURE 14.10 DNA Synthesis Proceeds in Two Directions from a Point of Origin. (a) A micrograph of a "replication bubble." **(b)** In most bacteria, chromosomes are circular and there is a single point of origin during DNA replication. **(c)** Most eukaryotes have linear chromosomes; each contains several to many points of origin for DNA synthesis.

● **EXERCISE** Circle the two replication forks in part (a).

strand of DNA present and then label the direction of DNA synthesis for each strand at both ends of the bubble.

A specific suite of proteins is responsible for recognizing sites where replication begins and opening the double helix at those points. These proteins are activated by the proteins responsible for initiating S phase in the cell cycle (see Chapter 11). Once a replication bubble opens, a different suite of enzymes takes over and initiates replication. The action takes place in the corners of each replication bubble—at a structure called the replication fork. A **replication fork** is a Y-shaped region where the parent DNA double helix is split into two single strands, which are then copied. How does the splitting event occur, and how are each of the resulting strands replicated?

How Is the Helix Opened and Stabilized?

Several key events occur at the point where the double helix opens. An enzyme called a **helicase** catalyzes the breaking of hydrogen bonds between deoxyribonucleotides. This reaction causes the two strands of DNA to separate. Proteins called **single-strand DNA-binding proteins** (**SSBPs**) attach to the separated strands and prevent them from snapping back into a double helix. In combination, then, the helicase and single-strand DNA-binding proteins open up the double helix and make both strands available for copying (**Figure 14.11**, step 1).

The "unzipping" process that occurs at the replication fork creates tension farther down the helix, however. To understand why, imagine what would happen if you started to pull apart the twisted strands of a rope. The untwisting movements at one end would force the intact section to rotate in response. If the intact end of the rope were fixed in place, though, it would eventually begin to coil on itself and kink in response to the twisting forces. This does not happen in DNA, because the twisting stress induced by helicase is relieved by proteins called topoisomerases. A **topoisomerase** is an enzyme that cuts and

rejoins the DNA downstream of the replication fork. Topoisomerases do this cutting and pasting in a way that undoes twists and knots.

Now, what happens once the DNA helix is open and has stabilized?

How Is the Leading Strand Synthesized?

The keys to understanding what happens at the start of DNA synthesis are to recall that DNA polymerase III works only in the $5' \to 3'$ direction and to recognize that to start synthesis, it requires both a 3' end *and* a single-stranded template. The single-stranded template dictates which deoxyribonucleotide should be added next, while a **primer**—which consists of a few nucleotides bonded to the template—provides a free 3' hydroxyl (–OH) group that can combine with an incoming dNTP to form a phosphodiester bond.

Once a primer is added to a single-stranded template, DNA polymerase III begins working in the $5' \to 3'$ direction and adds deoxyribonucleotides to complete the complementary strand.

Before DNA synthesis can get under way, then, an enzyme called **primase** has to synthesize a short stretch of ribonucleic acid (RNA) that acts as a primer for DNA polymerase. Primase is a type of **RNA polymerase**—an enzyme that catalyzes the polymerization of ribonucleotides to RNA (see Chapter 4 to review RNA's structure). Unlike DNA polymerases, primase

FIGURE 14.11 Synthesizing the Leading Strand during DNA Replication.

and other RNA polymerases do not require a primer. These enzymes can simply match ribonucleotides directly by complementary base pairing on single-stranded DNA. In this way, primase creates a primer for DNA synthesis.

Once the primer is in place, DNA polymerase III begins adding deoxyribonucleotides to the 3′ end of the new strand, in a sequence that is complementary to the template strand. As Figure 14.11 step 2 shows, the shape of DNA polymerase III grips the DNA strand during synthesis, a little like your hand clasping a rope. Catalysis takes place in the groove inside the enzyme, at an active site between the enzyme's "thumb" and "fingers." As DNA polymerase moves along the DNA molecule, a doughnut-shaped structure behind it, called the sliding clamp, holds the enzyme in place. The enzyme's product is called the **leading strand**, or **continuous strand**, because it leads into the replication fork and is synthesized continuously. ● If you understand leading-strand synthesis, you should be able to list the enzymes involved and for each one, predict the consequences if it were defective.

How Is the Lagging Strand Synthesized?

Synthesis of the leading strand is straightforward after an RNA primer is in place—DNA polymerase III chugs along, adding bases to the 3′ end of that strand. The enzyme moves into the replication fork, which "unzips" ahead of it. By comparison, events on the opposite strand are much more involved. To understand why, recall that the two strands in the DNA double helix are *antiparallel*—meaning they are parallel to one another but oriented in opposite directions. DNA polymerase works in only one direction, however, so if the DNA polymerase that is synthesizing the leading strand works into the replication fork, then a DNA polymerase must work away from the replication fork to synthesize the other strand in the 5′ → 3′ direction. The strand that is synthesized in the opposite direction of the replication fork is called the **lagging strand**, because it lags behind the fork. ● If you understand this concept, you should be able to draw a replication bubble in DNA, assign the 3′ and 5′ polarity of each strand, label the direction of DNA synthesis for each strand at both ends of the bubble, and indicate which of the four newly synthesized strands—two at each end of the bubble—are the leading strands and lagging strands.

The synthesis of the lagging strand starts when primase synthesizes a short stretch of RNA that acts as a primer. DNA polymerase III then adds bases to the 3′ end of the lagging strand. The key observation is that the enzyme moves away from the replication fork. But behind it, helicase continues to open the replication fork and expose new single-stranded DNA on the lagging strand.

These events create a paradox. New single-stranded template DNA is constantly appearing behind DNA polymerase III—in the direction *opposite* the direction of lagging-strand synthesis—as the helix continues to unzip during leading-strand synthesis.

FIGURE 14.12 Gaps Form in the Lagging Strand Gaps form in the lagging strand as helicase opens the double helix ahead of the DNA polymerase synthesizing the leading strand. The gaps occur because DNA polymerase can work only away from the replication fork—in the 5′ → 3′ direction—on the lagging strand.

To see this point, consider the state of the lagging strand at time 1 and time 2 in **Figure 14.12**. As DNA polymerase works into the replication fork in the 5′ → 3′ direction and away from the replication fork in the 5′ → 3′ direction, gaps appear in the lagging strand.

The puzzle posed by lagging-strand synthesis was resolved when Reiji Okazaki and colleagues tested a hypothesis called discontinuous replication. This hypothesis stated that once primase synthesizes an RNA primer on the lagging strand (step 1 in **Figure 14.13**), DNA polymerase III might synthesize short fragments of DNA along the lagging strand and that these fragments would later be linked together to form a continuous whole. The idea was that primase would add a primer to the newly exposed single-stranded DNA at intervals and that DNA polymerase III would then synthesize the lagging strand until it reached the fragment produced earlier.

To test this hypothesis, Okazaki's group set out to test a key prediction: Could they document the existence of short DNA fragments produced during replication? Their critical experiment was based on the pulse-chase strategy introduced in Chapter 7. Specifically, they added a short "pulse" of a radioactive deoxyribonucleotide to *E. coli* cells, followed by a large "chase" of nonradioactive deoxyribonucleotide. According to the discontinuous replication model, some of this radioactive deoxyribonucleotide should end up in short, single-stranded fragments of DNA.

As predicted, the researchers succeeded in finding these fragments when they purified DNA from the experimental cells and separated the molecules by centrifugation. A small number of labeled pieces of DNA, about 1000 base pairs long, were present. These short sections came to be known as **Okazaki fragments** (steps 2 and 3 of Figure 14.13). Subsequent work showed that Okazaki fragments in eukaryotes are just 100 to 200 base pairs long, though a chromosome typically contains millions of base pairs.

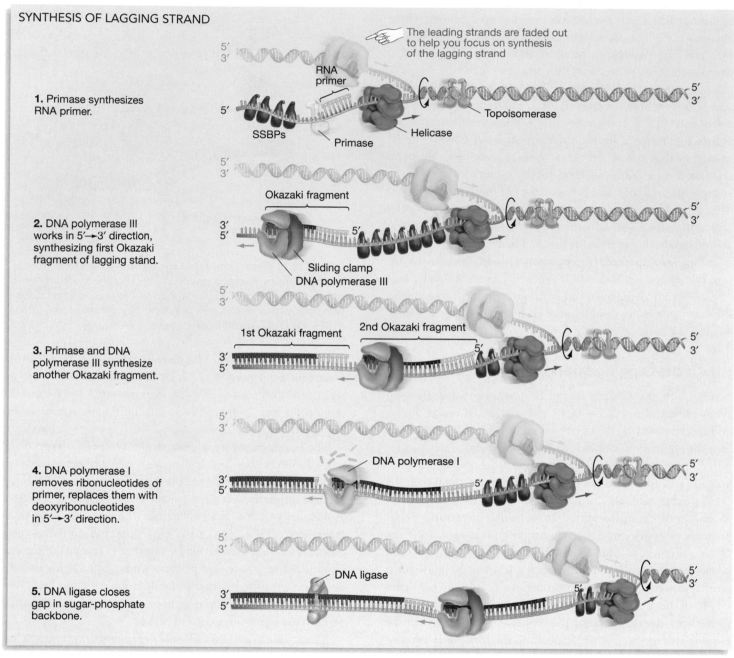

FIGURE 14.13 The Completion of DNA Replication.

How are Okazaki fragments connected into a continuous whole? As step 4 of Figure 14.13 shows, DNA polymerase I removes the RNA primer at the start of each fragment and fills in the appropriate deoxyribonucleotides. Finally, an enzyme called **DNA ligase** catalyzes the formation of a phosphodiester bond between the adjacent fragments (Figure 14.13, step 5). Because Okazaki fragments are synthesized independently and joined together later, the lagging strand is also called the **discontinuous strand**. ● If you understand lagging-strand synthesis, you should be able to draw what the two molecules resulting from DNA synthesis—starting from a single origin of replication—would look like if DNA ligase were defective.

In combination, then, the enzymes that open the replication fork and manage the synthesis of the leading and lagging strands succeed in producing a faithful copy of the original DNA molecule prior to mitosis or meiosis. Given the number of enzymes and proteins involved in the synthesis machinery, it is not surprising that it took 25 years for biologists to assemble the results summarized in **Table 14.1.**

It is also worth mentioning the following points:

- Although Figures 14.12 and 14.13 show the enzymes involved in DNA synthesis at different locations around the replication fork, in reality most are joined into one large multi-enzyme machine at the replication fork. This complex of proteins opens the double helix and synthesizes both the leading strand and the Okazaki fragments on the lagging strand. The lagging strand actually loops out and around one of the polymerases so that the entire machine can move together down the replication fork.

SUMMARY TABLE **14.1 Proteins Involved in DNA Synthesis**

Name	Structure	Function
Opening the helix		
Helicase		Catalyzes the breaking of hydrogen bonds between base pairs and the opening of the double helix
Single-strand DNA-binding proteins		Stabilize single-stranded DNA
Topoisomerase		Breaks and rejoins the DNA double helix to relieve twisting forces caused by the opening of the helix
Leading strand		
Primase		Catalyzes the synthesis of the RNA primer
DNA polymerase III		Extends the leading strand
Sliding clamp		Holds DNA polymerase in place during strand extension
Lagging strand		
Primase		Catalyzes the synthesis of the RNA primer on an Okazaki fragment
DNA polymerase III		Extends an Okazaki fragment
Sliding clamp		Holds DNA polymerase in place during strand extension
DNA polymerase I		Removes the RNA primer and replaces it with DNA
DNA ligase		Catalyzes the joining of Okazaki fragments into a continuous strand

• The basic elements of DNA synthesis were worked out in experiments on _E. coli_. Follow-up research has shown that the fundamental elements of this process are almost identical in eukaryotes. The exceptions are that (1) some of the enzymes at the replication fork in eukaryotes are larger and more complex than similar enzymes in bacteria; and (2) on the lagging strand, two different DNA polymerases do the work done by DNA polymerase III in bacteria. In almost every other detail, however, the critical components of DNA synthesis have been highly conserved over the course of evolution.

Most aspects of the model summarized in Figures 14.11 and 14.13 had emerged in the early 1980s, and subsequent decades saw major advances in our understanding of the detailed molecular structures of the major enzymes involved. But in addition to studying the replication process itself, biologists have been exploring other questions about DNA synthesis. This chapter closes with a look at two of these research areas: How are the ends of linear chromosomes replicated? And how can incorrect or damaged bases be repaired? As it turns out, both questions have practical implications—particularly for the origin of certain types of cancer.

 Web Animation at www.masteringbio.com
DNA Synthesis

Check Your Understanding

If you understand that...

• DNA synthesis begins at specific points of origin on the chromosome and then proceeds in both directions.

• A three-step process occurs at the replication fork.

(1) Helicase opens the double helix, SSBPs stabilize the exposed single strands, and topoisomerase prevents kinks downstream of the fork;

(2) DNA polymerase synthesizes the leading strand after primase has added an RNA primer; and

(3) a series of enzymes synthesizes the lagging strand.

• Lagging-strand synthesis cannot be continuous, because it moves away from the replication fork. In bacteria, enzymes called primase, DNA polymerase III, DNA polymerase I, and ligase work in sequence to synthesize Okazaki fragments and link them into a continuous whole.

You should be able to...

1) Draw a replication fork.

2) Label the leading and lagging strands and indicate their polarity.

3) Label primase, topoisomerase, single-stranded DNA binding proteins, the sliding clamp and DNA polymerase on each strand, and DNA ligase.

4) Explain the function of each structure involved.

14.4 Replicating the Ends of Linear Chromosomes

The circular DNA molecules in bacteria and archaea can be synthesized via the sequence of events diagrammed in Figures 14.11 and 14.13. Similarly, the leading and lagging strands of the linear DNA molecules found in eukaryotes are copied efficiently by the enzymes illustrated in those figures. But copying the ends of linear chromosomes is another story.

The region at the end of a linear chromosome is called a **telomere** (literally, "end-part"). **Figure 14.14** illustrates the problem that arises during the replication of telomeres. Steps 1 and 2 show that as the replication fork nears the end of a linear chromosome, DNA polymerase synthesizes the leading strand all the way to the end of the parent DNA template. As a result, leading-strand synthesis results in a normal copy of the DNA molecule. But the situation is different on the lag-

ging strand. As step 2 of Figure 14.14 illustrates, primase adds an RNA primer close to the tip of the chromosome. DNA polymerase III synthesizes the final Okazaki fragment on the lagging strand (step 3), and DNA polymerase I removes the primer. The DNA polymerase III is unable to add DNA near the tip of the chromosome, however, because there is not enough room for primase to add a new RNA primer (step 4). As a result, the single-stranded DNA that is left once the primer has been removed must stay single stranded.

The single-stranded DNA that remains after telomeres are copied is eventually degraded, which results in the shortening of the chromosome. If this process were to continue, every chromosome would shorten by 50–100 deoxyribonucleotides on average each time that DNA replication occurred prior to mitosis or meiosis. Over time, linear chromosomes would be expected to disappear completely.

CHROMOSOME SHORTENING DURING NORMAL DNA REPLICATION

Leading strand

DNA polymerase
Sliding clamp

End of chromosome

Lagging strand

SSBPs   Helicase

1. Helicase unwinds end of DNA helix (at end of chromosome).

2. DNA polymerase completes the leading strand. Primase synthesizes RNA primer at end of lagging strand.

RNA primer

3. DNA polymerase synthesizes the last Okazaki fragment in lagging strand.

DNA polymerase   Last Okazaki fragment

4. No DNA synthesis occurs after primer is removed (no free 3′ end for DNA polymerase); chromosome is shortened.

Unreplicated end

FIGURE 14.14 Telomeres Shorten during Normal DNA Replication. An RNA primer is added to the lagging strand near the end of the chromosome. Once the primer is removed, it cannot be replaced with DNA. As a result, the chromosome shortens.

Bacteria and archaea do not have this end-replication problem, because virtually all of these cells have a single, circular chromosome. How do eukaryotes maintain the integrity of their linear chromosomes? An answer emerged after two important discoveries were made:

1. Telomeres do not contain genes that code for products needed in the cell. Instead, they consist of short stretches of bases that are repeated over and over. In humans, for example, the base sequence TTAGGG is repeated thousands of times. (Because DNA and RNA sequences are always written 5′ → 3′, the base sequence on the complementary strand is CCCTAA rather than AATCCC.) Human telomeres consist of a total of about 10,000 deoxyribonucleotides with the sequence TTAGGGTTAGGGTTAGGG....

2. An interesting enzyme called telomerase is involved in replicating telomeres. **Telomerase** is remarkable because it catalyzes the synthesis of DNA from an RNA template. Further, the enzyme carries an RNA molecule with it that acts as a built-in template. Telomerase catalyzes the addition of deoxyribonucleotides to the ends of chromosomes. The deoxyribonucleotides that are added are complementary to the ribonucleotides in its built-in RNA template As a result, telomerase can add DNA onto the end of a chromosome and prevent it from getting shorter.

Figure 14.15 shows how this process works. Step 1 shows the unreplicated segment at the 3′ end of the lagging strand at the telomere. It forms a single-strand "overhang." In step 2, telomerase binds to this overhanging section of single-stranded parent DNA. Once it has bound, it begins catalyzing the addition of deoxyribonucleotides, in the 5′ → 3′ direction, to the overhang. In step 3, telomerase begins moving in the 5′ → 3′ direction and continues to catalyze the addition of deoxyribonucleotides. In this way, telomerase lengthens the overhang on the lagging strand. The normal machinery of DNA synthesis—primase, DNA polymerase, and ligase—then synthesizes the lagging strand in the 5′ → 3′ direction (see step 4). The result is that the lagging strand becomes slightly longer than it was originally.

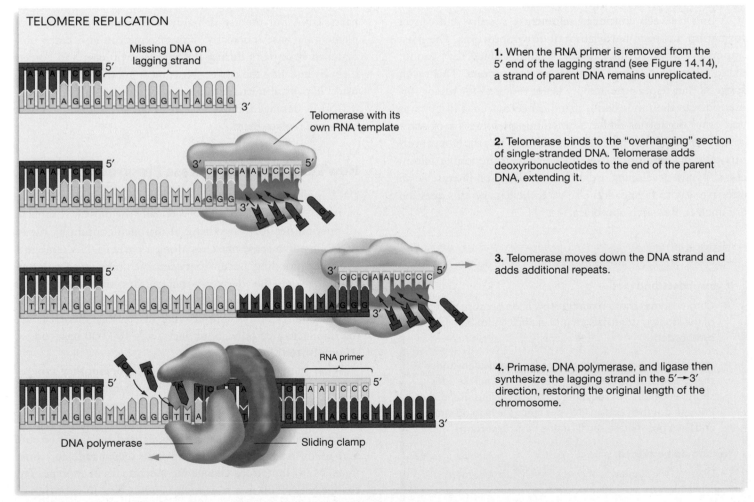

TELOMERE REPLICATION

Missing DNA on lagging strand

Telomerase with its own RNA template

RNA primer

DNA polymerase — Sliding clamp

1. When the RNA primer is removed from the 5′ end of the lagging strand (see Figure 14.14), a strand of parent DNA remains unreplicated.

2. Telomerase binds to the "overhanging" section of single-stranded DNA. Telomerase adds deoxyribonucleotides to the end of the parent DNA, extending it.

3. Telomerase moves down the DNA strand and adds additional repeats.

4. Primase, DNA polymerase, and ligase then synthesize the lagging strand in the 5′→3′ direction, restoring the original length of the chromosome.

FIGURE 14.15 Telomerase Prevents Shortening of Telomeres during Replication. By extending the number of repeated sequences in the 5′ → 3′ direction, telomerase provides room for enzymes to add an RNA primer to the lagging strand. DNA polymerase can then fill in the missing section of the lagging strand.

● **QUESTION** Would this telomerase work as well if its RNA template had a different sequence?

It is important to recognize, though, that telomerase is not active in most types of cells. In humans, for example, active telomerase is found primarily in the cells of reproductive organs—specifically, in the cells that eventually undergo meiosis and produce gametes. Cells that are not involved in gamete formation, or what biologists call **somatic cells**, normally lack telomerase. As predicted, the chromosomes of somatic cells gradually shorten with each mitotic division, getting progressively smaller as an individual grows and ages.

These observations inspired a pair of important hypotheses. The first was that telomere shortening causes cells to stop dividing and enter the non-dividing state called G_0 (see Chapter 11). The second was that if telomerase were mistakenly activated in a somatic cell, telomeres would fail to shorten. This would allow the cell to keep dividing and possibly contribute to uncontrolled growth and cancer.

To test the first hypothesis, biologists added functioning telomerase to human cells growing in vitro. The treated cells continued dividing long past the age when otherwise identical cells stop growing. These results have convinced most biologists that telomere shortening is involved in limiting how long cells remain in an actively growing state.

A link between continued telomerase activity and cancer formation has been harder to nail down, however. The most important work to date involves mutant mice that cannot make telomerase. Many of these develop tumors. This result suggests that telomerase activity is not required for tumor formation. On the other hand, many cancerous cells in humans and other organisms do have functioning telomerase or some other mechanism for maintaining telomere length. Noncancerous cells in these organisms do not have functioning telomerase. Would developing drugs that knock out telomerase be an effective way to fight cancer? To date, the data on this question are unclear. Research continues.

Check Your Understanding

If you understand that...

- Chromosomes shorten during replication because the end of the lagging strand lacks a primer and cannot be synthesized.

- Shortening is prevented in many cells—particularly those that produce sperm and egg—because telomerase adds short, repeated DNA sequences to the template strand. Primase can then add an RNA primer to the lagging strand, and DNA polymerase can fill in the missing sections.

You should be able to...

1) Diagram the sequence of events involved in telomere replication.

2) Add labels to indicate the enzymes involved and the polarity of each DNA strand.

14.5 Repairing Mistakes and Damage

DNA polymerases work fast. In yeast, for instance, each replication fork is estimated to move at a rate of about 50 bases per second. But the replication process is also astonishingly accurate. In organisms ranging from *E. coli* to animals, the error rate during DNA replication averages less than one mistake per *billion* deoxyribonucleotides. This level of accuracy is critical. Humans, for example, develop from a fertilized egg that has DNA containing over 6 billion deoxyribonucleotides. The DNA inside the fertilized egg is replicated over and over to create the trillions of cells that eventually make up the adult body. If more than one or two mutations occurred during each cell division cycle as a human grew, genes would be riddled with errors by the time the individual reached maturity. Genes that contain errors are often defective. Based on these observations, it is no exaggeration to claim that the accurate replication of DNA is a matter of life and death. Natural selection favors individuals with enzymes that copy DNA as exactly as possible.

How can the enzymes involved in replicating DNA be as precise as they are? The answer to this question has several parts. DNA polymerase is highly selective in matching complementary bases correctly. This enzyme can also catch mismatched base pairs during the synthesis process and correct them by inserting the correct base, meaning that it can "proofread." Finally, if mistakes remain after synthesis is complete or if DNA is damaged, repair enzymes can cut out the defective bases and replace them.

How Does DNA Polymerase Proofread?

DNA polymerases are selective because the correct base pairings (A-T and G-C) are energetically the most favorable of all possibilities for the pairing of nitrogen-containing bases. As DNA polymerase marches along a parent DNA template, hydrogen bonding occurs between incoming deoxyribonucleotides and the deoxyribonucleotides on the template strand. Because DNA polymerase is so selective in matching deoxyribonucleotides, the enzyme inserts the incorrect deoxyribonucleotide only about once every 100,000 bases added (**Figure 14.16a**).

What happens when a G-T, an A-C, or another type of mismatch occurs? An answer to this question emerged when researchers found mutants in *E. coli* with error rates that were 100 times greater than normal. Recall from Chapter 13 that mutants are individuals with traits caused by mutation and that mutation is a change in the gene responsible for those traits. Many mutations change the individual's phenotype. The change may result in a trait such as white eyes in fruit flies or an elevated mutation rate in *E. coli*. At the molecular level, a mutant phenotype usually results from a change in an enzyme or other type of protein.

(a) Mismatched bases

(b) DNA polymerase III can repair mismatches.

FIGURE 14.16 DNA Polymerase Can Proofread. (a) In bacteria, DNA polymerase adds an incorrect base to a growing strand of DNA about once in every 100,000 bases added. The result is a mismatch such as the pairing of A with C. **(b)** DNA polymerase can act as a 3′ → 5′ exonuclease, meaning that it can remove bases in that direction. The enzyme then adds the correct base.

In the case of *E. coli* cells with high mutation rates, biologists found that the mutation was localized to a particular portion of the the ends of DNA strands polymerase III enzyme, called the ε (epsilon) subunit. Further analyses showed that this subunit of the enzyme acts as an exonuclease—meaning an enzyme that removes deoxyribonucleotides from the ends of DNA strands (**Figure 14.16b**). The DNA polymerase III exonuclease activity removes deoxyribonucleotides in a 3′ to 5′ direction, and only if they are not hydrogen bonded to a base on the complementary strand. These results led to the conclusion that DNA polymerase III can **proofread**. If the wrong base is added during DNA synthesis, the enzyme pauses, removes the mismatched base that was just added, and then proceeds with synthesis.

Eukaryotic DNA polymerases have the same type of proofreading ability. Typically, proofreading reduces a DNA

polymerase's error rate to about 1×10^{-7} (one mistake per 10 million bases). Is this accurate enough? The answer is no. If DNA polymerase leaves a mismatched pair behind in the DNA sequence by mistake, a battery of enzymes springs into action to correct the problem. **Mismatch repair** occurs when mismatched bases are corrected after DNA synthesis is complete.

The proteins responsible for mismatch repair were discovered in the same way that the proofreading capability of DNA polymerase III was uncovered—by analyzing *E. coli* mutants. In this case, the mutants had normal DNA polymerase III but abnormally high mutation rates. The first mutant gene that caused a deficiency in mismatch repair was identified in the late 1960s and was called *mutS*. (The *mut* is short for "mutator.") By the late 1980s researchers had identified 10 proteins involved in the identification and repair of base-pair mismatches in *E. coli*. These proteins recognize the mismatched pair, remove a section of the newly synthesized strand that contains the incorrect base, and fill in the correct bases. The mismatch repair enzymes are like a copy editor who corrects typos that the author—DNA polymerase, in this case—did not catch.

Nucleotide Excision Repair

Even after DNA is synthesized and proofread and mismatches repaired, all is not well. Genes are under constant assault. Nucleotides are damaged by chemicals like the hydroxyl (OH) radicals produced during aerobic metabolism, the aflatoxin B1 found in moldy peanuts and corn, and the benzo[α]pyrene in cigarette smoke. Radiation is also dangerous. Ultraviolet (UV) light, for example, can cause a covalent bond to form between adjacent pyrimidine bases. The thymine-thymine pair illustrated in **Figure 14.17** is an example. This defect is called a thymine dimer. Thymine dimers create kinks in the secondary structure of DNA that stall the enzymes responsible for transcription and DNA synthesis. The kinks also stall the movement of the replication fork during DNA replication. If the damage is not repaired, the cell may die.

UV light ⟹ Kink

DNA strand with adjacent thymine bases

Thymine dimer

FIGURE 14.17 UV Light Damages DNA. When UV light strikes a section of DNA with adjacent thymines, the energy can break bonds within each base and result in the formation of bonds *between* them. The thymine dimer that is produced causes a kink in DNA.

● **QUESTION** Why are infrared wavelengths much less likely to damage DNA rather than UV?

NUCLEOTIDE EXCISION REPAIR

1. Enzymes detect an irregularity in DNA structure and cut the damaged strand.

Damaged bases

2. An enzyme excises nucleotides on the damaged strand.

3. DNA polymerase fills in the gap in the 5′→ 3′ direction.

4. DNA ligase links the new and old nucleotides.

Repaired damage

FIGURE 14.18 In Excision Repair, Defective Bases Are Removed and Replaced.

Cells have an efficient system to fix these types of problems, called **nucleotide excision repair** (**Figure 14.18**). As step 1 in the figure shows, the symmetry and regularity of DNA's secondary structure makes it possible for repair proteins to recognize thymine dimers and other types of damaged bases that produce an irregularity in the molecule. Once a damaged region is recognized, enzymes remove the single-stranded DNA around the damaged site. The presence of a DNA strand complementary to the damaged strand provides a template for resynthesis of the defective sequences. In this way, DNA's structure makes accurate repair possible. Because the molecule is repairable, DNA's structure supports its function in information storage and processing. But what happens when repair systems are defective?

Xeroderma Pigmentosum: A Case Study

Xeroderma pigmentosum (XP) is a rare autosomal recessive disease in humans. Individuals with this condition are extremely sensitive to ultraviolet (UV) light. Their skin develops lesions after even slight exposure to sunlight. In unaffected individuals, these kinds of lesions develop only after extensive exposure to UV light, X-rays, or other forms of high-energy radiation.

In 1968 James Cleaver proposed a connection between XP and DNA excision repair systems. He knew that in *E. coli*, mu-

tations in certain genes cause DNA nucleotide excision repair proteins to fail. Cells with these mutations have an increased sensitivity to radiation. Cleaver's hypothesis was that people with XP have similar mutations. He claimed that they are extremely sensitive to sunlight because they are unable to repair the damage that occurs when the nitrogen-containing bases of DNA absorb UV light.

Cleaver's hypothesized connection between DNA damage, faulty nucleotide excision repair, and XP turned out to be correct. Much of the work that he and other investigators did relied on the use of cell cultures (see Chapter 11). In this case, the researchers collected skin cells from people with XP and from people with a normal phenotype for excision repair. When these cell populations were grown in culture and exposed to increasing amounts of ultraviolet radiation, a striking difference emerged: Cells from xeroderma pigmentosum individuals died off much more rapidly than did unaffected cells (**Figure 14.19a**).

(a) Vulnerability of cells to UV light damage

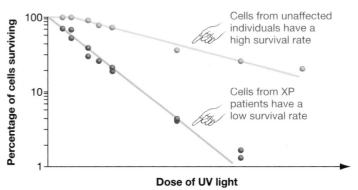

Cells from unaffected individuals have a high survival rate

Cells from XP patients have a low survival rate

(b) Ability of cells to repair damage

Damaged DNA is repaired in unaffected individuals

Repair is defective in XP patients

FIGURE 14.19 DNA Damage from UV Light Is Not Repaired Properly in Individuals with XP. (a) When cell cultures from unaffected individuals and from XP patients are irradiated with various doses of UV light, the percentage of cells that survive is strikingly different. **(b)** When cell cultures from unaffected individuals and from XP patients are irradiated with various doses of UV light and then fed radioactive thymidine, only unaffected individuals incorporate the labeled base.

● **QUESTION** Why are people who cultivate a skin tan likely to develop cancer? (Hint: Tanning is a response to UV light.)

The connection to excision repair systems was confirmed when Cleaver exposed cells from unaffected individuals and cells from XP individuals to various amounts of UV light, then fed the cells a radioactive deoxyribonucleotide to label DNA synthesized during the repair period. If repair is defective in the XP individuals, then their cells should incorporate virtually no radioactive deoxyribonucleotide into their DNA. Cells from unaffected individuals, in contrast, should incorporate large amounts of labeled deoxyribonucleotide into their DNA. As **Figure 14.19b** shows, this is exactly what happens. These data are consistent with the hypothesis that excision repair is virtually nonexistent in XP individuals.

More recently, genetic analyses of XP patients have shown that the condition can result from mutations in any one of seven genes. This result is not surprising in light of the large number of enzymes involved in repairing damaged DNA.

Finally, defects in the genes required for DNA repair are frequently associated with cancer. Individuals with xeroderma pigmentosum, for example, are 1000 to 2000 times more likely to get skin cancer than are individuals with intact excision repair systems. To explain this pattern, biologists suggest that if mutations in the genes involved in the cell cycle (see Chapter 11) go unrepaired, the cell may begin to grow in an uncontrolled manner. Tumor formation could result. Stated another way, if the overall mutation rate in a cell is elevated because of defects in DNA repair genes, then the mutations that trigger cancer become more likely. Research on "simple" model organisms can be extraordinarily fruitful. In this case, research on fundamental aspects of DNA replication and repair, using *Escherichia coli* as a model system, led directly to a major advance in understanding a form of cancer in humans.

Chapter Review

SUMMARY OF KEY CONCEPTS

● **Genes are made of DNA. When DNA is copied, each strand of a DNA double helix serves as the template for the synthesis of a complementary strand.**

By labeling DNA with ^{15}N or ^{14}N, Meselson and Stahl were able to validate the hypothesis that DNA replication is semiconservative. Each strand of a parent DNA molecule provides a template for the synthesis of a daughter strand, resulting in two complete DNA double helices.

You should be able to write a sequence of double-stranded DNA that is 10 bases long, separate the sequence, and write in the bases that are added during DNA replication. ●

● **DNA is synthesized only in the $5' \rightarrow 3'$ direction. When a DNA molecule is being copied, a large suite of specialized enzymes is involved in unwinding the double helix, continuously synthesizing the "leading strand" in the $5' \rightarrow 3'$ direction and synthesizing the "lagging strand" as a series of fragments that are then linked together.**

DNA synthesis is an enzyme-catalyzed reaction that takes place in one direction. DNA synthesis requires both a template and a primer sequence, and it takes place at the replication fork where the double helix is being opened.

Synthesis of the leading strand in the $5' \rightarrow 3'$ direction is straightforward, but synthesis of the lagging strand is more complex because DNA polymerase has to work away from the replication fork. By feeding *E. coli* cells a short pulse of radioactive deoxyribonucleotide, Okazaki and co-workers confirmed that short DNA fragments form on the lagging strand. These Okazaki fragments are primed by a short strand of RNA and are linked together after synthesis.

At the ends of linear chromosomes in eukaryotes, the enzyme telomerase adds short, repeated sections of DNA so that the lagging strand can be synthesized without shortening the chromosome. Telomerase is active in reproductive cells that eventually undergo meiosis. As a result, gametes contain chromosomes of normal length.

You should be able to compare and contrast the functions of the three polymerases introduced in this chapter: DNA polymerase, primase, and telomerase. ●

(MB) **Web Animation** at www.masteringbio.com
DNA Synthesis

● **Most mistakes that occur in DNA synthesis are repaired by specialized enzymes. If these repair enzymes are defective, the mutation rate increases. Mutations in many types of genes can lead to cancer.**

DNA replication is remarkably accurate because DNA polymerase proofreads and because mismatch repair enzymes remove incorrect bases once synthesis is complete and replace them with the correct sequence. In addition, DNA repair occurs after bases have been damaged by chemicals or radiation. Nucleotide excision repair systems cut out damaged portions of genes and replace them with correct sequences. Several types of human cancers are associated with defects in the genes responsible for DNA repair.

You should be able to explain the logical connections between failure of repair systems, increases in mutation rate, and high likelihood of cancer developing. ●

QUESTIONS

● Test Your Knowledge

1. The experiment that removed proteins, RNA, or DNA from *Streptococcus* extracts purported to show that DNA is the hereditary material. Why did it fail to convince many skeptics?
   a. It had no control treatments, so it was poorly designed.
   b. Genes had already been shown to be made of protein.
   c. The trait that was studied—virulence—is not genetic.
   d. Critics argued that the chemical treatments had not removed all of the protein present.

2. Which of the following is *not* a property of DNA polymerase?
   a. It catalyzes the addition of dNTPs only in the $5' \rightarrow 3'$ direction.
   b. It requires a primer to work.
   c. It is associated with a sliding clamp only on the leading strand.
   d. It can proofread because it has an exonuclease activity.

3. What is the function of topoisomerase?
   a. holding DNA polymerase steady as it moves down the leading or lagging strand
   b. opening the DNA helix at the replication fork
   c. stabilizing single strands of DNA, once the replication fork is open
   d. preventing kinks in DNA as the replication fork opens and unwinds

4. What is the function of primase?

   a. synthesis of the short section of double-stranded DNA required by DNA polymerase
   b. synthesis of a short section of RNA, complementary to single-stranded DNA
   c. closing the gap at the $3'$ end of DNA after excision repair
   d. removing primers and synthesizing a short section of DNA to replace them

5. Where and how are Okazaki fragments synthesized?
   a. on the leading strand, in a $5' \rightarrow 3'$ direction
   b. on the leading strand, in a $3' \rightarrow 5'$ direction
   c. on the lagging strand, in a $5' \rightarrow 3'$ direction
   d. on the lagging strand, in a $3' \rightarrow 5'$ direction

6. What does telomerase do?
   a. It adds a protein primer to the ends of linear chromosomes.
   b. It adds double-stranded DNA to the "blunt end" of a linear chromosome.
   c. It adds double-stranded DNA to the lagging strand at the end of a linear chromosome.
   d. It adds single-stranded DNA to the lagging strand at the end of a linear chromosome.

Test Your Knowledge answers: 1. d; 2. c; 3. d; 4. b; 5. c; 6. d

● Test Your Understanding

Answers are available at www.masteringbio.com

1. Researchers try to design experiments so that an experimental treatment shows the effect that one, and only one, condition or agent has on the phenomenon being studied. Examine Figure 14.2, and decide which treatments in the transformation study acted as experimental treatments and which acted as controls. Write a general statement explaining the role of control treatments in experimental design.

2. What does it mean to say that DNA replication is "bidirectional?"

3. Why is the synthesis of the lagging strand of DNA discontinuous? How is it possible for the synthesis of the leading strand to be continuous?

4. Explain how telomerase prevents linear chromosomes from shortening during replication.

5. List in chronological order the events that increase the accuracy of DNA replication. Indicate which events involve DNA polymerase III and which involve specialized repair enzymes.

6. Explain how the structure of DNA makes it relatively easy for proteins to recognize base-pair mismatches or damaged bases. How does DNA's secondary structure make it possible for damaged sections or incorrect bases to be removed and repaired?

● Applying Concepts to New Situations

Answers are available at www.masteringbio.com

1. If DNA polymerase III did not require a primer, which steps in DNA synthesis would differ from what is observed? Would any special enzymes be required to replicate telomeres? Explain your answers.

2. In the late 1950s Herbert Taylor grew bean root-tip cells in a solution of radioactive thymidine and allowed them to undergo one round of DNA replication. He then transferred the cells to a solution without the radioactive deoxyribonucleotide, allowed them to replicate again, and examined their chromosomes for the presence of radioactivity. His results are shown in the following figure, where red indicates a radioactive chromatid.
   a. Draw diagrams explaining the pattern of radioactivity observed in the sister chromatids after the first and second rounds of replication.
   b. What would the results of Taylor's experiment be if eukaryotes used a conservative mode of DNA replication?

Non-radioactive   Radioactive

1. One round of DNA replication in radioactive solution   **2.** Mitosis   **3.** One round of DNA replication in non-radioactive solution

3. The graph that follows shows the survival of four different *E. coli* strains after exposure to increasing doses of ultraviolet light. The wild-type strain is normal, but the other strains have a mutation in genes called *uvrA*, *recA*, or both *uvrA* and *recA*.

   a. Which strains are most sensitive to UV light? Which strains are least sensitive?

   b. What are the relative contributions of these genes to the repair of UV damage?

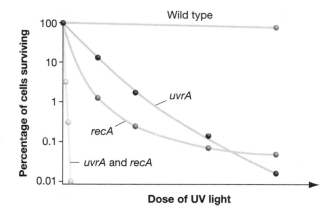

4. One widely used test to identify whether certain chemicals, such as pesticides or herbicides, might be carcinogenic (cancer causing) involves exposing bacterial cells to a chemical and recording whether the exposure leads to an increased mutation rate. In effect, this test equates cancer-causing chemicals with mutation-causing chemicals. Why is this an informative test?

15

How Genes Work

KEY CONCEPTS

- ● Most genes code for proteins.

- ● In cells, information flows from DNA to RNA to proteins. DNA is transcribed to messenger RNA by RNA polymerase, and then messenger RNA is translated to proteins by ribosomes.

- ● Each amino acid in a protein is specified by a group of three bases in RNA.

Each A, T, G, and C here represents a nitrogen-containing base in DNA. This chapter explores how biologists determined that genes are made of DNA and how a base sequence in DNA is translated into an amino acid sequence in a protein.

DNA has been called the blueprint of life. If an organism's DNA is like a set of blueprints, then its cells are like a construction site. How does the DNA inside each cell specify the types and amounts of lumber, nails, and cement that are needed as the cell grows and as its structure and function change? If the enzymes inside a cell are like construction workers, how does DNA organize them into a team that can build and maintain the cell, and perhaps remodel it when conditions change?

Mendel provided insights that made it possible to ask these questions. He discovered that particular alleles are associated with certain phenotypes and that alleles are transmitted faithfully from parent to offspring. The chromosome theory of inheritance established that genes are found in chromosomes, and the theory detailed how the movement of chromosomes during meiosis explains Mendel's results. The molecular era in biology began with the discovery that DNA is the hereditary material, and that DNA is a double helical structure containing sequences of four bases. Based on these early advances, it was clear that genes are made of DNA and that genes carry information—the instructions for making and maintaining an individual. But biologists still didn't know how the information in DNA is translated into action. How does an organism's genotype specify its phenotype?

This chapter introduces some of the most central ideas in all of biology—ideas that connected genotypes and phenotypes by

● Key Concept   Important Information   ● Practice It

revealing how genes work at the molecular level. Understanding how genes work, in turn, triggered a major transition in the history of biological science. Instead of thinking about genes solely in relation to their effects on eye color in fruit flies or on seed shape in garden peas, biologists could begin analyzing the molecular composition of both genes and their products. The molecular revolution in biology took flight.

15.1 What Do Genes Do?

Although in the early twentieth century biologists made tremendous progress in understanding how genes are inherited, an explicit hypothesis explaining what genes do did not appear until 1941. That year George Beadle and Edward Tatum published a series of breakthrough experiments on a bread mold called *Neurospora crassa*. Their research was inspired by an idea that was brilliant in its simplicity. As Beadle said: "One ought to be able to discover what genes do by making them defective." The idea was to knock out a gene by damaging it and then infer what the gene does by observing the phenotype of the mutant individual.

Today, alleles that do not function at all are called **knock-out mutants**, **null mutants**, or **loss-of-function mutants**. Creating and analyzing the effects of knock-out mutant alleles is still one of the most common research strategies in studies of gene function. But Beadle and Tatum were the pioneers.

To start their work, the researchers exposed a large number of *N. crassa* individuals to radiation. As Chapter 14 indicated, high-energy radiation damages the double-helical structure of DNA—often in a way that makes the affected gene nonfunctional. Then the researchers examined the mutant individuals. Eventually they succeeded in finding mutant *N. crassa* individuals that could not make specific compounds. For example, one of the mutants could not make pyridoxine, also called Vitamin B6, even though normal individuals can. Further, Beadle and Tatum showed that the inability to synthesize pyridoxine was due to a defect in a single gene, and that the inability to synthesize other molecules was due to defects in other genes.

These results inspired their **one-gene, one-enzyme hypothesis**. Beadle and Tatum proposed that the mutant *N. crassa* individual could not make pyridoxine because it lacked an enzyme required to synthesize the compound and that the lack of the enzyme was due to a genetic defect. ● Based on analyses of knock-out mutants, the one-gene, one-enzyme hypothesis claimed that genes contain the information needed to make proteins, many of which function as enzymes.

Three years later, Adrian Srb and Norman Horowitz published a rigorous test of the one-gene, one-enzyme hypothesis. These biologists focused on the ability of *N. crassa* individuals to synthesize the amino acid arginine. In the lab, normal cells of this bread mold grow well on a laboratory culture medium that lacks arginine. This is possible because *N. crassa* cells are able to synthesize their own arginine.

Previous work had shown that organisms synthesize arginine via a series of steps, or what biologists call a **metabolic pathway**. As **Figure 15.1** shows, compounds called ornithine and citrulline act as intermediates in the metabolic pathway leading to arginine. Further, specific enzymes are required to synthesize ornithine, convert ornithine to citrulline, and change citrulline to arginine. Srb and Horowitz hypothesized that specific genes in *N. crassa* cells are responsible for producing each of the three enzymes involved.

To test this idea, Srb and Horowitz used radiation to create a large number of mutant individuals. High-energy radiation is equally likely to damage DNA in any part of the organism's genome, however, and most organisms have thousands of genes. Of the many mutants that the biologists created, then, it was likely that only a handful contained knock-out mutations in the pathway for arginine synthesis.

To find the mutants they were looking for, the researchers performed what is now known as a genetic screen. A **genetic screen** is a technique for picking certain types of mutants out of many thousands of randomly generated mutants.

Srb and Horowitz began their screen by raising the irradiated cells on a medium that included arginine. Then they grew a sample of each type of cell on a medium that *lacked* arginine. If an individual could grow in the presence of arginine but failed to grow without arginine, they concluded that it couldn't make its own arginine. The biologists followed up by confirming that the offspring of these cells also had this defect. Based on these data, they were confident that they had isolated individuals with mutations in the genes for one or more of the enzymes shown in Figure 15.1.

To follow up on this result, the biologists grew the mutants on normal media that lacked arginine and was supplemented with nothing, or ornithine, or citrulline, or arginine. In each case, the media were solid and had a slanted surface to provide

Metabolic pathway for arginine synthesis: Precursor → Ornithine (Enzyme 1) → Citrulline (Enzyme 2) → Arginine (Enzyme 3)

FIGURE 15.1 There Are Several Steps in the Metabolic Pathway for Arginine Synthesis. Most organisms synthesize the amino acid arginine via the series of steps shown here. Each step is catalyzed by a different enzyme.

● **QUESTION** If a cell lacked enzyme 2 but got ornithine in its diet, could it still grow? Could it still grow if it got citrulline instead?

Experiment

Question: What do genes do?

Hypothesis: Each gene contains the information required to make a different enzyme.

Null hypothesis: Genes have nothing to do with making enzymes.

Experimental strategy: Knock out specific genes; see if the enzymes required for different steps in the pathway for synthesizing arginine are missing.

Experimental setup: Isolate mutant *N. crassa* that cannot synthesize arginine. Grow each type of mutant on normal medium with:

Neurospora crassa
Medium

No supplements
(no ornithine, citrulline or arginine)

Supplemented with ornithine only
(no citrulline or arginine)

Supplemented with citrulline only
(no ornithine or arginine)

Supplemented with arginine only
(no ornithine or citrulline)

Prediction: There will be three distinctive types of mutants, corresponding to defects in enzyme 1, enzyme 2, and enzyme 3 in the pathway for synthesizing arginine. Each type of mutant will be able to grow on different combinations of the four types of media.

Prediction of null hypothesis: There will not be distinctive types of mutants.

Results: There are three distinctive types of mutants, called *arg 1*, *arg 2*, and *arg 3*.

| | | Supplement type | | | |
		None	Ornithine only	Citrulline only	Arginine only
Mutant type	*arg 1*	no growth	GROWTH	GROWTH	GROWTH
	arg 2	no growth	no growth	GROWTH	GROWTH
	arg 3	no growth	no growth	no growth	GROWTH

Interpretation:

Precursor → Ornithine → Citrulline → Arginine

arg 1 cells lack enzyme 1   *arg 2* cells lack enzyme 2   *arg 3* cells lack enzyme 3

Conclusion: The one-gene, one-enzyme hypothesis is supported.

FIGURE 15.2 Experimental Support for the One-Gene, One-Enzyme Hypothesis. The association between specific genetic defects in *N. crassa* and specific deficits in the metabolic pathway for arginine synthesis convinced biologists that the one-gene, one-enzyme hypothesis was correct.

● **QUESTION** Why did they bother to do the "No supplements" treatment?

adequate room for cells to grow. As **Figure 15.2** shows, the results from these growth experiments were dramatic. Some of the mutant cells were able to grow in the presence of some of these compounds but not in the presence of others. More specifically, the mutants fell into three distinct classes, which the researchers called *arg1*, *arg2*, and *arg3*. As the Interpretation section of the figure shows, the data make sense only if each type of mutant lacked a different, specific step in a metabolic pathway because of a defect in a particular gene. Srb and Horowitz had documented a correlation between a specific

genetic defect and a defect at a specific point in a metabolic pathway. Their results convinced most investigators that the one-gene, one-enzyme hypothesis was correct.

Follow-up work showed that genes are responsible for all of the different types of proteins produced by cells—not just enzymes. Biologists finally understood what most genes do: They contain the instructions for making proteins.

MB **Web Animation** at **www.masteringbio.com**

The One-Gene One-Enzyme Hypothesis

15.2 The Central Dogma of Molecular Biology

How does a gene specify the production of a protein? After Beadle and Tatum's hypothesis had been supported in *N. crassa* and a variety of other organisms, this question became a central one. Part of the answer lay in understanding the molecular nature of the gene. Biochemists knew that the primary components of DNA were four nitrogen-containing bases: the pyrimidines thymine (abbreviated T) and cytosine (C), and the purines adenine (A) and guanine (G). They also knew that these bases were connected in a linear sequence by a sugar-phosphate backbone. Watson and Crick's model for the secondary structure of the DNA molecule, introduced in Chapter 4 and reviewed in Chapter 14, revealed that two strands of DNA are wound into a double helix, held together by hydrogen bonds between the complementary base pairs A-T and G-C.

Given DNA's structure, it appeared extremely unlikely that DNA directly catalyzed the reactions that produce proteins. Its shape was too regular to suggest that it could bind a wide variety of substrate molecules and lower the activation energy for chemical reactions. Instead, Crick proposed that the sequence of bases in DNA might act as a code. The idea was that DNA was *only* an information-storage molecule. The instructions it contained would have to be read and then translated into proteins.

Crick offered an analogy with Morse code. Morse code is a message-transmission system using dots and dashes to represent the letters of the alphabet. Crick's idea was that different combinations of bases could specify the 20 amino acids, just as different combinations of dots and dashes specify the 26 letters of the alphabet. In this way, simple dots and dashes can code for the complexity of information in a human language.

A particular stretch of DNA, then, could contain the information needed to specify the amino acid sequence of a particular enzyme. In code form, the tremendous quantity of information needed to build and run a cell could be stored compactly. This information could also be copied through complementary base pairing and transmitted efficiently from one generation to the next.

It soon became apparent, however, that the information encoded in the base sequence of DNA is not translated into the amino acid sequence of proteins directly. Instead, the link between DNA as information repository and proteins as cellular machines is indirect.

RNA as the Intermediary between Genes and Proteins

The first clue that biological information does not flow directly from DNA to proteins came from data on the structure of cells. In eukaryotic cells, DNA is enclosed within a membrane-

FIGURE 15.3 The Messenger RNA Hypothesis. In the cells of plants, animals, fungi, and other eukaryotes, most DNA is found only in the nucleus, but proteins are manufactured outside the nucleus, at ribosomes. Biologists proposed that the information coded in DNA is carried from inside the nucleus out to the ribosomes by messenger RNA (mRNA).

● **QUESTION** Bacteria and archaea do not have a nucleus. Based on the logic presented here, would they need to make an mRNA? Explain why or why not.

bound organelle called the nucleus (see Chapter 7). But the structures called ribosomes, where protein synthesis takes place, are outside the nucleus, in the cytoplasm. This observation began to make sense after François Jacob and Jacques Monod suggested that RNA molecules act as a link between genes and the protein-manufacturing centers. Jacob and Monod's hypothesis is illustrated in **Figure 15.3**. They predicted that short-lived molecules of RNA called **messenger RNA**, or **mRNA**, carry information from DNA to the site of protein synthesis. Messenger RNA is one of several distinct types of RNA in cells.

Follow-up research confirmed that the messenger RNA hypothesis is correct. One particularly important piece of evidence was the discovery of an enzyme that catalyzes the synthesis of RNA. This protein is called **RNA polymerase**, because it polymerizes ribonucleotides into strands of RNA. The key observation was that RNA polymerase synthesizes RNA molecules according to the information provided by the sequence of bases in a particular stretch of DNA. Unlike DNA polymerase, RNA polymerase does not require a primer to begin adding ribonucleotides to a growing strand of RNA.

Figure 15.4 provides experimental evidence on the transfer of information from DNA to mRNA. The experiment high-

Experiment

Question: How is information transferred from DNA to RNA?

Hypothesis: RNA is synthesized by complementary base pairing with DNA.

Null hypothesis: RNA is not synthesized by complementary base pairing with DNA.

Experimental setup: Create a reaction mix containing three elements:

Ribonucleotides   RNA polymerase

DNA strand with thymine as the only base

Prediction: An RNA strand will be produced containing only adenine (A).

Prediction of null hypothesis: No RNA strand will be produced, or an RNA strand with a random assortment of bases will be produced.

Results:

RNA strand with adenine as the only base

Conclusion: Information is transferred from DNA to RNA via complementary base-pairing.

FIGURE 15.4 Complementary Base Pairing Transfers Information from DNA to RNA. This experiment supported the hypothesis that RNA molecules are synthesized with a base sequence that is complementary to the corresponding DNA sequence.

lighted here began with a reaction mix containing three critical elements: (1) the enzyme RNA polymerase; (2) ribonucleotides containing the bases adenine (A), uracil (U), guanine (G), and cytosine (C); and (3) copies of a strand of synthetic DNA that contained deoxyribonucleotides in which the only base was thymine (T). (Recall from Chapter 4 that RNA contains the base uracil instead of thymine. If an RNA molecule binds to a DNA or RNA molecule, uracil forms complementary base pairs with adenine.) When researchers allowed the polymerization reaction to proceed, the only base in the RNA molecules that resulted was adenine. This result strongly supported the hypothesis that RNA polymerase synthesizes RNA according to the rules of complementary base pairing introduced in Chapter 4, because thymine pairs with adenine. In

this case, a sequence of DNA containing thymine as the only base directed the polymerization of a complementary RNA sequence containing adenine as the only base. Similar experiments showed that synthetic DNAs containing no bases other than cytosine result in the production of RNA molecules containing no bases other than guanine.

The Central Dogma

Once the mRNA hypothesis was accepted, Francis Crick articulated what became known as the central dogma of molecular biology. The **central dogma** summarizes the flow of information in cells. It simply states that DNA codes for RNA, which codes for proteins:

$$DNA \rightarrow RNA \rightarrow proteins$$

Crick's simple statement encapsulates much of the research reviewed in this chapter. ● DNA is the hereditary material. Genes consist of specific stretches of DNA that code for products used in the cell. The sequence of bases in DNA specifies the sequence of bases in an RNA molecule, which specifies the sequence of amino acids in a protein. In this way, genes ultimately code for proteins.

Many proteins function as enzymes that catalyze chemical reactions in the cell. Other proteins perform the types of roles introduced in earlier chapters: Motor proteins and contractile proteins move the cell itself or cellular cargo, structural proteins provide support for the cell or tracks for transporting cargo, peptide hormones carry signals from cell to cell, membrane transport proteins conduct specific ions or molecules across the plasma membrane, and antibodies and other immune system proteins provide defense by recognizing and destroying invading viruses and bacteria.

Biologists use specialized vocabulary to summarize the sequence of events encapsulated in the central dogma. For example, biologists say that DNA is transcribed to RNA. In everyday English, the word *transcription* simply means making a copy of information. The scientific use of the term is appropriate because DNA acts as a permanent record—an archive or blueprint containing the information needed to build and run the cell. This permanent record is copied, via **transcription**, to the short-lived form called mRNA. The information is then transferred to a new molecular form—a sequence of amino acids. In everyday English, the word *translation* refers to the transferring of information from one language to another. In biology, the synthesis of protein from mRNA is called **translation**. Translation is the transferring of information from one type of molecule to another—from the "language" of nucleic acids to the "language" of proteins.

Translation is also often referred to simply as protein synthesis. **Figure 15.5a** summarizes the relationship between transcription and translation, as well as the relationship among

(a) Information flows from DNA to RNA to proteins.

(b) DNA sequences define the genotype; proteins create the phenotype.

(c) Changes in the genotype may lead to changes in the phenotype.

FIGURE 15.5 The Flow of Information in the Cell. (a) DNA is the information archive in the cell. mRNA is a short-lived copy of the information in DNA, used to manufacture proteins. Most of the machinery and many of the structures in cells consist of proteins. **(b)** The central dogma revealed the connection between genotype and phenotype. **(c)** Alleles of genes differ in their DNA base sequence. As a result, they may produce proteins with different primary structures and thus different functions. In this way, different alleles may be associated with different phenotypes.

DNA, RNA, and proteins. The following diagram presents these terms and relationships in a condensed form:

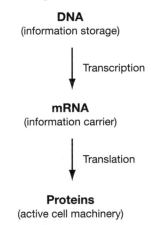

DNA
(information storage)

Transcription

mRNA
(information carrier)

Translation

Proteins
(active cell machinery)

According to the central dogma, an organism's genotype is determined by the sequence of bases in its DNA while its phenotype is a product of the proteins it produces (**Figure 15.5b**). To drive this point home, consider that the enzymes and other proteins encoded by genes are what make the "stuff" of the cell and dictate which chemical reactions occur inside. In garden peas, for example, the "gene for height" that Mendel studied actually codes for an enzyme involved in synthesizing a growth hormone. An important aspect of a pea plant's phenotype—its height—is determined in part by the DNA sequence at the gene for this hormone-producing enzyme.

Later work revealed that alleles of the same gene differ in their DNA sequence. As a result, the proteins produced by different alleles of the same gene frequently differ in their amino acid sequence (**Figure 15.5c**). If the primary structures of proteins vary, their functions are likely to vary as well. The dwarfed pea plants that Mendel studied, for example, have a DNA sequence that is different from the normal allele. The altered sequence is dysfunctional, meaning that the mutants have a knock-out allele at the gene for height. These individuals do not produce the enzyme required to synthesize growth hormone. As a result, their cells do not receive the message to continue growth and the plants are short. (Dwarfing varieties like these are actually preferred in many types of crop plants and fruit trees.) The variation in phenotypes that we observe depends in part on variations in DNA sequences.

The central dogma provided an important conceptual framework for the burgeoning field called molecular genetics, and it inspired a series of fundamental questions about how genes and cells work. But important modifications to the central dogma have occurred in the decades since Frances Crick first proposed it.

• Many genes code for RNA molecules that do not function as mRNAs and are not translated into proteins. Stated another way, there are many types of RNA in addition to mRNA that perform important functions in cells. Besides carrying information from DNA so translation can occur, there are RNAs that (1) help regulate which genes are transcribed at various times, (2) process mRNAs prior to translation, (3) transport the amino acids required to build proteins, and (4) catalyze the formation of peptide bonds during translation. For these genes, information flow would be diagrammed as simply DNA → RNA.

• The central dogma suggests that information flow occurs in only one direction—from DNA to RNA to proteins. RNA does not code for the production of other RNA or DNA, and proteins do not code for the production of RNA or DNA or other proteins. But there are important exceptions to these rules. In some cases, information flows from RNA back to DNA.

Check Your Understanding

If you understand that...

- Genes code for proteins, but they do so indirectly.
- The sequence of bases in DNA is used to produce messenger RNA (mRNA). The sequence of bases in an mRNA molecule is complementary to the DNA sequence of a gene.

You should be able to...

1) Draw a DNA double helix, writing out a sequence of bases on both strands and indicating the 5′ → 3′ polarity of each strand.

2) Choose one of the two strands of the double helixes as the template for RNA synthesis and write the sequence of the mRNA molecule that would be synthesized (in the 5′ → 3′ direction!) from that strand.

15.3 The Genetic Code

Once biologists understood the basic nature of information flow in the cell, the next challenge was to understand the final link between DNA and proteins. Exactly how does the sequence of bases in a strand of mRNA code for the sequence of amino acids in a protein? If this question could be answered, biologists would have cracked the **genetic code**—the rules that

specify the relationship between a sequence of nucleotides in DNA or RNA and the sequence of amino acids in a protein. Researchers from all over the world took up the challenge. A race was on.

How Long Is a Word in the Genetic Code?

The first step in cracking the genetic code was to determine how many bases make up a "word." In a sequence of mRNA, how long is a message that specifies one amino acid? Based on some simple logic, George Gamow suggested that each code word contains three bases. His reasoning derived from the observation that there are 20 amino acids commonly used in cells, and the hypothesis that each amino acid must be specified by a particular sequence of mRNA. Because there are only four different bases in ribonucleotides (A, U, G, and C), a one-base code could specify only four amino acids. Similarly, a two-base code could represent just 4 × 4, or 16, amino acids. ◯ A three-base code, though, could specify 4 × 4 × 4, or 64, different amino acids. As a result, a three-base code would provide more than enough messages to code for all 20 amino acids (**Figure 15.6**). This three-base code is known as a **triplet code**.

These simple calculations highlight an important point: Gamow's hypothesis predicted that the genetic code is redundant. That is, more than one triplet of bases might specify the same amino acid. The idea was that different three-base sequences in an mRNA—say, AAA and AAG—might code for the same amino acid—say, lysine. The group of three bases that specifies a particular amino acid is called a **codon**. According to the triplet code hypothesis, many of the 64

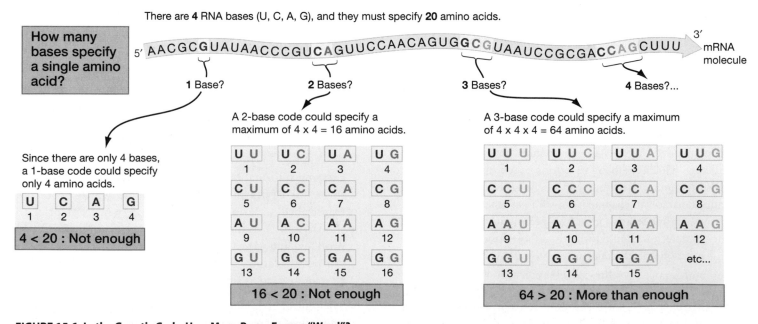

FIGURE 15.6 In the Genetic Code, How Many Bases Form a "Word"?

◯ **QUESTION** How many amino acids could be specified by a four-base code? Why did biologists conclude that a four-base code was extremely unlikely? Your answer should include an explanation for why a four-base code is unlikely to evolve.

Experiment

Question: In the genetic code, how many bases act as a codon and specify an amino acid?

Hypothesis: A codon is 3-bases long.

Alternative hypothesis: A codon is 1- or 2-bases long.

Experimental setup:

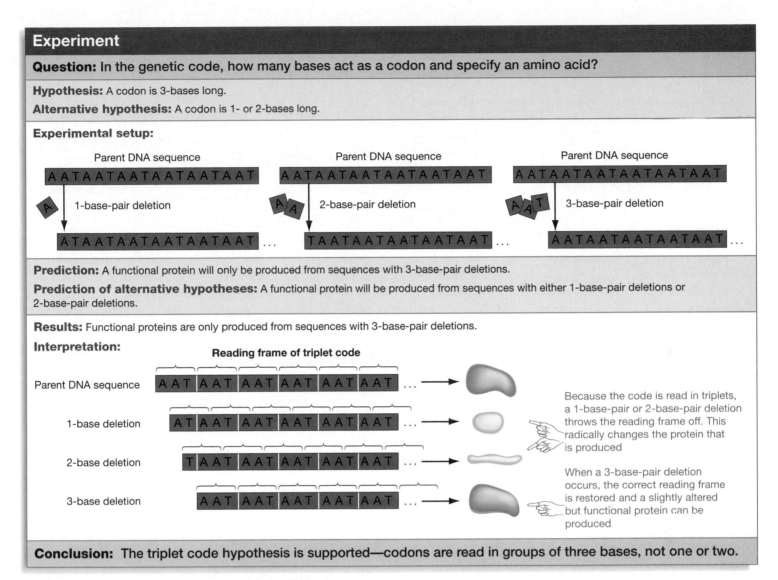

Prediction: A functional protein will only be produced from sequences with 3-base-pair deletions.

Prediction of alternative hypotheses: A functional protein will be produced from sequences with either 1-base-pair deletions or 2-base-pair deletions.

Results: Functional proteins are only produced from sequences with 3-base-pair deletions.

Interpretation:

Because the code is read in triplets, a 1-base-pair or 2-base-pair deletion throws the reading frame off. This radically changes the protein that is produced

When a 3-base-pair deletion occurs, the correct reading frame is restored and a slightly altered but functional protein can be produced

Conclusion: The triplet code hypothesis is supported—codons are read in groups of three bases, not one or two.

FIGURE 15.7 The Experiments That Confirmed the Triplet Code.

● **EXERCISE** Write a sequence of bases in a strand of DNA. Make vertical marks between every third base to separate codons and indicate the reading frame. At random, select a base in the sequence and delete it. Write the new sequence and indicate its reading frame. Circle the codons that have changed after the deletion. Do the same exercise when you delete two randomly chosen bases in the sequence, and when you delete three. Does the reading frame change when 3 bases are deleted?

codons that are possible actually specify the same amino acids.

Work by Francis Crick and Sydney Brenner confirmed that codons are three bases long. Their experiments used chemicals that caused an occasional addition or deletion of a base in DNA. As predicted for a triplet code, a one-base addition or deletion in the base sequence led to a loss of function in the gene being studied. This is because a single addition or deletion mutation throws the sequence of codons, or the **reading frame**, out of register. To understand how this works, consider the sentence "The fat cat ate the rat." The reading frame of this sentence is a three-letter word and a space. If the fourth letter in this sentence—the *f* in *fat*—were deleted but the reading frame stayed intact, the sentence would be transformed into "The atc ata tet her at." This is gibberish. When the reading frame in a DNA sequence is

thrown out of register by the addition or deletion of a base, the composition of each codon changes just like the letters in each word of the example sentence. The protein produced from the altered DNA sequence has a completely different sequence of amino acids. In terms of its normal function, this protein is gibberish.

Figure 15.7 summarizes some of the experiments showing that codons are three bases long. In the Experimental setup portion of the figure, notice that the parent DNA sequence (top line) shows the reading frame for a series of AAT codons. In the sequences just below this, you can see what happened when Crick and co-workers used chemicals to remove one, two, or three bases from the parent sequence. Now look at the Results section of the figure to see how each type of mutation affected the protein. The only time functional proteins were produced was when three bases were removed. In the sentence "The fat cat ate the rat," the combination of removing one letter from each of the first three words might result in "Tha tca ate the rat." Just as the altered sentence still conveys some meaning, genes with three deletion mutations were able to produce a functional protein.

The researchers interpreted these results as strong evidence in favor of the triplet code hypothesis. Most other biologists agreed. The results also launched a long, laborious, and ultimately successful effort to determine which amino acid is specified by each of the 64 codons.

How Did Researchers Crack the Code?

The initial advance in deciphering the genetic code came in 1961, when Marshall Nirenberg and Heinrich Matthaei created a method for synthesizing RNAs of known sequence. Their method was based on an enzyme called polynucleotide phosphorylase, which catalyzes the formation of phosphodiester bonds between any ribonucleotides available, in random order. By providing ribonucleotides whose only base was uracil (U) to a reaction mix containing this enzyme, Nirenberg and Matthaei were able to create a long polymer of uracil-containing ribonucleotides. These synthetic RNAs were added to an in vitro system for synthesizing proteins. The researchers analyzed the resulting amino acid chain and determined that it was polyphenylalanine—a polymer consisting of the amino acid phenylalanine.

This result could mean only one thing: The RNA triplet UUU codes for the amino acid phenylalanine. By complementary base pairing, it was clear that the corresponding DNA sequence would be AAA. This initial observation was followed by experiments with RNAs consisting of only A or C. RNAs consisting of only AAAAA.... produced polypeptides consisting of only lysine; poly-C RNAs (RNAs consisting of only CCCCC...) produced polypeptides composed entirely of proline.

Nirenberg and Philip Leder later devised a system for synthesizing specific codons. Once they had copies of specific codons, they performed a series of experiments in which they added each of the codons to a cell extract that included the 20 different amino acids, ribosomes, and other molecules required for protein synthesis. As Chapter 7 noted, ribosomes are the multimolecular machines where proteins are synthesized. Then the researchers determined which amino acid was bound to the ribosomes when a particular codon was present. For example, when the codon CAC was in the reaction mix, the amino acid histidine would bind to the ribosomes. This result confirmed that CAC codes for histidine. These ribosome-binding experiments allowed Nirenberg and Leder to determine which of the 64 codons coded for each of the 20 amino acids (**Figure 15.8**).

In addition to matching codons to amino acids, researchers discovered that certain codons are punctuation marks signaling "start of message" or "end of message." These codons relay information that the protein chain is complete or that

SECOND BASE

	U	C	A	G	
U	UUU UUC ⎤– Phenylalanine (Phe) UUA UUG ⎤– Leucine (Leu)	UCU UCC UCA UCG ⎤– Serine (Ser)	UAU UAC ⎤– Tyrosine (Tyr) UAA – **Stop codon** UAG – **Stop codon**	UGU UGC ⎤– Cysteine (Cys) UGA – **Stop codon** UGG – Tryptophan (Trp)	U C A G
C	CUU CUC CUA CUG ⎤– Leucine (Leu)	CCU CCC CCA CCG ⎤– Proline (Pro)	CAU CAC ⎤– Histidine (His) CAA CAG ⎤– Glutamine (Glu)	CGU CGC CGA CGG ⎤– Arginine (Arg)	U C A G
A	AUU AUC AUA ⎤– Isoleucine (Ile) AUG – Methionine (Met) **Start codon**	ACU ACC ACA ACG ⎤– Threonine (Thr)	AAU AAC ⎤– Asparagine (Asn) AAA AAG ⎤– Lysine (Lys)	AGU AGC ⎤– Serine (Ser) AGA AGG ⎤– Arginine (Arg)	U C A G
G	GUU GUC GUA GUG ⎤– Valine (Val)	GCU GCC GCA GCG ⎤– Alanine (Ala)	GAU GAC ⎤– Aspartic acid (Asp) GAA GAG ⎤– Glutamic acid (Glu)	GGU GGC GGA GGG ⎤– Glycine (Gly)	U C A G

(Left side label: **FIRST BASE**; Right side label: **THIRD BASE**)

FIGURE 15.8 The Genetic Code. To read the code, match the first base in an mRNA codon, in the red band on the left side, with the second base in a codon, in the blue band along the top, and the third base in a codon, in the green band on the right side. The 64 codons, along with the amino acid or stop (termination) signal that they specify, are given in the boxes. By convention, codons are always written in the 5′ → 3′ direction.

● **EXERCISE** Pick four codons at random. Next to each of them, write the three bases in DNA that specify the codon. (Write this DNA sequence in the 3′ → 5′ direction.)

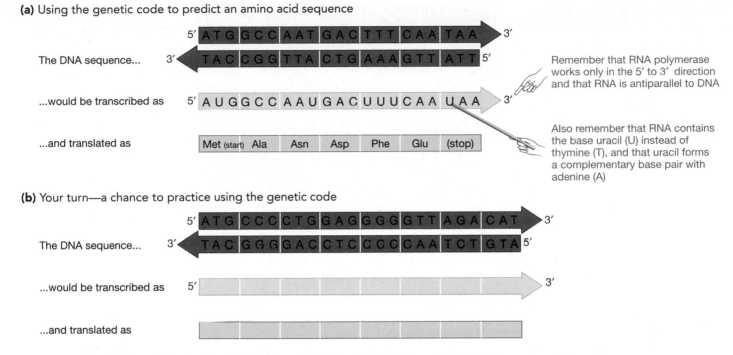

(a) Using the genetic code to predict an amino acid sequence

The DNA sequence...

5′ ATG GCC AAT GAC TTT CAA TAA 3′
3′ TAC CGG TTA CTG AAA GTT ATT 5′

Remember that RNA polymerase works only in the 5′ to 3′ direction and that RNA is antiparallel to DNA

...would be transcribed as   5′ AUG GCC AAU GAC UUU CAA UAA 3′

Also remember that RNA contains the base uracil (U) instead of thymine (T), and that uracil forms a complementary base pair with adenine (A)

...and translated as   Met (start) | Ala | Asn | Asp | Phe | Glu | (stop)

(b) Your turn—a chance to practice using the genetic code

The DNA sequence...

5′ ATG CCC CTG GAG GGG GTT AGA CAT 3′
3′ TAC GGG GAC CTC CCC CAA TCT GTA 5′

...would be transcribed as   5′ _____ 3′

...and translated as   _____

FIGURE 15.9 Using the Genetic Code to Predict Sequences. Biologists often have to predict RNA, DNA, or amino acid sequences from the types of data given here.

● **EXERCISE** Write an mRNA that codes for the amino acid sequence Ala-Asn-Asp-Phe-Glu that is different from the one given here. Indicate the mRNA's 5′ → 3′ polarity. Then write the double-stranded DNA that corresponds to this mRNA. Indicate the 5′ → 3′ polarity of both strands.

● **EXERCISE** Write the RNA sequence that is transcribed from the top DNA strand in part (a). Then write the amino acid sequence translated from that RNA. Do the same for the top strand in part (b).

protein synthesis should start at a given codon. There is one **start codon** (AUG), which codes for the amino acid methionine, and which also signals that protein synthesis should start at that point on the mRNA molecule. There are three **stop codons** (UAA, UAG, and UGA). The stop codons are also known as termination codons, because they signal that the protein is complete and end the translation process.

Using the genetic code, biologists are able to predict the amino acid sequences encoded by a particular DNA sequence (**Figure 15.9a**) and predict the mRNA sequence implied by a particular sequence of amino acids. ● If you understand how to read the genetic code, you should be able to identify the codons in Figure 15.5 and check to make sure that they are translated correctly. You should also be able to complete the exercises in **Figure 15.9b.**

The deciphering of the full genetic code, presented in Figure 15.8, is a tremendous achievement. It represents more than five years of work by several teams of researchers. As predicted, the genetic code is redundant. All amino acids except methionine and tryptophan are coded by more than one codon. The code is also unambiguous, however, because one codon never codes for more than one amino acid. Later work showed that the genetic code is also nearly universal: With a few minor exceptions, all codons specify the same amino acids in all organisms. **Box 15.1** explores the evolutionary implications of the universal code.

Check Your Understanding

If you understand that...

- The sequence of bases in mRNA forms a code in which particular combinations of three bases lead to the addition of a specific amino acid to the protein encoded by the gene.

- The genetic code is redundant because 64 combinations of bases occur, but only 20 amino acids and a stop "punctuation mark"—indicating the end of the coding sequence—need to be specified.

You should be able to...

1) Explain the consequences of mutations that occur in DNA and that change the codon ATA to one of the following codons: GTA, TTA, or GCA.

2) In each case, describe the effect on the resulting protein.

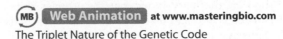

MB **Web Animation** at www.masteringbio.com
The Triplet Nature of the Genetic Code

BOX 15.1 The Evolution of the Genetic Code

Figure 15.10 is a simplified version of the tree of life that was introduced in Chapter 1. Recall that the three largest branches on this tree represent what biologists call the three domains of life: the Bacteria, the Archaea, and the Eukarya. Smaller branches on the tree represent groups of species within each of the three domains, such as the cyanobacteria, α-proteobacteria, land plants, fungi, and animals.

Virtually all organisms on this tree use the genetic code in Figure 15.8. Exceptions are few. For example, in the single-celled eukaryotes *Tetrahymena* and *Paramecium*, UAA and UAG code for glutamine instead of stop; and in the yeast *Candida cylindracea*, CUG codes for serine instead of leucine. In a few species other codons in addition to AUG initiate protein synthesis.

To explain why the genetic code is nearly universal, biologists hypothesize that the common ancestor of all species living today used this same code. This common ancestor is indicated at the base of the tree in Figure 15.10. The logic here is that if all species living today use the same genetic code, then it is reasonable to infer that their common ancestor did also. (For help reading trees, see **BioSkills 2**.)

Given that the genetic code has been in existence for a long time, the question arises, Is the code arbitrary? That is, is the genetic code random with respect to the ability of organisms to survive and produce offspring? A brief examination of Figure 15.8 suggests that the answer is no. Note that when several codons specify the same amino acid, the first two bases in those codons are almost always identical. This observation suggests that the code does not represent a random assemblage of bases, like letters drawn from a hat. In fact, extensive analysis has shown that, compared with randomly generated codes, the existing genetic code is structured in a way that efficiently minimizes the phenotypic effects of small changes in DNA and errors during translation. This result suggests that the genetic code has been honed by natural selection. Stated another way, it is likely that other genetic codes may have existed early in the history of life. But because the code in Figure 15.8 is particularly efficient, it emerged as the most successful version.

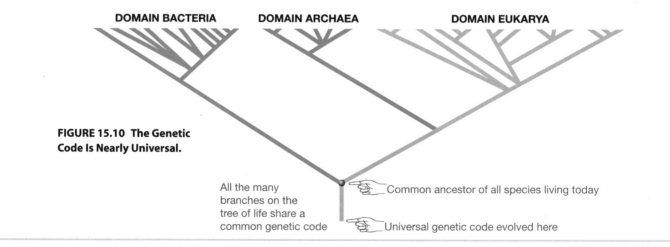

DOMAIN BACTERIA   DOMAIN ARCHAEA   DOMAIN EUKARYA

FIGURE 15.10 The Genetic Code Is Nearly Universal.

All the many branches on the tree of life share a common genetic code

Common ancestor of all species living today

Universal genetic code evolved here

Chapter Review

SUMMARY OF KEY CONCEPTS

● **Most genes code for proteins.**

The question "What do genes do?" was solved in a series of experiments on the bread mold *Neurospora crassa*. Researchers isolated mutants that cannot grow unless they are supplied with a specific amino acid and then showed that different mutants were unable to make different chemical precursors of the amino acid. The one-gene, one-enzyme hypothesis explains these results by proposing that most genes code for proteins and that most proteins act as enzymes that catalyze specific reactions in metabolic pathways.

You should be able to explain the molecular basis of an organism's genotype and phenotype—including why alleles of the same gene exist and why different alleles produce different phenotypes. ●

(MB) **Web Animation** at www.masteringbio.com
The One-Gene, One-Enzyme Hypothesis

● **In cells, information flows from DNA to RNA to proteins. DNA is transcribed to messenger RNA by RNA polymerase, and then messenger RNA is translated to proteins by ribosomes.**

Experiments confirmed that DNA does not code for proteins directly. Instead, mRNA molecules are transcribed from DNA and then translated into proteins. One-way flow of information, from DNA to RNA to proteins, is called the central dogma of molecular biology.

You should be able to explain why the structure of DNA supports its function as an information archive, why the structure of RNA supports its array of functions—including information carrier—and why the structure of proteins supports their function as active cell machinery.

● **Each amino acid in a protein is specified by a group of three bases in RNA.**

By synthesizing RNAs of known base composition and then observing how they functioned during the process of translation, researchers were able to unravel the genetic code. It is now established that the code is read in triplets and that the code is redundant—meaning that most of the 20 amino acids are specified by more than one codon.

You should be able to explain why some changes in DNA sequences do not change the corresponding protein—meaning, why certain changes in an organism's genotype do not change its phenotype. ●

 Web Animation at www.masteringbio.com
The Triplet Nature of the Genetic Code

QUESTIONS

● Test Your Knowledge

1. What does the one-gene, one-enzyme hypothesis state?
   a. Genes are composed of stretches of DNA.
   b. Genes are made of protein.
   c. Genes code for ribozymes.
   d. A single gene codes for a single protein.

2. Which of the following is not an important exception to the one-gene, one-enzyme hypothesis?
   a. Many genes code for messenger RNAs.
   b. Many proteins are not enzymes.
   c. Many genes code for RNAs that function directly in the cell.
   d. Many enzymes are composed of multiple polypeptides, encoded by different genes.

3. DNA's primary structure is made up of just four different bases, and its secondary structure is regular and highly stable. How can a molecule with these characteristics hold all of the information required to build and maintain a cell?
   a. The information is first transcribed, then translated.
   b. The messenger RNA produced from DNA has much more complex secondary structures, and thus holds much more information.
   c. A protein produced (indirectly) from DNA has much more complex primary and secondary structures, and thus holds much more information.
   d. The information in DNA is in code form.

4. Why did researchers suspect that DNA does not code for proteins directly?
   a. In eukaryotes, DNA is found inside the nucleus, but proteins are produced outside the nucleus.
   b. In prokaryotes, DNA and proteins are never found together.

   c. When DNA was damaged by ultraviolet radiation or other sources of energy, the proteins in the cell did not change accordingly.
   d. There are several distinct types of RNA, of which only one functions as messenger RNA.

5. Which of the following describes an important experimental strategy in deciphering the genetic code?
   a. comparing the amino acid sequences of proteins with the base sequence of their genes
   b. analyzing the sequence of RNAs produced from known genes
   c. analyzing mutants that changed the code
   d. examining the proteins produced when RNAs of known sequence were translated

6. Knock-out mutations often produce recessive alleles. When a normal allele and a knock-out allele are paired in a heterozygous individual, the normal copy of the gene is frequently able to produce enough functional protein to give a normal phenotype. Why are these facts important?
   a. They explain why some mutants are still able to produce functional proteins.
   b. They explain the phenomenon of dominance and recessiveness discovered by Mendel.
   c. They offer a serious challenge to the one-gene, one-enzyme hypothesis.
   d. They illustrate why the central dogma is considered central.

Test Your Knowledge answers: 1. d; 2. a; 3. d; 4. d; 5. a; 6. b

● Test Your Understanding

Answers are available at www.masteringbio.com

1. DNA is referred to as an "information-storage molecule." How can the base sequence of DNA store information?

2. Draw a hypothetical metabolic pathway involving a sequence of five substrates, five enzymes, and a product called BiolSciazine. Number the substrates 1–5, and label the enzymes A–E, in order. (For instance, enzyme A applies to the reaction between substrates 1 and 2.)
   • Suppose a mutation made the gene for enzyme C nonfunctional. What molecule would accumulate in the affected cells?
   • Suppose some individuals with a mutation affecting this metabolic pathway can survive if given substrate 5 in the diet. But they die even if given substrates 1, 2, 3, and 4. State a hypothe-

sis for which enzyme in the pathway is affected by this mutation.

3. Why did experiments with *Neurospora crassa* mutants support the one-gene, one-enzyme hypothesis?

4. When researchers discovered that a combination of three deletion mutations or three addition mutations would restore the function of a gene, most biologists were convinced that the genetic code was read in triplets. Explain the logic behind this conclusion.

5. Why is the genetic code redundant?

6. Explain how a single-base deletion disrupts the reading frame of a gene.

● Applying Concepts to New Situations

Answers are available at www.masteringbio.com

1. Recall that DNA and RNA are synthesized only in the 5′ → 3′ direction and that DNA and RNA sequences are always written in the 5′ → 3′ direction. Consider the following DNA sequence:

5′ T T G A A A T G C C C G T T T G G A G A T C G G G G T T A C A G C T A G T C A A A G 3′

3′ A A C T T T A C G G G C A A A C C A C T A G C C C A A T G T C G A T C A G T T T C 5′

   - Identify the bases in the bottom strand that code for start and stop codons. Write the mRNA sequence that would be transcribed between them if the bottom strand served as the template.

   - Write the amino acid sequence that would be translated from the mRNA sequence you just wrote.

2. What problems would arise if the genetic code contained only 22 codons—one for each amino acid, a start signal, and a stop signal? (Hint: When DNA is copied prior to mitosis or meiosis, random errors occur that change its primary base sequence.)

3. Scientists say that a phenomenon is a "black box" if they can describe it and study its effects but don't yet know the underlying mechanism that causes it. In what sense was genetics—meaning the transmission of heritable traits—a black box before the central dogma of molecular biology was understood?

4. One of the possibilities that researchers interested in the genetic code had to consider was that the code was overlapping, meaning that a single base could be part of more than one codon. Make a diagram showing how an overlapping code would work, assuming that each codon is three bases long.

www.masteringbio.com is also your resource for • Answers to text, table, and figure caption questions and exercises • Answers to *Check Your Understanding* boxes • Online study guides and quizzes • Additional study tools including the *E-Book for Biological Science* 3rd ed., textbook art, animations, and videos.

Transcription and Translation

16

A section of a chromosome inside a frog egg, where extensive transcription is taking place. Proteins involved in transcription are stained a fluorescent green.

KEY CONCEPTS

- After RNA polymerase binds DNA with the help of other proteins, it catalyzes the production of an RNA molecule whose base sequence is complementary to the base sequence of the DNA template strand.

- Some sections of an RNA are encoded by gene regions called exons, while others are encoded by gene regions called introns. During RNA processing, introns are removed and the ends of the RNA receive a cap and tail.

- Inside ribosomes, mRNAs are translated to proteins via intermediary molecules called transfer RNAs. Transfer RNAs carry an amino acid and have a three-base-pair anticodon, which binds to a three-base-long mRNA codon. The amino acid carried by the transfer RNA is then added to the growing protein via formation of a peptide bond.

- Mutations are random changes in DNA that may or may not produce changes in the phenotype.

Proteins are the stuff of life. They give cells their shape, control the chemical reactions that go on inside them, and regulate the import, export, and movement of materials. No one knows exactly how many different proteins can be made in your body's cells, but 100,000 is a reasonable guess. While some of those 100,000 different proteins may not be manufactured at all in certain types of cells, other proteins may be present in quantities ranging from millions of copies to fewer than a dozen. If even one of those types of protein is defective, the cell may sicken or even die.

A cell builds the proteins it needs from instructions encoded in its genome. Chapter 15 explored how the information archived in DNA is retrieved and converted into action in the form of proteins. The first step in the process is the transcription of a gene and the production of a messenger RNA, or

mRNA. The RNA message is a short-lived copy of the archived instructions in DNA. The sequence of ribonucleotides in the mRNA is then translated into a sequence of amino acids. This series of events defines the flow of information in the cell:

DNA
(information storage)

↓ Transcription

mRNA
(information carrier)

↓ Translation

Proteins
(product)

Key Concept   Important Information   ● Practice It

The discovery of the relationship between DNA and proteins was one of the great scientific advances of the twentieth century. But once the relationship between transcription and translation had been firmly established, biologists puzzled over how cells actually accomplish the feat of converting genetic information into protein. How do transcription and translation work? For example, how does RNA polymerase know where to start transcribing a gene and where to end? Once an RNA message is produced, how is the linear sequence of ribonucleotides translated into the linear sequence of amino acids in a protein?

This chapter explores the experiments that answered these questions. Besides providing a deeper understanding of how cells work, the results summarized here have had a significant impact on biomedicine and drug development. Many of the antibiotics prescribed today work by disrupting translation in bacteria that cause disease. In addition, research on transcription and translation set the stage for an explosion of work, introduced in Chapters 17 and 18, on how the expression of specific genes is regulated. Most genes are transcribed and translated only at certain times or in certain types of cells, and in carefully regulated amounts.

Let's begin by examining research on the mechanisms of transcription in bacteria and eukaryotes, and then delve into the mechanics of translation. The chapter concludes by examining how changes in DNA can lead to changes in protein structure and function.

16.1 Transcription in Bacteria

The first step in converting genetic information into proteins is the synthesis of a messenger RNA version of the instructions archived in DNA. Enzymes called RNA polymerases, introduced in Chapters 14 and 15, are responsible for synthesizing mRNA. Like the DNA polymerases introduced in Chapter 14, an RNA polymerase performs a template-directed synthesis in the $5' \rightarrow 3'$ direction. (Recall that $5'$ and $3'$ refer to carbons on the sugar subunits of ribonucleotides.) But unlike DNA polymerases, RNA polymerases do not require a primer to begin transcription.

Transcription occurs when RNA polymerase matches the base in a ribonucleotide triphosphate with the complementary base in a gene—a section of DNA that codes for a protein or RNA. Once a matching ribonucleotide is in place, RNA polymerase catalyzes the formation of a phosphodiester bond between the $3'$ end of the growing mRNA chain and the new ribonucleotide. As this matching-and-catalysis process continues, an RNA that is complementary to the gene is synthesized (**Figure 16.1**). Notice that only one of the two DNA strands is used as a template and transcribed, or "read," by RNA polymerase. The strand that is read by the enzyme is called the **template strand**; the other strand is called the **non-template strand**. The non-template strand is also known as the coding strand because its sequence matches the sequence of the RNA that is transcribed from the template strand and codes for a

FIGURE 16.1 Transcription Is the Synthesis of RNA from a DNA Template. The reaction catalyzed by RNA polymerase results in the formation of a phosphodiester bond between ribonucleotides. RNA polymerase produces an RNA strand whose sequence is complementary to the bases in the DNA template.

● **QUESTION** In which direction is RNA synthesized, $5' \rightarrow 3'$ or $3' \rightarrow 5'$? In which direction is the DNA template "read"?

polypeptide. A key difference is that RNA has uracil (U) rather than the thymine (T) found in the non-template or coding strand. An adenine (A) in the DNA template strand specifies a U in the complementary RNA strand.

Biologists reached these conclusions by studying RNA synthesis in cell-free, or in vitro, systems. Once the basic chemical reaction was understood, an entirely new set of questions arose:

- What does RNA polymerase look like?

- Are any other proteins or factors involved in transcription, or does RNA polymerase act alone?

- How does the enzyme know where to start transcription on the DNA template?

RNA Polymerase Structure and Function

To understand the structure of RNA polymerase, researchers employed the technique called X-ray crystallography introduced in **BioSkills 8**. This procedure allows biologists to obtain information about the three-dimensional structure of large, complex molecules.

The most recent models of bacterial RNA polymerase estimated from X-ray crystallographic data indicate that the enzyme is large and globular and has several prominent channels running through the interior. The enzyme's active site, where phosphodiester bonds form, is located where several of these channels intersect.

Because the structure of molecules correlates with their function, it is logical to predict that DNA fits into one of the enzyme's channels and that the two strands in the double helix separate inside the enzyme to expose a single-stranded template at the active site. But how do the enzyme and DNA come together in the first place?

Initiation: How Does Transcription Begin?

Soon after the discovery of RNA polymerase, researchers realized that the enzyme cannot initiate transcription on its own. Instead, a detachable protein subunit called sigma must bind to RNA polymerase before transcription can begin. RNA polymerase and sigma form what biologists call a **holoenzyme** ("whole enzyme"). A holoenzyme consists of a **core enzyme**, which contains the active site for catalysis, and other required proteins (**Figure 16.2**).

If RNA polymerase is the core enzyme of this holoenzyme, what does sigma do? When researchers mixed RNA polymerase, sigma, and DNA together, they found that the holoenzyme bound tightly to specific sections of DNA. These binding sites were named **promoters**, because they are sections of DNA where transcription begins. The discovery of promoters suggested that sigma's function is regulatory in nature. Sigma appeared to be responsible for guiding RNA polymerase to specific locations

FIGURE 16.2 The Three-Dimensional Structure of RNA Polymerase. Bacterial RNA polymerase is a globular enzyme. Sigma is a regulatory component, and the core enzyme contains the active site.

where transcription should begin. What is the nature of these specific locations? What do promoters look like, and what do they do?

David Pribnow offered an initial answer to these questions in the mid-1970s. When Pribnow analyzed the base sequence of promoters from a variety of bacteria and from viruses that infect bacteria, he found that the promoters were located on the non-template strand, were 40–50 base pairs long, and had a particular section that looked similar. This similar segment of DNA had a series of bases identical or similar to TATAAT. This six-base-pair sequence is now known as the −10 box, because it is centered about 10 bases from the point where RNA polymerase starts transcription. (DNA that is located in the direction RNA polymerase moves during transcription is said to be **downstream**; DNA in the opposite direction is said to be **upstream**. Thus, the −10 box is centered about 10 bases upstream from the transcription start site.) The place where transcription begins is called the +1 site. Soon after, researchers recognized that the sequence TTGACA occurred in these same promoters, centered about 35 bases upstream from the +1 site. This second key sequence is called the **−35 box**. Sequences that are inside the promoter but outside the −10 and −35 boxes vary widely among genes.

Follow-up work showed that transcription begins when sigma, as part of the holoenzyme complex, binds to the −35 and −10 boxes (**Figure 16.3**, step 1). This is a key point. Sigma, and not RNA polymerase, makes the initial contact with DNA that starts transcription in bacteria. This finding supports the hypothesis that sigma is a regulatory protein. Sigma tells RNA polymerase where and when to start synthesizing RNA.

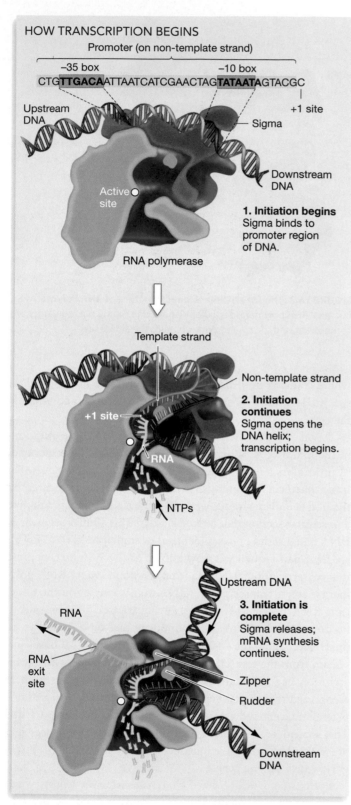

HOW TRANSCRIPTION BEGINS

Promoter (on non-template strand)

−35 box −10 box

CTG**TTGACA**ATTAATCATCGAACTAG**TATAATA**GTACGC

+1 site

Upstream DNA

Sigma

Active site

Downstream DNA

1. Initiation begins
Sigma binds to promoter region of DNA.

RNA polymerase

Template strand

Non-template strand

2. Initiation continues
Sigma opens the DNA helix; transcription begins.

+1 site

RNA

NTPs

Upstream DNA

3. Initiation is complete
Sigma releases; mRNA synthesis continues.

RNA

RNA exit site

Zipper

Rudder

Downstream DNA

FIGURE 16.3 In Bacteria, Sigma Plays a Key Role in Initiating Transcription. When sigma attaches to RNA polymerase, the holoenzyme is able to bind to the −35 box and the −10 box of a promoter. Once it has bound to a promoter, sigma opens the DNA helix and threads the template strand through the enzyme's active site. When transcription is under way, sigma disengages from RNA polymerase.

● **EXERCISE** Circle and label grooves in the enzyme that form exit sites for RNA and DNA.

Once sigma binds to a promoter, the DNA helix opens and creates two strands of single-stranded DNA. As shown in step 2 of Figure 16.3, the template strand is threaded through a channel that leads to the active site inside RNA polymerase. Monomers known as ribonucleoside triphosphates, or NTPs, enter a channel at the bottom of the enzyme and diffuse to the active site. NTPs are like the dNTPs introduced in Chapter 14's discussion of DNA synthesis.

When an incoming NTP pairs with a complementary base on the template strand of DNA, RNA polymerization begins. The reaction catalyzed by RNA polymerase is exergonic and spontaneous because NTPs have so much potential energy, owing to their three phosphate groups. As step 3 of Figure 16.3 shows, sigma is released once RNA synthesis is under way. The **initiation** phase of transcription is complete.

Recent work has shown that most bacteria have several types of sigma proteins, each with a distinctive structure and function. *Escherichia coli* has seven different sigma proteins, for example, while *Streptomyces coelicolor* has more than 60. Each of these proteins binds to promoters with slightly different DNA base sequences. Although all promoters have a −10 box and a −35 box, the sequences within each box vary somewhat. These variations in DNA sequences affect the ability of different sigma proteins to bind. As a result, each type of sigma protein allows RNA polymerase to bind to a different type of promoter and thus different kind of gene. The identity of the sigma protein in the RNA polymerase holoenzyme determines which types of genes will be transcribed. Controlling which sigma proteins are active is one of the ways that bacterial cells control which genes are expressed.

Elongation and Termination

During the **elongation** phase of transcription, RNA polymerase moves along the DNA template in the 3′ → 5′ direction of the template strand, synthesizing RNA in the 5′ → 3′ direction. In the interior of the enzyme, a group of projecting amino acids called the enzyme's zipper helps open the double helix at the upstream end, and a nearby group of amino acids called the rudder helps steer the template and non-template strands through channels inside the enzyme (see Figure 16.3, step 3). Meanwhile, the enzyme's active site catalyzes the addition of nucleotides to the 3′ end of the growing RNA molecule at the rate of about 50 nucleotides per second.

Note that during the elongation phase of transcription, all of the prominent channels or grooves in the enzyme are filled (Figure 16.3, step 3). Double-stranded DNA goes into and out of one groove; ribonucleoside triphosphates enter another; and the growing RNA strand exits to the rear. In this way, the enzyme's structure correlates closely with its function.

Transcription ends with a **termination** phase. In most cases, transcription stops when RNA polymerase reaches a stretch of DNA sequence that functions as a transcription termination signal. The bases that make up the termination signal code for

HOW TRANSCRIPTION ENDS

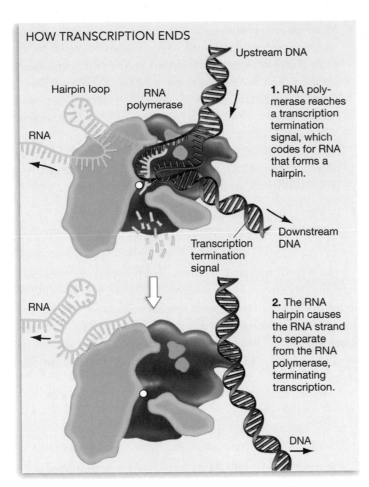

FIGURE 16.4 Transcription Terminates When a Hairpin Forms

1. RNA polymerase reaches a transcription termination signal, which codes for RNA that forms a hairpin.

2. The RNA hairpin causes the RNA strand to separate from the RNA polymerase, terminating transcription.

a stretch of RNA with an unusual property: As soon as it is synthesized, the RNA sequence folds back on itself and forms a short double helix that is held together by complementary base pairing. The secondary structure that results is called a hairpin (**Figure 16.4**). The formation of the hairpin structure is thought to disrupt the interaction between RNA polymerase and the RNA transcript, resulting in the physical separation of the enzyme and its product.

To summarize, transcription begins when sigma, as part of the holoenzyme complex, binds to the promoter at the start of a gene. Once binding occurs, RNA polymerase begins to synthesize mRNA by adding ribonucleotides that are complementary to the template strand in DNA. Transcription ends when a termination signal at the end of the gene leads to the formation of a hairpin in the mRNA, disrupting the transcription complex. If you understand these concepts, you should be able to (1) draw the RNA polymerase holoenzyme, (2) label sigma and the RNA polymerase core enzyme and describe their function, and (3) diagram the initiation, elongation, and termination phases of transcription in bacteria.

Web Animation at www.masteringbio.com

Transcription

16.2 Transcription and RNA Processing in Eukaryotes

The data reviewed in Section 16.1 established how transcription occurs in bacteria. How do these results compare with research on the mechanism of transcription in eukaryotes? What similarities and differences have come to light?

One of the first conclusions about transcription in eukaryotes is that it is similar to bacterial transcription in an important respect: RNA polymerase does not bind directly to promoter sequences by itself. Instead, proteins that came to be called **basal transcription factors** initiate eukaryotic transcription by matching the enzyme with the appropriate promoter region in DNA. In eukaryotes, the function of basal transcription factors is analogous to the function of the sigma proteins in bacteria, except that basal transcription factors interact with DNA independent of RNA polymerase.

The discovery of basal transcription factors emerged from studies with cell-free, or in vitro, systems. The basic approach was to purify RNA polymerase from human cells, add template DNA, and analyze the RNAs that were produced. But in early experiments, researchers found that RNA polymerase began copying at random locations on the template DNA instead of specifically at promoter regions. In addition, when RNA polymerase was alone, both strands of the DNA were transcribed instead of just one. This observation inspired the hypothesis that eukaryotic cells must contain sigma-like proteins that are required for normal transcription. To test this hypothesis, biologists added proteins isolated from human cells to the cell-free reaction system. By adding one or a few proteins at a time and recording which ones enabled RNA polymerase to bind to promoters and correctly transcribe the template strand, researchers were gradually able to characterize a suite of proteins required for proper initiation.

Follow-up work confirmed several important distinctions about how transcription works in bacteria and eukaryotes:

- In bacteria a single sigma protein binds to a promoter and initiates transcription, but in eukaryotes many basal transcription factors are required to initiate transcription. In eukaryotes, the machinery required to start transcription is complex.

- Eukaryotes have three distinct types of RNA polymerase (abbreviated RNA pol) instead of just one. As **Table 16.1** shows, RNA pol I, pol II, and pol III each transcribe a discrete class of RNA. Only RNA pol II transcribes the genes that code for proteins—meaning that it produces mRNA. RNA pol I makes the large RNA molecules that are found in ribosomes. RNA pol III manufactures one of the small RNAs that are found in ribosomes and the molecules called transfer RNAs, which are required for translation. **Box 16.1** explains how research on toxic mushrooms led to the discovery that eukaryotes have three different RNA polymerases.

TABLE **16.1** Eukaryotic RNA Polymerases

Name of Enzyme	Type of Gene Transcribed
RNA polymerase I (RNA pol I)	Genes that code for most of the large RNA molecules (rRNAs) found in ribosomes (see Section 16.3)
RNA polymerase II (RNA pol II)	Protein-coding genes (produce mRNAs)
RNA polymerase III (RNA pol III)	Genes that code for transfer RNAs (see Section 16.4), and genes that code for one of the small RNA molecules (rRNAs) found in ribosomes
RNA pol II and RNA pol III	RNA molecules found in snRNPs (see Section 16.2)

- Although eukaryotic genomes contain promoters that signal where transcription should begin, just as bacteria do, the promoters in eukaryotic DNA are much more diverse and complex than bacterial promoters. Many of the eukaryotic promoters recognized by RNA polymerase II include a unique sequence called the **TATA box**, located 30 base pairs upstream of the transcription start site. Some of the promoters recognized by pol II do not contain a TATA box, however. In addition, RNA pol I and pol III interact with entirely different promoters.

- In eukaryotes, transcription is followed by several important **RNA processing** steps that result in production of an mRNA that leaves the nucleus.

The overall message of this research is that the molecular mechanisms involved in transcription are much more complex in eukaryotes than in bacteria. Perhaps the most striking contrast between bacterial and eukaryotic gene expression emerged when researchers discovered that eukaryotic genes do not consist of one continuous DNA sequence that codes for a product,

BOX 16.1   Toxins and Transcription

Although wild mushrooms can be delicious, dining on certain species may be lethal. For example, in late summer and fall the reproductive structures of the death cap mushroom (*Amanita phalloides*) appear in forests, fields, and backyards in North America and Europe (**Figure 16.5**).

FIGURE 16.5 The Death Cap Mushroom. These mushrooms contain the toxin α-amanitin, which at low concentrations poisons the eukaryotic enzyme RNA polymerase II.

● **QUESTION** State a hypothesis to explain why the presence of α-amanitin increases the fitness of death cap mushrooms—meaning, their ability to survive and produce offspring.

The death cap is among the most poisonous of all mushrooms. Worldwide, it is thought to account for 90 percent of mushroom poisoning deaths. Eating a single death cap can kill a healthy adult.

The active agent in death caps is a toxin called α-amanitin. By happenstance, researchers who were investigating α-amanitin's mode of action made a crucial discovery about transcription in eukaryotes. The researchers observed that mRNA synthesis stopped when eukaryotic cells were treated with low concentrations of α-amanitin. The production of other types of RNAs—specifically, transfer RNAs (tRNAs) and ribosomal RNAs (rRNAs)—continued, however. At very high concentrations of the toxin, tRNA production also stopped but transcription of rRNA genes continued. These observations eventually led to the discovery that eukaryotes have three RNA polymerases, each with a distinct role in the transcription of mRNA, tRNA, and rRNA. Low concentrations of α-amanitin inhibit RNA polymerase II, which synthesizes

mRNAs. As a result, low concentrations of the toxin halt protein synthesis. High concentrations of this poison block RNA polymerase III, which synthesizes tRNAs. RNA polymerase I is not affected by α-amanitin at any concentration.

Toxins such as α-amanitin are useful in research on transcription and translation for the same reason that knock-out mutations, which completely disable genes, are useful in genetics. Toxins allow researchers to understand what happens when a molecule or a particular stage in a process does *not* work. This is similar, in concept, to learning how a car works by removing a part or a system and carefully recording the machine's response; or to learning how the brain works by studying the behavior of people whose brains have been damaged in specific, quantifiable ways by accidents. Toxins and knock-out mutations provide a molecular dissecting kit. They let biologists explore the intricacies of cell function by inactivating key components.

as do bacterial genes. Instead, the regions of eukaryotic genes that code for proteins come in pieces that are separated by hundreds or many thousands of intervening DNA bases. Although these intervening bases are part of the gene, they do not code for a product.

The Startling Discovery of Eukaryotic Genes in Pieces

When researchers realized that the protein-coding regions of eukaryotic genes are interrupted by stretches of noncoding DNA, it became clear that the nature of the gene and information processing are different in bacteria and eukaryotes. In eukaryotes, the conversion of information in DNA sequences to mRNA sequences does not occur directly. To make a functional mRNA, eukaryotic cells must dispose of certain sequences inside genes and then combine the separated coding sections into an integrated whole.

What sort of data would provoke such a startling claim? To answer this question, consider work that Phillip Sharp and colleagues carried out in the late 1970s to determine how DNA templates are transcribed. They began one of their experiments by heating DNA molecules enough to break the hydrogen bonds between complementary bases. This treatment separated the two strands. The single-stranded DNA was then incubated with the mRNA encoded by the sequence. The team's idea was to promote base pairing between the mRNA and the single-stranded DNA. Under these conditions, the research team expected the mRNA to form base pairs with the DNA sequences that act as the template for its synthesis. When the researchers examined the DNA-RNA hybrid molecules with the electron microscope, however, they observed the structure shown in **Figure 16.6a**. Instead of matching up exactly, parts of the DNA formed loops. As **Figure 16.6b** shows, Sharp and his coworkers interpreted these loops as stretches of nucleotides that are present in the DNA template strand but are *not* in the corresponding mRNA.

To make sense of these and other results, Sharp's group and a team headed by Richard Roberts proposed that there is not a one-to-one correspondence between the nucleotide sequence of a eukaryotic gene and its mRNA. By analogy, eukaryotic genes do not carry messages such as "Biology is my favorite course of all time." Instead, eukaryotic genes carry messages that read something like "BIOL$\tau\eta\epsilon\pi\rho o\tau\epsilon\iota\nu\chi o\delta\iota\nu\gamma'''$ $\rho\epsilon\gamma\iota o\nu\sigma o\phi\gamma\epsilon\nu\epsilon\sigma$OGY IS MY FAVOR$\alpha\rho\epsilon\iota\nu\tau\epsilon\rho\rho\upsilon\pi\tau\epsilon\delta\beta\psi'''$ $\nu o\nu\chi o\delta\iota\nu\gamma\delta$ITE COURSE OF$\alpha\nu\delta\eta\alpha\omega\epsilon\tau o\beta\epsilon\sigma\pi\lambda\iota\chi\epsilon\delta\tau o\gamma\epsilon\tau'''$ $\eta\epsilon\rho$ALL TIME." Here the sections of noncoding sequence are represented with Greek letters. They must be removed from the mRNA before it can carry an intelligible message to the translation machinery.

When it became clear that the genes-in-pieces hypothesis was correct, Walter Gilbert suggested that regions of eukaryotic genes that are part of the final mRNA be referred to as **exons** (because they are *ex*pressed) and the untranslated stretches as **introns** (because they are *int*ervening). Exons code for segments of functional proteins or RNAs; introns do not. Introns are sections of genes that are not represented in the final mRNA product. As a result, eukaryotic genes are much larger than their corresponding mature RNA transcripts (**Figure 16.6c**).

Exons, Introns, and RNA Splicing

The transcription of eukaryotic genes by RNA polymerase generates a **primary RNA transcript** that contains both the exon and intron regions (**Figure 16.7a**). As transcription proceeds, the introns are removed from the growing RNA strand by a process known as **splicing**. In this phase of information processing, pieces of the primary transcript are removed and the remaining segments are joined together. Splicing occurs while transcription is still under way and results in an RNA that contains an uninterrupted genetic message.

(a) Micrograph of DNA-RNA hybrid

(b) Interpretation of micrograph

Single-stranded DNA only

Single-stranded DNA base paired with mRNA

(c) Genes and RNA transcripts differ in length.

Intron   Exon

Size of gene (DNA)

Size of mature RNA transcript

FIGURE 16.6 The Discovery of Noncoding Regions of DNA. **(a)** Electron micrograph of single-stranded DNA that is bonded via complementary base pairing to the mRNA it encodes. **(b)** Researchers' interpretation of the photograph in part (a). The loops represent regions of DNA that do not have an equivalent sequence in the mRNA. These intervening regions are "extra" DNA compared with the sequences in the mRNA. **(c)** Because eukaryotic genes contain both introns and exons, these genes are much larger than the corresponding mature mRNA.

● **QUESTION** If introns did not exist, what would the micrograph in part (a) look like?

(a) Introns must be removed from RNA transcripts.

(b) snRNPs ARE THE EDITORS.

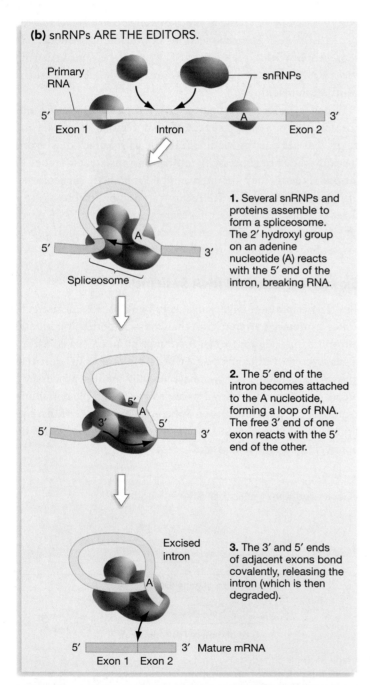

1. Several snRNPs and proteins assemble to form a spliceosome. The 2′ hydroxyl group on an adenine nucleotide (A) reacts with the 5′ end of the intron, breaking RNA.

2. The 5′ end of the intron becomes attached to the A nucleotide, forming a loop of RNA. The free 3′ end of one exon reacts with the 5′ end of the other.

3. The 3′ and 5′ ends of adjacent exons bond covalently, releasing the intron (which is then degraded).

FIGURE 16.7 Introns Are Spliced Out of the Original mRNA. (a) RNA polymerase produces a primary RNA transcript that contains exons and introns. Subsequent processing generates a mature transcript. **(b)** snRNPs are responsible for the splicing reactions that take place in the nucleus of eukaryotes.

Figure 16.7b provides more detail about how introns are removed from genes. In the case illustrated here, splicing is catalyzed by a complex of proteins and small RNAs known as **small nuclear ribonucleoproteins,** or **snRNPs** (pronounced "snurps"). The process begins when a snRNP binds to the 5′ exon-intron boundary. Once the initial snRNPs are in place, other snRNPs arrive to form a multipart complex called a **spliceosome.** Recent data have shown that the spliceosomes found in human cells contain about 145 different proteins and RNAs, making them the most complex molecular machines known. Once the spliceosome forms, the intron forms a loop with an adenine ribonucleotide at its base. The adenine participates in a reaction that cuts the loop out. A phosphodiester bond then links the exons on either side, forming a contiguous coding sequence. Splicing is now complete. In most cases, the excised intron is degraded to ribonucleotide monophosphates.

Current data suggest that both the cutting and rejoining reactions that occur during splicing are catalyzed by RNA molecules in the spliceosome. Recall from Chapter 4 that RNA molecules with catalytic ability are called ribozymes. In eukaryotes, ribozymes play a key role in the production of proteins.

Adding Caps and Tails to Transcripts

Intron splicing is not the only way that primary RNA transcripts are processed in eukaryotes. As soon as the 5′ end of a eukaryotic RNA emerges from RNA polymerase, enzymes add a structure called the **5′ cap** (**Figure 16.8**). The cap consists of the molecule 7-methylguanylate and three phosphate groups. In addition, an enzyme cleaves the 3′ end of most RNAs once transcription is complete, and another enzyme adds a long tract of 100–250 adenine nucleotides. This sequence is known as the **poly(A) tail.** These poly(A) tails are not encoded on the template strand. ⬤ With the addition of the cap and tail, processing of the primary RNA transcript is complete. The product is a mature mRNA. The molecule contains the coding sequence for a polypeptide flanked by sequences that are not destined to be translated. These 5′ and 3′ untranslated regions help stabilize the mature RNA and regulate its translation.

Not long after the caps and tails on eukaryotic mRNAs were described, evidence began to accumulate that they also protect mRNAs from degradation by ribonucleases and enhance the efficiency of translation. For example, experimental mRNAs that have a cap and a tail last longer when they are introduced into cells than do experimental mRNAs that lack a cap, a tail, or both a cap and a tail. The mRNAs with caps and tails also produce more proteins than do mRNAs without caps and tails.

Recent work has shown why the cap and tail are so critical to mRNAs. The 5′ cap serves as a recognition signal for the translation machinery, and the poly(A) tail extends the life span of an mRNA by protecting the message from degradation by ribonucleases in the cytosol.

FIGURE 16.8 In Eukaryotes, mRNAs Are Given a Cap and a Tail. Eukaryotic mRNAs have a cap consisting of a molecule called 7-methylguanylate (symbolized as m^7G) bonded to three phosphate groups; the tail is made up of a long series of adenine residues.

Splicing and the addition of caps and tails are major differences in how mRNAs are produced in bacteria versus eukaryotes. It is not the only important contrast, however. **Table 16.2** summarizes some of the similarities and differences observed in transcription in these two lineages.

Check Your Understanding

If you understand that...

- Transcription occurs when RNA polymerase catalyzes the $5' \rightarrow 3'$ synthesis of an mRNA molecule.
- RNA synthesis is driven by the potential energy stored in nucleotide triphosphates.
- The sequence of an RNA transcript is complementary to the sequence in a template DNA strand.

You should be able to...

1) Compare and contrast how transcription occurs in bacteria and eukaryotes.
2) Diagram the capping, splicing, and tail-addition steps that occur as primary RNA transcripts are processed in eukaryotes.

16.3 An Introduction to Translation

To synthesize a protein, the sequence of bases in a messenger RNA molecule is translated into a sequence of amino acids in a polypeptide. The genetic code specifies the relationship between the bases of a triplet codon in mRNA and the amino acid it codes for (see Chapter 15). But how are the amino acids assembled into a polypeptide according to the information in messenger RNA?

Studying how translation occurred in cell-free systems proved to be an extremely productive approach to answering this question. Once in vitro translation systems had been developed from human cells, *E. coli*, and a variety of other organisms, it became clear that the basic mechanisms of translation are fundamentally the same throughout the tree of life. The sequence of events that occurs during protein synthesis is similar in bacteria, archaea, and eukaryotes.

Ribosomes Are the Site of Protein Synthesis

The first question that biologists answered about translation concerned where it occurs. The answer was inspired by a simple observation: There is a strong, positive correlation between the number of small structures known as **ribosomes** in a given type of cell and the rate at which that cell synthesizes proteins. For example, immature human red blood cells divide rapidly, synthesize millions of copies of the protein hemoglobin, and contain large numbers of ribosomes. In contrast, the same cells at maturity have low rates of protein synthesis and very few ribosomes. Based on this correlation, investigators proposed that ribosomes are the site of protein synthesis in the cell.

TABLE **16.2 Comparing Transcription in Bacteria and Eukaryotes**

Aspect	Bacteria	Eukaryotes
RNA polymerase	One	Three; each produces a different class of RNA
Promoter structure	Typically contains a −35 box and a −10 box	Complex and variable; often includes a TATA box −30 from start of gene
Protein(s) involved in contacting promoter	Sigma; different versions of sigma bind to different promoters	Many basal transcription factors
RNA processing steps	None; translation occurs while transcription is still under way	Extensive; several processing steps occur in nucleus before RNA is exported to cytoplasm for translation: **1.** Enzyme-catalyzed addition of 5′ cap **2.** Splicing (intron removal) by spliceosome **3.** Enzyme-catalyzed addition of 3′ poly(A) tail

To test this hypothesis, Roy Britten and collaborators did a pulse-chase experiment similar in design to experiments introduced in Chapter 7. Recall that the goal of a pulse-chase experiment is to label a population of molecules as they are being produced. The location of the tagged molecules is then followed through time. The tagging is done by supplying a pulse of radioactive atoms that are incorporated into the molecule as it is being synthesized, followed by a chase of unlabeled atoms. Britten's group wanted to label amino acids with radioactive atoms and then track the molecules as translation occurred.

The biologists began by feeding a pulse of radioactive sulfate ($^{35}SO_4^{-2}$) to growing cultures of *E. coli*. They expected the cells to incorporate the radioactive sulfur into the amino acids methionine and cysteine—which contain sulfur (see Chapter 3)—and then into newly synthesized proteins. Fifteen seconds after adding the radioactive sulfate, the researchers "chased" the label by adding a large excess of nonradioactive sulfate to the culture medium. If the ribosome hypothesis was correct, then the radioactive signal should be associated with ribosomes for a short period of time (while they are making proteins). Later, all of the radioactivity should be found in proteins not in the ribosomes.

This is exactly what the researchers found. Soon after the pulse of labeled sulfate ended, the radioactive atoms were found in free amino acids or in ribosomes. Later, all of the radioactive atoms were found on completed proteins. Based on these data, biologists concluded that proteins are synthesized at ribosomes and then released.

About a decade after the ribosome hypothesis was confirmed, electron micrographs showed bacterial ribosomes in action (**Figure 16.9a**). The images confirmed that in bacteria, ribosomes attach to mRNAs and begin synthesizing proteins even before transcription is complete (**Figure 16.9b**). Transcription and translation can occur concurrently, or be coupled, in bacteria because there is no nuclear envelope to separate the two processes. Thus, transcription and translation are physically connected. But in eukaryotes, RNAs are processed in the nucleus and then mRNAs are exported to the cytoplasm (**Figure 16.10**). Once mRNAs are outside the nucleus, ribosomes attach to them and begin translation. In eukaryotes, transcription and translation are separated in time and space.

How Does an mRNA Triplet Specify an Amino Acid?

When an mRNA interacts with a ribosome, hereditary instructions encoded in nucleic acids are translated into a different chemical language—the amino acid sequences found in proteins. The discovery of the genetic code revealed that triplet codons in mRNA specify particular amino acids in a protein. How does this conversion happen?

(a) Bacterial ribosomes during translation

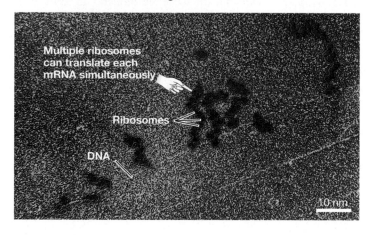

(b) In bacteria, transcription and translation are tightly coupled.

FIGURE 16.9 Transcription and Translation Occur Simultaneously in Bacteria. In bacteria, ribosomes attach to mRNA transcripts and begin translation while RNA polymerase is still transcribing the DNA template strand. Note that the colors in the micrograph in part (a) are artificial—they've been added to make the structures more visible.

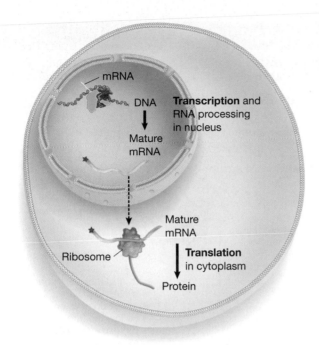

FIGURE 16.10 Transcription and Translation Are Separated in Space and Time in Eukaryotes. In eukaryotes, mRNA transcripts are processed inside the nucleus before they are exported to the cytoplasm. There, they are translated by ribosomes.

(a) Hypothesis 1: Amino acids interact directly with mRNA codons.

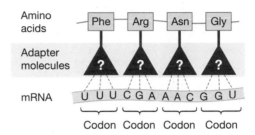

(b) Hypothesis 2: Adapter molecules hold amino acids and interact with mRNA codons.

FIGURE 16.11 Two Hypotheses for How mRNA Codons Interact with Amino Acids.

One early hypothesis was that mRNA codons and amino acids interact directly. The proposal was that the bases in a particular codon were complementary in shape or charge to the side group of a particular amino acid (**Figure 16.11a**). But Francis Crick pointed out that the chemistry involved didn't make sense. For example, how could the nucleic acid bases interact with a hydrophobic amino acid side group, which does not form hydrogen bonds?

Crick proposed an alternative hypothesis. As **Figure 16.11b** shows, he suggested that some sort of adapter molecule holds amino acids in place while interacting directly and specifically with a codon in mRNA via hydrogen bonding. In essence, Crick predicted the existence of a chemical go-between that produced a physical connection between the two types of molecules. As it turns out, Crick was right.

16.4 The Role of Transfer RNA

Crick's adapter molecule was discovered by accident. Biologists were trying to work out a cell-free protein synthesis system derived from mammalian liver cells and had discovered that ribosomes, mRNA, amino acids, ATP, and a molecule called guanosine triphosphate, or GTP, had to be present for translation to occur. (GTP is similar to ATP but contains guanosine instead of adenosine.) These results were logical—ribosomes provide the catalytic machinery, mRNAs contribute the message to be translated, amino acids are the building blocks of proteins,

and ATP and GTP supply potential energy to drive the endergonic polymerization reactions responsible for forming proteins. But in addition, a cellular fraction that contained a previously unknown type of RNA turned out to be indispensable. If this type of RNA is missing, protein synthesis does not occur. What is this mysterious RNA, and why is it essential to translation?

The novel class of RNAs eventually became known as **transfer RNA (tRNA)**. The role of tRNA in translation was a mystery until some researchers happened to add a radioactive amino acid—leucine—to an in vitro protein synthesis system. The treatment was actually done as a control for an unrelated experiment. To the researchers' amazement, some of the radioactive leucine attached to tRNA molecules. Follow-up experiments showed that the attachment of an amino acid to a tRNA requires an input of energy in the form of ATP. A tRNA molecule that becomes covalently linked to an amino acid is called an **aminoacyl tRNA**. More recent research has shown that enzymes called **aminoacyl tRNA synthetases** are responsible for catalyzing the addition of amino acids to tRNAs. For each of the 20 major amino acids, there is a different aminoacyl tRNA synthetase and one or more tRNAs. **Figure 16.12** summarizes how amino acids are transferred from aminoacyl tRNA synthetases to tRNAs.

What happens to the amino acid that is carried by an aminoacyl tRNA? Biologists who tracked the fate, over time, of radioactive leucine molecules that were attached to tRNAs discovered that the amino acids are transferred from aminoacyl tRNAs to proteins. The data supporting this conclusion are

HOW AMINO ACIDS ARE LOADED ONTO tRNAs

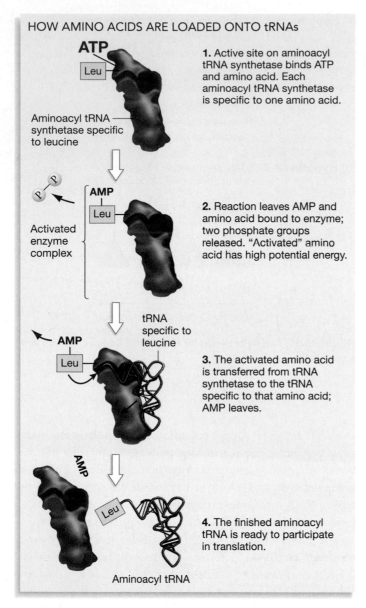

1. Active site on aminoacyl tRNA synthetase binds ATP and amino acid. Each aminoacyl tRNA synthetase is specific to one amino acid.

2. Reaction leaves AMP and amino acid bound to enzyme; two phosphate groups released. "Activated" amino acid has high potential energy.

3. The activated amino acid is transferred from tRNA synthetase to the tRNA specific to that amino acid; AMP leaves.

4. The finished aminoacyl tRNA is ready to participate in translation.

FIGURE 16.12 Aminoacyl tRNA Synthetases Load Amino Acids onto the Appropriate tRNA. An aminoacyl tRNA synthetase—in this case, for leucine—becomes activated in a reaction that requires ATP. Activated aminoacyl tRNA synthetase transfers an amino acid to a tRNA, producing an aminoacyl tRNA.

Experiment

Question: What happens to the amino acids attached to tRNAs?

Hypothesis: Aminoacyl tRNAs transfer amino acids to growing polypeptides.

Null Hypothesis: Aminoacyl tRNAs do not transfer amino acids to growing polypeptides.

Experimental setup:

1. Attach radioactive leucine molecules to tRNAs.

2. Add these aminoacyl tRNAs to in vitro translation system. Follow the fate of the radioactive amino acids.

Prediction: Radioactive amino acids will be found in proteins.

Prediction of null hypothesis: Radioactive amino acids will not be found in proteins.

Results:

Radioactive amino acids start attached to tRNA

Radioactive amino acids are rapidly incorporated into protein

Conclusion: Aminoacyl tRNAs transfer amino acids to growing polypeptides.

FIGURE 16.13 Experimental Evidence that Amino Acids Are Transferred from tRNAs to Proteins.

shown in **Figure 16.13**. The graph in the Results section of this figure shows that radioactive amino acids are lost from tRNAs and incorporated into polypeptides synthesized in ribosomes. These results inspired the use of transfer in tRNA's name. The experiment also confirmed that aminoacyl tRNAs act as the interpreter in the translation process. tRNAs are Crick's adapter molecules.

What Do tRNAs Look Like?

Transfer RNAs serve as chemical go-betweens that allow amino acids to interact with an mRNA template. But precisely how does the connection occur?

This question was answered by research on tRNA's molecular structure. The initial studies established the sequence of nucleotides in various tRNAs, or what is termed their primary structure. Transfer RNAs are relatively short, ranging from 75 to 85 nucleotides in length. When biologists studied the primary sequence closely, they noticed that certain parts of the molecules can form secondary structures. Specifically, some of the bases in the tRNA molecule can form hydrogen bonds with complementary bases in a different region of the same molecule. As a result, the entire tRNA molecule could assume the cloverleaf shape illustrated in **Figure 16.14a**. The stems in the cloverleaf are produced by complementary base pairing between different portions of the molecule running in antiparallel directions, while the loops consist of single-stranded RNA (see Chapter 4).

Two aspects of this secondary structure proved especially interesting. A CCA sequence at the 3′ end of each tRNA molecule offered a binding site for amino acids, while the triplet on a loop on the opposite end of the cloverleaf could serve as an anticodon. An **anticodon** is a set of three ribonucleotides that forms base pairs with the mRNA codon. **Figure 16.14b** depicts an early model of how the anticodon of a tRNA connects an mRNA codon with the appropriate amino acid.

This model had to be modified, however, when X-ray crystallography studies revealed the tertiary structure of tRNAs. Recall from Chapter 3 that the tertiary structure of a molecule is defined by the three-dimensional arrangement of its atoms and is usually a product of folding. According to the X-ray crystallography data, the cloverleaf structure folds over

to produce an L-shaped molecule (**Figure 16.14c**). All of the tRNAs in a cell have the same L-shaped structure, but each has a distinct anticodon and attached amino acid. The tertiary structure of tRNAs is important because it results in a precise separation between the anticodon and the attached amino acid. As we'll see, this separation is key to the positioning of the amino acid and the anticodon in the ribosome.

● If you understand the structure of tRNAs, you should be able to make a rough sketch of a tRNA with and without an amino acid attached, and explain the relationship between the anticodon of a tRNA and a codon in an mRNA.

How Many tRNAs Are There?

When research succeeded in characterizing all of the different types of tRNAs available in cells, a paradox arose. According to the genetic code introduced in Chapter 15, the 20 most common amino acids found in proteins are specified by 61 different mRNA codons. Instead of containing 61 different tRNAs with 61 different anticodons, though, most cells contain only

(a) Secondary structure of tRNA

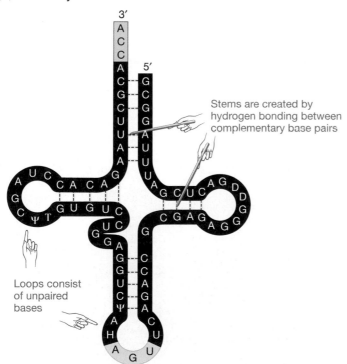

(b) Early model of aminoacyl tRNA function

(c) Revised model incorporating tertiary structure of tRNA

FIGURE 16.14 The Structure of Transfer RNA. (a) The secondary structure of tRNA resembles a cloverleaf. **(b)** If amino acids are attached to the 3′ end of a tRNA with an anticodon appropriate for that amino acid, then the tRNA anticodon will form a complementary base pair with the mRNA codon. **(c)** Recent data indicate that tRNAs have an L-shaped tertiary structure.

about 40. How can all 61 mRNA codons in the genome be translated with only two-thirds of the tRNAs required?

To resolve this paradox, Francis Crick proposed what is known as the **wobble hypothesis**. To understand his logic, recall from Chapter 15 that many amino acids are specified by more than one codon. Further, recollect that codons for the same amino acid tend to have the same nucleotides at the first and second positions but a different nucleotide at the third position. For example, both of the codons CAA and CAG code for the amino acid glutamine. (Codons are always written in the $5' \rightarrow 3'$ direction.) Surprisingly, experimental data have shown that a tRNA with an anticodon of GUU can base pair with both CAA and CAG in mRNA. (Anticodons are written in the $3' \rightarrow 5'$ direction.) The GUU anticodon matches the first two bases (C and A) in both codons, but the U in the third position forms a nonstandard base pair with a G in the CAG codon.

Crick proposed that inside the ribosome, tRNAs with nonstandard base pairing at the third position of the codon can still bind successfully to an anticodon. If so, it would allow a limited flexibility, or "wobbling," in the base pairing. According to the wobble hypothesis, a nonstandard base pair in the third position of a codon—such as G-U—is acceptable as long as it does not change the amino acid for which the codon codes. In this way, wobble in the third position of a codon allows just 40 or so tRNAs to bind to all 61 mRNA codons.

16.5 Ribosomes and the Mechanism of Translation

Protein synthesis occurs when the sequence of bases in an RNA message is translated into a sequence of amino acids in a polypeptide. The conversion begins when the anticodon of a tRNA that carries an amino acid binds to a codon in mRNA. The conversion is complete when a peptide bond forms between that amino acid and the growing polypeptide chain.

Both of these events occur inside a ribosome. Biologists have known since the 1930s that ribosomes contain a considerable amount of protein along with a great deal of **ribosomal RNA (rRNA)**. When high-speed centrifuges became available, researchers found that ribosomes could also be separated into two major substructures, called the large subunit and small subunit. Each subunit consists of a complex of RNA molecules and proteins. For example, the large subunit in *E. coli* contains two different rRNA molecules and 34 proteins. More recent work has shown that the small subunit holds the mRNA in place during translation and that peptide bond formation takes place in the large subunit.

The three-dimensional ribbon diagram of the ribosome in **Figure 16.15a** shows how all of the molecules required for translation fit together. Note that during protein synthesis, three distinct tRNAs are lined up inside the ribosome. All three are bound to their corresponding mRNA codon at the base of

(a) Ribbon model of ribosome during translation

tRNA in **E** site   tRNA in **P** site   tRNA in **A** site
(blue)   (green)   (red)

(b) Diagram of ribosome during translation

FIGURE 16.15 The Structure of the Ribosome. (a) During translation, three tRNAs line up side by side inside the ribosome. **(b)** Different events occur at each of the three sites inside the ribosome where tRNAs are found.

● **EXERCISE** Circle and label the active site in the ribosome in part (b).

the structure. The tRNA that is on the right in the figure, and colored red, carries an amino acid. This tRNA's position in the ribosome is called the A site—"A" for acceptor or aminoacyl. The tRNA that is in the middle (green) holds the growing polypeptide chain and occupies the P site, for peptidyl, inside the ribosome. (Think of "P" for peptide-bond formation.) The left-hand (blue) tRNA no longer has an amino acid attached and is about to leave the ribosome. It occupies the ribosome's E site—"E" for exit. **Figure 16.15b** summarizes what happens to each tRNA inside the ribosome. Because all tRNAs have similar secondary and tertiary structure, they all fit equally well in the A, P, and E sites.

⬤ The ribosome is a molecular machine that synthesizes proteins in a three-step sequence:

1. An aminoacyl tRNA diffuses into the A site; its anticodon binds to a codon in mRNA.

2. A peptide bond forms between the amino acid held by the aminoacyl tRNA in the A site and the growing polypeptide, which is held by a tRNA in the P site.

3. The ribosome moves ahead, and all three tRNAs move one position down the line. The tRNA in the E site exits; the tRNA in the P site moves to the E site; and the tRNA in the A site switches to the P site.

The protein that is being synthesized grows by one amino acid each time this three-step sequence repeats. The process occurs up to 20 times per second in bacterial ribosomes and about 2 times per second in eukaryotic ribosomes. Protein synthesis starts at the amino-end (N-terminus) of a polypeptide and proceeds to the carboxy end (C-terminus; see Chapter 3).

This introduction to how tRNAs, mRNAs, and ribosomes interact during protein synthesis leaves several key questions unanswered, however. How do mRNAs and ribosomes get together to start the process? Once protein synthesis is under way, how is peptide bond formation catalyzed inside the ribosome? And how does protein synthesis conclude when the ribosome reaches the end of the message? Let's consider each question in turn.

Initiation

To translate an mRNA, a ribosome must begin at a specific point in the message, translate the mRNA up to the message's termination codon, and then stop. Biologists call these three phases of protein synthesis **initiation**, **elongation**, and **termination**, respectively. One key to understanding initiation is to recall, from Chapter 15, that a start codon (usually AUG) is found near the 5′ end of all mRNAs and that it codes for the amino acid methionine. The presence of this start codon is an aspect of initiation that is common to both bacteria and eukaryotes.

Figure 16.16 shows how translation gets under way in bacteria. The process begins when a section of rRNA in the small subunit of a ribosome binds to a complementary sequence on an mRNA. The mRNA region is called the **ribosome binding site** or **Shine-Dalgarno sequence**, after the biologists who discovered it. The site is about six nucleotides upstream from the AUG start codon. It consists of all or part of the bases 5′-AGGAGGU-3′. The complementary sequence in the rRNA of the small subunit reads 3′-UCCUCCA-5′. The interaction between the small subunit and the message is mediated by proteins called **initiation factors** (Figure 16.16, step 1). In eukaryotes, initiation factors bind to the 5′ cap on mRNAs and guide it to the ribosome.

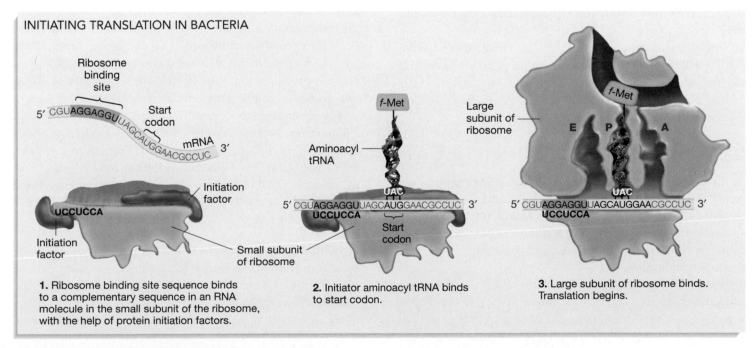

FIGURE 16.16 The Initiation Phase of Translation in Bacteria.

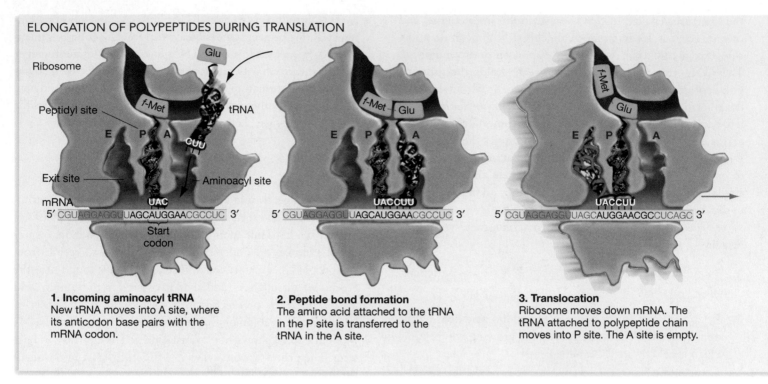

ELONGATION OF POLYPEPTIDES DURING TRANSLATION

1. Incoming aminoacyl tRNA
New tRNA moves into A site, where its anticodon base pairs with the mRNA codon.

2. Peptide bond formation
The amino acid attached to the tRNA in the P site is transferred to the tRNA in the A site.

3. Translocation
Ribosome moves down mRNA. The tRNA attached to polypeptide chain moves into P site. The A site is empty.

FIGURE 16.17 The Elongation Phase of Translation.

Once the Shine-Dalgarno sequence has attached to the small ribosomal subunit, an aminoacyl tRNA bearing a modified form of methionine called *N*-formylmethionine (abbreviated *f*-met) binds to the AUG start codon (Figure 16.16, step 2). In eukaryotes, this initial amino acid is normal methionine. Initiation is complete when the large subunit joins the complex (Figure 16.16, step 3). When the ribosome is completely assembled, the tRNA bearing *f*-met occupies the P site.

To summarize, initiation is a three-step process in bacteria: (1) The mRNA binds to the small subunit of the ribosome, (2) the initiator aminoacyl tRNA bearing f-Met binds to the start codon, and (3) the large subunit of the ribosome completes the complex.

Elongation

At the start of elongation, the E and A sites in the ribosome are empty of tRNAs. As a result, an mRNA codon is exposed at the base of the A site. As step 1 in **Figure 16.17** illustrates, elongation proceeds when an aminoacyl tRNA binds to the codon in the A site via complementary base pairing between anticodon and codon.

When a tRNA occupies both the P site and A site, the amino acids they hold are placed in the ribosome's active site. This is where peptide bond formation—the essence of protein synthesis—occurs. Peptide bond formation is considered one of the most important reactions that takes place in cells, because

manufacturing proteins is among the most fundamental of all cell processes. The question is, how does it happen?

Is the Ribosome an Enzyme or a Ribozyme? Because ribosomes contain both proteins and RNA, researchers had argued for decades over whether the active site consisted of protein or RNA. The debate was not resolved until the year 2000, when researchers completed three-dimensional models that were detailed enough to view the structure of the active site. These models confirmed that the active site consists entirely of ribosomal RNA. Based on these results, biologists are now convinced that protein synthesis is catalyzed by RNA. The ribosome is a **ribozyme**—not an enzyme.

The observation that protein synthesis is catalyzed by RNA is important, because it supports the RNA world hypothesis introduced in Chapter 4. Recall that proponents of this hypothesis claim that life began with RNA molecules and that the presence of DNA and proteins in cells evolved later. If the RNA world hypothesis is correct, then it is logical to observe that the production of proteins is catalyzed by RNA.

Moving Down the mRNA What happens after a peptide bond forms? Step 2 in Figure 16.17 shows that when peptide bond formation is complete, the polypeptide chain is transferred from the tRNA in the P site to the amino acid held by the tRNA in the A site. Step 3 shows the process called **translocation**, which occurs when the ribosome moves the mRNA through in the 5′ → 3′ direction. Translocation does several things: It moves

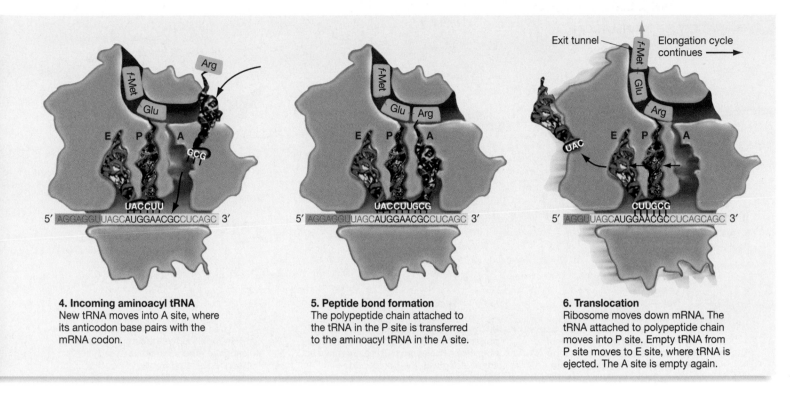

4. Incoming aminoacyl tRNA
New tRNA moves into A site, where its anticodon base pairs with the mRNA codon.

5. Peptide bond formation
The polypeptide chain attached to the tRNA in the P site is transferred to the aminoacyl tRNA in the A site.

6. Translocation
Ribosome moves down mRNA. The tRNA attached to polypeptide chain moves into P site. Empty tRNA from P site moves to E site, where tRNA is ejected. The A site is empty again.

the empty tRNA into the E site; it moves the tRNA containing the growing polypeptide into the P site; and it opens the A site and exposes a new mRNA codon. If the E site is occupied when translocation occurs, the tRNA there is ejected into the cytosol.

The three steps in elongation—(1) arrival of aminoacyl tRNA, (2) peptide bond formation, and (3) translocation—repeat down the length of the mRNA. By adding specific molecules to in vitro translation systems, researchers have confirmed that each elongation cycle depends on an input of energy from several GTP molecules as well as assistance from proteins called **elongation factors**. Further, recent three-dimensional models of ribosomes in various stages of the translation sequence show that the machine as a whole is highly dynamic during the process. The ribosome constantly changes shape as tRNAs come and go and catalysis and translocation occur. Its structure changes in conjunction with its multipart function: coordinating interactions between tRNAs and an mRNA, and catalyzing peptide bond formation.

Electron microscopy has also confirmed that once elongation is under way, additional ribosomes bind to the start site in the same message and initiate translation. As this process continues, strings of ribosomes called **polyribosomes** assemble along an mRNA. Each ribosome synthesizes a copy of the protein (**Figure 16.18**). Even though it may take a minute or more for a single ribosome to produce a large polypeptide, the presence of polyribosomes can quicken the overall pace of protein production. Polyribosomes are observed routinely in both bacteria and eukaryotes.

Termination

How does protein synthesis end? To answer this question, recall from Chapter 15 that the genetic code includes three stop codons: UAA, UAG, and UGA. In most cells, no aminoacyl tRNA has an anticodon that binds to these sequences. When translocation opens the A site and exposes one of the stop

FIGURE 16.18 Polyribosomes. In both bacteria and eukaryotes, many ribosomes synthesize proteins from the same mRNA at the same time. The resulting structures are called polyribosomes.

TERMINATION OF TRANSLATION

1. When translocation opens the A site and exposes one of the stop codons, a protein called a release factor fills the A site. The release factor catalyzes the hydrolysis of the bond linking the tRNA in the P site with the polypeptide chain.

2. The hydrolysis reaction frees the polypeptide, which is released from the ribosome. The empty tRNAs are released either along with the polypeptide or...

3. ...when the ribosome separates from the mRNA, and the two ribosomal subunits dissociate. The subunits are ready to attach to the start codon of another message and start translation anew.

FIGURE 16.19 The Termination Phase of Translation.

codons, a protein called a **release factor** fills the A site (**Figure 16.19**). Release factors are proteins that fit tightly into the A site because their size, shape, and electrical charge are tRNA-like. Release factors do not carry an amino acid, however. When a release factor occupies the A site, the active site catalyzes the hydrolysis of the bond linking the tRNA in the P site with the polypeptide chain. This reaction frees the polypeptide. The newly synthesized polypeptide is released from the ribosome, the ribosome separates from the mRNA, and the two ribosomal subunits dissociate. The subunits are ready to attach to the start codon of another message and start translation anew.

Post-Translational Modifications Proteins are not fully formed and functional when termination occurs. From earlier chapters, it should be clear that most proteins go through an extensive series of processing steps, collectively called post-translational modification, before they are ready to go to work in a cell:

- Recall from Chapter 3 that a protein's function depends on its shape and that a protein's shape depends on how it folds. Folding actually begins during elongation, long before termination occurs and ribosomes disassemble. Although it occurs spontaneously, in the sense that no energy input is required, folding is frequently speeded up by proteins called **molecular chaperones**. Recent data have shown that in some bacteria, chaperone proteins actually bind to the ribosome near the

"tunnel" where the growing polypeptide emerges from the ribosome. This result suggests that folding occurs as the polypeptide is emerging from the ribosome.

- Chapter 7 pointed out that many eukaryotic proteins are extensively modified after they are synthesized. For example, small chemical groups may be added to proteins in the organelles called the rough endoplasmic reticulum and the Golgi apparatus. Some proteins receive a sorting signal that serves as an address and ensures that the molecule will be carried to the correct location in the cell. Certain proteins are also augmented with sugar or lipid groups that are critical for normal functioning.

- Many proteins are altered by enzymes that add or remove a phosphate group. Phosphorylation (addition of phosphate) and dephosphorylation (removal of phosphate) of proteins were introduced in Chapter 9. Recall that because a phosphate group has two negative charges, adding or removing a phosphate group may cause major changes in the shape and chemical reactivity of proteins. These changes have a dramatic effect on the protein's activity—often switching it from an inactive state to an active state or vice versa.

The overall message is that the manufacture of a completely functional protein depends not only on translation in the ribosome but also on a wide array of other molecules and events that take place throughout the cell.

Check Your Understanding

If you understand that...

- Translation begins when (1) the ribosome binding site on an mRNA binds to an rRNA sequence in the small subunit of the ribosome, (2) the initiator aminoacyl tRNA binds to the start codon in the mRNA, and (3) the large subunit of the ribosome attaches to the small subunit to complete the ribosome.

- Translation elongation occurs when (1) an appropriate aminoacyl tRNA enters the A site, (2) a peptide bond forms between the amino acid held by that tRNA in the A site and the polypeptide held by the tRNA in the P site, and (3) the ribosome moves down the mRNA one codon.

- Translation ends when the ribosome reaches a stop codon.

You should be able to...

1) Diagram a ribosome during translation initiation, elongation, and termination.

2) Label each major structure in the diagrams (large and small ribosomal subunits, ribosome binding site and initiation factors, mRNA, E/P/A sites of ribosome, growing polypeptide, release factor).

3) Describe the function of each major structure.

(MB) Web Animation *BioFlix* at www.masteringbio.com
Translation; Protein Synthesis

16.6   What Is the Molecular Basis of Mutation?

This chapter has been focusing on how the information in DNA is transcribed to RNA and then translated into a functioning protein. Now the question is, what happens if the information in DNA changes? What are the consequences for the cell?

A **mutation** is any permanent change in an organism's DNA. It is a modification in a cell's information archive—a change in its genotype. The molecular machinery introduced in this chapter faithfully transcribes and translates mutations in DNA sequences. The result is the production of novel types of proteins. In this way, mutations may lead to changes in proteins and RNAs that affect the organism's phenotype.

Point Mutation

Figure 16.20 shows how a common type of mutation occurs. If DNA polymerase mistakenly inserts the wrong base as it synthesizes a new strand of DNA, and if proofreading by DNA polymerase and the mismatch repair system fail to correct the mismatched base before another round of DNA replication occurs, a change in the sequence of bases in DNA results. A single base change such as this is called a **point mutation**.

What happens when point mutations are transcribed and translated? To answer this question, consider the first point mutation ever described. This mutation occurs in the human gene for the protein hemoglobin. Hemoglobin is abundant in red blood cells, and it carries oxygen to tissues. You might recall from Chapter 3 that each hemoglobin protein is made up of four polypeptides—two copies of two different polypeptides. In adult humans, the two distinct polypeptides are α-globin and β-globin.

Figure 16.21a shows a small section of DNA sequence from the normal gene for β-globin along with the same sequence from the mutant allele found in individuals who suffer from sickle-cell disease. Note that the mutant form of β-globin has a thymine in place of an adenine at the second position in the sixth codon specified by this gene. A glance back at the genetic code in Chapter 15 confirms why this point mutation is significant: During protein synthesis, the mutant codon specifies valine instead of glutamic acid in the amino acid chain of β-globin. Point mutations that cause changes in the amino acid sequence of proteins are called **missense mutations** or **replacement mutations**. In the case of hemoglobin, the single

FIGURE 16.20 Unrepaired Mistakes in DNA Synthesis Lead to Point Mutations.

● **QUESTION** Why is it logical that the type of mutation illustrated here is termed a point mutation?

change in primary structure of β-globin causes the protein to crystallize when oxygen levels in the blood are low. When hemoglobin crystallizes, it causes red blood cells to become sickle shaped (**Figure 16.21b**). The misshapen cells get stuck in blood vessels, so nearby cells are starved for oxygen. The result is intense pain. Sickled cells also have a much shorter life span in the body than normal red blood cells, reducing overall red blood cell counts and making people anemic.

Missense mutations are often **deleterious**, meaning they reduce an individual's fitness—the ability to survive and reproduce. Missense mutations may also be beneficial, however. For example, sickle cell disease develops only in people who are homozygous for the sickling allele. But people who are heterozygous for the sickling allele have a fitness advantage in regions of the world where the life-threatening disease malaria is common. The advantageous effect is possible because the parasite that causes malaria infects red blood cells. If an individual has one normal hemoglobin allele and one sickling allele, red blood cells that become infected with the parasite tend to sickle while others function normally. The deformed cells are quickly destroyed by the body, eliminating the parasitic cells inside the blood cells. As a result, individuals who have one copy of the mutant gene have lower parasite loads and tend to be much healthier than people who have two normal copies of the gene. Although having two copies of the sickling allele is disadvantageous in all environments because it leads to sickle cell disease, having one copy of the sickling allele is advantageous in malaria-prone areas of the world.

In contrast to having a negative or positive effect on the organism, many point mutations have virtually no effect at all. To understand why, suppose that at the next site in the hemoglobin gene of Figure 16.21 (the third position in the same sixth codon), a thymine had been substituted for the cytosine. This point mutation would have no consequence. Why? Both GAA and GAG specify glutamic acid. In this case, the amino acid sequence of the gene product does not change even though the DNA sequence is altered due to mutation. This type of alteration in the base sequence is called a **silent mutation**. Silent mutations are said to be **neutral** in their effect on an individual's fitness. It is possible for point mutations to be silent and neutral with respect to fitness because of the redundancy in the genetic code.

The take-home message from this analysis of hemoglobin, sickle cell disease, and malaria is simple: Point mutations may be deleterious, beneficial, or neutral. Recent studies have shown that the vast majority of point mutations are neutral or slightly deleterious with respect to fitness. **Table 16.3** summarizes the types of point mutations that have been documented and reviews their consequences for the amino acid sequences of proteins.

Chromosome-Level Mutations

Besides documenting various types of point mutations, biologists study mutations involving larger-scale changes in the com-

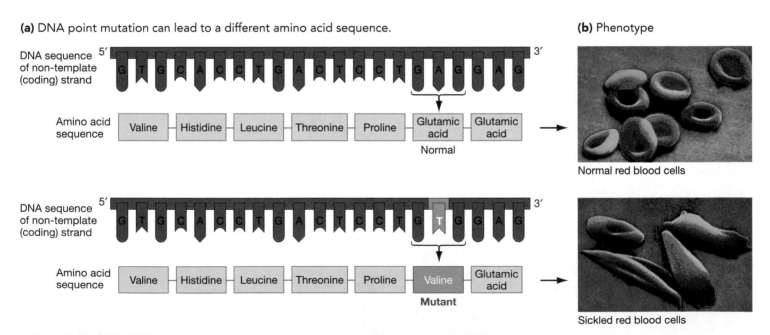

(a) DNA point mutation can lead to a different amino acid sequence.

(b) Phenotype

Normal red blood cells

Sickled red blood cells

FIGURE 16.21 Sickle-Cell Disease Results from a Point Mutation in the Gene for Hemoglobin. (a) These normal and mutant genes that code for hemoglobin differ by a single nucleotide. The sequences shown are just a small portion of the gene. **(b)** The mutation in part (a) changes the primary sequence of the gene product. The change in amino acid sequence causes hemoglobin molecules to crystallize when oxygen levels in the blood are low. As a result, red blood cells sickle and get stuck in small blood vessels.

● **EXERCISE** In part (a), add the base sequence of the template strand and label its 5′ and 3′ ends. Write out the base sequence of the mRNA transcribed from the template strand and label its 5′ and 3′ ends.

SUMMARY TABLE **16.3 Known Types of Point Mutations**

Name	Definition	Example	Consequence
		Original DNA sequence of non-template (coding strand) — **TAT TGG CTA GTA CAT** / Tyr — Trp — Leu — Val — His — Original polypeptide	
Silent	Change in nucleotide that does not change amino acid specified by codon	**TAC TGG CTA GTA CAT** / Tyr — Trp — Leu — Val — His	Change in genotype but no change in phenotype
Missense (Replacement)	Change in nucleotide that changes amino acid specified by codon	**TAT TGT CTA GTA CAT** / Tyr — **Cys** — Leu — Val — His	Change in primary structure of protein
Nonsense	Change in nucleotide that results in early stop codon	**TAT TGA CTA GTA CAT** / Tyr — **STOP**	Premature termination—polypeptide is truncated
Frameshift	Addition or deletion of a nucleotide	**TAT TCG GCT AGT ACA T** / Tyr — Ser — Ala — Ser — Thr	Reading frame is shifted (see Chapter 15)—massive missense

position of chromosomes. Chapter 12 introduced **polyploidy,** which is a change in the number of each type of chromosome present, and **aneuploidy,** which involves the addition or deletion of a chromosome. Polyploidy, aneuploidy, and other changes in chromosome number result from chance mistakes in the partitioning of chromosomes during meiosis or mitosis. In this sense they are similar to the random errors that DNA polymerase makes, resulting in point mutations.

But in addition to changes in overall chromosome number, the composition of individual chromosomes can change in important ways. For example, chromosome segments can become detached when accidental breaks in chromosomes occur. The segments may become flipped and rejoin—a phenomenon known as a chromosome **inversion**—or become attached to a different chromosome, causing a chromosome **translocation** (**Table 16.4**). These mutations affect the arrangement of genes on chromosomes. They are important because they influence the frequency of crossing over between chromosomes as well as how certain genes are turned on or off.

To summarize, point mutations and chromosome-level mutations are random changes in DNA that produce new genes, new alleles, and new traits. At the level of individuals, mutations can cause disease or death or lead to increases in fitness. At the level of populations, mutations furnish the heritable variation that Mendel and Morgan analyzed and that makes evolution possible.

SUMMARY TABLE **16.4 Some Types of Chromosome-Level Mutations**

Name	Description	Example	Consequence
Chromosome inversion	Chromosome segment flips and reattaches to same chromosome.	Genes — A B C D	Changes gene order along chromosome. Frequency of crossing over is reduced (when homologs synapse, inverted segments can't cross-over with non-inverted segments).
Chromosome translocation	Chromosome segment detaches and attaches to different chromosome.	A B C D / E F G H I	Leads to deletion or addition of chromosome segments. In many cases, expression of genes in translocated segments changes.

Chapter Review

SUMMARY OF KEY CONCEPTS

The instructions for making and operating a cell are archived in DNA, transcribed into messenger RNA, and then translated into protein.

● **After RNA polymerase binds DNA with the help of other proteins, it catalyzes the production of an RNA molecule whose sequence is complementary to the base sequence of the DNA template strand.**

RNA polymerase begins transcription by binding to promoter sequences in DNA. In bacteria, this binding occurs in conjunction with a regulatory protein called sigma. Sigma recognizes particular sequences within promoters that are centered 10 bases and 35 bases upstream from the start of the actual genetic message. These binding sites ensure that the correct DNA strand is transcribed in the correct direction. Eukaryotic promoters are more complex and variable than bacterial promoters are, and they interact with many more transcription factors.

You should be able to predict the consequences of a mutation that changes the sequence of nucleotides in a promoter. ●

(MB) Web Animation at www.masteringbio.com
Transcription

● **Some sections of an RNA are encoded by gene regions called exons, while others are encoded by gene regions called introns. During RNA processing, introns are removed and the ends of the RNA receive a cap and tail.**

Stretches of noncoding RNA called introns are spliced out by complex molecular machines called spliceosomes, a "cap" signal is added to the 5′ end, and a poly(A) tail is added to the 3′ end. The 5′ cap and the poly(A) tail serve as recognition signals for the translation machinery and protect the message from degradation by ribonucleases.

You should be able to draw the structure of a primary mRNA and a mature mRNA from a eukaryote, and contrast the two molecules with a bacterial mRNA. ●

● **Inside ribosomes, mRNAs are translated to proteins via intermediary molecules called transfer RNAs. Transfer RNAs carry an amino acid and have a three-base-pair anticodon, which binds to a three-base-long mRNA codon. The amino acid carried by the transfer RNA is then added to the growing protein via formation of a peptide bond.**

Experiments with radioactively labeled amino acids confirmed that ribosomes are the site of protein synthesis, and that transfer RNAs (tRNAs) serve as the chemical bridge between the RNA message and the polypeptide product. tRNAs have an L-shaped tertiary structure. One leg of the L contains the anticodon, which forms a base pair with the mRNA codon, while the other leg holds the amino acid appropriate for that codon. Because imprecise pairing—or "wobbling"—is allowed in the third position of a codon-anticodon pair, only about 40 different tRNAs are required to translate the 61 codons that code for amino acids.

Inside the ribosome, protein synthesis involves three steps: (1) an incoming aminoacyl tRNA occupies the A site; (2) the growing polypeptide chain is transferred from a peptidyl tRNA in the ribosome's P site to the amino acid bound to the tRNA in the A site, and a peptide bond is formed; and (3) the ribosome is translocated to the next codon on the mRNA, accompanied by ejection of the empty tRNA from the E site. Peptide bond formation is catalyzed by a ribozyme (RNA), not an enzyme (protein).

While translation is in progress, proteins fold into their three-dimensional conformation (tertiary structure), sometimes with the aid of chaperone proteins. Some proteins are targeted to specific locations in the cell by the presence of signal sequences, while others remain inactive until modified by phosphorylation or the removal of certain amino acids.

You should be able to create a concept map (see **BioSkills 6**) describing the events in translation, starting with the addition of amino acids to tRNAs and including the major steps in initiation, elongation, and termination. ●

(MB) Web Animation *BioFlix* at www.masteringbio.com
Translation; Protein Synthesis

● **Mutations are random changes in DNA that may or may not produce changes in the phenotype.**

Depending on the location and type of alteration in DNA and its impact on the resulting RNA or protein product, a mutation can be beneficial, deleterious, or neutral with respect to fitness. Mutations produce novel proteins and RNAs. They are the source of the heritable variation that makes evolution possible.

You should be able to explain why mutation is random. Your answer should be based on the molecular mechanisms responsible for point and chromosomal mutations. ●

QUESTIONS

● **Test Your Knowledge**

1. How did the A site of the ribosome get its name?
   a. It is where amino acids are affixed to tRNAs, producing aminoacyl tRNAs.
   b. It is where the amino group on the growing polypeptide chain is available for peptide bond formation.
   c. It is the site occupied by incoming aminoacyl tRNAs.
   d. It is surrounded by α-helices of ribosomal proteins.

2. How did the P site of the ribosome get its name?
   a. It is where the promoter resides.
   b. It is made up of protein.
   c. It is where peptidyl tRNAs reside.
   d. It is where a growing polypeptide chain is phosphorylated.

3. What is a molecular chaperone?
   a. a protein that recognizes the promoter and guides the binding of RNA polymerase
   b. a protein that activates or deactivates another protein by adding or removing a phosphate group
   c. a protein that is a component of the large ribosomal subunit and that assists with peptide bond formation
   d. a protein that helps newly translated proteins fold into their proper three-dimensional configuration

4. The three types of RNA polymerase found in eukaryotic cells transcribe different types of genes. What does RNA polymerase II produce?
   a. rRNAs   c. mRNAs
   b. tRNAs   d. spliceosomes

5. What is an anticodon?
   a. the part of an mRNA that signals translation termination
   b. the part of an mRNA that signals the start of translation
   c. the part of a tRNA that binds to a codon in mRNA
   d. the part of a tRNA that accepts an amino acid, via a reaction catalyzed by tRNA synthetase

6. What happens during a chromosome inversion?
   a. Pieces of different chromosomes break and swap positions.
   b. Homologs stay together during cell division, so a daughter cell is a chromosome short.
   c. Chromosome number doubles.
   d. A segment of a chromosome breaks, flips in orientation, and reattaches.

Test Your Knowledge answers: 1. c; 2. c; 3. d; 4. c; 5. c; 6. d

● Test Your Understanding

Answers are available at www.masteringbio.com

1. Explain the relationship among eukaryotic promoter sequences, basal transcription factors, and RNA polymerase. Explain the relationship among bacterial promoter sequences, sigma, and RNA polymerase.

2. According to the wobble rules, the correct amino acid can be added to a growing polypeptide chain even if the third base in an mRNA codon does not correctly match the corresponding base in a tRNA anticodon. How do the wobble rules relate to the redundancy of the genetic code?

3. Why does splicing occur in eukaryotic mRNAs? Where does it occur, and how are snRNPs involved?

4. Explain the sequence of events that occurs during translation as a protein elongates by one amino acid. At each step, specify what is happening in the ribosome's A site, P site, and E site.

5. What evidence supports the hypothesis that peptide bond formation is catalyzed by a ribozyme?

6. Explain why all point mutations change the genotype, but why only some point mutations change the phenotype.

● Applying Concepts to New Situations

Answers are available at www.masteringbio.com

1. The 5′ cap and poly(A) tail on eukaryotic mRNAs protect the message from degradation by ribonucleases. But why do ribonucleases exist? What function would an enzyme that destroys messages serve? Answer this question using the example of an mRNA for a hormone that causes human heart rate to increase.

2. The nucleotide shown below is called cordycepin.

If cordycepin triphosphate, which has three phosphate groups bonded to the 5′ hydroxyl group in the figure, is added to a cell-free transcription reaction, the nucleotide is added onto the growing RNA chain. This observation confirms that synthesis occurs in the 5′ → 3′ direction. Make a drawing, similar to Figure 16.1, showing why cordycepin triphosphate cannot be added to the 5′ end of an RNA.

3. Certain portions of the rRNAs in the large subunit of the ribosome are very similar in all organisms. To make sense of this result, Carl Woese suggests that the conserved sequences have an important functional role. His logic is that these conserved sequences are so important to cell function that any changes in the sequence cause death. Which specific portions of the ribosome would you expect to be identical or nearly identical in all organisms, and which would you expect to be more variable? Explain your logic.

4. Recent structural models show that α-amanitin inhibits translation by binding to a site inside RNA pol II, but not to the active site itself. Based on the model of bacterial RNA polymerase in Figure 16.2, predict where α-amanitin binds and why it inhibits transcription.

www.masteringbio.com is also your resource for • Answers to text, table, and figure caption questions and exercises • Answers to *Check Your Understanding* boxes • Online study guides and quizzes • Additional study tools including the *E-Book for Biological Science* 3rd ed., textbook art, animations, and videos.

17

Control of Gene Expression in Bacteria

KEY CONCEPTS

- Gene expression can be controlled at three levels: transcription, translation, or post-translation (protein activation).

- Changes in gene expression allow bacterial cells to respond to environmental changes.

- Transcriptional control can be negative or positive. Negative control occurs when a regulatory protein prevents transcription. Positive control occurs when a regulatory protein increases the transcription rate.

- Many regulatory proteins bind to specific sites in DNA. Because each type of regulatory protein has a distinct amino acid sequence, each type binds to different DNA sequences.

The red mounds in this micrograph are human intestinal cells; the yellow structures are bacteria. In the intestine, the nutrients available to bacteria constantly change. This chapter explores how changes in gene expression help bacteria respond to environmental changes.

Imagine waiting anxiously to hear the opening lines of a wonderfully melodic symphony, played by a renowned orchestra. The crowd applauds as the celebrated conductor comes onstage, then hushes as he takes the podium. He cocks the baton; the musicians raise their instruments. As the baton comes down, every instrument begins blaring a different tune at full volume. A tuba plays "Dixie"; a violinist renders "In-A-Gadda-Da-Vida"; and a cellist begins Mexico's national anthem. A snare drum lays down beats for Outkast's "Hey Ya" while the bass drum simulates the cannons in the "1812 Overture." Instead of music, there is pandemonium. The conductor staggers off stage, clutching his heart.

A cacophony like this would result if a bacterial cell "played" all its genes at full volume all the time. The *Escherichia coli* cells living in your gut right now have over 4300 genes. If all of those

genes were expressed at the fastest possible rate at all times, the *E. coli* cells would stagger off the stage, too. But this does not happen. Cells are extremely selective about which genes are expressed, in what amounts, and when.

This chapter explores how bacterial cells control the activity of their genes. **Gene expression** is said to occur when a protein or other gene product is synthesized and is active in the cell.

Understanding how bacterial cells regulate gene expression is a fundamental issue in biological science, for two reasons. First, bacteria are the most abundant organisms on Earth, and they occupy virtually every habitat known. In many cases, bacterial cells are able to grow and reproduce successfully because they respond rapidly to dramatic changes in temperature, pH, light, competitors, and nutrients. Changes in gene expression give bacteria the ability to cope with these types of environ-

○ Key Concept   Important Information   ● Practice It

mental change. Second, questions about bacterial gene expression have enormous practical significance. For example, biologists are trying to understand which genes are expressed when disease-causing bacteria colonize a person and start an infection. The goal of this research is to develop drugs that alter the expression of key genes in the disease-causing cells. Bacterial cells are also used commercially to manufacture important products such as human insulin and growth hormone. Efficient use of these cells depends on a solid understanding of how transcription and translation are regulated.

This chapter's goal is to introduce key concepts in the regulation of gene expression. Let's begin by reviewing some of the environmental challenges that bacterial cells face and then explore how these organisms meet them.

17.1 Gene Regulation and Information Flow

Bacteria are found in virtually every habitat on Earth, from boiling hot springs to alpine snowfields and from the open ocean to crevices miles underground. The millions of bacterial species existing today have evolved a bewildering variety of ways to solve a central problem of living: obtaining the materials and energy required for growth and reproduction. Although some bacteria specialize by using just one type of food, the vast majority of species are able to switch among several different sources of carbon and energy, according to which nutrients are available in the environment at the time. The fundamental question addressed in this chapter is *how* this switching occurs. Each type of nutrient requires a different membrane transport protein to bring the nutrient molecule into the cell and a different suite of enzymes to process it. How does a bacterial cell start transcribing certain genes and end the transcription of others, so that it can take advantage of alternative food sources? More generally, how do bacteria regulate gene expression so that a cell makes only the products that it needs?

As a case study, this chapter focuses on strategies observed in *E. coli*. These cells can use a wide array of carbohydrates to supply the carbon and energy they need. For example, *E. coli* are abundant in your intestinal tract and use carbohydrates found in your food. But as your diet changes from day to day, the availability of different sugars in your intestines varies. Precise control of gene expression gives *E. coli* the ability to respond to these changes in its environment and use the different sugars.

To appreciate why precise control over gene expression is so important to these cells, it is critical to realize that bacteria can be packed an inch thick along your intestinal walls. The organisms represent many different species that are competing for space and nutrients. For a cell to survive and reproduce in this environment, it must use resources efficiently—particularly resources that provide energy and carbon compounds. Using energy and carbon compounds to synthesize proteins that aren't needed would reduce a cell's ability to produce the proteins it *does* need, compete successfully for resources, and produce offspring.

Based on this reasoning, biologists predicted that the transcription and translation of individual genes in bacteria are triggered by specific signals from the environment, such as the presence of specific sugars. Did you drink milk at your last meal, or eat French fries and a candy bar? Each type of food contains different sugars. Each sugar should induce a different response by the *E. coli* cells in your intestine. Just as a conductor needs to regulate the orchestra's musicians, cells need to regulate which proteins they produce at any given time.

Mechanisms of Regulation—An Overview

Gene expression can be controlled at any step between the synthesis of RNA and the activation of the final gene product. Three steps occur as information flows from DNA to proteins, represented by arrows in the following expression:

$$\text{DNA} \rightarrow \text{mRNA} \rightarrow \text{protein} \rightarrow \text{activated protein}$$

The arrow from DNA to RNA represents transcription—the making of messenger RNA (mRNA). The arrow from RNA to protein represents translation, in which ribosomes read the information in mRNA and use that information to synthesize a protein. The arrow from protein to activated protein represents post-translational modifications—including folding, addition of carbohydrate or lipid groups, or perhaps phosphorylation.

How can a bacterial cell avoid producing proteins that are not needed at a particular time, and thus use resources efficiently? A look at the flow of information from DNA to protein suggests three possible mechanisms:

1. The cell could avoid making the mRNAs for particular enzymes. If there is no mRNA, then ribosomes cannot make the gene product. For example, various regulatory proteins affect the ability of RNA polymerase to bind to a promoter and initiate transcription. Genes that are controlled in this way are said to undergo **transcriptional control**:

$$\text{DNA} \xrightarrow{\times} \text{mRNA} \rightarrow \text{protein} \rightarrow \text{activated protein}$$

2. If the mRNA for an enzyme has been transcribed, the cell might have a way to prevent the mRNA from being translated into protein. Mechanisms that alter the length of time an mRNA survives before it is degraded by ribonucleases, that affect translation initiation, or that affect elongation factors and other proteins during the translation process are forms of **translational control**:

$$\text{DNA} \rightarrow \text{mRNA} \xrightarrow{\times} \text{protein} \rightarrow \text{activated protein}$$

3. Chapter 16 pointed out that some proteins are manufactured in an inactive form and have to be activated by chemical

FIGURE 17.1 Gene Expression Can Be Regulated at Three Levels. Although this chapter focuses on how regulatory molecules affect the ability of RNA polymerase to initiate transcription, translational and post-translational controls also occur in bacteria.

● **EXERCISE** Label which mode of regulation is the slowest in response time, and which is fastest. Label which is the most efficient and least efficient in resource use.

modification, such as the addition of a phosphate group. This type of regulation is **post-translational control**:

DNA → mRNA → protein ─✕→ activated protein

● Which of these three forms of control occur in bacteria? The short answer to this question is "all of the above." As **Figure 17.1** shows, many factors affect how much active protein is produced from a particular gene. Transcriptional control is particularly important due to its efficiency—it saves the most energy for the cell, because it stops the process at the earliest possible point. Translational control is advantageous because it allows a cell to make rapid changes in its array of proteins. Post-translational control is significant as well, and it provides the most rapid response of all three mechanisms. Among these mechanisms of gene regulation, there is a clear trade-off between the speed of response and the conservation of ATP, amino acids, and other resources. Transcriptional control is slow but efficient in resource use. Post-translational control is fast but energetically expensive.

Although this chapter focuses almost exclusively on mechanisms of transcriptional control in bacteria, it is important to keep in mind that both translational and post-translational controls occur in these organisms. ● It's also important to recognize that some genes—such as those that code for the enzymes required for glycolysis—are transcribed all the time, or **constitutively**. But the expression of other genes is regulated, meaning that they may be induced or repressed. Finally, it is critical to realize that gene expression is not an all-or-none proposition. Genes are not just "on" or "off"—instead, the level of expression is highly variable. Variation in gene expression allows cells to respond to changes in their environment. What factors determine how much a particular gene is expressed at a particular time?

Metabolizing Lactose—A Model System

As Chapters 13 through 16 have shown, many of the fundamental advances in genetics have been achieved through the analysis of various model systems. Studying the inheritance of seed shape in garden peas revealed the fundamental patterns of gene transmission. Exploring transcription in viruses and *E. coli* led to the discovery of RNA polymerase, transcription factors such as the sigma proteins, and promoters. In studies of gene regulation, the key model system has been the metabolism of the sugar lactose in *E. coli*.

Jacques Monod, François Jacob, and many colleagues introduced lactose metabolism in *E. coli* as a model system during the 1950s and 1960s. Although they worked with a single species of bacteria, their results had a profound effect on thinking about gene regulation in all organisms. Some details turned out to be specific to the *E. coli* genes responsible for lactose metabolism, but many of Monod and Jacob's results are universal.

Escherichia coli can use a wide variety of sugars for ATP production, via glycolysis and the electron transport chain or via fermentation. These sugars also serve as a raw material in the synthesis of amino acids, vitamins, and other complex compounds. Glucose is *E. coli*'s preferred carbon source, however, meaning that it is the source of energy and carbon atoms that the organism uses most efficiently. This observation is logical, because glycolysis begins with glucose and is the main pathway for the production of ATP. Lactose, the sugar found in milk, is also used by *E. coli*, but only when glucose supplies are depleted. Recall from Chapter 5 that lactose is a disaccharide made up of one molecule of glucose and one molecule of galactose.

To use lactose, *E. coli* must first transport the sugar into the cell. Once lactose is inside the cell, the enzyme β-galactosidase catalyzes a reaction that breaks the sugar down into glucose and galactose. The glucose released by this reaction is used directly via glycolysis; other enzymes convert the galactose to an intermediate in the glycolytic pathway.

In the early 1950s, biologists discovered that *E. coli* produces high levels of β-galactosidase only when lactose is present in the environment. If lactose is absent, little β-galactosidase is synthesized. Because lactose appears to induce the production of the enzyme that acts on lactose, researchers proposed that lactose itself regulates the gene for β-galactosidase. More formally, the hypothesis was that lactose acts as an inducer. An **inducer** is a molecule that stimulates the expression of a specific gene or genes.

In the late 1950s Jacques Monod investigated how the presence of glucose affects the regulation of the β-galactosidase gene. Would *E. coli* produce high levels of β-galactosidase when both glucose and lactose were present in the surrounding environment? As the experiment summarized in **Figure 17.2** shows, the answer was no. β-galactosidase is produced only when lactose is present and glucose is not present.

Monod teamed up with François Jacob to investigate exactly how lactose and glucose regulate the genes responsible for lactose metabolism. Research on how these genes are regulated opened a window on how genes in all organisms are controlled. Research on this system is still continuing, over 50 years later.

17.2 Identifying the Genes Involved in Lactose Metabolism

To understand how *E. coli* controls production of β-galactosidase and the membrane transport protein that brings lactose into the cell, Monod and Jacob first had to find the genes that code for these proteins. To do this, they employed the same tactic used in the pioneering studies of DNA replication, transcription, and translation reviewed in earlier chapters: They isolated and analyzed mutant individuals. In this case, their goal was to find *E. coli* cells that were *not* capable of metabolizing lactose. Cells that can't use lactose must lack either β-galactosidase or the lactose-transporter protein.

Finding mutants with respect to a particular trait is a two-step process. A researcher's first step is to generate a large number of individuals with mutations at random locations in their genomes. To produce mutant cells, Monod and others exposed *E. coli* populations to X-rays, UV light, or chemicals that damage DNA and increase mutation rates. The second step is to screen the mutants to find individuals with defects in the process or biochemical pathway in question—in this case, defects in lactose metabolism. A technique that allows researchers to identify individuals with a particular type of mutation is called a **genetic screen**.

Monod and colleagues were looking for cells that cannot grow in an environment that contains only lactose as an energy source. Normal cells grow well in this environment. How could the researchers select cells on the basis of *lack* of growth?

Experiment

Question: *E. coli* produces β-galactosidase when lactose is present. Does *E. coli* produce β-galactosidase when both glucose and lactose are present?

Hypothesis: *E. coli* does not produce β-galactosidase when glucose is present, even if lactose is present. (Glucose is the preferred food source.)

Alternative hypothesis: *E. coli* produces β-galactosidase whenever lactose is present, regardless of the presence or absence of glucose.

Experimental setup:

Prediction: β-Galactosidase will be produced only in treatment 3.

Prediction of alternative hypothesis: β-Galactosidase will be produced in treatments 2 and 3.

Results:

Conclusion: Glucose prevents the expression of the gene for β-galactosidase. The presence of lactose without glucose stimulates the expression of that gene.

FIGURE 17.2 Glucose Affects the Regulation of the β-Galactosidase Gene.

How Were the Genes Found?

Replica plating and growth on indicator plates were key techniques in the search for mutants with defects in lactose metabolism. **Replica plating** begins with spreading mutagenized bacteria on a plate that is filled with gelatinous agar containing glucose (**Figure 17.3**). This plate is known as the master plate. It is important that the mutant cells you want to identify are able to grow on this master plate. The bacteria are then allowed to grow, so that each cell produces a single colony. (A bacterial colony consists of a large number of identical cells that are descended from a single cell.) Next, a block covered with a piece of sterilized velvet is pressed onto the master plate. Because of the contact, cells from each colony on the master

FIGURE 17.3 Replica Plating Is a Technique for Identifying Mutant Cells. Here, replica plating is used to isolate mutant *E. coli* cells with a deficiency in lactose metabolism. There are two keys: (1) The location of colonies on the master plate and replica plate must match exactly, and (2) the media in the two plates must differ by just one component.

● **QUESTION** How would you alter this protocol to isolate mutant cells with a deficiency in the enzymes required to synthesize tryptophan?

plate are transferred to the velvet. Then the velvet is pressed onto a plate containing a medium that differs from that of the master plate by a single component. For instance, in Figure 17.3, step 3, the second medium has only lactose as a source of carbon and energy. Cells from the velvet stick to the plate's surface, producing an exact copy of the colonies on the master plate. This copy is called the replica plate. After these cells grow, an investigator can compare the colonies that thrive on the replica plate's medium with those on the master plate. In this case, colonies that grow on the master plate but are missing on the replica plate represent mutants deficient in lactose metabolism. By picking these particular colonies from the master plate, researchers build a collection of lactose mutants.

Monod also used an alternative strategy based on **indicator plates**, where mutants with metabolic deficiencies are observed directly. In this case, he added a compound that is acted on by β-galactosidase. The compound acts as an indicator molecule for the presence of functioning β-galactosidase because one of the molecules produced by the reaction is yellow. Colonies that stay white are unable to process the indicator molecule, meaning they have a defect in the β-galactosidase enzyme or its production.

Different Classes of Lactose Metabolism Mutants

The initial mutant screen yielded the three types of mutants summarized in **Table 17.1**. In one class, the mutant cells were unable to cleave the indicator molecule even if lactose was present inside the cells to induce production of the β-galactosidase protein. The investigators concluded that these mutants must lack a functioning version of the β-galactosidase protein—meaning the gene that encodes β-galactosidase is defective. This gene was designated *lacZ*, and the mutant allele *lacZ⁻*.

In the second class of mutants, the cells failed to accumulate lactose inside the cell. In normal cells the concentration of lactose is about 100 times that of lactose in the surrounding

TABLE **17.1 Three Distinct Types of Mutants in Lactose Metabolism of *E. coli***

Mutant Phenotype	Interpretation	Inferred Genotype
Cells cannot cleave indicator molecule even if lactose is present as an inducer.	No β-galactosidase; gene for β-galactosidase is defective. Call this gene *lacZ*.	*lacZ⁻*
Cells cannot accumulate lactose inside the cell.	No membrane protein (galactoside permease) required for import of lactose; gene for galactoside permease is defective. Call this gene *lacY*.	*lacY⁻*
Indicator molecule is cleaved even if lactose is absent (no inducer).	Constitutive expression of *lacZ* and *lacY*; gene for regulatory protein that shuts down *lacZ* and *lacY* is defective—it does not need to be induced by lactose. Call this gene *lacI*.	*lacI⁻*

FIGURE 17.4 Two Proteins Are Critical for _E. coli_ Cells to Use Lactose. For _E. coli_ cells to use lactose, the membrane protein galactoside permease must be present to bring the sugar into the cell. Then the enzyme β-galactosidase must be present to break lactose into its glucose and galactose subunits.

environment, but in the mutant cells lactose concentrations were much lower. To interpret this result, Jacob and Monod hypothesized that the mutant cells had defective copies of the membrane protein responsible for transporting lactose into the cell. This protein was identified and named galactoside permease; the gene that encodes it was designated _lacY_. **Figure 17.4** summarizes the functions of β-galactosidase and galactoside permease.

The third and most surprising class of mutants did not regulate expression of β-galactosidase and galactoside permease normally. For example, when these mutant cells were grown on lactose alone and the indicator molecule was added, they turned yellow just as normal cells do. But if they were grown on a medium that contained glucose but no lactose, they still turned yellow after the indicator molecule was added. Normal cells, in contrast, remain white when grown on glucose and given the indicator molecule. Normal cells do not produce β-galactosidase when glucose is present.

Cells that are abnormal because they produce a product at all times are called **constitutive** mutants. Unlike normal cells, their product is always part of the cell's constitution. The gene that mutated to produce constitutive β-galactosidase expression was named _lacI_. The use of "I" was appropriate because these

mutants did not need an inducer to express β-galactosidase or galactoside permease. Recall that in normal cells, the expression of these genes is induced by the presence of lactose. But in cells with a mutant form of _lacI_ (_lacI⁻_ mutants), gene expression occurred with or without lactose. This meant that _lacI⁻_ mutants have a defect in gene regulation. In these mutants, the gene remains "on" when it should be turned off.

Based on these observations, it is logical to infer that the normal product of the _lacI_ gene prevents the transcription of _lacZ_ and _lacY_ when lactose is absent. Because lactose acts as an inducer for the production of β-galactosidase, it is reasonable to expect that the _lacI_ gene or gene product interacts with lactose in some way. (Later work showed that the inducer is actually a derivative of lactose called _allolactose_. For the sake of historical accuracy and simplicity, however, this discussion refers to lactose itself as the inducer.)

Several Genes Are Involved in Metabolizing Lactose

Jacob and Monod had succeeded in identifying three genes involved in lactose metabolism: _lacZ_, _lacY_, and _lacI_. They had concluded that _lacZ_ and _lacY_ code for proteins involved in the metabolism and import of lactose, while _lacI_ is responsible for some sort of regulatory function. When lactose is absent, the _lacI_ gene or gene product shuts down the expression of _lacZ_ and _lacY_. But when lactose is present, the opposite occurred—transcription of _lacZ_ and _lacY_ was induced. ● If you understand the genes involved in lactose metabolism, you should be able to describe the specific function of _lacZ_ and _lacY_. You should also be able to describe the effect of the _lacI_ gene product when lactose is present versus absent, and explain why these effects are logical.

When Jacob and Monod followed up on these experiments by mapping the physical location of the three genes on _E. coli_'s circular chromosome, they found that the genes are close together (**Figure 17.5**). This was a crucial result, because it suggested that both _lacZ_ and _lacY_ might be controlled by _lacI_. Could one regulatory gene manage more than one protein-encoding gene? If so, how does _lacI_ actually work? And why do lactose and glucose have opposite effects on it?

FIGURE 17.5 The _lac_ Genes Are in Close Physical Proximity.

17.3 Mechanisms of Negative Control: Discovery of the Repressor

In principle, there are two ways that transcription can be regulated: via negative control or positive control. Negative control occurs when a regulatory protein binds to DNA and shuts down transcription (**Figure 17.6a**); positive control, also called **induction,** occurs when a regulatory protein binds to DNA and triggers transcription (**Figure 17.6b**). ● When a car is sitting at the curb, negative control occurs when the parking brake is set; positive control occurs when you step on the gas pedal. It turned out that the *lacZ* and *lacY* genes in *E. coli* are controlled by both a parking brake and a gas pedal. These genes are under both negative control and positive control.

The hypothesis that the *lacZ* and *lacY* genes might be under negative control originated with Leo Szilard in the late 1950s. Szilard suggested to Monod that the *lacI* gene codes for a product that represses transcription of the *lacZ* and *lacY* genes. Recall from Section 17.2 that in *lacI⁻* mutants, the production of β-galactosidase and galactoside permease is constitutive, even if lactose is not available. Szilard proposed that regulation fails because the lacI⁺ protein—which normally prevents enzyme

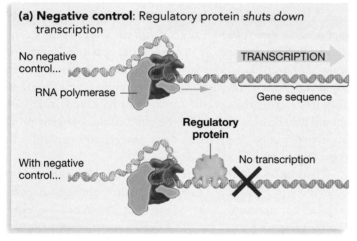

(a) Negative control: Regulatory protein *shuts down* transcription

No negative control...

RNA polymerase

TRANSCRIPTION

Gene sequence

Regulatory protein

With negative control...

No transcription

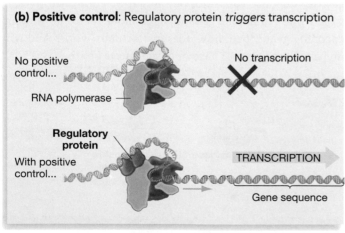

(b) Positive control: Regulatory protein *triggers* transcription

No positive control...

No transcription

RNA polymerase

Regulatory protein

With positive control...

TRANSCRIPTION

Gene sequence

FIGURE 17.6 Genes Are Regulated by Negative Control, Positive Control, or Both.

synthesis—is inactive. Stated another way, the *lacI* gene produces an inhibitor that exerts negative control over the *lacZ* and *lacY* genes. This transcription inhibitor was called a **repressor** and was thought to bind directly to DNA near or on the promoter for the *lacZ* and *lacY* genes (**Figure 17.7a**).

To explain how the presence of lactose triggers transcription in normal cells, Szilard and Monod proposed that lactose interacts with the repressor in a way that makes the repressor release from its binding site (**Figure 17.7b**). The idea was that lactose induces transcription by removing negative control. The repressor is the parking brake; lactose releases the parking brake. As **Figure 17.7c** shows, it was logical to observe that constitutive transcription occurs in *lacI⁻* mutants because a functional repressor is absent—the parking brake is broken.

To test these hypotheses, Jacob and Monod and co-workers created *E. coli* cells that had functioning copies of the genes for β-galactosidase and galactoside permease but lacked a functional gene for the repressor. As predicted, these cells made β-galactosidase all the time. But when these cells received a functioning copy of the repressor gene, β-galactosidase production declined and then stopped. This result supported the hypothesis that the repressor codes for a protein that shuts down transcription—that it is indeed the "parking brake" on transcription. But if an inducer such as lactose was added to the experimental cells, β-galactosidase activity continued. This result supports the hypothesis that lactose removes the repressor—it releases the parking brake. The *lacI* gene codes for a repressor protein that exerts negative control on *lacZ* and *lacY*. Lactose acts as an inducer by removing the repressor and ending negative control.

The *lac* Operon

Jacob and Monod summarized the results of their experiments with a comprehensive model of negative control that was published in 1961. In essence, their experimental data had confirmed the hypotheses illustrated in Figure 17.7. One of their key conclusions was that the genes for β-galactosidase and galactoside permease are controlled together. To capture this point, they coined the term **operon** for a set of coordinately regulated bacterial genes that are transcribed together into one mRNA. Logically enough, the group of genes involved in lactose metabolism was termed the *lac* operon. Later, a gene called *lacA* was found to be tightly linked to *lacY* and *lacZ* and part of the same operon. The *lacA* gene codes for the enzyme transacetylase. The enzyme's function is protective in nature. It catalyzes reactions that allow certain types of sugars to be exported from the cell when they are too abundant.

Three hypotheses are central to the Jacob-Monod model of *lac* operon regulation:

1. The *lacZ*, *lacY*, and *lacA* genes are adjacent and are transcribed into one mRNA initiated from the single promoter of the *lac* operon. As a result, the expression of the three genes is coordinated.

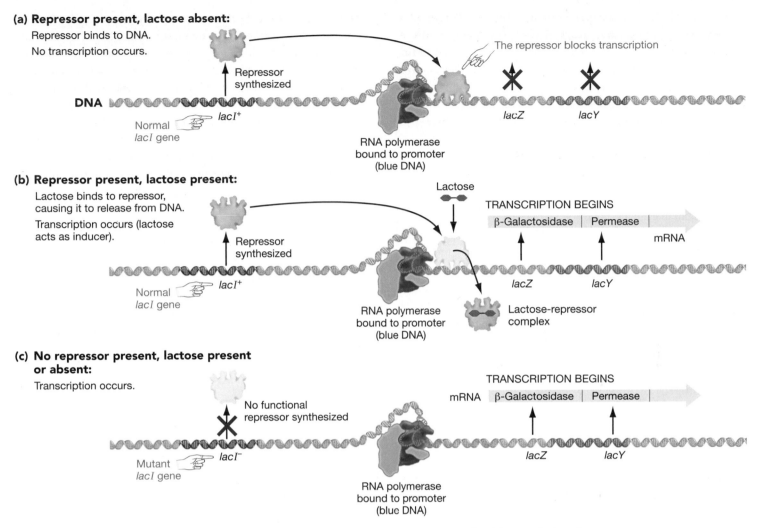

FIGURE 17.7 The Hypothesis of Negative Control. The negative-control hypothesis maintains that **(a)** transcription of genes involved in lactose use is normally blocked by a repressor molecule that binds to DNA on or near the promoter for *lacZ* and *lacY*, **(b)** lactose induces transcription of *lacZ* and *lacY* by interacting with the repressor, and **(c)** when a functional repressor is absent, transcription proceeds.

2. The repressor is a protein encoded by *lacI* that binds to DNA and prevents transcription of *lacZ*, *lacY*, and *lacA*. Jacob and Monod proposed that *lacI* is expressed constitutively and that the repressor binds to a section of DNA in the *lac* operon called the **operator**.

3. The inducer (lactose) interacts directly with the repressor by binding to it. As a result, the repressor changes shape in a way that causes it to drop off the DNA strand. Recall from Chapter 3 that this form of control over protein function is called **allosteric regulation**. When allosteric regula-

tion occurs, a small molecule binds directly to a protein and causes it to change its shape and activity. When the inducer binds to the repressor, negative control ends and transcription can proceed (**Figure 17.8**).

After their model of negative control was published, Jacob and Monod confirmed the existence of the operator by finding *E. coli* mutants that had normal forms of the repressor but that still expressed *lacZ* constitutively. In these new mutants, the repressor protein was unable to function because the nucleotide sequence of the operator was altered. In 1967 Walter Gilbert

FIGURE 17.8 Components of Negative Control in the *lac* Operon.

● **EXERCISE** Using small, colored bits of candy or paper, add RNA polymerase to the figure. Then add the repressor protein, and then add lactose. At each step, explain what happens after the molecule is added.

and Benno Müller-Hill were able to tag copies of the repressor protein with a radioactive atom. Their experiments showed that the repressor physically binds to the DNA sequences of the operator. This result confirmed that the operator is not a protein or an RNA product, but instead is part of the DNA sequence of the *lac* operon. **Box 17.1** explains how researchers eventually determined the operator's exact DNA sequence.

Why Has the *lac* Operon Model Been So Important?

Regulation of the *lac* operon provided an immensely important model system in genetics. Follow-up work has shown that numerous bacterial genes and operons are under negative control via repressor proteins. Even more important, the *lac* operon model introduced the idea that gene expression is regulated by physical contact between specific regions in regulatory proteins and specific regulatory sites in DNA. Publication of the *lac* operon model was a watershed event in the history of biological science. Its impact on thinking about gene regulation is hard to overestimate.

Besides confirming the existence of negative control, regulatory proteins, and regulatory sites in DNA, work on the *lac* operon offered an important example of post-translational control over gene expression. The key observation here was that the repressor protein is always present—because it is transcribed and translated constitutively at low levels. When a rapid change in *lac* operon activity is required, it does not occur via changes in the transcription or translation of the repressor. Instead, the activity of existing repressor proteins is altered. This turns out to be common. In virtually all cases, the activity of key regulatory proteins is controlled by post-translational modifications.

A New Twist on Negative Control: Comparing the *trp* and *lac* Operons

The success of the *lac* operon model inspired work on how a variety of other genes and operons are regulated in *E. coli*. Work on the *trp* operon (pronounced "trip") turned out to be particularly informative, because it extended and deepened the concept of negative control.

The *trp* operon, like the *lac* operon, produces a **polycistronic mRNA**—a message that contains more than one protein-encoding segment. In most cases, polycistronic mRNAs code for products devoted to a common end. The *lacZ* and *lacY* products are required for lactose metabolism, and the five genes in the *trp* operon in *E. coli* code for enzymes required for different steps in the synthesis of the amino acid tryptophan. Biologists interpret the polycistronic structure of many bacterial genes as an adaptation that increases the efficiency of gene expression. By binding to one promoter, regulatory proteins can control the expression of several genes that are required for the same process. Polycistronic mRNAs are common in bacteria but rare or nonexistent in most eukaryotes.

Like the *lac* operon, the *trp* operon is under negative control. There is a twist, however. Instead of coding for enzymes that perform catabolism—meaning that they break down a compound (see Chapter 9)—the *trp* operon codes for enzymes that perform anabolism. These enzymes synthesize a product. In this case, natural selection should favor cells where gene expression starts when the molecule is absent and is needed by the cell, and where transcription is reduced when the molecule is present. This is the opposite of what happens in the *lac* operon. How does negative control work in a case like this?

Early work on the *trp* operon established that it has an operator sequence that overlaps the promoter and a repressor protein that binds to the operator to prevent transcription—as in the *lac* operon. Experiments also showed that tryptophan binds to the repressor to regulate transcription, just as lactose binds to the *lac* operon's repressor to regulate transcription. But as **Figure 17.9a** shows, the *trp* repressor binds operator DNA and exerts negative control only when it is complexed with tryptophan. When tryptophan is absent, the repressor does not bind to the operator, and the operon is expressed constitutively (**Figure 17.9b**). This is exactly the opposite of how the presence of lactose affects the *lac* repressor (**Figure 17.10**). And it makes perfect sense for the *trp* operon: Bacterial cells will be more efficient, and have higher fitness, if the operon is transcribed when tryptophan levels are low but shut down when tryptophan is abundant.

After considering how the *lac* and *trp* operons are regulated in *E. coli*, it's easy to have new respect for "simple" forms of life. Bacteria may be small and unicellular, but they perform fantastically complex, finely controlled chemistry. Even the simplest of cells are marvelous machines.

(a) When tryptophan is present, transcription is blocked.

(b) When tryptophan is absent, transcription occurs.

FIGURE 17.9 The *trp* Operon Is under Negative Control. (a) When tryptophan is present, it binds to a repressor protein. The tryptophan-repressor complex binds to the operator and shuts down transcription, which ends the synthesis of tryptophan. **(b)** The *trp* operon contains five coding genes.

lac operon	trp operon
Catabolism (breakdown of lactose)	**Anabolism** (synthesis of tryptophan)

FIGURE 17.10 Comparing and Contrasting Negative Control of the *lac* and *trp* Operons. In contrast to the *lac* operon, negative control of the *trp* operon is released when tryptophan is absent.

Check Your Understanding

If you understand that...

- Negative control occurs when something must be taken away for transcription to occur.
- The *lac* operon repressor exerts negative control over three protein-coding genes by binding to the operator site in DNA, near the promoter.
- For transcription to occur in the *lac* operon, an inducer molecule (a derivative of lactose) must bind to the repressor, causing it to release from the operator.

You should be able to...

1) Explain why it is logical to observe that a molecule derived from lactose induces transcription of the *lac* operon.
2) Diagram the *lac* operon, showing the relative positions of the operator, the promoter, and the three protein-coding genes.
3) Show what is happening at the operon in the presence of lactose and in the absence of lactose.

BOX 17.1 DNA Footprinting

DNA footprinting is used to characterize DNA sequences that are bound by regulatory proteins (**Figure 17.11**). Researchers begin a footprinting study by obtaining DNA from a region of interest—the *lac* operon of *E. coli*, for example. Then they allow copies of the regulatory protein to bind to the DNA sequence being analyzed, and add a nuclease—an enzyme that cleaves DNA in random locations—to cut the DNA. Segments of DNA that are bound by the regulatory protein are protected from the enzyme and don't get cut. Segments of DNA that are not bound by the regulatory protein are not protected, meaning that they do get cut.

By analyzing the protected, uncut portion of the DNA segment, researchers can determine the exact nucleotide sequence of the regulatory site.

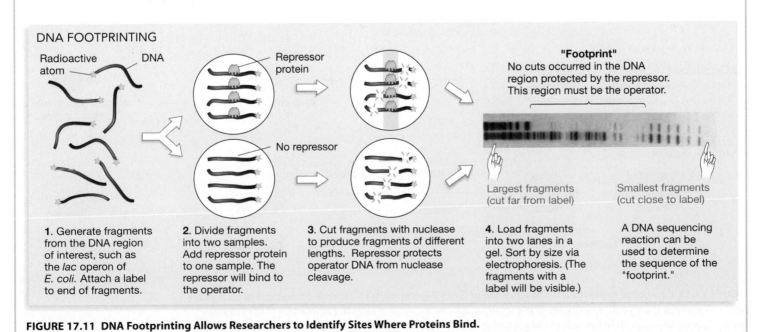

DNA FOOTPRINTING

1. Generate fragments from the DNA region of interest, such as the *lac* operon of *E. coli*. Attach a label to end of fragments.

2. Divide fragments into two samples. Add repressor protein to one sample. The repressor will bind to the operator.

3. Cut fragments with nuclease to produce fragments of different lengths. Repressor protects operator DNA from nuclease cleavage.

4. Load fragments into two lanes in a gel. Sort by size via electrophoresis. (The fragments with a label will be visible.)

A DNA sequencing reaction can be used to determine the sequence of the "footprint."

FIGURE 17.11 DNA Footprinting Allows Researchers to Identify Sites Where Proteins Bind.

17.4 Mechanisms of Positive Control: Catabolite Repression

The model of negative control over the *lac* operon, summarized in Figure 17.7, is elegant and successful in explaining experimental results. But it is not complete. After studying the model, you may think of an important question: Where does glucose fit in?

Transcription of the *lac* operon is drastically reduced when glucose is present in the environment—even when lactose is available to induce β-galactosidase expression. Given that glucose is *E. coli*'s preferred carbon source, this observation is logical. Glucose is produced when β-galactosidase cleaves lactose. When glucose is already present, it is not necessary for the cell to cleave lactose and produce still more glucose.

Biologists use the term **catabolite repression** to describe a situation like this. The term is appropriate because the hydrolysis of lactose into its glucose and galactose subunits is an example of catabolism (**Figure 17.12a**). In many cases, operons that encode catabolic enzymes are inhibited when the end product of the reaction, the catabolite, is abundant. Catabolite repression is a form of end-product inhibition. In the case of the *lac* operon, glucose is the catabolite. When glucose is abundant in the cell, transcription of the *lac* operon is decreased by catabolite repression (**Figure 17.12b**).

How does glucose prevent expression of the *lac* operon? An answer to this question began to emerge when researchers discovered a second major control element in the *lac* operon—an example of positive control. This regulatory switch consists of a DNA sequence known as the **CAP binding site**, which is located just upstream of the *lac* promoter, and a regulatory protein called **catabolite activator protein (CAP)**, which binds to this DNA sequence.

The CAP protein binds to the CAP binding site and triggers transcription of the *lac* operon. To understand how this happens, it's important to realize that not all promoters are created equal. Strong promoters allow efficient initiation of transcription by RNA polymerase; weak ones support much less efficient initiation of transcription. The *lac* promoter is weak. But when the CAP regulatory protein is bound to the CAP site just upstream of the *lac* promoter, the protein interacts with RNA polymerase in a way that allows transcription to begin much more frequently. CAP binding greatly strengthens the *lac* promoter. In this way, CAP exerts positive control of the *lac* operon. When CAP is active, transcription increases. If the inducer removes the parking brake, CAP can push the gas pedal to the floor.

Researchers also discovered that CAP, like the repressor protein, is allosterically regulated. CAP changes shape when the regulatory molecule **cyclic AMP (cAMP)** binds to it. Only when cAMP is attached to it can CAP bind to DNA (**Figure 17.13a**). The same type of allosteric regulation goes on with the repressor and lactose during negative control, except that the repressor binds to DNA only in the *absence* of the inducer. During positive control, the CAP-cAMP complex binds to the CAP binding site. As long as a repressor isn't bound to the operator, the complex increases the efficiency of transcription. If no cAMP is bound to CAP, then CAP has a conformation that does not allow binding to the CAP site (**Figure 17.13b**). cAMP is the green light that tells CAP to floor it.

How Does Glucose Influence Formation of the CAP-cAMP Complex?

Where does the elusive molecule glucose fit into this scheme? Glucose's role in positive control of the *lac* operon is indirect. More specifically, it is mediated by the influence that glucose has on the concentration of cAMP. Glucose levels outside the cell and cAMP levels inside the cell are inversely related. When extracellular glucose concentrations are high, intracellular cAMP concentrations are low; when extracellular glucose con-

(a) Catabolism

Large reactant molecule   Enzyme   Small product molecule (a catabolite)   +   Small product molecule (a catabolite)

(b) Catabolite repression of the *lac* operon

Lactose   β-Galactosidase required / No enzyme, no reaction   Galactose   +   Glucose

Inhibition of β-galactosidase synthesis

FIGURE 17.12 Catabolite Repression Is a Mechanism of Gene Regulation. (a) A generalized example of catabolism. **(b)** Catabolite repression occurs when one of the small product molecules represses the production of the enzyme responsible for the reaction. In the case of lactose metabolism, the production of β-galactosidase is suppressed when glucose is present.

(a) When cAMP is present:

cAMP binds to CAP and the cAMP-CAP complex binds to DNA at the CAP site. RNA polymerase binds the promoter efficiently. Transcription occurs frequently.

(b) When cAMP is absent:

CAP does not bind to DNA. RNA polymerase binds the promoter inefficiently. Transcription occurs rarely.

FIGURE 17.13 Positive Control of the *lac* Operon. (a) When glucose levels in an *E. coli* cell are low, cAMP is produced. cAMP then interacts with CAP to increase transcription of the *lac* operon. **(b)** When glucose is abundant, cAMP is rare in the cell and positive control does not occur.

centrations are low, intracellular cAMP concentrations are high. This seesaw is driven by the enzyme **adenylyl cyclase**, which produces cAMP from ATP (**Figure 17.14a**). Adenylyl cyclase's activity is inhibited by extracellular glucose. To see the consequences of this fact, imagine a situation in which glucose is abundant outside the cell (**Figure 17.14b**). In this state, adenylyl cyclase activity is low. Therefore, cAMP levels inside

the cell are low. CAP is not in a CAP-cAMP complex, so it does not have the conformation that allows it to bind to the CAP site and stimulate *lac* operon transcription. Conversely, when the extracellular concentration of glucose is low, there is an increase in the intracellular level of cAMP. In this case, the CAP-cAMP complex forms, binds to the CAP site, and allows RNA polymerase to initiate transcription efficiently. As long as

(a) Glucose inhibits the activity of the enzyme adenylyl cyclase, which catalyzes production of cAMP from ATP.

(b) The amount of cAMP and the rate of transcription of the *lac* operon are inversely related to the concentration of glucose.

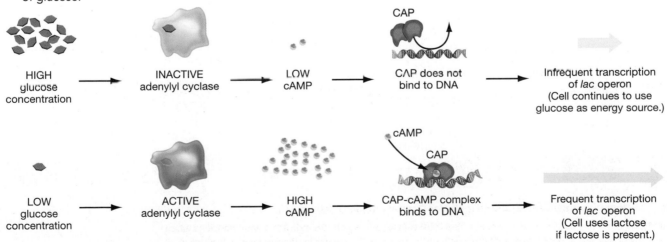

FIGURE 17.14 Cyclic AMP (cAMP) Is Synthesized When Glucose Levels Are Low.

FIGURE 17.15 An Overview of Lactose Use.

the repressor is not bound to the operator, transcription of the *lac* operon is initiated frequently and lactose can be employed as an alternative energy source.

It is important to note that the CAP-cAMP system influences many genes in addition to the *lac* operon. CAP sites are found adjacent to the promoters for several operons that are required for the metabolism of sugars other than glucose. When glucose levels fall and cAMP concentrations rise, the effect on gene regulation is similar to ringing an alarm bell. In response to the alarm, genes that encode enzymes required for the use of lactose, maltose, glycerol, and other food sources may be turned on—but only if the compound that a particular operon acts on is available to the cell. Conversely, cAMP levels are low when glucose supplies are adequate. In this case, the

genes for these enzymes are expressed only at low levels even when the appropriate metabolite is present.

Figure 17.15 summarizes how positive control and negative control combine to regulate the *lac* operon. **Figure 17.16** offers a similar summary in the form of a flowchart. The general message of these figures is that interactions among regulatory elements produce finely tuned control over gene expression. Because positive and negative control elements are superimposed, *E. coli* fully activates the genes for lactose metabolism only when lactose is available and when glucose is scarce or absent. In this way, control over gene expression increases the ability of these cells to compete, grow, and reproduce efficiently.

Web Animation at www.masteringbio.com

The *lac* Operon

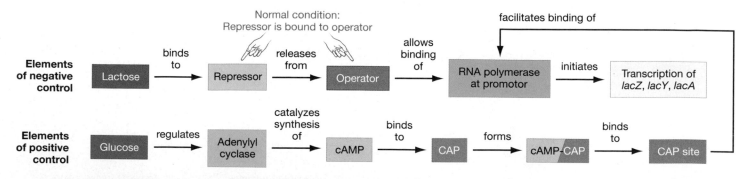

FIGURE 17.16 Negative and Positive Control in the *lac* Operon Interact: A Flowchart.

● EXERCISE Redraw this flowchart as a concept map (**BioSkills 6**), and add notes indicating what happens when concentrations of glucose and lactose are (1) glucose high/lactose low, (2) glucose low/lactose high, (3) glucose high/lactose high, and (4) glucose low/lactose low.

Check Your Understanding

If you understand that...

- Positive control occurs when something must be added for transcription to occur.
- CAP exerts positive control over the *lac* operon by binding to the CAP site and increasing the transcription rate.
- CAP can bind only when it is complexed with cAMP—a signaling molecule indicating that glucose levels are low.

You should be able to...

1) Diagram what the operon looks like when cAMP levels are low and when cAMP levels are high.

2) Explain how positive control and negative control work together to control the *lac* operon in the presence or absence of glucose and lactose.

17.5 The Operator and the Repressor—An Introduction to DNA-Binding Proteins

The insights that biologists gained from studying the *lac* operon proved to be relevant to many other genes and species. For example, transcription of bacterial and eukaryotic genes is either constitutive or controlled. When regulation occurs, it is negative, positive, or based on a combination of negative and positive factors. Negative control is based on repressor proteins, such as the *lacI* product, whereas positive control depends on transcription activator proteins, such as CAP. In both bacteria and eukaryotes, as in the *lac* operon, it is particularly common for genes to be regulated by a combination of several control systems. Catabolite repression is also extremely common.

Currently, biologists who are interested in gene regulation are trying to understand how DNA-binding proteins such as CAP and the repressor can control transcription. Precisely how do these proteins interact with DNA? How do small regulatory molecules such as cAMP and lactose induce changes in the conformation of these large DNA-binding proteins? Once again, investigators are using the *lac* operon as a model system that offers insights applicable to all cells.

Before taking a detailed look at the repressor and operator, let's step back and think about what is required for a DNA-binding protein to control transcription. First, the protein must bind to a specific DNA sequence associated with a specific gene or operon. Often, an investigator's first task is to identify what that sequence is. Second, the protein's structure must allow it to bind to DNA, interact with a regulatory molecule such as cAMP or lactose, and affect RNA polymerase. What structural features make each of these functions possible? As it turns out, the *lac* operon's operator DNA sequence and repressor protein structure have features that are common to many regulatory elements in bacteria and eukaryotes.

How Was the Operator Found?

To understand how the *lac* repressor controls transcription, a logical first step is to identify the DNA sequences it targets: the operator. By mapping mutants with defects in the operator, researchers were able to confirm that the target site is just downstream from the promoter. But what is the actual DNA sequence?

Footprinting experiments with the *lac* repressor yielded two new discoveries. First, the *lac* operon actually contains three sites where the repressor protein may bind. As **Figure 17.17a** shows, these sites are called O_1, O_2, and O_3. When the repressor protein is bound to DNA, the protein simultaneously binds O_1 and one of the two other operators. This finding required a modification of the original Jacob–Monod model, which proposed the existence of a single operator. Second, all three operators have a similar DNA sequence that has an unusual characteristic: It contains an axis of symmetry. To understand what an axis of symmetry in DNA is, look at the O_1 sequence shown in **Figure 17.17b**. Note that if the sequence were rotated 180°, it would be little changed. This type of symmetry is known as **twofold rotational symmetry** or **dyad symmetry**. Follow-up work showed that the nucleotide sequences recognized by many DNA-binding proteins have twofold rotational symmetry.

(a) Multiple binding sites for repressor.

"Upstream" O_3 O_1 O_2 "Downstream"

lacI promoter   *lacI*   CAP site   Promoter   Operator   *lacZ*   *lacY*   *lacA*

(b) Sequence at O_1 is symmetrical.

5′ **A A T T G T G A G C** G G A T A A C A A T A 3′
3′ T T A A C A C T C G C **C T A T T G T T A T** 5′

Axis of symmetry

FIGURE 17.17 Operator Sequences. (a) The *lac* operon contains three operator sequences, called O_1, O_2, and O_3, where the repressor can bind. A repressor simultaneously binds O_1 and either O_2 or O_3. **(b)** Many sequences targeted by DNA-binding proteins are symmetrical. In this example, all but three bases are identical on the two sides of the plane of symmetry.

Now that we've examined the operator, we need to consider the repressor. How does this DNA-binding protein recognize specific sequences along the double helix? In particular, why is twofold rotational symmetry important?

How Does the Repressor Bind to the Operator?

Like other proteins, the *lac* operon's repressor resembles a hand tool in one important respect: Both a hand tool and a repressor protein have distinct regions or parts. Consider a screwdriver, which consists of a handle, a shaft, and a blade. Each of these parts has a distinctive three-dimensional structure and function. Proteins also have regions or parts, called **domains**, that have a distinctive three-dimensional structure and function. Just as screwdrivers of various shapes and sizes all have a handle, shaft, and blade, it is common to observe the same domain in proteins of various shapes and sizes. A domain that is observed in many different proteins is known as a **motif**. Each individual screwdriver has a handle domain, a shaft domain, and a blade domain. As a group, screwdrivers are characterized by a handle motif, a shaft motif, and a blade motif. The handle and shaft motifs are also found in hammers and socket wrenches, but the blade motif is unique to screwdrivers. Hundreds of different motifs have been characterized in proteins.

When investigators analyzed repressor proteins from different bacteria, they discovered that many of the bacteria contained the domain illustrated in **Figure 17.18a**. This structure is a **helix-turn-helix motif**. As its name indicates, it consists of two α-helices connected by a short stretch of amino acids that forms a turn. If you hold your thumb and index finger apart, they would resemble a helix-turn-helix motif. Physical models constructed during the 1980s suggested that the helix-turn-helix region is the section of the repressor that binds to DNA. This hypothesis was later supported by structural studies based on X-ray crystallography (see **BioSkills 8**).

To see how a helix-turn-helix motif works, recall from Chapter 4 that the exterior of the DNA double helix has a minor groove and a major groove. A protein that has a helix-turn-helix domain can bind to DNA because one of the helices interacts with the sugar-phosphate backbone of a DNA strand while the other helix binds to the base pairs in the major groove (**Figure 17.18b**). The "index finger" in the helix-turn-helix motif fits in the major groove; the "thumb" hooks around the sugar-phosphate backbone. Interactions between the amino acids in the α-helices and the bases inside the major groove allow the finger and thumb to "grab" DNA. ● If you understand this concept, you should be able to predict what would happen to the interaction illustrated in Figure 17.18b if either the amino acids in the "index finger" or the bases in the major groove were changed.

Notice that DNA does not have to become single stranded for the helix-turn-helix domain of the repressor to bind to the

(a) Helix-turn-helix motif in in DNA-binding protein

(b) Recognition sequence of helix-turn-helix motif binds to DNA sequences in major groove

Interactions between amino acids in helix-turn-helix and bases in DNA

FIGURE 17.18 Helix-Turn-Helix Motifs in DNA-Binding Proteins.

● **QUESTION** The part of a helix-turn-helix motif that interacts directly with DNA is called the recognition sequence. Why?

operator. This observation holds for nearly all proteins that recognize specific sequences of double-stranded DNA.

The section of the helix-turn-helix motif that binds inside the major groove is called the recognition sequence. The amino acid sequence of this section is crucial because specific sequences of amino acids bind to specific sequences of DNA. Each type of regulatory protein with a helix-turn-helix motif has a unique sequence of amino acids in its recognition sequence. As a result, each of these regulatory proteins binds to a unique regulatory sequence in DNA. In many cases, precise control over gene expression is based on precise chemical interactions between the amino acid sequence in a recognition sequence and the DNA sequence in the major groove of a regulatory site.

Biologists used X-ray crystallography to document how the *lac* repressor binds to the DNA of the operator. Consider the structure of the repressor alone, shown in **Figure 17.19a**. Note that the protein consists of four identical polypeptide chains, and that four helix-turn-helix domains project from the main body of the molecule. When researchers analyzed the structure of the repressor while it was bound to the operator sequence, they confirmed that each of these helix-turn-helix motifs binds operator DNA inside the major groove. But when lactose molecules bind to the repressor, the helix-turn-helix motifs move in a way that disrupts their interaction with the operator **Figure 17.19b**. As a result, the repressor releases from the site. The parking brake is off, and transcription can begin.

(a) Structure of repressor protein

Helix-turn-helix

One polypeptide   Four polypeptides

(b) Interaction of inducer and repressor

O_3 O_1

DNA

Helix-turn-helix

Repressor protein

Inducer molecules

FIGURE 17.19 How Does the Repressor Interact with the Operator and the Inducer? (a) The repressor is a complex of four polypeptides, each a product of the *lacI* locus. **(b)** The repressor binds to the operator sequences only when the inducer is not present (top). When the inducer binds to the repressor, the protein changes shape and falls off the operator DNA (bottom).

What's Next?

Now that biologists understand the physical basis of negative control and induction, what does the future hold? One promising line of research is focused on how the activity of many genes and operons is coordinated. For example, suppose that an *E. coli* cell in your gut is faced with starvation. Does it adjust the activity of hundreds or even thousands of its genes in order to survive? If so, how are changes in gene expression coordi-nated throughout the genome? Because biologists have a fairly solid understanding of how individual genes and operons are regulated, the current challenge is to explore how multiple genes and operons are controlled in a way that increases the fitness of bacterial cells. Answering these questions may be important to our understanding of how certain bacteria are able to sustain in-fections that threaten human health, and how some bacterial species are able to detoxify pollutants by using them as food.

Chapter Review

SUMMARY OF KEY CONCEPTS

● **Gene expression can be controlled at three levels: transcription, translation, or post-translation (protein activation).**

Among the three levels of gene regulation, there is a trade-off be-tween the speed of response to changed conditions and the effi-cient use of resources. In post-translational control, for example, the change in gene expression is extremely fast but demands a con-siderable investment of resources to make the protein in question.

You should be able to predict whether the same three levels of control and the same trade-offs in speed versus cost occur in eukaryotes. ●

● **Changes in gene expression allow bacterial cells to respond to environmental changes.**

Transcription in bacteria is constitutive or inducible. Constitutive expression occurs in genes whose products are required at all times, such as genes that encode glycolytic enzymes. Expression of most genes is induced by environmental signals, however, so gene products are produced on demand.

Precise regulation over gene expression is important to bacterial cells. For example, some species switch between different food sources to take advantage of the nutrients currently available in their environment. Efficient use of nutrients is beneficial because cells are usually competing for access to resources. In the intestinal bacterium *E. coli*, for example, changes in gene expression allow cells to use lactose as a food source when glucose is not available.

You should be able to discuss whether changes in the environment inside a multicellular eukaryote—such as a drop in oxygen or glucose availability—would trigger changes in gene expression in its cells. ●

● **Transcriptional control can be negative or positive. Negative control occurs when a regulatory protein prevents transcription. Positive control occurs when a regulatory protein increases the transcription rate.**

Protein-coding genes involved in lactose metabolism are under both negative and positive control. Negative control occurs because a repressor protein is transcribed at low levels. The repressor binds to a DNA sequence—the operator—near the promoter of the protein-encoding genes. But when lactose is present, it binds to the repressor and induces a conformational change that causes the repressor to fall off the operator. Rapid transcription of the *lac* operon does not begin, however, unless glucose is also lacking. When glucose is scarce, cAMP levels rise in the cell, cAMP binds to the regulatory protein CAP, and the CAP-cAMP complex binds to a control sequence near the *lac* promoter. Binding by CAP-cAMP increases transcription and exerts positive control because it facilitates binding by RNA polymerase.

You should be able to extend this chapter's analogy for negative and positive control by identifying (1) how the operator, repressor, and inducer relate to a parking brake; and (2) how CAP, cAMP, and the CAP-binding site relate to a gas pedal. ●

(MB) **Web Animation** at www.masteringbio.com
The *lac* Operon

● **Many regulatory proteins bind to specific sites in DNA. Because each type of regulatory protein has a distinct amino acid sequence, each type binds to different DNA sequences.**

The regulatory proteins involved in negative control and positive control are often allosterically regulated, meaning that they have the ability to switch between two different conformations in response to binding a small molecule—often a nutrient or product affected by the gene being regulated. The change in conformation affects the protein's ability to bind to DNA.

You should be able to make a drawing that predicts how CAP interacts with the CAP-binding site in the presence or absence of cAMP and explain why a mutation that changed the binding-site sequence might increase or decrease CAP's ability to bind. ●

QUESTIONS

● Test Your Knowledge

1. *E. coli* expresses genes for glycolytic enzymes constitutively. Why?
   a. Their expression is controlled at three levels: transcriptional, translational, and post-translational.
   b. Cells always need fresh supplies of ATP, so glycolysis is occurring constantly.
   c. Transcription is activated only in response to signals indicating that ATP supplies are low.
   d. They do not need to be expressed when fermentation instead of cellular respiration is occurring.

2. Why are the genes involved in lactose metabolism considered to be an operon?
   a. They occupy adjacent locations on the *E. coli* chromosome.
   b. They have a similar function.
   c. They are all required for normal cell function.
   d. They are under the control of the same promoter.

3. In the *lac* operon, which regulatory molecule exerts negative control by inhibiting transcription?
   a. lactose (the inducer)
   b. the repressor protein
   c. the catabolite activator protein (CAP)
   d. cAMP

4. In the *lac* operon, which regulatory molecule is controlled at the post-translational level, by cAMP?
   a. lactose (the inducer)
   b. the repressor protein
   c. the catabolite activator protein (CAP)
   d. cAMP

5. What is catabolite repression?
   a. a mechanism that turns off the synthesis of enzymes responsible for catabolic reactions when the product is present
   b. a mechanism that turns off the synthesis of enzymes responsible for catabolic reactions when the product is absent
   c. repression that occurs because of allosteric changes in a regulatory protein
   d. repression that occurs because of allosteric changes in a DNA sequence

6. What is a helix-turn-helix motif?
   a. a protein domain involved in folding into the active conformation
   b. a protein domain involved in induction
   c. a protein domain involved in DNA binding
   d. a protein domain involved in catalysis

Test Your Knowledge answers: 1. b; 2. d; 3. b; 4. c; 5. a; 6. c

● Test Your Understanding

Answers are available at www.masteringbio.com

1. Explain the difference between positive and negative control over transcription. Why is it advantageous for the *lac* operon in *E. coli* to be under both positive control and negative control? What would happen if only negative control occurred? What would happen if only positive control occurred?

2. In *E. coli*, rising levels of cAMP can be a considered a starvation signal. Explain.

3. Explain the role of post-translational control in the *lac* operon.

4. CAP is also known as the cAMP-receptor protein. Why?

5. Explain how the *lac* repressor binds to DNA. Then explain the relationship between the twofold rotational symmetry of the operator and the four-part structure of the repressor protein.

6. The galactose released when β-galactosidase cleaves lactose enters the glycolytic pathway in *E. coli* once a series of enzyme-catalyzed reactions has converted the galactose to glucose-6-phosphate. Why, then, is glucose, not lactose, the preferred sugar in *E. coli*?

● Applying Concepts to New Situations

Answers are available at www.masteringbio.com

1. You are interested in using bacteria to metabolize wastes at an old chemical plant and convert them into harmless compounds. You find bacteria that are able to tolerate high levels of the toxic compounds toluene and benzene, and you suspect that it is due to the ability of the bacteria to break these compounds into less-toxic products. If that is true, these toluene- and benzene-resistant strains will be valuable for cleaning up toxic sites. How could you find out whether these bacteria have enzymes that allow them to metabolize toluene?

2. Assuming that the bacteria you examined in Applying Concepts to New Situations Problem 1 do have an enzymatic pathway to break down toluene, would you predict that the genes involved are constitutively expressed, under positive control, or under negative control? Why? What experiments could you conduct to test your hypothesis?

3. The *lacI* gene mutants produce β-galactosidase constitutively because no repressors are present to bind to the operator. Other repressor mutants have been isolated that are called *LacI*^{S} mutants. These repressor proteins continue to bind to the operator, even in the presence of the inducer. How would this mutation affect the function of the *lac* operon? Specifically, how well would *LacI*^{S} mutants do in an environment that has lactose as its sole sugar?

4. X-gal is a colorless, lactose-like molecule that can be split into two fragments by β-galactosidase. One of these product molecules is blue. The following photograph is a close-up of *E. coli* colonies growing in a medium that contains lactose.

Draw a line to three colonies whose cells have functioning copies of β-galactosidase. Draw a line to three colonies whose cells have mutations in the *lacZ* locus or in one of the genes involved in regulation of the *lacZ*. Suppose you could analyze the sequence of the β-galactosidase gene from each of the mutant colonies. How would these data help you distinguish which cells are structural mutants and which are regulatory mutants?

www.masteringbio.com is also your resource for • Answers to text, table, and figure caption questions and exercises • Answers to *Check Your Understanding* boxes • Online study guides and quizzes • Additional study tools including the *E-Book for Biological Science* 3rd ed., textbook art, animations, and videos.

18

Control of Gene Expression in Eukaryotes

KEY CONCEPTS

- Changes in gene expression allow eukaryotic cells to respond to changes in the environment and cause distinct cell types to develop.

- In eukaryotes, DNA is packaged with proteins into complex structures that must be opened before transcription can occur.

- In eukaryotes, transcription can be initiated only when specific proteins bind to the promoter and to regulatory sequences that may be close to the promoter or far from it.

- Alternative splicing allows a single gene to code for several different products.

- Once an mRNA is in the cytoplasm, gene expression is controlled by molecules that regulate the (1) life span of mRNAs, (2) efficiency of translation, and (3) activation or inactivation of protein products.

- Cancer can develop when mutations disable genes that regulate cell-cycle control genes.

This model shows what eukaryotic DNA looks like in the condensed state. The DNA has to be uncoiled before transcription can take place.

In bacteria, precise regulation of genes is essential for cells to respond to changes in their environment, as Chapter 17 indicated. In *Escherichia coli*, the genes that are required to import and cleave lactose are rapidly expressed only when the cells have to rely on lactose as a source of energy—when glucose is absent and lactose is present.

Unicellular eukaryotes face similar challenges. Consider the yeast *Saccharomyces cerevisiae*, which is used extensively in the production of beer, wine, and bread. In nature, the cells live on the skins of grapes and other fruits. In the wild, there is often wide variation in temperature and humidity. In addition, the sugars that are available to the cells change dramatically in type and concentration as the fruit they are living on

ripens, falls, and rots. For yeast cells to grow and reproduce efficiently, gene expression has to be modified in response to these environmental changes. Further, the offspring of a cell that lives on a grape may be blown to a nearby orchard and take up residence on an apple. Different genes are expressed in yeast cells that live on grapes versus those that live on apples.

The cells that make up multicellular eukaryotes face an even greater challenge. Consider that your body consisted of a single cell at one time—a fertilized egg. Now you have trillions of cells, each with a specialized structure and function. You have heart muscle cells, lung cells, nerve cells, skin cells, and so on. Even though these cells look and behave very differently, each

● Key Concept   Important Information   ○ Practice It

of them contains the same genes. Your bone cells are different from your blood cells not because they contain different genes, but because they express different genes. Your bone cells have blood cell genes—they just don't transcribe them.

How is it possible for the cells inside your body to express unique sets of genes? The answer is that your cells respond to changes in their environment, just as bacteria and unicellular eukaryotes do. But instead of responding to the presence of molecules such as glucose or lactose in the external environment, cells in a multicellular eukaryote respond to the presence of signals from other cells—from an internal environment. As a human being or an oak tree develops, cells that are located in different parts of the organism are exposed to different signals. As a result, they express different genes. ● **Differential gene expression** is responsible for creating different cell types, arranging them into tissues, and coordinating their activity to form the multicellular society we call an individual.

How does all this happen? Unit 4 explores the nature of the signals that trigger the formation of muscle, bone, leaf, and flower cells in multicellular organisms. Here we need to focus on what happens once a eukaryotic cell receives a signal. Let's begin by surveying the various ways that gene expression can be controlled and then plunge into a detailed look at key points in gene regulation. The chapter closes with a look at how defects in the control of gene expression can help trigger cancer.

18.1 Mechanisms of Gene Regulation— An Overview

Like bacteria, eukaryotes can control gene expression at the levels of transcription, translation, and post-translation (protein activation or inactivation). But as **Figure 18.1** shows, two additional levels of control occur in eukaryotes as genetic information flows from DNA to proteins. The first level involves the DNA-protein complex at the top of the figure. In eukaryotes, DNA is wrapped around proteins to create a protein-DNA complex called **chromatin**. Eukaryotic genes have promoters, just as bacterial genes do. Before transcription can begin in eukaryotes, DNA near the promoter must be released from tight interactions with proteins so that RNA polymerase can make contact with the promoter. To capture this point, biologists say that **chromatin remodeling** must occur prior to transcription.

The second level of regulation that is unique to eukaryotes involves **RNA processing**—the steps required to produce a mature, processed mRNA from a **primary RNA transcript**, the preliminary result of transcription. Recall from Chapter 16 that introns have to be spliced out of primary transcripts. In some cases, carefully orchestrated alterations in the splicing pattern occur. When the splicing events in a particular primary RNA transcript change, a different message emerges. The altered message, in turn, leads to the production of a different product.

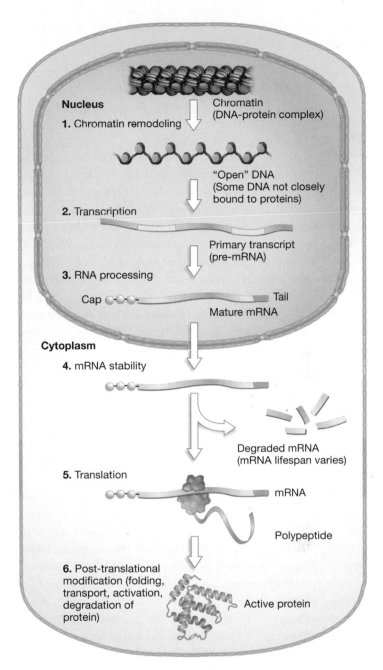

FIGURE 18.1 In Eukaryotes, Gene Expression Can Be Controlled at Many Different Steps. Numerous steps occur during protein synthesis in eukaryotes. Each step provides an opportunity for regulation of gene expression.

● **EXERCISE** Place and label a horizontal line such that mechanisms of transcriptional control fall above the line and mechanisms of post-transcriptional control fall below it.

Each of the six potential control points shown in Figure 18.1 is employed at certain times in a typical eukaryotic cell. This chapter explores all six—chromatin remodeling, transcription initiation, RNA processing, mRNA stability, translation, and post-translational modification of proteins.

To appreciate the breadth and complexity of gene regulation in eukaryotes, let's follow the series of events that occurs as an embryonic cell responds to a developmental signal. Suppose a molecule arrives that specifies the production of a muscle-specific protein. What happens next?

18.2 Eukaryotic DNA and the Regulation of Gene Expression

If the arrival of a signaling molecule from outside the cell is going to result in the transcription of a particular gene, the chromatin around the target gene must be drastically remodeled. To appreciate why, consider that a typical cell in your body contains about 6 billion base pairs of DNA. Lined up end to end, these nucleotide pairs would form a double helix about 2 m (6.5 feet) long. But the nucleus that holds this DNA is only about 5 μm in diameter—less than the thickness of this page. In eukaryotes, DNA is packed inside the nucleus so tightly that RNA polymerase can't access it. Part of this packing is done by supercoiling—meaning the DNA double helix is twisted on itself many times, just as in bacterial chromosomes (see Chapter 7). The supercoiled DNA found in bacteria does not need to be altered extensively before it can be transcribed, however. Eukaryotic DNA is different—it has to undergo a series of significant physical changes before transcription can take place. The reason is that supercoiling is just part of the packing system found in eukaryotes.

How Is Chromatin Structured?

The first data on the physical nature of eukaryotic DNA were published in the early 1900s, when chemical analyses established that eukaryotic DNA is intimately associated with proteins. Later work documented that the most abundant DNA-associated proteins belong to a group called the **histones**. In the 1970s electron micrographs like the one in **Figure 18.2a** revealed that the protein-DNA complex, or chromatin, has a regular structure. In some preparations for electron microscopy, chromatin actually looked like beads on a string. The "beads" came to be called **nucleosomes**. More details emerged in 1984 when researchers determined the three-dimensional structure of eukaryotic DNA by using X-ray crystallography (a technique introduced in **BioSkills 8**). The X-ray crystallographic data indicated that each nucleosome consists of DNA wrapped almost twice around a core of eight histone proteins. As **Figure 18.2b** indicates, a histone called H1 "seals" DNA to each set of eight nucleosomal histones. Between each pair of nucleosomes there is a "linker" stretch of DNA.

The intimate association between DNA and histones occurs in part because DNA is negatively charged and histones are positively charged. DNA has a negative charge because of its phosphate groups; histones are positively charged because they contain many lysine and/or arginine residues (see Chapter 3).

More recent work has shown that there is another layer of complexity in eukaryotic DNA. As **Figure 18.2c** indicates, H1 histones interact with each other and with histones in other

(a) Nucleosomes in chromatin

(b) Nucleosome structure

(c) In some cases, nucleosomes may be grouped into 30-nanometer fibers.

FIGURE 18.2 Chromatin Has Several Levels of Structure. (a) Electron micrograph of chromatin. **(b)** The nucleosome is the basic unit of chromosome structure. **(c)** Nucleosomes (shown as green spheres here) may become organized into 30-nm fibers, with H1 proteins forming a core in the center.

nucleosomes to produce a tightly packed structure. Based on its width, this structure is called the 30-nanometer fiber. (Recall that a nanometer is one-billionth of a meter and is abbreviated nm.) Often, 30-nm fibers are packed into still larger structures, though it is still not clear exactly what these higher-order structures look like.

A chromosome, then, is made up of chromatin—or DNA complexed with proteins—that has several layers of organization. DNA is wrapped around nucleosomes; nucleosomes are packed into 30-nm fibers, and 30-nm fibers are folded into structures that are still being studied.

The elaborate structure of chromatin does more than just package DNA into the nucleus. Chromatin structure has profound implications for the control of gene expression. To drive this point home, consider the 30-nm fiber illustrated in Figure 18.2c. If this tightly packed stretch of DNA contains a promoter, how can RNA polymerase bind to it and initiate transcription?

Evidence that Chromatin Structure Is Altered in Active Genes

Once the nucleosome-based structure of chromatin was established, biologists hypothesized that the close physical interaction between DNA and histones must be altered for RNA polymerase to make contact with DNA. More specifically, biologists hypothesized that a gene could not be transcribed until the chromatin near its promoter was remodeled. ● The idea was that the chromatin must be relaxed or decondensed for RNA polymerase to bind to the promoter. If so, then chromatin remodeling would represent the first step in the control of eukaryotic gene expression. Two types of studies have provided strong support for this hypothesis.

The first type of evidence comes from studies with the enzyme DNase, which cuts DNA at random locations. DNase cannot cut DNA efficiently if the molecule is tightly complexed with histones. As **Figure 18.3** shows, the enzyme works effectively only if DNA is in the "open" configuration. Harold Weintraub and Mark Groudine used this observation to test the hypothesis that the DNA of actively transcribed genes is in an open configuration. They performed the test by comparing chromatin structure in two genes in blood cells of chickens: the β-globin and ovalbumin genes. β-Globin is a protein that is part of the hemoglobin found in red blood cells; ovalbumin is a major protein of egg white. In blood cells, the β-globin gene is transcribed at high levels. The ovalbumin gene, in contrast, is not transcribed at all. After treating blood cells with DNase and then analyzing the state of the β-globin and ovalbumin genes, the researchers found that DNase cut up the β-globin gene much more readily than the ovalbumin gene. They interpreted this finding as evidence that chromatin in blood cells was in an open configuration at the β-globin gene but closed at the ovalbumin gene. Analogous studies using DNase on different genes and in different cell types yielded similar results.

The second type of evidence in support of the chromatin-remodeling hypothesis comes from studies of mutant brewer's yeast cells that do not produce the usual complement of histones. Researchers found that many yeast genes that are normally never transcribed are instead transcribed at high levels at all times in these mutant cells. To interpret this finding, biologists hypothesized that the lack of histone proteins prevented the assembly of normal chromatin. If the absence of normal histone-DNA interactions promotes transcription, then the presence of normal histone-DNA interactions must prevent it.

Taken together, the data suggest that the normal or default state of eukaryotic genes is to be turned off. This is a new mechanism of negative control—different from the repressor proteins introduced in Chapter 17. When DNA is wrapped into a 30-nm fiber, the parking brake is on. If so, then gene expression depends on chromatin being opened up in the promoter region. Some sort of positive control must be at work.

How Is Chromatin Altered?

Although much remains to be learned about how chromatin is opened up prior to transcription, researchers have made important progress recently. One key finding is that two major types of protein are involved in modifying chromatin structure. One group of proteins creates structures called chromatin-remodeling complexes, which reshape chromatin through a series of reactions that are dependent on ATP. It is not known exactly how these ATP-dependent enzymes work, however. The second major type of chromatin-modifying protein works by adding small molecules such as acetyl (CH_3COOH) or methyl (CH_3) groups to histones. These processes are known as **acetylation** and **methylation**, respectively.

The **histone acetyl transferases (HATs)** are some of the best-studied enzymes that modify chromatin through acetylation or methylation. HATs acetylate the positively charged lysine residues in histones. When a HAT adds an acetyl group to selected histones, the number of positive charges on the histones is reduced. The result is less electrostatic attraction between histones and the negatively charged DNA. The association between nucleosomes and DNA is loosened and the chromatin decondenses.

Condensed chromatin (many histones)

Open chromatin

Treatment with DNase I

Condensed chromatin

Degraded DNA

FIGURE 18.3 DNase Assay for Chromatin Structure. DNase is an enzyme that cuts DNA at random locations. It cannot cut intact condensed chromatin.

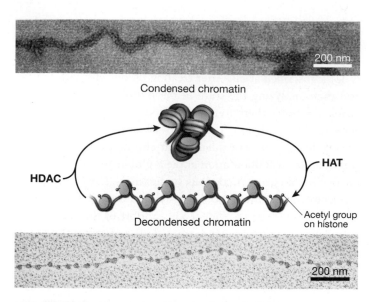

FIGURE 18.4 Chromatin Decondensation and Condensation Are Driven by Histone Acetyl Transferases and Histone Deacetylases. Histone acetyl transferases (HATs) cause chromatin to decondense, and histone deacetylases (HDACs) cause it to condense.

QUESTION Are HATs and HDACs elements in positive control or in negative control?

As **Figure 18.4** shows, chromatin is "recondensed" by a group of enzymes called **histone deacetylases (HDACs)**, which remove the acetyl groups added by HATs. Histone deacetylase activity reverses the effects of acetylation, returning chromatin to its default condensed state. If HATs are an on switch for transcription, HDACs are the off switch.

Chromatin Modifications Can Be Inherited

Work on chromatin remodeling has one fundamental message: The state of the histone proteins that are complexed with DNA is fundamental to determining whether transcription can occur. Acetylation of histones is usually associated with positive control—meaning, activation of genes. Methylation can be correlated with either activation or inactivation, depending on which histones are altered and where the methyl groups occur on the protein. Methylation can set the parking brake or release it, depending on conditions.

The pattern of chemical modifications occurring on histones varies from one cell type to another. For example, suppose you analyzed the same gene in a cell that was destined to give rise to muscle and in a cell that is a precursor to part of the brain. The gene would likely have a completely different pattern of acetylation and methylation in the two cell types. Just as important, some or most these chemical modifications are passed on to daughter cells when the muscle- and brain-associated cells divide. This is a key observation, because it means that daughter cells inherit patterns of gene expression from their

parent cells. This is an example of **epigenetic inheritance**, or patterns of inheritance that are not due to differences in gene sequences. Muscle cells are different from brain cells in part because they inherited different types of modified histones—not different types of genes.

Now the question is, what happens once a section of DNA is opened up by chromatin remodeling and exposed to RNA polymerase? In bacteria, the protein sigma binds to RNA polymerase and allows it to contact the promoter. Then transcription begins if any required positive regulators, such as CAP, are in place and negative regulators, such as the repressor, are absent. Does anything similar happen in eukaryotes?

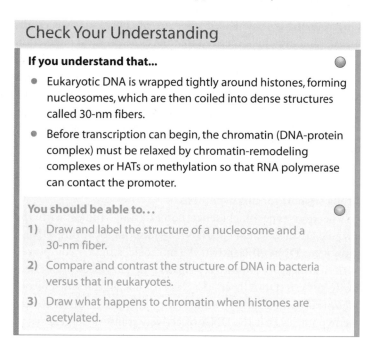

Check Your Understanding

If you understand that...

- Eukaryotic DNA is wrapped tightly around histones, forming nucleosomes, which are then coiled into dense structures called 30-nm fibers.

- Before transcription can begin, the chromatin (DNA-protein complex) must be relaxed by chromatin-remodeling complexes or HATs or methylation so that RNA polymerase can contact the promoter.

You should be able to...

1) Draw and label the structure of a nucleosome and a 30-nm fiber.

2) Compare and contrast the structure of DNA in bacteria versus that in eukaryotes.

3) Draw what happens to chromatin when histones are acetylated.

18.3   Regulatory Sequences and Regulatory Proteins

Chapter 16 introduced the **promoter**—the site in DNA where RNA polymerase binds to initiate transcription. Based on their position and function, eukaryotic promoters are similar to bacterial promoters. Most eukaryotic promoters are located just upstream of the point where RNA polymerase begins transcription, and all have a highly conserved element analogous to the −35 box and −10 box in bacterial promoters. For example, many genes that are transcribed by RNA polymerase II contain a specific base sequence, known as a **TATA box**, where a sigma-like protein binds and allows the enzyme to contact DNA. Recall from Chapter 16 that bacterial promoters may vary in sequence and bind different types of sigma proteins. Eukaryotic genes also have promoters that vary in sequence, but all eukaryotic promoters are bound by the same protein: the **TATA-binding protein** (TBP).

If eukaryotic genes have promoters that interact with the same promoter-binding protein, how can transcription be controlled? The answer lies in interactions between regulatory sequences other than the promoter and regulatory proteins other than TBP. **Regulatory sequences** are sections of DNA that are involved in controlling the activity of genes, similar to the CAP site and operators introduced in Chapter 17. Eukaryotic regulatory proteins, which change gene activity when they bind to regulatory sites, are analogous to *E. coli*'s CAP and the repressor protein analyzed in Chapter 17.

As it turns out, some regulatory sequences found in eukaryotes are similar to those observed in bacteria. Others are radically different. Let's take a closer look.

Some Regulatory Sequences Are Near the Promoter

The first regulatory sequences in eukaryotic DNA were discovered in the late 1970s, when Yasuji Oshima and co-workers developed an important model system to study control of gene expression in eukaryotes. Much as François Jacob and Jacques Monod had focused on lactose metabolism in *E. coli*, Oshima's team set out to understand how yeast cells control the metabolism of the sugar galactose. When galactose is absent, yeast cells produce tiny quantities of the enzymes required to metabolize it. But when galactose is present, transcription of the genes encoding these enzymes increases by a factor of 1000.

Oshima and co-workers focused their early work on mutants that were unable to use galactose. The team's first major result was the discovery of mutant cells that failed to produce any of the five enzymes required for galactose metabolism, even if galactose was present. To interpret this observation, they hypothesized that the cells had a knock-out (loss-of-function) mutation that completely disabled a regulatory protein. Like CAP, this hypothesized protein was thought to exert positive control over five genes—the five that coded for the enzymes involved in galactose metabolism. Although not on the same chromosome, the five genes appeared to be regulated together. The team called the hypothesized regulatory protein GAL4 and the gene that encodes it *GAL4*. (In *S. cerevisiae*, the names of genes are written in full caps and italicized. As in other species, gene names are italicized but protein names are not.)

This hypothesis was supported when other researchers isolated the regulatory protein and found that it has a DNA-binding domain, analogous to the helix-turn-helix motif introduced in Chapter 17. The domain binds to a short stretch of DNA located just upstream from the promoter for the five genes that GAL4 regulates. The location and structure of this regulatory sequence are comparable to those of the CAP binding site in the *lac* operon of *E. coli*. Similar regulatory sequences have now been found in a wide array of eukaryotic genes and species. Because such sequences are located close to

FIGURE 18.5 Promoter-Proximal Elements Regulate the Expression of Some Eukaryotic Genes. All eukaryotic genes have promoters, the sites where RNA polymerase initially is in contact with the gene. Exons and introns are not drawn to scale throughout this chapter. They are typically very large compared with promoters and promoter-proximal (regulatory) elements.

the promoter and bind regulatory proteins, they are termed **promoter proximal elements**.

As **Figure 18.5** shows, promoter-proximal elements are just upstream from the promoter and the gene's start site; the exons and introns that make up the actual gene are farther downstream. Unlike the promoter itself, promoter-proximal elements have sequences that are unique to specific genes. In this way, they furnish a mechanism for eukaryotic cells to exert precise control over transcription.

The discovery of positive control and promoter-proximal elements provided a satisfying parallel between gene regulation in bacteria and in eukaryotes. This picture changed, however, when researchers discovered a new class of eukaryotic DNA regulatory sequences—sequences unlike anything in bacteria.

Some Regulatory Sequences Are Far from the Promoter

Susumu Tonegawa and colleagues made a discovery that may rank as the most startling in the history of research on gene expression. Tonegawa's group was exploring how human immune system cells regulate the genes involved in the production of antibodies. **Antibodies** are proteins that bind to specific sites on other molecules. In your immune system, antibodies bind to viruses and bacteria and mark them for destruction by other cells. The antibody gene that the researchers were working with is broken into many introns and exons. Introns are DNA sequences spliced out of the primary mRNA transcript; exons are regions of eukaryotic genes that are included in the mature RNA once splicing is complete. The biologists used techniques that will be introduced in Chapter 19 to place copies of an intron in new locations and found that, when they placed the intron close to a gene, the gene's transcription rate increased. Based on this observation, Tonegawa and co-workers hypothesized that the intron contained some sort of regulatory sequence.

Question: Can a regulatory sequence in DNA be located far from the promoter?

Hypothesis: A regulatory sequence exists far from the promoter of an antibody-producing gene—in an intron.

Alternative hypothesis: Regulatory sequences are located close to the promoter—not in introns or in other distant sequences.

Experimental setup:

Antibody-producing gene

Promoter   Exon   Intron   Exon

1. An antibody-producing gene in humans has a large intron.

Intron deletion

2. Use enzymes to cut out specific sections of the intron. (Several sections were cut out—only one example is shown.)

3. Use an enzyme to put DNA fragments back together.

1   2   3   4

Normal human antibody-producing cell

Control mouse cell (no gene inserted)

Intact gene (as in cell 1) in mouse cell

Modified genes in mouse cell

4. Introduce intact or modified genes into mouse cells, which do not normally produce this protein. Different types of modified genes are missing different portions of the intron.

Prediction: Transcription of antibody mRNA will be reduced when part of the intron is deleted.

Prediction of alternative hypothesis: Transcription of antibody mRNA will not be affected by deleting part of the intron.

Results:

	1	2	3	4	
	Normal human antibody-producing cell	Control mouse cell (no gene inserted)	Intact gene (as in cell 1) in mouse cell	Modified genes in mouse cell	Genes that have a specific section of intron deleted
mRNA levels:	HIGH	Not detectable	HIGH	Not detectable	are not transcribed

Conclusion: The deleted part of the intron must contain a regulatory sequence that is required for transcription. Thus, regulatory sequences in DNA may be located far from the promoter—in this case, in an intron.

FIGURE 18.6 Evidence that Enhancers Are Required for Transcription.

● **QUESTION** Why did the researchers assess mRNA production in human cells that did *not* receive a modified or intact gene?

To test this hypothesis, the biologists performed what is now considered a classic experiment (**Figure 18.6**). The protocol was simple in concept: Starting with a human antibody-producing gene that included an intron flanked by two exons (step 1 of Figure 18.6), the team used enzymes to cut out several different specific pieces of the intron (step 2) and to ligate (link) the resulting sections of the gene back together (step 3). Each of the modified genes was missing a different section of the intron. Then the team inserted the various versions of the gene into mouse cells, which do not normally contain the antibody-producing gene. Some mouse cells received normal copies of the gene; others received copies of the gene missing different portions of the intron; still others received no gene (step 4). Finally, the researchers analyzed the mRNAs that were produced and compared them with mRNA from a normal human cell and from mouse cells that received the normal gene. If a regulatory sequence was located inside the intron as

they predicted, then some of the modified genes should lack that sequence and fail to transcribe the antibody gene.

As Figure 18.6 shows, some of the modified copies of the gene were not transcribed at all. Based on these results, Tonegawa and co-workers proposed that the intron contains a regulatory sequence that is required for transcription to occur.

This result was remarkable for two reasons: (1) The regulatory sequence was thousands of bases away from the promoter, and (2) it was downstream of the promoter instead of upstream. Regulatory elements that are far from the promoter are termed **enhancers**. Follow-up work has shown that enhancers occur in all eukaryotes and that they have several key characteristics:

● Enhancers can be more than 100,000 bases away from the promoter. They can be located in introns or in untranscribed 5′ or 3′ sequences flanking the gene (see **Figure 18.7**). Researchers have yet to find enhancers located in exons.

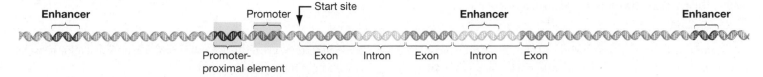

FIGURE 18.7 **Enhancers Are Far from the Genes They Regulate.** Eukaryotic genes often have more than one enhancer. Enhancers can be located either 5′ or 3′ to the gene they regulate or in introns and can be located tens of thousands of base pairs away from the promoter. All of the colored sections of the DNA strand shown here are part of the same gene.

● **QUESTION** How does the structure of this typical eukaryotic gene compare with the structure of a bacterial operon?

- Like promoter-proximal elements, many types of enhancers exist. Different enhancers are associated with different genes.

- Enhancers can work even if their normal 5′ → 3′ orientation is flipped.

- Enhancers can work even if they are moved to a new location in the vicinity of the gene, on the same chromosome.

Enhancers are regulatory sequences that are unique to eukaryotes. When regulatory proteins bind to enhancers, transcription begins. Enhancers are a gas pedal—an element in positive control. In addition, eukaryotic genomes contain regulatory sequences that are similar in structure to enhancers but opposite in function. These sequences are **silencers**. When regulatory proteins bind to silencers, transcription is shut down. Silencers are a brake—an element in negative control.

The discovery of enhancers and silencers expanded the catalog of regulatory sites known in organisms and inspired researchers to reconsider the nature of the gene. Biologists now define the **gene** as a section of DNA that codes for a functional polypeptide or RNA molecule along with the regulatory sequences required for expression.

Once enhancers and silencers had been characterized, it was clear that they represented a type of regulatory sequence that worked very differently than promoter-proximal elements did. For example, there was good evidence that GAL4 and other types of regulatory proteins bind to promoter-proximal elements and interact directly with TBP or RNA polymerase. These interactions stabilize binding and promote transcription. But how could a regulatory sequence that is distant from the promoter help initiate transcription?

What Role Do Regulatory Proteins Play?

An experiment that followed up on Tonegawa's work supported the hypothesis that enhancers are binding sites for proteins that regulate transcription. As **Figure 18.8** shows, researchers cut the enhancer sequence from an antibody-producing gene and spliced it into copies of the gene that codes for the protein β-globin. When this modified gene was inserted into antibody-producing cells, which normally do not express the gene for β-globin, the β-globin mRNA was produced. This result suggests that antibody-producing cells contain some factor that interacts specifically with the antibody-producing gene's enhancer to induce

FIGURE 18.8 **Evidence that Enhancers Are Involved in Tissue-Specific Expression of Genes**

transcription. Follow-up work showed that the factor is a regulatory protein. Enhancers are regulatory sequences where regulatory proteins bind.

By analyzing mutant yeast, fruit flies, and roundworms that have defects in the expression of particular genes, biologists have identified a large number of regulatory proteins that bind to enhancers and silencers. This research program has supported one of the most general conclusions about gene regulation in eukaryotes: In multicellular species, different types of cells express different genes because they contain different regulatory proteins. The regulatory proteins, in turn, are produced in response to signals that arrive from other cells early in embryonic development. For example, a signaling molecule might arrive at a cell early in development and trigger the production of regulatory proteins that are specific to muscle cells. Because the regulatory proteins bind to specific enhancers and silencers and promoter-proximal elements, they trigger the production of muscle-specific proteins (**Figure 18.9**). If no "become a muscle cell" signal arrives, then no muscle-specific regulatory proteins are produced and no muscle-specific gene expression takes place.

⬤ To summarize, differential gene expression is based on the production of specific regulatory proteins. Eukaryotic genes are turned on when specific regulatory proteins bind to enhancers and promoter-proximal elements; the genes are turned off when regulatory proteins bind to silencers. Distinctive regulatory proteins are what make a muscle cell a muscle cell and a bone cell a bone cell.

How do the interactions between regulatory proteins and DNA sequences control transcription? As we'll see in the next section, the initiation of transcription is a marvelously complex process in eukaryotes. It requires the coordinated activity of many types of proteins interacting with each other and a variety of DNA sequences.

Check Your Understanding

If you understand that...   ⬤

- Eukaryotic genes have regulatory sequences called promoter-proximal elements close to their promoters.

- Eukaryotic genes also have regulatory sequences called enhancers or silencers far from their promoters.

You should be able to...   ⬤

1) Explain the logic of the experiment that led to the discovery of an enhancer in an intron.

2) Compare and contrast regulatory sequences in the DNA of bacteria to those in eukaryotes.

3) Describe how gene expression is dependent on the production of specific regulatory proteins in eukaryotes.

EXTRACELLULAR SIGNALS TRIGGER CELL-SPECIFIC GENE EXPRESSION.

Extracellular signals

Receptor protein in membrane

1. Signal arrives at cell with message: "Become a muscle cell."

Plasma membrane

Intracellular signals

2. Signal transduction results in production of intracellular signal.

Cytoplasm

Regulatory proteins

3. Regulatory proteins are produced or activated in response to intracellular signal.

Nuclear envelope

Promoter-proximal element

Promoter

4. Regulatory proteins bind to regulatory sites in DNA, triggering expression of muscle-cell-specific genes.

TRANSCRIPTION

Enhancer

RNA polymerase

Exon   Intron   Exon   Intron   Exon

Enhancer

Gene for muscle-specific protein

FIGURE 18.9 Why Are Certain Proteins Produced Only in Certain Types of Cells? Signaling molecules from outside the cell trigger the production or activation of cell-specific regulatory proteins, which then influence transcription by binding to enhancers, silencers, or promoter proximal elements in DNA. (See Chapter 8 to review how signal transduction occurs.)

18.4 Transcription Initiation

Many questions remain about how transcription is initiated in eukaryotes. It is clear, however, that two broad classes of regulatory proteins interact with regulatory sequences at the start of transcription: (1) regulatory transcription factors and (2) basal transcription factors. **Regulatory transcription factors** are proteins that bind to enhancers, silencers, or promoter-proximal elements. These transcription factors are responsible for the expression of particular genes in particular cell types and at particular stages of development. **Basal transcription factors**, in contrast, interact with the promoter and are not restricted to particular cell types. Basal transcription factors must be present for transcription to occur, but they do not provide much in the way of regulation. TBP, for example, is a basal transcription factor that is common to all genes. Other basal transcription factors are specific to promoters recognized by RNA polymerase I, II, or III. ● If you understand this concept, you should be able to compare and contrast the regulatory and basal transcription factors found in muscle cells versus nerve cells.

In addition, proteins called **coactivators** are involved in starting transcription. Coactivators do not bind to DNA. Instead, they link the proteins involved in initiating transcription—regulatory transcription factors and basal transcription factors. Their activity helps explain how regulatory sites can be so far from the transcription start site.

Figure 18.10 summarizes a current model for how transcription is initiated in eukaryotes. The first step is the binding of regulatory transcription factors to DNA; these factors recruit chromatin-remodeling complexes and histone acetyl transferases (HATs). The result is chromatin remodeling, a loosening of the chromatin structure. Once the chromatin-remodeling

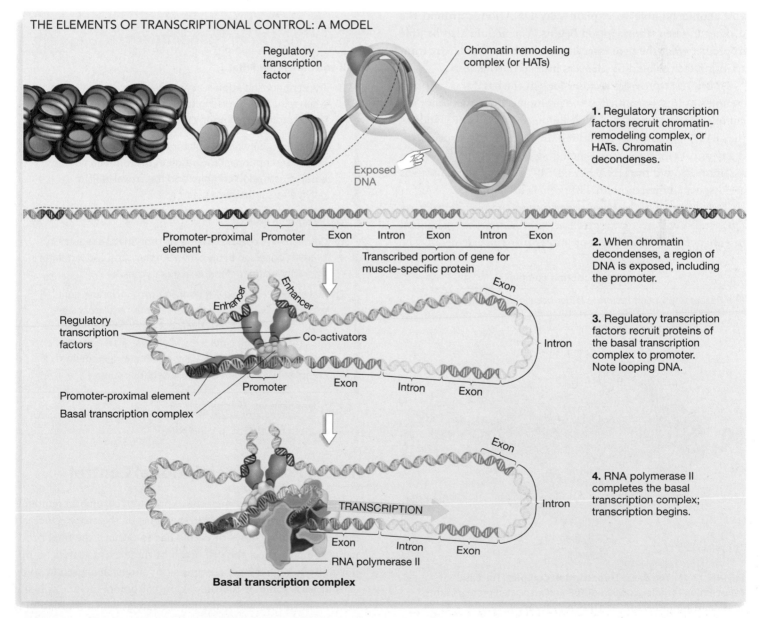

FIGURE 18.10 The Elements of Transcriptional Control: A Model. According to the currently accepted model, transcription is initiated through a series of steps.

complexes and HATs are in place, they open a broad swath of chromatin that includes the promoter region (step 2).

Once chromatin's grip is relaxed, other regulatory transcription factors can bind to enhancers or promoter-proximal elements. When these regulatory transcription factors are bound to DNA, they interact with basal transcription factors. Because enhancers are located far from the promoter, the regulatory transcription factors are normally far from the basal transcription factors. For the two sets of proteins to make contact, DNA has to loop out and away from the promoter, as shown in Figure 18.10 (step 3). When all of the basal transcription factors have assembled at the promoter in response to interactions with regulatory transcription factors and coactivators, they form a multi-protein machine called the **basal transcription complex**. The basal transcription complex then "recruits" RNA polymerase II so that transcription can begin (step 4).

● If you understand transcriptional control in eukaryotes, you should be able to explain why DNA loops around the promoter when transcription begins. You should also be able to predict what the promoter looks like when regulatory transcription factors bind to silencers instead of enhancers.

Figure 18.11 provides a closer look at the basal transcription complex as it is assembled at the promoter. Construction of the complex begins when the TATA-binding protein (TBP) binds to the TATA box in the promoter. In a multistep process, as many as 60 other proteins—including RNA polymerase—then assemble around the DNA-bound TBP. Some of these proteins are associated directly with TBP; others are not.

The assembly of the basal transcription complex depends on interactions with regulatory transcription factors that are bound to enhancers, silencers, and promoter-proximal elements. The

result is a large, multimolecular machine that is positioned at the start site and able to start transcription. Compared with the situation in bacteria, where just 3 to 5 proteins may interact at the promoter to initiate transcription, the state of affairs in eukaryotes is remarkably complicated.

Currently, biologists are focused on understanding exactly how regulatory and basal transcription factors interact to control the formation of the basal transcription complex. Given the number of different proteins involved, progress may be slow. Progress is important, however, for transcription initiation lies at the heart of gene expression. Careful regulation of transcription is critical not only to the development of embryos but also to the daily life of eukaryotes. Right now, cells throughout your body are starting and stopping the transcription of specific genes in response to signals from nearby and distant cells. As the environment inside and outside your body continually changes, your cells continually change which genes are being transcribed.

Check Your Understanding

If you understand that...

● Transcription initiation is a multistep process that begins when regulatory transcription factors bind to DNA and recruit proteins that open chromatin.

● Interactions between regulatory transcription factors and basal transcription factors result in the formation of the basal transcription complex and the arrival of RNA polymerase at the gene's start site.

You should be able to...

1) Compare and contrast the sigma proteins and regulatory proteins observed in bacteria with the basal and regulatory transcription factors found in eukaryotes.

2) Make a concept map with boxed terms including gene expression, bacteria, eukaryotes, sigma, regulatory sites, regulatory proteins, coactivators, promoter, promoter-proximal element, enhancers, silencers, basal transcription complex, TBP, RNA polymerase, chromatin remodeling complex/HATs, positive control, negative control.

(MB) **Web Animation** at www.masteringbio.com
Transcription Initiation in Eukaryotes

18.5 Post-Transcriptional Control

In the process of regulating gene expression, chromatin remodeling and transcription are just the start of the story. Once an mRNA is made, a series of events has to occur if the final product is going to affect the cell. Each of these events offers an opportunity to regulate gene expression, and each is used in some cells at least some of the time. Control points include splicing mRNAs in various ways, altering the rate at which translation is initiated, and modifying the life span of mRNAs and proteins after translation has occurred. Let's consider each in turn.

Basal transcription complex

Basal transcription factors associated with TBP
Other basal transcription factors
TATA
TBP
RNA polymerase II
Promoter
Start site

FIGURE 18.11 The Basal Transcription Complex. The basal transcription complex consists of TBP and associated proteins, other basal transcription factors, and RNA polymerase II.

● **QUESTION** How does this situation compare with the situation at bacterial promoters?

Alternative Splicing of mRNAs

Introns are spliced out of primary RNA transcripts while the message is still inside the nucleus. The mRNA that results from splicing consists of sequences that are encoded by exons and are protected by a cap on the 5′ end and a long poly(A) tail on the 3′ end. Recall from Chapter 16 that splicing is accomplished by the molecular machines called **spliceosomes.** What that chapter did not mention, however, is that splicing provides an opportunity for the regulation of gene expression.

⬤ During splicing, changes in gene expression are possible because selected exons, as well as introns, may be removed. As a result, the same primary RNA transcript can yield mature, processed mRNAs with several different combinations of transcribed exons. This is important. If the sequence of ribonucleotides in the mature mRNAs varies, then the polypeptides translated from those mature mRNAs will likewise vary. When the same primary RNA transcript is spliced in different ways to produce different mature mRNAs and thus different proteins, **alternative splicing** is said to occur.

To see how alternative splicing works, consider the muscle-cell protein tropomyosin. The tropomyosin gene is expressed in both skeletal muscle and smooth muscle, which are distinct muscle types. (Skeletal muscle is responsible for moving your bones; smooth muscle lines many parts of your gut and certain blood vessels. The two types of muscle are composed of distinct types of muscle cells.) As **Figure 18.12a** shows, the primary transcript from the tropomyosin gene contains 14 exons. In skeletal muscle and smooth muscle cells, different subsets of the 14 exons are spliced together to produce two different messages for translation (**Figure 18.12b**). Each mature mRNA contains information from a different combination of exons. As a result of alternative splicing, the tropomyosin proteins found in these two cell types are distinct. One of the reasons skeletal muscle and smooth muscle are different is that they contain different types of tropomyosin.

Alternative splicing is controlled by proteins that bind to mRNAs in the nucleus and interact with spliceosomes. When cells that are destined to become skeletal muscle or smooth muscle are developing, they receive signals leading to the production of specific proteins that are active in the regulation of splicing. Instead of transcribing different versions of the tropomyosin gene, the cells splice the same primary RNA transcript but in different ways.

Before the importance of alternative splicing was widely appreciated, a gene was considered to be a nucleotide sequence that encodes a single protein or RNA. Based on this view, estimates for the number of genes in the human genome were typically in the range of 60,000 to 100,000. Once the complete human genome sequence was available, however, researchers realized that we actually have fewer than 20,000 genes. Even though our genomes contain a relatively low number of genes, recent data indicate that at least 35 percent of human genes undergo alternative splicing. Thus, the number of different proteins that your cells can produce is believed to be between 100,000 and 1 million.

Thanks to results like these, genes now have to be thought of as DNA sequences that are capable of directing the production of one or more related polypeptides or RNAs. The current record holder for the number of mRNAs derived from one gene is the *Dscam* gene in the fruit fly *Drosophila melanogaster*. The products of this gene are involved in guiding growing nerve cells within the embryo. Because the primary transcript is spliced into about 38,000 distinct forms of mRNA, the *Dscam* gene can produce about 38,000 different products. Alternative splicing ranks as a major mechanism in the control of gene expression in multicellular eukaryotes. ⬤ If you understand alternative splicing, you should be able to explain why it does not occur in bacteria, comment on its relationship to the evolution of the nuclear envelope, and indicate where it occurs in Figure 18.1.

mRNA Stability and RNA Interference

Once splicing is complete and processed mRNAs are exported to the cytoplasm, new regulatory mechanisms come into play. For example, it has long been known that the life span of an mRNA in the cell can vary. The mRNA for casein—the major protein in milk—is produced in the mammary gland tissue of female mammals. Normally, many of these mRNAs persist in the cell for just

(a) Tropomyosin gene

Intron Intron Intron

Exon Exon Exon Exon

(b) Processed mRNAs

Some exons are specific to tropomyosin in skeletal or smooth muscle; some exons are common to both muscle types

Skeletal muscle

Smooth muscle

FIGURE 18.12 Alternative Splicing Produces More than One Mature mRNA from the Same Gene. (a) The tropomyosin gene has a large series of exons and introns. The purple exons are found in skeletal muscle tropomyosin, and the blue exons are found in smooth muscle tropomyosin. The orange exons are found in the tropomyosin of both types of muscle cells. **(b)** In skeletal and smooth muscle cells, alternative splicing leads to the production of distinct mature mRNAs for the tropomyosin protein.

an hour, and little casein protein is produced. But when a female mouse is lactating, regulatory molecules help the mRNAs persist almost 30 times longer—leading to a huge increase in the production of casein protein. In the case of casein, mRNA stability is associated with changes in the length of its poly (A) tail.

In many cases, the life span of an mRNA is controlled by tiny, single-stranded RNA molecules that bind to complementary sequences in the mRNA. Once part of an mRNA becomes double stranded in this way, specific proteins degrade the mRNA or prevent it from being translated into a polypeptide. This phenomenon is known as **RNA interference**. How does it work?

As **Figure 18.13** shows, RNA interference begins when RNA polymerase transcribes DNA sequences that code for an unusual product—a small RNA molecule that doubles back on itself to form a hairpin (step 1). Hairpin formation occurs because pairs of sequences within the RNA transcript are complementary. Some of the RNA is trimmed by enzymes in the nucleus, then the double-stranded segment that remains is exported to the cytoplasm (Figure 18.13, step 2). In the cytoplasm, the double-stranded RNA sequence is cut by another enzyme into molecules that are typically about 22 nucleotides long. One of these short RNA strands is taken up by a group of proteins called the RNA-induced silencing complex, or RISC (Figure 18.3, step 3). The single-stranded RNA held by the RISC is a called a **microRNA** (**miRNA**). Once it is part of a RISC, the miRNA binds to its complementary sequences in a target RNA (step 4). If the match between an miRNA and an mRNA is perfect, an enzyme in the RISC destroys the mRNA by cutting it in two (step 5). In effect, tight binding by an miRNA is a "kiss of death" for the mRNA. If the match isn't perfect, the mRNA is not destroyed. Instead, its translation is inhibited. In both cases, miRNAs are responsible for interfering with mRNAs—hence the phenomenon known as RNA interference.

The first papers on RNA interference were published in the mid-1990s, and the first miRNAs were characterized in 2001. In the short time since then, research on miRNAs and RNA interference has virtually exploded. Current data suggest that a typical animal or plant species has at least 500–1000 sequences that code for miRNAs. Because many miRNAs regulate more than one mRNA, it's estimated that some 20–30 percent of all animal and plant genes are regulated by these tiny molecules.

Although a great deal remains to be learned about what miRNAs do and how they are regulated, their discovery has had a tremendous impact on how biologists think about gene regulation in eukaryotes. Just when researchers thought they were starting to get a handle on how eukaryotic genes are controlled, RNA interference came along and added an entirely new layer of complexity. ○ If you understand RNA interference, you should be able to label where it occurs in Figure 18.1.

How Is Translation Controlled?

Stopping or starting translation is a common mechanism for controlling gene expression. Although RNA interference usu-

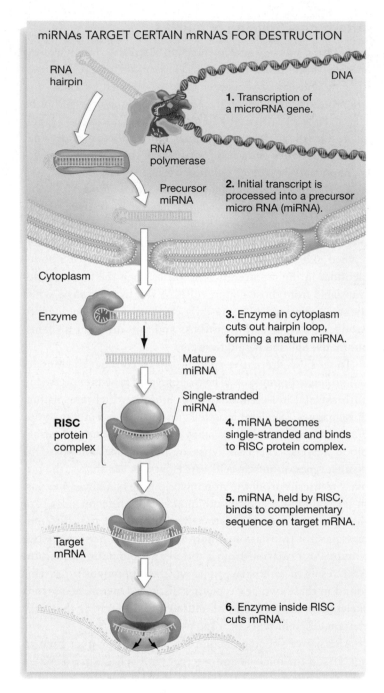

FIGURE 18.13 RNA Interference: Micro RNAs Target Certain mRNAs for Destruction.

ally operates at the level of the mRNA—after RNA processing is complete, but independently of the translation machinery—many of the small RNAs responsible for RNA interference disrupt translation directly. In other cases, mechanisms that do not involve miRNAs are responsible for controlling the timing and rate of translation. The best-studied examples depend on regulatory proteins that bind to mRNAs or the ribosome.

• In many species of animals, eggs are loaded with mRNAs that are not translated until fertilization occurs. The proteins produced from these messages are involved in directing the early development of the embryo, so they are not needed until fertilization is complete. Translation of the mRNAs in the egg is prevented when a regulatory protein binds to them, or when the mRNA's cap or tail is modified.

- The overall rate of translation in a cell may slow or stop in response to a sudden increase in temperature or infection by a virus. The slowdown occurs because regulatory proteins add a phosphate group to a protein that is part of the ribosome. You might recall from earlier chapters that phosphorylation frequently leads to changes in the shape and chemical reactivity of proteins. In the case of the phosphorylated ribosomal protein, the shape change slows or prevents translation. For the cell, this dramatic change in gene expression can mean the difference between life and death. If the danger is due to a sudden increase in ambient temperature that disrupts protein folding, shutting down translation prevents the production of improperly folded polypeptides; if the insult is a virus, the cell avoids manufacturing viral proteins.

Observations like these are a reminder that gene expression can be regulated at multiple points: at the level of chromatin structure, transcription initiation, RNA processing, mRNA life span, and translation. The sixth and last level occurs after translation is complete. In many cases, proteins are not activated as soon as they are produced.

Post-Translational Control

Control of gene expression may continue even after translation occurs and a protein product is complete. Recall from Chapter 17 that in bacteria, mechanisms of post-translational regulation are important because they allow the cell to respond to new conditions rapidly. The same is true for eukaryotes. Instead of waiting for transcription, RNA processing, and translation to occur, the cell can respond to altered conditions by quickly activating or inactivating existing proteins. Regulatory mechanisms occurring late in the flow of information from DNA to RNA to protein involve a trade-off between speed and resource use, however, because transcription and translation require energy and materials.

As an example of post-translational control, consider a group of regulatory transcription factors called **signal transducers and activators of transcription (STATs)**. STATs are common in the cytoplasm of white blood cells in mammals, where they normally reside as single polypeptide chains. In this form, the STATs are inactive. But a chain of events is triggered when a signaling molecule binds to a receptor on the cell surface (**Figure 18.14**). When the receptor is activated by the signal, it adds a phosphate to a STAT polypeptide, triggering the formation of a **dimer** (a unit made of two parts). The activated STAT dimer then moves into the nucleus, binds to an enhancer, and activates transcription of genes that trigger cell growth and division. In this way, STATs are key players in a signal

FIGURE 18.14 STATs Activate Transcription after They Are Phosphorylated. Signal transduction triggers a phosphorylation event that causes two STAT polypeptide chains to join and form an activated dimer. Activated STAT proteins move to the nucleus, bind to DNA, and activate the transcription of selected target genes.

transduction pathway required for your white blood cells to respond to invading bacteria, viruses, or other threats.

Phosphorylation is a common mechanism of post-translational control over gene expression, particularly in signal transduction pathways (see Chapter 8). It is only one of several post-translational control mechanisms that have been documented, however. The activity of a protein may also be modified by folding or by enzymes that cleave off a portion of the molecule or degrade the entire structure. But no matter what the mechanism, post-translational control is correlated with particularly rapid changes in gene expression.

A 50-Year Retrospective: How Does Gene Expression in Bacteria Compare with That in Eukaryotes?

For about five decades, biologists have been studying how gene expression is controlled. Almost as soon as they knew that information in DNA is transcribed into RNA and then translated into proteins, researchers began asking questions about how that flow of information is regulated. Early studies on bacteria established the existence of promoters, the structure and function of regulatory proteins and regulatory sequences, and the principles of positive control and negative control. More recent work on eukaryotes has highlighted the importance of chromatin structure, alternative splicing, and RNA interference.

What are the fundamental differences in how gene expression is regulated in bacteria and eukaryotes? Biologists point to four primary differences, two of which involve levels of control that exist in eukaryotes but not in bacteria:

1. *Packaging* The chromatin structure of eukaryotic DNA must be opened in order for TBP, the basal transcription complex, and RNA polymerase to gain access to genes and initiate transcription. A key insight here is that, because eukaryotic DNA is packaged so tightly, the default state of transcription in eukaryotes is off. In contrast, the default state of transcription in bacteria, which lack histone proteins and have freely accessible promoters, is on. Chromatin structure provides a mechanism of negative control that does not exist in bacteria.

2. *Alternative splicing* Prior to translation, primary transcripts in eukaryotes must be spliced—an event that is extremely rare in bacteria and archaea. The fundamental consequence is that the one-to-one correspondence between the number of genes and the number of gene products observed in bacteria and archaea is not seen in eukaryotes. Instead, each eukaryotic gene may code for one to thousands of distinct products.

3. *Complexity* Transcriptional control is much more complex in eukaryotes than in bacteria. The function of sigma proteins in bacteria is analogous to the role of the basal transcription complex in eukaryotes. Likewise, the function of CAP, the repressor, and other regulatory proteins is analogous to the role of regulatory transcription factors in eukaryotes. But the sheer number of eukaryotic proteins involved in regulating transcription—and the complexity of their interactions—dwarfs those in bacteria.

4. *Coordinated expression* In bacteria, genes that are involved in the same cellular response are organized into operons controlled by a single promoter. Because their polycistronic mRNAs are translated together, several proteins are produced in a coordinated fashion. In contrast, operons are rare in eukaryotes. In these organisms, genes that are physically scattered can be expressed at the same time because a single set of regulatory transcription factors can trigger the transcription of several genes. For example, muscle-specific genes found on several different chromosomes can be transcribed in response to the same muscle-specific regulatory transcription factor. In this way, eukaryotes coordinate the expression of functionally related genes.

To date, biologists do not have a good explanation for why gene expression is so much more complex in unicellular eukaryotes than it is in bacteria and archaea. All unicellular organisms have to respond to environmental changes in an appropriate way. It is easier to generate a logical hypothesis to explain why gene expression is complex in multicellular eukaryotes. In these organisms, cells have to differentiate as an individual develops. Changes in gene expression are responsible for the differentiation of muscle cells, bone cells, leaf cells, and flower cells in response to signals from other cells. The need for each cell type to have a unique pattern of gene expression may explain why control of gene expression is so much more complex in multicellular eukaryotes than in bacteria.

One of the two great frontiers in research on gene expression is to understand how developmental signals produce cell-specific gene expression in multicellular organisms. The other major frontier is to understand how certain defects in gene regulation result in uncontrolled cell growth and the suite of diseases called cancer.

18.6 Linking Cancer with Defects in Gene Regulation

Normal regulation of gene expression results in the orderly development of an embryo and appropriate responses to environmental changes. Abnormal regulation of gene expression, in contrast, can lead to developmental abnormalities and diseases such as cancer. Hundreds of distinct cancers exist. These diseases are enormously variable regarding the tissues they affect, their rate of progression, and their outcome. Because the underlying defects, symptoms, and consequences are so diverse, cancer is not a single disease but a family of related diseases. Cancers are related because they all originate in uncontrolled cell growth. For a cancer to become dangerous, however, two other things have to happen: The rapidly growing cells must metastasize, meaning that some cells leave their point of origin and invade other tissues (see Chapter 11), and they must stimulate the growth of blood vessels that supply them with nutrients.

Here let's consider the first step in cancer formation, and the question of what causes uncontrolled cell growth. The short answer is that each type of cancer is caused by a different set of genetic defects that lead to uncontrolled cell growth. The longer and more accurate answer is based on two key observations. First, recall from Chapter 11 that cancer results from defects in the proteins that control the cell cycle. Second, cancer is associated with mutations that alter key genes. For example, Chapter 14 linked the cancer-causing disorder xeroderma pigmentosum with defects in DNA repair. This association supports the hypothesis that increased mutation rates are involved in triggering cancer. The hypothesis is also supported by the observation that people who have been exposed to **mutagens**—

that is, radiation or chemicals that induce mutation—have an increased risk of developing cancer. For example, cancer rates were extremely high in the tens of thousands of people who were exposed to massive doses of radiation from the atomic bombs dropped on Hiroshima and Nagasaki, Japan, at the end of World War II. The link between mutation and cancer is so strong that suspected **chemical carcinogens** (cancer-causing compounds) are first tested by their ability to cause mutations in bacteria.

The fundamental question of cancer biology comes into sharper focus once the role of cell-cycle regulators and mutation is acknowledged. Now we can ask: Which genes, when mutated, disrupt the cell cycle and trigger uncontrolled cell growth?

● Intensive research over the past two decades has shown that many cancers are associated with mutations in regulatory transcription factors. These mutations lead to cancer when they affect one of two classes of genes: (1) genes that stop or slow the cell cycle, and (2) genes that trigger cell growth and division by initiating specific phases in the cell cycle.

Genes that stop or slow the cell cycle are called **tumor suppressor** genes. Their products prevent the cell cycle from progressing unless specific signals indicate that conditions are right for moving forward with mitosis and cell division. If a mutation disrupts normal function of a tumor suppressor gene, then a key "brake" on the cell cycle is eliminated.

Genes that encourage cell growth by triggering specific phases in the cell cycle are called **proto-oncogenes** (literally, "first-cancer-genes"). In normal cells, proto-oncogenes are required to initiate each phase in the cell cycle. They are active only when conditions are appropriate for growth, however. In cancerous cells, defects in the regulation of proto-oncogenes cause these genes to stimulate growth at all times. In cases such as this, a mutation has converted the proto-oncogene into an **oncogene**—an allele that promotes cancer development.

To gain a deeper understanding of how defects in gene expression can lead to cancer, consider research on the gene that is most often defective in human cancers. The gene is called **p53** because the protein it codes for has a molecular weight of approximately 53 kilodaltons (53,000 daltons, or 53,000 amu—see Chapter 2). DNA sequencing studies have revealed that mutant, nonfunctional forms of *p53* are found in over half of all human cancers. The *p53* gene codes for a regulatory transcription factor.

What is the link between a loss of p53 protein activity and cancer? A key observation suggested an answer to this question: When researchers exposed normal, noncancerous human cells to UV radiation, levels of p53 protein increased markedly. Recall from Chapter 14 that UV radiation damages DNA. Follow-up studies confirmed that there is a close correlation between DNA damage and the amount of p53 in a cell. In addition, analyses of the protein's structure showed that it contains a DNA-binding domain.

These observations inspired the hypothesis that p53 is a transcription factor that serves as the master brake on the cell cycle. In this model, p53 is activated after DNA damage occurs. The activated protein binds to the enhancers of genes that arrest the cell cycle. Once these genes are activated, the cell has time to repair its DNA before continuing to grow and divide.

Recent research has shown that this model of p53 function is correct in almost every detail. For example, consider the results of studies based on X-ray crystallography. The three-dimensional models generated by this technique confirmed that p53 binds directly to DNA (**Figure 18.15**). In addition, researchers who mapped the location of mutations that make p53 defective found that virtually all of the cancer-causing mutations were located in p53's DNA-binding site (see the amino acids highlighted in Figure 18.15). This observation supports the hypothesis that defective forms of the protein can't bind to enhancers. Further, investigators have documented that one of the genes induced by p53 encodes a protein that prevents cell-cycle regulatory proteins from triggering M (mitosis) phase.

Recent research has also shown that when a cell's DNA is extensively damaged and cannot be repaired, p53 activates the transcription of genes that cause the cell to take its own life via apoptosis (see Chapter 11). But if mutations in the *p53* gene make the protein product inactive, then damaged cells are not shut down or killed. They continue to move through the cell

Cancer-causing amino acid mutations occur in regions involved in DNA binding

FIGURE 18.15 p53 Is a DNA-Binding Protein That Acts as a Tumor Suppressor. Ribbon model showing the DNA-binding domain of p53. In cancer patients, the amino acids highlighted in yellow often differ from the normal versions. The molecule shown in red and pink is DNA.

● **QUESTION** Why would substituting one amino acid for another at one of the highlighted locations affect a protein's ability to bind to DNA? What happens to a cell when p53 cannot bind to DNA efficiently?

cycle, except now they are likely to contain many mutations because of the DNA damage they have sustained. If these mutations create oncogenes, the cells have taken a key step on the road to cancer.

To summarize, *p53* functions as a tumor suppressor gene. It prevents cancer onset by stopping the cell cycle when DNA is damaged. When *p53* is functioning normally, mutations that produce oncogenes are repaired or eliminated before uncontrolled cell growth can begin. The role of *p53* in preventing cancer is so fundamental that biologists call this gene "the guardian of the genome." Currently, research on the p53 protein is forging ahead on two fronts: Biologists are striving to identify more of the genes that are regulated by this protein and to find molecules that could act as anticancer drugs by mimicking p53's shape and activity.

Check Your Understanding

If you understand that...

- Cancer is associated with mutations and with loss of control over the cell cycle.
- Uncontrolled cell growth may occur when a protein that activates the cell cycle is activated constitutively because of a mutation in a regulatory gene.
- Uncontrolled cell growth may occur when tumor suppressor genes stop functioning, due to mutation.

You should be able to...

Explain why knock-out (loss-of-function) mutations in *p53* occur in so many cancers.

Chapter Review

SUMMARY OF KEY CONCEPTS

● **Changes in gene expression allow eukaryotic cells to respond to changes in the environment and cause distinct cell types to develop.**

In a multicellular eukaryote, different types of cells are radically different in size, shape, and function—even though they have the same DNA. Cells are different not because they have different genes but because they express different genes.

In embryos, cells begin expressing specific genes in response to signals from other cells. Because different cells receive different signals, one cell might begin expressing muscle-specific genes while a nearby cell expresses bone-specific genes. Once cells mature, they continue to receive signals that are released by other cells and that carry information about changes in the environment. In response, changes in gene expression may occur.

In both embryonic cells and mature cells, gene expression is regulated at several levels. The transcription of specific genes may be initiated or repressed, mRNAs may be spliced in different ways to produce a different product, the life span of specific mRNAs may be extended or shortened, and the life span or activity of particular proteins may be altered.

You should be able to explain why muscle-specific genes are expressed in muscle cells while nearby nerve-specific genes are not. ●

● **In eukaryotes, DNA is packaged with proteins into complex structures that must be opened before transcription can occur.**

Eukaryotic DNA is wrapped around histone proteins to form a bead-like nucleosome, which is then coiled into 30-nm fibers and higher-order chromatin structures. Transcription cannot be initiated until the interaction between DNA and histones in chromatin is relaxed. These changes depend on the acetylation or methylation of histones and the action of molecular machines called chromatin-remodeling complexes.

You should be able to draw what a gene looks like when chromatin is in a condensed versus uncondensed state. ●

● **In eukaryotes, transcription can be initiated only when specific proteins bind to the promoter and to regulatory sequences that may be close to the promoter or far from it.**

Regulatory transcription factors are proteins that bind to regulatory sequences called enhancers and silencers, which are often located at a distance from the gene in question, or to promoter-proximal sequences near the start of the coding sequence. The first regulatory transcription factors that bind to DNA recruit proteins that loosen the histones' grip on the gene, making the promoter accessible to basal transcription factors. Interactions between regulatory and basal transcription factors lead to the formation of the basal transcription complex. Once this large, multi-protein complex is intact, RNA polymerase is recruited to the site and transcription begins.

You should be able to draw what a eukaryotic gene looks like when it is being transcribed. Label enhancers, promoter proximal elements, the promoter, regulatory transcription factors, basal transcription factors, and RNA polymerase. ●

(MB) **Web Animation** at www.masteringbio.com
Transcription Initiation in Eukaryotes

● **Alternative splicing allows a single gene to code for several different products.**

Once a message has been transcribed, several other regulatory events come into play. Alternative splicing allows a single gene to produce more than one mRNA and more than one protein. It is regulated by proteins that interact with the spliceosome.

You should be able to explain why humans have so few genes. ●

● Once an mRNA is in the cytoplasm, gene expression is controlled by molecules that regulate the (1) life span of mRNAs, (2) efficiency of translation, and (3) activation or inactivation of protein products.

RNA interference occurs when tiny strands of RNA, called microRNAs (miRNAs), bind to mRNAs in company with the protein complex called RISC and mark the mRNAs for degradation or when small RNAs inhibit translation. Once translation occurs, proteins may be activated or inactivated by the addition or removal of a phosphate group or other events.

You should be able to draw the interaction between an miRNA, RISC, and an mRNA. ●

● Cancer can develop when mutations disable genes that regulate cell-cycle control genes.

The cell cycle is controlled by specific genes. But if mutations alter transcription factors that control the cell cycle, then uncontrolled cell growth and tumor formation may result. For example, the regulatory transcription factor *p53* is responsible for stopping the cell cycle when DNA is damaged. If *p53* is mutated in a way that prevents its protein product from binding to DNA, then the cell cycle is not arrested and the damaged DNA is not repaired—leading to mutations.

You should be able to explain why cancer is common in people who have been exposed to high levels of radiation, and why it is more common in older people than younger people. ●

QUESTIONS

● Test Your Knowledge

1. What is chromatin?
   a. the protein core of the nucleosome, which consists of histones
   b. the 30-nm fiber
   c. the DNA-protein complex found in eukaryotes
   d. the histone *and* non-histone proteins in eukaryotic nuclei

2. What is a tumor suppressor?
   a. a gene associated with tumor formation when its product does not function
   b. a gene associated with tumor formation when its product functions normally
   c. a gene that accelerates the cell cycle and leads to uncontrolled cell growth
   d. a gene that codes for a transcription factor involved in tumor formation

3. Which of the following statements about enhancers is correct?
   a. They contain a unique base sequence called a TATA box.
   b. They are located only in 5′-flanking regions.
   c. They are located only in introns.
   d. They are found in a variety of locations and are functional in any orientation.

4. In eukaryotes, why are certain genes expressed only in certain types of cells?
   a. Different cell types contain different genes.
   b. Different cell types have the same genes but different promoters.
   c. Different cell types have the same genes but different enhancers.
   d. Different cell types have different regulatory transcription factors.

5. What is alternative splicing?
   a. the phosphorylation events that lead to different types of post-translational regulation
   b. mRNA processing events that lead to different combinations of exons being spliced together
   c. folding events that lead to proteins with alternative conformations
   d. action by regulatory proteins that leads to changes in the life span of an mRNA

6. What types of proteins bind to promoter-proximal elements?
   a. the basal transcription complex
   b. the basal transcription complex plus RNA polymerase
   c. basal transcription factors
   d. regulatory transcription factors

Test Your Knowledge answers: 1. c; 2. a; 3. d; 4. d; 5. b; 6. d

● Test Your Understanding

Answers are available at www.masteringbio.com

1. Compare and contrast (a) enhancers and the CAP site; (b) promoter-proximal elements and the *lac* operon operator; (c) basal transcription factors and sigma; and (d) eukaryotic promoters and bacterial promoters.

2. Why does chromatin need to be "remodeled" for transcription to occur? Why was it logical to observe that DNA-protein interactions can be loosened by the addition of acetyl groups to histones?

3. Compare and contrast (a) enhancers and silencers; (b) promoter-proximal elements and enhancers; (c) transcription factors and coactivators; and (d) regulatory transcription factors and basal transcription factors.

4. Explain the relationship between complementary base pairing and RNA interference. Suggest a way that miRNAs that are complementary to viral RNAs could be used as drugs.

5. Now that the human genome has been sequenced, researchers estimate that it contains fewer than 20,000 genes. Previously, most investigators expected that the human genome held at least 100,000 genes. Explain how alternative splicing can explain the discrepancy between the observed and predicted numbers of genes.

6. Explain why mutations in regulatory proteins and regulatory sequences can lead to loss of control over the cell cycle and the development of cancer. Include a specific example.

● Applying Concepts to New Situations

1. Histone proteins have been extremely highly conserved during evolution. The histones found in fruit flies and humans, for example, are nearly identical in amino acid sequence. Offer an explanation for this observation. (Hint: What are the consequences of a mutation in a histone?)

2. Cancers are most common in tissues where cell division is common, such as blood cells and cells in the lining of the lungs or gut. Why is this observation logical?

3. Levels of p53 protein in the cytoplasm increase after DNA damage. Design an experiment to determine whether this increase is due to increased transcription of the *p53* gene or to activation of preexisting p53 proteins via a post-translational mechanism such as phosphorylation.

4. Why are individuals who experience severe sunburn or deep tanning more likely to develop skin cancer?

www.masteringbio.com is also your resource for • Answers to text, table, and figure caption questions and exercises • Answers to *Check Your Understanding* boxes • Online study guides and quizzes • Additional study tools including the *E-Book for Biological Science* 3rd ed., textbook art, animations, and videos.

Analyzing and Engineering Genes

19

A mouse that has been genetically engineered to express a gene from jellyfish. The jellyfish gene codes for a protein that naturally luminesces, or emits light. The gene's product is called green fluorescent protein, or GFP.

KEY CONCEPTS

● The discovery of enzymes that cut DNA at specific locations, along with enzymes that piece DNA segments back together, gave biologists the ability to move genes from one location to another.

● To analyze a gene, biologists have to obtain many identical copies of it. This can be done by inserting the gene into a bacterial cell that copies the gene as it grows, or by performing a polymerase chain reaction.

● Once researchers have many identical copies of a gene, its sequence of bases can be determined by dideoxy sequencing.

● To find the genes associated with a particular trait, researchers start with a genetic map. If certain markers on the map are found only in individuals that have a distinctive phenotype, the gene responsible for that phenotype is likely to be near that marker.

● Researchers are attempting to insert genes into humans to cure genetic diseases. Efforts to insert genes into plants to provide them with novel traits, such as the ability to withstand insect attacks, have been much more successful.

The molecular revolution in biological science began when researchers confirmed that DNA is the hereditary material and understood the molecule's secondary structure. But when biologists discovered how to remove DNA sequences from an organism, manipulate them, and insert them into different individuals, the molecular revolution took off. Efforts to manipulate DNA sequences in organisms are often referred to as genetic engineering. The goal of this chapter is to introduce some of the key techniques and conceptual issues involved in analyzing and engineering DNA, using a case history approach.

Genetic engineering became possible with the discovery of bacterial enzymes that cut DNA at specific sites and that paste DNA sequences together. Although cutting and pasting DNA sequences sounds simple in concept, these new molecular tools were extremely powerful. Biologists no longer had to rely solely on controlled breeding experiments to change the genetic characteristics of individuals. Instead, they could quickly mix and match specific DNA sequences in the lab. Efforts to manipulate genes usually result in novel combinations of genes along chromosomes, or genetic recombination (see Chapter 12).

● Key Concept   Important Information   ● Practice It

389

Appropriately enough, techniques used to engineer genes are often referred to as **recombinant DNA technology**.

The enzymes used to cut and paste DNA were initially discovered and used during the 1960s and 1970s. Over the subsequent two decades, researchers followed up by developing systems for transferring recombinant genes into various types of organisms and by working out techniques for controlling the expression of the introduced DNA. Since then, research has concentrated on applying these techniques to solve problems in medicine, industry, and agriculture. The goals of genetic engineering are to (1) improve our understanding of how genes work and (2) advance **biotechnology**—the manipulation of organisms to create products or cure disease.

To introduce the techniques that biologists are using to analyze and engineer genes, we're going to view them in action. More specifically, we'll consider a series of case histories that introduce the basic techniques in molecular biology in the context of solving problems. Our first example represents one of the initial efforts to use recombinant DNA technology to cure an inherited disease in humans. Section 19.1 focuses on how the search for a cure for pituitary dwarfism relied on techniques for identifying genes, copying them, moving them into a new host organism, and then controlling their expression. Later sections introduce techniques biologists use to produce many identical copies of a gene, determine a gene's sequence, and find the gene or genes associated with a particular disease or other trait. The chapter closes with a look at how genetic engineering has been applied to humans and crop plants in the hope of curing devastating genetic diseases and increasing the quality and quantity of food available to people in poverty-stricken areas of the globe.

In addition to exploring the techniques used to manipulate DNA, it is essential to consider the ethical, economic, ecological, and political issues involved. Manipulating the genetic makeup of organisms raises ethical concerns and has the potential to create harm. Gene therapy—the introduction of normal alleles to cure genetic defects—and the release of genetically modified plants are under intense scrutiny in the press. Lawmaking bodies around the world are debating legislation to regulate both efforts. What are the potential perils and benefits of introducing recombinant genes into human beings, food plants, and other organisms? This question represents one of the great ethical challenges of the twenty-first century, and is a recurrent theme in this chapter.

19.1 Using Recombinant DNA Techniques to Manufacture Proteins: The Effort to Cure Pituitary Dwarfism

To understand the basic techniques and tools that are used by genetic engineers, let's consider the effort to treat pituitary dwarfism in humans. The pituitary gland is a structure at the base of the mammalian brain that produces several important molecules, including a protein that stimulates growth. The molecule, which was found to be just 191 amino acids long, was named growth hormone. The gene that codes for this protein is called *GH1*.

Once growth hormone was discovered, researchers immediately suspected that at least some forms of inherited dwarfism might be due to a defect in the GH1 protein. This hypothesis was confirmed when it was established that people with certain types of dwarfism produce little growth hormone or none at all. These people have defective copies of *GH1* and exhibit pituitary dwarfism, type I.

By studying the pedigrees of families in which dwarfism was common, several teams of researchers established that pituitary dwarfism, type I, is an autosomal recessive trait (see Chapter 13). Stated another way, affected individuals have two copies of the defective allele. People who have only one defective allele are carriers of the trait, meaning that they can transmit the defective allele to their offspring but are not themselves affected. Affected individuals grow more slowly than average people, reach puberty from two to ten years later than average, and are short in stature as adults—typically no more than 120 cm (4 feet) tall.

Why Did Early Efforts to Treat the Disease Fail?

Once the molecular basis of pituitary dwarfism was understood, physicians began treating the disease with injections of growth hormone. This approach was inspired by the spectacular success that physicians had achieved in treating type I diabetes mellitus. Diabetes mellitus is due to a deficiency of the peptide hormone insulin, and clinicians had been able to alleviate the disease's symptoms by injecting patients with insulin from pigs. Early trials showed that people with pituitary dwarfism could be treated successfully with growth hormone therapy, but only if the protein came from humans. Growth hormones isolated from pigs, cows, or other animals were ineffective. Until the 1980s, however, the only source of human growth hormone was pituitary glands dissected from human cadavers. Up to 20,000 pituitary glands had to be collected from cadavers to supply enough growth hormone to treat the population of affected individuals. As a result, the drug was extremely scarce and expensive.

Meeting demand turned out to be the least of the problems with growth hormone therapy, however. To understand why, recall from Chapter 3 that infectious proteins called prions can cause degenerative brain disorders in mammals. Kuru and Creutzfeldt–Jakob disease are prion diseases that affect humans. Certain prion diseases, including some forms of Creutzfeldt–Jakob disease, are hereditary. In most cases, however, people and other animals become infected with prion diseases by directly ingesting prion proteins in food. When physicians found that some of the children treated with human growth hormone developed Creutzfeldt–Jakob disease in their teens and twenties, the physicians realized that the supply of growth hormone was contaminated with a prion protein from the brains of the cadavers supplying the hormone. In 1984, the use of growth hormone isolated from cadavers was banned.

Using Recombinant DNA Technology to Produce a Safe Supply of Growth Hormone

To replace natural sources of growth hormone, researchers turned to genetic engineering. The idea was to insert fully functional copies of human *GH1* into the bacterium *Escherichia coli*; the hope was that huge quantities of recombinant bacterial cells could be grown. If so, then the recombinant cells would produce uncontaminated growth hormone in sufficient quantities to meet demand at an affordable price.

To meet this goal, investigators had to find *GH1*, obtain many copies of the gene, and insert them into *E. coli* cells. Their ability to do this hinged on using three of the most basic tools in molecular biology. Let's consider each in turn.

Using Reverse Transcriptase to Produce cDNAs

An enzyme called reverse transcriptase is responsible for a major exception to the central dogma of molecular biology: It allows information to flow from RNA to DNA. **Reverse transcriptase can do this because it catalyzes the synthesis of DNA from an RNA template.**

DNA that is produced from RNA is called **complementary DNA**, or **cDNA**. Although reverse transcriptase initially produces a single-stranded cDNA, it is also capable of synthesizing the complementary strand to yield a double-stranded DNA. In many cases, however, researchers add a primer to single-stranded cDNAs and use DNA polymerase to synthesize the second strand (**Figure 19.1**).

Reverse transcriptase played a key role in the search for the growth hormone gene because *GH1* is actively transcribed in cells from the pituitary gland. To get the gene hunt started, then, researchers isolated mRNAs from pituitary glands and used reverse transcriptase to reverse-transcribe the mRNAs to cDNAs. When this step was completed, researchers had a solution containing double-stranded cDNAs that corresponded to each gene that is actively expressed in pituitary cells. Their next

move? Isolating each of these cDNAs and obtaining many identical copies of them. Efforts to produce many identical copies of a gene are referred to as **DNA cloning**. If a researcher says that she has cloned a gene, it means that she has isolated it and produced many identical copies.

How Are Plasmids Used in Cloning?

In many cases, researchers can clone a gene by inserting it into a small, circular DNA molecule called a **plasmid**. You might recall from Chapter 7 that plasmids are common in bacterial cells but are physically distinct from the bacterial chromosome. They are not required by the cell for normal growth and reproduction, and most replicate independently of the chromosome. Some plasmids carry genes for antibiotic resistance or other traits that increase the cell's ability to grow in a particular environment. In natural populations of bacteria, plasmids are sometimes copied and transferred from one cell to another bacterial cell during conjugation (see Chapter 12). Plasmids have also been discovered in the nuclei of certain unicellular eukaryotes, including baker's yeast.

Researchers quickly realized that if they could splice a loose piece of DNA into a plasmid and then insert the modified plasmid into a bacterial cell, the engineered plasmid would be replicated and passed on to daughter cells as the bacterium grew and divided. If this recombinant bacterium were placed in a nutrient broth and allowed to grow overnight, billions of copies of the plasmid DNA would be produced in the resulting bacterial cells. When plasmids are used to carry DNA from another source, the plasmid is called a **cloning vector**, or simply a **vector.** Vector DNA can carry foreign DNA and make several copies of it. Plasmids are often used as vectors.

Biologists could harvest the recombinant genes by breaking the bacteria open, isolating all of the DNA, and then separating the plasmids from the main chromosomes. This is a promising strategy for obtaining many identical copies of a gene, but how can a gene be inserted into a plasmid in the first place?

Using Restriction Endonucleases and DNA Ligase to Cut and Paste DNA

To cut a gene out for later insertion into a cloning vector, researchers use an enzyme called a restriction

FIGURE 19.1 Reverse Transcriptase Catalyzes the Synthesis of DNA from RNA. The single-stranded DNA produced by reverse transcriptase is complementary to the RNA template. The cDNA can be made double stranded by reverse transcriptase or DNA polymerase. DNA polymerase requires a primer.

endonuclease. Restriction endonucleases are bacterial enzymes that cut DNA molecules at specific base sequences.

Most of the 400 known restriction endonucleases cut DNA only at sites that form palindromes. In English, a word or sentence is a palindrome if it reads the same way backward as it does forward. "Madam, I'm Adam" is an example. In biology, a stretch of double-stranded DNA forms a palindrome if the $5' \rightarrow 3'$ sequence of one strand is identical to the $5' \rightarrow 3'$ sequence on the antiparallel, complementary strand.

To insert the pituitary gland cDNAs into plasmids, researchers performed the series of steps outlined in **Figure 19.2**. They began with a plasmid containing a palindromic sequence that is cut by a specific restriction endonuclease. Notice in step 1 of the figure that the $5' \rightarrow 3'$ sequence on one strand of the endonuclease recognition site reads GAATTC and that the $5' \rightarrow 3'$ sequence on the complementary strand also reads GAATTC. Then they attached the same palindromic sequence to the ends of each cDNA in their sample (step 2).

The researchers' next step was to cut the recognition sites at the ends of each cDNA with a restriction endonuclease called EcoRI. (The name stands for *Escherichia coli* restriction I, because it was the first restriction endonuclease discovered in *E. coli*.) As the top part of step 3 in Figure 19.2 shows, however, EcoRI did not break its recognition sequence neatly in two. Like most restriction endonucleases, it made a staggered cut in the palindrome. The result was DNA fragments that had **sticky ends**. The ends of the fragments are said to be sticky because the single-stranded bases on one fragment are complementary to the single-stranded bases on the other fragment. As a result, the two ends will tend to pair up and hydrogen bond to one another.

The creation of sticky ends in DNA is significant. ● If restriction sites in different DNA sequences are cut with the same restriction endonuclease, the presence of sticky ends allows the resulting fragments to be spliced together via complementary base pairing. This is the essence of recombinant DNA technology—the ability to create novel combinations of DNA sequences by cutting specific sequences and pasting them into new locations.

Step 4 of Figure 19.2 demonstrates this aspect of genetic engineering. The base sequences of the sticky ends of the plasmid are shown in black, while the base sequences of the sticky ends of the cDNA are shown in red. As the figure shows, the complementary base pairs in the pairs of sticky ends bind to each other via complementary base pairing. Researchers then used **DNA ligase**—the enzyme introduced in Chapter 14 that connects Okazaki fragments during DNA replication—to seal the recombinant pieces of DNA together at the arrows marked in green (step 5).

At the end of this cutting and pasting step in the search for the growth hormone gene, researchers had a set of recombinant plasmids. Each contained a different cDNA made from a human pituitary gland mRNA.

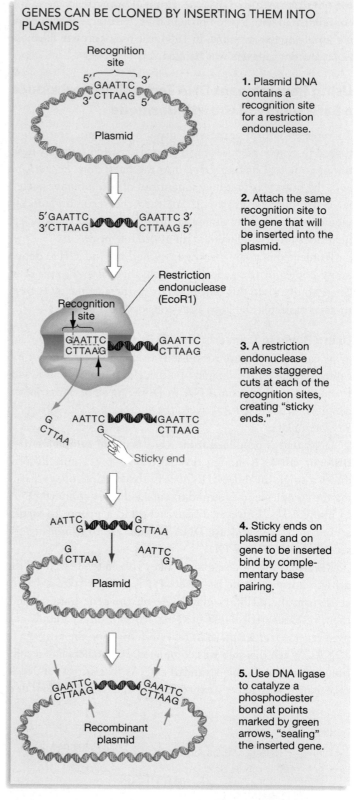

GENES CAN BE CLONED BY INSERTING THEM INTO PLASMIDS

1. Plasmid DNA contains a recognition site for a restriction endonuclease.

2. Attach the same recognition site to the gene that will be inserted into the plasmid.

3. A restriction endonuclease makes staggered cuts at each of the recognition sites, creating "sticky ends."

4. Sticky ends on plasmid and on gene to be inserted bind by complementary base pairing.

5. Use DNA ligase to catalyze a phosphodiester bond at points marked by green arrows, "sealing" the inserted gene.

FIGURE 19.2 Genes Can Be Cloned into Plasmids. To clone a gene, researchers often start by inserting the sequence into a plasmid. Once the gene is in the plasmid, the plasmid can be introduced into bacterial cells and grown to produce many identical copies of the gene.

● **QUESTION** Some restriction endonucleases make even cuts instead of staggered cuts. Are they useful for cloning?

Transformation: Introducing Recombinant Plasmids into Bacterial Cells Plasmids serve as a vector—meaning, a vehicle for transferring recombinant genes to a new host. ○ If a recombinant plasmid can be inserted into a bacterial or yeast cell, the foreign DNA will be copied and transmitted to new cells as the host cell grows and divides. In this way, researchers can obtain millions or billions of copies of specific genes. Although viruses are used as vectors in some applications (see Section 19.5), plasmids are often the vector of choice in genetic engineering.

How can recombinant plasmids be introduced into cells? Cells that take up DNA from the environment and incorporate it into their genomes are said to undergo **transformation**. To transform bacterial cells with a plasmid, researchers increase the permeability of the cell's plasma membranes using a specific chemical treatment or an electrical shock. Typically, just a single plasmid enters the cell during this treatment. The resulting cells are spread out on plates in a layer thin enough to ensure that each cell is physically isolated. The cells are then allowed to grow into colonies containing millions of identical cells.

Producing a cDNA Library **Figure 19.3** summarizes the steps covered thus far in the hunt for the growth hormone gene. The result, shown in step 5, is a collection of transformed bacterial cells. Each of the cells contains a plasmid with one cDNA from a pituitary gland mRNA. A collection of genes like this, each of which is inserted into a vector, is called a **DNA library**. If the genes are cDNAs from a particular cell type or tissue, the library is called a **cDNA library**. If the genes are fragments of DNA that collectively represent the entire genome of an individual, the library is called a **genomic library**. In each case, the library is made up of cloned genes—each gene present can be produced in large quantity and isolated in pure form. ○ If you understand this concept, you should be able to make a diagram showing the steps involved in making a genomic library, starting with DNA from your own cells and using the restriction endonuclease EcoRI to cut the genome into fragments.

DNA libraries are important because they give researchers a way to store information from a particular cell type or genome in a form that is accessible. But like a college or community library, a DNA library isn't very useful unless there is a way to retrieve specific pieces of information. At your college library, you use call numbers or computer searches to retrieve a particular book or article. How do you go about retrieving a particular gene from a DNA library? For example, how did researchers find the growth hormone gene in the cDNA library of the human pituitary gland?

How Is a DNA Library Screened? Molecular biologists are often faced with the task of finding one specific gene in a large

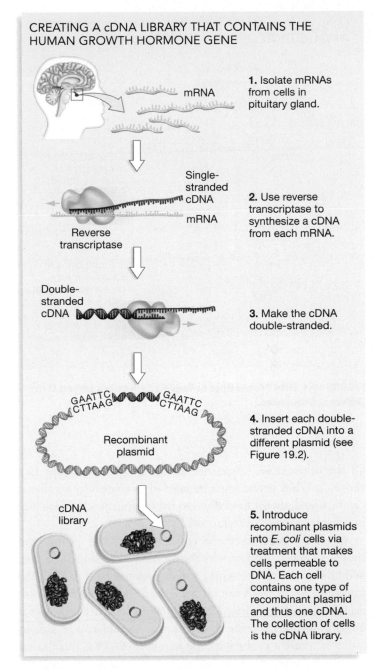

CREATING A cDNA LIBRARY THAT CONTAINS THE HUMAN GROWTH HORMONE GENE

1. Isolate mRNAs from cells in pituitary gland.

2. Use reverse transcriptase to synthesize a cDNA from each mRNA.

3. Make the cDNA double-stranded.

4. Insert each double-stranded cDNA into a different plasmid (see Figure 19.2).

5. Introduce recombinant plasmids into *E. coli* cells via treatment that makes cells permeable to DNA. Each cell contains one type of recombinant plasmid and thus one cDNA. The collection of cells is the cDNA library.

FIGURE 19.3 Complementary DNA (cDNA) Libraries Represent the mRNAs in a Cell. The hunt for growth hormone gene began with the creation of a cDNA library from pituitary-gland cells. Because growth hormone is produced in this tissue, the cDNA for growth hormone should exist somewhere in the library.

○ **QUESTION** Would each type of cDNA in the library be represented just once? Why or why not?

collection of DNA fragments. To do this, you have to have a **probe**—a marked copy of the molecule you are looking for. A DNA probe, for example, is a single-stranded fragment of a labeled, known gene that binds to a single-stranded complementary sequence in the sample of DNA being analyzed. By binding to the target sequence, it marks it—distinguishing it from

USING A DNA PROBE TO FIND A TARGET SEQUENCE IN A COLLECTION OF MANY DNA SEQUENCES

Labeled probe

1. Single-stranded DNA probe has a label that can be visualized.

2. Expose probe to collection of single-stranded DNA sequences.

3. Probe binds to complementary sequences in target DNA—and only to that DNA. Target DNA is now labeled and can be isolated.

FIGURE 19.4 DNA Probes Bind to Specific Sequences among Many Different Sequences.

all the other DNA fragments in the sample. As **Figure 19.4** shows, a DNA probe must be labeled in some way so that it can be found after it has bound to the complementary sequence in the large sample of fragments.

● If you understand the concept of a DNA probe, you should be able to explain why the probe must be single stranded and labeled in order to work; why it does not bind to all of the fragments in the DNA sample being analyzed, or even to a random assortment of those DNA fragments; and instead, why it binds to just one specific fragment. You should also be able to indicate where a probe with the sequence AATGC (recall that sequences are always written 5′ to 3′) will bind to a target DNA with the sequence TTTTACCCATTTACGATTGGCCT (again written 5′ to 3′).

To find an appropriate probe for the human growth hormone gene, researchers began by using the genetic code to infer the approximate DNA sequence of *GH1*. This was possible because the sequence of amino acids in the polypeptide was known. Thus, the researchers could infer which mRNA codon and DNA sequences coded for each amino acid. You did similar exercises in Chapter 15. But recall from Chapter 15 that the genetic code is redundant, so it is possible for more than one codon to code for an amino acid. As a result, the researchers' inferred sequence for the growth hormone gene was actually a set of related sequences.

The next step was to synthesize many copies of a short, single-stranded stretch of DNA that was complementary to

SCREENING a cDNA LIBRARY TO FIND THE GROWTH HORMONE GENE

1. Grow *E. coli* cells containing plasmids on many plates. Each colony contains a different cDNA.

2. Lay a filter on each plate, then remove. Some cells from each colony stick to filters.

3. Treat bacteria with chemicals to break open cells and make DNAs single stranded.

Labeled probe

4. Probe filters with labeled DNA (short sequence inferred from amino acid sequence of growth hormone).

5. The labeled probe DNA binds to its complementary sequence in the cDNA library.

E. coli containing growth hormone gene

6. On original plates, find colony of *E. coli* cells that contains growth hormone gene. Sample cells, grow, and analyze.

FIGURE 19.5 Finding Specific Genes by Probing a cDNA Library. Once bacteria with the growth hormone cDNA were isolated from the cDNA library, researchers could produce virtually unlimited quantities of the cDNA.

the inferred sequence. Because these molecules would bind to single-stranded fragments from the actual gene via complementary base pairing, they could act as a probe. In this case, researchers attached a radioactive atom to the probe.

Figure 19.5 shows how researchers used this probe to find the plasmid in the cDNA library that contained *GH1*. In a similar vein, **Box 19.1** (page 398) explains how DNA, RNA, or protein probes are used to find particular molecules in a collection of molecules separated on a gel by electrophoresis.

As predicted, the radioactively labeled growth hormone probe bound to its complementary sequence in the cDNA library. The probe had identified the recombinant cell that contained the human growth hormone.

Mass-Producing Growth Hormone   To accomplish their goal of producing large quantities of the human growth hormone, the investigators used recombinant DNA techniques to transfer the growth hormone cDNA to a new plasmid. The plasmid in question was one containing a promoter sequence, like those introduced in Chapter 17, that is recognized by *E. coli*'s RNA polymerase holoenzyme. The recombinant plasmids were then introduced into *E. coli* cells.

The *E. coli* cells that resulted from these steps contained a gene for human growth hormone attached to an *E. coli* promoter. As a result, the transformed *E. coli* cells began to transcribe and translate the human growth hormone gene. (Note that because the cDNAs are synthesized from mature mRNAs, which do not contain introns, cDNAs do not contain introns either. Thus, bacterial cells can translate an mRNA transcribed from a cDNA directly—no splicing is required.)

The human growth hormone accumulated in the cells and could be isolated and purified. Bacterial cells containing the human growth hormone gene are now grown in huge quantities. These cells have proved to be a safe and reliable source of the human growth hormone protein. The effort to cure pituitary dwarfism using recombinant DNA technology was a spectacular success.

Ethical Concerns over Recombinant Growth Hormone

As supplies of growth hormone increased, physicians used it in treating not only people with pituitary dwarfism but also children who were short, although they had no actual growth hormone deficiency. Even though the treatment requires several injections per week until adult stature is reached, growth hormone therapy was popular because it often increased the height of these children by a few centimeters. In essence, the drug was being used as a cosmetic—a way to improve appearance in cultures where height is deemed attractive. But should a parent request growth hormone treatment for genetically normal children to change their appearance? If short people are discriminated against in a culture, is a medical treatment a better solution than education and changes in attitudes? And what if parents wanted a tall child to be even taller to enhance her potential success as, say, a basketball player? Currently, the U.S. Food and Drug Administration has approved the use of human growth hormone for only the shortest 1.2 percent of children, who are projected to reach adult heights of less than 160 cm (5′3″) in males and 150 cm (4′11″) in women.

Growth hormone has also been found to enhance the maintenance of bone density and muscle mass. As a result, it has become a popular performance-enhancing drug for athletes. Part of its popularity stems from the fact that it is virtually undetectable in the drug tests currently administered by governing bodies. Should athletes be able to enhance their physical skills by taking hormones or other types of drugs? And is the drug safe at the dosages being used by athletes? These types of questions are being debated by physicians and researchers, agencies that govern sports, and legislative bodies. Even though these questions are still unanswered, it is clear that, while solving one important problem, recombinant DNA technology created others. Throughout this chapter, an important theme is that genetic engineering has costs that must be carefully weighed against its benefits.

Check Your Understanding

If you understand that...   ○

- The essence of recombinant DNA technology is to cut DNA into fragments with a restriction endonuclease, paste specific sequences together via complementary base pairing of sticky ends and the action of DNA ligase, and insert the resulting recombinant genes into a bacterial (or yeast) cell so that the genes are expressed.

You should be able to...   ●

1) Make a diagram showing how certain restriction endonucleases create DNA fragments with sticky ends.

2) Make a diagram showing how DNA ligase can be used to splice sequences with sticky ends into a plasmid or other type of DNA sequence.

3) Explain how a cDNA library is constructed.

4) Make a diagram showing how a DNA library is probed with a known sequence.

 Web Animation   at www.masteringbio.com

Producing Human Growth Hormone

19.2 Another Approach to Cloning: The Polymerase Chain Reaction

Inserting a gene into a bacterial plasmid is one method for cloning DNA. The polymerase chain reaction is another. ○ The **polymerase chain reaction** (**PCR**) is an in vitro DNA synthesis reaction in which a specific section of DNA is replicated over and over to amplify the number of copies of that sequence. It's a technique for generating many identical copies of a particular section of DNA.

Although PCR is much faster and technologically easier than cloning genes into a DNA library, there is a catch: PCR is possible only when a researcher already has some information about

DNA sequences near the gene in question. Sequence information is required because to do a polymerase chain reaction, you have to start with short lengths of single-stranded DNA that match sequences on either side of the gene of interest. These short segments act as primers for the synthesis reaction.

As **Figure 19.6a** shows, the primer sequences must be complementary to bases on either side of the gene in question. One primer is complementary to a sequence on one strand upstream of the target DNA—the sequence of interest; the other primer is complementary to a sequence on the other strand, downstream of the target DNA. If the target DNA molecule is made single stranded, then the primers will bond or anneal to their complementary sequences as shown in **Figure 19.6b**. You might recall that DNA polymerase cannot work unless a sequence is primed in this way. Once the primers are bound, DNA polymerase can extend each strand in the 5′ to 3′ direction.

As step 1 of **Figure 19.7** shows, a PCR procedure begins with a reaction mix containing an abundant supply of the four deoxynucleoside triphosphates (dNTPs; see Chapter 14), copies of the template DNA—that is, a DNA sample that includes the gene of interest—copies of the two primers, and an enzyme called *Taq* polymerase. *Taq* polymerase is a DNA polymerase found in the bacterium *Thermus aquaticus*, which was originally discovered in a hot spring inside Yellowstone National Park, Wyoming. *Taq* polymerase is the enzyme of choice in PCR because the technique depends on heating the reaction mix, and because *Taq* polymerase is heat stable. Although most DNA polymerases are destroyed at high temperature, *Taq* polymerase continues to function normally even when heated to 95°C.

(a) PCR primers must be located on either side of the target sequence, on opposite strands.

(b) When target DNA is single stranded, primers bind and allow DNA polymerase to work.

FIGURE 19.6 The Polymerase Chain Reaction (PCR) Depends on Having Specific Primers. (a) The orange and yellow sequences indicate a set of single-stranded primers, which bracket the region of DNA to be amplified. To do PCR, a researcher must have enough sequence information to be able to design the appropriate primers. **(b)** The primers bind to single-stranded target DNA as shown.

● **EXERCISE** Indicate where DNA polymerase would begin to work on each strand and add arrows indicating the direction of DNA synthesis.

THE POLYMERASE CHAIN REACTION IS A WAY TO PRODUCE MANY IDENTICAL COPIES OF A SPECIFIC GENE

1. Start with a solution containing template DNA, synthesized primers, and an abundant supply of the four dNTPs.

dNTPs   Primers

2. Denaturation Heating leads to denaturation of the double-stranded DNA.

3. Primer annealing At cooler temperatures, the primers bind to the template DNA by complementary base pairing.

One cycle

4. Extension During incubation, *Taq* polymerase uses dNTPs to synthesize complementary DNA strand, starting at the primer.

5. Repeat cycle of three steps (2–4) again, doubling the copies of DNA.

6. Repeat cycle again, up to 20–30 times, to produce millions of copies of template DNA.

FIGURE 19.7 The Polymerase Chain Reaction (PCR) Is a Method for Producing Many Copies of a Specific Sequence. Each PCR cycle (denaturation, primer annealing, and extension) results in a doubling of the number of copies of the sequence between and including the primers.

● **EXERCISE** Draw the events that produced the sequences in step 5.

As step 2 in Figure 19.7 shows, PCR gets under way when the reaction mix is heated to 94°C. At this temperature, the double-stranded template DNA denatures. This means that the two DNA strands separate, forming single-stranded templates. Then the mixture is allowed to cool to 50–60°C. In this temperature range, some of the denatured DNA strands re-form double helices. But some of the primers bond, or anneal, to complementary portions of the single-stranded template DNA (step 3). This step is called *primer annealing*. The reaction mix is then heated to 72°C. At this temperature, *Taq* polymerase synthesizes the complementary DNA strand from the dNTPs, starting at the primer. This step is called *extension* (step 4). The temperature changes required in each step are now automated by PCR machines.

The denaturation, primer annealing, and extension steps constitute a single PCR cycle. If one copy of the template sequence existed in the sample originally, then there are two copies at the end of the first cycle (see step 4 in Figure 19.7). The two copies present at the start of the second cycle then act as templates for another round of denaturation, primer annealing, and extension, and four copies of the target gene exist at the end of cycle number 2 (step 5). As the cycle repeats, the amount of template sequence doubles each time (step 6). Doubling occurs because each newly synthesized segment of DNA serves as a template in the subsequent cycle. Starting with a single copy, successive cycles result in the production of 2, 4, 8, 16, 32, 64, 128, 256 copies, and so on. A total of n cycles generates 2^n copies; so in just 20 cycles, one sequence can be amplified to over a million copies. By performing up to 30 cycles, researchers obtain enormous numbers of copies of the template sequence.

PCR in Action: Studying Fossil DNA

To understand why PCR is so valuable, consider a recent study by Svante Pääbo and colleagues. These biologists wanted to analyze DNA recovered from the 30,000-year-old bones of a fossilized human of the species *Homo neanderthalensis*. The individual involved was actually the first Neanderthal fossil ever discovered, from the Neander valley of Germany. Pääbo's goal was to determine the sequence of bases in the ancient DNA, compare it with DNA from modern humans (*Homo sapiens*), and analyze how similar the two species are. If some modern humans have sequences that are identical or almost identical to the base sequences found in Neanderthals, it would support the hypothesis that some of us inherited at least some of our DNA directly from a Neanderthal ancestor. That could happen only if *Homo sapiens* and *Homo neanderthalensis* interbred while they coexisted in Europe about 30,000 years ago.

The Neanderthal bone was so old, however, that most of the DNA in it had degraded into tiny fragments. The biologists could recover only a minute amount of DNA that was still in moderate-sized pieces. But two fortunate events occurred: (1) The Neanderthal DNA sample included a few fragments of the gene region that Pääbo's team wanted to study, and (2) the researchers were able to design primers that bracketed this gene region, based on the sequence of highly conserved sections of the same gene from *Homo sapiens*. Using PCR, the researchers were able to produce millions of copies of a specific DNA section from a Neanderthal who died over 30,000 years ago. Upon analyzing these sequences, the team found that they are highly distinct from the same gene segment found in modern humans. Subsequent work with DNA from seven other Neanderthal fossils, from locations throughout Europe, gave the same result. These data support the hypothesis that Neanderthals never interbred with modern humans—even though they lived in the same areas of Europe at the same time.

PCR has been used to study other fossil DNAs, with the current record holder for oldest DNA to be amplified via PCR coming from 17-million-year-old magnolia trees. But PCR is useful in a wide array of other situations as well. Forensic biologists, who use biological analyses to help solve crimes, clone DNA from tiny drops of blood or hair gathered at crime scenes. The copied DNA can then be analyzed to identify victims or implicate perpetrators. Genetic counselors, who advise pregnant couples on how likely their offspring are to suffer from inherited diseases, can use PCR to find out if an embryo being carried by a client has alleles associated with deadly illness. For example, it is possible to obtain a small sample of the embryo's cells, isolate the DNA, add primers on either side of the Huntington's disease gene, and clone enough copies to analyze the alleles and let the parents know if their offspring is at risk for the illness.

Because the complete genomes of a wide array of organisms have now been sequenced, researchers can find appropriate primer sequences to use in cloning almost any target gene, in a wide array of organisms, via PCR. The polymerase chain reaction is now one of the most basic and widely used techniques in molecular biology.

Check Your Understanding

If you understand that...

- PCR is a technique for amplifying a specific region of DNA into millions of copies, which can then be sequenced or used for other types of analyses.

You should be able to...

1) Explain the steps in a PCR cycle.

2) Write down the sequence of a DNA strand 50 base pairs long, then design 20-base-pair-long primers that would allow you to amplify the segment via PCR.

3) Explain why "chain reaction" is an appropriate part of PCR's name.

(MB) **Web Animation** at www.masteringbio.com

The Polymerase Chain Reaction

BOX 19.1 Southern Blotting

Southern blotting was named after its inventor, Edwin Southern, and it is among the most basic techniques in molecular biology. It is a multistep procedure that allows researchers to identify and characterize specific genes within an organism's genome. If you have a DNA sample from an individual and wish to find a specific gene in that sample, Southern blotting is a way to do it.

The first step in Southern blotting is to obtain DNA from the cells of the organisms being studied and digest it with restriction endonucleases (**Figure 19.8**, step 1). The DNA fragments generated in this way can then be separated by gel electrophoresis (steps 2 and 3). Once the fragments are sorted by size, they are treated with an alkaline solution that breaks the hydrogen bonds between base pairs, resulting in the formation of single-stranded DNA (step 4). Next the fragments of single-stranded DNA are transferred from the gel to a piece of nitrocellulose or nylon filter, using the blotting technique illustrated in step 5.

The product of these steps is a series of single-stranded DNA fragments, sorted by size and permanently bound to the filter. In some cases, these fragments represent the entire genome of the organism being studied. To find a certain gene in this collection of fragments, a researcher must have a DNA sequence that is complementary to some region of that gene. Recall that such a sequence is called a probe. The

SOUTHERN BLOTTING: ISOLATING AND FINDING A TARGET DNA IN A LARGE COLLECTION OF DNA FRAGMENTS

1. Restriction endonucleases cut DNA sample into fragments of various lengths. Each type of restriction endonuclease cuts a specific sequence of DNA.

2. A sample consists of all the DNA fragments of various lengths. The sample is loaded into a gel for electrophoresis.

3. During **electrophoresis**, a voltage gel separates DNA fragments by size. Small fragments run faster.

FIGURE 19.8 Southern Blotting Protocol. Southern blotting is a technique for locating a specific gene in a DNA sample containing many genes. In essence, the investigator cuts the DNA of interest into fragments, separates the fragments by size using gel electrophoresis, and then probes the fragments to find a specific gene.

19.3 Dideoxy DNA Sequencing

Once researchers have cloned a gene from a DNA library or via PCR, determining the gene's base sequence is usually one of the first things they want to do. Understanding a gene's sequence is valuable for a variety of reasons:

- Sequence differences among alleles can be analyzed to understand why some versions of the gene function better or differently than others do. The Huntington's disease allele, for example, has a different base sequence from alleles of the same gene that don't cause disease.

- Once a gene's sequence is known, it is straightforward to infer the amino acid sequence of its product from the genetic code. Understanding a protein's primary structure, in turn, is often helpful in inferring its function.

- A gene sequence can be compared with sequences of genes that have the same function in other species. These comparisons are often interesting. For example, bacteria, yeast, and humans are about as different as organisms can be; yet all three contain DNA polymerase genes with sections that are nearly identical in base sequence. Biologists explain this similarity by hypothesizing that the bacteria inhabiting your gut, the yeast in the bread you eat, and you are descended from a common ancestor that had DNA polymerase with the same sequence. Genes that are similar due to descent from a common ancestor are said to be **homologous**.

probe DNA is labeled—typically with a radioactive atom or fluorescent marker—and made single stranded by heating. The probe is then added to a solution bathing the filter (step 6). During incubation, the labeled probe binds to the fragment or fragments on the filter that have complementary base pairs. This step is called **nucleic acid hybridization.** In this way, the probe identifies the gene of interest.

To visualize which fragments hybridized with a radioactively labeled probe, a researcher lays X-ray film over the filter. As step 7 of Figure 19.8 shows, radioactive emissions from the probe DNA expose the film. The black band that results identifies the target gene. Fluorescent markers can be visualized and photographed under the appropriate wavelengths of light.

A variation on Southern blotting is based on separating RNAs via gel electrophoresis, transferring them to a filter paper, and probing them with a single-stranded and radioactively labeled DNA probe. This technique is used to identify the RNA fragments produced by a particular gene. It is called **Northern blotting,** in a lighthearted tribute to the protocol from which it was derived. The variation called **Western blotting** involves separating proteins via electrophoresis and then probing the resulting filter with an antibody probe that binds to the protein of interest. The use of antibodies in research is explored in detail in later chapters.

4. The DNA fragments are treated to make them single stranded.

5. Blotting. An alkaline solution wicks up through the gel into blotting paper. DNA fragments from the gel are carried to the filter, where they are permanently bound.

6. Hybridization with labeled probe. The filter is put into a solution containing labeled probe DNA. The probe binds to DNA fragments containing complementary sequences.

7. Visualize fragments bound by probe. Fluorescence or autoradiography (see **BioSkills 7**) is used to find label.

• Gene sequences can be compared to infer how closely related various species are. For example, analyzing DNA sequences isolated from fossil bones allowed researchers to evaluate whether our species, *Homo sapiens,* ever interbred with the species of human called *Homo neanderthalensis,* Neanderthals.

Given that sequence data are important, how can biologists obtain them? In 1975 Frederick Sanger developed **dideoxy sequencing** as a clever variation on the basic in vitro DNA synthesis reaction to determine exact base sequences. But saying "clever" is an understatement. ◯ Sanger had to link three important insights, summarized in **Figure 19.9,** to make his sequencing strategy work:

1. Sanger chose the name "dideoxy" because the method uses monomers called dideoxyribonucleoside triphosphates, or ddNTPs, along with deoxyribonucleoside triphosphates (dNTPs). These ddNTPs are identical to the dNTPs found in DNA, except that they lack a hydroxyl group at their 3' carbon (Figure 19.9a). Sanger realized that if a ddNTP were added to a growing DNA strand, it would terminate synthesis. Why? A ddNTP has no hydroxyl group available on its 3' carbon to link to the 5' carbon on an incoming dNTP monomer. As a result, DNA polymerization stops once a ddNTP is added. Four types of ddNTPs are used in dideoxy sequencing, each named according to whether it contains adenine (ddATP), thymine (ddTTP), cytosine (ddCTP), or guanine (ddGTP).

DIDEOXY SEQUENCING

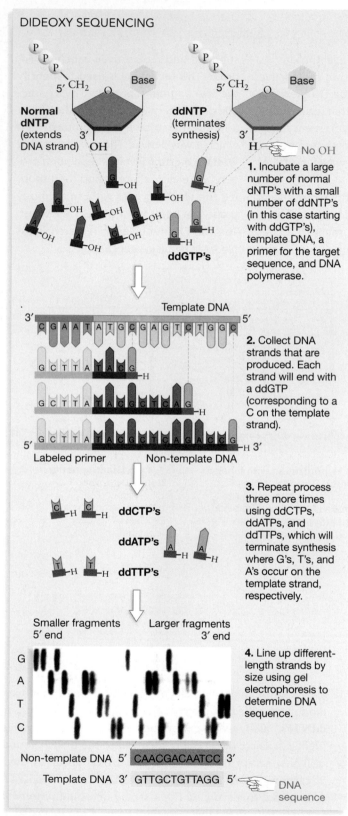

1. Incubate a large number of normal dNTP's with a small number of ddNTP's (in this case starting with ddGTP's), template DNA, a primer for the target sequence, and DNA polymerase.

2. Collect DNA strands that are produced. Each strand will end with a ddGTP (corresponding to a C on the template strand).

3. Repeat process three more times using ddCTPs, ddATPs, and ddTTPs, which will terminate synthesis where G's, T's, and A's occur on the template strand, respectively.

4. Line up different-length strands by size using gel electrophoresis to determine DNA sequence.

FIGURE 19.9 Dideoxy Sequencing Is a Method for Determining the Base Sequence of DNA.

● **EXERCISE** Starting at the far left end of the gel pictured in step 4 write down the entire sequence shown. Hint: The sequence starts with GG.

● **QUESTION** Why is it necessary to use a labeled DNA primer?

2. Sanger linked this property of ddNTPs to a second fundamental insight. Suppose, he reasoned, that a labeled primer could be attached to a template DNA, and that a large number of copies of this template DNA with a labeled primer attached could be added to a reaction mix. Suppose further that this population of labeled templates were incubated with DNA polymerase, the four dNTPs, and a small amount of ddGTP. Under these conditions, an array of different daughter strands would be synthesized. Each of these resulting daughter strands would be a different length, with each distinct DNA fragment ending with a ddGTP. To understand why, consider that DNA polymerase would synthesize a complementary strand from each labeled template in the reaction mix. The synthesis of each one of these complementary strands would start at the same point—the primer. Because there are many dGTPs and relatively few ddGTPs in the reaction mix, dGTPs are usually incorporated opposite C's on the template strand as DNA works its way along the template strand. But occasionally, one of the few ddGTPs present will be incorporated into the growing strand, opposite a C in the template. Which C in the template strand got paired with a ddGTP would be random. The addition of the ddGTP at this location would stop further elongation. Because this would happen in each template strand present, the overall reaction would produce a collection of newly synthesized strands that vary in length, corresponding to the location of each C in the template strand (Figure 19.9b). Analogous reactions done using ddTTP, ddATP, and ddCTP would give the distances between successive A's, T's, and G's, respectively.

3. Finally, Sanger realized that when the fragments produced by the four reactions are lined up by size, they reveal the sequence of bases in the template DNA. To line up fragments in order of size, he separated them using gel electrophoresis. As step 4 in Figure 19.9 shows, the sequence can then be read directly from the resulting gel.

● If you understand the essence of a DNA sequencing reaction, you should be able to consider a template strand with the sequence AGGCTACCATTCTGCTGAAG (remember, this is written 5′ to 3′) and write out, in the 5′ to 3′ direction, the sequence of each of the five fragments that would be produced in a sequencing reaction that includes ddGTP. (You should also be able to do the same exercise for reaction mixes containing ddCTP, ddATP, and ddTTP. There are five fragments from this template for each reaction mix, for a total of 20 from the four reactions.)

Several important innovations have enormously improved the speed and affordability of dideoxy sequencing. For example, researchers traditionally attached a radioactive isotope to the primer in the reaction mix to label the fragments of DNA produced by a sequencing reaction. This meant that each se-

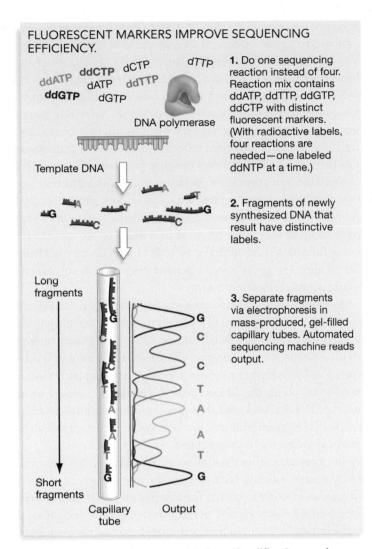

FLUORESCENT MARKERS IMPROVE SEQUENCING EFFICIENCY.

1. Do one sequencing reaction instead of four. Reaction mix contains ddATP, ddTTP, ddGTP, ddCTP with distinct fluorescent markers. (With radioactive labels, four reactions are needed—one labeled ddNTP at a time.)

Template DNA

2. Fragments of newly synthesized DNA that result have distinctive labels.

Long fragments

3. Separate fragments via electrophoresis in mass-produced, gel-filled capillary tubes. Automated sequencing machine reads output.

Short fragments

Capillary tube   Output

FIGURE 19.10 Use of Fluorescent Markers Simplifies Sequencing. Researchers made DNA sequencing more efficient by attaching distinctive fluorescent markers to the ddATP, ddGTP, ddCTP, and ddTTP used in sequencing reactions. Fluorescent markers allow researchers to use all four ddNTPs in a single reaction mix instead of having to perform four separate reactions. In addition, fluorescent markers are safer than radioactive labels and can be read by a machine.

● **EXERCISE** State another advantage of this protocol compared with sequencing methods that rely on radioactive markers.

quencing experiment required four reactions—one with each of the four ddNTPs. Now fluorescent markers are bonded to the dideoxyribonucleoside triphosphates (ddNTPs) used in a sequencing reaction—with a different color attached to each ddNTP. As **Figure 19.10** shows, the switch to fluorescent markers was important for two reasons: (1) Each stretch of DNA could be sequenced with one dideoxy reaction instead of four separate ones, and (2) machines were developed to detect the fluorescence produced by each fragment and read the output of the sequencing reaction. In addition, instead of separating the fragments generated by sequencing reactions through electrophoresis in hand-prepared gels that are poured between glass plates, researchers now perform the electrophoresis step with mass-produced, gel-filled capillary tubes.

Dideoxy sequencing ranks among the greatest of all technological advances in the history of biological science. Its impact is comparable to the development of the light microscope, the electron microscope, and recombinant gene technology.

Check Your Understanding

If you understand that...

● Dideoxy DNA sequencing allows researchers to determine the sequence of bases present in a length of DNA.

You should be able to...

1) Explain why a dideoxy sequencing reaction mix has to contain all four ddNTPs.

2) Explain why the ddNTPs have to be labeled when fragments are separated in capillary tubes.

3) Explain why few of the labeled ddNTPs have to be present relative to the number of unlabeled dNTPs.

4) Read a base sequence off a sequencing gel.

19.4 Finding Genes by Mapping: The Huntington's Disease Story

For almost 100 years after Mendel's work launched the field of genetics, the nature of the gene was a black box. Then the confirmation of DNA as the genetic material and the discovery of the double helix changed everything. It launched an explosive era of work on the nature of DNA synthesis and the central dogma of molecular biology—including cracking the genetic code and investigating the mechanisms responsible for controlling transcription and translation.

Although work on the regulation of gene expression continues at a rapid clip, the field of genetics has also become focused on efforts to find and characterize individual genes—to understand the connection between genotype and phenotype as explicitly and directly as possible. Efforts to engineer genes, for example, depend on having copies of specific genes in hand. The question is, How do researchers find the genes associated with certain traits in the first place? How do you find the gene responsible for seed shape in peas, or white eyes in flies, or the mutS DNA repair enzyme in *E. coli*?

As an example of how this type of gene hunt is done, let's consider the first successful effort ever conducted in humans—the search for the gene associated with Huntington's disease.

How Was the Huntington's Disease Gene Found?

Huntington's disease is a rare but devastating illness. Typically, affected individuals first show symptoms between the ages of 35 and 45. At onset, an individual appears to be clumsier than normal and tends to develop small tics and abnormal movements. As the disease progresses, uncontrollable movements become more pronounced. Eventually the affected individual twists and writhes involuntarily. Personality and intelligence are also affected—to the extent that the early stage of this disease is sometimes misdiagnosed as the brain disorder schizophrenia. The illness may continue to progress for 10 to 20 years and is eventually fatal.

Because Huntington's disease appeared to run in families, physicians suspected that it was a genetic disease. Recall from Chapter 13 that an analysis of pedigrees from families affected by Huntington's disease suggested that the trait was due to a single, autosomal dominant allele. To test this hypothesis, researchers set out to identify the gene or genes involved and to document that one or more genes are altered in affected individuals. Reaching this goal took over 10 years of intensive effort.

The search for the Huntington's disease gene was led by Nancy Wexler, whose mother had died of the disease. If the trait was indeed due to an autosomal dominant allele, it meant that there was a 50 percent chance that Wexler had received the allele from her mother and would begin to show symptoms when she reached middle age.

Using Genetic Markers To locate the gene or genes associated with a particular phenotype, such as a disease, researchers start with a **genetic map,** also known as a **linkage map** or **meiotic map** (see Chapter 13). Recall that in fruit flies and other organisms, the relative positions of genes on the same chromosome can be determined by analyzing the frequency of recombination between pairs of genes. In contrast to a genetic map, a **physical map** of the genome records the absolute position—in numbers of base pairs—along a chromosome. Techniques for creating physical maps are introduced briefly in Chapter 20.

A genetic map is valuable in gene hunts because it contains **genetic markers**—genes or other genetic loci or sites that have known locations. Each genetic marker provides a landmark at a position along a chromosome that is known relative to other markers.

To understand how genetic markers can be used to locate the positions of unknown genes, suppose that you knew the position of a hair-color gene in humans relative to other genetic markers. Suppose further that various alleles of this gene contributed to the development of black hair, red hair, blond hair, and brown hair in the family that you were studying. This variation in phenotype associated with the marker is crucial. To be useful in a gene hunt, a genetic marker has to be **polymorphic**, meaning that the phenotype associated with the marker varies. In our hypothetical example, hair color is a polymorphic genetic marker.

Now suppose that the genetic disease called cystic fibrosis is common in the family you are studying. Then consider what it means if your data suggest that individuals who have cystic fibrosis almost always have black hair—even though they were just as likely as unaffected individuals to have any other inher-

ited trait observed in the family, such as the presence or absence of a widow's peak or detached earlobes. ● If you observe that a certain marker and a certain phenotype are almost always inherited together, it is logical to conclude that the genes involved are physically close to each other on the same chromosome—meaning that they are closely linked. If they were not closely linked, then crossing over between them would be common and they would *not* be inherited together. Based on these data and this logic, you could infer that the gene for cystic fibrosis is very close to the hair-color gene. ● If you understand this concept, you should be able to explain why a gene hunt can be successful only if one of the genetic markers on the existing map happens to be in very close physical proximity to the gene that's being hunted. You should also be able to explain why it's helpful to hunt for genes using a genetic map with many genetic markers rather than only a few.

In essence, then, disease-gene hunting in humans boils down to this: Researchers have to find a large number of closely related people who are affected and unaffected, and then attempt to locate a genetic marker that almost always occurs in the affected individuals but not in the unaffected people. If such a marker is found, the disease gene is almost guaranteed to be nearby. The markers used in a gene hunt are mapped on their species' chromosomes, so this analysis tells researchers the physical location of a disease gene.

In organisms other than humans, researchers don't have to study a large, existing family where the disease or trait in question exists in order to search for associations between a phenotype and a known genetic marker. Instead, they can create a large family by arranging experimental crosses between individuals with different versions of the trait in question. Then they can look for associations between genetic markers in the offspring of the crosses and the presence of that trait in those offspring. The ability to make experimental crosses makes gene hunts in these organisms much more efficient than in humans.

It's also important to note that the types of genetic markers used in gene hunts have changed over time. Today, researchers often have a large catalog of genetic markers available, including the particularly abundant markers known as **single nucleotide polymorphisms** (**SNPs**, pronounced "snips"). A SNP is a site in DNA where some individuals in the population have different bases. For example, some might have an A at a certain site while others have a C. That site would qualify as a SNP. But in the late 1970s and early 1980s, when biologists were searching for the Huntington's gene, the best genetic markers available were restriction sites, the short stretches of DNA where restriction endonucleases cut the double helix. These sequences are also known as restriction endonuclease recognition sites. The sites that Wexler's team used were polymorphic—in the family they studied, some individuals had sites where cuts occurred; but in other individuals, the DNA sequences at the same site were different and no cuts occurred. Thus, each restriction site was either present in an individual or not present—just as an

Chromosome of
diseased individual

Genetic marker
restriction sites
absent

Defective
Huntington's gene
(disease allele)

Close physical association between
recognition site and defective allele.
Genetic markers at other locations are
equally likely to be found in affected
and unaffected individuals

Chromosome of
healthy individual

Genetic marker
restriction sites
present

Normal
Huntington's
gene

FIGURE 19.11 Genetic Markers and Disease Alleles Are Inherited Together if They Are Closely Linked. To find a gene, researchers look for an association between a phenotype and a genotype. The phenotype is associated with the gene of interest, and the genotype is a mapped genetic marker.

individual might have an A instead of a C at a certain SNP or a yellow versus black hair-color allele. Wexler's team was looking for restriction sites that are almost always present in diseased individuals but absent in healthy individuals, or absent in diseased individuals but present in healthy people (**Figure 19.11**).

Using a Pedigree Once a genetic map containing many genetic markers has been assembled, biologists need help from families affected by an inherited disease to find the gene in question. Recall that the fundamental goal is to find a genetic marker that is almost always inherited along with the disease-causing allele. Gene hunts are more likely to be successful if large families are involved. Large sample sizes minimize the probability that researchers will observe an association between one or more markers and the disease just by chance—not because they are closely linked. Wexler's Huntington's disease team was fortunate

to find a large, extended family affected with the disease living along the shores of Lake Maracaibo, Venezuela.

From historical records, the researchers deduced that the Huntington's disease allele was introduced to this family by a European sailor or trader who visited the area in the early 1800s. When family members agreed to participate in the study, there were over 3000 of his descendants living in the area. One hundred of these people had been diagnosed with Huntington's disease. To help in the search for the gene, family members agreed to donate skin or blood samples for DNA analysis and to furnish information on who was related to whom.

This research resulted in the pedigree shown in **Figure 19.12**. The diagram includes information about the disease phenotype and the particular genetic markers observed in each family member. The combinations of markers called A, B, C, and D turned out to be particularly important. These

FIGURE 19.12 Certain Genetic Markers Are Inherited with the Huntington's Disease Gene. Part of a pedigree from seven generations (I–VII) of a large extended family. Individuals affected by Huntington's disease inherited the defective allele from a common ancestor. The letters A–D represent combinations of restriction sites found in each individual—in essence, each is a different allele of the same genetic marker. Two letters are indicated for each individual because humans are diploid.

● **QUESTION** In generation VI, an individual with the genotype AC does not have Huntington's disease. How is this possible?

four combinations of restriction sites represent alleles. The key finding is that the C allele was almost always found in diseased individuals. Apparently, the English sailor who introduced the Huntington's disease allele also had the C allele—this particular combination of restriction sites—in his DNA. The C allele and the Huntington's disease gene are so close together that recombination between them—which would put an A or B or D allele next to the Huntington's disease allele in his descendants—has been extremely rare.

From the existing genetic map of humans, Wexler's team knew that the C restriction sites were on chromosome 4. Eventually the team succeeded in narrowing down the location of the Huntington's disease gene to a region about 500,000 base pairs long. Because the haploid human genome contains over 3 billion base pairs, this was a huge step in focusing the search for the gene.

Pinpointing the Defect

Once the general location of the Huntington's disease gene was known, biologists looked in that region for exons that encode a functional mRNA. Then they sequenced exons from diseased and normal individuals, compared the data, and pinpointed specific bases that differed between the two groups of individuals. In this way, dideoxy sequencing played a key role in the gene hunt.

When this analysis was complete, the research team found that individuals with Huntington's disease have an unusual number of CAG codons near the 5′ end of a particular gene. CAG codes for glutamine. Healthy individuals have 11–25 copies of the CAG codon, while affected individuals have 42 or more copies. Although many genetic diseases are caused by single base changes that alter the amino acid sequence of a protein, diseases caused by the expansion of a particular codon repeat have also been observed.

When the Huntington's disease research team confirmed that the increase in the CAG codon was always observed in affected individuals, the team concluded that the long search for the Huntington's disease gene was over. They named the newly discovered gene *IT15* and its protein product huntingtin. In both affected and normal individuals, the huntingtin protein is involved in the early development of nerve cells. It is only later in life that mutant forms of the protein cause disease.

What Are the Benefits of Finding a Disease Gene?

Finding the genes associated with specific traits has helped biologists gain a better understanding of hundreds of issues, ranging from why some pea plants are dwarfed to how the repressor shuts down the *lac* operon. In general, how have efforts to find disease genes improved human health and welfare? More specifically, how has the effort to locate the Huntington's disease gene helped researchers and physicians understand and treat the illness? Biomedical researchers point to three major benefits of successful disease-gene hunts.

Improved Understanding of the Phenotype

Once a disease gene is found and its sequence is known, researchers can usually figure out why its product causes disease. In the case of *IT15*, autopsies of Huntington's patients had shown that their brains actually decrease in size due to the death of neurons, and that the brain tissue contains clumps, or aggregates, of the huntingtin protein. These aggregates are thought to be a direct consequence of the changes in the number of CAG repeats in the *IT15* gene and thus the number of glutamine residues in the protein. Long stretches of polyglutamine (a polymer of the amino acid glutamine) are known to result in the formation of protein aggregates. The leading hypothesis to explain Huntington's disease proposes that a gradual buildup of the huntingtin protein aggregates triggers neurons to undergo apoptosis, or programmed cell death.

These results helped explain why Huntington's disease is pleiotropic. Patients suffer from abnormal movements *and* personality changes because neurons from throughout the brain are being killed. The results also help explain why the disease takes so long to appear, and why it is progressive: The defective huntingtin proteins take time to build up to deadly levels, but then continue to increase over time. Finally, understanding the molecular mechanism responsible for the illness explained why the disease allele is dominant. One copy of the defective gene is enough to produce fatal concentrations of polyglutamine aggregates.

Therapy

Once the sequence of a gene is known, researchers can better understand the structure and function of its protein product. Information on a protein's mode of action is extremely helpful in developing drugs or other treatments.

The search for new therapies for Huntington's disease began with an effort to introduce the defective allele into mice, using the types of genetic engineering techniques discussed in Section 19.5. These mice are called **transgenic** (literally, "across-genes") because they have alleles that have been modified by genetic engineering. Transgenic mice that produce defective versions of the huntingtin protein develop tremors and abnormal movements, exhibit higher-than-normal levels of aggression toward litter and cage mates, and experience a loss of neurons in the brain. The transgenic mice get a version of Huntington's disease. Laboratory animals with disease symptoms that parallel those of a human disease are said to provide an **animal model** of the disease.

Animal models are valuable in disease research, because they can be used to test potential treatments before investigators try them on human patients. Now that an animal model for Huntington's disease is available, research groups have begun testing drugs that appear to prevent or reduce the aggregation of the huntingtin protein. Other biologists are attempting to design drugs that prevent the death of cells containing aggregated huntingtin proteins.

Genetic Testing   When the Huntington's gene was found, biologists used the information to develop a test for the presence of the defective allele. The test involves obtaining a DNA sample from an individual and using the polymerase chain reaction to amplify the chromosome region that contains the CAG repeats responsible for the disease. If the number of CAG repeats is 35 or less, the individual is considered to be normal. Forty or more repeats results in a positive diagnosis for Huntington's.

As more becomes known about the human genome, more and more tests like this will be developed for diseases that have some genetic component. What types of genetic testing are done currently?

- *Carrier Testing*   Before starting their own family, people from families affected by a genetic disease frequently want to know whether they carry the allele responsible. That is especially true for diseases, such as cystic fibrosis (CF), that are due to recessive alleles. If only one of the prospective parents has the allele, then none of the children they have together should develop CF. But if both prospective parents carry the allele, then each child they produce has a 25 percent chance of having CF. Carrier testing can determine whether an individual is a carrier for a genetic disease.

- *Prenatal Testing*   Suppose that two parents, both carrying the CF allele, decide to have children but do not want to pass along that allele. Once the mother is pregnant, a physician can obtain fetal cells early in gestation. The cells can be cultured and DNA isolated. The CF allele can then be amplified by PCR and sequenced. Based on the test results, the couple may choose to continue or terminate a pregnancy.

- *Adult Testing*   Because the symptoms of Huntington's disease do not appear until middle age, many affected people have married and had children by the time they develop symptoms. But people with a family history of Huntington's disease can now be tested as young adults. If they learn that they have the defective allele, they might make different decisions about starting their own families and passing on the illness. And Huntington's disease isn't the only trait that appears in adulthood. For example, 5–10 percent of the women who develop breast cancer have inherited a faulty gene. To date, more than five genes are known to create a predisposition to breast cancer. Adult women with a family history of the illness can be tested for the presence of these mutant genes. If they have a susceptibility allele, they can be tested for the onset of cancer more frequently than normal. They may also be able to make changes in their diet or activity that make them less likely to actually develop the illness.

Recent work on the Huntington's disease gene highlights just how far the field of molecular genetics has come in just a few short decades. Before the molecular revolution, the molecular nature of the gene was not well understood. Now, researchers can find and sequence the alleles associated with a wide array of traits in humans and other organisms and place them in other organisms. **Table 19.1** summarizes some of the most important tools and techniques that made these advances possible.

Along with any other type of change, however, the advances emerging from gene hunts create new problems that must be addressed. Like the misuse of human growth hormone produced with recombinant DNA technology, successful gene hunts can create serious moral and legal dilemmas.

Ethical Concerns over Genetic Testing

Genetic testing raises controversial ethical issues. For example, some individuals maintain that it is morally wrong to terminate any pregnancy, even if the fetus involved is guaranteed to be born with a debilitating genetic disease. Also, consider Nancy Wexler's position soon after the discovery of *IT15*. Would you choose to be tested for the defective allele and risk finding out that you were almost certain to develop Huntington's disease? Should physicians agree to test people for genetic diseases that have no cure? Should it be legal for health insurance companies to test clients for genetic diseases that will require expensive treatments so that the companies can refuse coverage?

These questions are still being debated by political and religious leaders, health-care workers, philosophers, and members of the general public. Difficult as they are, the issues raised by genetic testing pale in comparison with some of the questions raised by the effort to create transgenic humans.

Check Your Understanding

If you understand that...
- Genes for particular traits can be located if they are closely linked to a known genetic marker and are thus inherited along with the marker.

You should be able to...

Draw a pedigree showing a close association between a particular genetic illness and a specific genetic marker.

19.5   Can Gene Therapy Cure Inherited Diseases in Humans? Research on Severe Immune Disorders

For biomedical researchers interested in curing inherited diseases such as Huntington's, sickle-cell anemia, and cystic fibrosis, the

SUMMARY TABLE **19.1** **Some Common Tools and Techniques Used in Genetic Engineering**

		Tools	
Name	**Description**	**How Used**	**Example or Illustration**
Reverse transcriptase	Enzyme that catalyzes synthesis of a complementary DNA (cDNA) from an RNA template	Making cDNAs used in constructing a genetic library (many other applications).	cDNA / RNA
Restriction endonucleases	Enzyme that cuts DNA at a specific sequence—often a palindromic sequence that is 5 base pairs long.	Allows researchers to cut DNA at specific locations. Cuts in palindromic sites create "sticky ends."	G AATTC / CTTAA G
DNA ligase	An enzyme that catalyzes the formation of a phosphodiester bond between nucleotides on the same DNA strand.	Ligates (joins) sequences that were cut with a restriction endonuclease. Gives researchers the ability to splice fragments of DNA together.	G AATTC / C TTAAG
Plasmids	Small, extrachromosomal loops of DNA found in many bacteria and in some yeast.	After a target gene is inserted into a plasmid, the recombinant plasmid serves as a vector for transferring the gene into a bacterial or yeast cell, so the gene can be cloned.	GAATTC / CTTAAG ... GAATTC / CTTAAG
Taq polymerase	DNA polymerase from the bacterium *Thermus aquaticus*. Catalyzes synthesis of DNA from a primed DNA template; remains stable at 95°C.	Responsible for the "primer extension" step in the polymerase chain reaction. Heat stability allows enzyme to be active even after denaturation step of PCR cycle at 94°C.	
Single nucleotide polymorphisms (SNPs)	Sites in DNA where the identity of the base varies among individuals in a population.	One of many polymorphic types of DNA sequences that are useful in creating the genetic maps required for gene hunts.	Sequence of some individuals ...GATTCTTATTAACTACGC... ...GATTCTTCTTAACTACGC... Sequence of other individuals

ultimate goal is to replace or augment defective copies of the gene with normal alleles. This approach to treatment is called **gene therapy**.

For gene therapy to succeed, two crucial steps must be completed. First, the "wild-type" allele for the gene in question and its regulatory sequences must be sequenced and understood. This allele codes for a product associated with a healthy phenotype. Second, there must be a method for introducing a copy of the wild-type allele into affected individuals. The DNA has to be introduced in a way that ensures expression of the gene in the correct tissues, in the correct amount, and at the correct time. If the defective allele is dominant, then the introduction step may be even more complicated: In at least some cases, the introduced allele must physically replace or block the expression of the existing dominant allele.

Chapter 18 discussed aspects of gene regulation in humans and other eukaryotes; here, we focus on techniques for introducing DNA into humans. As a case study, let's consider the first successful application of gene therapy in humans.

How Can Novel Alleles Be Introduced into Human Cells?

Section 19.1 reviewed how recombinant DNA sequences are packaged into plasmids and taken up by *E. coli* cells. Humans and other mammals lack plasmids, however, and their cells do not take up foreign DNA in response to chemical or electric treatments as efficiently as bacterial cells do.

How can foreign genes be introduced efficiently into human cells? To date, researchers have focused on packaging

SUMMARY TABLE **19.1** **Some Common Tools and Techniques Used in Genetic Engineering** *(continued)*

	Techniques		
Name	**Description**	**How Used**	**Example or Illustration**
Recombinant DNA technology ("genetic engineering")	Taking a copy of a gene from one individual and placing it in the genome of a different individual (often of a different species).	Many applications, including DNA cloning, gene therapy (see Section 19.5), and biotechnology (see Sections 19.1 and 19.6).	
Genetic library	A collection of all DNA sequences present in a particular source. The library consists of individual DNA fragments that are isolated and inserted into a plasmid or other vector, so they can be cloned.	cDNA libraries allow researchers to catalog the genes being expressed in a particular cell type. Genomic libraries allow researchers to archive all the DNA sequences present in a genome. Libraries can be screened to find a particular target gene.	
DNA Probing/Screening a genetic library	Use of a labeled, known DNA fragment to hybridize (via complementary base pairing) with a collection of unlabeled, unknown fragments.	Allows a researcher to find a particular DNA sequence in a large collection of sequences.	
Polymerase chain reaction	Requires known primer sequences on either side of a target gene. Reaction is based on many cycles of primer annealing, primer extension, and DNA denaturation	Produces many identical copies of a target sequence. A shortcut method for DNA cloning.	
Dideoxy sequencing	In vitro DNA synthesis reaction that includes dideoxyribonucleotide monomers (ddNTPs).	Determining the base sequence of a gene or other section of DNA.	
Genetic mapping	Creation of a map showing the relative positions of genes or specific DNA sequences on chromosomes. Done by analyzing the frequency of recombination between sequences (see Chapter 13).	Many applications, including use of mapped genetic markers to find unknown genes associated with diseases or other distinctive phenotypes.	

● **EXERCISE** Fill in the "Example or Illustration" cells in the Techniques column of the table.

foreign DNA into viruses for transport into human cells. A viral infection begins when a virus particle enters or attaches to a host cell and inserts its genome into that host cell (see Chapter 35). In some cases the viral DNA becomes integrated into a host-cell chromosome. As a result, viruses that infect human cells can be used as vectors to carry engineered alleles into the chromosomes of target cells. Potentially, the alleles delivered by the virus could be expressed and produce a product capable of curing a genetic disease.

Currently, the vector of choice in gene therapy is **retroviruses**. Retroviruses are viruses that have an RNA genome. The genomes include the gene for the enzyme reverse transcriptase, which catalyzes the production of a double-stranded DNA molecule from a single-stranded RNA template.

(Recall from Section 19.1 that researchers use reverse transcriptase to produce cDNAs.) When a retrovirus infects a human cell, reverse transcriptase catalyzes the production of a DNA copy of the viral genome. Other viral enzymes catalyze the insertion of the viral DNA into a host-cell chromosome. If human genes can be packaged into a retrovirus, then the virus is capable of inserting the human alleles into a chromosome in a target cell (**Figure 19.13**).

Although viruses would appear to be ideal vectors for the delivery of normal human alleles, there are serious problems associated with their use. Viruses usually cause disease. The retroviruses include the human immunodeficiency virus (HIV), which causes AIDS, as well as an array of viruses that cause cancer in humans and other animals. Using these agents for

USING ENGINEERED VIRUSES TO INTRODUCE ALLELES INTO HUMAN CELLS

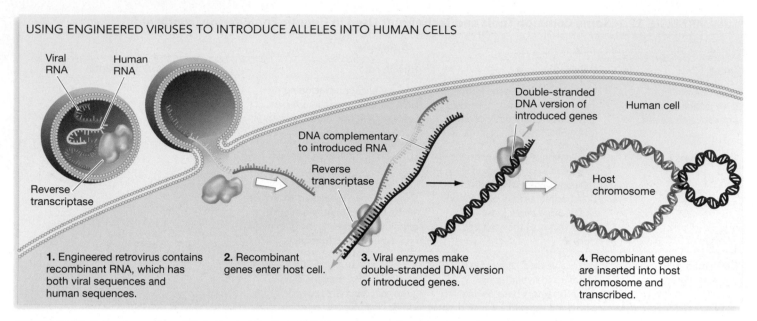

Viral RNA   Human RNA

Reverse transcriptase

DNA complementary to introduced RNA

Reverse transcriptase

Double-stranded DNA version of introduced genes   Human cell

Host chromosome

1. Engineered retrovirus contains recombinant RNA, which has both viral sequences and human sequences.

2. Recombinant genes enter host cell.

3. Viral enzymes make double-stranded DNA version of introduced genes.

4. Recombinant genes are inserted into host chromosome and transcribed.

FIGURE 19.13 Retroviruses Insert Their Genes into Host-Cell Chromosomes. Retroviruses can be used as vectors to carry specific genes into cells and to insert these genes into the chromosomes of target cells.

● **QUESTION** What happens if the recombinant DNA is inserted in the middle of a gene that is critical to the cell's normal function?

gene therapy requires that sequences responsible for causing disease be inactivated or removed from their genomes. Even if they have been inactivated, the altered particles may be able to recombine with viral DNA that already exists in the individual being treated and lead to the formation of a new infectious strain. In addition, the presence of viral proteins triggers a response by the immune system that can cause dangerous side effects during treatment. Finally, if viral genes happen to insert themselves in a position that disrupts the function of an important gene in the target cell, the consequences can be serious. Despite these risks, retroviruses are still the best vectors currently available for human gene therapy. ● If you understand this concept, you should be able to add a row called "Retroviruses" to the "Tools" section of Table 19.1 and fill in the "Description," "How Used," and "Illustration" columns.

Using Gene Therapy to Treat X-Linked Immune Deficiency

In 2000 a research team reported the successful treatment of a disease by means of gene therapy. The illness is called **severe combined immunodeficiency (SCID)**. Children who are born with SCID lack a normal immune system and are unable to fight off infections.

The type of SCID treated by the group is designated SCID-X1, because it is caused by mutations in a gene on the X chromosome. The gene codes for a receptor protein necessary for the development of immune system cells called T cells. T cells develop in bone marrow, from undifferentiated cells that divide continuously.

Traditionally, physicians have treated SCID-X1 by keeping the patient in a sterile environment, isolated from any direct human contact until the patient could receive bone-marrow tissue transplanted from a close relative (**Figure 19.14**). In most bone-marrow transplants, the T cells that the patient needs are produced by the transplanted bone-marrow cells and allow the individual to live normally. In some cases, though, no suitable donor is available. Could gene therapy cure this disease by furnishing functioning copies of the defective gene?

After extensive testing suggested that their treatment plan was safe and effective, the research team gained approval to

FIGURE 19.14 A "Bubble Child". Children who are born with SCID cannot fight off bacterial or viral infections. As a result, such children must live in a sterile environment.

ONE APPROACH TO GENE THERAPY IN HUMANS

SCID patient

1. Stem cells are isolated from the patient's bone marrow and grown in vitro.

2. Recombinant retroviruses carry the normal allele into host cells.

3. Cells that are expressing normal alleles are isolated and implanted into patient.

FIGURE 19.15 A Gene Therapy Protocol. For gene therapy to work, copies of a normal allele have to be introduced into a patient's cells and be expressed.

treat 10 boys, each less than 1 year old, who had SCID-X1 but no suitable bone-marrow donor. To implement gene therapy, the researchers removed bone marrow from each patient and collected the cells that produce mature T cells (**Figure 19.15**). The researchers then infected the cells with an engineered retrovirus that carried the normal receptor gene. Cells that were producing normal receptor protein were then isolated and transferred back into the patients.

Within four months after reinsertion of the transformed marrow cells, nine of the boys had normal levels of functioning T cells. These patients were removed from germ-free isolation rooms and began residing at home, where they grew and developed normally. During a routine check-up 30 months after gene therapy, however, the research team discovered that two of the boys had developed a type of cancer characterized by unchecked growth of T cells. Follow-up analyses of their bone-marrow cells showed that in both boys, a viral-borne receptor gene had been inserted near the gene for a regulatory transcription factor that triggers the growth of T cells. The viral sequences apparently acted as an enhancer and led to constitutive expression of the transcription factor. As this book goes to press, the two boys have responded to cancer chemotherapy and are healthy.

The tenth boy to receive gene therapy never succeeded in producing T cells at all. For unknown reasons, his recombinant stem cells failed to function normally when they were transplanted back into his bone marrow. Fortunately, physicians were later able to find a bone marrow donor whose cells were a close enough match to the boy's cells to make a successful transplant possible.

Ethical Concerns over Gene Therapy

Throughout the history of medicine, efforts to test new drugs, vaccines, and surgical protocols have always carried a risk for the patients involved. Gene therapy experiments are no different. The researchers who run gene therapy trials must explain the risks clearly and make every effort to minimize them. The initial report on the development of cancer in the boys who received gene therapy for SCID-X1 concluded with the following statement: "We have proposed ... a halt to our trial until further evaluation of the causes of this adverse event and a careful reassessment of the risks and benefits of continuing our study of gene therapy." Although recombinant gene therapy still holds promise for the treatment of a wide variety of devastating inherited diseases, fulfilling that promise is almost certain to require many years of additional research and testing, as well as the refinement of legal and ethical guidelines.

When recombinant DNA technology first became possible, many researchers thought they would live to see most or all of the serious inherited diseases in humans cured by gene therapy. Several decades later, that optimism is tempered. In humans, gene therapy is still highly experimental, extremely expensive, and intensely controversial.

19.6 Biotechnology in Agriculture: The Development of Golden Rice

Progress in human gene therapy has been slow, but progress in transforming crop plants with recombinant genes has been breathtakingly rapid. In the United States alone, genetically modified or transgenic strains of 15 major crop species are now approved for commercial use. Farmers growing maize (corn) in the United States currently have over 15 distinct transgenic varieties to choose from. In Mississippi, 97 percent of the land area devoted to cotton is now planted with recombinant strains. Globally, about 87.2 million hectares (192 million acres) of transgenic crops were grown in 2006. The use of genetically modified plants has become a major element of modern agriculture.

Recent efforts to develop transgenic plants have focused on three general objectives:

1. *Reducing losses to herbivore damage* The most popular strategy for reducing losses caused by plant-eating organisms is based on introducing the gene for a naturally occurring insecticide. For example, researchers have transferred a gene from the bacterium *Bacillis thuringiensis* into corn; the presence of the "Bt toxin" encoded by this gene protects the plant from corn borers and other caterpillar pests.

2. *Reducing competition with weeds* Competition with weeds is usually reduced by introducing a gene whose product makes the crop plant resistant to an herbicide (a chemical that destroys plants). For example, soybeans have been genetically engineered for resistance to the herbicide glyphosate. Soybean fields with the engineered strain can be sprayed with glyphosate to kill weeds without harming the soybeans.

3. *Improving the quality of the product consumed by people* Some crop plants, including soybeans and canola plants, have been engineered to produce a higher percentage of unsaturated fatty acids relative to saturated fatty acids (see Chapter 6). Saturated fatty acids can contribute to heart disease, so crops with less of them are healthier to eat.

To gain a better understanding of genetic engineering in plants, let's take a detailed look at efforts to produce transgenic rice with improved nutritional qualities.

Rice as a Target Crop

Although almost half the world's population depends on rice as its staple food, this grain is an extremely poor source of certain vitamins and essential nutrients. For example, rice contains no vitamin A. This is a serious issue. In addition to causing blindness in 250,000 Southeast Asian children each year, vitamin A deficiency can make children more susceptible to diarrhea, respiratory infections, and childhood diseases such as measles.

Humans and other mammals synthesize vitamin A from a precursor molecule known as β-carotene (beta-carotene). β-Carotene belongs to a family of plant pigments called the carotenoids, which were introduced in Chapter 10. Carotenoids are orange, yellow, and red and are especially abundant in carrots. Rice plants synthesize β-carotene in their chloroplasts but not in the part of the seed that is eaten by humans. Could genetic engineering produce a strain of rice that synthesized β-carotene in the carbohydrate-rich seed tissue called endosperm, which humans do eat? A research team set out to answer this question. If successful, their research could help to solve an important global health problem.

Synthesizing β-Carotene in Rice

The research team began their work by searching for compounds in rice endosperm that could serve as precursors for the synthesis of β-carotene. They found that maturing rice endosperm contains a molecule called geranyl geranyl diphosphate (GGPP), which is an intermediate in the synthetic pathway that leads to the production of carotenoids. As **Figure 19.16** shows, three enzymes are required to produce β-carotene from GGPP. If genes that encode these enzymes could be introduced into rice plants along with regulatory sequences that would trigger their synthesis in endosperm, the researchers could produce a transgenic strain of rice that would contain β-carotene.

Fortunately, genes that encode two of the required enzymes had already been isolated from daffodils, and the gene for the third enzyme had been purified from a bacterium. Because the sequences had been inserted into plasmids and grown in bacteria, many copies were available for manipulation. To each of the coding sequences in the plasmids, biologists added the promoter region from an endosperm-specific protein. This segment included a regulatory site that would promote transcription of the recombinant sequences in endosperm cells.

To develop transgenic rice strains that are capable of producing β-carotene, the three sets of sequences had to be inserted into rice plants. As you know from reading this chapter, introducing recombinant DNA is fairly straightforward into *E. coli* but difficult in humans. How are foreign genes introduced into plants?

The *Agrobacterium* Transformation System

Agrobacterium tumefaciens is a bacterium that infects plants. Plant tissues infected with this parasite form a tumorlike growth called a **gall**. Such plants are said to have crown gall disease (**Figure 19.17a**). When researchers looked into how

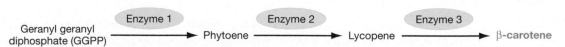

FIGURE 19.16 Synthetic Pathway for β-Carotene. GGPP is a molecule found in rice seeds. Three enzymes are required to produce β-carotene from GGPP.

(a) Plant with crown gall disease

(b) *AGROBACTERIUM* INDUCES GALL FORMATION.

Agrobacterium cell

Ti plasmid

Main chromosome

T-DNA

1. *Agrobacterium* cells contain a Ti (*tumor-inducing*) plasmid.

Host-cell chromosomes

Plant cell nucleus

2. A section of DNA from the Ti plasmid, called T-DNA, incorporates into the chromosomes of plant cells infected by the bacterium.

3. When transcribed, Ti genes induce the affected cell to begin growing and dividing. The resulting gall encloses an increasing number of *Agrobacterium* cells.

Agrobacterium cells

FIGURE 19.17 Agrobacterium Infections Introduce Foreign Genes into a Host-Cell Chromosome. (a) The gall, or tumorlike growth at the bottom of this plant's stem, is due to an infection by the bacterium *Agrobacterium*. **(b)** Ti plasmids of *Agrobacterium* cells induce gall formation.

these infections occur, they found that a plasmid carried by the *Agrobacterium* cells, called a **Ti** (tumor-inducing) **plasmid**, plays a key role (**Figure 19.17b**, step 1). Ti plasmids contain several functionally distinct sets of genes. One set encodes products that allow the bacterium to bind to the cell walls of a host. Another set, referred to as the virulence genes, encodes

the proteins required to transfer part of the Ti DNA, called T-DNA (transferred DNA), into the interior of the plant cell. The T-DNA then travels to the plant cell's nucleus and integrates into its chromosomal DNA (step 2). When transcribed, T-DNA induces the infected cell to grow and divide. The result is the formation of a gall that encloses a growing population of *Agrobacterium* cells (step 3).

Researchers soon realized that the Ti plasmid offers an efficient way to introduce recombinant genes into plant cells. Follow-up experiments confirmed that recombinant genes could be added to the T-DNA that integrates into the host chromosome, that the gall-inducing genes could be removed, and that the resulting sequence is efficiently transferred and expressed in its new host plant.

Using the Ti Plasmid to Produce Golden Rice

To generate a strain of rice that produces all three enzymes needed to synthesize β-carotene in endosperm, the researchers exposed embryos to *Agrobacterium* cells containing genetically modified Ti plasmids (**Figure 19.18**). The transformed plants were then grown in the greenhouse. When the transgenic individuals had matured and set seed, the researchers found that some rice grains contained so much β-carotene that they appeared yellow. This "golden rice" is compared with unmodified

GENETIC ENGINEERING OF Ti PLASMIDS

Tumor-inducing genes

T-DNA

1. Start with normal Ti plasmids.

2. Remove tumor-inducing genes.

Promoter

Genes for three enzymes

3. Add genes for enzymes required for β-carotene synthesis along with promoter that will be activated in endosperm.

FIGURE 19.18 "Golden Rice" Is a Transgenic Strain. The construction of a Ti plasmid to produce a strain of rice capable of synthesizing β-carotene in the endosperm of their seeds.

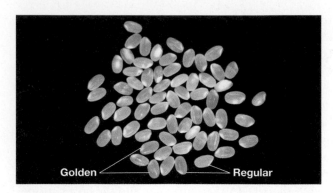

FIGURE 19.19 Comparing Regular Rice and Golden Rice Chemical analyses confirmed that the golden color in transformed seeds is due to large quantities of β-carotene.

rice in **Figure 19.19**. Testing is under way to determine whether the transgenic rice contains enough β-carotene to relieve vitamin A deficiency in children.

While the production of golden rice in the laboratory is promising, there is still much to do before the advance has an impact on public health. The transgenic rice must be grown successfully under normal field conditions. In addition, most or all of the rice strains grown in different parts of the world will need to be crossed with transgenic stock to acquire the appropriate genes. Finally, the transgenic strains must be made available to poor farmers at an affordable price.

In addition to the possibility of alleviating vitamin A deficiency, transgenic strains of rice may be able to ease other nutritional problems in rice-eating people—specifically, protein deficiency and iron deficiency. Early attempts to develop transgenic rice strains with augmented amounts of stored protein and available iron have been promising.

As the examples reviewed in this chapter show, recent advances in agricultural and medical biotechnology promise to increase the quality and quantity of food and to alleviate at least some inherited diseases that plague humans. Each solution offered by genetic engineering tends to introduce new issues, however. For example, researchers and consumer advocates have expressed concerns about the increased numbers and types of genetically modified foods. During this debate, biology students and others who are well informed about the techniques and issues involved will be an important source of opinions and information for the community.

Chapter Review

SUMMARY OF KEY CONCEPTS

● **The discovery of enzymes that cut DNA at specific locations, along with enzymes that piece DNA segments back together, gave biologists the ability to move genes from one location to another.**

Genetic engineering is based on inserting alleles from one organism into another organism. Biologists were able to manipulate genes after the discovery of restriction endonucleases, DNA ligase, and plasmids. Restriction endonucleases allow researchers to cut DNA at specific locations and insert it into plasmids or other vectors with the help of DNA ligase. In many cases, the engineered alleles are modified by the addition of certain types of promoters or other regulatory sequences.

You should be able to make a flowchart illustrating the steps in cutting a gene out of one location and pasting it into another location. ●

● **To analyze a gene, biologists have to obtain many identical copies of it. This can be done by inserting the gene into a bacterial cell that copies the gene as it grows, or by performing a polymerase chain reaction.**

The gene for the human growth hormone was cloned after it was isolated by introducing it into a plasmid that was taken up by *E. coli* cells. There the gene was expressed. Thus, genetic engineering and DNA cloning have resulted in large populations of recombinant *E. coli* cells that produce large, safe supplies of human growth hormone at an affordable price.

The most common alternative to cloning genes in plasmids is an in vitro DNA synthesis reaction called the polymerase chain reaction (PCR). The PCR technique is based on primers that bracket a target stretch of DNA. Using the heat-stable form of DNA polymerase called *Taq* polymerase, a single target DNA sequence can be amplified to millions of identical copies.

You should be able to compare the advantages and disadvantages of cloning genes via plasmids versus PCR. ●

(MB) Web Animation at www.masteringbio.com
Producing Human Growth Hormone; The Polymerase Chain Reaction

● **Once researchers have many identical copies of a gene, its sequence of bases can be determined by dideoxy sequencing.**

The most common way to analyze the gene copies produced by plasmid-based cloning or PCR is to sequence them by the dideoxy sequencing method. This is an in vitro synthesis reaction that employs dideoxyribonucleotides to stop DNA replication at each base in the sequence. By running the resulting DNA fragments out on a gel, the sequence of nucleotides in a gene can be determined.

You should be able to explain why the use of fluorescent markers and gel-filled capillary tubes increased the efficiency of sequencing studies. ●

● **To find the genes associated with a particular trait, researchers start with a genetic map. If certain markers on the map are found only in individuals that have a distinctive phenotype, the gene responsible for that phenotype is likely to be near that marker.**

To find the gene associated with Huntington's disease, investigators analyzed a large number of genetic markers and a large pedigree of an affected family. Their goal was to find a marker that was inherited along with the allele responsible for the disease. Once this strategy pinpointed the general area where the gene was located, biologists sequenced DNA from the region to determine exactly where the defect occurred.

You should be able to create a genetic map for a hypothetical organism and explain how you could use it to find a gene associated with body size. ●

● **Researchers are attempting to insert genes into humans to cure genetic diseases. Efforts to insert genes into plants to provide them with novel traits, such as the ability to produce important nutrients, have been much more successful.**

Once genes are located and characterized, they can be introduced into other individuals or species in an effort to change their traits. Genetic transformation can occur in several ways, depending on the species involved. In humans, recombinant DNA must be introduced by viruses. Because introducing foreign genes into humans is difficult, and because complex ethical and safety issues are involved, progress in human gene therapy has been slow. Genetic engineering of crop plants is much easier technically because certain bacteria that infect plants have plasmids that integrate their genes into the host-plant genome. By adding recombinant alleles to these plasmids, researchers have been able to introduce alleles that improve the nutritional quality of crops, make them resistant to herbicides, or allow them to produce insecticides.

You should be able to explain what would have to happen for recombinant gene technology to become as feasible in humans as it is in plants. ●

QUESTIONS

● Test Your Knowledge

1. What do restriction endonucleases do?
   a. They cleave bacterial cell walls and allow viruses to enter the cells.
   b. They join pieces of DNA by catalyzing the formation of phosphodiester bonds between them.
   c. They cut stretches of DNA at specific sites known as recognition sequences.
   d. They act as genetic markers in the chromosome maps used in gene hunts.

2. What is a plasmid?
   a. an organelle found in many bacteria and certain eukaryotes
   b. a circular DNA molecule that in some cases replicates independently of the main chromosome(s)
   c. a type of virus that has a DNA genome and that infects certain types of human cells, including lung and respiratory tract tissue
   d. a type of virus that has an RNA genome, codes for reverse transcriptase, and inserts a cDNA copy of its genome into host cells

3. When present in a DNA synthesis reaction, a ddNTP molecule is added to the growing chain. No further nucleotides can be added afterward. Why?
   a. There are not enough dNTPs available.
   b. A ddNTP can be inserted at various locations in the sequence, so fragments of different length form—each ending with a ddNTP.
   c. The 5' carbon on the ddNTP lacks a hydroxyl group, so no phosphodiester bond can form.
   d. The 3' carbon on the ddNTP lacks a hydroxyl group, so no phosphodiester bond can form.

4. Once the gene that causes Huntington's disease was found, researchers introduced the defective allele into mice to create an animal model of Huntington's disease. Why was this model valuable?
   a. It allowed them to test potential drug therapies without endangering human patients.
   b. It allowed them to study how the gene is regulated.
   c. It allowed them to make large quantities of the huntingtin protein.
   d. It allowed them to study how the gene was transmitted from parents to offspring.

5. To begin the hunt for the human growth hormone gene, researchers created a cDNA library from cells in the pituitary gland. What did this library contain?
   a. only the sequence encoding growth hormone
   b. DNA versions of all the mRNAs in the pituitary-gland cells
   c. all of the coding sequences in the human genome, but no introns
   d. all of the coding sequences in the human genome, including introns

6. What does it mean to say that a genetic marker and a disease gene are closely linked?
   a. The marker lies within the coding region for the disease gene.
   b. The sequence of the marker and the sequence of the disease gene are extremely similar.
   c. The marker and the disease gene are on different chromosomes.
   d. The marker and the disease gene are in close physical proximity and tend to be inherited together.

Test Your Knowledge answers: 1. c; 2. b; 3. d; 4. a; 5. b; 6. d

● Test Your Understanding

Answers are available at www.masteringbio.com

1. Explain how restriction endonucleases and DNA ligase are used to insert foreign genes into plasmids and create recombinant DNA. Make a drawing that shows why sticky ends are sticky, and that identifies the exact location where DNA ligase catalyzes a key reaction.

2. Explain the function of a vector in genetic engineering. Then explain how recombinant plasmids are introduced into *E. coli* cells, how viruses are used to transport genes into human cells, and how the Ti plasmid in *Agrobacterium* cells is used to transport genes into plants. List the attributes of a "perfect" vector.

3. What is a cDNA library? How is one created? Would you expect the cDNA library from a human muscle cell to be different from the cDNA library from a human nerve cell in the same individual? Explain why or why not.

4. What are genetic markers, and how are they used to create a genetic map? Explain how researchers combine analyses of human pedigrees and genetic markers to narrow down the location of disease genes.

5. Researchers added the promoter sequence from an endosperm-specific gene to the Ti plasmids used in creating golden rice. Why was this step important? Comment on the roles of promoter and enhancer sequences in genetic engineering in eukaryotes.

6. List the molecules that are required for a typical PCR reaction, and explain their role in the primer-annealing, primer-extension, and denaturation steps of a PCR cycle. Then list the function of each molecule required for the replication of DNA in vivo. For each molecule on the in vivo list, explain why it is or is not required for the corresponding in vitro reaction.

● Applying Concepts to New Situations

Answers are available at www.masteringbio.com

1. Suppose you had a large amount of sequence data, similar to the data that Nancy Wexler's team had in the region of the Huntington's disease gene, and that you knew that genes in the species being considered typically contain about 1500 bases. How would you use the genetic code (see Chapter 15) and information on the structure of promoters (see Chapters 17 and 18) to locate the precise location of one more genes in your sequence?

2. Should human gene therapy be restricted to somatic cells, or should individuals be able to alter their germ-line cells (meaning that they would alter the alleles they pass on to their offspring)? Should gene therapy be approved for disease-causing alleles only, or should parents also be able to pay to transform their children with alleles associated with height, intelligence, hair color, eye color, athletic performance, musical ability, or similar traits?

3. Describe similarities between how researchers screen a DNA library and how they perform a genetic screen—for example, for mutant *E. coli* cells that cannot metabolize lactose (see Chapter 17).

4. A friend of yours is doing a series of PCR reactions and comes to you for advice. She purchased three sets of primers, hoping that one set would amplify the template sequence shown here. (The dashed lines in the template sequence stand for a long sequence of bases.) None of the three primer pairs produced any product DNA, however.

	Primer a	Primer b
Primer Pair 1:	5′ GTCCAGC 3′ &	5′ CCTGAAC 3′
Primer Pair 2:	5′ GGACTTG 3′ &	5′ GCTGCAC 3′
Primer Pair 3:	5′ GTCCAGG 3′ &	5′ CAAGTCC 3′

Template

5′ ATTCGGACTTG — — — GTCCAGCTAGAGG 3′

3′ TAAGCCTGAAC — — — CAGGTCGATCTCC 5′

a. Explain why each primer pair didn't work. Indicate whether both primers are at fault or just one primer is the problem.

b. Your friend doesn't want to buy new primers. She asks you whether she can salvage this experiment. What do you tell her to do?

Genomics

20

Output from an automated genome sequencing machine, representing about 48,000 bases from the human genome. Each vertical stripe represents the sequence of a stretch of DNA.

KEY CONCEPTS

- Once a genome has been completely sequenced, researchers use a variety of techniques to identify which sequences code for products and which act as regulatory sites.

- In bacteria and archaea, there is a positive correlation between the number of genes in a species and the species' metabolic capabilities. Gene transfer between species is also common.

- In eukaryotes, genomes are dominated by sequences that have little to no effect on the fitness of the organism.

- Data from genome sequencing projects are now being used in the development of new drugs and vaccines, and to search for alleles associated with inherited diseases.

The first data sets describing the complete DNA sequence, or **genome**, of humans were published in February 2001. These papers were immediately hailed as a landmark in the history of science. In just 50 years, biologists had gone from not understanding the molecular nature of the gene to knowing the molecular makeup of every gene present in our species.

Leading up to the 2001 papers and since then, the multinational effort called the **Human Genome Project** has produced a wide array of data on the locations and functions of genes and other types of DNA sequences found in humans. It's important to recognize, though, that research on *Homo sapiens* is part of a much larger and ongoing effort to sequence genomes from an array of other eukaryotes, hundreds of bacteria, and dozens of archaea. The effort to sequence, interpret, and compare whole genomes is referred to as **genomics**. The pace of research in this field is nothing short of explosive.

The arrival of genomics has also triggered the development of the field called functional genomics. Genomics supplies a list of the genes present in an organism; **functional genomics** asks when those genes are expressed and how their products interact. This effort is important. Besides providing new insights into how gene expression is coordinated as an embryo develops into a college student or an oak tree, functional genomics is helping researchers identify new drug targets in disease-causing species, design new vaccines, and speed efforts to locate the genes responsible for inherited diseases.

As an introductory biology student, you are part of the first generation trained in the genome era. Genomics is revolutionizing several fields within biological science and will almost certainly be an important part of your personal and professional life. Let's look at what genomics is, how and why it's being done, and what is being learned.

● Key Concept   Important Information   ● Practice It

415

20.1   Whole-Genome Sequencing

Genomics has moved to the cutting edge of research in biology largely because a series of technological advances—including the fluorescent markers and gel-filled capillary tubes introduced in Chapter 19—have increased the speed of DNA sequencing and driven down the expense. Thanks largely to increased automation, the cost of sequencing DNA has declined by a factor of two every year and a half since the human genome project began in 1988. Researchers have now established factory-style DNA sequencing centers, each containing dozens of automated sequencing machines, at over 20 locations in the United States, the United Kingdom, Germany, France, Japan, and China. Some of these laboratories employ dozens of biologists and can conduct 100,000 sequencing reactions daily.

As data became less expensive and faster to obtain, the pace of whole-genome sequencing accelerated. The result is that an almost mind-boggling number of sequences are now being generated. As this book goes to press, the primary international repositories for DNA sequence data contain over 146 *billion* nucleotides. (By way of comparison, a haploid human genome contains about 3 billion bases.) The size of this database is increasing by about 20 percent every year (**Figure 20.1**).

How Are Complete Genomes Sequenced?

Genomes range in size from a few million base pairs to several billion. But a single sequencing reaction can analyze only about 1000 base pairs. How do investigators break a genome into sequencing-sized pieces, and then figure out how the thousands or millions of pieces go back together?

Most recent genome sequencing projects answer this question with a whole-genome shotgun approach, or simply **shotgun sequencing**. In shotgun sequencing, a genome is broken up into a set of overlapping fragments that are small enough to be sequenced. Using the regions of overlap, the sequenced fragments are then put back into the correct order.

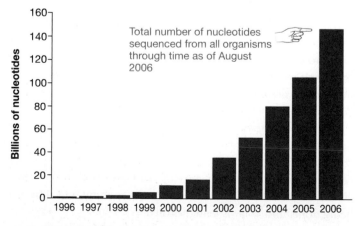

FIGURE 20.1 The Total Number of Bases Sequenced Is Growing Rapidly.

As step 1 of **Figure 20.2** shows, shotgun sequencing begins by using high-frequency sound waves, or sonication, to break up a genome into pieces about 160 kilobases (kb) long (1 kb = 1000 bases). Next, each 160-kb piece is inserted into a plasmid called a **bacterial artificial chromosome (BAC).** BACs are able to replicate large segments of DNA. Using transformation techniques introduced in Chapter 19, each BAC is then inserted into a different *Escherichia coli* cell, creating what researchers call a BAC library (step 2). A BAC library is a genomic library: a set of all the DNA sequences in a particular genome, split into small segments and inserted into cloning a vector (see Chapter 19). By separating the cells in a BAC library and then allowing each cell to grow into a large colony, researchers can isolate large numbers of each 160-kb fragment.

Once a research group has many copies of each 160-kb fragment, the DNA is again broken into fragments—but this time, segments that are about 1000 base pairs long (step 3). These small fragments are then inserted into plasmids and placed inside bacterial cells (step 4). In this way, a genome is broken down into two manageable levels: 160-kb fragments in BACs and 1-kb segments in plasmids. The plasmids are copied many times as the bacterial cells grow into a large population. Large numbers of each 1000-base-pair fragment are then available for sequencing reactions (step 5).

Once the 1000-base-pair fragments from each 160-kb BAC clone are sequenced, computer programs analyze regions where the ends of each 1000-base-pair segment overlap (step 6). Overlaps occur because there were many copies of each 160-kb segment, and each was fragmented randomly by sonication. The computer mixes and matches segments from a single BAC clone until an alignment consistent with all available data is obtained. Then the ends of each BAC are analyzed in a similar way (step 7). The goal is to arrange each 160-kb segment in its correct position along the chromosome, based on regions of overlap.

In essence, the shotgun strategy consists of breaking a genome into tiny fragments, sequencing the fragments, and then putting the sequence data back into the correct order. ● If you understand shotgun sequencing, you should be able to explain why it is essential for regions of overlap to exist between fragments that are adjacent to each other on a chromosome.

Once complete genome sequences became available, databases that could hold completed sequence information had to be created and managed in a way that made the raw data and a variety of annotations available to the international community of researchers. These sequence databases also had to be searchable, so that investigators could evaluate how similar newly discovered genes were to genes that had been studied previously.

Because the amount of data involved is so large, the computational challenges involved in genomics are formidable. Thus far, sophisticated algorithms and continually improving computer hardware have allowed researchers to keep pace with the rate of data acquisition. The vast quantity of data generated by

SHOTGUN SEQUENCING A GENOME

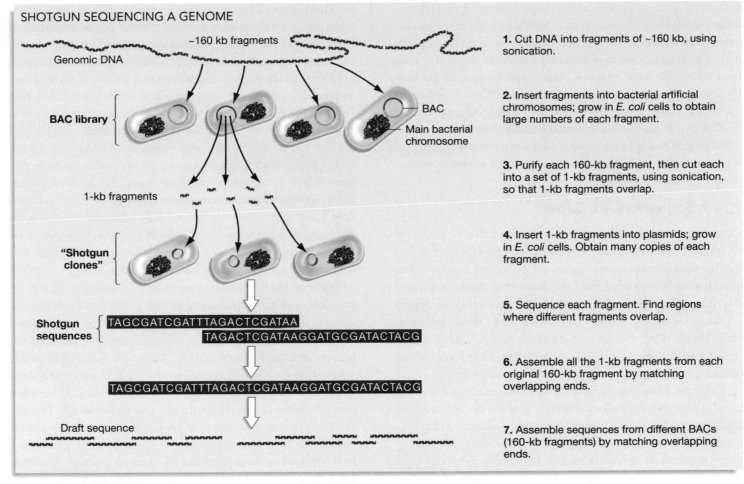

1. Cut DNA into fragments of ~160 kb, using sonication.

2. Insert fragments into bacterial artificial chromosomes; grow in *E. coli* cells to obtain large numbers of each fragment.

3. Purify each 160-kb fragment, then cut each into a set of 1-kb fragments, using sonication, so that 1-kb fragments overlap.

4. Insert 1-kb fragments into plasmids; grow in *E. coli* cells. Obtain many copies of each fragment.

5. Sequence each fragment. Find regions where different fragments overlap.

6. Assemble all the 1-kb fragments from each original 160-kb fragment by matching overlapping ends.

7. Assemble sequences from different BACs (160-kb fragments) by matching overlapping ends.

FIGURE 20.2 Shotgun Sequencing Breaks Large Genomes into Many Short Segments. Shotgun sequencing is a multistep process. A genome is initially fragmented into 160-kb sections via sonication and cloned into bacterial artificial chromosomes (BACs). Each 160-kb section is then cut into 1-kb fragments that can be cloned into plasmids, grown to a large number of copies, and sequenced.

QUESTION Why is "shotgun" an appropriate way to describe this sequencing strategy?

genome sequencing centers has made **bioinformatics**—the effort to manage, analyze, and interpret information in biology—a key to continued progress in the field.

Which Genomes Are Being Sequenced, and Why?

The first genome to be sequenced from an organism—not a virus—came from a bacterium that lives in the human upper respiratory tract. This bacterium, *Haemophilus influenzae*, has one circular chromosome and a total of 1,830,138 base pairs of DNA. This genome was small enough to sequence completely with a reasonable amount of time and money, given the technology available in the early 1990s. *H. influenzae* was an important research subject because it causes earaches and respiratory tract infections in children; what's more, one particular strain is capable of infecting the membranes surrounding the brain and spinal cord, causing meningitis.

Publication of the *H. influenzae* genome in 1995 was quickly followed by complete genomes sequenced from an assortment of bacteria and archaea. The first eukaryotic genome, from the yeast *Saccharomyces cerevisiae*, was finished in 1996. After that breakthrough, complete genome sequences were published from a variety of protists, plants, and animals. To date, complete genomes have been sequenced from over 510 bacterial species, 47 archaeal species, and 52 eukaryotic species. Incomplete genome sequence data are available from more than 200 other species.

Most of the organisms that have been selected for whole-genome sequencing cause disease or have other interesting biological properties. For example, genomes of bacteria and archaea that inhabit extremely hot environments have been sequenced in the hopes of discovering enzymes useful for high-temperature industrial applications and understanding how proteins can work under those conditions. Other bacteria and archaea were chosen for sequencing because they do interesting chemistry, such as producing methane (natural gas; CH_4) or other compounds. In some cases, researchers hope that these organisms might act as an important source of commercial products. The rice genome was sequenced because rice is the main food source for most humans. Finally, species such as the fruit fly *Drosophila melanogaster*, the roundworm *Caenorhabditis elegans*, the house mouse *Mus musculus*, and the mustard plant *Arabidopsis thaliana* were analyzed because they serve as model organisms in biology and because data from well-studied organisms promised to help researchers interpret the human genome.

Which Sequences Are Genes?

Obtaining raw sequence data is just the beginning of the effort to understand a genome. As researchers point out, raw sequence data are analogous to the parts list for a house. The parts list reads something like "windowwabeborogovestaircase-doorjubjub …," however, because it has no punctuation. Where do the genes for "window," "staircase," and "door" start and end? Are the segments that read "wabeborogove" and "jubjub" important in gene regulation, or are they simply spacers or other types of sequences that have no function at all?

The most basic task in annotating or interpreting a genome is to identify which bases constitute genes—the segments of DNA that code for an RNA or a protein product and that regulate their production. In bacteria and archaea, identifying genes is relatively straightforward. It is much more difficult, however, in eukaryotes.

Identifying Genes in Bacterial and Archaeal Genomes

Biologists begin with computer programs that scan the sequence of a genome in both directions. These programs identify each reading frame that is possible on the two strands of the DNA. (Recall from Chapter 15 that a reading frame is the sequence in which codons are read.) With codons consisting of three bases, three reading frames are possible on each strand, for a total of six possible reading frames (**Figure 20.3**). Because randomly generated sequences contain a stop codon about one in every 20 codons on average, a long stretch of codons that lacks a stop codon is a good indication of a coding sequence. The computer program highlights any "gene-sized" stretches of sequence that lack a stop codon but are flanked by a stop codon and a start codon. Because polypeptides range in size from a few dozen amino acids to many hundreds of amino acids, gene-sized stretches of sequence range from several hundred bases to thousands of bases. ◯ In addition, the computer programs look for sequences typical of promoters, operators, or other regulatory sites. DNA segments that are identified in this way are called **open reading frames**, or **ORFs**.

Once an ORF is found, a computer program compares its sequence with the sequences of known genes from well-studied species. If the ORF appears to be a gene that has not yet been described in any other species, further research is required before it can actually be considered a gene. A "hit," in contrast, means that the ORF shares a significant amount of sequence with a known gene from another species. Similarities between genes in different species are usually due to **homology**. If genes are homologous, it means they are similar because they are related by descent from a common ancestor. Homologous genes have similar base sequences and the same or a similar function. For example, consider the genes introduced in Chapter 14 that code for enzymes involved in repairing mismatches in DNA. Recall that the mismatch repair genes in *E. coli*, yeast, and humans are extremely similar in structure, DNA sequence, and function. To explain this similarity, biologists hypothesize that the common ancestor of all cells living today had mismatch re-

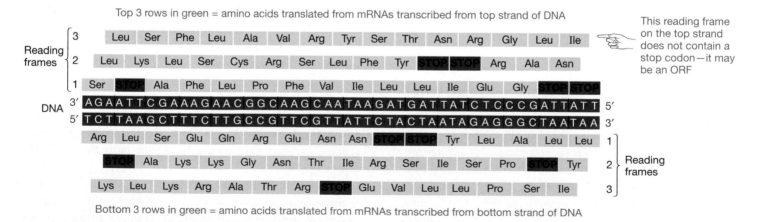

FIGURE 20.3 Open Reading Frames May Be the Locations of Genes. Computer programs scan the three possible reading frames on each strand of DNA and use the genetic code to translate each codon. A long stretch of codons that contains a start codon but lacks a stop codon may be an open reading frame (ORF)—a possible gene.

◯ **QUESTION** To predict the mRNA codons that would be produced by a particular reading frame, a computer analyzes the DNA in the 3′ to 5′ direction. Why?

pair genes—thus, the descendants of this ancestral species also have versions of these genes.

Based on this logic, researchers can confirm that an ORF is actually a gene by finding that it is homologous to a known gene. They can also analyze the product that would be produced by an ORF, and see if it conforms to a known gene. Unfortunately, finding and analyzing genes by identifying ORFs does not work well in eukaryotes.

Identifying Genes in Eukaryotic Genomes Mining eukaryotic sequence data for genes is complicated by two observations: Coding regions are broken up by introns, and the vast majority of eukaryotic DNA does not actually code for a product. In the human genome, for example, it is estimated that less than 2 percent of the DNA present actually codes for proteins, tRNAs, ribosomal RNAs, or other types of products. Finding a coding region in eukaryotic DNA is like finding a diamond in a huge pile of rock. To do so, researchers pursue several strategies:

- Computer programs can be written to search for sequences that are homologous to known genes. If a stretch of bases in the newly sequenced genome is similar to the sequence of a known gene, then researchers infer that it codes for a product whose function is similar to the function of the known gene.

- As Chapter 19 showed, investigators can isolate mRNAs from the organism being studied and then use enzymes to make the complementary DNAs (cDNAs). If the sequence of these cDNAs is determined, then a computer program can scan the genome sequence and pinpoint where each of the cDNAs is located. This approach allows researchers to identify genes that are expressed in certain cell types—the tissues where the original mRNA was found.

- To identify genes that have no known function, computers compare the genomes of closely related species and highlight sequences that are similar. Sequences that are shared by closely related species are hypothesized to be located in the protein-coding or regulatory regions of genes. The logic behind this claim runs as follows: Sequences that are part of a gene are expected to change much more slowly over time than sequences that are not actually part of a gene. Gene sequences change slowly over time because most gene products work less efficiently when they change randomly by mutation. Thus, it is logical to expect that natural selection eliminates most mutations in genes and that genes should change slowly over time. But changes in sequences that do not code for products or regulate gene expression do not affect the organism's phenotype. Mutations in these regions are much less likely to be eliminated by natural selection, so they change relatively rapidly over time.

Although each of these gene-finding strategies has been productive, it will probably be many years before biologists are convinced they have identified all of the coding regions in even a single eukaryotic genome. As that effort continues, though, researchers are analyzing the data and making some remarkable observations. Let's first consider what genome sequencing has revealed about the nature of bacterial and archaeal genomes and then move on to eukaryotes. Is the effort to sequence whole genomes paying off?

(MB) **Web Animation** at www.masteringbio.com
Human Genome Sequencing Strategies

20.2 Bacterial and Archaeal Genomes

By the time you read this paragraph, the genomes of over 600 bacterial and archaeal species will have been sequenced. In addition to this impressive array of different species, complete genome sequences are now available for several strains of the same bacterial species. For example, researchers have sequenced the genome of a laboratory population of *Escherichia coli*—derived from the harmless strain that lives in your gut—as well as the genome of a form that causes severe disease in humans. As a result, researchers can now compare the genomes of closely related cells that have different ways of life.

This section focuses on a simple question: Based on data published between 1995 and 2007, what general observations have biologists been able to make about the nature of bacterial and archaeal genomes?

The Natural History of Prokaryotic Genomes

In a sense, biologists who are working in genomics can be compared to the naturalists of the eighteenth and nineteenth centuries. These early biologists explored the globe, collecting the plants and animals they encountered. Their goal was to describe what existed. Similarly, the first task of a genome sequencer is to catalog what is in a genome—specifically, the number, type, and organization of genes. Several interesting conclusions can be drawn from relatively straightforward observations about the data obtained thus far:

- In bacteria, there is a general correlation between the size of a genome and the metabolic capabilities of the organism. For example, most parasites have much smaller genomes than nonparasitic organisms do. **Parasites** live off a host and thus reduce the host's fitness—its ability to produce offspring. Some of the smallest genomes are found in parasitic bacteria from the genus *Mycoplasma*. These bacteria live and multiply inside host cells. *Mycoplasma pneumoniae*, for example, parasitizes lung cells and causes pneumonia in humans. *Mycoplasma* lack the enzymes required to manufacture many essential compounds. Instead they acquire almost all of their nutrients from their hosts. In contrast, the genomes of nonparasitic strains of the bacteria *E. coli* and *Pseudomonas aeruginosa* are 8 to 10 times larger.

Their genes code for enzymes that synthesize virtually every molecule needed by the cell. Based on this observation, it is not surprising that *E. coli* is able to grow under a wide variety of environmental conditions. Using similar logic, researchers hypothesize that the large genome of *P. aeruginosa* explains why it is able to occupy a wide array of soil types, including marine and marshy habitats, as well as human tissues, where it can cause illness.

- Biologists still do not know the function of many of the genes that have been identified. Although *E. coli* probably qualifies as the most intensively studied of all organisms, the function of over 30 percent of its genes is unknown.

- There is tremendous genetic diversity among bacteria and archaea. About 15 percent of the genes in each species' genome appear to be unique. That is, about one in six genes in one of these species is found nowhere else.

- Redundancy among genes is common. For instance, the genome of E. coli has 86 pairs of genes whose DNA sequences are nearly identical—meaning that the proteins they produce are nearly alike in structure and presumably in function. Although the significance of this redundancy is unknown, biologists hypothesize that slightly different forms of the same protein are produced in response to slight changes in environmental conditions.

- Multiple chromosomes are more common than anticipated. Several species of bacteria and archaea have two different circular chromosomes instead of one. And at least some bacteria have linear chromosomes.

- Many species contain the small, extrachromosomal DNA molecules called plasmids. Recall from Chapter 19 that plasmids contain a small number of genes, though not genes that are absolutely essential for growth. In many cases, plasmids can be exchanged between cells of the same species or even of different species (see Chapter 12).

⚪ Perhaps the most surprising observation of all is that in many bacterial and archaeal species, a significant proportion of the genome appears to have been acquired from other, often distantly related, species. In some bacteria and archaea, 15–25 percent of the genetic material appears to be "foreign." This is a remarkable claim. What evidence backs up the assertion that prokaryotes acquire DNA from other species? How could this happen, and what are the consequences?

Evidence for Lateral Gene Transfer

Biologists use two general criteria to support the hypothesis that sequences in bacterial or archaeal genomes originated in another species: (1) when stretches of DNA are much more similar to genes in distantly related species than to those in closely related species and (2) when the proportion of G-C base pairs to A-T base pairs in a particular gene or series of genes is markedly different from the base composition of the rest of the genome. In many cases, the proportion of G-C bases in a genome is characteristic of a genus or species.

How can genes move from one species to another? In at least some cases, plasmids appear to be responsible. For example, most of the genes that are responsible for conferring resistance to antibiotics are found on plasmids. Researchers have documented the transfer of plasmid-borne antibiotic-resistance genes between very distantly related species of disease-causing bacteria. In some cases, genes from plasmids become integrated into the main chromosome of a bacterium, resulting in genetic recombination (see Chapter 12). The movement of DNA from one species to another species is called **lateral gene transfer** (**Figure 20.4**).

Some biologists hypothesize that lateral gene transfer also occurs via transformation—when bacteria and archaea take up raw pieces of DNA from the environment, perhaps in the course of acquiring other molecules. This may have occurred in the bacterium *Thermotoga maritima*, which occupies the high-temperature environments near deep-sea vents. Almost 25 percent of the genes in this species are extremely closely related to genes found in archaea that live in the same habitats. The archaea-like genes occur in distinctive clusters within the *T. maritima* genome, which supports the hypothesis that the sequences were transferred in large pieces from an archaean to the bacterium.

Similar types of direct gene transfer are hypothesized to have occurred in the bacterium *Chlamydia trachomatis*. This organism is a major cause of blindness in humans from Africa and Asia; it also causes chlamydia, the most common sexually transmitted bacterial disease in the United States. The *C. trachomatis* genome contains 35 genes that resemble eukaryotic genes in structure. Because *C. trachomatis* lives inside the cells that it parasitizes, the most logical explanation for this observation is that the bacterium occasionally takes up DNA directly from its host cell, resulting in a eukaryote-to-bacterium transfer.

In addition to being transferred between species by means of plasmids or DNA fragments, genes can be transported by viruses. For example, investigators who compared the sequences of laboratory and pathogenic (disease-causing) strains of *E. coli* found that the pathogenic cells have almost 1400 "extra" genes. Compared with the rest of the genome, most of these genes have a distinctive G-C to A-T ratio. Many are also extremely similar to sequences isolated from viruses that infect *E. coli*. Based on these observations, most researchers support the hypothesis that at least some of the disease-causing genes in *E. coli* were brought in by viruses.

To summarize, mutation and genetic recombination within species are not the only source of genetic variation in bacteria and archaea. Over the course of evolution, lateral gene transfer has been an important source of new genes and allelic diversity in these domains. This insight would not have been possible without data from whole-genome sequencing. Have efforts to sequence eukaryotic genomes led to similar types of insights?

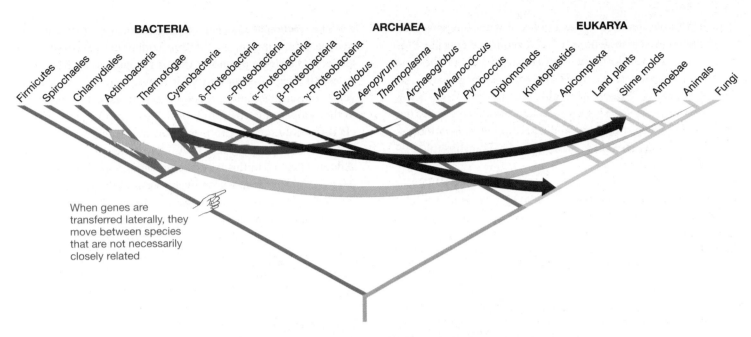

BACTERIA **ARCHAEA** **EUKARYA**

When genes are transferred laterally, they move between species that are not necessarily closely related

FIGURE 20.4 Lateral Gene Transfer Is Movement of DNA between Species. Lateral gene transfer can occur between very distantly related organisms. For tips on how to read an evolutionary tree like this, see **BioSkills 2**.

⦿ **EXERCISE** Beside this tree, list the mechanisms responsible for lateral gene transfer.

Check Your Understanding

If you understand that...

- The major surprise that came out of genome sequencing projects involving bacteria and archaea was the extent and importance of lateral gene transfer—that is, the movement of DNA from one species to another.

You should be able to...

1) Summarize how researchers infer that lateral gene transfer is responsible for the presence of a particular DNA sequence.

2) Summarize evidence that the size of a prokaryotic genome is correlated with the organism's metabolic capabilities.

20.3 Eukaryotic Genomes

Sequencing eukaryotic genomes presents two daunting challenges. The first is sheer size. Compared with the genomes of bacteria and archaea, which range from 580,070 base pairs in *Mycoplasma genitalium* to over 6.3 million base pairs in *Pseudomonas aeruginosa*, eukaryotic genomes are even larger. The haploid genome of *Saccharomyces cerevisiae* (baker's yeast), a unicellular eukaryote, contains over 12 million base pairs. The roundworm *Caenorhabditis elegans* has a genome of 97 million base pairs; the fruit-fly genome contains 180 million base pairs; the mustard plant *Arabidopsis thaliana*'s genome has 130 million base pairs; and humans, rats, mice, and cattle contain roughly 3 billion base pairs each.

The second great challenge in sequencing eukaryotic genes is coping with noncoding sequences that are repeated many times. ⦿ Many eukaryotic genomes are dominated by repeated DNA sequences that occur between genes and do not code for products used by the organism. These repeated sequences pose serious problems in aligning and interpreting sequence data. What are they? If such sequences don't code for a product, why do they exist?

Natural History: Types of Sequences

In many eukaryotic genomes, the exons and regulatory sequences associated with genes make up a relatively small percentage of the genome. Recall from Section 20.1 that in humans, protein-coding sequences constitute less than 2 percent of the total genome while repeated sequences account for well over 50 percent. In contrast, over 90 percent of a bacterial or archaeal genome consists of genes—DNA sequences that code for a product needed by the cell and regulate its transcription.

When noncoding and repeated sequences were discovered, they were initially considered "junk DNA" that was nonfunctional and probably unimportant and uninteresting. But subsequent work has shown that many of the repeated sequences observed in eukaryotes are actually derived from sequences known as transposable elements. **Transposable elements** are segments of DNA that are capable of moving from one location to another, or transposing, in a genome. They are similar to viruses, except that viruses leave a host cell that they have infected and find a new cell to infect. In contrast, transposable elements never leave their host cell—they simply make copies of themselves and move to new locations in the genome. Transposable elements are passed from parents to offspring, generation after generation, because they are part of the genome.

Transposable elements are examples of what biologists call selfish genes. A selfish gene is a DNA sequence that survives and reproduces but does not increase the fitness of the host genome. Transposable elements and viruses are classified as parasitic because it takes time and resources to copy them along with the rest of the genome, and because they can disrupt gene function when they move and insert in a new location. As a result, they decrease their host's fitness. Transposable elements are genomic parasites.

How Do Transposable Elements Work? Transposable elements come in a wide variety of types and spread through genomes in a variety of ways. Different species—fruit flies, yeast, and humans, for example—contain distinct types of transposable elements.

As an example of how these selfish genes work, let's consider a well-studied type called a **long interspersed nuclear element (LINE)** that is found in humans and other eukaryotes. Because LINEs are so similar to the retroviruses, introduced in Chapter 19, biologists hypothesize that they are derived from them evolutionarily. Your genome contains tens of thousands of LINEs, each between 1000 and 5000 bases long.

An active LINE contains all the sequences required for it to make copies of itself and insert them into a new location in the genome (**Figure 20.5**, step 1): a gene that codes for the enzyme reverse transcriptase, a gene that codes for the enzyme integrase, and a single promoter that is recognized by RNA polymerase II (step 2). After a LINE is transcribed to an mRNA, reverse transcriptase and integrase are synthesized by ribosomes in the cytoplasm (steps 3 and 4). Reverse transcriptase makes a cDNA version of the LINE mRNA, and integrase inserts the newly synthesized LINE DNA into a new location in the genome (step 5). In this way, the parasitic sequence reproduces (step 6). If the transposition event occurs in reproductive cells that go on to form eggs or sperm, the copied LINE will be passed on to offspring. If the LINE happens to insert itself inside a gene or a regulatory sequence, it causes a mutation that is almost certain to reduce the host's fitness.

Most of the LINEs observed in the human genome do not actually function, however, because they don't contain a promoter or the genes for either reverse transcriptase or integrase. To make sense of this observation, researchers hypothesize that the insertion process illustrated in steps 6 and 7 of Figure 20.5 is usually disrupted in some way. Analyses of the human genome have revealed that only a handful of our LINEs appear to be complete and potentially active.

Virtually every prokaryotic and eukaryotic genome examined to date contains at least some transposable elements. They vary

FIGURE 20.5 Transposable Elements Spread within a Genome.

widely in type and number, however, and bacterial and archaeal genomes have relatively few transposable elements compared to most eukaryotes studied thus far. This observation has inspired the hypothesis that bacteria and archaea either have efficient means of removing parasitic sequences or can somehow thwart insertion events. To date, however, this hypothesis has yet to be tested rigorously.

Research on transposable elements and lateral gene transfer has revolutionized how biologists view the genome. Many genomes are riddled with parasitic sequences, and others have undergone radical change in response to lateral gene transfer events. Genomes are much more dynamic and complex than previously thought. Their size and composition can change dramatically over time.

Repeated Sequences and DNA Fingerprinting In addition to containing repeated sequences from transposable elements, eukaryotic genomes have several thousand loci called simple tandem repeats (STRs). These are small sequences repeated one after another down the length of a chromosome. There are two major classes of STRs: Repeating units that are just 1 to 5 bases long are known as **microsatellites** or **simple sequence repeats**; repeating units that are 6 to 500 bases long are called **minisatellites** or **variable number tandem repeats** (**VNTRs**). Both types of repeated sequences make up 3 percent of the human genome. The most common type of microsatellite is a repeated stretch of the dinucleotide AC, giving the sequence ACACACAC.... Microsatellite sequences are thought to originate when DNA polymerase skips or mistakenly adds extra bases during replication; the origin of minisatellites is still unclear.

Soon after these sequences were first characterized, Alec Jeffreys and co-workers established that microsatellite and minisatellite loci are "hypervariable," meaning that they vary among individuals much more than any other type of sequence does. **Figure 20.6** illustrates one hypothesis for why microsatellites and minisatellites have so many different alleles: These highly repetitive stretches often align incorrectly when homologous chromosomes synapse and cross over during prophase of meiosis I. Instead of lining up in exactly the same location, the two chromosomes pair in a way that matches up bases in different repeated segments. Due to this misalignment, **unequal crossover** occurs. Chromosomes produced by unequal crossover contain different numbers of repeats. The key observation is that if a particular microsatellite or minisatellite locus has a unique number of repeats, it represents a unique allele. Each allele has a unique length. As with any allele, microsatellite and minisatellite alleles are transmitted from parents to offspring.

Misalignment or errors by DNA polymerase are so common at these loci that, in most eukaryotes, the genome of virtually every individual has at least one new allele. This variation in repeat number among individuals is the basis of DNA fingerprinting. **DNA fingerprinting** refers to any technique for identifying individuals based on the unique features of their genomes. Because microsatellite and minisatellite loci vary so much among individuals, they are now the loci of choice for DNA fingerprinting. To fingerprint an individual, researchers obtain a DNA sample and perform the polymerase chain reaction (PCR), using primers that flank a region containing an STR. Once many copies of the region are available, they can be analyzed to determine the number of repeats present (**Figure 20.7a**). Primers are now available

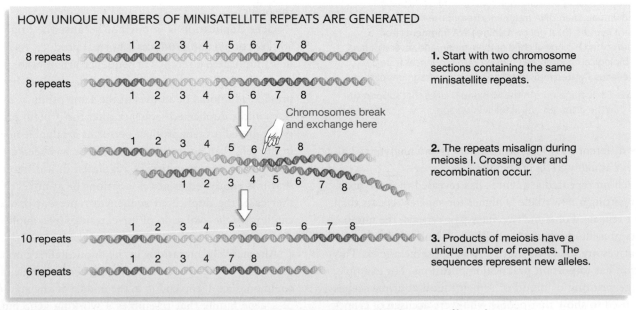

HOW UNIQUE NUMBERS OF MINISATELLITE REPEATS ARE GENERATED

8 repeats   1 2 3 4 5 6 7 8

8 repeats

1 2 3 4 5 6 7 8

1. Start with two chromosome sections containing the same minisatellite repeats.

Chromosomes break and exchange here

1 2 3 4 5 6 7 8

1 2 3 4 5 6 7 8

2. The repeats misalign during meiosis I. Crossing over and recombination occur.

10 repeats   1 2 3 4 5 6 5 6 7 8

6 repeats   1 2 3 4 7 8

3. Products of meiosis have a unique number of repeats. The sequences represent new alleles.

FIGURE 20.6 Unequal Crossover Changes the Numbers of Simple Sequence Repeats. The alignment of homologous chromosomes during prophase of meiosis I is driven by sequence similarity between homologs. Because simple sequence repeats are so similar, they are likely to misalign during synapsis.

(a) Using PCR to amplify minisatellite and microsatellite loci

(b) A gel showing minisatellite sequences from unrelated and related individuals

Lane sources:

X: An unrelated individual

M: A mother

B: A boy the mother claims is her own

U: Undisputed children of the mother

FIGURE 20.7 DNA Fingerprinting Can Be Used to Identify Fathers.
(a) If a minisatellite locus contains different numbers of repeats in different individuals, then DNA fragments from those individuals will have different lengths. **(b)** A gel containing DNA fragments from a mother, a purported biological child of the woman, and several of her confirmed biological children was probed with sequences from a minisatellite locus. Related individuals tend to share fragment patterns.

● **EXERCISE** Circle fragments in the M, B, and U lanes that support the hypothesis that the disputed boy actually is M's son.

for many different STR loci, so researchers can analyze the alleles present at many STRs efficiently.

Research on repeated sequences has revealed that the probability of getting a new allele is higher for shorter repeats than for longer repeats. For some two-base-pair repeats, the number of repeats present changes so quickly over time that only very close relatives are likely to share any of the same alleles. This observation has important practical implications. For example, DNA fingerprinting of blood or semen found at crime scenes has been used to show that people who were accused of crimes were actually innocent. DNA fingerprinting has also been used as evidence to convict criminals or assign paternity in birds,

humans, and other species that have well-characterized microsatellite or minisatellite sequences (**Figure 20.7b**).

Now that we've reviewed the characteristics of some particularly prominent types of noncoding sequences in eukaryotes, it's time to consider the nature of the coding sequences in these genomes. Let's start with the most basic question of all: Where do eukaryotic genes come from?

Gene Duplication and the Origin of Gene Families

In eukaryotes, the major source of new genes is the duplication of existing genes. Biologists infer that genes have been duplicated recently when they find groups of similar genes clustered along the same chromosome. The genes are usually similar in structural aspects, such as the arrangement of exons and introns, and in their base sequence. The degree of sequence similarity among these clustered genes varies. In the genes that code for ribosomal RNAs (rRNAs) in vertebrates, the sequences are virtually identical—meaning that each individual has many exact copies of the same gene. In other cases, though, the proportion of bases that are identical is 50 percent or less.

Within a species, genes that are extremely similar to each other in structure and function are considered to be part of the same **gene family**. Genes that make up gene families are hypothesized to have arisen from a common ancestral sequence through gene duplication. When **gene duplication** occurs, an extra copy of a gene is added to the genome.

The most common type of gene duplication results from crossover during meiosis. As **Figure 20.8** shows, gene-sized segments of chromosomes can be deleted or duplicated if homologous chromosomes misalign during prophase of meiosis I and an unequal crossover occurs. The duplicated segments resulting from unequal crossover are arranged in tandem—one after the other.

Gene duplication is important because the original gene is still functional and produces a normal product. As a result, the new, duplicated stretches of sequence are redundant. In some cases the duplicated genes retain their original function and provide additional quantities of the same product. But if mutations in the duplicated sequence alter the protein product, and if the altered protein product performs a valuable new function in the cell, then an important new gene has been created. The duplicated gene may also be regulated in a different way, so that it is expressed in novel locations or at novel times. In either case, the duplicated sequences represent new genes and can lead to the evolution of novel traits. Gene duplication produces new genes and creates gene families.

Alternatively, mutations in the duplicated region may make expression of the new gene impossible. For example, a mutation could produce a stop codon in the middle of an exon. A member of a gene family that resembles a working gene but does not code for a functional product, due to early stop codons or other defects, is called a **pseudogene**. Pseudogenes have no function.

FIGURE 20.8 **Unequal Crossover Changes the Numbers of Genes along a Chromosome.** If unequal crossover occurs as diagrammed here, the resulting chromosomes contain one gene fewer than the original chromosome or an additional copy of a gene.

As an example of a gene family, consider the human globin genes diagrammed in **Figure 20.9**. These genes code for proteins that form part of hemoglobin—the oxygen-carrying molecule in your red blood cells. Analyzing the globin genes illustrates several important points about gene families. In humans, the globin gene family contains several pseudogenes, along with several genes that code for oxygen-transporting proteins. The various coding genes in the family serve slightly different functions. For example, some genes are active only in the fetus or the adult. Follow-up work showed that oxygen is much more likely to bind to the proteins encoded by the fetal genes compared to the proteins expressed in adults. As a result, oxygen is able to move from the mother's blood to the fetus's blood (see Chapter 44).

In addition to the gene duplication events resulting from unequal crossover, the entire complement of chromosomes may be duplicated due to a mistake in either mitosis or meiosis. In this case, the resulting cell contains double the normal complement of chromosomes. Recall from Chapter 12 that species with duplicated chromosome complements are said to be **polyploid**. When polyploidy occurs, every gene in the duplicated genome is redundant. As a result, each gene may experience mutations that lead to new functions or to the loss of function and creation of a pseudogene.

By comparing how many copies of gene families occur in eukaryotes that have completely sequenced genomes, researchers have concluded that a whole-genome duplication occurred early in the evolution of vertebrates. They further conclude that another genome duplication occurred early in the evolution of the ray-finned fish—a lineage of over 24,000 living species, including familiar groups such as the trout, tuna, and guppies. Genome duplication has also been a particularly important source of new genes in plants.

Insights from the Human Genome Project

The human genome is rapidly becoming the most intensively studied of all eukaryotic genomes. In many or even most cases, researchers are gaining insights into how the human genome works by comparing it to the genomes of other

FIGURE 20.9 **Gene Families Are Clusters of Closely Related Genes.** Genes within the globin family. Red segments represent functioning genes, and yellow segments are pseudogenes. The members of a gene family are arranged one after the other, or in tandem. Most of these genes are expressed at different times during development.

● **EXERCISE** Suppose that during prophase of meiosis I, the β locus on one chromosome aligned with the $\psi\beta2$ locus on another chromosome, and then crossing over occurred just to their left. Draw the chromosomes that would result.

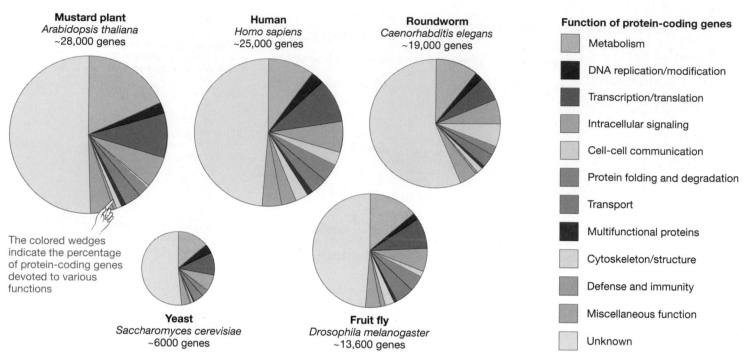

FIGURE 20.10 Comparing Gene Functions in Various Eukaryotes.

● **QUESTION** In the organisms pictured here, about what percentage of the genes present have unknown function?

species. **Figure 20.10**, for example, shows the relative proportion of genes devoted to various functions in humans and four other species of eukaryotes. A careful look at this figure should convince you that humans have a particularly large proportion of their genome devoted to immunity (defense against bacteria, viruses, and other parasites), and that humans and the roundworm *Caenorhabditis elegans* have a larger percentage of their genome devoted to cell-cell signaling than do other eukaryotes studied thus far. But even a quick look at these diagrams carries an important message: No one knows the function of most genes found in humans and other eukaryotes.

Although it is clear that a great deal remains to be learned about the human genome, two important questions have emerged from the early studies. Let's consider each of them in turn.

Why Do Humans Have So Few Genes? Of all observations about the nature of eukaryotic genomes, perhaps the most striking is that organisms with complex morphology and behavior do not appear to have particularly large numbers of genes. **Table 20.1** indicates the estimated number of genes found in selected eukaryotes. Notice that the total number of genes in *Homo sapiens*, which is considered a particularly complex organism, is not that much higher than the total number of genes in fruit flies. Gene number in humans is about the same as in roundworms, mice, rats, puffer fish, and chickens and substantially less than in rice and the weedy mus-

tard plant *Arabidopsis thaliana*. Before the human genome was sequenced, many biologists expected that humans would have at least 100,000 genes. We now know that we have only a fifth of that number—perhaps less.

How can this be? In prokaryotes there is a correlation between genome size, gene number, a cell's metabolic capabilities, and the cell's ability to live in a variety of habitats. Similarly, it is logical to observe that plants have exceptionally large numbers of genes because they synthesize so many different and complex molecules from just carbon dioxide, nitrate ions, phosphate ions, and other simple nutrients. The idea is that large numbers of genes enable plants to produce large numbers of enzymes. But why isn't there a stronger correlation between gene number and morphological and behavioral complexity in animals?

The leading hypothesis focuses on **alternative splicing**. Recall from Chapter 18 that the exons of a particular gene can be spliced in ways that produce distinct mature mRNAs. As a result, a single eukaryotic gene can code for multiple transcripts and thus multiple proteins. The alternative-splicing hypothesis claims that at least certain multicellular eukaryotes do not need enormous numbers of distinct genes. Instead, alternative splicing creates different proteins from the same gene. The alternative forms might be produced at different developmental stages or in response to different environmental conditions.

In support of the alternative-splicing hypothesis, researchers have analyzed the mRNAs produced by human genes and have estimated that each gene produces an average of slightly more

TABLE **20.1 Number of Genes in Selected Eukaryotes**

Species	Description	Genome Size (Millions of Base Pairs)	Estimated Number of Genes
Saccharomyces cerevisiae	Baker's and brewer's yeast; a unicellular fungus; an important model organism in biochemistry and genetics	12	6000
Plasmodium falciparum	Single-celled, parasitic eukaryote; causes malaria in humans	30	6500
Drosophila melanogaster	Fruit fly; an important model organism in genetics and developmental biology	180	13,600
Caenorhabditis elegans	A roundworm; an important model organism in developmental biology	97	19,000
Canis familiarus	Domestic dog	2410	19,300
Gallus gallus	Chicken	1050	20,000–23,000
Homo sapiens	Humans	3000	20,000
Rattus norvegicus	Norway rat; an important model organism in physiology and behavior	2750	21,000
Mus musculus	House mouse; an important model organism in genetics and developmental biology	2500	~30,000
Arabidopsis thaliana	A mustard plant; an important model organism in genetics and developmental biology	119	28,000
Oryza sativa	Rice	389	37,500

than three distinct transcripts. If this result is valid for the rest of the genome, the actual number of different proteins that can be produced is more than triple the gene number. Humans may have fewer than 20,000 genes according to current estimates, but these genes may have the ability to produce 100,000 different transcripts. Researchers are currently trying to assess whether alternative splicing is this frequent in other eukaryotes, as well.

How Can the Human and Chimp Genomes Be So Similar?

Comparing the numbers of genes found in humans and in mice created a paradox that may be resolved by the alternative-splicing hypothesis. Comparing the base sequences of genes in humans and in other species has created an analogous paradox.

Here is the issue: At the level of base sequences, human beings and chimpanzees are 98.8 percent identical on average. Of the homologous genes analyzed in humans and chimps, 29 percent are identical in amino acid sequence; the average difference between homologous proteins is just two amino acids. If humans and chimps are so similar genetically, why do they appear to be so different in their morphology and behavior?

The leading hypothesis to resolve this paradox focuses on the importance of regulatory genes and regulatory sequences. Recall from Chapter 18 that a **regulatory sequence** is a section of DNA involved in controlling the activity of other genes; it may be a promoter, a promoter-proximal element, an enhancer, or a silencer. The term **structural gene**, in contrast, refers to a sequence that codes for a tRNA, rRNA, protein, or other type of product. **Regulatory genes** code for regulatory transcription factors that alter the expression of specific genes.

To resolve the sequence-similarity paradox, biologists propose that even though many structural genes in closely related species, such as humans and chimps, are identical or nearly identical, regulatory sequences and regulatory genes might have important differences between the two species. Suppose that the structural gene for human growth hormone and chimp growth hormone are identical in base sequence. But if changes in transcription factors, enhancers, or promoters change the pattern of expression of that gene—perhaps turning it on later and longer in humans than in chimps—then height and other characteristics will change even though the structural gene is the same. Based on current analyses, biologists suggest that the human genome contains about 3000 different regulatory transcription factors. Subtle mutations in these proteins and the regulatory sites that they bind to could have a significant effect on gene expression and thus on the phenotype.

The regulatory hypothesis is certainly logical, and it is consistent with data suggesting that regulation of alternative splicing underlies phenotypic complexity in *Homo sapiens* and other large vertebrates. It may be true that most of the genetic changes responsible for the rapid evolution of humans over the past 5 million years have been due to changes in regulatory genes and sequences and alternative splicing rather than to changes in structural genes. To date, however, there are no specific examples of changes in the regulatory sequences responsible for the phenotypic differences observed between humans and chimps or other closely related species. The regulatory hypothesis still needs to be tested rigorously.

Check Your Understanding

If you understand that...

- Eukaryotic genomes are riddled with parasitic sequences that do not contribute to the fitness of the organism.

- Simple repeated sequences are also common in eukaryotic genomes.

- In eukaryotes, many of the coding sequences are organized into families of genes with related functions.

You should be able to...

1) Explain why transposable elements are considered selfish genes.

2) Explain why simple sequence repeats make DNA fingerprinting possible.

3) Explain how unequal crossover leads to duplicated sequences.

4) Explain why researchers hypothesize that human evolution has been dominated by alternative splicing and other types of regulatory changes.

20.4 Functional Genomics and Proteomics

To explain the impact of genomics on the future of biological science, Eric Lander has compared the sequencing of the human genome to the establishment of the periodic table of the elements in chemistry. Once the periodic table was established and validated, chemists focused on understanding how the elements combine to form molecules. Similarly, biologists now want to understand how the elements of the human genome combine to produce an individual.

In essence, a genome sequence is a parts list. Once that list is assembled, researchers delve deeper to understand how genes interact to produce an organism. Let's explore some of the ways in which researchers use whole-genome data to answer fundamental questions about how organisms work.

What Is Functional Genomics?

For decades, biologists have worked at understanding how and when individual genes are expressed. Research on the *lac* operon and *trp* operon, reviewed in Chapter 17, are typical of this effort. But now, with complete catalogs of the genes present in a variety of organisms whose genomes have been sequenced, researchers can ask how and when *all* of the genes in an organism are expressed. These types of large-scale analyses of gene expression are sometimes called functional genomics. The research is motivated by the realization that gene products do not exist in a vacuum. Instead, groups of RNAs and proteins act together to respond to environmental challenges such as extreme heat or

drought. Similarly, distinct groups of genes are transcribed at different stages as a multicellular eukaryote grows and develops.

One of the most basic tools used in functional genomics is called a microarray. A **DNA microarray** consists of a large number of single-stranded DNAs that are permanently affixed to a glass slide. For example, the slide pictured in **Figure 20.11** contains thousands of spots, each of which contains single-stranded DNA from a unique exon found in the human genome.

To do an experiment with a DNA microarray, researchers follow the protocol outlined in **Figure 20.12**. The first step is to isolate the mRNAs being produced by two contrasting types of cells. In this example, the control cells are functioning at normal temperature. The other cells, in contrast, have been exposed to high temperatures. The goal of this experiment is to compare genes that are expressed during normal cell activity with those expressed under heat stress.

Once they've purified mRNAs from the two populations of cells, investigators use reverse transcriptase to make a single-stranded cDNA version of each RNA in the samples (see Chapter 19). In addition to the four standard dNTPs, one of the DNA building blocks used in this reaction carries a fluorescent label. The label used for normal cells glows green, while the label chosen for the heat-stressed cells glows red.

The labeled cDNAs can then be used to probe the microarray. As Chapter 19 noted, a probe allows an investigator to find a particular molecule in a sample containing many different molecules. In this case, the labeled cDNAs will bind to the single-stranded DNAs on the plate by complementary base pairing. Out of all the exons in the genome, then, only the exons that are being expressed will be labeled. In our example, genes that are expressed under normal conditions will be labeled green, while those expressed during heat stress will be labeled red. If one of

FIGURE 20.11 DNA Microarrays Represent Every Gene in a Genome. To create a DNA microarray, investigators spot thousands of short, single-stranded DNA sequences from coding sequences onto a glass plate. The DNAs typically represent every exon in the genome of a particular species.

PROTOCOL FOR A MICROARRAY EXPERIMENT

Normal temperature

High temperature

1. Use reverse transcriptase to prepare single-stranded cDNA from mRNA of control cells and treatment cells (e.g., cells grown at normal and high temperature).

cDNA
mRNA

cDNA probes

2. Use labeled nucleotides in reaction, so fluorescent green label is added to control cDNA and fluorescent red label to treatment cDNA.

3. Probe a microarray with the labeled cDNAs. If probe cDNA sequence is complementary to the sequence in an exon fragment on the plate, it will bind and label that spot.

Microarray

4. Shine laser light on one spot at a time to induce fluorescence. Analyze the pattern of hybridization between the two cDNAs and the DNA on the microarray.

Microarray computer output:

Green spots indicate genes transcribed in **control cells**

Yellow spots indicate genes transcribed equally in **both cells**

Dark spots indicate low gene expression

Red spots indicate genes transcribed in **treatment cells**

FIGURE 20.12 DNA Microarrays Are Used to Study Changes in Gene Expression. By probing a microarray with labeled cDNAs synthesized from mRNAs, researchers can identify which coding sequences are being transcribed. Here mRNAs from cells growing at normal temperature are green, while mRNAs from cells growing at high temperature are red.

● **QUESTION** If every spot on a microarray represents a different exon, how could an experiment like this document the existence of alternative splicing?

the exons in the microarray is expressed under both sets of conditions, then both green- and red-labeled cDNAs will bind to that spot and make it appear yellow. In this way, a microarray lets researchers study the expression of thousands of genes at a time. As a result, they can identify which sets of genes are expressed in concert under specific sets of conditions.

Once a microarray has been used, the bound cDNA probes can be removed. The original DNAs remain in place, so the slide may then be reused to assess gene expression in a different type of cell, or in the same cell type under different conditions. Researchers can use microarrays to establish which genes are transcribed in different organs and tissues, during cancerous growth, or in response to changes in environmental conditions, such as starvation, the presence of a toxin, or a viral infection. ● If you understand the concept of how microarrays are used, you should be able to design an experiment that uses a DNA microarray to compare the genes expressed in brain cells versus liver cells of an adult human.

What Is Proteomics?

The Greek root –*ome*, meaning "all," inspired the term *genome*. Similarly, biologists use the term **transcriptome** in referring to the complete set of genes that are transcribed in a particular cell, and **proteome** in referring to the complete set of proteins that are produced. **Proteomics**, it follows, is the large-scale study of protein function. Proteomic studies begin by identifying the proteins present in a cell or organelle; then, researchers attempt to determine the locations and interactions of proteins and document how they change through time or compare with other cells.

Proteomics can be thought of as a branch of functional genomics. Instead of studying individual proteins or how two proteins might interact, biologists can study all of the proteins present at once. One approach to studying protein-protein interactions is similar to the use of DNA microarrays, except that large numbers of proteins, rather than DNA sequences, are affixed on a glass plate. This microarray of proteins is then treated with an assortment of proteins produced by the same organism. These proteins are labeled with a fluorescent or radioactive tag. If any labeled proteins bind to the proteins in the microarray, the two molecules may also interact in the cell. In this way, researchers hope to identify proteins that physically bind to one another—like the G proteins and associated enzymes introduced in Chapter 8, or the cyclin and Cdk molecules introduced in Chapter 11. Microarray technology is allowing biologists to study protein-protein interactions on a massive scale.

20.5 Can Genomics Help Improve Human Health and Welfare?

With the advent of microarray technology, the "periodic tables" provided by genome projects are having an important impact on research into gene expression and protein-protein interactions.

But the governments and corporations that fund genome projects have underwritten the expense primarily because of the potential benefits for improving human health and welfare. In this respect, is genomics living up to its promise?

Although large amounts of genome sequence data have been available for only a few years, early indications are that the investment may indeed pay off with substantial advances in biomedicine. Let's consider how whole-genome data are informing the development of new drugs and vaccines, and then look at a project focused on searching for the alleles associated with inherited diseases.

Identifying Potential Drug Targets

Currently, dozens of whole-genome sequencing projects are focused on species that cause disease in humans or the livestock and crops that we depend on. As each of these projects is completed, biologists begin comparing the genomes of pathogenic strains with closely related species that are harmless. The goal is to achieve a much more detailed understanding of the genetic basis of **virulence**—the tendency for a parasite to harm its host.

One of the first tasks biologists undertake is to identify genes that occur only in pathogens and that may be required for virulence. ● Typically, biologists are finding that "virulence genes" code for proteins that allow parasitic cells to adhere to host cells, produce enzymes that break down host-cell walls or membranes, or secrete toxins that poison host-cell enzymes. Identifying these genes is important because it gives investigators targets for the development of new drugs. If drugs that knock out the protein products of these genes can be formulated, they would inhibit disease-causing cells while leaving closely related but helpful species unharmed. For example, comparing the genomes of benign strains of *E. coli* with *E. coli* strains that cause food poisoning allowed researchers to find virulence genes in the pathogenic strain. These genes code for proteins that poison cells in the human intestine and cause severe diarrhea. If drugs that neutralize these proteins can be developed, they might provide an effective treatment for food poisoning.

Having the complete gene catalog from pathogenic species is giving biomedical researchers new targets for drug development and new possibilities for therapy. This work is particularly urgent, because disease-causing bacteria continue to evolve resistance to many of the antibiotics currently in use (see Chapter 24).

Designing Vaccines

Although efforts to exploit genome data in drug design are still in their infancy, genomics has already inspired important advances in vaccine development. In essence, researchers are testing proteins that are identified by genome sequencing to see if the molecules stimulate the immune system enough to function as vaccines.

To illustrate how this work is proceeding, consider recent research on the bacterium *Neisseria meningitidis*. This species is a major cause of meningitis and blood infections in children

Experiment

Question: Could vaccines be developed from the products of newly discovered genes?

Hypothesis: Some of the genes discovered through genome sequencing code for proteins that can be used in vaccines.

Null hypothesis: None of the genes discovered through genome sequencing code for proteins that can be used in vaccines.

Experimental setup:

1. Isolate open reading frames (ORFs) from pathogen genome sequence.

2. Introduce ORFs into *E. coli* cells.

3. Isolate proteins that result from transcription and translation.

4. Inject proteins into mice. As a control, inject only the solution used to suspend the proteins.

Prediction: Some proteins will elicit an immune response similar to that elicited by vaccines. The control solution will not elicit an immune response.

Prediction of null hypothesis: No injections (control or treatments) will elicit an immune response similar to that elicited by vaccines.

Results:	Number of mice with immune response:	Number of mice with no immune response:
Strong immune responses 7		343

Conclusion: The seven proteins that elicited an immune response are potential vaccine components. Further research is needed to test their safety and effectiveness.

FIGURE 20.13 Newly Discovered Proteins Can Be Tested for Vaccine Development. Because all potential genes are identified after whole-genome sequencing, virtually all proteins produced by a pathogen can be tested for their ability to provoke an immune response and act as a vaccine.

and was one of the first bacteria to have its genome sequenced. Although antibiotics can treat *N. meningitidis* infections effectively, the organism grows so quickly that it often injures or even kills the victim before a diagnosis can be made and drugs administered. As a result, biomedical researchers have been interested in developing a vaccine that would prime the immune system and allow children to ward off infections.

Vaccine development has been difficult in this case, however. As Chapter 49 will explain in detail, the immune system usually responds to molecules on the outer surface of bacteria or viruses. When an immune system cell recognizes one of these surface molecules as foreign, the invading bacterium or viral particle is destroyed. This recognition step by the immune system takes time, however, and is speeded up by vaccination. Vaccines contain surface molecules from a bacterium or virus. Even though the vaccine is harmless, the immune system cells go through their normal recognition sequence. In this way, ingesting a vaccine alerts or primes the immune system. If an actual infection occurs later, the cells are ready to spring into action and destroy the pathogens before they can grow and cause disease.

Unfortunately, *N. meningitidis* is covered with a polysaccharide that is identical to a compound found on the surface of brain cells. Immune system cells normally do not attack compounds found on the body's own cells, so a vaccine composed of the *N. meningitidis* polysaccharide would elicit no response.

To circumvent this problem, biologists analyzed the genome sequence of *N. meningitidis* and tested 600 open reading frames for the ability to encode vaccine components. The researchers inserted the 600 DNA sequences into *E. coli* cells, following the steps shown in **Figure 20.13**. Later they succeeded in isolating 350 different *N. meningitidis* proteins from the transformed cells. The biologists injected these proteins into mice and then analyzed whether an immune response occurred. Their results show that seven of the proteins tested evoked a strong immune response and represent promising vaccine components. Follow-up work is now under way to determine whether one or more of these proteins could act as a safe and effective vaccine in humans.

Finding Genes Associated with Inherited Disease: The HapMap Project

Chapter 19 explained how gene hunters analyzed genetic markers to find the allele responsible for Huntington's disease. Data from human genome sequencing have made this approach to finding disease genes much more powerful.

Because DNA from many individuals was used as source material during the Human Genome Project, and because overlapping segments of genes were routinely sequenced, researchers were able to identify 1.42 million sites where single bases vary among individuals. Where you might have a "C" at a particular site, others may have a "T." Recall that these variable sites are called **single nucleotide polymorphisms,** or **SNPs,** and that they can serve as genetic markers—mapped sites in the genome that vary among individuals.

By sequencing DNA samples from dozens of people representing four ethnic groups from widely dispersed geographic areas, a recent research effort called the HapMap Project has extended this initial catalog of SNPs to a current total of 10 million sites that vary among humans. HapMap is short for haplotype mapping. A **haplotype** is the set of alleles found on a single chromosome or chromosome segment. By mapping SNPs, researchers hope to be able to determine the haplotype of any individual at any of their chromosomes.

For disease gene hunters, this new catalog of SNPs is an enormously powerful resource. ● The fundamental idea is to compare the haplotypes of individuals who have an inherited disease with the haplotypes of unaffected people. If certain SNPs are extremely common in affected individuals but rare or absent in unaffected individuals, it is likely that those SNPs are near or even within a gene that contributes to the disease.

The possibility of analyzing the inheritance of millions of polymorphic sites all over the genome—instead of just a few hundred widely scattered ones like those used in the Huntington's disease gene hunt—makes it much more likely that researchers can locate genes associated with illnesses such as Alzheimer's disease, bipolar disorder, diabetes, rheumatoid arthritis, and cardiovascular disease. The database promises to be particularly important in understanding the genetic basis of diseases that involve many different genes, instead of a single allele as Huntington's disease does. As biologists continue their efforts to annotate the human genome, it's very likely that they will be able to track down the genes responsible for many or even most inherited diseases.

Chapter Review

SUMMARY OF KEY CONCEPTS

● Once a genome has been completely sequenced, researchers use a variety of techniques to identify which sequences code for products and which act as regulatory sites.

Recent technical advances have allowed investigators to sequence DNA much more rapidly and cheaply than before, resulting in a flood of genome data. Researchers annotate genome sequences by finding genes and determining their function. To identify genes in bacteria and archaea, researchers use computers to scan the genome for start and stop codons that are in the same reading frame and that are separated by gene-sized stretches of sequence. Finding such open reading frames (ORFs) is difficult in eukaryotes, because exons are interrupted by introns and because most eukaryotic DNA does not code for a product. One approach to finding eukaryotic genes is to analyze the sequences of complementary DNAs (cDNAs) synthesized from mRNAs and then match these sequences to DNA found in the genome itself. Sequences that are highly conserved among species are also hypothesized to indicate the locations of genes.

You should be able to describe how a research group that discovered a gene for coat color in mice would determine whether a homologous gene existed in the human genome. ●

(MB) **Web Animation** at www.masteringbio.com
Human Genome Sequencing Strategies

● **In bacteria and archaea, there is a positive correlation between the number of genes in a species and the species' metabolic capabilities. Gene transfer between species is also common.**

Species of bacteria and archaea are usually targeted for whole-genome sequencing because they cause disease or have interesting metabolic abilities. In these groups, the size of an organism's genome and its morphological complexity or biochemical capabilities are correlated. Parasites tend to have small genomes; organisms that live in a broad array of habitats or that use a wide variety of nutrients tend to have larger genomes. Many of the genes identified in bacteria and archaea still have no known function, however, and a significant percentage of them are extremely similar to other genes in the same genome. Another generalization about prokaryotic genomes is that genes are frequently transferred laterally, or between species. Lateral gene transfer appears to be common in genes responsible for causing disease.

You should be able to describe two mechanisms responsible for lateral gene transfer in bacteria and archaea. ●

● **In eukaryotes, genomes are dominated by sequences that have little to no effect on the fitness of the organism.**

Compared with prokaryotic genomes, eukaryotic genomes are large and contain a high percentage of transposable elements, repeated sequences, and other noncoding sequences. There is no obvious correlation between morphological complexity and gene number in eukaryotes, although the number of distinct transcripts produced may be much larger than the actual gene number in certain species as a result of alternative splicing. Gene duplication and polyploidy have been the most important sources of new genes in eukaryotes.

You should be able to explain what biologists mean when they refer to "junk DNA," and whether these sequences lack function and are uninteresting, as originally proposed. ●

● **Data from genome sequencing projects are now being used in the development of new drugs and vaccines, and to search for alleles associated with inherited diseases.**

The availability of whole-genome sequences is inspiring new research programs. Biologists are affixing exons or proteins to microarrays in order to study changes in gene expression or protein-protein interactions. In addition, the availability of whole-genome data has allowed investigators to find new drug targets, new proteins that may serve as vaccine candidates, and new genetic markers that should aid in the hunt for alleles associated with human disease.

You should be able to explain the difference between studies of gene expression in single genes versus microarrays, and expression of individual proteins versus microarrays. ●

QUESTIONS

● **Test Your Knowledge**

1. What is an open reading frame?
   a. a gene whose function is already known
   b. a DNA section that is thought to code for a protein because it is similar to a complementary DNA (cDNA)
   c. a DNA section that is thought to code for a protein because it has a start codon and a stop codon flanking hundreds of base pairs
   d. any member of a gene family

2. What best describes the logic behind shotgun sequencing?
   a. Break the genome into tiny pieces. Sequence each piece. Use overlapping ends to assemble the pieces in the correct order.
   b. Start with one end of each chromosome. Sequence straight through to the other end of the chromosome.
   c. Use a variety of techniques to identify genes and ORFs. Sequence these segments—not the noncoding and repeated sequences.
   d. Break the genome into pieces. Map the location of each piece. Then sequence each piece.

3. What are minisatellites and microsatellites?
   a. small, extrachromosomal loops of DNA that are similar to plasmids
   b. parts of viruses that have become integrated into the genome of an organism
   c. incomplete or "dead" remains of transposable elements in a host cell
   d. short and simple repeated sequences in DNA

4. What is the leading hypothesis to explain the paradox that large, morphologically complex eukaryotes such as humans have relatively small numbers of genes?
   a. lateral transfer of genes from other species
   b. alternative splicing of mRNAs
   c. polyploidy, or the doubling of the genome's entire chromosome complement
   d. expansion of gene families through gene duplication

5. What evidence do biologists use to infer that a gene is part of a gene family?
   a. Its sequence is exactly identical to that of another gene.
   b. Its structure—meaning its pattern of exons and introns—is identical to that of a gene found in another species.
   c. Its composition, in terms of percentage of A-T and G-C pairs, is unique.
   d. Its sequence, structure, and composition are similar to those of another gene in the same genome.

6. What is a pseudogene?
   a. a coding sequence that originated in a lateral gene transfer
   b. a gene whose function has not yet been established
   c. a polymorphic gene—meaning that more than one allele is present in a population
   d. a gene whose sequence is similar to that of functioning genes but does not produce a functioning product

Test Your Knowledge answers: 1. c; 2. a; 3. d; 4. b; 5. d; 6. d

● Test Your Understanding

Answers are available at www.masteringbio.com

1. Explain how open reading frames are identified in the genomes of bacteria and archaea. Why is it more difficult to find open reading frames in eukaryotes?

2. Why is the observation that parasitic organisms tend to have relatively small genomes logical?

3. Review how a LINE sequence transmits a copy of itself to a new location in the genome. Why are LINEs and other repeated sequences referred to as "genomic parasites"?

4. How does DNA fingerprinting work? Stated another way, how does variation in the size of microsatellite and minisatellite loci allow investigators to identify individuals?

5. Researchers can create microarrays of short, single-stranded DNAs that represent many or all of the exons in a genome. Explain how these microarrays are used to document changes in the transcription of genes over time or in response to environmental challenges.

6. Explain the concept of homology and how identifying homologous genes helps researchers identify the function of unknown genes. Are duplicated sequences that form gene families homologous? Explain.

● Applying Concepts to New Situations

Answers are available at www.masteringbio.com

1. Parasites lack genes for many of the enzymes found in their hosts. Most parasites, however, have evolved from free-living ancestors that had large genomes. Based on these observations, W. Ford Doolittle claims that the loss of genes in parasites represents an evolutionary trend. He summarizes his hypothesis with the quip "use it or lose it." What does he mean?

2. According to eyewitness accounts, communist revolutionaries executed Nicholas II, the last czar of Russia, along with his wife and five children, the family physician, and several servants. Many decades after this event, a grave purporting to hold the remains of the royal family was identified. Biologists were asked to analyze DNA from each adult and juvenile skeleton and determine whether the bodies were indeed those of several young siblings, two parents, and several unrelated adults. If the grave was authentic, describe how similar the DNA fingerprints of each skeleton would be relative to the fingerprints of other individuals in the grave.

3. The human genome contains a gene that encodes a protein called syncytin. This gene is expressed in placental cells during pregnancy.

The syncytin gene is nearly identical in DNA sequence to a gene in a virus that infects humans. In this virus, the syncytin-like gene codes for a protein found in the virus's outer envelope. State a hypothesis to explain the similarity between the two genes.

4. A recent study used microarrays to compare the patterns of expression of genes that are active in the brain, liver, and blood of chimpanzees and humans. Although the overall patterns of gene expression were similar in the liver and blood of the two species, expression patterns were strikingly different in the brain. How does this study relate to the hypothesis that most differences between humans and chimps involve changes in gene regulation?

www.masteringbio.com is also your resource for • Answers to text, table, and figure caption questions and exercises • Answers to *Check Your Understanding* boxes • Online study guides and quizzes • Additional study tools including the *E-Book for Biological Science* 3rd ed., textbook art, animations, and videos.

21

Principles of Development

KEY CONCEPTS

- ● As an embryo develops, cells have to proliferate, move or expand in a directed manner, become specialized, and interact with other cells.

- ● Cells become specialized because they express different genes, not because they contain different genes.

- ● Cells interact continuously during development via cell-cell signals.

- ● Early signals trigger a cascade of effects, including the production of transcription factors and other signals. Interactions among these signals, transcription factors, and regulatory DNA sequences cause cells to differentiate as development proceeds.

- ● The evolution of new body sizes, shapes, and structures is possible when mutations alter genes responsible for development.

A young fish, hatching from an egg. This chapter introduces the processes responsible for transforming a fertilized egg into an individual with specialized cells, tissues, and organs.

In 1859 Charles Darwin referred to the origin of species as "the mystery of mysteries." At that time the most urgent task confronting biologists was to explain how species come to be. Today, however, Darwin's theory of evolution by natural selection explains most of our fundamental questions about how new species arise and how they change through time. What question qualifies as the current mystery of mysteries in biological science? Although there are many candidates, one of the most compelling is the question addressed in this unit: How does a multicellular individual develop from a single cell—the fertilized egg?

It's important to pause for a moment and think about the magnitude of this problem. For example, at one time you consisted of a single cell. If you had been able to watch your own development, you would have seen that cell divide rapidly and form a ball of tiny, identical-looking cells. At that point, the fertilized egg had given rise to an **embryo**—a young, developing organism. After continued cell division, large groups of these cells would begin moving into the interior of the embryo, in a series of coordinated but mysterious-looking movements. Cell division would continue rapidly. After a week or two the embryo would elongate, and a recognizable head and tail portion would appear. Then tiny precursors of your vertebrae would be visible, along with your eyes. Eventually, buds would emerge and go on to form your limbs. As development continued, that embryo would eventually become you.

Biologists who have watched this process in humans or other organisms never cease to marvel at it. The question is, how does all the growth and body formation happen? In many species, the adult body contains trillions of cells and

● Key Concept   Important Information   ● Practice It

dozens of different types of highly organized tissues and organs. How does a fully formed individual that can move, eat, and reproduce develop from a tiny and formless-looking mass of cells?

People have pondered this question for centuries. Using light microscopes, early investigators observed the development of fertilized eggs and documented the events that occur as an individual grows from an embryo to an adult. Based on these studies, the physical pattern of development appeared to be highly variable among species. You've already thought for a moment about the events of early human development. But consider that a fertilized egg from a burr oak tree takes about four months to grow into an **embryo** that becomes encased in an acorn. After spending the winter in a dormant state, the embryo grows into a seedling. The organism will continue to grow and develop new leaves, roots, flowers, and seeds every year for the next 250–300 years. At another extreme, a fertilized egg of the fruit fly *Drosophila melanogaster* takes just a day to develop into a caterpillar-like larva that can move about and feed. About five days later, the larva stops feeding and forms a cocoon-like, stationary pupa. In another four days, an adult fly emerges from the pupal case. The adult flies about, feeds, courts, and mates—starting the cycle anew.

The variation in developmental patterns observed among species adds another layer of complexity to today's mystery of mysteries. Biologists have to explain the fundamental mystery of how highly integrated individuals, both large and small, can develop from a single cell. What's more, they have to make sense of all the variability that exists among species in the rate and pattern of development.

Fortunately, a century's worth of research has culminated in one of the great insights of contemporary biology: A few fundamental principles are common to all developmental sequences observed in multicellular organisms. Development is like jazz, where different members of a band play variations on the same theme. If you understand the theme and how chords and scales are put together, you can understand what each soloist is doing and how they are doing it. Although many critically important elements remain to be worked out, the discovery of the general principles underlying development has unified the vast variation observed in how embryos grow. And it has given biologists a framework for explaining how a single cell can give rise to a complex, multicellular individual.

The goal of this chapter is to explore the common principles that underlie development. As you study this material, you'll need to draw on what you've already learned about cell-cell interactions, the regulation of gene expression, and a host of other topics. Part of the excitement surrounding work on development is that it draws on insights from biochemistry, cell biology, genetics, and evolution. Developmental biology is among the most synthetic fields in all of biological science. Let's delve in.

21.1 Four Essential Developmental Processes

At the level of the cell, four general processes occur throughout development. Cells divide, move or expand in a directed way, begin to express certain genes rather than others, and signal to each other about where they are, what they are doing, and what type of cell they are becoming (**Table 21.1**). We'll consider each of these four processes in turn, and then go on to consider more specific issues about cell specialization and cell-cell interactions.

Cell Proliferation and Programmed Cell Death

For an organism to grow and develop, its cells have to proliferate—they have to divide and make more cells. This statement seems obvious. Perhaps less obvious is an equally important point: For an individual to develop from an undifferentiated mass of cells, the location, timing, and extent of cell division have to be tightly controlled.

Chapter 11 introduced the processes of mitosis and cytokinesis, which are responsible for cell proliferation in eukaryotes. That chapter also introduced the stages of the cell cycle and how they are controlled. You might recall that cells initiate mitosis in response to a regulatory protein complex called mitosis-promoting factor (MPF), that each stage of the cell cycle has checkpoints that are carefully regulated, and that cells continue to grow or stop growing in response to what biologists like to call "social controls"—meaning, signals from other cells. During development, all of these layers of regulation interact to determine when, where, and how much each embryonic cell proliferates.

What Chapter 11 did not mention is that death is also a highly regulated aspect of life for some cells. Programmed cell death in animals, or what biologists call **apoptosis** (literally, "falling away"), occurs as certain tissues and organs take shape. As the human hand and foot form, for example, cells that are initially present between the fingers and toes must die in order for separate digits to form. In plants, programmed cell death allows flower petals to fall after pollination has occurred, and for leaves or entire stems to be lost in autumn.

In these and many other cases, cell death is a normal part of development—it is carefully timed and regulated. **Box 21.1** introduces the genetic mechanisms responsible for apoptosis and highlights how defects in the normal patterns of programmed cell death can lead to abnormal development and disease.

To summarize, complex structures develop in multicellular organisms due in part to the growth of some cells and the regulated loss of others. Together, cell proliferation and programmed cell death are essential to the emergence of complex structures from the initial mass of embryonic cells.

SUMMARY TABLE **21.1 Four Essential Developmental Processes**

Cell proliferation and programmed cell death		Cells divide by mitosis and cytokinesis. The timing, location, and amount of cell division and cell death is regulated.
Cell movement or differential expansion		Cells can move past one another within a block of animal cells, causing drastic shape changes within the embryo.
		Certain animal cells can break away from a block of cells and migrate to new locations.
		Plant cells can divide in certain planes and expand in specific directions, causing dramatic changes in shape.
Cell differentiation		Undifferentiated cells specialize at specific times and places in a stepwise fashion. Cells that do not undergo differentiation are called stem cells in animals. Many plant cells are capable of de-differentiating.
Cell-cell interactions		Embryonic cells grow, move, or differentiate in response to signals from other cells.

Cell Movement and Expansion

In addition to growing and dividing (and dying), many animal cells have to move for normal development to occur. Some of the most dramatic cell movements occur early in animal development, after rapid cell divisions have produced a mass of similar-looking cells. During a process called **gastrulation**, cells in different parts of the mass rearrange themselves into three distinctive types of embryonic tissues, which then give rise to the skin, gut, and other basic parts of the body (see Chapter 22). Throughout development, certain animal cells break away from their original sites and migrate to new locations in the embryo. There they give rise to many different types of cells, including germ cells (sperm or eggs), pigment-containing cells, precursors of blood cells, or certain nerve cells. If any of these cell movements is inhibited, or if migrating cells end up in the wrong place or move at the wrong time, the embryo is likely to be deformed or die before development is complete.

Plant cells, in contrast, are encased in stiff cell walls and do not move. Instead, plant cells control how the cleavage plane is oriented during cell division and the direction of cell growth that subsequently takes place. This directionality in cell division and growth results in the proper formation of straight and branched stems, leaf blades, and other structures.

Cell Differentiation

For development to occur, many cells have to proliferate, some cells have to die, and other cells have to move to new locations or grow in specific directions. In addition, it is essential for most cells to undergo **differentiation**—the process of becoming a specialized type of cell. As you developed, some of your embryonic

cells differentiated to form muscle cells that contract and relax, while others became nerve cells that conduct electrical signals throughout the body. As an oak tree develops, some cells secrete thickened walls and transport water as part of a woody stem, while others become flattened and secrete the waxy coating found on the surface of leaves. In most multicellular eukaryotes, the fertilized egg gives rise to hundreds of distinctive cell types.

Differentiation may be the most basic process in the development of a multicellular organism, and it probably qualifies as the most intensively studied. In addition to the more detailed analysis upcoming in Sections 21.2 and 21.3, it's important to note several general points about differentiation.

- *Differentiation is a progressive, step-by-step process.* Cells initially are determined to become a certain cell type—meaning they are committed to a specific developmental pathway. Only later do they become differentiated—that is, they actually begin to look and behave like a specific cell type.

- *Some cells do not become specialized adult cells.* Undifferentiated cells are common in embryos, but even adult organisms have populations of cells that retain the ability to divide and give rise to an array of specialized cell types. Animal cells with this ability are called **stem cells**. The products of animal stem cell division can differentiate to help heal wounds, replace skin and blood cells that die, or perform other functions that maintain the body in good working order. In plants, populations of undifferentiated cells called **meristems** give rise to the flowers, roots, leaves, and other structures that develop throughout life.

- *Many or even most plant cells are capable of "de-differentiating"*—meaning they can change their structure and func-

BOX 21.1 Apoptosis: Programmed Cell Death

How does programmed cell death occur? An answer to this question emerged from studies of the roundworm *Caenorhabditis elegans*. This species is a popular model organism in developmental biology because it has a complex array of organs and tissues but only about a thousand cells. In addition, the cells are transparent (**Figure 21.1**). As a result, biologists are able to identify individual cells in both embryos and adults. Thanks to these features, biologists know the fate of each cell in the embryo at every stage of development.

As a *C. elegans* individual matures, 131 of its 1090 original somatic cells—cells that aren't involved in gamete formation—undergo apoptosis. To explore how this process occurs, Hillary Ellis and Robert Horvitz set out to identify mutants that do not undergo the normal pattern of cell deaths. You might recall from Unit 3 that this is a common research strategy in genetics. To find the genes responsible for a

trait, researchers look for mutants that do not have the trait.

Ellis and Horvitz's initial work uncovered two genes that are essential for apoptosis. If a worm carries mutations that inactivate either gene, cells that would normally die survive. The researchers proposed that the genes are part of a genetic program that executes apoptosis—in short, that they were cell-suicide genes.

By searching databases of known DNA sequences, biologists found genes in mice that are similar to the worm genes. To test the hypothesis that these genes and their products are important in mouse development, a research team used genetic engineering techniques to produce mice in which both copies of one of their cell-suicide genes were disrupted. In contrast to the normal embryo in **Figure 21.2a**, the mutant embryos like the one in **Figure 21.2b** exhibited a severe malformation of the brain. The defect occurred because cells that would normally die early in development survived.

As an aside, it's interesting to note that the mutant mouse embryos had normal digit separation, even though they had abnormal brains. Based on this observation, the researchers suggested that other cell-suicide genes are involved in programmed cell death during the formation of mouse toes.

Are the same genes found in humans? The answer is yes. Genes that are homologous to the worm and mouse cell-suicide genes have been identified in the human

(a) Mouse embryo with normal cell suicide genes   **(b)** Mouse embryo with defective cell suicide gene

FIGURE 21.2 Defects Occur When Programmed Cell Death Fails. (a) A normal mouse embryo; **(b)** a mouse embryo that has two defective alleles of a cell suicide gene.

● **QUESTION** If the defective version of the cell suicide gene is recessive, what would the embryo in part (b) look like if it were heterozygous instead of homozygous?

genome. (Recall from earlier chapters that homologous genes are similar in structure and function because they are derived from the same gene in a common ancestor.) Normal apoptosis is important in the development of human embryos, and abnormal apoptosis has been implicated in certain diseases of adults. For example, inappropriate activation of programmed cell death is involved in some neurodegenerative diseases, including ALS (Lou Gehrig's disease). Work on worm genes opened a new frontier in developmental biology, inspiring new insights into some devastating human illnesses.

50 µm

FIGURE 21.1 *Caenorhabditis elegans* **Is an Important Model Organism.** *C. elegans* has a defined number of transparent cells. Each cell can be identified and followed during development.

---

tion, even after they have specialized. For example, if a branch of a cedar tree droops down low enough to make contact with the soil, cells in the branch will de-differentiate and then re-differentiate to form root cells. A fully formed root will grow where a branch initially made contact with the ground.

Cell-Cell Interactions

Chapter 8 introduced the topic of cell-cell interactions by examining the extracellular matrix found between cells. That

chapter also explored how adjacent cells are attached and exchange materials and how cells respond to signals from other cells. During development, the most important cell-cell interactions involve sending and receiving signals. You might recall from Section 8.3 that when a signal arrives at the surface of a cell, its message is transduced to a new molecular form and amplified inside the receiving cell. Chapter 8 concluded by noting that cells change their activity in response to the signals they receive. What the chapter did not explain is that cell-cell signals are essential for changing cell activity during development.

As Section 21.3 will show, embryonic cells grow, move, or differentiate in response to signals from other cells. In many cases, the signal transduction pathways introduced in Chapter 8 trigger the production of the transcription factors introduced in Chapter 18. As a result, the arrival of cell-cell signals changes patterns of gene expression and thus the embryonic cell's structure and behavior.

21.2 The Role of Differential Gene Expression in Development

Differential gene expression is the essence of cell differentiation during development. The muscle cells in your body are different from your nerve cells because they express different genes. The same is true for plants. The water-transport cells in an oak tree are different from its leaf-surface cells because they express different genes and produce different proteins.

If you think about these statements, you'll realize that they have to be true. The only way that cells can have different structures and functions is if they contain different molecules. What is less obvious is whether cells express different genes because they contain different genes, or whether all the cells in a body contain the same genes but express only a specialized subset.

As it turns out, researchers found this question relatively simple to answer for plants. You've already learned, for example, that cells from the stem of a cedar tree can de-differentiate to form roots. This differentiation would be possible only if the cells from a branch continue to contain the genes required by root cells. Gardeners and farmers have known for centuries that with many plant species, complete new individuals can be produced from a small section of a root or shoot. Based on observations like this, researchers strongly suspected that all plant cells contain the same genes—meaning that they are genetically equivalent. This result was confirmed in the 1950s when biologists were able to grow entire tobacco plants or carrots from a single, differentiated cell taken from an adult (see Chapter 39).

In contrast to plants, the issue of genetic equivalence was extremely difficult to resolve for animals. Although serious experimental work on the problem began in the 1950s, the question was not answered until the late 1990s.

Are Differentiated Animal Cells Genetically Equivalent?

Early experiments on genetic equivalence in animals focused on transferring nuclei from diploid cells in frog tadpoles or adults into unfertilized eggs whose nuclei had been removed. At least some of these transplanted nuclei were able to direct the development of tadpoles successfully. These results provided strong evidence that all mature cells are genetically equivalent.

In 1997 Ian Wilmut and colleagues reinforced this conclusion through nuclear transfer experiments in sheep. As shown in **Figure 21.3**, these researchers removed mammary-gland cells

CLONING A SHEEP

Mammary-cell donor sheep   Egg-cell donor sheep

1. Start with two female sheep. Each will donate one cell.

Mammary cells   Egg cell

2. Culture mammary-gland cells. Remove nucleus from egg cell.

3. Fuse the mammary-gland cell to enucleated egg cell.

Fused cell

4. Egg cell now contains nucleus from mammary-gland cell.

Early embryo

5. Grow in culture. Embryo begins development.

Surrogate mother

6. Implant early embryo in uterus of third sheep.

Cloned sheep "Dolly"

7. Embryo develops normally, resulting in lamb that is genetically identical to mammary-cell donor.

This result supports the hypothesis that mature cells contain all the genes in the genome.

FIGURE 21.3 Mammals Can Be Cloned by Transplanting Nuclei from Mature Cells. The lamb that resulted from the nuclear transfer experiment shown here resembled the breed from which the donor *nucleus* came, not the breed of the egg donor or the surrogate mother. DNA testing also showed that the lamb was genetically identical to the individual that donated the nucleus.

from a 6-year-old pregnant female and grew them in culture (steps 1 and 2). Later they fused these cells with enucleated eggs—eggs whose nuclei had been removed. This procedure resulted in egg cells that contained nuclei from mammary-gland cells (steps 3 and 4). As the upper drawings show, the eggs came from a black-faced breed of sheep, while the donor nuclei came from a white-faced breed. After developing in culture, the resulting embryos were implanted in the uteri of surrogate mothers

(steps 5 and 6). In one of several hundred such transfer attempts, a white-faced lamb named Dolly was born (step 7). Genetically, Dolly was a clone of the white-faced donor of the mammary-gland cell. (Recall from Chapter 12 that a **clone** is a genetically exact copy of a parent.) Dolly grew into a fertile adult and, by normal mating, produced her own lamb named Bonnie (**Figure 21.4**). In 1998 other research groups reported similar results in mice and cows; since then, cats, horses, a monkey, and individuals of several other species have been cloned (but see **Box 21.2**).

⬤ Taken together, work on cloning plants and animals has shown that in most cases, the process of cellular differentiation does not involve changes in the genetic makeup of cells. Instead, it is based on differential gene expression. There are some important exceptions to this rule, however. For example, small stretches of DNA are rearranged in certain immune system cells in humans and other mammals, late in development. As a result, many immune cells are genetically unique. Chapter 49 explains how this happens.

What Is the Most Important Level of Control over Gene Expression?

As Chapter 18 emphasized, eukaryotic cells control gene expression at several different levels. A gene can be regulated by turning transcription on or off (or up or down), alternative splicing of mRNAs, selective destruction of mRNAs, speeding up or slowing down translation, and activating or deactivating proteins after they are translated. All of these processes occur during development. But which is most important?

FIGURE 21.4 Dolly with an Offspring of Her Own.

BOX 21.2 Human Cloning?

Biologists can harvest nuclei from the somatic cells of mice, cows, and sheep and use the nuclei to produce embryos. These nuclear transfer experiments produce offspring that are genetically identical to the individual providing the donor nucleus. A group of genetically identical organisms are clones, so the process of creating individuals by transferring nuclei into enucleated eggs is called cloning. Can humans be cloned?

Before addressing the technical feasibility of cloning humans, let's consider a more practical issue. Why would anyone want to clone a human? Proponents of this technology contend that cloning by nuclear transfer would allow otherwise infertile couples to have offspring that are genetically related—in fact, identical—to

one of the parents. Opponents of this technology are concerned that, in an attempt at immortality, dictators or wealthy eccentrics could finance clones of themselves in perpetuity.

Even benign uses of human cloning create ethical dilemmas that are yet to be resolved. What is the effect on a family of having children who are genetically identical to a parent? Would cloning be limited to individuals with "desirable" characteristics? Given the number and seriousness of the questions involved, several nations have proposed a moratorium on all research related to cloning humans.

Besides the legal and ethical issues involved, significant technical hurdles need to be overcome before human cloning would be feasible. Of the species

that have been successfully cloned, monkeys and mice are most closely related to humans. Less than 3 percent of nuclear transfers result in viable mice, however. Most result in aborted pregnancies; some produce individuals that die soon after birth. Currently, the success rate is similar in cloning sheep and cows, and deformed newborns have been produced. Most people regard a failure rate this high as unacceptable for humans, especially given the high likelihood of producing grossly deformed offspring.

As cloning techniques improve, however, human cloning may become technologically feasible. Meanwhile, the nations of the world are debating the merits of the technology and deciding whether it will be regulated or banned outright.

The answer is transcriptional control. To understand why, ask yourself whether a muscle cell should produce mRNAs or proteins that are specifically required by nerve cells. If it did, it would also have to produce microRNAs that disable nerve-cell mRNAs or regulatory proteins that keep nerve-cell proteins inactivated. It is much more logical to predict that muscle cells transcribe only genes required by muscle cells. This is exactly what researchers have found. Transcription is the fundamental level of control in differential gene expression during development. In eukaryotes, transcription is controlled primarily by the presence of **regulatory transcription factors** that influence chromatin remodeling and bind to promoter-proximal elements, enhancers, silencers, or other regulatory sites in DNA (see Chapter 18).

This simple insight is extremely important. To understand differentiation, researchers have to understand how and why regulatory transcription factors vary among cells.

Check Your Understanding

If you understand that...

- Differentiation occurs because embryonic cells express distinctive subsets of genes, not because they contain different genes.

- Differential gene expression is predominantly based on transcriptional control.

- The essence of differentiation is the production of distinctive sets of transcription factors.

You should be able to...

1) Explain the evidence for genetic equivalence in both plant and animal cells.

2) Explain why it is logical to observe that the key difference between developing cell types is the regulatory transcription factors that they contain.

21.3 What Triggers Differential Gene Expression?

To understand development, you have to think like a cell. Suppose you were one of the hundreds or thousands of cells in a developing animal embryo. Your fate—whether you ended up as part of an arm or a kidney, and whether you differentiated into a nerve cell or a blood-vessel cell—would depend on your location in four dimensions: time (meaning the stage of development that is currently occurring) plus the three spatial dimensions of the body axis. The three body axes are illustrated in **Figure 21.5**.

1. One axis runs **anterior** (toward the head) to **posterior** (toward the tail).

(a) The three body axes observed in humans and other animals...

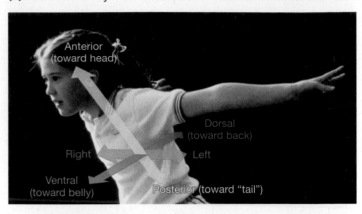

(b) ...are initially established in embryos (such as this mouse).

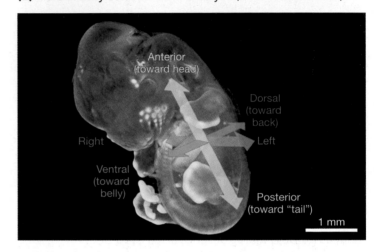

FIGURE 21.5 Most Animals Have Three Major Body Axes.

2. One axis runs **ventral** (toward the belly) to **dorsal** (toward the back).

3. One axis runs left to right.

   Cells "know" where they are in time and space because they are constantly interacting via cell-cell signals. In effect, much of development is organized by signals that cells send and receive. These signals activate transcription factors that turn specific genes on or off. As development proceeds, the distinctive suite of genes that are activated at successive stages determines the fate of each cell. If you understand this concept, you should be able to explain (1) how a cell could "know" where it is in relation to the three body axes and (2) how the same cell could receive different information early versus late in development.

   Let's consider how this process happens, beginning with one of the first developmental signals ever discovered. Although you'll be analyzing what happens as a fruit-fly embryo develops, keep an important point in mind: Principles that were discovered in fruit flies are relevant to virtually all multicellular organisms studied to date—from mustard plants to humans.

Master Regulators Set Up the Major Body Axes

Biologists refer to the events that determine the spatial organization of an embryo as **pattern formation**. If a molecule signals that a target cell is in the embryo's head, or tail, or dorsal side, or ventral side, that molecule is involved in pattern formation. As it turns out, certain early signals act as master regulators that set up the general anterior-posterior, dorsal-ventral, and left-right axes of an embryo. Then a network of genes activated by these master regulators sends signals with more specific information about the location of cells in space. The process occurs again and again: As development proceeds, a series of signals arrive and activate genes that specify finer and finer control over what a cell becomes.

The Discovery of *bicoid* The insights you just reviewed first began to emerge from work on the fruit fly *Drosophila melanogaster*—the model organism that has been so important in genetics (see Chapter 13). Because fruit flies produce large numbers of offspring rapidly, researchers could survey laboratory populations for rare mutant embryos in which the normal spatial relationships among cells are disrupted. A *Drosophila* larva that lacked a head, for example, was likely to have a mutation in a gene that helps pattern the head. Stated another way, biologists took a genetic approach to studying development. This approach has turned out to be extraordinarily productive. For example, studies of mutant fruit flies identified cell-cell signals that also direct pattern formation in mice, humans, and other animals.

The effort to dissect pattern formation in *Drosophila* began in the 1970s, when Christiane Nüsslein-Volhard and Eric Wieschaus exposed adult flies to treatments that cause mutations and then examined embryos or larvae descended from these individuals for body plan defects. After intensive effort, Nüsslein-Volhard and Wieschaus were able to identify over 100 genes that play fundamental roles in pattern formation.

One of the most dramatic mutations found by Nüsslein-Volhard and Wieschaus alters the normal arrangement of segments in the embryo. A **segment** is a distinct region of an animal body that is repeated along its length. In fly embryos, segments are grouped in three broad regions: head (anterior), thoracic (middle), and abdominal (posterior); these segments are pictured in **Figure 21.6a**. Embryos with the dramatic mutation were missing all of the segments and other structures normally found in the anterior end (**Figure 21.6b**). These had been replaced by duplicated posterior structures. The gene responsible for this phenotype was therefore called *bicoid*, meaning "two tailed." Based on its phenotype, Nüsslein-Volhard and Wieschaus suspected that the *bicoid* gene's product must provide positional information. The hypothesis was that the *bicoid* gene coded for a signal that tells cells where they are located along the anterior-posterior body axis.

When they mated flies that carried the mutant *bicoid* allele, Nüsslein-Volhard and Wieschaus found that the mutant allele was autosomal and recessive. More importantly, they found an interesting pattern in the inheritance and expression of this gene. Suppose that *b* stands for the mutant *bicoid* allele. The researchers found that some *bb* offspring are normal—even though they had two copies of the defective allele. In contrast, all mothers with the *bb* genotype looked normal but produced offspring that were all deformed. To interpret this finding, Nüsslein-Volhard and Wieschaus hypothesized that the *bicoid* gene is not expressed in embryos. Instead, it is expressed in mothers—as eggs are being formed. Mothers who were *bb* could not add any normal bicoid product to their eggs, resulting in offspring that couldn't develop normally. This hypothesis meant that *bicoid* mRNA or protein had to be present in the egg. Nüsslein-Volhard and Wieschaus thought they had found a master regulator that mothers loaded into their eggs.

Where Is the *bicoid* Product Found? To determine how the maternally endowed *bicoid* product works, Nüsslein-Volhard and colleagues cloned and sequenced the gene using

(a) A normal fruit-fly embryo

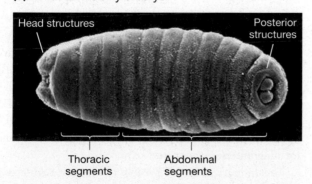

Head structures

Posterior structures

Thoracic segments

Abdominal segments

(b) A *bicoid* mutant

Posterior structures

Posterior structures

Abdominal segments

FIGURE 21.6 Pattern-Formation Mutants Have Misshapen Bodies. (a) Normal embryos have distinct head, thorax, and abdominal regions. **(b)** In *bicoid* mutants, embryos have no head or thoracic segments. Instead they have duplicated sets of the posteriormost regions, making them "two tailed."

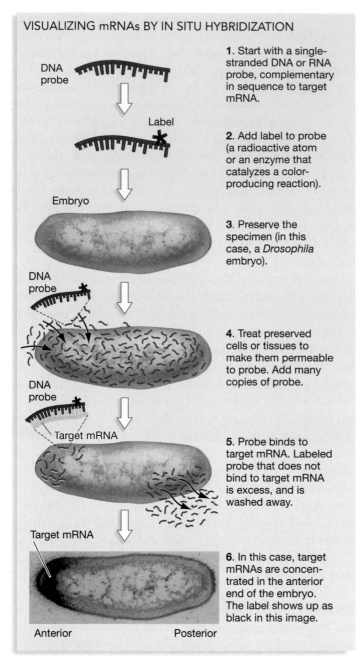

VISUALIZING mRNAs BY IN SITU HYBRIDIZATION

DNA probe

1. Start with a single-stranded DNA or RNA probe, complementary in sequence to target mRNA.

Label

2. Add label to probe (a radioactive atom or an enzyme that catalyzes a color-producing reaction).

Embryo

3. Preserve the specimen (in this case, a *Drosophila* embryo).

DNA probe

4. Treat preserved cells or tissues to make them permeable to probe. Add many copies of probe.

DNA probe

Target mRNA

5. Probe binds to target mRNA. Labeled probe that does not bind to target mRNA is excess, and is washed away.

Target mRNA

6. In this case, target mRNAs are concentrated in the anterior end of the embryo. The label shows up as black in this image.

Anterior   Posterior

FIGURE 21.7 In Situ Hybridization Allows Researchers to Pinpoint the Location of Specific mRNAs. The micrograph in the last step shows the location of mRNAs from the *bicoid* gene in a fruit fly egg.

● **QUESTION** In situ hybridization is usually used to identify cells that are expressing a particular gene. Why is this method valid?

techniques introduced in Chapter 19. Then they used a technique called **in situ** ("in place") **hybridization** to find where *bicoid* mRNAs are located in embryos. As **Figure 21.7** shows, researchers perform an in situ hybridization by adding a label to single-stranded copies of DNA or RNA molecules that are complementary in sequence to the mRNA they want to find. In this case, the probes were designed to bind to *bicoid*

mRNA inside the embryo—marking the location of the mRNAs.

In situ hybridization is enormously important in developmental biology because it allows researchers to pinpoint where and when specific genes are expressed. It is a way of visualizing how gene expression varies among cells in space. By doing in situ hybridizations with embryos of different ages, researchers can also track how gene expression changes through time. Differential gene expression is the essence of development, and in situ hybridization is an important way to document it.

When Nüsslein-Volhard's group treated eggs and early embryos with labeled copies of a probe that bound to *bicoid* RNA, they found that the mRNA is highly localized at the anterior end (see step 6 in Figure 21.7). Their next task was to determine where and when the Bicoid protein is produced. To do this, they made Bicoid protein from *bicoid* DNA (step 1 in **Figure 21.8**) and generated antibodies that bind specifically to the Bicoid protein (step 2). Recall that antibodies are proteins that bind to specific segments of a molecule. They attached a fluorescent compound to the Bicoid antibodies to mark the location of the Bicoid protein in embryos (step 3). These experiments indicated that Bicoid protein was produced from mRNAs localized at the anterior end. The protein then diffused away from the anterior end of the embryo. As the "Results" photograph in Figure 21.8 indicates, a steep concentration gradient is produced: The protein is abundant in the anterior end but declines to progressively lower concentrations in the posterior end.

To pull these observations together, Nüsslein-Volhard and co-workers hypothesized that high concentrations of Bicoid protein lead to the formation of anterior structures such as the head, with progressively lower concentrations giving rise to thoracic segments and the first abdominal segments. Absence of Bicoid, in contrast, contributes to formation of posterior structures. This hypothesis provided an explanation for the original *bicoid* mutants that the researchers had observed. Larvae that lack *bicoid* mRNA have posterior structures in their anterior end.

How Does Bicoid Work? DNA sequencing studies provided important insights into how the Bicoid protein acts as a master regulator. The *bicoid* gene contains sequences that are typical of regulatory transcription factors (see Chapter 18). Based on this observation, biologists hypothesized that the Bicoid protein must enter nuclei, bind to regulatory sites in DNA, and change the expression of specific genes (**Figure 21.9**). Follow-up work showed that Bicoid turns on genes responsible for forming anterior structures.

To summarize, Bicoid protein is a master regulator. Because it is present in a concentration gradient, it provides cells with information about their position along the anterior-posterior

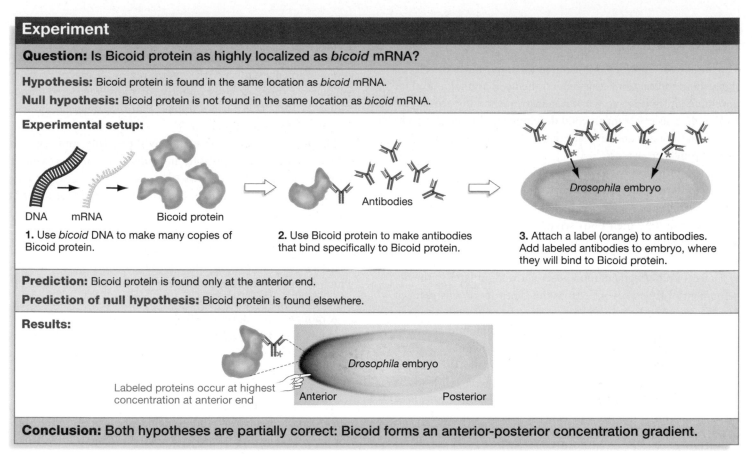

Experiment

Question: Is Bicoid protein as highly localized as *bicoid* mRNA?

Hypothesis: Bicoid protein is found in the same location as *bicoid* mRNA.

Null hypothesis: Bicoid protein is not found in the same location as *bicoid* mRNA.

Experimental setup:

DNA mRNA Bicoid protein

Antibodies

Drosophila embryo

1. Use *bicoid* DNA to make many copies of Bicoid protein.

2. Use Bicoid protein to make antibodies that bind specifically to Bicoid protein.

3. Attach a label (orange) to antibodies. Add labeled antibodies to embryo, where they will bind to Bicoid protein.

Prediction: Bicoid protein is found only at the anterior end.

Prediction of null hypothesis: Bicoid protein is found elsewhere.

Results:

Drosophila embryo

Labeled proteins occur at highest concentration at anterior end

Anterior   Posterior

Conclusion: Both hypotheses are partially correct: Bicoid forms an anterior-posterior concentration gradient.

FIGURE 21.8 The *bicoid* mRNA Is Localized, and Bicoid Protein Forms a Concentration Gradient. Bicoid protein forms an anterior-posterior concentration gradient.

body axis. Other initial signals are found in concentration gradients that tell cells where they are along the left-right and dorsal-ventral axes. Thanks to this information, cells have a general sense of where they are early in development. As cell division continues and the embryo grows and changes, how do the descendants of these cells get more specific information about where they are and what they should become?

MB **Web Animation** at **www.masteringbio.com**
Early Pattern Formation in *Drosophila*

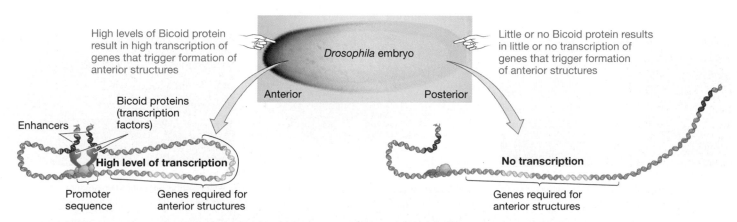

High levels of Bicoid protein result in high transcription of genes that trigger formation of anterior structures

Little or no Bicoid protein results in little or no transcription of genes that trigger formation of anterior structures

Drosophila embryo

Anterior   Posterior

Bicoid proteins (transcription factors)

Enhancers

High level of transcription

No transcription

Promoter sequence

Genes required for anterior structures

Genes required for anterior structures

FIGURE 21.9 Differential Gene Expression Is Based on the Presence or Absence of Regulatory Transcription Factors. Bicoid protein is a regulatory transcription factor. It binds to DNA and triggers the expression of specific genes—those involved in the formation of anterior structures. Because Bicoid is present in a gradient, different genes are expressed in the anterior of the embryo versus the posterior.

Regulatory Genes Provide Increasingly Specific Positional Information

The initial work on *bicoid* illustrated the importance of cell-cell signals and interactions as a general theme in animal and plant development. Follow-up work was instrumental in focusing attention on a second fundamental developmental principle, common to both plants and animals: that differentiation is a progressive, step-by-step process.

Along with the "two-posteriored monsters" they named *bicoid* mutants, Nüsslein-Volhard and Wieschaus found an array of embryos that had normal anterior-posterior patterning but defects in how their body segments became organized later in development. These mutants had defective **segmentation genes**.

Defects in the segmentation genes showed up later in the development of fly embryos because the genes act downstream of *bicoid*—meaning later in the developmental sequence. Research on the relationships among these genes helped illuminate one of the great general principles of development: ⦿ Regulatory genes act in a sequence that provides progressively detailed information on where cells are in time and space. One signal leads to another, and signals interact with other signals to tell cells how they should differentiate. This is another case where detailed work on fruit flies established principles that hold for other organisms as well. Let's take a closer look.

What Do Segmentation Genes Do? Researchers who first identified the segmentation genes noticed that they are expressed in sequence. Sequences called gap genes are expressed first. Experiments with antibodies to gap gene proteins showed that the genes are expressed in broad regions along the head-to-tail axis (**Figure 21.10a**), suggesting that they define the general position of segments in the anterior, middle, or posterior part of the body. Pair-rule genes are expressed next. In situ hy-

bridization experiments showed that their mRNAs occur in alternating bands (**Figure 21. 10b**). This pattern of expression suggests that pair-rule genes demarcate the boundaries of individual segments within each general region. Segment polarity genes are expressed subsequently, in more restricted bands (**Figure 21.10c**). This pattern implies that they delineate boundaries within individual segments.

The general observation is that gap, pair-rule, and segment polarity genes are expressed in order and in increasingly restricted regions. In this way, the anterior to posterior axis set up by the Bicoid signal is progressively divided into what becomes a segmented body.

These data also suggested that the segmentation genes might interact directly. Specifically, investigators hypothesized that the gap genes may activate the transcription of pair-rule genes, which in turn may trigger the expression of other pair-rule genes and segment polarity genes.

Although research on interactions among segmentation gene products is continuing, an enormous amount of progress has already been made. One of the key early discoveries was that the Bicoid protein directly activates transcription of a gap gene. This gap gene is not expressed until the Bicoid protein binds to its regulatory DNA sequences. As a result, the gap gene's mRNA and protein are found in high concentration in the anterior part of the embryo, where the Bicoid protein is abundant.

Subsequent work showed that this and other gap genes encode transcription factors that regulate the expression of pair-rule genes. Each gap gene product regulates the production of a defined set of pair-rule proteins. In addition, several of the gap genes' products regulate the expression of other gap genes or even their own expression. The pair-rule genes activated by the gap genes also encode regulatory transcription factors, and some of the genes they regulate are segment polarity genes. In this way, cells along the anterior-posterior axis of the embryo have their fate determined in a step-by-step process.

(a) Early in development: Gap genes

(b) Later: Pair-rule genes

(c) Still later: Segment polarity genes

Each stripe indicates the position of a group of segments

Each stripe indicates the position of a segment boundary

Each stripe indicates the position of regions within individual segments

FIGURE 21.10 A Series of Genes Demarcates Body Segments in Fruit Flies. Antibody-staining and in situ hybridizations show the location of segmentation gene products in fly embryos. The embryos in **(a)** and **(b)** were stained for two different gene products. The embryo in **(c)** was stained for one gene product.

⦿ **QUESTION** From (a) to (c), these photographs show embryos at progressively later stages of development. Why is the sequence logical?

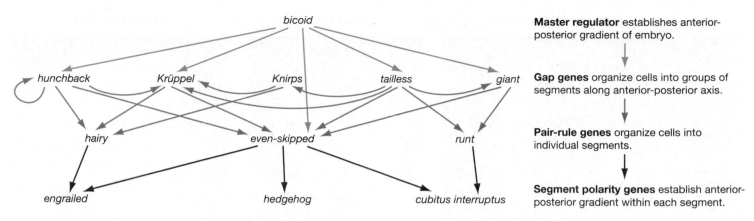

FIGURE 21.11 Segmentation Genes Regulate Each Other. The arrows indicate a small subset of the known interactions among the segmentation genes and their products. In each case, the outcome depends on the concentration of each protein involved. In reality, 9 gap genes, 8 pair-rule genes, and 9 segment polarity genes have been described thus far. Interactions among all their gene products are still being worked out. It's important to realize that analogous networks occur along the dorsal-ventral and right-left axes of an embryo.

Regulatory Genes Form Cascades, and Are Networked

As **Figure 21.11** shows, the interactions among *bicoid* and the segmentation genes can be described as a hierarchy called a **regulatory cascade**. The *bicoid* gene, gap genes, pair-rule genes, and segment polarity genes each define a level in the cascade. The cascade is triggered by the initial concentration of Bicoid protein and is later modified by other signals. For example, some of the segment polarity genes code for proteins that are secreted from the cell. These proteins bind to receptors on nearby cells, which modify their gene expression in response.

To illustrate how a regulatory cascade works, consider a cell that happens to be located at one end of a fly embryo. Very early in development, the Bicoid protein that binds to its DNA acts as a signal that means "You're in the anterior end." In response, the cell begins expressing genes that are expressed only in anterior cells. So do its descendants. Later, a gap gene product signals "You're part of the anterior segments," triggering other changes in transcription. Still later, pair-rule and segment polarity gene products send signals that specify, "Here's the boundary of the head segment, and here's the boundary of its anterior part." Step by step, the cell's progeny get increasingly specific information about where they are. In response, they become progressively committed to becoming brain cells or other components of the head segment. ● If you understand the relationship between regulatory cascades and segmentation, you should be able to describe the sequence of events that occur in a cell in the posterior of a fly embryo, where a different regulatory protein—analogous to Bicoid—is expressed.

The full scope of interactions occurring in this and other regulatory cascades are marvelously complex and are still being worked out. But several general take-home messages are clear, and they are just as relevant to maize, mice, and other eukaryotes as they are to flies:

● Differential gene expression results from regulatory cascades that start early in development: Master regulators

trigger the production of other regulatory signals and transcription factors, which trigger production of another set of signals and regulatory proteins, and so on.

● Because the identity and concentration of signals and transcription factors vary along the three major body axes, cells in different locations receive unique positional information.

● Each level in a regulatory cascade provides more specific information about where a cell is.

● As regulatory cascades proceed over time, a cell's fate becomes more and more finely determined.

Homeotic Genes: What Are They, and Where Do They Fit In?

Once gap genes, pair-rule genes, and segment polarity genes have established the identity of each segment along the anterior-posterior axis of a fly embryo, the regulatory cascade continues with the activation of **homeotic genes**. Segmentation genes establish the boundaries of each segment; homeotic genes identify which segment is which. More specifically, homeotic genes trigger the development of structures that are appropriate to each type of segment, such as antennae, wings, or legs.

The homeotic genes were discovered when researchers found adult fruit flies with body parts in the wrong place. For example, a series of homeotic mutations can transform thoracic segment number 3 in flies into thoracic segment 2. Instead of bearing the pair of small stabilizer structures called halteres normally found on thoracic segment 3, the transformed segment bears a pair of wings. The mutant has four wings instead of two (**Figure 21.12**). This type of replacement of one structure by another is termed **homeosis** ("like-condition"). Homeosis occurs when cells get incorrect information about where they are in the body.

Studies conducted during the 1970s and 1980s initially identified eight genes in the *Drosophila* genome that lead to

Normal fruit fly

Homeotic mutant

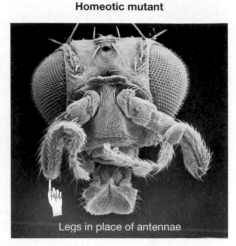

Antennae

Haltere

Wings in place of halteres

Legs in place of antennae

FIGURE 21.12 Homeotic Mutants in *Drosophila* Have Structures in the Wrong Locations. Among the most spectacular homeotic mutants in fruit flies are individuals with wings growing where small, stabilizing structures called halteres should be, or legs growing where antennae should be.

homeosis when they are defective. As **Figure 21.13a** shows, the eight genes are found in two clusters on the same chromosome. Together, these eight genes became known as the homeotic complex, or ***Hox* genes.** Once body segments are established in a *Drosophila* embryo, the *Hox* genes are expressed in a

(a) The location of *Hox* genes on the fly chromosome correlates with their pattern of expression in fly embryos.

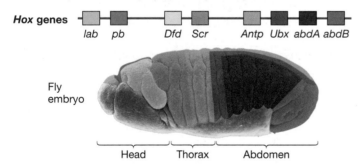

Hox genes

lab pb   Dfd Scr   Antp Ubx abdA abdB

Fly
embryo

Head   Thorax   Abdomen

(b) The location of *Hox* genes on the mouse chromosome correlates with their pattern of expression in mouse embryos.

1   2   3   4   5   6   7   8   9   10 11 12 13

Hox genes

Mouse
embryo

FIGURE 21.13 Organization and Expression of *Hox* Genes.

● **QUESTION** Suppose someone claimed that the anterior-posterior expression of *Hox* genes in a mouse embryo correlated with their positions along the length of the chromosome, as well as with the expression of the homologous genes in flies. Do these data support these assertions or contradict them?

distinctive pattern along the anterior-posterior axis. These genes code for transcription factors that trigger the production of segment-specific structures. Subsequent work revealed that not all *Hox* genes cause homeosis when they are defective.

Although it was exciting to establish that homeotic genes form the next link in the regulatory cascade responsible for building a fruit fly, that is not the reason they qualify as one of the most intensively studied groups of genes in developmental biology. Homeotic genes helped illuminate several other key principles of development.

Cell-Cell Signals and Regulatory Genes Are Evolutionarily Conserved

After homeotic genes were discovered and characterized in fruit flies, researchers began looking for similar genes in other animals. The results have been spectacular. Investigators have found that clusters of *Hox* genes occur in virtually every animal examined to date, including frogs, crustaceans (crabs and their relatives), birds, various types of worms, mice, and humans. Although the number of *Hox* genes varies widely among species, their chromosomal organization is similar to that of the fly homeotic genes shown in Figure 21.13a.

Recent studies of *Hox* genes have shown that they are expressed along the head-to-tail axis of the mouse embryo in the same sequence as in fruit flies (**Figure 21.13b**). In addition, experiments have shown that when mouse *Hox* genes are altered by mutation, defects in pattern formation result. Based on these data, biologists conclude that in flies, mice, humans, and most other animals, *Hox* genes play a key role in identifying the position of cells along the head-to-tail axis of the body. This conclusion was supported in dramatic fashion when researchers in William McGinnis's lab introduced the *Hoxb6* gene from mice into fruit-fly eggs. The *Hoxb6* gene in mice is similar in structure and sequence to the *Antp* gene of flies. Because it was introduced without its normal regulatory sequences, the *Hoxb6* gene was expressed throughout the treated fly embryos. The resulting larvae had defects identical to those observed in naturally occurring fly mutants, in which the *Antp* gene is mistakenly expressed throughout the embryo. This is a stunning result: A mouse allele not only affected the development of a fly but also mimicked the effect of a specific fly allele.

To interpret these observations, biologists hypothesize that the genes in *Hox* complexes of animals are homologous—meaning that they are similar because they are descended from genes in a common ancestor. If this hypothesis is correct, it means that the first *Hox* genes arose before the origin of animals. For the past billion years, *Hox* gene products have been helping to direct the development of animals.

Hox genes are by no means an isolated example of signals or regulatory proteins that direct development and are evolutionarily conserved. A well-studied gene called *Pax* has a similar structure and function in many animals—it is involved in the formation of eyes in species ranging from fruit flies to humans. Similarly, homologs of certain segment polarity genes have been found in a wide array of animals, where they perform similar cell-cell signaling functions during development.

The take-home message from these studies is that at least some of the molecular mechanisms of pattern formation have been highly conserved during animal evolution. The discovery of these shared mechanisms is one of the most significant results to have emerged from animal development studies to date. Although animal bodies are spectacularly diverse in size and shape, the underlying mechanisms responsible for their development are similar. What varies among species is less a matter of which genes are present and more a matter of when, where, and in what quantity similar genes are expressed.

Common Signaling Pathways Are Used in Many Contexts

The observation that key regulatory transcription factors and signals are conserved during evolution qualifies as a general principle of animal and plant development. But a related observation also qualifies as a general principle: During development, the same regulatory transcription factors and cell-cell signals are used in a variety of contexts. Regulatory proteins and signals are not only conserved, they are reused. For example, *Hox* genes are involved in the development of an array of animal structures, including appendages like your legs and arms, in addition to their role in defining the anterior-posterior axis of the animal body. But the most spectacular example of this phenomenon may be the *wingless* genes, abbreviated *Wnt* and pronounced "wint."

In humans and other mammals, the genome contains a family of 15 *Wnt* genes. Each of these genes codes for proteins responsible for cell-cell signaling during development. More specifically, they are involved in the regulatory cascade involved in setting up the anterior-posterior axis in the embryo—much like one of the roles of the *Wnt* genes of fruit flies. But in addition, signals from mammalian *Wnt* genes are involved in

the formation of back muscles, the midbrain region, the limbs, the gonads (testes or ovaries), hair follicles, parts of the intestine, and structures inside the kidney, among other things. To capture this point, biologists like to say that multicellular organisms have a tool kit of common signals, signal transduction pathways, and regulatory proteins that are used over and over during development. The common tool kit can direct the development of dramatically different structures because its tools are deployed at different times and in different locations.

To create a human analogy for this developmental principle, consider the signal called a pinch. When you were little, a pinch on the cheek from your grandmother meant something very different from a pinch on the ear from your mother or a pinch on the arm from an older sibling. It also means something different from a pinch you might receive from a close friend, now. In human communication and in development, the context in which a signal is sent and received—its location, timing, and intensity—is crucial.

Check Your Understanding

If you understand that...

- Cell-cell signals trigger the production of specific sets of transcription factors that change gene expression in receiving cells.

- Different cells express different genes because they receive different sets of signals, and because they produce different sets of transcription factors in response.

You should be able to...

1) Explain why the gradient of Bicoid protein delivers information about where cells are along the anterior-posterior axis of a fly embryo.

2) Describe how the cascade of transcription factors triggered by Bicoid leads to the gradual differentiation of segments along the anterior-poster axis of a fly embryo.

21.4 Changes in Developmental Pathways Underlie Evolutionary Change

For an embryo to develop, cells have to proliferate, move, differentiate, and interact. Differentiation is caused by differential gene expression. It results from signals that tell cells where they are in time and space, and which trigger a complex cascade of interacting transcription factors.

If any of these processes are disrupted, the embryo is likely to die. But if one of these processes is modified in some slight way, the affected structure is likely to have a different size or shape or activity. As a result, the embryo will develop new features and the adult will have a novel phenotype.

Once biologists began working out the regulatory signals and cascades introduced earlier in the chapter, they realized that

FIGURE 21.14 Some Snakes Have Rudimentary Hindlimbs. This claw, of a South African python, is attached to a rudimentary femur and hip.

the genetic changes altering these developmental processes must be the foundation of evolutionary change. For example, the increase in body size that has occurred during human evolution must have resulted from mutations that altered the signals, regulatory sequences in DNA, or transcription factors that are involved in the amount and timing of cell proliferation throughout the body.

A research field called evolution and development, or **evo-devo**, is focused on understanding how changes in developmentally important genes have led to the evolution of new phenotypes such as the limbs found in tetrapods (amphibians, reptiles, mammals) and arthropods (crabs, insects, millipedes). As an example of how this work is done, let's consider an instance of limb *loss*.

An array of analyses has shown that snakes are clearly tetrapods and that they are most closely related to lizards.

Although some snakes do not develop any sort of forelimb or hindlimb at all, boas and pythons have tiny pelvic (hip) bones and a rudimentary femur (thigh bone) and claw, as **Figure 21.14** shows. The ancestors of all snakes had four functional legs along with feet and toes, however. How did snakes lose their limbs? What developmental changes were responsible for this dramatic evolutionary event?

Biologists were able to answer this question because the cell-cell signals and regulatory cascades involved in the development of the tetrapod limb are reasonably well understood. For example, researchers knew that two *Hox* genes, *Hoxc6* and *Hoxc8*, are expressed together in vertebrae that form ribs, but that *Hoxc6* is expressed without *Hoxc8* in the region that gives rise to the forelimbs (**Figure 21.15a**). In snakes, however, they found that *Hoxc6* and *Hoxc8* are expressed together throughout the part of the trunk where forelimbs should form (**Figure 21.15b**). These data suggest that a change in the regulation of *Hoxc8*—specifically, where the gene is expressed—led to the evolutionary loss of the forelimb. Hindlimb loss, in contrast, appears to be due to defects in a signaling molecule encoded by a gene called *Sonic hedgehog*. This gene product is homologous to that of the fruit fly's *hedgehog* gene, introduced in Section 21.3. In snakes, hindlimb loss is due to a genetic change in part of the signaling cascade that sets up the anterior-posterior body axis.

In the case of both forelimb and hindlimb loss, biologists can point to alterations in a specific transcription factor or cell-cell signal that explains why an important evolutionary change occurred. Changes in a regulatory cascade led to changes in the adult body of a lizard-like ancestor. Snakes have been legless ever since.

(a) Pattern of gene expression in tetrapods with forelimbs.

(b) Pattern of gene expression in snakes (no forelimbs).

FIGURE 21.15 Changes in Homeotic Gene Expression Led to Limb Loss in Snakes. *Hoxc6* and *Hoxc8* code for transcription factors. If both transcription factors are present in a population of cells along the anterior-posterior axis of a tetrapod, genes that lead to the formation of vertebrae and ribs are activated. But if only *Hoxc6* is present, then genes that lead to the formation of forelimbs are activated. Contrasting patterns of gene expression are responsible for **(a)** the development of forelimbs in most tetrapods and **(b)** the lack of forelimbs in snakes.

Chapter Review

SUMMARY OF KEY CONCEPTS

● As an embryo develops, cells have to proliferate, move or expand in a directed manner, become specialized, and interact with other cells.

Four essential processes underlie development: (1) Cells have to grow and divide in a regulated manner as the body develops, and some cells have to undergo programmed cell death. (2) In animals, masses of cells move in a coordinated way early in development; later, smaller populations of cells move to specific locations in the body. Plant cells do not move, but normal development depends on precise control over the plane of cell division and directed expansion of cells. (3) Cells undergo a step-by-step process that leads to differentiation, or the production of a specialized cell type. (4) Cell-cell signals provide a constant flow of information about where cells are in space and time.

You should be able to predict how the development of an oak seedling would be affected by a drug that (1) reduces the rate of cell division by half, or (2) prevents cells from expanding after mitosis, or (3) obstructs the first step in the sequence that leads to flower formation, or (4) blocks the signal that sets up the shoot-to-root axis. ●

● Cells become specialized because they express different genes, not because they contain different genes.

In animals and plants, cells differentiate due to differential gene expression, not differential loss of genes. Cells of various cell types are specialized because they contain transcription factors that activate certain genes and not others. Differentiation occurs in a step-by-step manner because cell-cell signals trigger a cascade of effects that lead to certain transcription factors being present in the cell over time.

You should be able to explain why chromatin structure in a muscle cell is different from chromatin structure in a nerve cell. ●

● Cells interact continuously during development via cell-cell signals.

Cells "know" where they are in the body and how far along development has progressed because they produce and receive a steady stream of signals. These signaling molecules operate through signal transduction pathways that alter gene expression—primarily transcription. Cells begin expressing a distinc-

tive suite of proteins and differentiate because they receive a distinctive suite of signals.

You should be able to explain why development is based on transcriptional control rather than RNA processing, RNA stability, translation, or protein modification. ●

● Early signals trigger a cascade of effects, including the production of transcription factors and other signals. Interactions among these signals, transcription factors, and regulatory DNA sequences cause cells to differentiate as development proceeds.

In fruit flies, a master regulator called Bicoid is responsible for setting up the overall anterior-posterior axis of the body. Bicoid triggers the production of transcription factors encoded by the gap genes, which activate the expression of pair-rule genes, which stimulate the transcription of segment polarity genes, which regulate the expression of homeotic genes. As this cascade continues, cell-cell signals from genes like *hedgehog* and *wingless* moderate the array of transcription factors produced in cells throughout the body. Complex interactions among these signals and transcription factors are responsible for differentiation.

You should be able to explain why *bicoid* mutants die as embryos, while most homeotic mutants can develop into adults. ●

 at www.masteringbio.com
Early Pattern Formation in *Drosophila*

● The evolution of new body sizes, shapes, and structures is possible when mutations alter the genes responsible for development.

If the cell-cell signals, transcription factors, and regulatory DNA sequences that are active in development change, then the adult phenotype is likely to change as a result. Variation in phenotypes is the basis of evolutionary change. In several cases, researchers have been able to pinpoint changes in cell-cell signals, transcription factors, or regulatory DNA sequences that were responsible for major events in evolution, such as the loss of limbs in snakes.

You should be able to explain the significance of the four-winged homeotic mutant in Figure 21.12, in terms of efforts to explain the genetic mechanisms responsible for insect evolution. (The ancestors of fruit flies had two pairs of full-sized wings that functioned in flight.) ●

QUESTIONS

● **Test Your Knowledge**

1. What is apoptosis?
   a. an experimental technique that biologists use to kill specific cells
   b. programmed cell death that is required for normal development
   c. a pathological condition observed only in damaged or diseased organisms
   d. a developmental mechanism unique to the roundworm *C. elegans*

2. What is the function of stem cells in adult mammals?
   a. Some of their daughter cells remain as stem cells and continue to divide throughout life.
   b. They give rise to hair, fingernails, and other structures that grow throughout life.
   c. They produce compounds that stem blood loss from wounds.
   d. They produce cells that differentiate to replace dead or damaged cells.

3. Why are in situ hybridizations such a valuable tool for studying development?
   a. They identify the location of specific mRNAs, and so provide a picture of differential gene expression.
   b. They allow researchers to understand how cell-cell signals and regulatory transcription factors interact.
   c. They provide data on homology—the presence of similar genes in different species.
   d. They can be done with RNA or DNA probes.

4. Why is it logical to observe that differentiation is based on transcriptional control, versus translational or post-translational control?
   a. There is a trade-off between speed and efficiency at the various levels of control over gene expression.
   b. Differentiation is based on differential gene expression.
   c. Differentiation is usually permanent, so cells should not produce proteins that will never be used.
   d. Transcriptional control is the only level of control that can be carefully regulated.

5. What is a homeotic mutant?
   a. an individual with a structure located in the wrong place
   b. an individual with an abnormal head-to-tail axis
   c. in flies, an individual that is missing segments; in *Arabidopsis*, an individual that is missing a hypocotyl or other embryonic structure
   d. an individual with double the normal number of structures or segments

6. Why are regulatory cascades in development also referred to as networks?
   a. The cascade is not linear—within and between levels in the cascade, various gene products interact to form a network.
   b. Signals as well as transcription factors are involved in the cascade.
   c. Each transcription factor and signal in the cascade has a limited life span.
   d. Each step in the cascade takes time, so the steps in the cascade are sequential.

Test Your Knowledge answers: 1. b; 2. d; 3. a; 4. c; 5. a; 6. a

Test Your Understanding

Answers are available at www.masteringbio.com

1. How did researchers go about looking for molecules that have important roles in establishing the anterior-posterior body axis and body segments in *Drosophila*?

2. What evidence suggests that at least some of the molecular mechanisms responsible for pattern formation have been highly conserved over the course of animal evolution?

3. Why is it significant that many of the genes involved in development encode regulatory transcription factors?

4. Explain the connection between the existence of regulatory genes and signaling cascades and the observation that differentiation is a step-by-step process.

5. Explain the logic behind using nuclear transplant experiments to test the hypothesis that all animal cells in the body are genetically equivalent.

6. Explain why cell proliferation and death, cell movement or growth, cell differentiation, and cell-cell interactions are considered the four fundamental processes in development.

Applying Concepts to New Situations

Answers are available at www.masteringbio.com

1. Recent research has shown that the products of two different *Drosophila* genes are required to keep *bicoid* mRNA concentrated at the anterior end of the egg. In individuals with mutant forms of these proteins, *bicoid* mRNA diffuses farther toward the posterior pole than it normally does. Predict what effect these mutations will have on segmentation of the larva.

2. In 1992 David Vaux and colleagues used genetic engineering technology to introduce an active human gene for a protein that inhibits apoptosis into embryos of the roundworm *C. elegans*. When the team examined the embryos, they found that cells that normally undergo programmed cell death survived. What is the significance of this observation?

3. Suppose that physicians implanted human stem cells into the brains of patients with Parkinson's disease, in the area of the brain where the patients had lost many nerve cells. What would have to happen for the stem cells to differentiate into functioning nerve cells?

4. According to the data presented in this chapter, segmentation genes set up the segmented body plan of a fruit fly while homeotic genes trigger the production of segment-specific structures such as antennae, legs, and wings. Sketch the body of a fruit fly next to the body of a spider, centipede, and earthworm. Generate hypotheses to explain the differences between the bodies of fruit flies and the other three species, based on changes in the segmentation genes and homeotic genes.

www.masteringbio.com is also your resource for • Answers to text, table, and figure caption questions and exercises • Answers to *Check Your Understanding* boxes • Online study guides and quizzes • Additional study tools including the *E-Book for Biological Science* 3rd ed., textbook art, animations, and videos.

An Introduction to Animal Development

22

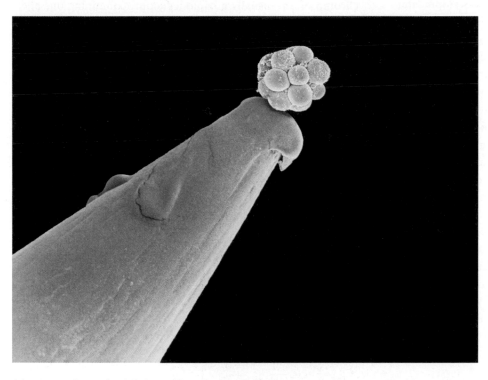

A human embryo, about 3 days old, on the head of a pin.

KEY CONCEPTS

- In most animals, fertilization depends on specific interactions between proteins on the plasma membranes of sperm and egg.

- The earliest cell divisions divide the fertilized egg into a mass of cells. The fate of these cells depends on the cytoplasmic determinants they contain, the signals they receive, or a combination of both.

- Early in development, cells undergo massive, coordinated movements to form distinct tissue layers. Each of these layers gives rise to distinct groups of tissues and organs in the adult.

- As development proceeds, the formation of specialized organs and other structures depends on cell-cell signals, cell proliferation, cell movements, and differentiation. Differentiation is complete when cells express tissue-specific proteins.

W hen the physician, researcher, and writer Lewis Thomas considered how a human being develops from a fertilized egg, he could only marvel: "You start out as a single cell derived from the coupling of a sperm and an egg, this divides into two, then four, then eight, and so on, and at a certain stage there emerges a single cell which will have as all its progeny the human brain. The mere existence of that cell should be one of the great astonishments of the earth. People ought to be walking around all day, all through their waking hours, calling to each other in endless wonderment, talking of nothing except that cell. It is an unbelievable thing, and yet there it is, popping neatly into its place amid the jumbled cells of every one of the several billion human embryos around the planet, just as if it were the easiest thing in the world to do."[1]

Many biologists share Thomas's feelings. The experience of watching a fertilized egg go through early development inspires awe and wonder. In animals, the fertilized egg undergoes a rapid series of cell divisions. The resulting cells begin a series of extensive, tightly coordinated movements, and the embryo is dramatically reorganized. Just a few hours or days later, a recognizable creature with internal organs, a head, eyes, back, belly, and other

[1]Lewis Thomas, *The Medusa and the Snail* (New York: Viking Press, 1979), 156. Thomas was exercising some poetic license here. The brain actually arises from a group of cells in the embryo rather than from a single cell.

● Key Concept   Important Information   ● Practice It

body parts appears. In many cases, these structures continue to grow and develop for years after hatching or birth occurs.

Understanding how and why these events happen is among the greatest of all challenges facing biologists today. Chapter 21 introduced this challenge by reviewing the basic principles that unify animal and plant development. The most fundamental message of Chapter 21 was that cell-cell signals cause specific sets of transcription factors to be produced in various cells throughout the embryo, resulting in differential gene expression and differentiation.

This chapter's goal is to apply the general developmental principles to understand the sequence of events that occurs as an animal embryo develops. We begin by examining gametogenesis, the process responsible for producing gametes. Recall from Chapter 12 that **gametes** are haploid reproductive cells. In animals and many other organisms, male gametes are called **sperm** and female gametes are called **eggs**. As **Figure 22.1** shows, gametogenesis is followed by fertilization, or the union of a sperm and an egg. In animals, a fertilized egg undergoes a series of rapid cell divisions called cleavage, forming a mass of cells known as a blastula. The cells then move and rearrange themselves in the process of gastrulation. During this phase of development, the embryo is referred to as a gastrula. These early events in development are analyzed in Sections 22.1 through 22.3. They set the stage for the development of specialized tissues and organs during the process of organogenesis, introduced in Section 22.4.

The events summarized in Figure 22.1 represent the cycle of life in animals. As you study each step in this cycle, remember to keep an eye out for the cell-cell signals and regulatory proteins that allow developing cells to divide, move, interact, and differentiate—to do the things required for them to "pop into place" and contribute to the formation of a new individual.

22.1 Gametogenesis

The development of an offspring begins with **gametogenesis**, or the formation of gametes—sperm and egg—in the reproductive organs of adult organisms. The DNA and cytoplasm in these reproductive cells are the initial components of the new individual. Both sperm and egg contribute an equal number of chromosomes—usually a haploid genome containing one allele of each gene—to the offspring. But because egg cells are routinely hundreds or thousands of times larger than sperm cells, an egg contributes much more cytoplasm than does a sperm.

The sequence of mitotic and meiotic cell divisions leading up to the production of sperm and eggs are discussed in Chapter 40 (plant reproduction) and Chapter 48 (animal reproduction). Here, let's focus on the structure of the mature reproductive cells. Understanding how these specialized cells are put together will lay the groundwork for addressing questions about fertilization and early development.

Sperm Structure and Function

Sperm cells begin to develop after meiosis has resulted in the production of a haploid nucleus. As a mammalian sperm cell matures, the four main compartments shown in **Figure 22.2** develop: the head, neck, midpiece, and tail.

- The head region contains the nucleus and an enzyme-filled structure called the **acrosome**. The chromosomes inside the

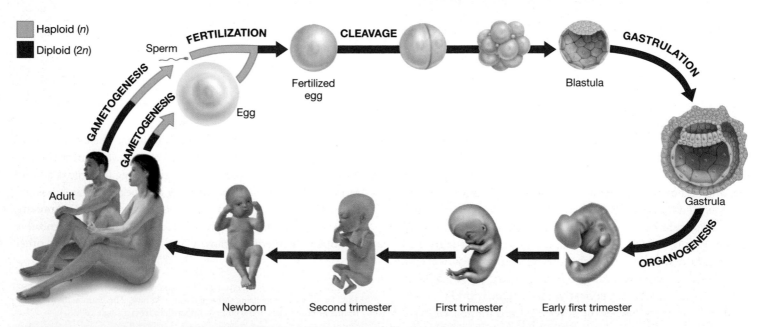

FIGURE 22.1 Development Proceeds in Ordered Phases. In animals, the development of a new individual starts with the formation of gametes and continues with fertilization and early cell divisions (cleavage), resulting in a blastula. Gastrulation then rearranges the blastula into a gastrula. Organogenesis follows and leads to the formation of adult tissues and structures.

Head
Neck
Midpiece
Tail

Acrosome
Nucleus
Centriole
Plasma membrane
Vacuole (not present in all sperm)
Mitochondria
0.5 µm
Flagellum

FIGURE 22.2 Animal Sperm Structure. The morphology of human sperm is typical of many animal species.

● QUESTION If you were examining the gametes of an animal that you'd never seen before, how would you decide which cells are sperm and which are eggs?

nucleus are condensed into an even more densely packed arrangement than in other cells. The acrosome at the tip of the sperm is derived from vesicles produced by the Golgi apparatus (see Chapter 7). The enzymes stored in the acrosome allow the sperm to penetrate barriers that exist around the egg.

- The neck encloses a **centriole**. You might recall from Chapter 11 that animal centrioles play a key role in the formation of the spindle apparatus, which moves chromosomes to daughter cells during mitosis. The fertilized egg inherits one centriole from the sperm cell and one centriole from the egg.

- The midpiece is packed with **mitochondria** (see Chapter 9). Sperm swim long distances to reach an egg. This movement is powered by ATP molecules synthesized in the mitochondria of the midpiece.

- The tail region consists of a **flagellum**—the long structure composed of microtubules and surrounded by the plasma membrane that was introduced in Chapter 7. Sperm cells are simple enough to be characterized as "DNA with a propeller." The "propeller" is the flagellum, which arises from the centriole early in sperm-cell maturation. The flagellum powers movement by whipping back and forth.

Sperm are race cars—stripped down, streamlined, souped-up cells that are specialized for racing other sperm to the egg. Eggs, in comparison, are like semitrailers—hulking, locked, immobile storage containers that are packed with valuable merchandise.

Egg Structure and Function

Animal gametes are easy to tell apart: Sperm are small and motile; eggs are large and nonmotile. Eggs are large mainly because they contain the nutrients required for the embryo's early development. But in addition to nutrients, the eggs of many species contain key developmental regulatory molecules called **cytoplasmic determinants**—like the *bicoid* mRNA introduced in Chapter 21—that control the early events of development.

The quantity of nutrients present in the egg varies widely among species. In mammals, embryos start to obtain nutrition through a maternal organ called the placenta within a week or two after fertilization. Thus, the egg only has to supply nutrients for early development and is relatively small. But in species where females lay eggs directly into the environment, stores in the egg are the *only* source of nutrients until organs have formed and a larva or juvenile hatches and begins to feed. In these species, the nutrients required for early development are provided by **yolk**—a fat- and protein-rich cytoplasm that is loaded into egg cells as they mature. Yolk may be present as one large mass or as many small granules.

Yolk and the cytoplasmic determinants present in some species are not the only specialized material or structure found inside animal eggs. Many eggs also contain organelles called **cortical granules**, which are small vesicles filled with enzymes that are activated during fertilization. As the egg matures, cortical granules are synthesized in the Golgi apparatus, transported to the cell surface, and localized just under the inner surface of the plasma membrane.

Just outside the plasma membrane of eggs, a fibrous, mat-like sheet of glycoproteins called the **vitelline envelope** forms and surrounds the egg (**Figure 22.3**). In the eggs of humans and other mammals, this structure is unusually thick and is called the **zona pellucida**. In some species, a large gelatinous matrix known as a jelly layer surrounds the vitelline envelope to further enclose the egg.

These sperm are digesting their way through the jelly layer

Jelly layer
Vitelline envelope
Plasma membrane
Cortical granules
Nucleus
Yolk granules

FIGURE 22.3 Sea Urchin Eggs Are Covered by a Jelly Layer. Animals that lay their eggs in water frequently surround their eggs with a jelly layer, a gelatinous coat.

The egg is loaded with rich stores of nutrients and guarded by one or more layers of protective material. How does a sperm cell penetrate these coatings to reach the egg's plasma membrane and fertilize the egg?

22.2 Fertilization

Fertilization seems like a simple process: A sperm cell fuses with an egg cell to form a diploid cell known as a **zygote** (a fertilized egg). Upon reflection, though, you begin to appreciate that the process is extraordinarily complex. For fertilization to take place, sperm and egg cells must be in the same place at the same time. Then they must recognize and bind to each other. Next they have to fuse—even though most of the other cells in the body do not fuse with cells they contact. In most species, fusion must also be limited to a single sperm so that the zygote does not receive extra chromosomes. Finally, the fusion of the two gametes has to trigger the onset of development. This complexity has made fertilization a fascinating research topic. The contact that takes place between sperm and egg at fertilization qualifies as the best studied of all cell-to-cell interactions.

Fertilization in Sea Urchins

Research on fertilization began in earnest early in the twentieth century, when biologists started to study the sperm-egg interaction in sea urchins. Sea urchins are marine organisms that are closely related to sea stars and sand dollars. Like most aquatic animals, they shed their gametes into the surrounding water and fertilization occurs externally (**Figure 22.4**). To maximize the probability that sperm and egg will meet, male and female sea urchins secrete huge quantities of gametes. As a result, this organism provides researchers with large numbers of gametes that can be studied in a test tube or culture dish filled with seawater. Sea urchins continue to be an intensively studied

FIGURE 22.4 Many Aquatic Species Have External Fertilization. Sea urchins release their gametes into the ocean. Fertilization and early development take place in the open water.

● **EXERCISE** Sea urchins cannot see. Suggest a hypothesis to explain how it is possible for many sea urchins to release gametes at the same time.

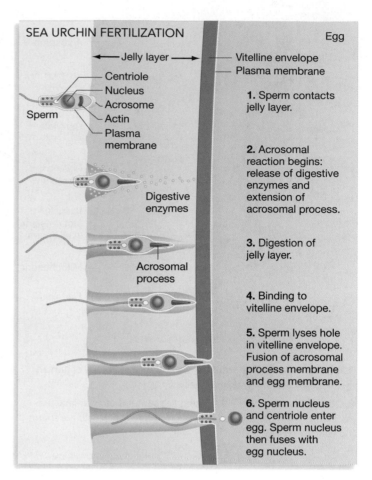

FIGURE 22.5 The Fertilization Sequence in Sea Urchins Is Based on Cell-Cell Interactions.

model system in research on fertilization and other aspects of development.

In sea urchins, the jelly layer surrounding the egg contains a molecule that attracts sperm. The attractant is a small peptide that diffuses away from the egg and into the surrounding seawater. Sperm respond to the attractant by swimming toward areas where its concentration is high. The same type of interaction between sperm and eggs has been observed in a wide variety of animals that lay their eggs in seawater.

When a sea urchin sperm and egg meet, the head of the sperm initially encounters the jelly layer of the egg cell (**Figure 22.5**). To reach the plasma membrane of the egg cell, the sperm head must digest its way through the egg's jelly layer and vitelline envelope. This process begins with the **acrosomal reaction**, which is triggered by contact between the sperm's head and the jelly layer. In the first part of this reaction, the contents of the acrosome—the structure at the tip of the sperm's head—are released. The enzymes that diffuse away digest a pathway through the egg's jelly layer so the sperm can reach the vitelline envelope of the egg. The second part of the acrosomal reaction involves polymerization of actin into microfilaments that form the acrosomal process, a protrusion that extends from the sperm's head until it makes contact with the vitelline envelope. Finally, the plasma membranes of the egg and sperm fuse. The sperm nucleus, mitochondria, and centriole enter the egg—though the sperm mitochondria later disintegrate—and the sperm and egg nuclei fuse to form the zygote nucleus. Fertilization is complete.

In 1910 Frank Lillie recognized that an important question was hidden in this sequence of events: What prevents cross-species fertilization and the production of dysfunctional hybrid offspring? After all, in many habitats sperm and eggs from a particular sea urchin species float in seawater along with eggs and sperm from other sea urchin species and many other organisms. The effort to answer this question blossomed into an extraordinarily productive line of research.

How Do Gametes from the Same Species Recognize Each Other?

Lillie was the first researcher to identify a substance on the surface of egg cells that appeared to be involved in binding sperm. He called the compound *fertilizin* and showed that if the purified molecule were experimentally applied to a group of sperm, it caused the cells to clump together. This result supported the hypothesis that fertilizin is involved in sperm-egg coupling, because it implied that clumps of fertilizin molecules were binding to sperm, causing them to clump. Lillie also showed that clumping occurred only when fertilizin from the eggs of a particular sea urchin species was combined with sperm from the same species. Each species appeared to have its own version of fertilizin.

Based on these observations, Lillie proposed that fertilizin on the surface of an egg interacted with a substance on sperm in a lock-and-key fashion. Further, Lillie suggested that this interaction was a product of natural selection (see Chapter 1), because it increased the probability of fertilization occurring between sperm and eggs of the same species.

Decades passed before Lillie's hypothesis was extended and tested, however. Then, in the 1970s, Victor Vacquier and co-workers succeeded in identifying a protein on the head of sea urchin sperm that binds to the surface of sea urchin eggs in a species-specific manner. They called this protein *bindin*. Follow-up work showed that the bindin proteins from even very closely related species are distinct. As a result, bindin should ensure that a sperm binds only to eggs from the same species. The next question was, To what does bindin bind? If bindin acts as a key, what acts as the lock?

To address these questions, Kathleen Foltz and William Lennarz hypothesized that fertilizin is the "lock" protein, that it is found on the surface of sea urchin eggs, and that it acts as a receptor for bindin. To test these hypotheses, they set out to isolate the bindin receptor. More specifically, they set out to isolate the part of fertilizin that is exposed on the outside of the egg. They predicted that this region of the protein interacts with the bindin on sperm.

Figure 22.6 illustrates Foltz and Lennarz's experimental approach. They began by treating the surface of sea urchin eggs

---

Experiment

Question: If bindin is the protein on the surface of sperm that acts like a key, what is the "lock"?

Hypothesis: The lock is a receptor protein on the surface of sea urchin eggs. Bindin binds to this receptor.

Null hypothesis: The lock is not a receptor protein on the surface of sea urchin eggs. Bindin does not bind to this receptor.

Experimental setup:

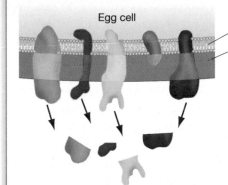

Egg cell

Plasma membrane

Vitelline envelope

1. Use protease to release protein fragments from egg surface.

2. Isolate each type of protein fragment and see which ones bind to sperm and to isolated bindin molecules.

Prediction: One of the protein fragments isolated from egg surface will bind to sperm and to isolated bindin molecules.

Prediction of null hypothesis: The protein fragments isolated from egg surface will not bind to sperm or to isolated bindin molecules.

Results:

A fragment from the egg protein fertilizin binds to bindin protein on sperm—*if* sperm is from same species as egg

Bindin

Plasma membrane

Sperm cell

Conclusion: Fertilizin is a "lock" protein on the surface of eggs that binds to bindin (the "key") in a species-specific manner.

FIGURE 22.6 The Egg-Cell Receptor for Sperm Was Isolated and Characterized. By treating sea urchin egg cells with a protease, researchers isolated the fragment of the egg-cell receptor that binds to sperm.

---

with a **protease**—an enzyme that cleaves peptide bonds. When the investigators isolated the protein fragments that were released from the egg surface, they found one that bound to sperm and to isolated bindin molecules. Further, this binding occurred in a species-specific manner. A protein fragment from the eggs of one species bound to sperm of its own species, but did not bind to sperm of different species. Based on these observations, the biologists claimed that they had found the outward-facing portion of the egg-cell receptor protein for sperm.

These experiments provided convincing evidence that an egg receptor for sperm exists on the surface of sea urchin eggs. Foltz and Lennarz succeeded in isolating the compound Lillie had named fertilizin. In doing so, they provided important support for Lillie's lock-and-key hypothesis. ● During sea urchin fertilization, species-specific bindin molecules on sperm interact with species-specific fertilizin receptors on the surface of the egg. This interaction is required for the plasma membranes of sperm and egg to fuse. As a result, cross-species fertilization is rare.

Why Does Only One Sperm Enter the Egg? Early in the history of studies on sea urchin fertilization, researchers noticed that only one sperm succeeded in fertilizing the egg, even when dozens or even hundreds of sperm were clustered around the vitelline envelope. From the standpoint of producing a viable offspring, this observation was logical. If multiple fertilization, or **polyspermy**, occurred, the resulting zygote would have more than two copies of each chromosome. Sea urchin embryos with more than two copies of each chromosome die. If many sperm are present, how is polyspermy avoided?

Research over the past 75 years has revealed a wide array of mechanisms that block polyspermy in various animal species. In sea urchins, for example, fertilization results in the erection of a physical barrier to sperm entry. As **Figure 22.7a** shows, the entry of a sperm causes calcium ions (Ca^{2+}) to be

released from storage areas inside the egg. A wave of ions starts at the point of sperm entry and propagates throughout the egg. In response to this dramatic increase in Ca^{2+} concentration, a series of events occurs in the egg. For example, the cortical granules located just inside the egg cell's plasma membrane respond to the Ca^{2+} signal by fusing with the membrane and releasing their contents to the exterior. The contents include proteases that digest the exterior-facing fragment of the egg-cell receptor for sperm, which prevents any new sperm from binding to the egg surface. In addition, ions and other compounds released by the cortical granules accumulate between the egg cell's plasma membrane and the vitelline envelope. Because these solutes are at high concentration, they cause water to flow by osmosis into the space between the plasma membrane and vitelline envelope (see Chapter 6). The influx of water causes the envelope matrix to lift away from the cell and form a **fertilization envelope** (**Figure 22.7b**). The fertilization envelope, in turn, keeps additional sperm from contacting the egg's plasma membrane.

To summarize, a century of research on sea urchin fertilization has illuminated one of the most important cell-to-cell interactions in nature. The interaction between sperm and egg

(a) A WAVE OF Ca^{2+} SPREADS FROM THE SITE OF SPERM ENTRY.

Sperm enters egg here

Ca^{2+}

Ca^{2+}

A wave of calcium ions starts at the point of sperm entry and propagates across the egg

(b) THE FERTILIZATION ENVELOPE LIFTS AND BLOCKS EXCESS SPERM.

Sperm enters egg here

Excess sperm

Fertilization envelope

1. Egg is covered with sperm. One sperm enters.

2. Fertilization envelope begins to lift and clear away excess sperm.

3. Fertilization envelope expands across egg. When complete, all excess sperm are cleared away.

FIGURE 22.7 A Physical Barrier Erected after Fertilization Prevents Polyspermy. (a) During fertilization, a wave of Ca^{2+} begins at the point of sperm entry and spreads throughout the egg. The white dots are from a reagent that reacts with calcium ions. **(b)** In response to increased Ca^{2+} concentrations, a fertilization envelope rises and clears away excess sperm.

triggers a series of remarkable events, including the acrosomal reaction and mechanisms for blocking polyspermy.

Fertilization in Mammals

Although sea urchins have acted as a productive model system, an important question remains: How relevant are these findings to other species? In particular, how similar is sea urchin fertilization to events in humans and other mammals?

The most obvious difference between sea urchin and mammal fertilization is that the process occurs internally in mammals, not externally as in sea urchins. As a result, most female mammals actively choose a mate prior to the sperm-egg interaction. This simple difference means that a species-specific interaction between sperm and egg is much less important in mammals and other species with internal fertilization than it is in sea urchins and other species with external fertilization.

Internal fertilization also makes mammalian fertilization much more difficult to study than fertilization in sea urchins and similar species. With the advent of in vitro fertilization (IVF) methods, however, biologists have finally acquired the ability to study mammalian fertilization under laboratory conditions. What have they learned?

In humans and other mammals, the acrosomal reaction occurs after the sperm's head reaches the equivalent of the sea urchin vitelline envelope—the gelatinous extracellular matrix known as the **zona pellucida**. The enzymes released from the acrosome digest the glycoproteins that make up this protective layer. As a result, the sperm's head is able to reach the egg cell's plasma membrane and fuse with it.

Although as yet there is no convincing evidence for a bindin-like protein on the sperm head in mammals, molecules around and in the zona pellucida interact extensively with sperm. To search for these molecules in mouse eggs, researchers analyzed the three glycoproteins found in the zona pellucida. They found that one of the three glycoproteins binds to the heads of sperm. The interaction between the egg-cell glycoprotein and molecules on the surface of the sperm head causes sperm cells to stick to the zona pellucida. The interaction also triggers the acrosomal reaction.

Because male mammals typically introduce tens of millions of sperm during a single mating, sea urchin and mammal eggs share the problem of preventing polyspermy. Mammal eggs do not produce a fertilization envelope, but enzymes that are released from cortical granules perform a function similar to the sea urchin enzymes that destroy the egg-cell receptor for sperm. More specifically, the mammalian enzymes modify the egg cell glycoprotein in a way that releases sperm bound to the zona pellucida and prevents binding by additional sperm.

Once the sperm nucleus enters the egg, the two haploid nuclei fuse to form a diploid zygote. Development can begin.

Check Your Understanding

If you understand that...

- Fertilization is based on specific interactions between proteins in or around the plasma membranes of animal sperm and egg cells.

You should be able to...

1) Describe the protein-protein interactions that allow sea urchin and mammalian sperm to bind to eggs.

2) Predict the consequences of mutations that change the amino acid sequence of bindin or fertilizin.

3) Explain why it is important for sea urchin and mammal eggs to block polyspermy.

22.3 Cleavage

Cleavage refers to the rapid cell divisions that take place in animals after fertilization. Cleavage is the first step in **embryogenesis**—the process by which a single-celled zygote becomes a multicellular embryo.

In most animals, cleavage partitions the egg cytoplasm without additional cell growth taking place. The zygote simply divides into two cells, then four cells, then eight, and so on, without concurrent growth overall. The process is fast—the fastest cell divisions recorded over an individual's lifetime. In frogs, cleavage produces 37,000 cells in under two days. When cleavage occurs in fruit flies, mitosis can occur every 10 minutes; about 50,000 cells are produced in half a day. The key point about cleavage is that the cytoplasm present in the egg is divided into a larger and larger number of smaller and smaller daughter cells.

The cells that are created by cleavage divisions are called **blastomeres** (literally, "bud-part"). When cleavage is complete, the embryo consists of a mass of blastomere cells called a **blastula** ("little-sprout, bud").

Variation in Cleavage Patterns

The principles of cleavage are the same in all animals: The egg cytoplasm is divided to form many cells that constitute a blastula. As **Figure 22.8** shows, however, the exact pattern of cleavage varies widely among species. In groups that undergo radial cleavage, cells divide at right angles to each other so that they form tiers, as shown in Figure 22.8a. In species that have spiral cleavage, cells divide at oblique angles to each other so that they pile up as shown in Figure 22.8b. (Notice that in these two parts of the figure, a part of the original egg cytoplasm is highlighted in color. This is done to illustrate a key point about cleavage: Certain parts of the egg cytoplasm end up in certain blastomeres.)

(a) Radial cleavage: Cells divide at right angles to each other.

(b) Spiral cleavage: Cells divide at oblique angles to each other.

(c) Discoidal cleavage: A disk of small cells forms on top of a large cell containing the yolk.

Yolk

(d) Superficial cleavage: Mitosis is not immediately followed by cytokinesis, resulting in a multinucleate embryo.

Cells

Nucleus

Yolk

Fertilized egg   Multinucleate embryo   Blastula

FIGURE 22.8 Cleavage Patterns Vary Widely among Species. During cleavage, cells divide but growth does not occur. Just some of the known cleavage patterns are shown here. **(a)** Radial cleavage is typical in echinoderms such as sea urchins. **(b)** Spiral cleavage is typical in segmented worms and mollusks. **(c)** Discoidal cleavage occurs in fishes, lizards, birds, and other species with extremely large amounts of yolk. **(d)** Superficial cleavage occurs in insects and produces multinucleate embryos.

The radial and spiral patterns are not the only variations on the theme of cleavage, however. For example, birds, fish, and other species have eggs with large, membrane-bound structures filled with yolk. In these animals, cleavage does not split the egg completely. Instead, it produces a mound of cells around the yolk or on top of it, as shown in Figure 22.8c. This pattern is called discoidal cleavage. Fruit flies and other insects have another distinctive cleavage pattern, called superficial cleavage. As Figure 22.8d shows, mitosis is not followed by cytokinesis early in cleavage. Instead, a large number of nuclei become scattered through the cytoplasm. Later, the nuclei migrate to the periphery of the embryo and plasma membranes form around them. **Box 22.1** analyzes the events that occur during cleavage in humans.

Although the cleavage patterns in Figure 22.8c are logical to observe in eggs with large, dense yolk deposits—which cannot be divided up efficiently during cytokinesis—the adap-tive significance of the radial, spiral, and insect cleavage patterns is not well understood. Biologists still do not know if the variations in cleavage patterns affect fitness in any way or help individuals adapt to different environments—even though most of the patterns were described well over 100 years ago.

What Role Do Cytoplasmic Determinants Play?

The essence of cleavage is that the egg cytoplasm gets divided into many daughter cells. The key to understanding the process, then, is not to memorize the differences between spiral and radial and other cleavage patterns, but to analyze what is happening to the egg cytoplasm during all types of cleavage.

To drive this point home, think back to the Bicoid protein introduced in Chapter 21. Bicoid is a transcription factor

BOX 22.1 What Happens during Cleavage in Humans?

In humans, a special type of cleavage called rotational cleavage occurs when the fertilized egg is in a part of the female reproductive tract called the fallopian tube, or oviduct (**Figure 22.9**). The **fallopian tube, (oviduct)** is a structure that connects reproductive organs called the ovary and the uterus. The **ovary** is the organ in which the egg matures, and the **uterus** is the organ in which the embryo develops. Fertilization occurs near the ovary, and cleavage occurs as the embryo travels down the length of the fallopian tube toward the uterus. Cleavage results in a specialized type of blastula called a **blastocyst** ("sprout-bag"), which has two major populations of cells. The exterior of the blastocyst is a thin-walled, hollow structure called the **trophoblast** ("feeding-sprout"). The interior of the trophoblast contains a cluster of cells called the **inner cell mass** (**ICM**). After arriving at the uterus, the blastocyst implants into the uterine wall and a structure called the placenta begins to form. The **placenta** is an organ derived from a mixture of maternal cells and trophoblast cells. It allows nutrients and wastes to be exchanged between the mother's blood and the embryo's blood. The ICM, in contrast, contains the cells that undergo gastrulation and develop into the embryo. Later events in human pregnancy and birth are introduced in Chapter 48.

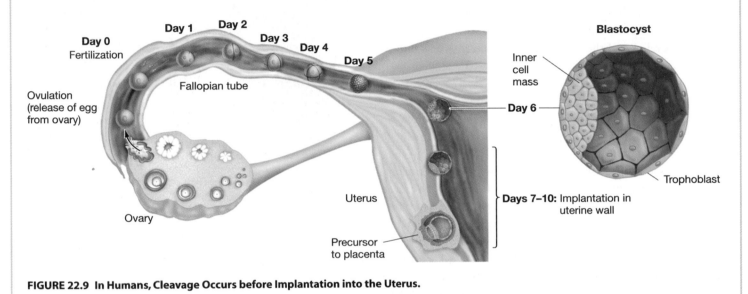

FIGURE 22.9 In Humans, Cleavage Occurs before Implantation into the Uterus.

found in an anterior-posterior concentration gradient. When the nuclei shown in Figure 22.8d form in the early stages of fruit-fly cleavage, only those in the anterior end of the egg are exposed to a high concentration of Bicoid. To extend this observation, consider what would happen if the cytoplasm that is highlighted in color in Figures 22.8a and b contained a key signal or transcription factor—a master regulator analogous to Bicoid. Only the colored cells in the figure would have this molecule.

A molecule that exists in eggs and helps direct early development is called a cytoplasmic determinant. *Bicoid* mRNA is a cytoplasmic determinant; the color highlighting in Figure 22.8a and b represents a cytoplasmic determinant. These examples underscore a key point: ● Cytoplasmic determinants are found in specific locations within the egg cytoplasm, so they end up in specific populations of blastomeres. As a result, only certain regulatory gene cascades are triggered in certain blastomeres. By dividing the egg cytoplasm to precisely distribute cytoplasmic determinants to certain cells, cleavage initiates the step-by-

step process that, in combination with signals received from other cells, results in the differentiation of cells.

Check Your Understanding

If you understand that...

- Cleavage is a series of rapid cell divisions that result in formation of a blastula.

- During cleavage, the egg cytoplasm is divided.

- Because certain regulatory signals and transcription factors are found at specific locations in the egg cytoplasm of most species, different blastomeres end up with different sets of cytoplasmic determinants.

You should be able to...

Explain how the asymmetric distribution of cytoplasmic determinants in blastomeres affects the eventual differentiation of muscle cells, nerve cells, and other cell types.

22.4 Gastrulation

As cleavage nears completion, cell division slows. During the next phase of the developmental sequence, cell proliferation stops being the most important developmental process. Instead, cell movement becomes primary. During **gastrulation**, extensive and highly organized cell movements radically rearrange the embryonic cells into a structure called the gastrula.

Research on this phase of development started in the 1920s, with efforts to document the movement of individual cells during newt and frog gastrulation. These early experiments were based on soaking tiny blocks of agar (a gelatinous compound) with a nontoxic dye. The dyed blocks were then pressed against the surface of blastula-stage embryos, so that a small number of blastomeres became marked with dye. By allowing marked embryos to develop and then examining them at intervals during gastrulation, researchers were able to follow the movement of cells.

Subsequent research has shown that the pattern of gastrulation varies among species of animals almost as much as cleavage patterns do. But like cleavage, the variation in the details is underlain by a common outcome: Gastrulation results in the formation of embryonic tissue layers. A **tissue** is an integrated set of cells that function as a unit. Most adult animals have a wide array of tissues, including muscle tissue, connective tissue, and nerve tissue. But most early animal embryos have three primary tissue layers: (1) ectoderm ("outside skin"), (2) mesoderm ("middle skin"), and (3) endoderm ("inner skin"). These embryonic tissues are called **germ layers** because they give rise to the organs and tissues of the adult. As **Figure 22.10** shows, embryonic **ectoderm** forms the outer covering of the adult body and the nervous system; **mesoderm** gives rise to muscle, most internal organs, and connective tissues such as bone and cartilage; and **endoderm** produces the lining of the digestive tract or gut, along with some of the associated organs.

Figure 22.11 shows how the cell movements of gastrulation result in the formation of the three embryonic tissue layers in a frog embryo. The drawings are based on experiments with dyed blastomeres, and they identify blastomeres according to the type of embryonic tissue they will become. Cells that will become ectoderm are shown in blue; cells destined to form mesoderm are pink, and cells that will form endoderm are colored yellow. These cell fates are determined by cell-cell signals, transcription factors, and cytoplasmic determinants in the blastomeres.

As you start to examine the steps of this figure, notice that the blastula contains a fluid-filled interior space called the **blastocoel** (step 1). In many animal species, the blastula is hollow or partially hollow like this. As gastrulation begins, cells begin moving into this space through an invagination—similar to the indentation that forms when you push a finger

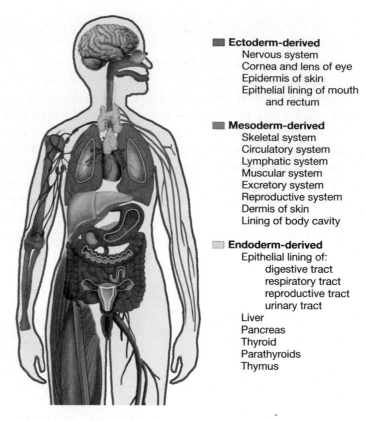

FIGURE 22.10 The Three Embryonic Tissues Give Rise to Different Adult Tissues and Organs.

● **QUESTION** In a gastrula, mesoderm is sandwiched between ectoderm on the outside of the embryo and endoderm on the inside. Relate this observation to the positions of adult tissues and organs derived from mesoderm, ectoderm, and endoderm.

■ **Ectoderm-derived**
Nervous system
Cornea and lens of eye
Epidermis of skin
Epithelial lining of mouth
and rectum

■ **Mesoderm-derived**
Skeletal system
Circulatory system
Lymphatic system
Muscular system
Excretory system
Reproductive system
Dermis of skin
Lining of body cavity

■ **Endoderm-derived**
Epithelial lining of:
digestive tract
respiratory tract
reproductive tract
urinary tract
Liver
Pancreas
Thyroid
Parathyroids
Thymus

into a balloon (step 2). In frogs this invagination is slit-like and eventually forms the **blastopore**. Cells from the periphery continue to move to the interior of the embryo through the blastopore, forming the initial part of the gut or digestive tract (step 3). When gastrulation is complete, the ectoderm, mesoderm, and endoderm cells are arranged in three distinct layers (step 4).

The other major outcome of gastrulation is that the major body axes become visible. In frogs, for example, the blastopore becomes the anus, and the region just above the blastopore (as drawn in Figure 22.11) becomes the dorsal or back side of the embryo. In this way, the anterior-posterior and dorsal-ventral axes of the body become apparent. These axes were partially determined early in development, by Bicoid-like cytoplasmic determinants that were stored in the egg and partitioned into blastomeres during cleavage.

As a comparison with frog gastrulation, the series of micrographs in **Figure 22.12** shows the sequence of events during gastrulation in a sea urchin. ● If you understand the major concepts of gastrulation, you should be able to (1) describe the cell movements taking place in panels 2 and 3 of Figure 22.12, (2) label the three embryonic tissue layers and

DURING GASTRULATION, EMBRYONIC TISSUES FORM DISTINCT LAYERS.   ■ Ectoderm   ■ Mesoderm   □ Endoderm

Cross section
— Blastocoel
— Blastopore

Start of gut

Whole embryo
Blastopore

Blastopore

1. Different regions of the frog blastula contain cytoplasmic determinants (signals or transcription factors) that determine their fate during gastrulation.

2. Gastrulation begins with the formation of an opening—the blastopore—that extends into the embryo. Cells from the surface move into the interior through the blastopore.

3. The blastocoel shrinks as the surface cells continue to move inward, forming the three embryonic tissue layers.

4. The three embryonic tissue layers are formed, ready for organogenesis. The blastopore (future anus in frogs) surrounds a plug of yolk cells.

FIGURE 22.11 Precise Cell Movements Occur during Gastrulation.

● **EXERCISE** In frogs, the initial opening of the blastopore is on what will become the dorsal side of the larva and adult. The blastopore becomes the anus of the larva and adult. On the last drawing, mark where the head, tail, back, and belly will develop.

the anterior-posterior axis visible at the end of gastrulation (panel 5), and (3) label a type of cell movement that occurs in sea urchin gastrulation but not frog gastrulation—the migration of individual cells from the outer cell layer to the interior of the embryo.

Current research on gastrulation is focused on understanding how cells move in such an organized way—on the mechanism responsible for the movement itself as well as how cells navigate. But like the embryo itself, we'll move on to the next phase in the developmental sequence—forming tissues, organs, and other basic structures as the body takes shape.

 Web Animation at www.masteringbio.com
Early Stages of Animal Development

Check Your Understanding

If you understand that...

● Gastrulation consists of extensive cell movements that reorganize the embryonic cells and result in the formation of embryonic germ layers.

● Once gastrulation is complete, the major body axes are visible and organs and other structures can begin to form.

You should be able to...

1) Name the three embryonic germ layers.

2) Describe the tissues and organs that each of these layers gives rise to in the adult.

FIGURE 22.12 The Sequence of Events during Gastrulation Varies among Species. These micrographs show stages in sea urchin gastrulation. Notice that some cells break off from the inner surface of the embryo's outer cell layer and migrate into the interior individually.

22.5 Organogenesis

Organogenesis ("organ-origin") is the process of tissue and organ formation that begins once gastrulation is complete and the embryonic germ layers are in place. During organogenesis, cells proliferate and become differentiated—meaning that they become a specialized cell type. Differentiated cells have a distinctive structure and function because they express a distinctive suite of genes.

As Chapter 21 noted, differentiation is a progressive, step-by-step process. Cells become specified to a fate long before they are differentiated—meaning, before they begin producing products that are specific to a certain cell type. This observation is logical, given that an irreversible commitment to become a particular cell type is the end point of a long and complex sequence of events, mediated by the cascades of interacting cell-cell signals and regulatory transcription factors introduced in Chapter 21.

To explore how organogenesis takes place, let's consider how muscle tissue forms and then how muscle cells become differentiated. What causes undifferentiated mesodermal cells to form muscle tissue and begin producing the muscle-specific proteins that make breathing, walking, and turning the pages of textbooks possible?

Organizing Mesoderm into Somites: Precursors of Muscle, Skeleton, and Skin

Figure 22.13 illustrates some of the key events that occur as organogenesis begins in frogs, humans, and other vertebrates. In this cross section of a frog embryo, the dorsal (back) side of the individual is up and the ventral (belly) side is at the bottom. You should already be familiar with the ectoderm, mesoderm, and endoderm layers shown in step 1, along with the early gut cavity. But you should also note that a new structure has appeared in the dorsal mesoderm—a rod-like element called the **notochord**. This structure is shown in cross section in the figure and runs the length of the anterior-posterior axis of the embryo, just under the dorsal surface. The notochord is unique to the group of animals called the **chordates**, which includes humans and other vertebrates. In some species of chordates, the notochord is a long-lasting structure that functions as a simple internal skeleton—it stiffens the body and makes efficient swimming movements possible. But in species like frogs and humans, the notochord is transient. It appears only in embryos. As organogenesis proceeds, many of the cells in the notochord undergo apoptosis—programmed cell death.

What does the notochord do in these species? The answer is that it functions as a key organizing element during organogenesis. As step 2 in Figure 22.13 shows, cells in or near the notochord produce signaling molecules that induce the ectoderm on the dorsal (back) side of the embryo to fold. More specifically, signals from the notochord trigger massive changes in cytoskeletal elements inside each of the dorsal ectodermal cells. As the cytoskeleton is reorganized, the ectodermal cells extend in length and then constrict at their dorsal end and expand at their ventral end. The constriction above and expansion below makes the sheet of cells fold upward. As folding continues, a structure called the **neural tube** forms (step 3). The neural tube is the precursor of the brain and spinal cord. Besides producing signals that direct formation of the neural tube, the notochord furnishes physical support as the ectodermal cells fold into their final configuration.

Once the neural tube forms, organogenesis continues when nearby mesodermal cells become organized into blocks of tissue called **somites**. Somites form on both sides of the neural tube, down the length of the body (**Figure 22.14**). Somite formation is

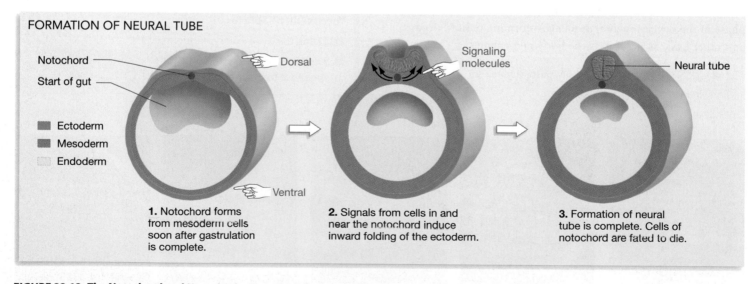

1. Notochord forms from mesoderm cells soon after gastrulation is complete.

2. Signals from cells in and near the notochord induce inward folding of the ectoderm.

3. Formation of neural tube is complete. Cells of notochord are fated to die.

FIGURE 22.13 The Notochord and Neural Tube Form Early in Organogenesis. In vertebrates, the notochord forms from mesoderm cells soon after gastrulation is complete. Molecules produced in or near the notochord induce the formation of the neural tube and other structures along the dorsal (back) side of the embryo.

(a) Surface view of somites

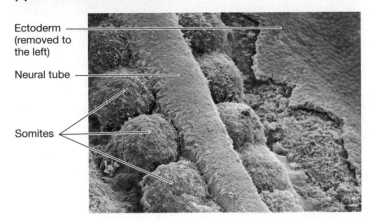

Ectoderm (removed to the left)

Neural tube

Somites

(b) Cross section of somites

Ectoderm

Neural tube

Somite

Notochord

Endoderm

FIGURE 22.14 Somites Form on Both Sides of the Neural Tube. Somites are made of mesodermal cells.

a response to changes in the cell adhesion molecules that keep mesodermal cells attached to each other (see Chapter 8).

Somites are transient structures, just as the notochord is. But by marking cells and following them over time, researchers have found that the mesodermal cells in a somite are destined to form a variety of structures. As development proceeds, somite cells do not die as notochord cells do. Instead, they break away as distinct populations that migrate to their final location in the developing embryo and continue to proliferate. These cell movements are critical to organogenesis. Cells from somites build the vertebrae and ribs, the deeper layers of the skin that covers the back, and the muscles of the back, body wall, and limbs. As **Figure 22.15** shows, the destiny of a somite cell depends on its position within the somite at a particular time during organogenesis.

By transplanting cells from one location to another early after somite formation, researchers have found that initially any cell in a somite can become any of the somite-derived elements of the body. For example, biologists transplanted cells from various locations to the side of the somite farthest from the neural tube. All the transplants eventually became committed to form limb muscles (**Figure 22.16**). Cells transplanted later in develop-

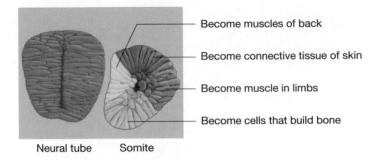

Become muscles of back

Become connective tissue of skin

Become muscle in limbs

Become cells that build bone

Neural tube   Somite

FIGURE 22.15 Cells in the Somite Give Rise to an Array of Tissues. Each somite eventually breaks up into distinct populations of cells, each with a different fate. Within a somite, the position of a cell determines its fate.

ment, however, failed to become the type of cell associated with their new position. Instead, the transplanted cells differentiated into the cell types they would normally form in their original position, even though they were now in a new and inappropriate location.

As the somite matures, then, somite cells become irreversibly determined—meaning that they will eventually differentiate into a specific cell type—based on their location within the somite. Experiments like this illustrate the general observation that a cell's fate becomes fixed in a step-by-step process.

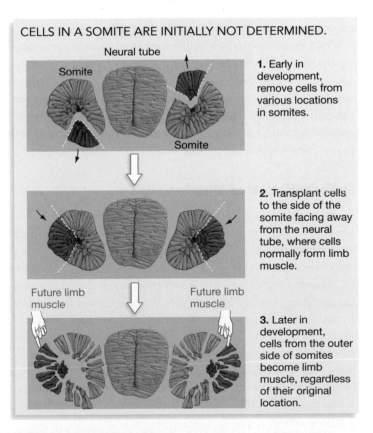

CELLS IN A SOMITE ARE INITIALLY NOT DETERMINED.

Neural tube

Somite

Somite

1. Early in development, remove cells from various locations in somites.

2. Transplant cells to the side of the somite facing away from the neural tube, where cells normally form limb muscle.

Future limb muscle

Future limb muscle

3. Later in development, cells from the outer side of somites become limb muscle, regardless of their original location.

FIGURE 22.16 Experimental Evidence for Progressive Differentiation of Mesodermal Cells in Somites. By transplanting cells early in somite formation to new locations in this structure, researchers showed that the cells are not determined initially.

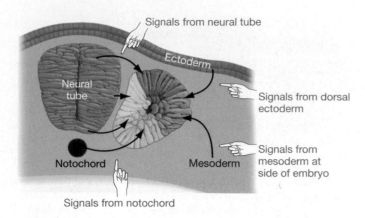

FIGURE 22.17 Somite Cells Differentiate in Response to Signals from Nearby Tissues. Somite cells become determined in response to signals from the neural tube, notochord, and nearby ectoderm and mesoderm.

Recent studies of this process, called **determination**, have shown that distinct populations of cells in a somite differentiate in response to distinct local combinations of cell-cell signals. **Figure 22.17** illustrates how these signals diffuse away from cells in the notochord, the neural tube, and nearby ectoderm and mesoderm to act on specific populations of target cells in the somite. These unique combinations of cell-cell signals direct the movement of somite cells when the structure breaks up and trigger the production of transcription factors required for the expression of muscle-, bone-, or skin-specific proteins.

Organogenesis begins with the formation of embryonic structures such as the notochord, neural tube, and somites. It continues with the formation of bones, muscles, skin, and other organs and tissues found in a larval or juvenile individual. At this point, the cells that make up these structures begin expressing tissue-specific proteins. Let's take a closer look at how this crucial step in organogenesis happens, using muscle cells as an example.

Differentiation of Muscle Cells

In a somite, why do cells located farthest from the neural tube become committed to producing muscle? Harold Weintraub and colleagues answered this question by experimenting with cells called myoblasts. A myoblast is a cell that is determined to become a muscle cell but has not yet begun producing muscle-specific proteins. Specifically, Weintraub and co-workers hypothesized that myoblasts contain at least one regulatory protein that commits them to their fate. Their idea was that myoblasts begin producing this muscle-determining protein after they receive an appropriate set of cell-cell signals from nearby tissues. In effect, they were looking for a regulatory transcription factor in myoblasts that dictates, "I will become a muscle cell."

Figure 22.18 outlines how the biologists went about searching for this hypothetical protein. They began by isolating mRNAs from myoblasts and using reverse transcriptase to con-

Experiment

Question: Why do cells on the outside of somites become committed to producing muscle in response to the signals they receive?

Hypothesis: Cell-cell signals trigger production of a regulatory protein (or proteins) that commits myoblasts to their fate.

Null hypothesis: Myoblasts do not contain a regulatory protein that commits them to their fate.

Experimental setup:

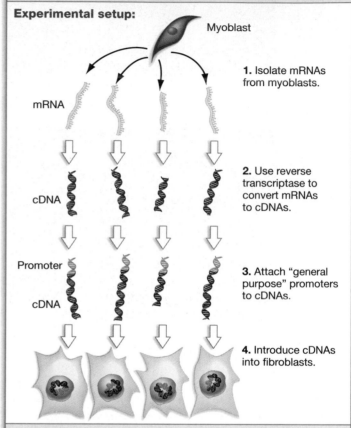

1. Isolate mRNAs from myoblasts.

2. Use reverse transcriptase to convert mRNAs to cDNAs.

3. Attach "general purpose" promoters to cDNAs.

4. Introduce cDNAs into fibroblasts.

Prediction: One of the myoblast-derived cDNAs will convert fibroblasts into cells that produce muscle-specific proteins.

Prediction of null hypothesis: None of the myoblast-derived cDNAs will convert fibroblasts into cells that produce muscle-specific proteins.

Results:

Muscle-like cell

One of the fibroblasts begins producing muscle-specific proteins

Conclusion: Cells on the outside of somites contain a regulatory protein (later called MyoD) that commits them to differentiate into muscle.

FIGURE 22.18 A Gene That Causes Muscle-Cell Differentiation.

● **QUESTION** Why did the researchers have to attach a "general purpose" promoter to the cDNAs?

vert the mRNAs to cDNAs (steps 1 and 2). Because myoblasts transcribe genes required for a muscle cell to function, these cDNAs represented muscle-specific genes. Then the biologists attached a type of promoter to the cDNAs that would ensure the genes' expression in any type of cell (step 3). Finally, they introduced the recombinant genes into non-muscle cells called fibroblasts and monitored the development of the transformed cells (step 4). Just as predicted, one of the myoblast-derived cDNAs converted fibroblasts into muscle-like cells. Follow-up experiments showed that the same gene could convert pigment cells, nerve cells, fat cells, and liver cells into cells that produced muscle-specific proteins.

Weintraub's group called the protein product of this gene **MyoD**, for *my*oblast *d*etermination. Follow-up work showed that the *MyoD* gene encodes a regulatory transcription factor and that the MyoD protein binds to enhancer elements located upstream of muscle-specific genes. In addition, the MyoD protein activates expression of the *MyoD* gene. This was a key observation because it meant that once *MyoD* is turned on, it triggers its own expression—meaning that the gene continues

to be transcribed. Other researchers have found that genes closely related to *MyoD* are also required for the differentiation of muscle cells.

To put this specific example into context, think back to the sequence of events occurring in early development and to the principles of development introduced in Chapter 21. ● To understand why the muscle cells in your bicep became muscle cells, you have to start with an egg inside your mother. After fertilization triggered the onset of cleavage, certain cells in the blastocyst began producing signals that triggered regulatory gene cascades—changes in gene expression that led specific cells to become mesoderm in your back. During gastrulation, these cells moved into position. Early in organogenesis, signals from the notochord and nearby cells induced the production of MyoD and other muscle-determining proteins in certain populations of cells from somites. In response, these target cells were committed to becoming muscle and moved into your upper arm as it formed. Later, the MyoD-containing cells began expressing muscle-specific proteins. All of these steps made it possible for you to move your arms after birth. The rest, as they say, is history.

Chapter Review

SUMMARY OF KEY CONCEPTS

● **In most animals, fertilization depends on specific interactions between proteins on the plasma membranes of sperm and egg.**

Animal development begins with the formation of gametes in parents. Sperm cells contribute a haploid genome and a centriole to the embryo. Eggs, in contrast, contribute a large amount of cytoplasm to the embryo in addition to a haploid genome and a centriole. The egg cytoplasm usually contains cytoplasmic determinants and a large supply of nutrients.

Fertilization is the best known of all cell-cell interactions, primarily because of observational and experimental studies on sea urchin eggs and sperm. When a sea urchin sperm contacts the jelly layer surrounding the egg, enzymes released from the sperm digest the jelly layer and bindin proteins on the sperm head attach to sperm-binding receptors on the egg. Once the sperm nucleus enters the egg, a fertilization envelope lifts off the egg cell's plasma membrane and protects against multiple fertilizations.

You should be able to draw what the "lock and key" interaction between bindin and fertilizin might look like, and explain why interactions between specific amino acids in the two proteins would cause binding to occur. ●

● **The earliest cell divisions divide the fertilized egg into a mass of cells. The fate of these cells depends on the cytoplasmic determinants they contain, the signals they receive, or a combination of both.**

Early embryonic development begins with cleavage—a series of cell divisions that divide the egg cytoplasm into a large number

of cells. During cleavage, an array of signals and regulatory transcription factors are apportioned to different blastomeres. As a result, different blastomeres have different fates.

You should be able to describe how Bicoid mRNA is partitioned into cells during cleavage in a fruit-fly embryo. ●

● **Early in development, cells undergo massive, coordinated movements to form distinct tissue layers. Each of these layers gives rise to distinct groups of tissues and organs in the adult.**

Once cleavage is complete, the embryo consists of a mass of cells. Studies with dyes and other tools have shown that embryonic cells undergo dramatic movements during gastrulation. These movements arrange the embryonic tissues into endoderm, mesoderm, and ectoderm layers and make the back-to-belly and head-to-tail axes of the body visible.

(MB) **Web Animation** at www.masteringbio.com
Early Stages of Animal Development

● **As development proceeds, the formation of specialized organs and other structures depends on cell-cell signals, cell proliferation, cell movements, and differentiation. Differentiation is complete when cells express tissue-specific proteins.**

Tissues, organs, and other structures form during the developmental phase called organogenesis. Early in vertebrate organogenesis, cells in a transient structure called the notochord release cell-cell signals that induce the formation of a neural tube—precursor to the brain and spinal cord. Blocks of mesodermal cells, called somites, form next to the neural tube. In response to cell-cell signals from the notochord, neural tube, and other nearby tissues, cells in the somites move to new positions, proliferate, and begin expressing tissue-specific proteins.

You should be able to describe how the regulatory gene and signaling cascades described in Chapter 21 relate to the step-by-step process of differentiation described in this chapter. ●

QUESTIONS

● Test Your Knowledge

1. How are the vitelline envelope of sea urchins and the zona pellucida of mammals similar?
   a. Both are gelatinous coats that protect the egg.
   b. Both hold cortical granules, which block polyspermy.
   c. Both hold stores of Ca^{2+}, which block polyspermy.
   d. Both are an extracellular matrix that sperm bind to and digest.

2. What happens during the acrosomal reaction?
   a. Bindin binds to the egg-cell receptor for sperm.
   b. The sperm and egg cells' plasma membranes fuse.
   c. Enzymes that digest the egg's jelly layer are released, and microfilaments in the tip of the sperm head polymerize to form a protrusion.
   d. Cortical granules fuse with the plasma membrane, triggering formation of the fertilization envelope.

3. What happens during cleavage?
   a. The neural tube—precursor of the spinal cord and brain—forms.
   b. Basal and apical cells—precursors of the suspensor and embryo, respectively—form.
   c. The fertilized egg divides without growth occurring, forming a mass of cells.
   d. Massive movements of cells make the primary body axes visible and organize the three embryonic tissues.

4. What happens during gastrulation?
   a. The neural tube—precursor of the spinal cord and brain—forms.
   b. Basal and apical cells—precursors of the suspensor and embryo, respectively—form.
   c. The fertilized egg divides without growth occurring, forming a ball of cells.
   d. Massive movements of cells make the primary body axes visible and organize the three embryonic tissues.

5. In animals, which adult tissues and organs are derived from ectoderm?
   a. lining of the digestive tract and associated organs
   b. blood, heart, kidney, bone, and muscle
   c. nerve cells and skin
   d. blastopore and blastocoel

6. During organogenesis in vertebrates, which of the following does *not* occur?
   a. establishment of the anterior-posterior axis
   b. differentiation of cells
   c. movement of cells into new positions
   d. extensive cell-cell signaling

Test Your Knowledge answers: 1.d; 2.c; 3.c; 4.d; 5.c; 6.a

● Test Your Understanding

Answers are available at www.masteringbio.com

1. Draw a typical animal sperm and egg and label the major structures. List the contributions to offspring that are made by a sperm versus an egg.

2. Why is it logical to observe that sperm-egg interactions are extremely species specific in sea urchins but much less so in mammals?

3. Blastomeres look identical. Explain why they are not.

4. Describe three elements in the cascade of interacting cell-cell signals and regulatory transcription factors that lead to differentiation of muscle cells in human embryos.

5. Why are ectoderm, mesoderm, and endoderm called germ layers?

6. Give an example of how each of the four fundamental developmental processes—cell proliferation, cell movements, cell-cell signals, and differentiation—plays a role in animal development.

● Applying Concepts to New Situations

Answers are available at www.masteringbio.com

1. At the level of the amino acid, explain why bindin on the sperm of the sea urchin *Strongylocentrotus purpuratus* attaches to the egg-receptor for sperm on eggs of *S. purpuratus*, but why bindin from other species of *Strongylocentrotus* cannot.

2. A molecule called α-amanitin inhibits translation. When researchers treat fruit-fly eggs with this compound, the early stages of cleavage proceed normally after fertilization. What can you conclude from this experiment?

3. In the marine organisms called sea squirts, eggs contain a yellow pigment. Blastomeres that contain this yellow pigment become muscle cells. State a hypothesis to explain this observation.

4. Would a molecule that bound to the glycoproteins in the human zona pellucida serve as an effective contraceptive? Explain why or why not.

www.masteringbio.com is also your resource for • Answers to text, table, and figure caption questions and exercises • Answers to *Check Your Understanding* boxes • Online study guides and quizzes • Additional study tools including the *E-Book for Biological* Science 3rd ed., textbook art, animations, and videos.

An Introduction to Plant Development

23

The red cells are an *Arabidopsis thaliana* embryo at the 8-cell stage; the purple cells are part of the seed coat. This chapter explores how plant embryos develop into adults that continue to grow throughout life and reproduce.

KEY CONCEPTS

- Several aspects of plant development contrast sharply with animals: Plants develop continuously, commit certain cells to gamete production late in development, and produce gametes by mitosis.

- In flowering plants, double fertilization results in the production of a zygote and a nutritive tissue that supports embryogenesis.

- Embryogenesis results in the formation of the major body axes and three types of embryonic tissue.

- Vegetative development is based on meristems where cell division occurs throughout life, producing cells that go on to differentiate.

- When a meristem is converted from vegetative to reproductive development, regulatory transcription factors are activated that control the position and identity of floral organs.

About the year 2700 B.C., perhaps less than a century before the first pyramids were constructed in ancient Egypt, a remarkable thing came to pass on a mountainside in southwestern North America. A sperm cell and an egg cell from a bristlecone pine tree fused to form a zygote. The offspring produced by that blessed event has been growing and developing ever since—for over 4700 years.

● Unlike many animals, plants continue to grow and develop throughout their lives, whether that life lasts two weeks or thousands of years. As a dramatic example of this continuous development, the entire aboveground shoot system of many plant species dies each fall and grows back the next spring. One hallmark of plant development is that the essential developmental processes of cell proliferation and expansion, differentiation, and cell-cell interaction occur throughout life.

In addition, most plant cells retain the ability to de-differentiate—to stop producing a certain suite of tissue-specific proteins and begin producing proteins typical of a different type of tissue. Under the right circumstances, for example, a photosynthetic cell in a plant stem can de-differentiate and give rise to cells that function as storage cells in a root. But none of your muscle cells or nerve cells will ever give rise to liver cells. Certain aspects of animal development can be flexible, but animal development is rigid and inflexible in comparison with plants. This extraordinary flexibility is another hallmark of plant development.

The striking differences existing between plant and animal development should not be surprising, given their very different ways of life. Most animals move around to find food and mates while plants stand their ground, make their own food, and use

● Key Concept   Important Information   ● Practice It

FIGURE 23.1 *Arabidopsis thaliana* **Is an Important Model Organism in Developmental Biology.**

● **QUESTION** What are the attributes of a model organism? In what sense is *Arabidopsis* a "model"?

water, wind, or insects to carry their gametes. In addition, multicellularity evolved independently in the two lineages. Chapter 8 reviewed evidence to back the claim that multicellularity has evolved many times during the history of life. That chapter also analyzed some consequences of multicellularity evolving independently in plants and animals, including striking differences in the chemical composition of plant versus animal extracellular matrices and the mechanisms of cell-cell contact and adhesion.

The goal of this chapter is to explore the sequence of events occurring as a plant develops from a fertilized egg into an adult

that can reproduce, and that continues to grow and develop new tissues and organs throughout its life. The discussion in this chapter focuses on the flowering plants called eudicots (see Chapter 30), because they are the most species-rich and abundant group of plants. In many cases our model organism will be a small flowering plant called *Arabidopsis thaliana*, which qualifies as the best-studied of all plants (**Figure 23.1**). *Arabidopsis* has been a successful model organism in genetics and developmental biology because it is relatively easy to grow in the greenhouse, produces large numbers of offspring, and completes its entire life cycle in as little as six weeks. Even though their life span is short and their body size is small, individuals are complex enough to be interesting to study.

A plant's life story begins with gamete formation and fertilization (**Figure 23.2**). In *Arabidopsis* and other flowering plants, an egg is fertilized inside a protective, womb-like structure called the **ovule**. Development continues inside the ovule with **embryogenesis** (literally, "embryo-origin"). In many plant species embryogenesis ends with the maturation of the ovule into a **seed**—a structure that contains the embryo and a supply of nutrients surrounded by a protective coat. The embryo can remain in a nongrowing state inside the seed for months, years, or in some cases, centuries. When conditions are favorable, however, the seed undergoes **germination**—meaning that it resumes growth—to form a seedling. Organogenesis continues throughout life, forming the three vegetative organs: roots, leaves, and stems. **Vegetative organs** are the nonreproductive portions of the plant body. Later, cells in the stem will become converted to reproductive structures, initiating gamete production and the sexual phase of the plant life cycle.

A common theme should emerge as you study the steps in this sequence. Despite numerous differences between plant and animal development, both are underlain by the common prin-

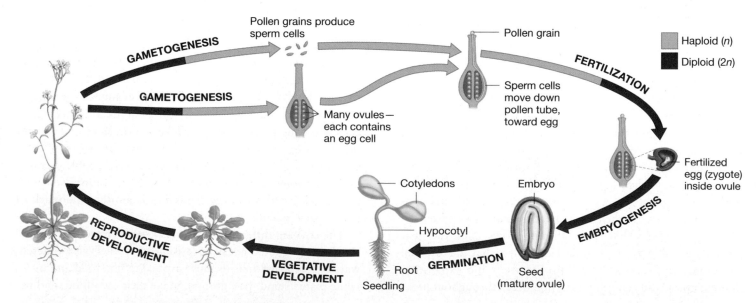

FIGURE 23.2 An Overview of Development in *Arabidopsis*. This chapter traces the events in the life of a plant, from fertilization and embryogenesis to vegetative and reproductive development.

ciples described in Chapter 21. As plant cells proliferate, they divide in precise orientations and expand in specific directions. In some cases the results are analogous to those achieved by the cell movements that occur during animal development. Developing plant cells also communicate constantly via cell-cell signals, and they differentiate due to specific combinations of cell-cell signals and regulatory transcription factors. This unity of underlying principles is one of the great results emerging from research on plant developmental biology.

23.1 Pollination and Fertilization

The story of pollination and fertilization starts with the production of a pollen grain. Inside a mature flower, diploid cells undergo meiosis to form haploid cells that give rise to pollen grains. More specifically, the haploid cells divide by mitosis to form a two-celled structure, which becomes encased in a tough, protective coat. As **Figure 23.3a** shows, the resulting **pollen grains** are tiny, multicellular individuals. One of the cells inside the pollen grain will give rise to a sperm cell that fertilizes the egg.

Pollen grains are carried to a mature flower by wind, water, insects, bats, birds, or some other agent. If the pollen grains land on a structure inside the flower called the stigma, one of the best-studied cell-cell interactions in all of plant biology takes place. For the pollination and fertilization process to continue successfully, proteins on the surface of the pollen grain have to interact with proteins on the surface of the stigma (**Figure 23.3b**). Like bindin attaching to the egg-cell receptor for sperm in sea urchins (see Chapter 22), these interactions prevent fertilization from occurring between members of different species. In many cases, the interactions also prevent self-fertilization—the union of a sperm and egg from the same individual (see Chapter 40 for details on these cell-cell interactions).

If the protein-protein interactions on the surface of the stigma are appropriate, the pollen begins to grow and a **pollen tube** extends toward the egg (**Figure 23.3c**). Recent research suggests that this growth is guided by cell-cell signals released from the egg at the base of the female reproductive structure, or **carpel**. Before the pollen tube is completed, one of the cells inside the pollen grain divides by mitosis to form two sperm cells.

If you think about that last sentence for a moment, you will realize that this aspect of gametogenesis is strikingly different from the pattern in animals. ● In plants, sperm and egg cells are produced from haploid cells via mitosis—not from diploid cells via meiosis, as in animals. This aspect of the plant life cycle is explored in detail in Chapter 30; the details of plant gametogenesis are analyzed in detail in Chapter 40.

Another dramatic contrast with animal development occurs when the pollen tube reaches the base of the carpel. ● The two sperm cells pass through the wall of the ovule and enter a structure that contains the egg cell and a maternal cell that in many species contains two haploid nuclei. One sperm fuses with the egg to form the diploid zygote, while the other sperm

(a) Pollen grains are microscopic but multicellular.

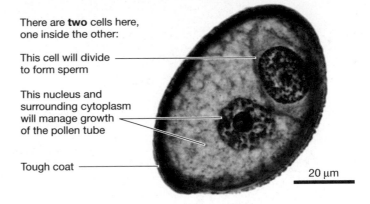

There are **two** cells here, one inside the other:

This cell will divide to form sperm

This nucleus and surrounding cytoplasm will manage growth of the pollen tube

Tough coat

20 μm

(b) Pollen grains interact with the stigma.

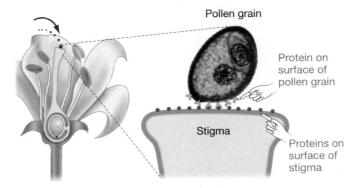

Pollen grain

Protein on surface of pollen grain

Stigma

Proteins on surface of stigma

(c) Plant sperm move to the egg through a pollen tube.

Germinated pollen grain

Pollen tube

Sperm cells

Pollen tube nucleus

Ovule

20 μm

FIGURE 23.3 The Fertilization Process Begins When Pollen and Stigma Interact. In many species, sperm cells are produced by mitosis after the pollen grain germinates and the pollen tube begins growing toward the egg.

● **QUESTION** How does the interaction between pollen and stigma in angiosperms compare to the interaction between bindin and fertilizin in sea urchins?

(a) How does double fertilization occur?

(b) Products of double fertilization in a wheat seed

Ovule

Pollen tube carries two sperm (*n*) to ovule

One sperm fuses with two maternal nuclei to form triploid cell (3*n*)

One sperm fuses with egg to form zygote (2*n*)

Double fertilization

Seed

Endosperm (3*n*)

Embryo (2*n*)

1 mm

FIGURE 23.4 Double Fertilization Leads to the Formation of Endosperm. (a) In flowering plants, a sperm (*n*) fuses with two or more haploid maternal nuclei near the egg to form a cell that gives rise to endosperm tissue. In many species, endosperm cells are 3*n*. **(b)** Endosperm is a nutritive tissue packed with proteins and carbohydrates, and fats or oils.

fuses with the cell with two maternal haploid nuclei to form a triploid (3*n*) cell. This event is known as double fertilization (**Figure 23.4a**).

The triploid cell divides repeatedly by mitosis to form a triploid nutritive tissue called endosperm. **Endosperm** ("inside-seed") provides the proteins, carbohydrates, and fats or oils required for embryonic development, **seed germination**, and early seedling growth. In species with large seeds, the endosperm grows into a sizeable nutrient reservoir as the ovule matures (**Figure 23.4b**). When you eat wheat, rice, corn, or other grains, you are eating primarily endosperm. Functionally, endosperm is analogous to the yolk found in animal eggs.

Once double fertilization is complete, the stage is set for embryogenesis—the early development of a new individual.

Check Your Understanding

If you understand that...

- Fertilization is based on specific interactions between proteins on the surfaces of plant pollen grains and stigmas.
- In flowering plants, double fertilization results in the production of a zygote and a (usually) triploid cell that grows into a nutritive tissue.

You should be able to...

1) Explain why protein-protein interactions between pollen grains and stigmas are important to the individual's ability to reproduce. Your answer should consider what might happen if the protein-protein interactions did *not* occur.

2) Describe when meiosis and mitosis occur during the development of a sperm cell in *Arabidopsis*.

23.2 Embryogenesis

In flowering plants like *Arabidopsis,* embryogenesis takes place inside the ovule as the seed is maturing. In essence, embryogenesis produces a tiny, simplified precursor to a mature plant. The process is equivalent to the cleavage, gastrulation, and organogenesis phases of early animal development introduced in Chapter 22.

Let's first look at the sequence of events, which was worked out through careful observation of different-aged embryos. Then we can consider more research on the genetics of embryogenesis—studies that are focused on finding and characterizing the genes that make this elaborate process work.

What Happens during Embryogenesis?

Figure 23.5 illustrates the key events in embryogenesis, using *Arabidopsis* as a model organism. After fertilization, the zygote undergoes a highly asymmetric cell division. The cells resulting from this initial division are unlike in size, content, and fate. The bottom, or basal cell, is large and is dominated by an extensive vacuole. It gives rise to a column of cells called the suspensor, which anchors the embryo as it develops. The small cell above the basal cell, called the apical cell, is rich in cytoplasm and gives rise to the mature embryo.

The asymmetries in the basal and apical cell help establish one of the primary axes of the plant body: the **apical-basal axis.** **Apex** refers to the tip; **basal** refers to the base or foundation.

As Figure 23.5 shows, the basal cell divides perpendicular to the apical-basal axis, giving rise to the suspensor. Only one cell in the suspensor—the one closest to the apical cell—contributes

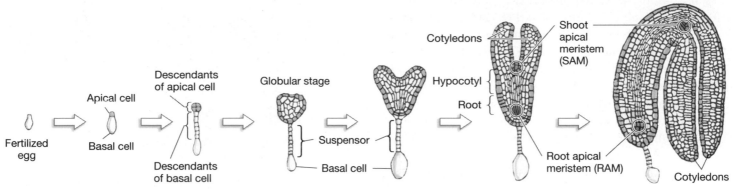

FIGURE 23.5 The Events of Embryogenesis. Embryogenesis takes place inside the developing seed.

● **EXERCISE** Label the apical-basal and radial axes on the globular-stage embryo.

cells to the embryo and thus to the mature adult. The apical cell, in contrast, divides both perpendicularly to the apical-basal axis and parallel to it. These divisions give rise to a simple ball of cells at the tip of the suspensor. The embryo is said to be in the globular stage, and it includes distinctive-looking cells that cover the exterior. Thus, there is now a difference between cells in the interior of the embryo and cells on the surface. In this way, the second major body axis becomes visible during the globular stage. The radial axis extends from the interior of the body out to the exterior.

These initial events in embryogenesis illustrate two general, and critically important, points about plant development: (1) The shape of the body depends on how the planes of cell division are oriented and on the direction of subsequent cell expansion, and (2) the fate of a plant cell can be summed up in the old quip about the three keys to success in real estate—"location, location, location." Starting with the initial division that creates the apical and basal cells, plant cells differentiate based on where they are in the body.

As the ball of cells continues to grow and develop, the initial leaves, or **cotyledons**, begin to take shape. The cotyledons are connected to the developing root by a stem-like structure called the **hypocotyl** (**Figure 23.6**). Together, the cotyledons and hypocotyl make up the **shoot**, which will become the aboveground portion of the body. The shoot system functions in photosynthesis and reproduction. The **root**, in contrast, forms the belowground portion of the body. The root system anchors the individual and functions as a water- and nutrient-gathering structure.

Once the apical-basal and radial axes are established and as the cotyledons, hypocotyl, and root begin to take shape, groups of cells called the **shoot apical meristem (SAM)** and **root apical meristem (RAM)** form (Figure 23.5). A **meristem** consists of undifferentiated cells that divide repeatedly, with some of their daughter cells becoming specialized cells. Meristem cells are analogous to the stem cells found in animals, except that plant meristems are much more flexible. The root meristem can

form all the underground portions of the plant, and the shoot meristem can form all of the aerial portions, including reproductive structures. Throughout a plant's life, meristematic tissues continue to produce embryonic cells that can differentiate into adult tissues and structures.

Note that all of this growth and development takes place without the aid of the cell migration seen in gastrulation in animals. Because plant cells have stiff cell walls, they do not move. For the cotyledons and other embryonic structures to take shape, then, cell divisions must occur in precise orientations, and the resulting cells must exhibit differential growth—meaning that some cells grow larger than others.

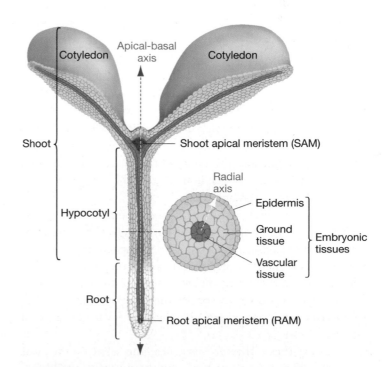

FIGURE 23.6 Embryogenesis Establishes the Two Body Axes and Three Embryonic Tissues. The apical-basal axis is the long axis of the individual; the radial axis extends from the center toward the exterior.

In addition to establishing the two body axes, early development in *Arabidopsis* produces three embryonic tissues. As Figure 23.6 shows, the embryo contains an **epidermis** ("over-skin"). This is an outer covering of specialized cells that protects the individual. Inside the epidermal layer of cells is **ground tissue**, a mass of cells that may later differentiate into cells that are specialized for photosynthesis, food storage, or other functions. The **vascular tissue** in the center of the plant will eventually differentiate into specialized cells that transport food and water between root and shoot. In the embryo, the three tissue systems are arranged in a radial pattern. These embryonic tissues make up the radial axis and are analogous to the ectoderm, mesoderm, and endoderm of developing animals (see Chapter 22).

FIGURE 23.7 Mutant *Arabidopsis* Embryos with Misshapen Bodies. Researchers have identified *Arabidopsis* mutant individuals missing certain defined sections of the body along the apical-basal axis.

● **QUESTION** How do these mutants compare and contrast with Bicoid mutants in fruit flies?

Which Genes and Proteins Set Up Body Axes?

The genetic approach to development that was pioneered with research on *Drosophila melanogaster* has also proven to be a powerful way of studying plant embryogenesis. Although the specific genes involved are different in plants than in animals, the basic logic of how genes direct the earliest events in development is similar in both.

Research on the genetics of early development in plants was pioneered by Gerd Jürgens and colleagues. This research group set out to identify genes that are transcribed in the zygote or embryo of *Arabidopsis* and that are responsible for establishing the apical-basal axis of the plant body. It's no surprise that this effort was similar to the work on anterior-posterior pattern formation in *Drosophila* introduced in Chapter 21. Jürgens had participated in the work with flies.

The biologists' initial goal was to identify individuals with developmental defects at the seedling stage. More specifically, they were looking for mutants that lacked particular regions along the apical-basal axis of the body. The team succeeded in finding several bizarre-looking mutants (**Figure 23.7**). Individuals called apical mutants lacked the first leaves, or cotyledons. Some individuals lacked the embryonic stem, or hypocotyl, and were named central mutants. Individuals dubbed basal mutants lacked both hypocotyls and roots.

To interpret these results, the researchers suggested that each type of *Arabidopsis* mutant had a defect in a different gene and that each gene played a role in specifying the position of cells along the apical-basal axis of the body. They hypothesized that these genes are analogous to the segmentation genes of fruit flies, which specify the identity of cells within well-defined regions along the anterior-posterior axis of insects.

What are these *Arabidopsis* genes, and what do they do? To answer these questions, consider the gene responsible for the mutants lacking hypocotyls and roots. This gene has been cloned and sequenced and named *MONOPTEROS*. Because its DNA sequence indicates that its protein product has a DNA-binding domain, *MONOPTEROS* is hypothesized to encode a transcription factor that regulates the activity of target genes. The MONOPTEROS protein, in turn, is manufactured in response to signals from **auxin**. Auxin is a cell-to-cell signal molecule that is produced in the shoot apical meristem and transported toward the basal parts of the individual. (For more detail on auxin's prominent role as a long-distance signal, or **hormone**, see Chapter 39). The concentration of auxin along the apical-basal axis of a plant forms a concentration gradient that provides positional information, not unlike the Bicoid concentration gradient found in fruit-fly embryos (see Chapter 22).

The take-home message of these results is that the auxin signal is part of a regulatory cascade that triggers the production of MONOPTEROS and other regulatory transcription factors specific to cells in the developing hypocotyl and roots. Although the genes and proteins involved in forming the cotyledons, hypocotyl, and root of *Arabidopsis* are not yet understood in as much detail as the segmentation genes of *Drosophila* are, several important similarities are clear. Both developmental pathways are based on cell-to-cell signals and regulatory cascades that result in the step-by-step specification of a cell's position and fate.

Many questions remain about embryogenesis in *Arabidopsis*. How is auxin production turned on as the shoot apical meristem first begins to form in embryos? Once production of the *MONOPTEROS* gene product begins, what target genes are affected? What genes other than *MONOPTEROS* are found in the regulatory cascade responsible for development along the apical-basal axis? Are any of them active in adults as well? These questions present a host of interesting challenges to current researchers.

Check Your Understanding

If you understand that...

- Early embryonic development results in the formation of the apical-basal and radial axes of the plant body.
- The early structures of the shoot and root systems form along the body's apical-basal axis.
- The genes responsible for setting up the body axes are currently the focus of intense research.

You should be able to...

1) Give an example of how an asymmetric cell division affects the fate of daughter cells.

2) Give an example of how variation in the plane of cell division affects the fate of daughter cells.

3) Predict the effect on the *MONOPTEROS* gene of adding auxin to embryonic root cells.

4) Predict the effect of adding a drug that blocks auxin transport away from the shoot apical meristem in embryos.

23.3   Vegetative Development

For a plant to thrive, it has to adjust to constantly changing conditions. Consider just one such condition—the availability of light. You might recall from Chapter 10 that plants use wavelengths in the blue and red portions of the spectrum to drive photosynthesis. Now suppose that you are an oak tree with a life expectancy of 300 years. The quality and quantity of light that your leaves receive depends on where you happened to germinate. Are you growing on flat ground? With a southern exposure in full sun? With a northern exposure rarely exposed to full sun? In addition, from the time you emerge from an acorn to the time of your death, the light you receive will depend on changes in climate and weather as well as on your size relative to the size and proximity of plants that compete with you for light and shade. Finally, the leaves in your bottommost branches experience a different light regime from the leaves at your apex.

How do plants cope with all this variation in their living conditions? Unlike most animals, they don't move around until they find a place that suits their requirements. Instead, a large part of the answer is the continuous growth and development of roots, stems, and leaves. If an oak tree is heavily shaded on one side, it stops growing in that direction and extends branches on the other side. If it is heavily shaded on all sides, its growth is directed upward. ● This constant adjustment to changing environmental conditions is possible because of the meristems that exist at the tips of shoots and roots.

Meristems Are Responsible for Continuous Growth and Development

When embryonic development is complete, the basic body axes are established and the initial structures in the root and shoot systems have formed. For the rest of the individual's life, further development is driven by the meristems. As **Figure 23.8a** shows, meristematic tissue exists at each tip in the shoot and root sys-

(a) Apical meristems are located at specific points throughout the body.

Shoot meristems

Root meristems

(b) Close-up of a shoot apical meristem

Developing leaves

Rapidly dividing, undifferentiated meristematic cells

Cells differentiating into **ground tissue**

Cells differentiating into **vascular tissue**

Cells differentiating into **epidermal tissue**

0.1 mm

FIGURE 23.8 Meristems Are "Growing Points" Where Development Takes Place.
(a) Each tip in the root and shoot system contains a meristem. The individual can grow in any direction that a meristem exists. **(b)** When meristem cells divide, the daughter cells either remain undifferentiated and continue to function as meristem cells or differentiate into new epidermal, ground, or vascular cells.

tems. As a result, the individual is capable of growing in any direction aboveground or belowground, depending on conditions.

Figure 23.8b provides a close-up view of a shoot apical meristem, or SAM. The cells within the meristem are small and undifferentiated. Within each meristem, the rate and direction of cell growth are dictated by cell-cell signals that are produced in response to environmental cues like the arrival of spring, the presence of abundant water, and the amount of light striking the plant. Just below the meristem, daughter cells produced by mitosis and cytokinesis in the meristem initially differentiate into epidermal, ground, or vascular tissue. Eventually these cells will differentiate into specialized cell types.

Careful microscopy allowed biologists to tease out the sequence of events that occur as meristems grow, and intense research continues to explore how interactions between auxin and other cell-cell signals influence the fate of cells produced by meristems. But recently researchers have also turned to the question of which genes respond to these cell-cell signals. Specifically, which genes and gene products direct the formation of specific structures during vegetative development? Let's consider one particularly well-studied example—the genetic control of shape in developing leaves.

Which Genes and Proteins Determine Leaf Shape?

Leaves can be induced to grow by applying tiny amounts of auxin to a SAM. Based on this observation, it appears clear that the initiation of a leaf depends on the concentration of auxin in parts of a SAM, although other types of cell-cell signals are undoubtedly involved as well. Once a leaf begins to grow, the next key event in its development is the formation of three axes: the proximal-distal, lateral, and upper-lower axes shown in **Figure 23.9**. Proximal is toward the main body while distal is away from the main body; the lateral axis runs from the middle of a leaf toward its margin. The amount and direc-

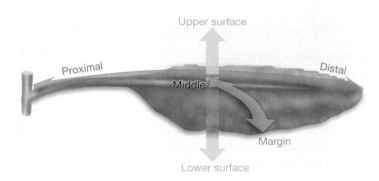

FIGURE 23.9 Leaves Have Three Axes. Overall, the plant body has just two axes: apical-basal and radial. Every leaf has the three axes shown here.

● **QUESTION** How do the body axes in plants compare and contrast with the body axes in animals?

tion of growth along these axes determines the shape of the leaf.

Recent research has begun to identify the genes responsible for specifying these three leaf axes. For example, analyses of mutant individuals in *Arabidopsis* and other species has shown that a gene called *PHANTASTICA* (abbreviated *PHAN*) is critical in setting up the upper-lower axis of leaves. The protein product of *PHAN* has a DNA-binding domain and acts as a regulatory transcription factor. *PHAN* triggers the expression of genes that cause cells to form the upper surface of leaves and suppresses transcription of genes required for forming the lower leaf surface. It is part of a regulatory cascade that begins with auxin and other cell-cell signals and ends with the growth of a normal-shaped leaf.

Recent research on tomatoes also suggests that changes in *PHAN* expression may underlie at least some of the evolutionary changes observed in leaf shape. Leaf shape varies widely among species (see Chapter 36). Simple leaves consist of a single blade, but as **Figure 23.10a** shows, tomatoes have compound leaves—a single leaf blade divided into smaller units

(a) Normal leaves—compound

(b) Transgenic leaves (reduced *PHAN* expression)—simple or palmately compound

FIGURE 23.10 Changes in Genes That Establish Leaf Axes Can Change Leaf Shape. (a) Normal tomato leaves consist of a series of leaflets. **(b)** When the expression of *PHAN* is reduced experimentally, leaves become cup shaped (left) or have leaflets arising from a single point. [Reprinted by permission from *Nature* 424 (July 24, 2003), p. 439, by Kim, M. et al. ©2004 Macmillan Publishers Ltd. Photographs supplied by Dr. Neelima Sinah.]

● **QUESTION** Suppose that *PHAN* were overexpressed. What type of leaf do you predict would form?

called leaflets. Other species have palmately compound leaves, meaning that leaflets radiate from a single point.

To explore whether changes in *PHAN* might have a role in the evolution of various leaf shapes, a team of biologists used techniques introduced in Chapter 19 to create transgenic tomato plants. In these individuals, the *PHAN* gene product was blocked to a moderate or large extent. As **Figure 23.10b** shows, leaf shape in the transgenic individuals changed dramatically. Some of the individuals had cup-shaped (simple) leaves, while others were palmately compound. Although it is still uncertain why changing the specification of the upper-lower leaf axis affects overall shape, it appears clear that *PHAN* expression plays an important role.

These results have inspired the hypothesis that at least some evolutionary changes in leaf size and shape are due to mutations that created new alleles of genes that regulate *PHAN* expression. Alleles that resulted in lowered *PHAN* expression might lead to simple leaves with a single blade; alleles that increased the extent of *PHAN* expression might result in compound leaves like those of normal tomatoes. By experimentally altering the genes that regulate development, researchers are beginning to understand the genetic changes leading to novel types of leaves. If this research continues to be productive, biologists will have another example of how changes in regulatory pathways that direct development underlie evolutionary change (see Chapter 21).

Check Your Understanding

If you understand that...

- Vegetative development is continuous, and vegetative growth can occur in all directions.
- Vegetative growth and development occur in ways that increase an individual's ability to survive and produce offspring in the face of changing environmental conditions.
- Continuous development is possible in plants because meristematic tissue is present at the tips of each component of the root and shoot system.

You should be able to...

1) Make a sketch of a flowering plant and label the locations of SAMs and RAMs. Make another sketch of what the same plant would look like after extensive growth, given that a cow defecated near its left side.

2) Sketch the three axes of a leaf, and give an example of how changes in *PHAN* expression can result in changes in overall leaf shape.

23.4 Reproductive Development

Among the many startling contrasts observed in animal and plant development, one of the most important concerns the development of reproductive tissues and organs. In animals, the cells that give rise to sperm and egg cells are set aside or sequestered early in development. These **germ cells** migrate to the ovaries or testes once the reproductive organs have developed. As a result, the cells that give rise to animal gametes undergo relatively few rounds of mitosis—perhaps 20 to 50—prior to meiosis and gametogenesis.

○ In contrast, plants do not have germ cells that are set aside early in development. Instead, flowering and gametogenesis occur when a SAM converts from vegetative development to reproductive development. As a result, meristematic cells that have divided thousands of times can give rise to the reproductive organs of plants, and eventually to sperm and eggs. Because mutations occur during each cell cycle, plants generate much more genetic variation via mutation than animals do. Although biologists have only begun to explore the consequences of this fact, research on the mechanisms responsible for the formation of reproductive structures has been intense. To introduce this huge body of work, let's consider some highlights of research on *Arabidopsis*.

The Floral Meristem and the Flower

When specialized proteins in *Arabidopsis* sense that days are getting longer and the temperature is favorable, they trigger the production of signals that convert SAMs from vegetative to reproductive development. A **floral meristem** is a modified SAM that produces flowers containing the reproductive organs instead of vegetative structures. The genes that take part in the regulatory cascade responsible for the maturation of a floral meristem are now well characterized.

As an *Arabidopsis* flower develops, the floral meristem produces four organs arranged as shown in **Figure 23.11**: (1) sepals, (2) petals, (3) stamens, and (4) carpels. Each of these organs is a modified leaf. **Sepals** are located around the outside of the flower and help protect it. In some species, sepals are also colorful and function in attracting pollinators. Inside the sepals is a whorl or circular arrangement of **petals**, which enclose the male and female reproductive organs. If insects or other animals pollinate the species in question, the petals may be colored to help

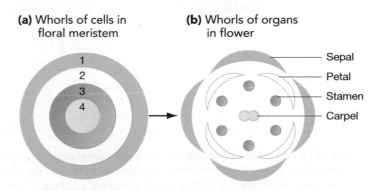

(a) Whorls of cells in floral meristem

(b) Whorls of organs in flower

Sepal
Petal
Stamen
Carpel

FIGURE 23.11 Flowers Are Composed of Four Organs. (a) In a floral meristem, whorls of cells produce **(b)** the four whorls of floral organs.

advertise the reproductive structures. The pollen-producing organs, or **stamens**, are located in a whorl inside the petals. In the center of the entire structure are egg-producing reproductive organs, or **carpels**. The question is, How does the floral meristem produce these four organs in the characteristic pattern of whorls within whorls?

The first hint of an answer came in the late 1800s, when researchers realized that several types of mutant flowering plants—including some that were popular garden plants—were homeotic. In the mutant individuals, one kind of floral organ was replaced by another, even though the total number and relative position of each organ did not change. For example, one homeotic mutant had flowers with sepals, petals, another ring of petals, and carpels instead of having sepals, petals, stamens, and carpels. The mutant phenotype was similar to the transformation of segments later observed in *Drosophila* homeotic mutants (see Chapter 21).

Just as an analysis of homeotic mutants in fruit flies triggered a breakthrough in understanding the genetic control of body axis formation and segmentation in animals, an analysis of homeotic floral mutants in *Arabidopsis* triggered a breakthrough in understanding the genetic control of flower structure in plants.

Which Genes Control the Development of Flowers?

Over 100 years after floral homeotic mutants were first described, Elliot Meyerowitz and colleagues assembled a large collection of *Arabidopsis* individuals with homeotic mutations in flower structure. The researchers' goal was to identify and characterize the genes responsible for specifying the four floral organs.

Meyerowitz's group found that the mutants could be sorted into the three general classes shown in **Figure 23.12**. The classes were distinguished by the type of homeotic transformation that occurred. Some mutants had only carpels and stamens; others had only sepals and carpels; still others had only petals and sepals. The key observation was that each type of mutant lacked the elements found in two of the four whorls.

What was going on? To begin searching for an answer, the biologists hypothesized that each of the three classes of homeotic mutants was due to a defect in a single gene. Meyerowitz then reasoned that if three genes are responsible for setting up the pattern of a flower, then the mutants suggested a hypothesis for how the three gene products interact. Because he referred to the three hypothetical genes as *A*, *B*, and *C*, his hypothesis is called the ABC model.

Three basic ideas underlie the ABC model. The first is that each of the three genes involved is expressed in two adjacent whorls. The second idea is that if each gene is expressed in two adjacent whorls, then a total of four different combinations of gene products can occur. The final idea is that each of these four combinations of gene products triggers the development of a different floral organ. Specifically, Meyerowitz proposed that (1) the A protein alone causes cells to form sepals, (2) a combination of A and B proteins sets up the formation of petals, (3) B and C combined specify stamens, and (4) the C protein alone designates cells as the precursors of carpels.

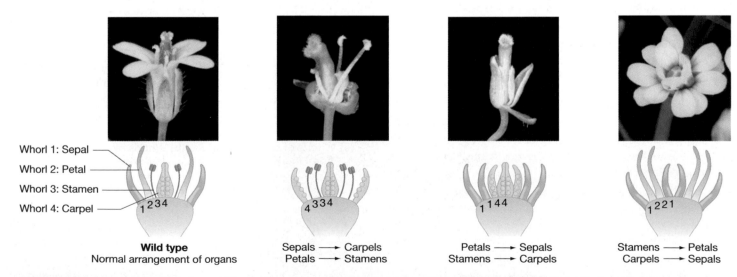

Whorl 1: Sepal
Whorl 2: Petal
Whorl 3: Stamen
Whorl 4: Carpel

1 2 3 4   4 3 3 4   1 1 4 4   1 2 2 1

Wild type
Normal arrangement of organs

Sepals → Carpels
Petals → Stamens

Petals → Sepals
Stamens → Carpels

Stamens → Petals
Carpels → Sepals

FIGURE 23.12 Homeotic Mutants in *Arabidopsis* Flowers Have Structures in the Wrong Locations. The photos show a normal (wild type) individual and three types of mutants. In each type of homeotic floral mutant, two adjacent whorls of organs are transformed into different organs. Sepals and petals, petals and stamens, or stamens and the carpel are transformed. The numbers indicate the wild-type whorls from outermost (1) to innermost (4).

(a) The ABC model

Idea 1: Three genes pattern the flower. Each gene is expressed in two adjacent whorls.

Idea 2: A total of four different combinations of gene products occurs.

Idea 3: Each of the four combinations of gene products triggers the development of a different floral organ.

Assumption: Presence of gene A product inhibits expression of gene C. Presence of gene C product inhibits expression of gene A.

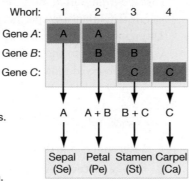

FIGURE 23.13 The ABC Model for Genetic Control of Flower Development. The ABC model is a hypothesis to explain why three types of homeotic mutants exist.

● **QUESTION** Why is it necessary to propose that the *A* gene product inhibits expression of the *C* gene, and that the *C* gene product inhibits expression of the *A* gene?

(b) Testing the ABC model

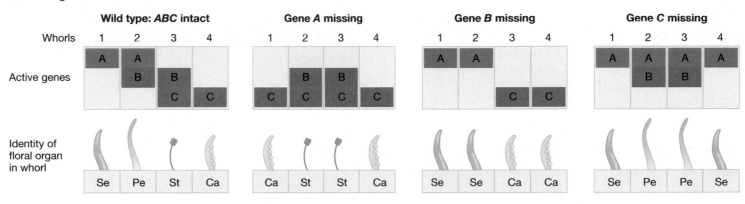

Does this model explain how the three classes of homeotic mutants occur? The answer is yes, if two additional elements are added to the model: The presence of the A protein inhibits the production of the C protein, and the presence of the C protein inhibits the production of the A protein (**Figure 23.13a**). Then the patterns of gene expression and mutant phenotypes in **Figure 23.13b** correspond. For example, if the *A* gene is disabled by mutation, then it no longer inhibits the expression of the *C* gene and all cells produce the C protein. As a result, cells in the outermost whorl express only C protein and develop into carpels, while cells in the whorl just to the inside produce B and C proteins and develop into stamens. ● If you understand the ABC model, you should be able to explain why biologists did not hypothesize that four genes are involved—one that specifies each of the four floral organs—and why they proposed that each gene involved is expressed in two adjacent whorls.

Although the model is plausible and appeared to explain the data, it needed to be tested directly. To accomplish this, Meyerowitz and co-workers mapped the genes responsible for the mutant phenotypes and identified the appropriate DNA

sequences. Once they had isolated the genes, they were able to obtain and use single-stranded DNAs to perform in situ hybridizations (see Chapter 21). The goal was to document the pattern of expression of the *A*, *B*, and *C* genes and see if that pattern corresponded to the model's predictions. As anticipated, the mRNAs for each of the three genes showed up in the sets of whorls predicted by the model. The *A* gene is expressed in the outer two whorls, the *B* gene is expressed in the middle two whorls, and the *C* gene is expressed in the inner two whorls.

This result strongly supported the validity of the ABC model. Just as different combinations of *Hox* gene products specify the identity of fly segments, different combinations of floral identity genes specify the parts of a flower.

Similarities with principles of animal development did not end there. When Meyerowitz and others analyzed the DNA sequences of the floral organ identity genes, they discovered that all three genes contained a segment that coded for a long sequence of 58 amino acids, called the **MADS box**, that binds to DNA. Based on this observation, the researchers hypothesized that the floral genes encode regulatory transcription factors that bind to enhancers or other regulatory sequences and trigger the expression of genes required for sepal, petal, carpel, and stamen formation. The discovery highlighted remarkable similarities between the *MADS-box* genes and the homeotic genes found in *Drosophila* and other animals. The *Hox* genes of animals have a region called the homeobox that is functionally similar to the MADS-box: It encodes a DNA-binding domain in Hox proteins. Both *Hox* and *MADS-box* gene products are parts of regulatory cascades that lead to the specification of important structures. *Hox* genes regulate the expression of genes responsible for forming limbs or other structures in specific parts of the body; *MADS-box* genes regulate the expression of genes responsible for forming flowers. Both sets of genes tell cells where they are in the body and can produce homeotic mutants if they do not work properly. Although different genes are involved, the logic of how to put a multicellular body together is similar in plants and animals.

Biologists are currently working to identify the genes targeted by the ABC proteins, just as they are working to identify the genes controlled by the *Hox* products in animals. Eventually, researchers hope to understand the complete regulatory cascade from initiation of the floral meristem to the expression of proteins that are specific to petals or other floral organs. Ultimately, the goal is to explore how changes in this cascade could have led to the evolution of the spectacular diversity of flowers observed today.

Check Your Understanding

If you understand that...
- Reproductive development begins when SAMs are converted to floral meristems.
- Development of the four floral organs depends on the expression of regulatory transcription factors encoded by the *A*, *B*, and *C* class genes.

You should be able to...
1) Describe a mechanism whereby an A protein could inhibit expression of a *C* gene.
2) Compare and contrast the *MADS-box* genes of plants with the *Hox* genes of animals.

Chapter Review

SUMMARY OF KEY CONCEPTS

● **Several aspects of plant development contrast sharply with that of animals: Plants develop continuously, commit certain cells to gamete production late in development, and produce gametes by mitosis.**

The striking differences between plant and animal development are not surprising, given their different ways of life and the observation that multicellularity evolved independently in the two groups. Surprising similarities exist in plant and animal development, however: (1) Fertilization is mediated by protein-protein interactions between cells produced by each parent, (2) the body axes are established early in development, (3) three types of embryonic tissues give rise to the array of adult tissues and organs, (4) homeotic genes exist in both lineages, (5) meristem cells and stem cells are similar in function, and (6) development is based on cell-cell signals that trigger and modulate cascades of regulatory transcription factors.

You should be able to explain the similarities between auxin and Bicoid; and between the epidermal, vascular, and ground tissues of embryonic plants and the endoderm, mesoderm, and ectoderm of animals. ●

● **In flowering plants, double fertilization results in the production of a zygote and a nutritive tissue that supports embryogenesis.**

The pollination and fertilization process begins with interactions between proteins on the surface of pollen grains and proteins on the surface of the stigma and continues with growth of the pollen tube and production of sperm cells. In flowering plants, one sperm fertilizes the egg to form a zygote while a second sperm fuses with a (usually) diploid cell to form a triploid cell. The triploid cell goes on to produce the nutritive tissue called endosperm.

You should be able to diagram the relationship between the zygote and endosperm nucleus right after double fertilization, and explain the relationship between an embryo and endosperm in a mature wheat seed. ●

● **Embryogenesis results in the formation of the major body axes and three types of embryonic tissue.**

The earliest cell divisions in embryogenesis establish the apical-basal and radial axes of the individual. Later, the embryonic structures called the cotyledons, hypocotyl, and root develop and the embryonic epidermal, ground, and vascular tissues form.

You should be able to sketch a seedling and label the apical-basal and radial axes, cotyledons, hypocotyl, and root. ●

● **Vegetative development is based on meristems where cell division occurs throughout life, producing cells that go on to differentiate.**

Plants can grow continuously because meristematic tissue exists at each tip in the root and shoot system. Continuous growth and development allows plants to respond to environmental conditions that change throughout their life.

You should be able to explain why plants don't have just one SAM and one RAM, at their apex and base, in terms of their ability to grow and reproduce. ⬤

⬤ **When a meristem is converted from vegetative to reproductive development, regulatory transcription factors are activated that control the position and identity of floral organs.**

Meristems that undergo vegetative development can be converted to reproductive development as plants respond to changes in day length or other environmental cues. Once a floral meristem is established, combinations of regulatory transcription factors encoded by *A*, *B*, and *C* class genes interact to produce the flower's sepals, petals, stamens, and carpels.

You should be able to explain why individuals with defective alleles at the *B* gene are considered homeotic mutants. ⬤

QUESTIONS

⬤ Test Your Knowledge

1. What is the fate of the two cells found inside pollen grains prior to germination?
   a. One cell directs development of the pollen tube; the other gives rise to sperm cells via mitosis.
   b. One cell directs development of the pollen tube; the other gives rise to sperm cells via meiosis.
   c. One cell fertilizes the egg; the other fuses with a diploid cell to form triploid endosperm.
   d. One cell initiates germination by interacting with proteins on the surface of the stigma; the other gives rise to sperm cells.

2. Which of the following does *not* represent a contrast between plant and animal development?
   a. Under certain conditions, plant cells can "de-differentiate" readily.
   b. The fate of a cell is determined in part by its location in the embryo.
   c. Germ cells are set aside early in development.
   d. Plant cells do not move.

3. What evidence suggests that auxin concentrations help determine where leaves form near SAMs?
   a. Auxin is produced in SAMs and transported from there toward the root.
   b. Auxin is present in a concentration gradient, with higher concentrations apically and lower concentrations basally.
   c. Auxin concentrations are relatively constant along the radial axis of the body, and leaves form along the radial axis.
   d. Addition of small quantities of auxin to a SAM can induce leaf development.

4. Which of the following does *not* occur during embryogenesis?
   a. Formation of the radial axis.
   b. Production of the suspensor.
   c. Formation of the cotyledons and hypocotyl.
   d. Formation of the lateral and proximal-distal axes.

5. When does the apical-basal axis first become apparent?
   a. When the epidermal, ground, and vascular tissues form.
   b. When the cotyledons, hypocotyl, and root form.
   c. When the first cell division produces the apical cell and basal cell.
   d. During the globular stage, when the suspensor is complete.

6. What evidence suggests that changes in the way that the *PHAN* gene is expressed could be partly responsible for evolutionary changes in leaf shape?
   a. If *PHAN* gene expression is manipulated experimentally, individuals produce leaf types found in different species.
   b. Experiments have shown that *PHAN* plays a role in establishing the upper-lower surface axis in leaves.
   c. Sequencing studies and other data have shown that *PHAN* encodes a regulatory transcription factor.
   d. All plant species surveyed to date have a gene homologous to *PHAN*.

Test Your Knowledge answers: 1. a; 2. b; 3. d; 4. d; 5. c; 6. a

⬤ Test Your Understanding

Answers are available at www.masteringbio.com

1. Make a set of sketches showing the sequence of events occurring between the time a pollen grain lands on a stigma and fertilization occurs. Note where important protein-protein interactions occur, and indicate how cell-cell signals allow sperm and egg to communicate. Explain what is double about double fertilization.

2. Sketch the two major body axes found in plants and the three axes present in a leaf. Compare and contrast these axes with the three body axes of animals.

3. Compare and contrast the stem cells of animals with the meristems of plants. How are these cells similar? How are they different?

4. In what sense are the epidermal, ground, and vascular tissues produced in the SAMs and RAMs of a 300-year-old oak tree "embryonic"?

5. When in situ hybridization experiments documented where *A*, *B*, and *C* class genes were expressed in developing *Arabidopsis* flowers, it was considered strong support for the ABC model. Explain why. Also, explain the significance of the observation that the *A*, *B*, and *C* class genes all encode a MADS-box domain.

6. Give an example of how each of the four fundamental developmental processes—cell proliferation, cell expansion, cell-cell signals, and differentiation—plays a role in plant development.

Applying Concepts to New Situations

1. Transcription factors with MADS boxes also occur in fungi and animals. But the *MADS-box* genes of fungi and animals are not active in the formation of complex structures like flowers or segments, and they do not produce homeotic mutations when they fail to function properly. State a hypothesis to explain why the function of homologous DNA sequences like the *MADS-box* genes can be so different in plants versus these other groups.

2. When growth occurs in an unlimited or unrestricted way, it is said to be indeterminate. But when growth is of limited duration and then stops, it is said to be determinant. Which process is observed in vegetative development and which in reproductive development? Explain your logic.

3. Convergent evolution is said to occur when similar structures or processes are observed in different species, but when the similar structures and processes are not due to homology—meaning, not due to shared ancestry. Stated another way, convergence is due to the independent evolution of similar traits. Explain why the existence of homeotic mutants in both plants and animals could be considered a form of convergent evolution. In this case, what would cause convergence?

4. Make a sketch showing the locations of SAM and RAM in a young oak tree. Suppose that a concrete sidewalk gets poured on the left side of the tree, preventing water penetration below. A building goes up on the right side, cutting off light from that direction. But after a few years, an underground water pipe just next to the building begins to leak, providing abundant water year round. Draw the expected locations of SAM and RAM in the same tree 50 years after your initial sketch. Explain your logic.

www.masteringbio.com is also your resource for • Answers to text, table, and figure caption questions and exercises • Answers to *Check Your Understanding* boxes • Online study guides and quizzes • Additional study tools including the *E-Book for Biological Science* 3rd ed., textbook art, animations, and videos.

Reading Graphs

Graphs are the most common way to report data, for a simple reason. Compared to reading raw numerical values in the form of a table or list, a graph makes it much easier to understand what the data mean.

Learning how to read and interpret graphs is one of the most basic skills you'll need to acquire as a biology student. As when learning piano or soccer or anything else, you need to understand a few key ideas to get started and then have a chance to practice—a *lot*—with some guidance and feedback.

Getting Started

To start reading a graph, ask yourself three questions:

1. *What do the axes represent?* As **Figure BS1.1a** shows, the horizontal axis of a graph is also called the *x*-axis or the abscissa. The vertical axis of a graph is also called the *y*-axis or the ordinate. Each axis represents a quantity that varies over a range of values. These values are indicated by the ticks and labels on the axis. In our example, the *x*-axis represents the width of broccoli stalks, in centimeters, while the *y*-axis represents the number of individuals in the sample. Note that the units being represented on each axis should *always* be clearly labeled or implied.

    To create a graph, researchers plot the independent variable on the *x*-axis and the dependent variable on the *y*-axis. The *independent* and *dependent* terms are appropriate because the values on the *y*-axis depend on the *x*-axis values. In our example, the researchers wanted to show the distribution of flowering stalk sizes in a large sample of individuals. Thus, the number of individuals present depended on the size of the flowering stalk. (This is similar to the way you would plot a graph showing the heights of students in your class.)

    In many graphs in biology, the independent variable is either time or the various treatments used in an experiment. In these cases, the *y*-axis records how some quantity changes as a function of time or as the outcome of the treatments applied to the experimental cells or organisms. The value on the *y*-axis depends on the value on the *x*-axis, but not vice versa.

(a) Read the axes—what's being plotted?

(b) Look at the bars or data points—what do they represent?

This bar means that there were 8 broccoli plants in the sample with a flowering stalk between 2.5 cm and 3.0 cm wide

(c) What's the punchline?

There are a few very narrow-stalked and a few very wide-stalked individuals in this sample, but most individuals have intermediate stalks about 2.0 cm wide

FIGURE BS1.1 Histograms Are a Common Way to Graph Data.

● **EXERCISE** Draw what this histogram would look like if almost all the individuals in the sample had flowering stalks between 3.0 and 3.5 cm wide.

● **EXERCISE** Suppose you measured the width of flowering stalks in a large sample of individuals over time—measuring the same individuals every week for 10 weeks. Draw a scatter plot predicting how the average width of flowering stalks might change over time.

2. *What do the bars or the data points represent?* Most of the graphs in this text are histograms like the one shown here, or bar charts—which are similar to histograms, but plot data that have discrete values instead of a continuous range of values—or scatter plots, where individual data points are plotted. Once you've read the axes, you need to figure out what each bar in the bar chart or each data point in the scatter plot represents and how the researchers determined it. In our broccoli stalk example, the bars in **Figure BS1.1b** represent the number of individuals in the sample with flowering stalks of a certain width. There was one individual with a stalk between 0 and 0.5 cm wide, two individuals with a stalk between 0.5 and 1.0 cm wide, and so on.

3. *What is the overall trend or message?* Look at the data as a whole, and figure out what they mean. **Figure BS1.1c** suggests an interpretation of the broccoli stalk example. If a bar chart plots values from different treatments in an experiment, ask yourself if these values are the same or different. If a scatter plot shows how some quantity changes over time, ask yourself if that quantity is increasing, decreasing, or staying the same.

Getting Practice

Working with this text will give you lots of practice with reading graphs—they appear in almost every chapter. In many cases we've put a little pointing hand, like your instructor's hand at the whiteboard, with a label that suggests an interpretation or draws your attention to an important point on the graph. In other cases, you should be able to figure out what the data mean on your own or with the help of other students or your instructor.

Make an effort to develop this skill. One of the most common complaints from professors who teach upper-level courses is, "My students can't read graphs!" When you become competent and confident at interpreting graphs, your performance will improve in this course, in subsequent courses, and on professional and graduate school admission tests.

Reading a Phylogenetic Tree

Phylogenetic trees show the evolutionary relationships among species, just as a genealogy shows the relationships among people in your family. They are unusual diagrams, however, and it can take practice to interpret them correctly.

To understand how evolutionary trees work, consider **Figure BS2.1**. Notice that a phylogenetic tree consists of branches, nodes, and tips. Branches represent populations through time. Nodes (also called forks) occur where an ancestral group splits into two or more descendant groups (see point B in Figure BS2.1). If more than two descendant groups emerge from a node, the node is called a polytomy (see node C). Tips (also called terminal nodes) are the tree's endpoints, which represent groups living today or a dead end—a branch ending in extinction. The names at the tips can represent species or larger groups such as mammals or conifers. Recall from Chapter 1 that a taxon (plural: *taxa*) is any named group of organisms. A taxon could be a single species, such as *Homo sapiens*, or a large group of species, such as Primates. Groups that occupy adjacent branches on the tree are called sister taxa.

The phylogenetic trees used in this text are all rooted—meaning the bottom, or most basal, node on the tree is the most ancient. To determine where the root on a tree occurs, bi-

ologists include one or more out-group species when they are collecting data to estimate a particular phylogeny. An out-group is a taxonomic group that is known to have diverged prior to the rest of the taxa in the study. In Figure BS2.1, "Taxon 1" is an out-group to the monophyletic group consisting of taxa 2–6. A monophyletic group consists of an ancestral species and all of its descendants. The root of a tree is placed between the out-group and the monophyletic group being studied. This position in Figure BS2.1 is node A.

Understanding monophyletic groups is fundamental to reading and estimating phylogenetic trees. Monophyletic groups may also be called lineages or clades and can be identified using the "one-snip test": If you cut any branch on a phylogenetic tree, all of the branches and tips that fall off represent a monophyletic group. Using the one-snip test, you should be able to convince yourself that the monophyletic groups on a tree are nested. In Figure BS2.1, for example, the monophyletic group comprising node A and taxa 1–6 contains a monophyletic group consisting of node B and taxa 2–6, which includes the monophyletic group represented by node C and taxa 4–6.

To put all these new terms and concepts to work, consider the phylogenetic tree in **Figure BS2.2**, which shows the relationships

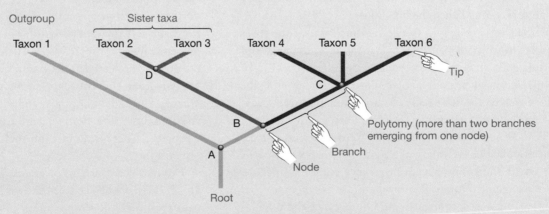

FIGURE BS2.1 Phylogenetic Trees Have Roots, Branches, Nodes, and Tips.
● **EXERCISE** Circle all nine monophyletic groups present.

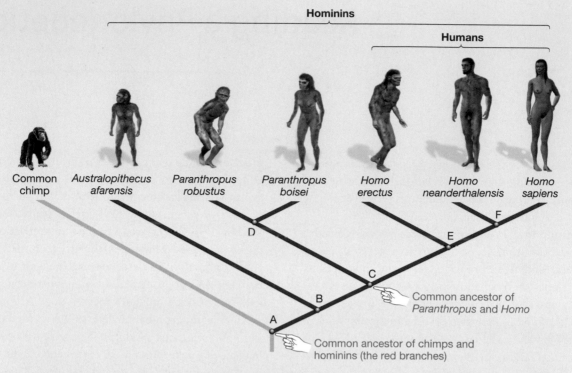

FIGURE BS2.2 An Example of a Phylogenetic Tree. A phylogenetic tree showing the relationships of species in the monophyletic group called hominins.

● **EXERCISE** All of the hominins walked on two legs—unlike chimps and all of the other primates. Add a mark on the phylogeny to show where upright posture evolved, and label it "origin of walking on two legs." Circle and label a pair of sister species. Circle the monophyletic group called hominins. Label an out-group to the monophyletic group called humans (species in the genus *Homo*).

between common chimpanzees and six human and humanlike species that lived over the past 5–6 million years. Chimps functioned as an out-group in the analysis that led to this tree, so the root was placed at node A. The branches marked in red identify a monophyletic group called the hominins.

To practice how to read a tree, put your finger at the tree's root and begin to work your way up. At node A, the ancestral population split into two descendant populations. One of these populations eventually evolved into today's chimps; the other gave rise to the six species of hominins pictured. Now continue moving your finger up the tree until you hit node C. It should make sense to you that at this splitting event, one descendant population eventually gave rise to two *Paranthropus* species, while the other became the ancestor of humans—species in the genus *Homo*. If multiple branches emerge from a node, creating a polytomy, it would be because the populations involved split from one another so quickly that it is not possible to tell which split off earlier or later.

As you study Figure BS2.2, you should consider a couple of important points. First, there are many equivalent ways of

drawing this tree. For example, this version shows *Homo sapiens* on the far right. But the tree would be identical if the two branches emerging from node E were rotated 180°, so that the species appeared in the order *Homo sapiens, Homo neanderthalensis, Homo erectus*. Trees are read from root to tips, not from left to right. Second, no species on any tree is any higher or lower than any other. Groups that branch off close to the root are referred to as basal, and groups that branch off further from the root are more derived. *Australopithecus afarensis* is a basal hominin and *Homo erectus* is a more derived hominin, but both species are at tips—*Homo erectus* is not higher. There is, in fact, no such thing as a higher or lower organism.

Figure BS2.3 presents a chance to test your tree-reading ability. Five of the six trees shown in this diagram are identical in terms of the evolutionary relationships they represent. One differs. You should be able to answer the question posed in the figure's caption and then, on tree (a), circle three nested monophyletic groups and label a node, a branch, a tip, and the out-group.

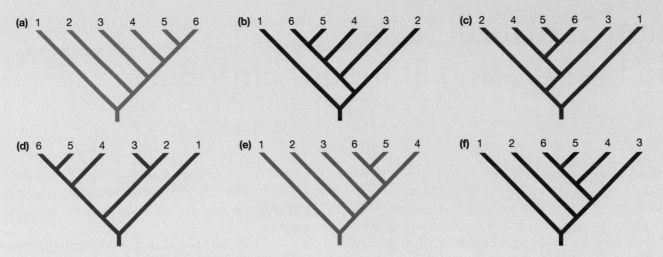

FIGURE BS2.3 Alternative Ways of Drawing the Same Tree.

◉ QUESTION Five of these six trees describe exactly the same relationships among taxa 1 through 6. Identify the tree that is different from the other five.

Answer: The unlike tree is (d).

Using Statistical Tests and Interpreting Standard Error Bars

When biologists do an experiment, they collect data on individuals in a treatment group and a control group (or several such comparison groups). Then they want to know whether the individuals in the two (or more) groups are different. For example, Chapter 1 introduced an experiment in which researchers fed three types of fruits to two different types of fruit-eating organisms—cactus mice and birds called thrashers—and measured how much of each type of fruit the animals ate.

Figure BS3.1 graphs the data for cactus mice and thrashers. The treatments are plotted on the *x*-axes as hackberry (H), non-pungent chilies (NP), and pungent chilies (P). The heights of the bars on the graphs indicate the average percentage of each type of fruit consumed by the 5 mice and 10 thrashers tested. The thin "I-beams" on each bar indicate the standard error of each average. The standard error is a quantity that indicates the uncertainty in the calculation of an average. For example, if two of the mice ate all of the hackberry offered, two ate none, and one ate half, then your estimate of how much an average mouse consumes would be 50 percent. The standard error of that average would be very large, though, because the amounts that the mice ate ranged from 0 to 100 percent. In contrast, if two of the mice ate 49 percent of the hackberry, two ate 51 percent, and one ate half, the average would still be 50 percent but the standard error would be very small. In effect, the standard error quantifies how confident you are that the average you've calculated is a good estimate of the true average—the average you'd observe if you tested all cactus mice in the world under the conditions used in the experiment.

Once they had calculated these averages and standard errors, the biologists wanted to know the answers to two questions: Do cactus mice eat all three types of fruit equally? And, do thrashers eat all three types of fruit equally?

After looking at the data, you might conclude that cactus mice ate different amounts of the fruits offered—specifically, that they ate much more hackberry than non-pungent chilies or pungent chilies. You might be less sure about suggesting that thrashers ate less hackberry and more of the non-pungent and pungent chilies. But how could you come to conclusions like this rigorously, instead of subjectively?

The answer is to use a statistical test. The first step in a statistical test is to specify the null hypothesis, which is that there is no difference among the groups. The second step is to calculate a test statistic, which is a number that characterizes the size of the difference among the groups. In this case, the test statistic compares the actual differences in amount of fruit consumed to the differences predicted by the null hypothesis. The third step is to determine the probability of getting a test statistic as large as the one calculated just by chance. The answer comes from a reference distribution—a mathematical function

FIGURE BS3.1 Standard Error Bars Indicate the Uncertainty in an Average.

that specifies the probability of getting various values of the test statistic if the null hypothesis is correct. (If you take a statistics course, you'll learn which reference distributions are relevant to different types of data.)

You are very likely to see small differences among treatment groups just by chance—even if no differences actually exist. If you flipped a coin 10 times, for example, you are unlikely to get exactly five heads and five tails, even if the coin is fair. A reference distribution tells you how likely you are to get each of the possible outcomes of the 10 flips if the coin is fair, just by chance.

In this case, the reference distribution indicated that if the null hypothesis of no actual fruit preference is correct, you would see differences as large as those observed for mice only 0.8 percent of the time just by chance. For thrashers, though, you would expect to see differences as large as those observed in the experiment 22 percent of the time, just by chance, if the

null hypothesis is correct. By convention, biologists consider a difference among treatment groups to be statistically significant if you have less than a 5 percent probability of observing it just by chance. Based on this convention, the researchers were able to claim that the null hypothesis is not correct for cactus mice—they really do prefer hackberries over chilies. But the null hypothesis is correct for thrashers—the data indicate that they show no preference among the three fruits.

It is likely that you'll be doing actual statistical tests early in your undergraduate career. To use this text, though, you only need to be aware of what statistical testing does. And you should take care to inspect the standard error bars on graphs in this book. As a *very* rough rule of thumb, averages often turn out to be significantly different, according to an appropriate statistical test, if two times the standard errors do not overlap.

Reading Chemical Structures

If you haven't had much chemistry yet, learning basic biological chemistry can be a challenge. One of the stumbling blocks is simply being able to read chemical structures efficiently and understand what they mean. This skill will come much easier once you have a little notation under your belt and you understand some basic symbols.

Atoms are the basic building blocks of everything in the universe, just as cells are the basic building blocks of your body. Every atom has a 1- or 2-letter symbol. The following table shows the symbols for most of the atoms you'll encounter in this book. You should memorize these. The table also offers details on how the atoms form bonds as well as how they are represented in visual models.

When atoms attach to each other by covalent bonding, a molecule forms. Biologists have a couple of different ways of representing molecules—you'll see each of these in the book and in class. A molecular formula like those in **Figure BS4.1a** simply lists the atoms present in a molecule, with subscripts indicating how many of each atom are present. If the formula has no subscript, only one of that type of atom is present. A methane (natural gas) molecule, for example, can be written as CH_4 because it consists of one carbon atom and four hydrogen atoms. Structural formulas like those in **Figure BS4.1b** show which atoms in the molecule are bonded to each other, with each bond indicated by a dash. The structural formula for methane indicates that each of the four hydrogen atoms forms one covalent bond with carbon, and that carbon makes a total of four covalent bonds. Note that single covalent bonds are symbolized by a single dash; double bonds are indicated by two dashes.

	Methane	Ammonia	Water	Oxygen
(a) Molecular formulas:	CH_4	NH_3	H_2O	O_2

(b) Structural formulas:

(c) Ball-and-stick models:

(d) Space-filling models:

FIGURE BS4.1 Molecules Can Be Represented in Several Different Ways.

⬤ **EXERCISE** Carbon dioxide consists of a carbon atom that forms a double bond with each of two oxygen atoms, for a total of four bonds. It is a linear molecule. Write carbon dioxide's molecular formula, then draw its structural formula, a ball-and-stick model, and a space-filling model.

Even simple molecules have distinctive shapes, because different atoms make covalent bonds at different angles. Ball-and-stick and space-filling models show the geometry of the bonds accurately. In a ball-and-stick model, a stick is used to represent each covalent bond (see **Figure BS4.1c**). In space-filling models, the atoms are simply stuck onto each other in their proper places (see **Figure BS4.1d**).

Atom	Symbol	Number of Bonds It Can Form	Standard Color Code*
Hydrogen	H	1	white
Carbon	C	4	black
Nitrogen	N	3	blue
Oxygen	O	2	red
Sodium	Na	1	(not used in this text)
Magnesium	Mg	2	(not used in this text)
Phosphorus	P	5	orange or purple
Sulfur	S	2	yellow
Calcium	Ca	2	(not used in this text)

*In ball-and-stick or space-filling models.

To learn more about a molecule when you look at a chemical structure, ask yourself three questions:

1. *Is the molecule polar—meaning that some parts are more negatively or positively charged than others?* Molecules that contain nitrogen or oxygen atoms are often polar, because these atoms have such high electronegativity (see Chapter 2). This trait is important because polar molecules dissolve in water.

2. *Does the structural formula show atoms that might participate in chemical reactions?* For example, are there charged atoms or amino ($-NH_2$) or carboxyl ($-COOH$) groups that might act as a base or an acid?

3. *In ball-and-stick and especially space-filling models of large molecules, are there interesting aspects of overall shape?* For example, is there a groove where a protein might bind to DNA, or a cleft where a substrate might undergo a reaction in an enzyme?

Using Logarithms

You have probably been introduced to logarithms and logarithmic notation in algebra courses, and you will encounter logarithms at several points in this course. Logarithms are a way of working with powers—meaning, numbers that are multiplied by themselves one or more times. Scientists use exponential notation to represent powers. For example,

$$a^x = y$$

means that if you multiply a by itself x times, you get y. In exponential notation, a is called the base and x is called the exponent. The entire expression is called an exponential function.

What if you know y and a, and you want to know x? This is where logarithms come in.

$$x = \log_a y$$

This equation reads, x is equal to the logarithm of y to the base a. Logarithms are a way of solving exponential functions. They are important because so many processes in biology (and chemistry and physics, for that matter) are exponential in nature. To understand what's going on, you have to describe the process with an exponential function and then use logarithms to work with that function.

Although a base can be any number, most scientists use just two bases when they employ logarithmic notation: 10 and e. Logarithms to the base 10 are so common that they are usually symbolized in the form log y instead of $\log_{10} y$. A logarithm to the base e is called a natural logarithm and is symbolized ln (pronounced *EL-EN*) instead of log. You write "the natural logarithm of y" as ln y. The base e is an irrational number (like π) that is approximately equal to 2.718. Like 10, e is just a number. But both 10 and e have qualities that make them convenient to use in biology (and chemistry, and physics).

Most scientific calculators have keys that allow you to solve problems involving base 10 and base e. For example, if you know y, they'll tell you what log y or ln y are—meaning that they'll solve for x in our example above. They'll also allow you to find a number when you know its logarithm to base 10 or base e. Stated another way, they'll tell you what y is if you know x, and y is equal to e^x or 10^x. This is called taking an antilog. In most cases, you'll use the inverse or second function button on your calculator to find an antilog (above the log or ln key).

To get some practice with your calculator, consider the equation

$$10^2 = 100$$

If you enter 100 in your calculator and then press the log key, the screen should say 2. The logarithm tells you what the exponent is. Now press the antilog key while 2 is on the screen. The calculator screen should return to 100. The antilog solves the exponential function, given the base and the exponent.

If your background in algebra isn't strong, you'll want to get more practice working with logarithms—you'll see them frequently during your undergraduate career. Remember that once you understand the basic notation, there's nothing mysterious about logarithms. They are simply a way of solving exponential functions, which describe what happens when something is multiplied by itself a number of times—like cells that divide and then divide again and then again.

Using logarithms will also come up when you are studying something that can have a huge range of values, like the concentration of hydrogen ions in a solution or the intensity of sound that the human ear can detect. In cases like this, it's convenient to express the numbers involved as exponents. Using exponents makes a large range of numbers smaller and more tractable. For example, instead of saying that hydrogen ion concentration in a solution can range from 1 to 10^{-14}, the pH scale allows you to simply say that it ranges from 1 to 14. Instead of giving the actual value, you're expressing it as an exponent. It just simplifies things.

Making Concept Maps

A concept map is a graphical device for organizing and expressing what you know about a topic. It has two main elements: (1) concepts that are identified by words or short phrases and placed in a box or circle, and (2) labeled arrows that physically link two concepts and explain the relationship between them. The concepts are arranged hierarchically on a page, with the most general concepts at the top and the most specific ideas at the bottom.

The combination of a concept, a linking word, and a second concept is called a proposition. Good concept maps also have cross-links—meaning, labeled arrows that connect different elements in the hierarchy, as you read down the page.

Concept maps were initially developed by Joseph Novak in 1972 and have proven to be an effective studying and learning tool. They can be particularly valuable if constructed by a group, or when different individuals exchange and critique concept maps they have created independently. Although concept maps vary widely in quality and can be graded using objective criteria, there are many equally valid ways of making a high-quality concept map on a particular topic.

When you are asked to make a concept map in this text, you will usually be given at least a partial list of concepts to use. As an example, suppose you were asked to create a concept map on experimental design and were given the following concepts: results, predictions, control treatment, experimental treatment, controlled (identical) conditions, conclusions, experiment, hypothesis to be tested, null hypothesis. One possible concept map is shown in **Figure BS6.1**.

Dr. Doug Luckie, who uses concept maps extensively in his teaching at Michigan State University, points out that good concept maps have four qualities.

- They exhibit an organized hierarchy, indicating how each concept on the map relates to larger and smaller concepts.

- The concept words are specific—not vague.

- The propositions are accurate.

- There is cross-linking between different elements in the hierarchy of concepts.

As you practice making concept maps, go through these criteria and use them to evaluate your own work, as well as the

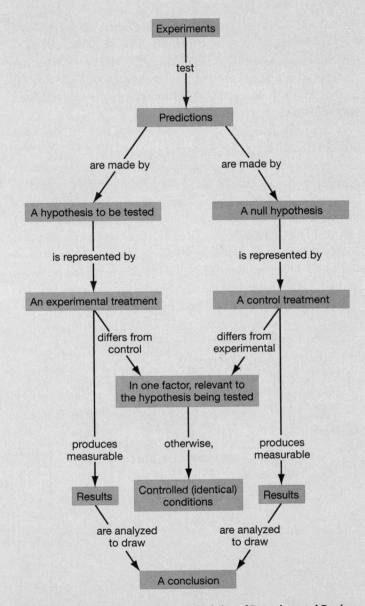

FIGURE BS6.1 A Concept Map on Principles of Experimental Design.

work of fellow students. For more about concept maps, see the website for this text. It has additional information as well as links to concept-mapping software developed by Dr. Luckie.

Using Electrophoresis to Separate Molecules

In molecular biology, the standard technique for separating and analyzing proteins and nucleic acids is called gel electrophoresis or, simply, electrophoresis (literally "electricity-moving"). You may be using electrophoresis in a lab for this course, and you will certainly be analyzing data derived from electrophoresis in this text.

The principle behind electrophoresis is fairly simple. Both proteins and nucleic acids carry a charge. As a result, these molecules move when placed in an electric field. Negatively charged molecules move toward the positive electrode (the positive end of the field), and positively charged molecules move toward the negative electrode (the negative end). To separate a mixture of macromolecules so that each can be isolated and analyzed, researchers place the sample in a gelatinous substance. The "gel" consists of long molecules that form a matrix of fibers. The presence of the fibers keeps molecules in the sam-

ple from moving around randomly, but the gelatinous matrix also has pores through which the molecules can pass. When an electrical field is applied across the gel, the molecules in the well move through the gel toward an electrode. Molecules that are smaller or more highly charged for their size move faster than do larger or less highly charged molecules. As they move, the molecules separate by size and by charge.

Figure BS7.1 shows the electrophoresis setup used in an experiment investigating how RNA molecules polymerize, described in Chapter 4. Step 1 shows how investigators loaded samples of macromolecules, taken on different days during the experiment, into cavities or "wells" at the top of the gel slab. In this and many other cases, the researchers also filled a well with a sample containing fragments of known size, called a size standard or "ladder." In step 2 the researchers immersed the gel in a solution that conducts electricity and applied a voltage

FIGURE BS7.1 Macromolecules Can Be Separated via Gel Electrophoresis.

● **QUESTION** DNA and RNA run toward the positive electrode. Why are these molecules negatively charged?

across the gel. After the samples had run down the gel for some time (step 3), they removed the electric field. By then, molecules of different size and charge had separated from one another. In this case, small RNA molecules had reached the bottom of the gel. Above them were larger RNA molecules, which had run more slowly.

Once molecules have been separated in this way, they have to be detected. Often, proteins or nucleic acids can be stained or dyed. In this case, however, the researchers had attached a radioactive atom to the monomers used in the experiment, so they could visualize the resulting polymers by laying X-ray film over the gel. Because radioactive emissions expose film, a black dot appears wherever a radioactive atom is located in the gel. This technique for visualizing macromolecules is called autoradiography. If the samples are loaded into a rectangular well, as is commonly done, molecules of a particular size and charge form a band in an autoradiograph.

The autoradiograph that resulted from the polymerization experiment is shown in **Figure BS7.2**. The samples, taken on days 2, 4, 6, 8, and 14 of the experiment, are labeled along the bottom. The far right lane contains macromolecules of known size; this lane is used to estimate the size of the molecules in the experimental samples. The bands that appear in each sample lane represent the different polymers that had formed. Darker bands contain more radioactive marker, indicating the presence of many radioactive molecules. Lighter bands contain fewer molecules.

To read a gel, you look for (1) the presence or absence of bands in some lanes—meaning, some experimental samples—versus others, and (2) contrasts in the darkness of the bands present—meaning, differences in the amount of molecule present. For example, several conclusions can be drawn from the data in Figure BS7.2. First, a variety of polymers formed at each

FIGURE BS7.2 Autoradiography Is a Technique for Visualizing Macromolecules. The molecules in a gel can be visualized in a number of ways. In this case, the RNA molecules in the gel exposed an X-ray film because they had radioactive atoms attached. When developed, the film is called an autoradiograph.

● **EXERCISE** Add labels to the photograph that read: "Top of gel—large molecules" and "Bottom of gel—small molecules."

stage. After the second day, for example, polymers from 12 to 18 monomers long had formed on the clay particles used in this experiment. Second, the overall length of polymers produced increased with time. At the end of the fourteenth day, most of the RNA molecules were between 20 and 40 monomers long.

Observing Microscopic Structures and Processes

With the unaided eye, it is not possible to see a lot of biology. Biologists have to use microscopes to study small multicellular organisms, individual cells, and the contents of cells. And to understand what individual macromolecules or multimolecular machines like ribosomes look like, researchers use data from a technique called X-ray crystallography.

You'll probably use dissecting microscopes and compound light microscopes to view specimens during your labs for this course, and throughout this text you'll be seeing images generated from other types of microscopy and from X-ray crystallographic data. One of the fundamental skills you'll be acquiring as an introductory student, then, is a basic understanding of how these techniques work. The key is to recognize that each approach for visualizing microscopic structures has strengths and weaknesses. As a result, each technique is appropriate for studying certain types or aspects of cells or molecules.

Trends in Microscopy: Increasing Magnification and Clarity

If you use a dissecting microscope during labs, you'll recognize that it works by magnifying light that bounces off a whole specimen—often a live organism. You'll be able to view the specimen in three dimensions, which is why these instruments are sometimes called stereomicroscopes, but the maximum magnification is only about 20 to 40 times normal size ($20\times$ to $40\times$).

To view smaller objects, you'll probably use a compound microscope. Compound microscopes magnify light that is passed through a specimen. The instruments used in introductory labs are usually capable of $400\times$ magnifications; the most sophisticated compound microscopes available can achieve magnifications of about $2000\times$. This is enough to view individual bacterial or eukaryotic cells and see large structures inside cells, like condensed chromosomes (see Chapter 11). To prepare a specimen for viewing under a compound light microscope, the tissues or cells are usually sliced to create a thin enough section for light to pass through efficiently. The section is then dyed to increase contrast and make structures visible. In many cases, different types of dyes are used to highlight different types of structures.

Until the 1950s, the compound microscope was the biologist's only tool for viewing cells directly. But the invention of the electron microscope provided a new way to view speci-mens. Two basic types of electron microscopy are now available: one that allows researchers to examine cross sections of cells at extremely high magnification, and one that offers a view of surfaces at somewhat lower magnification.

Transmission Electron Microscopy (TEM)

The transmission electron microscope is an extraordinarily effective tool for viewing cell structure at high magnification. TEM forms an image from electrons that pass through a specimen, just as a light microscope forms an image from light rays that pass through a specimen.

Biologists who want to view a cell under a transmission electron microscope begin by "fixing" the cell, meaning that they treat it with a chemical agent that stabilizes the cell's structure and contents while disturbing them as little as possible. Then they permeate the cell with an epoxy plastic that stiffens the structure. Once this epoxy hardens, the cell can be cut into extremely thin sections with a glass or diamond knife. Finally, the sectioned specimens are impregnated with a metal—often lead. (The reason for this last step is explained shortly.)

Figure BS8.1a outlines how the transmission electron microscope works. A beam of electrons is produced by a tungsten filament at the top of a column and directed downward. (All of the air is pumped out of the column, so that the electron beam isn't scattered by collisions with air molecules.) The electron beam passes through a series of lenses and through the specimen. The lenses are actually electromagnets, which alter the path of the beam much like a glass lens in a dissecting or compound microscope bends light. The lenses magnify and focus the image on a screen at the bottom of the column. There the electrons strike a coating of fluorescent crystals, which emit visible light in response—just like a television screen. When the microscopist moves the screen out of the way and allows the electrons to expose a sheet of black-and-white film, the result is a micrograph—a photograph of an image produced by microscopy.

The image itself is created by electrons that pass through the specimen. If no specimen were in place, all the electrons would pass through and the screen (and micrograph) would be uniformly bright. Unfortunately, cell materials by themselves would also appear fairly uniform and bright. This is because an atom's ability to deflect an electron depends on its mass. In

(a) Transmission electron microscopy: High magnification of cross sections

Tungsten filament (source of electrons)

Condenser lens

Specimen

Objective lens

Projector lens

Image on fluorescent screen

2 μm

Cross section of insect eye

(b) Scanning electron microscopy: Lower magnification of surfaces

20 μm

Surface view of insect eye

FIGURE BS8.1 There Are Two Basic Types of Electron Microscopy.

turn, an atom's mass is a function of its atomic number. The hydrogen, carbon, oxygen, and nitrogen atoms that dominate biological molecules have low atomic numbers. This is why cell biologists must saturate cell sections with lead solutions. Lead has a high atomic number and scatters electrons effectively. Different macromolecules take up lead atoms in different amounts, so the metal acts as a "stain" that produces contrast. With TEM, areas of dense metal scatter the electron beam most, producing dark areas in micrographs.

The advantage of TEM is that it can magnify objects up to 250,000×—meaning that intracellular structures are clearly visible. The downsides are that researchers are restricted to observing dead, sectioned material, and they must take care that the preparation process does not distort the specimen.

Scanning Electron Microscopy (SEM)

The scanning electron microscope is the most useful tool biologists have for looking at the surfaces of cells. Materials are prepared for scanning electron microscopy by coating their surfaces with a layer of metal atoms. To create an image of this surface, the microscope scans the surface with a narrow beam of electrons. Electrons that are reflected back from the surface or that are emitted by the metal atoms in response to the beam then strike a detector. The signal from the detector controls a second electron beam, which scans a TV-like screen and forms an image magnified up to 50,000 times the object's size. Because SEM records shadows and highlights, it pro-

vides images with a three-dimensional appearance (**Figure BS8.1b**). It cannot magnify objects nearly as much as TEM can, however.

Studying Live Cells and Real-Time Processes

Until the 1960s, it was not possible for biologists to get clear, high-magnification images of living cells. But a series of innovations over the past several decades has made it possible to observe organelles and subcellular structures in action.

The development of video microscopy, where the image from a light microscope is captured by a video camera instead of by an eye or a film camera, proved revolutionary. It allowed specimens to be viewed at higher magnification, because video cameras are more sensitive to small differences in contrast than are the human eye or still cameras. It also made it easier to keep live specimens functioning normally, because the increased light sensitivity of video cameras allows them to be used with low illumination, so specimens don't overheat. And when it became possible to digitize video images, researchers began using computers to remove out-of-focus background material and increase image clarity.

Another important innovation was the use of a fluorescent molecule called green fluorescent protein, or GFP, which allows researchers to tag specific molecules or structures and follow their movement over time. GFP is naturally synthesized in jellyfish that fluoresce, or emit light. By affixing GFP molecules

(a) Conventional fluorescence image of single cell

(b) Confocal fluorescence image of same cell

25 μm

FIGURE BS8.2 Confocal Microscopy Provides Sharp Images of Living Cells. (a) The conventional image of this mouse intestinal cell is blurred, because it results from light emitted by the entire cell. **(b)** The confocal image is sharp, because it results from light emitted at a single plane inside the cell.

to another protein and then inserting it into a cell, investigators can follow the protein's fate over time and even videotape its movement. For example, researchers have videotaped GFP-tagged proteins being transported from the rough ER through the Golgi apparatus and out to the plasma membrane. This is cell biology: the movie.

Visualizing Structures in 3-D

The world is three-dimensional. To understand how microscopic structures and macromolecules work, it is essential to understand their shape and spatial relationships. Consider three techniques currently being used to reconstruct the 3-D structure of cells, organelles, and macromolecules.

- Confocal microscopy is carried out by mounting cells that have been treated with one or more fluorescing tags on a microscope slide and then focusing a beam of ultraviolet light at a specific depth within the specimen. The fluorescing tag emits visible light in response. A detector for this light is then set up at exactly the position where the emitted light comes into focus. The result is a sharp image of a precise plane in the cell being studied (**Figure BS8.2**). By altering the focal plane, a researcher can record images from an array of depths in the specimen; a computer can then be used to generate a 3-D image of the cell.

- Electron tomography uses a transmission electron microscope to generate a 3-D image of an organelle or other subcellular structure. The specimen is rotated around a single axis, with the researcher taking many "snapshots." The individual images are then pieced together with a computer. This technique has provided a much more accurate view of mitochondrial structure than was possible using traditional TEM (see Chapter 7).

- X-ray crystallography, or X-ray diffraction analysis, is the most widely used technique for reconstructing the 3-D structure of molecules. As its name implies, the procedure is based on bombarding crystals of a molecule with X-rays. X-rays are scattered in precise ways when they interact with the electrons surrounding the atoms in a crystal, producing a diffraction pattern that can be recorded on X-ray film or other types of detectors (**Figure BS8.3**). By varying the orientation of the X-ray beam as it strikes a crystal and documenting the diffraction patterns that result, researchers can construct a map representing the density of electrons in the crystal. By relating these electron-density maps to information about the primary structure of the nucleic acid or protein, a 3-D model of the molecule can be built. Virtually all of the molecular models used in this book were built from X-ray crystallographic data.

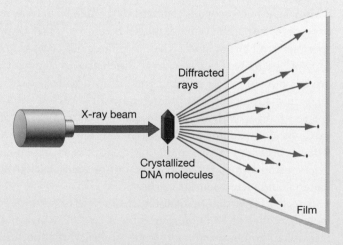

Diffracted rays

X-ray beam

Crystallized DNA molecules

Film

The patterns are determined by the structure of the molecules within the crystal

FIGURE BS8.3 X-Ray Crystallography. When crystallized molecules are bombarded with X-rays, the radiation is scattered in distinctive patterns. The photograph at the right shows an X-ray film that recorded the pattern of scattered radiation from DNA molecules.

Combining Probabilities

In several cases in this text, you'll need to combine probabilities from different events in order to solve a problem. One of the most common applications is in genetics problems. For example, Punnett squares work because they are based on two fundamental rules of probability, called the "both-and rule" and the "either-or" rule. Each rule pertains to a distinct situation.

The both-and rule applies when you want to know the probability that two or more independent events occur together. Let's use the rolling of two dice as an example. What is the probability of rolling two sixes? These two events are independent, because the probability of rolling a six on one die has no effect on the probability of rolling a six on the other die. (In the same way, the probability of getting a gamete with allele R from one parent has no effect on the probability of getting a gamete with allele R from the other parent. Gametes fuse randomly.) The probability of rolling a six on the first die is 1/6. The probability of rolling a six on the second die is also 1/6. The probability of rolling a six on both die, then, is $1/6 \times 1/6 = 1/36$. In other words, if you rolled two die 36 times, on average you would expect to roll two sixes once. It should make sense that the both-and rule is also called the multiplication rule or product rule. In the case of a cross between two parents heterozygous at the R gene, the probability of getting allele R from the father is 1/2 and the probability of getting R from the mother is 1/2. Thus, the probability of getting both alleles and creating an offspring with genotype RR is $1/2 \times 1/2 = 1/4$.

The either-or rule, in contrast, applies when you want to know the probability of an event happening when there are several ways for the event to occur. In this case, the probability that the event will occur is the sum of the probabilities of each way that it can occur. For example, suppose you wanted to know the probability of rolling either a one or a six when you toss a die. The probability of drawing each is 1/6, so the probability of getting one or the other is $1/6 + 1/6 = 1/3$. (The either-or rule is also called the addition rule or sum rule.) If you rolled a die three times, on average you'd expect to get a one or a six once. In the case of a cross between two parents heterozygous at the R gene, the probability of getting an R allele from the father and an r allele from the mother is $1/2 \times 1/2 = 1/4$. Similarly, the probability of getting an r allele from the father and an R allele from the mother is $1/2 \times 1/2 = 1/4$. Thus, the combined probability of getting the Rr genotype in either of the two ways is $1/4 + 1/4 = 1/2$.

Questions

1. Suppose that four students each toss a coin. What is the probability of four "tails"?
2. After a single roll of a die, what is the probability of getting either a two, a three, or a six?

Answers: 1. $1/2 \times 1/2 \times 1/2 \times 1/2 = 1/16$; 2. $1/6 + 1/6 + 1/6 = 1/2$

Glossary

abiotic Not alive (e.g., air, water, and soil). Compare with **biotic**.

ABO blood types Genetically determined classes of human blood that are distinguished by the presence or absence of specific glycolipids on the surface of red blood cells. Also called *ABO blood groups*.

aboveground biomass The total mass of living plants in an area, excluding roots.

abscisic acid (ABA) A plant hormone that inhibits cell elongation and stimulates leaf shedding and dormancy.

abscission In plants, the normal (often seasonal) shedding of leaves, fruits, or flowers.

abscission zone The region at the base of a petiole that thins and breaks during dropping of leaves.

absorption In animals, the uptake of ions and small molecules derived from food across the lining of the intestine and into the bloodstream.

absorption spectrum The amount of light of different wavelengths absorbed by a pigment. Usually depicted as a graph of light absorbed versus wavelength. Compare with **action spectrum**.

acclimation, acclimatization Gradual physiological adjustment of an organism to new environmental conditions that occur naturally or as part of a laboratory experiment.

acetylation Addition of an acetyl group (CH_3COO^-) to a molecule.

acetylcholine (ACh) A neurotransmitter, released by nerve cells at neuromuscular junctions, that triggers contraction of muscle cells. Also used as a neurotransmitter between neurons.

acetyl CoA A molecule produced by oxidation of pyruvate (the final product of glycolysis) in a reaction catalyzed by pyruvate dehydrogenase. Can enter the Krebs cycle and also is used as a carbon source in the synthesis of fatty acids, steroids, and other compounds.

acid Any compound that gives up protons or accepts electrons during a chemical reaction or that releases hydrogen ions when dissolved in water.

acid-growth hypothesis The hypothesis that auxin triggers elongation of plant cells by inducing the synthesis of proton pumps whose activity makes the cell wall more acidic, leading to expansion of the cell wall and an influx of water.

acoelomate An animal that lacks an internal body cavity (coelom). Compare with **coelomate** and **pseudocoelomate**.

acquired immune deficiency syndrome (AIDS) A human disease characterized by death of immune system cells (in particular helper T cells and macrophages) and subsequent vulnerability to other infections. Caused by the human immunodeficiency virus (HIV).

acquired immune response See **acquired immunity**.

acquired immunity Immunity to a particular pathogen or other antigen conferred by antibodies and activated B and T cells following exposure to the antigen. Is characterized by specificity, diversity, memory, and self-nonself recognition. Compare with **innate immunity**.

acrosomal reaction A set of events occurring in a sperm cell upon encountering an egg cell, including release of acrosomal enzymes and formation of the acrosomal process, which helps the sperm cell reach the egg.

acrosome A caplike structure, located on the head of a sperm cell, that contains enzymes capable of dissolving the outer coverings of an egg.

ACTH See **adrenocorticotropic hormone**.

actin A globular protein that can be polymerized to form filaments. Actin filaments are part of the cytoskeleton and constitute the thin filaments in skeletal muscle cells.

actin filament A long fiber, about 7 nm in diameter, composed of two intertwined strands of polymerized actin protein; one of the three types of cytoskeletal fibers. Involved in cell movement. Also called a *microfilament*. Compare with **intermediate filament** and **microtubule**.

action potential A rapid, temporary change in electrical potential across a membrane, from negative to positive and back to negative. Occurs in cells, such as neurons and muscle cells, that have an excitable membrane.

action spectrum The relative effectiveness of different wavelengths of light in driving a light-dependent process such as photosynthesis. Usually depicted as a graph of some measure of the process versus wavelength. Compare with **absorption spectrum**.

activation energy The amount of energy required to initiate a chemical reaction; specifically, the energy required to reach the transition state.

active site The portion of an enzyme molecule where substrates (reactant molecules) bind and react.

active transport The movement of ions or molecules across a plasma membrane or organelle membrane against an electrochemical gradient. Requires energy (e.g., from hydrolysis of ATP) and assistance of a transport protein (e.g., pump).

adaptation Any heritable trait that increases the fitness of an individual with that trait, compared with individuals without that trait, in a particular environment.

adaptive radiation Rapid evolutionary diversification within one lineage, producing numerous descendant species with a wide range of adaptive forms.

adenosine diphosphate (ADP) A molecule consisting of adenine, a sugar, and two phosphate groups. Addition of a third phosphate group produces adenosine triphosphate (ATP).

adenosine triphosphate (ATP) A molecule consisting of adenine, a sugar, and three phosphate groups that can be hydrolyzed to release energy. Universally used by cells to store and transfer energy.

adenylyl cyclase An enzyme that can catalyze the formation of cyclic AMP (cAMP) from ATP. Involved in controlling transcription of various operons in prokaryotes and in some eukaryotic signal-transduction pathways.

adhesion The tendency of certain dissimilar molecules to cling together due to attractive forces. Compare with **cohesion**.

adipocyte A fat cell.

adipose tissue A type of connective tissue whose cells store fats.

ADP See **adenosine diphosphate**.

adrenal glands Two small endocrine glands that sit above each kidney. The outer portion (cortex) secretes several steroid hormones; the inner portion (medulla) secretes epinephrine and norepinephrine.

adrenaline See **epinephrine**.

adrenocorticotropic hormone (ACTH) A peptide hormone, produced and secreted by the anterior pituitary, that stimulates release of steroid hormones (e.g., cortisol, aldosterone) from the adrenal cortex.

adventitious root A root that develops from a plant's shoot system instead of from the plant's root system.

aerobic Referring to any metabolic process, cell, or organism that uses oxygen as an electron acceptor. Compare with **anaerobic**.

afferent division The part of the nervous system, consisting mainly of sensory neurons, that transmits information about the internal and external environment to the central nervous system. Compare with **efferent division**.

agar A gelatinous mix of polysaccharides that is commonly used to grow cells on solid media.

age class All the individuals of a specific age in a population.

age structure The proportion of individuals in a population that are of each possible age.

age-specific fecundity The average number of female offspring produced by a female in a certain age class.

agglutination Clumping together of cells, typically caused by antibodies.

aggregate fruit A fruit (e.g., raspberry) that develops from a single flower that has many separate carpels. Compare with **multiple** and **simple fruit**.

AIDS See **acquired immune deficiency syndrome**.

albumen A solution of water and protein (particularly albumins), found in amniotic eggs, that nourishes the growing embryo. Also called *egg white*.

albumin A class of large proteins found in plants and animals, particularly in the albumen of eggs and in blood plasma.

alcohol fermentation Catabolic pathway in which pyruvate produced by glycolysis is converted to ethanol in the absence of oxygen.

aldosterone A hormone produced in the adrenal cortex that stimulates the kidney to conserve salt and water and promotes retention of sodium.

alimentary canal See **digestive tract**.

allele A particular version of a gene.

allopatric speciation The divergence of populations into different species by physical isolation of populations in different geographic areas. Compare with **sympatric speciation**.

allopatry Condition in which two or more populations live in different geographic areas. Compare with **sympatry**.

allopolyploidy (adjective: allopolyploid) The state of having more than two full sets of chromosomes (polyploidy) due to hybridization between different species. Compare with **autopolyploidy.**

allosteric regulation Regulation of a protein's function by binding of a regulatory molecule, usually to a specific site distinct from the active site, causing a change in the protein's shape.

alpha(α)-helix A protein secondary structure in which the polypeptide backbone coils into a spiral shape stabilized by hydrogen bonds between atoms.

alternation of generations A life cycle involving alternation of a multicellular haploid stage (gametophyte) with a multicellular diploid stage (sporophyte). Occurs in most plants and some protists.

alternative splicing In eukaryotes, the splicing of primary RNA transcripts from a single gene in different ways to produce different mature mRNAs and thus different polypeptides.

altruism Any behavior that has a cost to the individual (such as lowered survival or reproduction) and a benefit to the recipient. See **reciprocal altruism.**

alveolus (plural: alveoli) One of the tiny air-filled sacs of a mammalian lung.

ambisense virus A virus whose genome contains both positive-sense and negative-sense sequences.

amino acid A small organic molecule with a central carbon atom bonded to an amino group ($-NH_3$), a carboxyl group ($-COOH$), a hydrogen atom, and a side group. Proteins are polymers of 20 common amino acids.

aminoacyl tRNA A transfer RNA molecule that is covalently bound to an amino acid.

aminoacyl tRNA synthetase An enzyme that catalyzes the addition of a particular amino acid to its corresponding tRNA molecule.

ammonia (NH_3) A small molecule, produced by the breakdown of proteins and nucleic acids, that is very toxic to cells. Is a strong base that gains a proton to form the ammonium ion (NH_4^+).

amnion The membrane within an amniotic egg that surrounds the embryo and encloses it in a protective pool of fluid (amniotic fluid).

amniotes A major lineage of vertebrates (Amniota) that reproduce with amniotic eggs. Includes all reptiles (including birds) and mammals.

amniotic egg An egg that has a watertight shell or case enclosing a membrane-bound water supply (the amnion), food supply (yolk sac), and waste sac (allantois).

amoeba Any unicellular protist that lacks a cell wall, is extremely flexible in shape, and moves by means of pseudopodia.

amoeboid motion See **cell crawling.**

amphibians A lineage of vertebrates many of whom breathe through their skin and feed on land but lay their eggs in water; represent the earliest tetrapods. Include frogs, salamanders, and caecilians.

amphipathic Containing hydrophilic and hydrophobic elements.

amylase Any enzyme that can break down starch by catalyzing hydrolysis of the glycosidic linkages between the glucose residues.

amyloplasts Dense, starch-storing organelles that settle to the bottom of plant cells and that may be used as gravity detectors.

anabolic pathway Any set of chemical reactions that synthesizes larger molecules from smaller ones. Generally requires an input of energy. Compare with **catabolic pathway.**

anadromous Having a life cycle in which adults live in the ocean (or large lakes) but migrate up freshwater streams to breed and lay eggs.

anaerobic Referring to any metabolic process, cell, or organism that uses an electron acceptor other than oxygen, such as nitrate or sulfate. Compare with **aerobic.**

anaphase A stage in mitosis or meiosis during which chromosomes are moved to opposite ends of the cell.

anatomy The study of the physical structure of organisms.

androgens A class of steroid hormones that generally promote male-like traits (although females have some androgens, too). Secreted primarily by the gonads and adrenal glands.

aneuploidy (adjective: aneuploid) The state of having an abnormal number of copies of a certain chromosome.

angiosperm A flowering vascular plant that produces seeds within mature ovaries (fruits). The angiosperms form a single lineage. Compare with **gymnosperm.**

animal A member of a major lineage of eukaryotes (Animalia) whose members typically have a complex, large, multicellular body, eat other organisms, and are mobile.

animal model Any disease that occurs in a nonhuman animal and has many parallels to a similar disease of humans. Studied by medical researchers in hope that findings may apply to human disease.

anion A negatively charged ion.

annelids Members of the phylum Annelida (segmented worms). Distinguished by a segmented body and a coelom that functions as a hydrostatic skeleton. Annelids belong to the lophotrochozoan branch of the protostomes.

annual Referring to a plant whose life cycle normally lasts only one growing season—less than one year. Compare with **perennial.**

anoxygenic Referring to any process or reaction that does not produce oxygen. Photosynthesis in purple sulfur and purple nonsulfur bacteria, which does not involve photosystem II, is anoxygenic. Compare with **oxygenic.**

antenna (plural: antennae) A long appendage that is used to touch or smell.

antenna complex Part of a photosystem, containing an array of chlorophyll molecules and accessory pigments, that receives energy from light and directs the energy to a central reaction center during photosynthesis.

anterior Toward an animal's head and away from its tail. The opposite of posterior.

anterior pituitary The part of the pituitary gland containing endocrine cells that produce and release a variety of peptide hormones in response to other hormones from the hypothalamus. Compare with **posterior pituitary.**

anther The pollen-producing structure at the end of a stamen in flowering plants (angiosperms).

antheridium (plural: antheridia) The sperm-producing structure in most land plants except angiosperms.

anthropoids One of the two major lineages of primates, including apes, humans, and all monkeys. Compare with **prosimians.**

antibiotic Any substance, such as penicillin, that can kill or inhibit the growth of bacteria.

antibody An immunoglobulin protein, produced by B cells, that can bind to a specific part of an antigen, tagging it for attack by the immune system. All antibody molecules have a similar Y-shaped structure and, in their monomer form, consist of two identical light chains and two identical heavy chains.

anticodon The sequence of three bases (triplet) in a transfer RNA molecule that can bind to a mRNA codon with a complementary sequence.

antidiuretic hormone (ADH) A peptide hormone, secreted from the posterior pituitary gland, that stimulates water retention by the kidney. Also called *vasopressin.*

antigen Any foreign molecule, often a protein, that can stimulate a specific response by the immune system.

antigen presentation Process by which small peptides, derived from ingested particulate antigens (e.g., bacteria) or intracellular antigens (e.g., viruses in infected cell) are complexed with MHC proteins and transported to the cell surface where they are displayed and can be recognized by T cells.

antiparallel Describing the opposite orientation of the strands in a DNA double helix with one strand running in the $5' \rightarrow 3'$ direction and the other in the $3' \rightarrow 5'$ direction.

antiporter A carrier protein that allows an ion to diffuse down an electrochemical gradient, using the energy of that process to transport a different substance in the opposite direction *against* its concentration gradient. Compare with **symporter.**

antiviral Any drug or other agent that can kill or inhibit the transmission or replication of viruses.

anus In a multicellular animal, the end of the digestive tract where wastes are expelled.

aorta In terrestrial vertebrates, the major artery carrying oxygenated blood away from the heart.

aphotic zone Deep water receiving no sunlight. Compare with **photic zone.**

apical Toward the top. In plants, at the tip of a branch. In animals, on the side of an epithelial layer that faces the environment and not other body tissues. Compare with **basal.**

apical bud A bud at the tip of a stem, where growth occurs to lengthen the stem.

apical dominance Inhibition of lateral bud growth by the apical meristem at the tip of a plant branch.

apical meristem A group of undifferentiated plant cells, at the tip of a stem or root, that is responsible for primary growth. Compare with **lateral meristem.**

apomixis The formation of mature seeds without fertilization occurring; a type of asexual reproduction.

apoplast In plant roots, a continuous pathway through which water can flow, consisting of the porous cell walls of adjacent cells and the intervening extracellular space. Compare with **symplast.**

apoptosis Series of genetically controlled changes that lead to death of a cell. Occurs frequently during embryological development and later may occur in response to infections or cell damage. Also called *programmed cell death.*

aquaporin A type of channel protein through which water can move by osmosis across a plasma membrane.

arbuscular mycorrhizal fungi (AMF) Fungi whose hyphae enter the root cells of their host plants.

Archaea One of the three taxonomic domains of life consisting of unicellular prokaryotes distinguished by cell walls made of certain polysaccharides not found in bacterial or eukaryotic cell walls, plasma membranes composed of unique isoprene-containing phospholipids, and ribosomes and RNA polymerase similar to those of eukaryotes. Compare with **Bacteria** and **Eukarya.**

archegonium (plural: archegonia) The egg-producing structure in most land plants except angiosperms.

arteriole One of the many tiny vessels that carry blood from arteries to capillaries.

artery Any thick-walled blood vessel that carries blood (oxygenated or not) under relatively high pressure away from the heart to organs throughout the body. Compare with **vein.**

arthropods Members of the phylum Arthropoda. Distinguished by a segmented body; a hard, jointed exoskeleton; paired appendages; and an extensive body cavity called a hemocoel. Arthropods belong to the ecdysozoan branch of the protostomes.

articulation A movable point of contact between two bones of a skeleton. See **joint.**

artificial selection Deliberate manipulation by humans, as in animal and plant breeding, of the genetic composition of a population by allowing only individuals with desirable traits to reproduce.

ascocarp A large, cup-shaped reproductive structure produced by some ascomycete fungi. Contains many microscopic asci, which produce spores.

ascomycete See **sac fungus.**

ascus (plural: asci) Specialized spore-producing cell found at the ends of hyphae in sac fungi (ascomycetes).

asexual reproduction Any form of reproduction resulting in offspring that are genetically identical to the parent. Includes binary fission, budding, and parthenogenesis. Compare with **sexual reproduction.**

asymmetric competition Ecological competition between two species in which one species suffers a much greater fitness decline than the other. Compare with **symmetric competition.**

atomic mass unit (amu) A unit of mass equal to 1/12 the mass of one carbon-12 atom; about the mass of 1 proton or 1 neutron. Also called *dalton.*

atomic number The number of protons in the nucleus of an atom, giving the atom its identity as a particular chemical element.

ATP See **adenosine triphosphate.**

ATP synthase A large membrane-bound protein complex in chloroplasts, mitochondria, and some bacteria that uses the energy of protons flowing through it to synthesize ATP. Also called F_oF_1 *complex.*

atrioventricular (AV) node A region of the heart between the right atrium and right ventricle where electrical signals from the atrium are slowed briefly before spreading to the ventricle. This delay allows the ventricle to fill with blood before contracting. Compare with **sinoatrial (SA) node.**

atrium (plural: atria) A thin-walled chamber of the heart that receives blood from veins and pumps it to a neighboring chamber (the ventricle).

autocrine Relating to a chemical signal that affects the same cell that produced and released it.

autoimmunity A pathological condition in which the immune system attacks self cells or tissues of an individual's own body.

autonomic nervous system The part of the peripheral nervous system that controls internal organs and involuntary processes, such as stomach contraction, hormone release, and heart rate. Includes parasympathetic and sympathetic nerves. Compare with **somatic nervous system.**

autophagy The process by which damaged organelles are surrounded by a membrane and delivered to a lysosome to be destroyed.

autopolyploidy (adjective: autopolyploid) The state of having more than two full sets of chromosomes (polyploidy) due to a mutation that doubled the chromosome number.

autoradiography A technique for detecting radioactively labeled molecules separated by gel electrophoresis by placing an unexposed film over the gel. A black dot appears on the film wherever a radioactive atom is present in the gel. Also can be used to locate labeled molecules in fixed tissue samples.

autosomal inheritance The inheritance patterns that occur when genes are located on autosomes rather than on sex chromosomes.

autosome Any chromosome that does not carry genes involved in determining the sex of an individual.

autotroph Any organism that can synthesize reduced organic compounds from simple inorganic sources such as CO_2 or CH_4. Most plants and some bacteria and archaea are autotrophs. Also called *primary producer.* Compare with **heterotroph.**

auxin Indoleacetic acid, a plant hormone that stimulates phototropism and some other responses.

avirulence (*avr*) loci Genes in pathogens encoding proteins that trigger a defense response in plants. Compare with **resistance (R) loci.**

axillary bud A bud that forms in the angle between a leaf and a stem and may develop into a lateral (side) branch. Also called *lateral bud.*

axon A long projection of a neuron that can propagate an action potential and transmit it to another neuron.

axon hillock The site in a neuron where an axon joins the cell body and where action potentials are first triggered.

axoneme A structure found in eukaryotic cilia and flagella and responsible for their motion; composed of two central microtubules surrounded by nine doublet microtubules (9 + 2 arrangement).

background extinction The average rate of low-level extinction that has occurred continuously throughout much of evolutionary history. Compare with **mass extinction.**

Bacteria One of the three taxonomic domains of life consisting of unicellular prokaryotes distinguished by cell walls composed largely of peptidoglycan, plasma membranes similar to those of eukaryotic cells, and ribosomes and RNA polymerase that differ from those in archaeans or eukaryotes. Compare with **Archaea** and **Eukarya.**

bacterial artificial chromosome (BAC) An artificial version of a bacterial chromosome that can be used as a cloning vector to produce many copies of large DNA fragments.

bacteriophage Any virus that infects bacteria.

baculum A bone inside the penis usually present in mammals with a penis that lacks erectile tissue.

bark The protective outer layer of woody plants, composed of cork cells, cork cambium, and secondary phloem.

baroreceptors Specialized nerve cells in the walls of the heart and certain major arteries that detect changes in blood pressure and trigger appropriate responses by the brain.

basal Toward the base. In plants, at the base of a branch where it joins the stem. In animals, on the side of an epithelial layer that abuts underlying body tissues. Compare with **apical.**

basal body A structure of nine pairs of microtubules arranged in a circle at the base of eukaryotic cilia and flagella where they attach to the cell. Structurally similar to a centriole.

basal lamina A thick, collagen-rich extracellular matrix that underlies most epithelial tissues (e.g., skin) in animals.

basal metabolic rate (BMR) The total energy consumption by an organism at rest in a comfortable environment. For aerobes, often measured as the amount of oxygen consumed per hour.

basal transcription complex A large multiprotein structure that assembles near the promoter of eukaryotic genes and initiates transcription. Composed of basal transcription factors, TATA-binding protein, coactivators, and RNA polymerase.

basal transcription factor General term for proteins, present in all cell types, that bind to eukaryotic promoters and help initiate transcription. Compare with **regulatory transcription factor.**

base Any compound that acquires protons or gives up electrons during a chemical reaction or accepts hydrogen ions when dissolved in water.

basidiomycete See **club fungus.**

basidium (plural: basidia) Specialized spore-producing cell at the ends of hyphae in club fungi.

basilar membrane The membrane in the vertebrate cochlea on which the bottom portion of hair cells sit.

basolateral Toward the bottom and sides. In animals, the side of an epithelial layer that faces other body tissues and not the environment.

Batesian mimicry A type of mimicry in which a harmless or palatable species resembles a dangerous or poisonous species. Compare with **Müllerian mimicry.**

B cell A type of leukocyte that matures in the bone marrow and, with T cells, is responsible for acquired immunity. Produces antibodies and also functions in antigen presentation. Also called *B lymphocyte.*

B-cell receptor (BCR) An immunoglobulin protein (antibody) embedded in the plasma membrane of mature B cells and to which antigens bind.

behavior Any action by an organism.

benign tumor A mass of abnormal tissue that grows slowly or not at all, does not disrupt surrounding tissues, and does not spread to other organs. Benign tumors are not cancers. Compare with **malignant tumor.**

benthic Living at the bottom of an aquatic environment.

benthic zone The area along the bottom of an aquatic environment.

beta(β)-pleated sheet A protein secondary structure in which the polypeptide backbone folds into a sheetlike shape stabilized by hydrogen bonding.

bilateral symmetry An animal body pattern in which there is one plane of symmetry dividing the body into a left side and a right side. Typically, the body is long and narrow, with a distinct head end and tail end. Compare with **radial symmetry**.

bilaterian A member of a major lineage of animals (Bilateria) that are bilaterally symmetrical at some point in their life cycle, have three embryonic germ layers, and have a coelom. All protostomes and deuterostomes are bilaterians.

bile A complex solution produced by the liver, stored in the gall bladder, and secreted into the intestine. Contains steroid derivatives called bile salts that are responsible for emulsification of fats during digestion.

binary fission See **fission** (1).

binomial nomenclature A system of naming species by using two-part Latinized names composed of a genus name and a species name (e.g., *Homo sapiens,* humans; *Canis familiaris,* dogs). Always italicized, with genus name capitalized.

biodiversity The diversity of life considered at three levels: genetic diversity (variety of alleles in a population, species, or group of species); species diversity (variety and relative abundance of species present in a certain area); and ecosystem diversity (variety of communities and abiotic components in a region).

biogeochemical cycle The pattern of circulation of an element or molecule among living organisms and the environment.

biogeography The study of how species and populations are distributed geographically.

bioinformatics The field of study concerned with managing, analyzing, and interpreting biological information, particularly DNA sequences.

biological species concept The definition of a species as a population or group of populations that are reproductively isolated from other groups. Members of a species have the potential to interbreed in nature to produce viable, fertile offspring but cannot produce viable, fertile hybrid offspring with members of other species. Compare with **morphological** and **phylogenetic species concept.**

bioluminescence The emission of light by a living organism.

biomass The total mass of all organisms in a given population or geographical area; usually expressed as total dry weight.

biome A large terrestrial ecosystem characterized by a distinct type of vegetation and climate.

bioremediation The use of living organisms, usually bacteria or archaea, to degrade environmental pollutants.

biotechnology The application of biological techniques and discoveries to medicine, industry, and agriculture.

biotic Living, or produced by a living organism. Compare with **abiotic.**

bipedal Walking primarily on two legs.

bipolar cell A cell in the vertebrate retina that receives information from one or more photoreceptors and passes it to other bipolar cells or ganglion cells.

bivalves A lineage of molluscs that have two shells, such as clams and mussels.

bladder A mammalian organ that holds urine until it can be excreted.

blade The wide, flat part of a plant leaf.

blastocoel Fluid-filled cavity in the blastula of many animal species.

blastocyst Specialized type of blastula in mammals. A spherical structure composed of trophoblast cells on the exterior and a cluster of cells (the inner cell mass), which fills part of the interior space.

blastomere A small cell created by cleavage divisions in early animal embryos.

blastopore A small opening (pore) in the surface of an early vertebrate embryo, through which cells move during gastrulation.

blastula In vertebrate development, a hollow ball of cells (blastomere cells) that is formed by cleavage of a zygote and immediately undergoes gastrulation. See **blastocyst.**

blood A type of connective tissue consisting of red blood cells and leukocytes suspended in a fluid portion called plasma.

blood pressure See **diastolic blood pressure** and **systolic blood pressure.**

body mass index (BMI) A mathematical relationship of weight and height used to assess obesity in humans. Calculated as weight (kg) divided by the square of height (m²).

body plan The basic architecture of an animal's body, including the number and arrangement of limbs, body segments, and major tissue layers.

bog A wetland that has no or almost no water flow, resulting in very low oxygen levels and acidic conditions.

Bohr shift The rightward shift of the oxygen-hemoglobin dissociation curve that occurs with decreasing pH. Results in hemoglobin being more likely to release oxygen in the acidic environment of exercising muscle.

bone A type of vertebrate connective tissue consisting of living cells and blood vessels within a hard extracellular matrix composed of calcium phosphate ($CaPO_4$) and small amounts of calcium carbonate ($CaCO_3$) and protein fibers.

bone marrow The soft tissue filling the inside of long bones containing stem cells that develop into red blood cells and leukocytes throughout life.

Bowman's capsule The hollow, double-walled cup-shaped portion of a nephron that surrounds a glomerulus in the vertebrate kidney.

braincase See **cranium.**

brain stem The most posterior portion of the vertebrate brain, connecting to the spinal cord and responsible for autonomic body functions such as heart rate, respiration, and digestion.

branch (1) A part of a phylogenetic tree that represents populations through time. (2) Any extension of a plant's shoot system.

bronchiole One of the small tubes in mammalian lungs that carry air from the bronchi to the alveoli.

bronchus (plural: bronchi) In mammals, one of a pair of large tubes that lead from the trachea to each lung.

brown adipose tissue A specialized form of adipose tissue whose cells contain a high density of mitochondria as well as stored fats and that can produce extra body heat. Found in some mammals.

bryophytes Members of several phyla of green plants that lack vascular tissue including liverworts, hornworts, and mosses. Also called *nonvascular plants.*

budding Asexual reproduction via outgrowth from the parent that eventually break free as an independent individual; occurs in yeasts and some invertebrates.

buffer A substance that, in solution, acts to minimize changes in the pH of that solution when acid or base is added.

bulbourethral glands In male mammals, small paired glands at the base of the urethra that secrete an alkaline mucus (part of semen), which lubricates the tip of the penis and neutralizes acids in the urethra during copulation. In humans, also called *Cowper's glands.*

bulk flow The directional movement of a substantial volume of fluid due to pressure differences, such as movement of water through plant phloem and movement of blood in animals.

bundle-sheath cell A type of cell found around the vascular tissue (veins) of plant leaves.

C₃ photosynthesis The most common form of photosynthesis in which atmospheric CO_2 is used to form 3-phosphoglycerate, a three-carbon sugar.

C₄ photosynthesis A variant type of photosynthesis in which atmospheric CO_2 is first fixed into four-carbon sugars, rather than the three-carbon sugars of classic C₃ photosynthesis. Enhances photosynthetic efficiency in hot, dry environments, by reducing loss of oxygen due to photorespiration.

cadherin Any of a class of cell-surface proteins involved in cell adhesion and important for coordinating movements of cells during embryological development.

callus In plants, a mass of undifferentiated cells that can generate roots and other tissues necessary to create a mature plant.

Calorie A unit of energy often used to measure the energy content of food. Also called *kilocalorie.*

Calvin cycle In photosynthesis, the set of light-independent reactions that use NADPH and ATP formed in the light-dependent reactions to drive the fixation of atmospheric CO_2 and reduction of the fixed carbon, ultimately producing sugars. Also called *carbon fixation* and *light-independent reactions.*

calyx All of the sepals of a flower.

CAM See **crassulacean acid metabolism.**

cambium (plural: cambia) See **lateral meristem.**

Cambrian explosion The rapid diversification of animal body types that began about 543 million years ago and continued for approximately 40 million years.

camera eye The type of eye in vertebrates and cephalopods, consisting of a hollow chamber with a hole at one end (through which light enters) and a sheet of light-sensitive cells against the opposite wall.

cAMP See **cyclic AMP.**

cancer General term for any tumor whose cells grow in an uncontrolled fashion, invade nearby tissues, and spread to other sites in the body.

canopy The uppermost layers of branches in a forest (i.e., those fully exposed to the Sun).

5′ cap A chemical grouping, consisting of 7-methylguanylate and three phosphate groups, that is added to the 5′ end of newly transcribed messenger RNA molecules.

CAP binding site A DNA sequence upstream of certain prokaryotic operons to which catabolite activator protein can bind, increasing gene transcription.

capillarity The tendency of water to move up a narrow tube due to surface tension, adhesion, and cohesion.

capillary One of the numerous small, thin-walled blood vessels that permeate all tissues and organs, and allow exchange of gases and other molecules between blood and body cells.

capillary bed A thick network of capillaries.

capsid A shell of protein enclosing the genome of a virus particle.

carapace In crustaceans, a large platelike section of the exoskeleton that covers and protects the cephalothorax (e.g., a crab's "shell").

carbohydrate Any of a class of molecules that contain a carbonyl group, several hydroxyl groups, and several to many carbon-hydrogen bonds. See **monosaccharide** and **polysaccharide**.

carbon cycle, global The worldwide movement of carbon among terrestrial ecosystems, the oceans, and the atmosphere.

carbon fixation See **Calvin cycle**.

carbonic anhydrase An enzyme that catalyzes the formation of carbonic acid (H_2CO_3) from carbon dioxide and water.

cardiac cycle One complete heartbeat cycle, including systole and diastole.

cardiac muscle The muscle tissue of the vertebrate heart. Consists of long branched fibers that are electrically connected and that initiate their own contractions; not under voluntary control. Compare with **skeletal** and **smooth muscle**.

carnivore (adjective: carnivorous) An animal whose diet consists predominantly of meat. Most members of the mammalian taxon Carnivora are carnivores. Some plants are carnivorous, trapping and killing small animals, then absorbing nutrients from the prey's body. Compare with **herbivore** and **omnivore**.

carotenoid Any of a class of accessory pigments, found in chloroplasts, that absorb wavelengths of light not absorbed by chlorophyll; typically appear yellow, orange, or red. Includes carotenes and xanthophylls.

carpel The female reproductive organ in a flower. Consists of the stigma, to which pollen grains adhere; the style, through which pollen grains move; and the ovary, which houses the ovule. Compare with **stamen**.

carrier A heterozygous individual carrying a normal allele and a recessive allele for an inherited trait; does not display the phenotype of the trait but can pass the recessive gene to offspring.

carrier protein A membrane protein that facilitates diffusion of a small molecule (e.g., glucose) across the plasma membrane by a process involving a reversible change in the shape of the protein. Also called *carrier* or *transporter*.

carrying capacity (K) The maximum population size of a certain species that a given habitat can support.

cartilage A type of vertebrate connective tissue that consists of relatively few cells scattered in a stiff matrix of polysaccharides and protein fibers.

Casparian strip In plant roots, a waxy layer containing suberin, a water-repellent substance, that prevents movement of water through the walls of endodermal cells, thus blocking the apoplastic pathway.

catabolic pathway Any set of chemical reactions that breaks down larger, complex molecules into smaller ones, releasing energy in the process. Compare with **anabolic pathway**.

catabolite activator protein (CAP) A protein that can bind to the CAP binding site upstream of certain prokaryotic operons, facilitating binding of RNA polymerase and stimulating gene expression.

catabolite repression A type of positive transcriptional control in which the end product of a catabolic pathway inhibits further transcription of the gene encoding an enzyme early in the pathway.

catalysis (verb: catalyze) Acceleration of the rate of a chemical reaction due to a decrease in the free energy of the transition state, called the activation energy.

catalyst Any substance that increases the rate of a chemical reaction without itself undergoing any permanent chemical change.

catecholamines A class of small compounds, derived from the amino acid tyrosine, that are used as hormones or neurotransmitters. Include epinephrine, norepinephrine, and dopamine.

cation A positively charged ion.

cation exchange In botany, the release (displacement) of cations, such as magnesium and calcium from soil particles, by protons in acidic soil water. The released cations are available for uptake by plants.

CD4 A membrane protein on the surface of some T cells in humans. CD4+ T cells can give rise to helper T cells.

CD8 A membrane protein on the surface of some T cells in humans. CD8+ T cells can give rise to cytotoxic T cells.

Cdk See **cyclin-dependent kinase**.

cDNA See **complementary DNA**.

cDNA library A set of cDNAs from a particular cell type or stage of development. Each cDNA is carried by a plasmid or other cloning vector and can be separated from other cDNAs. Compare with **genomic library**.

cecum A blind sac between the small intestine and the colon. Is enlarged in some species (e.g., rabbits) that use it as a fermentation vat for digestion of cellulose.

cell A highly organized compartment bounded by a thin, flexible structure (plasma membrane) and containing concentrated chemicals in an aqueous (watery) solution. The basic structural and functional unit of all organisms.

cell body The part of a neuron that contains the nucleus and where incoming signals are integrated. Also called the *soma*.

cell crawling A form of cellular movement involving actin filaments in which the cell produces bulges (pseudopodia) that stick to the substrate and pull the cell forward. Also called *amoeboid motion*.

cell cycle Ordered sequence of events in which a eukaryotic cell replicates its chromosomes, evenly partitions the chromosomes to two daughter cells, and then undergoes division of the cytoplasm.

cell-cycle checkpoint Any of several points in the cell cycle at which progression of a cell through the cycle can be regulated.

cell division Creation of new cells by division of pre-existing cells.

cell-mediated (immune) response The type of immune response that involves generation of cytotoxic T cells from CD8+ T cells. Defends against pathogen-infected cells, cancer cells, and transplanted cells. Compare with **humoral (immune) response**.

cell membrane See **plasma membrane**.

cell plate A double layer of new plasma membrane that appears in the middle of a dividing plant cell; ultimately divides the cytoplasm into two separate cells.

cell sap An aqueous solution found in the vacuoles of plant cells.

cell theory The theory that all organisms are made of cells and that all cells come from preexisting cells.

cellular respiration A common pathway for production of ATP, involving transfer of electrons from compounds with high potential energy (often NADH and FADH$_2$) to an electron transport chain and ultimately to an electron acceptor (often oxygen).

cellulose A structural polysaccharide composed of β-glucose monomers joined by β-1,4-glycosidic linkages. Found in the cell wall of algae, plants, bacteria, fungi, and some other groups.

cell wall A protective layer located outside the plasma membrane and usually composed of polysaccharides. Found in algae, plants, bacteria, fungi, and some other groups.

Cenozoic era The most recent period of geologic time, beginning 65 million years ago. during which mammals became the dominant vertebrates and angiosperms became the dominant plants. Also called *Age of Mammals*.

central dogma The long-accepted hypothesis that information in cells flows in one direction: DNA codes for RNA, which codes for proteins. Exceptions are now known (e.g., retroviruses).

central nervous system (CNS) The brain and spinal cord of vertebrate animals. Compare with **peripheral nervous system (PNS)**.

centriole One of two small cylindrical structures, structurally similar to a basal body, found together within the centrosome near the nucleus of a eukaryotic cell.

centromere Constricted region of a replicated chromosome where the two sister chromatids are joined and the kinetochore is located.

centrosome Structure in animal and fungal cells, containing two centrioles, that serves as a microtubule-organizing center for the cell's cytoskeleton and for the mitotic spindle during cell division.

cephalization The formation of a distinct anterior region (the head) where sense organs and a mouth are clustered.

cephalochordates One of the three major chordate lineages (Cephalochordata), comprising small, mobile organisms that live in marine sands; also called *lancelets* or *amphioxi*. Compare with **urochordates** and **vertebrates**.

cephalopods A lineage of molluscs including the squid, octopuses, and nautiluses. Distinguished by large brains, excellent vision, tentacles, and a reduced or absent shell.

cerebellum Posterior section of the vertebrate brain that is involved in coordination of complex muscle movements, such as those required for locomotion and maintaining balance.

cerebrum The most anterior section of the vertebrate brain. Divided into left and right hemispheres and four lobes: parietal lobe, involved in complex decision making (in humans); occipital lobe, receives and interprets visual information; parietal lobe, involved in integrating sensory and motor functions; and temporal lobe, functions in memory, speech (in humans), and interpreting auditory information.

cervix The narrow passageway between the vagina and the uterus of female mammals.

chaetae (singular: chaeta) Bristle-like extensions found in some annelids.

channel protein A membrane protein that facilitates diffusion of a specific ion or small molecule across a plasma membrane through a central pore in the protein. Includes water channels (aquaporins) and ion channels. Also called simply *channel*.

character displacement The tendency for the traits of similar species that occupy overlapping ranges to change in a way that reduces interspecific competition.

chelicerae A pair of clawlike appendages found around the mouth of certain arthropods called chelicerates (spiders, mites, and allies).

chemical bond An attractive force binding two atoms together. Covalent bonds, ionic bonds, and hydrogen bonds are types of chemical bonds.

chemical carcinogen Any chemical that can cause cancer.

chemical energy The potential energy stored in covalent bonds between atoms.

chemical equilibrium A dynamic but stable state of a reversible chemical reaction in which the forward reaction and reverse reactions proceed at the same rate, so that the concentrations of reactants and products remain constant.

chemical evolution The theory that simple chemical compounds in the ancient atmosphere and ocean combined by spontaneous chemical reactions to form larger, more complex substances, eventually leading to the origin of life and the start of biological evolution.

chemical reaction Any process in which one compound or element is combined with others or is broken down; involves the making and/or breaking of chemical bonds.

chemiosmosis An energetic coupling mechanism whereby energy stored in an electrochemical proton gradient (proton-motive force) is used to drive an energy-requiring process such as production of ATP.

chemokine Any of several chemical signals that attract leukocytes to a site of tissue injury or infection.

chemolithotroph An organism that produces ATP by oxidizing inorganic molecules with high potential energy such as ammonia (NH_3) or methane (CH_4). Also called *lithotroph*. Compare with **chemoorganotroph.**

chemoorganotroph An organism that produces ATP by oxidizing organic molecules with high

potential energy such as sugars. Also called *organotroph*. Compare with **chemolithotroph.**

chemoreceptor A sensory cell or organ specialized for detection of specific molecules or classes of molecules.

chemotaxis Movement toward or away from a certain chemical.

chiasma (plural: chiasmata) The X-shaped structure formed during meiosis by crossing over between non-sister chromatids in a pair of homologous chromosomes.

chitin A structural polysaccharide composed of N-acetylglucosamine monomers joined end to end by β-1,4-glycosidic linkages. Found in cell walls of fungi and many algae, and in external skeletons of insects and crustaceans.

chitons A lineage of marine mollusc that have a protective shell formed of eight calcium carbonate plates.

chlorophyll Any of several closely related green pigments, found in chloroplasts and photosynthetic protists, that absorb light during photosynthesis.

chloroplast A chlorophyll-containing organelle, bounded by a double membrane, in which photosynthesis occurs; found in plants and photosynthetic protists. Also the location of amino acid, fatty acid, purine, and pyrimidine synthesis.

choanocyte A specialized flagellated feeding cell found in choanoflagellates (protists that are the closest living relatives of animals) and sponges (the oldest animal phylum).

cholecystokinin A peptide hormone secreted by cells in the lining of the small intestine. Stimulates the secretion of digestive enzymes from the pancreas and of bile from the liver and gallbladder.

chordates Members of the phylum Chordata, deuterostomes distinguished by a dorsal hollow nerve cord, pharyngeal gill slits, a notochord, a dorsal hollow nerve cord, and a post-anal tail. Include vertebrates, cephalochordata, and urochordata.

chromatid One of the two identical strands composing a replicated chromosome that is connected at the centromere to the other strand.

chromatin The complex of DNA and proteins, mainly histones, that compose eukaryotic chromosomes. Can be highly compact (heterochromatin) or loosely coiled (euchromatin).

chromatin remodeling The process by which the DNA in chromatin is unwound from its associated proteins to allow transcription or replication. May involve chemical modification of histone proteins or reshaping of the chromatin by large multi-protein complexes in an ATP-requiring process.

chromosome Gene-carrying structure consisting of a single long molecule of DNA and associated proteins (e.g., histones). Most prokaryotic cells contain a single, circular chromosome; eukaryotic cells contain multiple noncircular (linear) chromosomes located in the nucleus.

chromosome inversion See **inversion.**

chromosome painting A technique for producing high-resolution karyotypes by "painting" chromosomes with fluorescent tags that bind to particular regions of certain chromosomes. Also called *spectral karyotyping*.

chromosome theory of inheritance The principle that genes are located on chromosomes and that patterns of inheritance are determined by the behavior of chromosomes during meiosis.

chromosome translocation See **translocation** (2).

chylomicron A ball of protein-coated lipids used to transport the lipids through the bloodstream.

chytrid A member of a paraphyletic group of fungi (Chytridiomycota) that produce motile spores and gametes (both have flagella similar to those in sperm cells). Also called *chytridiomycete*.

cilium (plural: cilia) One of many short, filamentous projections of some eukaryotic cells containing a core of microtubules. Used to move the cell and/or to move fluid or particles along a stationary cell. See **axoneme.**

circadian clock An internal mechanism found in most organisms that regulates many body processes (sleep-wake cycles, hormonal patterns, etc.) in a roughly 24-hour cycle.

circulatory system The system in animals responsible for moving oxygen, carbon dioxide, and other materials (hormones, nutrients, wastes) around the body.

cisternae (singular: cisterna) Flattened, membrane-bound compartments that make up the Golgi apparatus.

clade See **monophyletic group.**

cladistic approach A method for constructing a phylogenetic tree that is based on identifying the unique traits of each monophyletic group. Compare with **phenetic approach.**

class In Linnaeus' system, a taxonomic category above the order level and below the phylum level.

Class I MHC protein An MHC protein that is present on the plasma membrane of virtually all nucleated cells and functions in presenting antigen to $CD8^+$ T cells.

Class II MHC protein An MHC protein that is present only on the plasma membrane of dendritic cells, macrophages, and B cells and functions in presenting antigen to $CD4^+$ T cells.

classical conditioning A type of learning in which an animal learns to associate two stimuli, so that a response originally given to just one stimulus can be evoked by the second stimulus as well.

cleavage In animal development, the series of rapid mitotic cell divisions, with little cell growth, that produces successively smaller cells and transforms a zygote into a multicellular blastula, or blastocyst in mammals.

cleavage furrow A pinching-in of the plasma membrane that occurs as cytokinesis begins in animal cells and deepens until the cytoplasm is divided.

climate The prevailing long-term weather conditions in a particular region.

climax community The stable, final community that develops from ecological succession.

clitoris A small rod of erectile tissue in the external genitalia of female mammals. Is formed from the same embryonic tissue as the male penis and has a similar function in sexual arousal.

cloaca An opening to the outside used by the excretory and reproductive systems in many nonmammalian vertebrate but few vertebrates.

clonal expansion Rapid cell division by a T cell or B cell in response to its binding the specific antigen recognized by receptors on the cell surface. Produces a large population of descendant cells all specific for that particular antigen.

clonal-selection theory The dominant explanation of the development of acquired immunity in vertebrates. According to the theory, the immune system retains a vast pool of inactive lymphocytes, each with a unique receptor for a unique antigen. Lymphocytes that encounter their antigens are stimulated to divide (selected and cloned), producing daughter cells that combat infection and confer immunity.

clone (1) An individual that is genetically identical to another individual. (2) A lineage of genetically identical individuals or cells. (3) As a verb, to make one or more genetic replicas of a cell or individual.

cloning vector A plasmid or other agent used to transfer recombinant genes into cultured host cells. Also called simply *vector*.

closed circulatory system A circulatory system in which the circulating fluid (blood) is confined to blood vessels and flows in a continuous circuit. Compare with **open circulatory system.**

club fungus A member of a lineage of fungi (Basidiomycota) that produce complex reproductive structures (e.g., mushrooms, puffballs) containing specialized spore-forming cells (basidia). Also called *basidiomycete.*

cnidocyte A specialized stinging cell found in cnidarians (e.g., jellyfish, corals, and anemones) that is used in capturing prey.

coactivator Any regulatory protein that help initiate transcription by bringing together the necessary transcription factors but does not itself bind to DNA.

cochlea The organ of hearing in the inner ear of mammals, birds, and crocodilians. A coiled, fluid-filled tube containing specialized pressure-sensing neurons (hair cells) that detect sounds of different pitches.

coding strand See **non-template strand.**

codominance An inheritance pattern in which heterozygotes exhibit both of the traits seen in either kind of homozygous individual.

codon A sequence of three nucleotides in DNA or RNA that codes for a certain amino acid or that initiates or terminates protein synthesis.

coefficient of relatedness (r) A measure of how closely two individuals are related. Calculated as the probability that an allele in two individuals is inherited from the same ancestor.

coelom An internal, usually fluid-filled, body cavity that is lined with mesoderm.

coelomate An animal that has a true coelom. Compare with **acoelomate** and **pseudocoelomate.**

coenocytic Containing many nuclei and a continuous cytoplasm through a filamentous body, without the body being divided into distinct cells. Some fungi are coenocytic.

coenzyme A small organic molecule that is a required cofactor for an enzyme-catalyzed reaction. Often donates or receives electrons or functional groups during the reaction.

coenzyme A (CoA) A nonprotein molecule that is required for many cellular reactions involving transfer of acetyl groups ($-COCH_3$).

coenzyme Q A nonprotein molecule that shuttles electrons between membrane-bound complexes in the mitochondrial electron transport chain. Also called *ubiquinone* or *Q.*

coevolution A pattern of evolution in which two interacting species reciprocally influence each other's adaptations over time.

cofactor A metal ion or small organic compound that is required for an enzyme to function normally. May be bound tightly to an enzyme or associate with it transiently during catalysis.

cognition The mental processes involved in recognition and manipulation of facts about the world, particularly to form novel associations or insights.

cohesion The tendency of certain like molecules (e.g., water molecules) to cling together due to attractive forces. Compare with **adhesion.**

cohesion-tension theory The theory that water movement upward through plant vascular tissues is due to loss of water from leaves (transpiration), which pulls a cohesive column of water upward.

cohort A group of individuals that are the same age and can be followed through time.

coleoptile A modified leaf that covers and protects the stems and leaves of young grasses.

collagen A fibrous, pliable, cable-like glycoprotein that is a major component of the extracellular matrix of animal cells. Various subtypes differ in their tissue distribution.

collecting duct In the vertebrate kidney, a large straight tube that receives filtrate from the distal tubules of several nephrons. Involved in the regulated reabsorption of water.

collenchyma cell In plants, an elongated cell with cell walls thickened at the corners that provides support to growing plant parts; usually found in strands along leaf veins and stalks. Compare with **parenchyma cell** and **sclerenchyma cell.**

colon The portion of the large intestine where feces are formed by compaction of wastes and reabsorption of water.

colony An assemblage of individuals. May refer to an assemblage of semi-independent cells or to a breeding population of multicellular organisms.

commensalism (adjective: commensal) A symbiotic relationship in which one organism (the commensal) benefits and the other (the host) is not harmed. Compare with **mutualism** and **parasitism.**

communication In ecology, any process in which a signal from one individual modifies the behavior of another individual.

community All of the species that interact with each other in a certain area.

companion cell In plants, a cell in the phloem that is connected via numerous plasmodesmata to adjacent sieve-tube members. Companion cells provide materials to maintain sieve-tube members and function in the loading and unloading of sugars into sieve-tube members.

compass orientation A type of navigation in which movement occurs in a specific direction.

competition In ecology, the interaction of two species or two individuals trying to use the same limited resource (e.g., water, food, living space). May occur between individuals of the same species (intraspecific competition) or different species (interspecific competition).

competitive exclusion principle The principle that two species cannot coexist in the same ecological niche in the same area because one species will out-compete the other.

competitive inhibition Inhibition of an enzyme's ability to catalyze a chemical reaction via a nonreactant molecule that competes with the substrate(s) for access to the active site.

complementary base pairing The association between specific nitrogenous bases of nucleic acids stabilized by hydrogen bonding. Adenine pairs only with thymine (in DNA) or uracil (in RNA), and guanine pairs only with cytosine.

complementary DNA (cDNA) DNA produced in the laboratory using an RNA transcript as a template and reverse transcriptase; corresponds to a gene but lacks introns. Also produced naturally by retroviruses.

complementary strand A newly synthesized strand of RNA or DNA that has a base sequence complementary to that of the template strand.

complement system A set of proteins that circulate in the bloodstream and can form holes in the plasma membrane of bacteria, leading to their destruction.

complete metamorphosis See **holometabolous metamorphosis.**

compound eye An eye formed of many independent light-sensing columns (ommatidia); occurs in arthropods. Compare with **simple eye.**

concentration gradient Difference across space (e.g., across a membrane) in the concentration of a dissolved substance.

condensation reaction A chemical reaction in which two molecules are joined covalently with the removal of an $-OH$ from one and an $-H$ from another to form water. Also called a *dehydration reaction.* Compare with **hydrolysis.**

conduction (1) Direct transfer of heat between two objects that are in physical contact. Compare with **convection.** (2) Transmission of an electrical impulse along the axon of a nerve cell.

cone In conifers (e.g., pines, cedars, and spruces), the reproductive structure in which microspores and macrospores are formed.

cone cell A photoreceptor cell with a cone-shaped outer portion that is particularly sensitive to bright light of a certain color. Also called simply *cone.* Compare with **rod cell.**

conjugation The process by which DNA is exchanged between unicellular individuals. Occurs in bacteria, archaea, and some protists.

connective tissue An animal tissue consisting of scattered cells in a liquid, jellylike, or solid extracellular matrix. Includes bone, cartilage, tendons, ligaments, and blood.

conservation biology The effort to study, preserve, and restore threatened populations, communities, and ecosystems.

constant (C) region The portion of an antibody's light chains or heavy chains that has the same amino acid sequence in the antibodies produced by every B cell of an individual. Compare with **variable (V) region.**

constitutive Always occurring; always present. Commonly used to describe enzymes and other proteins that are synthesized continuously or mutants in which one or more genetic loci are constantly expressed due to defects in gene control.

constitutive defense A defensive trait that is always manifested even in the absence of a predator or pathogen. Also called *standing defense.* Compare with **inducible defense.**

consumer See **heterotroph.**

continuous strand See **leading strand.**

control In a scientific experiment, a group of organisms or samples that do not receive the experimental treatment but are otherwise identical to the group that does.

convection Transfer of heat by movement of large volumes of a gas or liquid. Compare with **conduction.**

convergent evolution The independent evolution of analogous traits in distantly related organisms due to adaptation to similar environments and a similar way of life.

cooperative binding The tendency of the protein subunits of hemoglobin to affect each other's oxygen binding such that each bound oxygen molecule increases the likelihood of further oxygen binding.

copulation The act of transferring sperm from a male directly into a female's reproductive tract.

coral reef A large assemblage of colonial marine corals that usually serves as shallow-water, sunlit habitat for many other species as well.

co-receptor Any membrane protein that acts with some other membrane protein in a cell interaction or cell response.

cork cambium One of two types of lateral meristem, consisting of a ring of undifferentiated plant cells found just under the cork layer of woody plants; produces new cork cells on its outer side. Compare with **vascular cambium.**

cork cell A waxy cell in the protective outermost layer of a woody plant.

corm A rounded, thick underground stem that can produce new plants via asexual reproduction.

cornea The transparent sheet of connective tissue at the very front of the eye in vertebrates and some other animals. Protects the eye and helps focus light.

corolla All of the petals of a flower.

corona The cluster of cilia at the anterior end of a rotifer.

corpus callosum A thick band of neurons that connects the two hemispheres of the cerebrum in the mammalian brain.

corpus luteum A yellowish structure in an ovary that secretes progesterone. Is formed from a follicle that has recently ovulated.

cortex (1) The outermost region of an organ, such as the kidney or adrenal gland. (2) In plants, a layer of ground tissue found outside the vascular bundles and pith of a plant stem.

cortical granules Small enzyme-filled vesicles in the cortex of an egg cell. Involved in formation of the fertilization envelope after fertilization.

corticosterone A steroid hormone, produced and secreted by the adrenal cortex, that increases blood glucose and prepares the body for stress. The major glucocorticoid hormone in most reptiles, birds, and many mammals.

corticotropin-releasing hormone (CRH) A peptide hormone, produced and secreted by the hypothalamus, that stimulates the anterior pituitary to release ACTH.

cortisol A steroid hormone, produced and secreted by the adrenal cortex, that increases blood glucose and prepares the body for stress. The major glucocorticoid hormone in some mammals. Also called *hydrocortisone.*

cotransporter A transmembrane protein that facilitates diffusion of an ion down its previously established electrochemical gradient and uses the energy of that process to transport some other substance, in the same or opposite direction, *against* its concentration gradient. See **antiporter** and **symporter.**

cotyledon The first leaf, or seed leaf, of a plant embryo. Used for storing and digesting nutrients and/or for early photosynthesis.

countercurrent exchanger In animals, any anatomical arrangement that allows the maximum transfer of heat or a soluble substance from one fluid to another. The two fluids must be flowing in opposite directions and have a heat or concentration gradient between them.

covalent bond A type of chemical bond in which two atoms share one or more pairs of electrons. Compare with **hydrogen bond** and **ionic bond.**

cranium A bony, cartilaginous, or fibrous case that encloses and protects the brain of vertebrates. Forms part of the skull. Also called *braincase.*

crassulacean acid metabolism (CAM) A variant type of photosynthesis in which CO_2 is stored in organic acids at night when stomata are open and then released to feed the Calvin cycle during the day when stomata are closed. Helps reduce water loss and oxygen loss by photorespiration in hot, dry environments.

cristae (singular: crista) Sac-like invaginations of the inner membrane of a mitochondrion. Location of the electron transport chain and ATP synthase.

critical period See **sensitive period.**

Cro-Magnon A prehistoric European population of modern humans (*Homo sapiens*) known from fossils, paintings, sculptures, and other artifacts.

crossing over The exchange of segments of non-sister chromatids between a pair of homologous chromosomes that occurs during meiosis I.

cross-pollination Pollination of a flower by pollen from another individual, rather than by self-fertilization. Also called *crossing.*

crustaceans A lineage of arthropods that includes shrimp, lobster, and crabs. Many have a carapace (a platelike portion of the exoskeleton) and mandibles for biting or chewing.

culture In cell biology, a collection of cells or a tissue growing under controlled conditions, usually in suspension or on the surface of a dish on solid growth medium.

cup fungus See **sac fungus.**

Cushing's disease A human endocrine disorder caused by loss of feedback inhibition of cortisol on ACTH secretion. Characterized by high ACTH and cortisol levels and wasting of body protein reserves.

cuticle A protective coating secreted by the outermost layer of cells of an animal or a plant.

cyanobacteria A lineage of photosynthetic bacteria formerly known as blue-green algae. Likely the first life-forms to carry out oxygenic photosynthesis.

cyclic AMP (cAMP) Cyclic adenosine monophosphate; a small molecule, derived from ATP, that is widely used by cells in signal transduction and transcriptional control.

cyclic photophosphorylation Path of electron flow during the light-dependent reactions of photosynthesis in which photosystem I transfers excited electrons back to the electron transport chain of photosystem II, rather than to $NADP^+$. Also called *cyclic electron flow.* Compare with **Z scheme.**

cyclin One of several regulatory proteins whose concentrations fluctuate cyclically throughout the cell cycle.

cyclin-dependent kinase (Cdk) Any of several related protein kinases that are active only when bound to a cyclin. Involved in control of the cell cycle.

cytochrome c (cyt c) A soluble iron-containing protein that shuttles electrons between membrane-bound complexes in the mitochondrial electron transport chain.

cytokines A diverse group of autocrine signaling proteins, secreted largely by cells of the immune system, whose effects include stimulating leukocyte production, tissue repair, and fever. Generally function to regulate the intensity and duration of an immune response.

cytokinesis Division of the cytoplasm to form two daughter cells. Typically occurs immediately after division of the nucleus by mitosis or meiosis.

cytokinins A class of plant hormones that stimulate cell division and retard aging.

cytoplasm All of the contents of a cell, excluding the nucleus, and bounded by the plasma membrane.

cytoplasmic determinant A regulatory transcription factor or signaling molecule that is distributed unevenly in the cytoplasm of the egg cells of many animals and that directs early pattern formation in an embryo.

cytoplasmic streaming The directed flow of cytosol and organelles that facilitates distribution of materials within some large plant and fungal cells. Occurs along actin filaments and is powered by myosin.

cytoskeleton In eukaryotic cells, a network of protein fibers in the cytoplasm that are involved in cell shape, support, locomotion, and transport of materials within the cell. Prokaryotic cells have a similar but much less extensive network of fibers.

cytosol The fluid portion of the cytoplasm.

cytotoxic T cell An effector T cell that destroys infected cells and cancer cells. Is descended from an activated CD8$^+$ T cell that has interacted with antigen on an infected cell or cancer cell. Also called *cytotoxic T lymphocyte (CTL)* and *killer T cell.* Compare with **helper T cell.**

dalton (Da) See **atomic mass unit.**

Darwinian fitness See **fitness.**

day-neutral plant A plant whose flowering time is not affected by the relative length of day and night (the photoperiod). Compare with **long-day** and **short-day plant.**

dead space Portions of the air passages that are not involved in gas exchange with the blood, such as the trachea and bronchi.

deciduous Describing a plant that sheds leaves or other structures at regular intervals (e.g., each fall)

decomposer See **detritivore.**

decomposer food chain An ecological network of detritus, decomposers that eat detritus, and predators and parasites of the decomposers.

definitive host The host species in which a parasite reproduces sexually. Compare with **intermediate host.**

dehydration reaction See **condensation reaction.**

deleterious In genetics, referring to any mutation, allele, or trait that reduces an individual's fitness.

demography The study of factors that determine the size and structure of populations through time.

denaturation (verb: denature) For a macromolecule, loss of its three-dimensional structure and biological activity due to breakage of hydrogen bonds and disulfide bonds, usually caused by treatment with excess heat or extreme pH conditions.

dendrite   A short extension from a neuron's cell body that receives signals from other neurons.

dendritic cell   A type of leukocyte that ingests and digests foreign antigens, moves to a lymph node, and presents the antigens displayed on its membrane to CD4$^+$ T cells.

density dependent   In population ecology, referring to any characteristic that varies depending on population density.

density independent   In population ecology, referring to any characteristic that does not vary with population density.

deoxyribonucleic acid (DNA)   A nucleic acid composed of deoxyribonucleotides that carries the genetic information of a cell. Generally occurs as two intertwined strands, but these can be separated. See **double helix**.

deoxyribonucleoside triphosphate (dNTP)   A monomer that can be polymerized to form DNA. Consists of deoxyribose, a base (A, T, G, or C), and three phosphate groups; similar to a nucleotide, but with two more phosphate groups.

deoxyribonucleotide   See **nucleotide**.

depolarization   Change in membrane potential from its resting negative state to a less negative or a positive state; a normal phase in an action potential. Compare with **hyperpolarization**.

deposit feeder   An animal that eats its way through a food-containing substrate.

dermal tissue system   The tissue forming the outer layer of an organism. In plants, also called *epidermis;* in animals, forms two distinct layers: *dermis* and *epidermis*.

descent with modification   The phrase used by Darwin to describe his hypothesis of evolution by natural selection.

desmosome   A type of cell-cell attachment structure, consisting of cadherin proteins, that bind the cytoskeletons of adjacent animal cells together. Found where cells are strongly attached to each other. Compare with **gap junction** and **tight junction**.

determination   In embryogenesis, progressive changes in a cell that commit it to a particular cell fate. Once a cell is fully determined, it can differentiate only into a particular cell type (e.g., liver cell, brain cell).

detritivore   An organism whose diet consists mainly of dead organic matter (detritus). Various bacteria, fungi, and protists are detritivores. Also called *decomposer.*

detritus   A layer of dead organic matter that accumulates at ground level or on seafloors and lake bottoms.

deuterostomes   A major lineage of animals that share a pattern of embryological development, including radial cleavage, formation of the anus earlier than the mouth, and formation of the coelom by pinching off of layers of mesoderm from the gut. Includes echinoderms and chordates. Compare with **protostomes**.

developmental homology   A similarity in embryonic form, or in the fate of embryonic tissues, that is due to inheritance from a common ancestor.

diabetes insipidus   A human disease caused by defects in the kidney's system for conserving water. Characterized by production of large amounts of dilute urine.

diabetes mellitus   A human disease caused by defects in insulin production (type I) or the response of cells to insulin (type II). Characterized by abnormally high blood glucose levels and large amounts of glucose-containing urine.

diaphragm   An elastic, sheetlike structure. In mammals, the muscular sheet of tissue that separates the chest and abdominal cavities. Contracts and moves downward during inhalation, expanding the chest cavity.

diastole   The portion of the heartbeat cycle during which the atria or ventricles of the heart are relaxed. Compare with **systole**.

diastolic blood pressure   The force exerted by blood against artery walls during relaxation of the heart's left ventricle. Compare with **systolic blood pressure.**

dicot   Any plant that has two cotyledons (embryonic leaves) upon germination. The dicots do not form a monophyletic group. Also called *dicotyledonous plant.* Compare with **eudicot** and **monocot.**

dideoxy sequencing   A laboratory technique for determining the exact nucleotide sequence of DNA. Relies on the use of dideoxynucleotide triphosphates (ddNTPs), which terminate DNA replication.

diencephalon   The part of the mammalian brain that relays sensory information to the cerebellum and functions in maintaining homeostasis.

differential centrifugation   Procedure for separating cellular components according to their size and density by spinning a cell homogenate in a series of centrifuge runs. After each run, the supernatant is removed from the deposited material (pellet) and spun again at progressively higher speeds.

differential gene expression   Expression of different sets of genes in cells with the same genome. Responsible for creating different cell types.

differentiation   The process by which a relatively unspecialized cell becomes a distinct specialized cell type (e.g., liver cell, brain cell) usually by changes in gene expression. Also call *cell differentiation.*

diffusion   Spontaneous movement of a substance from a region of high concentration to one of low concentration (i.e., down a concentration gradient).

digestion   The physical and chemical breakdown of food into molecules that can be absorbed into the body of an animal.

digestive tract   The long tube that begins at the mouth and ends at the anus. Also called *alimentary canal, gastrointestinal(GI) tract,* or the *gut.*

dihybrid cross   A mating between two parents that are heterozygous for both of the two genes being studied.

dimer   An association of two molecules, which may be identical (homodimer) or different (heterodimer).

dioecious   Describing an angiosperm species that has male and female reproductive structures on separate plants. Compare with **monoecious.**

diploblast   (adjective: diploblastic) An animal whose body develops from two basic embryonic cell layers—ectoderm and endoderm. Compare with **triploblast.**

diploid   (1) Having two sets of chromosome s ($2n$). (2) A cell or an individual organism with two sets of chromosomes, one set inherited from the maternal parent and one set from the paternal parent. Compare with **haploid.**

directional selection   A pattern of natural selection that favors one extreme phenotype with the result that the average phenotype of a population changes in one direction. Generally reduces overall genetic variation in a population.

direct sequencing   A technique for identifying and studying microorganisms that cannot be grown in culture. Involves detecting and amplifying copies of certain specific genes in their DNA, sequencing these genes, and then comparing the sequences with the known sequences from other organisms.

disaccharide   A carbohydrate consisting of two monosaccharides (sugar residues) linked together.

discontinuous strand   See **lagging strand.**

discrete trait   An inherited trait that exhibits distinct phenotypic forms rather than the continuous variation characteristic of a quantitative traits such as body height.

dispersal   The movement of individuals from their place of origin (birth, hatching) to a new location.

disruptive selection   A pattern of natural selection that favors extreme phenotypes at both ends of the range of phenotypic variation. Maintains overall genetic variation in a population. Compare with **stabilizing selection.**

distal tubule   In the vertebrate kidney, the convoluted portion of a nephron into which filtrate moves from the loop of Henle. Involved in the regulated reabsorption of sodium and water. Compare with **proximal tubule.**

disturbance   In ecology, any event that disrupts a community, usually causing loss of some individuals or biomass from it.

disturbance regime   The characteristic disturbances that affect a given ecological community.

disulfide bond   A covalent bond between two sulfur atoms, typically in the side groups of some amino acids (e.g., cysteine). Often contributes to tertiary structure of proteins.

DNA   See **deoxyribonucleic acid.**

DNA cloning   Any of several techniques for producing many identical copies of a particular gene or other DNA sequence.

DNA fingerprinting   Any of several methods for identifying individuals by unique features of their genomes. Commonly involves using PCR to produce many copies of certain simple sequence repeats (microsatellites) and then analyzing their lengths.

DNA footprinting   A technique used to identify stretches of DNA that are bound by particular regulatory proteins.

DNA library   See **cDNA library** and **genomic library.**

DNA ligase   An enzyme that joins pieces of DNA by catalyzing formation of a phosphodiester bond between the pieces.

DNA microarray   A set of single-stranded DNA fragments, representing thousands of different genes, that are permanently fixed to a small glass slide. Can be used to determine which genes are expressed in different cell types, under different conditions, or at different developmental stages.

DNA polymerase   Any enzyme that catalyzes synthesis of DNA from deoxyribonucleotides.

domain   (1) A section of a protein that has a distinctive tertiary structure and function. (2) A taxonomic category, based on similarities in basic cellular biochemistry, above the kingdom level. The three recognized domains are Bacteria, Archaea, and Eukarya.

dominant   Referring to an allele that determines the phenotype of a heterozygous individual. Compare with **recessive.**

dopamine   A catecholamine neurotransmitter that functions mainly in a part of the mammalian brain involved with muscle control. Also functions as a hypothalamic inhibitory hormone that inhibits release of prolactin from the interior pituitary; also called *prolactin-inhibiting hormone (PIH)*.

dormancy   A temporary state of greatly reduced, or no, metabolic activity and growth in plants or plant parts (e.g., seeds, spores, bulbs, and buds).

dorsal   Toward an animal's back and away from its belly. The opposite of ventral.

double fertilization   An unusual form of reproduction seen in flowering plants, in which one sperm nucleus fuses with an egg to form a zygote and the other sperm nucleus fuses with two polar nuclei to form the triploid endosperm.

double helix   The secondary structure of DNA, consisting of two antiparallel DNA strands wound around each other.

downstream   In genetics, the direction in which RNA polymerase moves along a DNA strand. Compare with **upstream.**

Down syndrome   A human developmental disorder caused by trisomy of chromosome 21.

dyad symmetry   A type of symmetry in which an object can be superimposed on itself if rotated 180°. Occurs in some regulatory sequences of DNA. Also called *two-fold rotational symmetry.*

dynein   Any one of a class of motor proteins that use the chemical energy of ATP to "walk" along an adjacent microtubule. Dyneins are responsible for bending of cilia and flagella, play a role in chromosome movement during mitosis, and can transport certain organelles.

early endosome   A small membrane-bound vesicle, formed by endocytosis, that is an early stage in the formation of a lysosome.

ecdysone   An insect hormone that triggers either molting (to a larger larval form) or metamorphosis (to the adult form), depending on the level of juvenile hormone.

ecdysozoans   A major lineage of protostomes (Ecdysozoa) that grow by shedding their external skeletons (molting) and expanding their bodies. Includes arthropods, insects, crustaceans, nematodes, and centipedes. Compare with **lophotrochozoans.**

echinoderms   A major lineage of deuterostomes (Echinodermata) distinguished by adult bodies with five-sided radial symmetry, a water vascular system, and tube feet. Includes sea urchins, sand dollars, and sea stars.

echolocation   The use of echoes from vocalizations to obtain information about locations of objects in the environment.

ecology   The study of how organisms interact with each other and with their surrounding environment.

ecosystem   All the organisms that live in a geographic area, together with the nonliving (abiotic) components that affect or exchange materials with the organisms; a community and its physical environment.

ecosystem diversity   The variety of biotic components in a region along with abiotic components, such as soil, water, and nutrients.

ecosystem services   Alterations of the physical components of an ecosystem by living organisms, especially beneficial changes in the quality of the atmosphere, soil, water, etc.

ectoderm   The outermost of the three basic cell layers in most animal embryos; gives rise to the outer covering and nervous system. Compare with **endoderm** and **mesoderm.**

ectomycorrhizal fungi (EMF)   Fungi whose hyphae form a dense network that covers their host plant's roots but do not enter the root cells.

ectoparasite   A parasite that lives on the outer surface of the host's body.

ectotherm   An animal that does not use internally generated heat to regulate its body temperature. Compare with **endotherm.**

effector   Any cell, organ, or structure with which an animal can respond to external or internal stimuli. Usually functions, along with a sensor and integrator, as part of a homeostatic system.

efferent division   The part of the nervous system, consisting primarily of motor neurons, that carries commands from the central nervous system to the body.

egg   A mature female gamete and any associated external layers (such as a shell). Larger and less mobile than the male gamete. In animals, also called *ovum.*

ejaculation   The release of semen from the copulatory organ of a male animal.

ejaculatory duct   A short duct connecting the vas deferens to the urethra, through which sperm move during ejaculation.

elastic   Referring to a structure (e.g., lungs) with the ability to stretch and then spring back to its original shape.

electrical potential   Potential energy created by a separation of electric charges between two points. Also called *voltage.*

electric current   A flow of electrical charge past a point. Also called *current.*

electrocardiogram (EKG)   A recording of the electrical activity of the heart, as measured through electrodes on the skin.

electrochemical gradient   The combined effect of an ion's concentration gradient and electrical (charge) gradient across a membrane that affects the diffusion of ions across the membrane.

electrolyte   Any compound that dissociates into ions when dissolved in water. In nutrition, refers to the major ions necessary for normal cell function.

electromagnetic spectrum   The entire range of wavelengths of radiation extending from short wavelengths (high energy) to long wavelengths (low energy). Includes gamma rays, x-rays, ultra-violet, visible light, infrared, microwaves, and radio waves (from short to long wavelengths).

electron acceptor   A reactant that gains an electron and is reduced in a reduction-oxidation reaction.

electron carrier   Any molecule that readily accepts electrons from and donates electrons to other molecules.

electron donor   A reactant that loses an electron and is oxidized in a reduction-oxidation reaction.

electronegativity   A measure of the ability of an atom to attract electrons toward itself from an atom to which it is bonded.

electron microscope   See **scanning electron microscope** and **transmission electron microscope.**

electron shell   A group of orbitals of electrons with similar energies. Electron shells are arranged in roughly concentric layers around the nucleus of an atom, with electrons in outer shells having more energy than those in inner shells. Electrons in the outermost shell, the valence shell, often are involved in chemical bonding.

electron transport chain (ETC)   Any set of membrane-bound protein complexes and smaller soluble electron carriers involved in a coordinated series of redox reactions in which the potential energy of electrons transferred from reduced donors is successively decreased and used to pump protons from one side of a membrane to the other.

electroreceptor   A sensory cell or organ specialized to detect electric fields.

element   A substance, consisting of atoms with a specific number of protons, that cannot be separated into or broken down to any other substance. Elements preserve their identity in chemical reactions.

elongation   (1) The process by which messenger RNA lengthens during transcription. (2) The process by which a polypeptide chain lengthens during translation.

elongation factors   Proteins involved in the elongation phase of translation, assisting ribosomes in the synthesis of the growing peptide chain.

embryo   A young developing organism; the stage after fertilization and zygote formation.

embryo sac   The female gametophyte in flowering plants that exhibit alternation of generations.

embryogenesis   The process by which a single-celled zygote becomes a multicellular embryo.

embryophyte   A plant that nourishes its embryos inside its own body. All land plants are embryophytes.

emergent vegetation   Any plants in an aquatic habitat that extend above the surface of the water.

emerging disease   Any infectious disease, often a viral disease, that suddenly afflicts significant numbers of humans for the first time; often due to changes in the host species for a pathogen or host population movements.

emigration   The migration of individuals away from one population to other populations. Compare with **immigration.**

emulsification   (verb: emulsify) The dispersion of fat into an aqueous solution. Usually requires the aid of an amphipathic substance such as a detergent or bile salts, which can break large fat globules into microscopic fat droplets.

endangered species   A species whose numbers have decreased so much that it is in danger of extinction throughout all or part of its range.

endemic species A species that lives in one geographic area and nowhere else.

endergonic Referring to a chemical reaction that requires an input of energy to occur and for which the Gibbs free-energy change (ΔG) > 0. Compare with **exergonic**.

endocrine Relating to a chemical signal (hormone) that is released into the bloodstream by a producing cell and acts on a distant target cell.

endocrine gland A gland that secretes hormones directly into the bloodstream or interstitial fluid instead of into ducts. Compare with **exocrine gland**.

endocrine system All of the glands and tissues that produce and secrete hormones into the bloodstream.

endocytosis General term for any pinching off of the plasma membrane that results in the uptake of material from outside the cell. Includes phagocytosis, pinocytosis, and receptor-mediated endocytosis. Compare with **exocytosis**.

endoderm The innermost of the three basic cell layers in most animal embryos; gives rise to the digestive tract and organs that connect to it (liver, lungs, etc.). Compare with **ectoderm** and **mesoderm**.

endodermis In plant roots, a cylindrical layer of cells that separates the cortex from the vascular tissue.

endomembrane system A system of organelles in eukaryotic cells that performs most protein and lipid synthesis. Includes the endoplasmic reticulum (ER), Golgi apparatus, and lysosomes.

endoparasite A parasite that lives inside the host's body.

endophyte (adjective: endophytic) A fungus that lives inside the aboveground parts of a plant in a symbiotic relationship. Compare with **epiphyte**.

endoplasmic reticulum (ER) A network of interconnected membranous sacs and tubules found inside eukaryotic cells. See **rough** and **smooth endoplasmic reticulum**.

endoskeleton Bony and/or cartilaginous structures within the body that provide support. Examples are the spicules of sponges, the plates in echinoderms, and the bony skeleton of vertebrates. Compare with **exoskeleton**.

endosome See **early** and **late endosome**.

endosperm A triploid ($3n$) tissue in the seed of a flowering plant (angiosperm) that serves as food for the plant embryo. Functionally analogous to the yolk in some animal eggs.

endosymbiont An organism that lives in a symbiotic relationship inside the body of its host.

endosymbiosis theory The theory that mitochondria and chloroplasts evolved from prokaryotes that were engulfed by host cells and took up a symbiotic existence within those cells, a process termed primary endosymbiosis. In some eukaryotes, chloroplasts originated by secondary endosymbiosis, that is, by engulfing a chloroplast-containing protist and retaining its chloroplasts.

endotherm An animal whose primary source of body heat is internally generated heat. Compare with **ectotherm**.

endothermic Referring to a chemical reaction that absorbs heat. Compare with **exothermic**.

energetic coupling In cellular metabolism, the mechanism by which energy released from an exergonic reaction (commonly, hydrolysis of ATP) is used to drive an endergonic reaction.

energy The capacity to do work or to supply heat. May be stored (potential energy) or available in the form of motion (kinetic energy).

enhancer A regulatory sequence in eukaryotic DNA that may be located far from the gene it controls or within introns of the gene. Binding of specific proteins to an enhancer enhances the transcription of certain genes.

enrichment culture A method of detecting and obtaining cells with specific characteristics by placing a sample, containing many types of cells, under a specific set of conditions (e.g., temperature, salt concentration, available nutrients) and isolating those cells that grow rapidly in response.

entropy (S) A quantitative measure of the amount of disorder of any system, such as a group of molecules.

envelope, viral A membrane-like covering that encloses some viruses and their capsid coats, shielding them from attack by the host's immune system.

environmental sequencing The inventory of all the genes in a community or ecosystem by sequencing, analyzing, and comparing the genomes of the component organisms.

enzyme A protein catalyst used by living organisms to speed up and control biological reactions.

epicotyl In some embryonic plants, a portion of the embryonic stem that extends above the cotyledons.

epidemic The spread of an infectious disease throughout a population in a short time period. Compare with **pandemic**.

epidermis The outermost layer of cells of any multicellular organism.

epididymis A coiled tube wrapped around the testis in reptiles, birds, and mammals. The site of the final stages of sperm maturation and storage.

epigenetic inheritance Pattern of inheritance involving differences in phenotype that are not due to changes in the nucleotide sequence of genes.

epinephrine A catecholamine hormone, produced and secreted by the adrenal medulla, that triggers rapid responses relating to the fight-or-flight response. Also called *adrenaline*.

epiphyte (adjective: epiphytic) A nonparasitic plant that grows on trees or other solid objects and is not rooted in soil.

epithelium (plural: epithelia) An animal tissue consisting of sheet-like layers of tightly packed cells that lines an organ, a duct, or a body surface. Also called *epithelial tissue*.

epitope A small region of a particular antigen to which an antibody, B-cell receptor, or T-cell receptor binds.

equilibrium curve See **oxygen-hemoglobin equilibrium curve**.

equilibrium potential The membrane potential at which there is no net movement of a particular ion into or out of a cell.

ER signal sequence A short amino acid sequence that marks a polypeptide for transport to the endoplasmic reticulum where synthesis of the polypeptide chain is completed and the signal sequence removed. See **signal recognition particle**.

erythrocyte See **red blood cell**.

erythropoietin (EPO) A peptide hormone, released by the kidney in response to low blood oxygen levels, that stimulates the bone marrow to produce more red blood cells.

esophagus The muscular tube that connects the mouth to the stomach.

essential amino acid An amino acid that an animal cannot synthesize and must obtain from the diet. May refer specifically to one of the eight essential amino acids of adult humans: isoleucine, leucine, lysine, methionine, phenylalanine, threonine, tryptophan, and valine.

essential nutrient Any chemical element, ion, or compound that is required for normal growth, reproduction, and maintenance of a living organism and that cannot be synthesized by the organism.

ester linkage The covalent bond formed by a condensation reaction between a carboxyl group ($-COOH$) and a hydroxyl group ($-OH$). Ester linkages join fatty acids to glycerol to form a fat or phospholipid.

estradiol The major estrogen produced by the ovaries of female mammals. Stimulates development of the female reproductive tract, growth of ovarian follicles, and growth of breast tissue.

estrogens A class of steroid hormones, including estradiol, estrone, and estriol, that generally promote female-like traits. Secreted by the gonads, fat tissue, and some other organs.

estrous cycle A female reproductive cycle, seen in all mammals except Old World monkeys and apes (including humans), in which the uterine lining is reabsorbed rather than shed in the absence of pregnancy, and the female is sexually receptive only briefly during mid-cycle (estrus). Compare with **menstrual cycle**.

estuary An environment of brackish (partly salty) water where a river meets the ocean.

ethylene A gaseous plant hormone that induces fruits to ripen, flowers to fade, and leaves to drop.

eudicot A member of a monophyletic group (lineage) of angiosperms that includes complex flowering plants and trees (e.g., roses, daisies, maples). All eudicots have two cotyledons, but not all dicots are members of this lineage. Compare with **dicot** and **monocot**.

Eukarya One of the three taxonomic domains of life consisting of unicellular organisms (most protists, yeast) and multicellular organisms (fungi, plants, animals) distinguished by a membrane-bound cell nucleus, numerous organelles, and an extensive cytoskeleton. Compare with **Archaea** and **Bacteria**.

eukaryote A member of the domain Eukarya; an organism whose cells contain a nucleus, numerous membrane-bound organelles, and an extensive cytoskeleton. May be unicellular or multicellular. Compare with **prokaryote**.

eutherians A lineage of mammals (Eutheria) whose young develop in the uterus and are not housed in an abdominal pouch. Also called *placental mammals*.

evaporation The energy-absorbing phase change from a liquid state to a gaseous state. Many organisms evaporate water as a means of heat loss.

evo-devo Research field focused on how changes in developmentally important genes have led to the evolution of new phenotypes.

evolution (1) The theory that all organisms on Earth are related by common ancestry and that they have changed over time, predominantly via natural selection. (2) Any change in the genetic characteristics of a population over time, especially, a change in allele frequencies.

excitable membrane A plasma membrane that is capable of generating an action potential. Neurons, muscle cells, and some other cells have excitable membranes.

excitatory postsynaptic potential (EPSP) A change in membrane potential, usually depolarization, at a neuron dendrite that makes an action potential more likely.

exergonic Referring to a chemical reaction that can occur spontaneously, releasing heat and/or increasing entropy, and for which the Gibbs free-energy change (ΔG) < 0. Compare with **endergonic**.

exocrine gland A gland that secretes some substance through a duct into a space other than the circulatory system, such as the digestive tract or the skin surface. Compare with **endocrine gland**.

exocytosis Secretion of intracellular molecules (e.g., hormones, collagen), contained within membrane-bounded vesicles, to the outside of the cell by fusion of vesicles to the plasma membrane. Compare with **endocytosis**.

exon A region of a eukaryotic gene that is translated into a peptide or protein. Compare with **intron**.

exoskeleton A hard covering secreted on the outside of the body, used for body support, protection, and muscle attachment. Examples are the shell of molluscs and the outer covering (cuticle) of arthropods. Compare with **endoskeleton**.

exothermic Referring to a chemical reaction that releases heat. Compare with **endothermic**.

exotic species A nonnative species that is introduced into a new area. Exotic species often are competitors, pathogens, or predators of native species.

expansins A class of plant proteins that actively increase the length of the cell wall when the pH of the wall falls below 4.5.

exponential population growth The accelerating increase in the size of a population that occurs when the growth rate is constant and density independent. Compare with **logistic population growth**.

extensor A muscle that pulls two bones farther apart from each other, as in the extension of a limb or the spine. Compare with **flexor**.

extinct Said of a species that has died out.

extracellular digestion Digestion that takes place outside of an organism, as occurs in many fungi that make and secrete digestive enzymes.

extracellular matrix (ECM) A complex meshwork of proteins (e.g., collagen, fibronectin) and polysaccharides secreted by animal cells and in which they are embedded.

extremophile A bacterium or archaean that thrives in an "extreme" environment (e.g., high-salt, high-temperature, low-temperature, or low-pressure).

F_1 generation First filial generation. The first generation of offspring produced from a mating (i.e., the offspring of the parental generation).

facilitated diffusion Movement of a substance across a plasma membrane down its concentration gradient with the assistance of transmembrane carrier proteins or channel proteins.

facilitation In ecological succession, the phenomenon in which early-arriving species make conditions more favorable for later-arriving species. Compare with **inhibition** and **tolerance**.

facultative aerobe Any organism that can perform aerobic respiration when oxygen is available to serve as an electron acceptor but can switch to fermentation when it is not.

FAD/FADH$_2$ Oxidized and reduced forms, respectively, of flavin adenine dinucleotide. A nonprotein electron carrier that functions in the Krebs cycle and oxidative phosphorylation.

fallopian tube A narrow tube connecting the uterus to the ovary in humans, through which the egg travels after ovulation. Site of fertilization and cleavage. In nonhuman animals, called *oviduct*.

family In Linnaeus' system, a taxonomic category above genus and below order. In animals, family names usually end in the suffix *-idae*; Canidae (dogs, wolves and foxes) and Felidae (cats) are both in the order Carnivora.

fat A lipid consisting of three fatty acid molecules joined by ester linkages to a glycerol molecule. Also called *triacylglycerol* or *triglyceride*.

fatty acid A lipid consisting of a hydrocarbon chain bonded to a carboxyl group (−COOH) at one end. Used by many organisms to store chemical energy; a major component of animal and plant fats.

fauna All the animals characteristic of a particular region, period, or environment.

feather A specialized skin outgrowth, composed of β-keratin, present in all birds and only in birds. Used for flight, insulation, display, and other purposes.

feces The waste products of digestion.

fecundity The average number of female offspring produced by a single female in the course of her lifetime.

feedback inhibition A type of metabolic control in which high concentrations of the product of a metabolic pathway inhibit one of the enzymes early in the pathway. A form of negative feedback.

fermentation Any of several metabolic pathways that make ATP by transferring electrons from a reduced compound such as glucose to a final electron acceptor other than oxygen. Allows glycolysis to proceed in the absence of oxygen.

ferredoxin In photosynthetic organisms, an iron- and sulfur-containing protein in the electron transport chain of photosystem I. Can transfer electrons to the enzyme NADP$^+$ reductase, which catalyzes formation of NADPH.

fertilization Fusion of the nuclei of two haploid gametes to form a zygote with a diploid nucleus.

fertilization envelope A physical barrier that forms around a fertilized egg in amphibians and some other animals. Formed by an influx of water under the vitelline membrane.

fetal alcohol syndrome A condition, marked by hyperactivity, severe learning disabilities, and depression, thought to be caused by exposure of an individual to high blood alcohol concentrations during embryonic development.

fetus In live-bearing animals, the unborn offspring after the embryonic stage, which usually are developed sufficiently to be recognizable as belonging to a certain species. In humans, from 9 weeks after fertilization until birth.

fiber In plants, a type of elongated sclerenchyma cell that provides support to vascular tissue. Compare with **sclereid**.

fibronectin An abundant protein in the extracellular matrix that binds to other ECM components and to integrins in plasma membranes; helps anchor cells in place. Numerous subtypes are found in different tissues.

Fick's law of diffusion A mathematical relationship that describes the rates of gas exchange in animal respiratory systems.

fight-or-flight response Rapid physiological changes that prepare the body for emergencies. Includes increased heart rate, increased blood pressure, and decreased digestion.

filament Any thin, threadlike structure, particularly (1) the threadlike extensions of a fish's gills or (2) the slender stalk that bears the anthers in a flower.

filter feeder See **suspension feeder**.

filtrate Any fluid produced by filtration, in particular the fluid (pre-urine) in the nephrons of vertebrate kidneys.

filtration A process of removing large components from a fluid by forcing it through a filter. Occurs in a renal corpuscle of the vertebrate kidney, allowing water and small solutes to pass from the blood into the nephron.

finite rate of increase (λ) The rate of increase of a population over a given period of time. Calculated as the ending population size divided by the starting population size. Compare with **intrinsic rate of increase**.

first law of thermodynamics The principle of physics that energy is conserved in any process. Energy can be transferred and converted into different forms, but it cannot be created or destroyed.

fission (1) A form of asexual reproduction in which a prokaryotic cell divides to produce two genetically similar daughter cells by a process similar to mitosis of eukaryotic cells. Also called *binary fission*. (2) A form of asexual reproduction in which an animal splits into two or more individuals of approximately equal size; common among invertebrates.

fitness The relative ability of an individual to produce viable offspring compared with other individuals in the same population. Also called *Darwinian fitness*.

fitness trade-off See **trade-off**.

fixed action pattern (FAP) Highly stereotyped behavior pattern that occurs in a certain invariant way in a certain species. A form of innate behavior.

flaccid Limp as a result of low internal pressure (e.g., a wilted plant leaf). Compare with **turgid**.

flagellum (plural: flagella) A long, cellular projection that undulates (in eukaryotes) or rotates (in prokaryotes) to move the cell through an aqueous environment. See **axoneme**.

flatworms Members of the phylum Platyhelminthes. Distinguished by a broad, flat, unsegmented body that lacks a coelom. Flatworms belong to the lophotrochozoan branch of the protostomes.

flavin adenine dinucleotide See **FAD/FADH$_2$**.

flexor A muscle that pulls two bones closer together, as in the flexing of a limb or the spine. Compare with **extensor**.

floral meristem A group of undifferentiated plant cells that can give rise to the four organs making up a flower.

florigen In plants, a protein hormone that is synthesized in leaves and transported to the shoot apical meristem where it stimulates flowering.

flower In angiosperms, the part of a plant that contains reproductive structures. Typically includes a calyx, a corolla, and one or more stamens and/or carpels. See **perfect** and **inperfect flower.**

fluid feeder An animal that feeds by sucking or mopping up liquids such as nectar, plant sap, or blood.

fluid-mosaic model The widely accepted hypothesis that the plasma membrane and organelle membranes consist of proteins embedded in a fluid phospholipid bilayer.

fluorescence The spontaneous emission of light from an excited electron falling back to its normal (ground) state.

follicle An egg cell and its surrounding ring of supportive cells in a mammalian ovary.

follicle-stimulating hormone (FSH) A peptide hormone, produced and secreted by the anterior pituitary, that stimulates (in females) growth of eggs and follicles in the ovaries or (in males) sperm production in the testes.

follicular phase The first major phase of a menstrual cycle during which follicles grow and estrogen levels increase; ends with ovulation.

food Any nutrient-containing material that can be consumed and digested by animals.

food chain A relatively simple pathway of energy flow through a few species, each at a different trophic level, in an ecosystem. Might include, for example, a primary producer, a primary consumer, a secondary consumer, and a decomposer. Compare with **food web.**

food vacuole A membrane-bound organelle containing food ingested by a cell via phagocytosis.

food web Any complex pathway along which energy moves among many species at different trophic levels of an ecosystem.

foot One of the three main parts of the mollusc body; a muscular appendage, used for movement and/or burrowing into sediment.

forebrain One of the three main regions of the vertebrate brain; includes the cerebrum, thalamus, and hypothalamus. Compare with **hindbrain** and **midbrain.**

fossil Any trace of an organism that existed in the past. Includes tracks, burrows, fossilized bones, casts, etc.

fossil record All of the fossils that have been found anywhere on Earth and that have been formally described in the scientific literature.

founder effect A change in allele frequencies that often occurs when a new population is established from a small group of individuals (founder event) due to sampling error (i.e., the small group is not a representative sample of the source population).

free energy See **Gibbs free-energy change.**

free radical Any substance containing one or more atoms with an unpaired electron. Unstable and highly reactive.

frequency The number of wave crests per second traveling past a stationary point. Determines the pitch of sound and the color of light.

fronds The large leaves of ferns.

fruit In flowering plants (angiosperms), a mature, ripened plant ovary (or group of ovaries), along with the seeds it contains and any adjacent fused parts. See **aggregate, multiple,** and **simple fruit.**

fruiting body A structure formed in some prokaryotes, fungi, and protists for spore dispersal; usually consists of a base, a stalk, and a mass of spores at the top.

functional genomics The study of how a genome works, that is, when and where specific genes are expressed and how their products interact to produce a functional organism.

functional group A small group of atoms bonded together in a precise configuration and exhibiting particular chemical properties that it imparts to any organic molecule in which it occurs.

fundamental niche The ecological space that a species occupies in its habitat in the absence of competitors. Compare with **realized niche.**

fungi A lineage of eukaryotes that typically have a filamentous body (mycelium) and obtain nutrients by absorption.

fungicide Any substance that can kill fungi or slow their growth.

G₁ phase The phase of the cell cycle that constitutes the first part of interphase before DNA synthesis (S phase).

G₂ phase The phase of the cell cycle between synthesis of DNA (S phase) and mitosis (M phase); the last part of interphase.

gall A tumorlike growth that forms on plants infected with certain bacteria or parasites.

gallbladder A small pouch that stores bile from the liver and releases it as needed into the small intestine during digestion of fats.

gametangium (plural: gametangia) (1) The gamete-forming structure found in all land plants except angiosperms. Contains a sperm-producing antheridium and an egg-producing archegonium. (2) The gamete-forming structure of some chytrid fungi.

gamete A haploid reproductive cell that can fuse with another haploid cell to form a zygote. Most multicellular eukaryotes have two distinct forms of gametes: egg cells (ova) and sperm cells.

gametogenesis The production of gametes (eggs or sperm).

gametophyte In organisms undergoing alternation of generations, the multicellular haploid form that arises from a single haploid spore and produces gametes. A female gametophyte is commonly called an *embryo sac;* a male gametophyte, a *pollen grain.* Compare with **sporophyte.**

ganglion cell A neuron in the vertebrate retina that collects visual information from one or several bipolar cells and sends it to the brain via the optic nerve.

gap junction A type of cell-cell attachment structure that directly connects the cytoplasms of adjacent animal cells, allowing passage of water, ions, and small molecules between the cells. Compare with **desmosome** and **tight junction.**

gastrin A hormone produced by cells in the stomach lining in response to the arrival of food or to a neural signal from the brain. Stimulates other stomach cells to release hydrochloric acid.

gastrointestinal (GI) tract See **digestive tract.**

gastropods A lineage of mollusc distinguished by a large muscular foot and a unique feeding structure, the radula. Include slugs and snails.

gastrulation The process by which some cells on the outside of a young embryo move to the interior of the embryo, resulting in the three distinct germ layers (endoderm, mesoderm, and ectoderm).

gated channel A channel protein that opens and closes in response to a specific stimulus, such as the binding of a particular molecule or a change in the electrical charge on the outside of the membrane.

gel electrophoresis A technique for separating molecules on the basis of size and electric charge, which affect their differing rates of movement through a gelatinous substance in an electric field.

gemma (plural: gemmae) A small reproductive structure that is produced in some liverworts during the gametophyte phase and can grow into mature gametophyte.

gene A section of DNA (or RNA, for some viruses) that encodes information for building one or more related polypeptides or functional RNA molecules along with the regulatory sequences required for its transcription.

gene duplication The formation of an additional copy of a gene, typically by misalignment of chromosomes during crossing over. Thought to be an important evolutionary process in creating new genes.

gene expression Overall process by which the information encoded in genes is converted into an active product, most commonly a protein. Includes transcription and translation of a gene and in some cases protein activation.

gene family A set of genetic loci whose DNA sequences are extremely similar. Thought to have arisen by duplication of a single ancestral gene and subsequent mutations in the duplicated sequences.

gene flow The movement of alleles between populations; occurs when individuals leave one population, join another, and breed.

gene-for-gene hypothesis The hypothesis that there is a one-to-one correspondence between the resistance (*R*) loci of plants and the avirulence (*avr*) loci of pathogenic fungi; particularly, that *R* genes produce receptors and *avr* genes produce molecules that bind to those receptors.

gene pool All of the alleles of all of the genes in a certain population.

generation The average time between a mother's first offspring and her daughter's first offspring.

gene therapy The treatment of an inherited disease by introducing normal alleles.

genetic bottleneck A reduction in allelic diversity resulting from a sudden reduction in the size of a large population (population bottleneck) due to a random event.

genetic code The set of all 64 codons and the particular amino acids that each specifies.

genetic correlation A type of evolutionary constraint in which selection on one trait causes a change in another trait as well; may occur when the same gene(s) affect both traits.

genetic diversity The diversity of alleles in a population, species, or group of species.

genetic drift Any change in allele frequencies due to random events. Causes allele frequencies to drift up and down randomly over time, and eventually can lead to the fixation or loss of alleles.

genetic homology Similarities in DNA sequences or amino acid sequences that are due to inheritance from a common ancestor.

genetic map An ordered list of genes on a chromosome that indicates their relative distances from each other. Also called a *linkage map* or *meiotic map*. Compare with **physical map**.

genetic marker A genetic locus that can be identified and traced in populations by laboratory techniques or by a distinctive visible phenotype.

genetic model A set of hypotheses that explain how a certain trait is inherited.

genetic recombination A change in the combination of genes or alleles on a given chromosome or in a given individual. Also called *recombination*.

genetics The field of study concerned with the inheritance of traits.

genetic screen Any of several techniques for identifying individuals with a particular type of mutation. Also called a *screen*.

genetic variation (1) The number and relative frequency of alleles present in a particular population. (2) The proportion of phenotypic variation in a trait that is due to genetic rather than environmental influences in a certain population in a certain environment.

genitalia External copulatory organs.

genome All of the hereditary information in an organism, including not only genes but also other non-gene stretches of DNA.

genomic library A set of DNA segments representing the entire genome of a particular organism. Each segment is carried by a plasmid or other cloning vector and can be separated from other segments. Compare with **cDNA library**.

genomics The field of study concerned with sequencing, interpreting, and comparing whole genomes from different organisms.

genotype All of the alleles of every gene present in a given individual. May refer specifically to the alleles of a particular set of genes under study. Compare with **phenotype**.

genus (plural: genera) In Linnaeus' system, a taxonomic category of closely related species. Always italicized and capitalized to indicate that it is a recognized scientific genus.

geologic time scale The sequence of eons, epochs, and periods used to describe the geologic history of Earth.

geometric isomer A molecule that shares the same molecular formula as another molecule but differs in the arrangement of atoms or groups on either side of a double bond or ring structure. Compare with **optical isomer** and **structural isomer**.

germ cell In animals, any cell that can potentially give rise to gametes. Also called *germ-line cells*.

germination The process by which a seed becomes a young plant.

germ layer In animals, one of the three basic types of tissue formed during gastrulation; gives rise to all other tissues. See **endoderm, mesoderm,** and **ectoderm**.

germ theory of disease The theory that infectious diseases are caused by bacteria, viruses, and other microorganisms.

gestation The duration of embryonic development from fertilization to birth in those species that have live birth.

gibberellins A class of hormones found in plants and fungi that stimulate growth. Gibberellic acid is one of the major gibberellins.

Gibbs free-energy change (ΔG) A measure of the change in potential energy and entropy that occurs in a given chemical reaction. $\Delta G < 0$ for spontaneous reactions and >0 for nonspontaneous reactions.

gill Any organ in aquatic animals that exchanges gases and other dissolved substances between the blood and the surrounding water. Typically, a filamentous outgrowth of a body surface.

gill arch In aquatic vertebrates, curved region of tissue between the gills. Gills are suspended from the gill arches.

gill filament In fish, one of the many long, thin structures that extend from gill arches into the water and across which gas exchange occurs.

gill lamella (plural: gill lamellae) One of hundreds to thousands of sheetlike structures, each containing a capillary bed, that makes up a gill filament.

gland An organ whose primary function is to secrete some substance, either into the blood (endocrine gland) or into some other space such as the gut or skin (exocrine gland).

glia Collective term for several types of cells in nervous tissue that are not neurons and do not conduct electrical signals but provide support, nourishment, and electrical insulation and perform other functions. Also called *glial cells*.

global warming A sustained increase in Earth's average surface temperature.

glomerulus (plural: glomeruli) (1) In the vertebrate kidney, a ball-like cluster of capillaries, surrounded by Bowman's capsule, at the beginning of a nephron. (2) In the brain, a ball-shaped cluster of neurons in the olfactory bulb.

glucagon A peptide hormone produced by the pancreas in response to low blood glucose. Raises blood glucose by triggering breakdown of glycogen and stimulating gluconeogenesis. Compare with **insulin**.

glucocorticoids A class of steroid hormones, produced and secreted by the adrenal cortex, that increase blood glucose and prepare the body for stress. Include cortisol and corticosterone. Compare with **mineralocorticoids**.

gluconeogenesis Synthesis of glucose from non-carbohydrate sources (e.g., proteins and fatty acids). Occurs in the liver in response to low insulin levels and high glucagon levels.

glucose Six-carbon monosaccharide whose oxidation in cellular respiration is the major source of ATP in animal cells.

glyceraldehyde-3-phosphate (G3P) The phosphorylated three-carbon compound formed as the result of carbon fixation in the first step of the Calvin cycle.

glycerol A three-carbon molecule that forms the "backbone" of phospholipids and most fats.

glycogen A highly branched storage polysaccharide composed of α-glucose monomers joined by 1,4- and 1,6-glycosidic linkages. The major form of stored carbohydrate in animals.

glycolipid Any lipid molecule that is covalently bonded to a carbohydrate group.

glycolysis A series of 10 chemical reactions that oxidize glucose to produce pyruvate and ATP. Used by all organisms as part of fermentation or cellular respiration.

glycoprotein Any protein with one or more covalently bonded carbohydrate groups.

glycosidic linkage The covalent bond formed by a condensation reaction between two sugar monomers; joins the residues of a polysaccharide.

glycosylation Addition of a carbohydrate group to a molecule.

glyoxisome Specialized type of peroxisome found in plant cells and packed with enzymes for processing the products of photosynthesis.

goblet cell A cell in the stomach lining that secretes mucus.

goiter A pronounced swelling of the thyroid gland; usually caused by a deficiency of iodine in the diet.

Golgi apparatus A eukaryotic organelle, consisting of stacks of flattened membranous sacs (cisternae), that functions in processing and sorting proteins and lipids destined to be secreted or directed to other organelles. Also called *Golgi complex*.

gonad An organ that produces reproductive cells, such as a testis or an ovary.

gonadotropin-releasing hormone (GnRH) A peptide hormone, produced and secreted by the hypothalamus, that stimulates release of FSH and LH from the anterior pituitary.

G protein Any of various peripheral membrane proteins that bind GTP and function in signal transduction. Binding of a signal to its receptor triggers activation of the G protein, leading to production of a second messenger or initiation of a phosphorylation cascade.

grade In taxonomy, a group of species that share a position in an inferred evolutionary sequence of lineages but that are not a monophyletic group. Also called a *paraphyletic group*.

Gram-negative Describing bacteria that look pink when treated with a Gram stain. These bacteria have a cell wall composed of a thin layer of peptidoglycan and an outer phospholipid layer.

Gram-positive Describing bacteria that look purple when treated with a Gram stain. These bacteria have cell walls composed of a thick layer of peptidoglycan.

Gram stain A dye that distinguishes the two general types of cell walls found in bacteria. Used to routinely classify bacteria as Gram-negative or Gram-positive.

granum (plural: grana) In chloroplasts, a stack of flattened, membrane-bound vesicles (thylakoids) where the light reactions of photosynthesis occur.

gravitropism The growth or movement of a plant in a particular direction in response to gravity.

grazing food chain The ecological network of herbivores and the predators and parasites that consume them.

great apes See **hominids**.

green algae A paraphyletic group of photosynthetic organisms that contain chloroplasts similar to those in green plants. Often classified as protists, green algae are the closest living relatives of land plants and form a monophyletic group with them.

greenhouse gas   An atmospheric gas that absorbs and reflects infrared radiation, so that heat radiated from Earth is retained in the atmosphere instead of being lost to space.

green plant   A member of a lineage of eukaryotes that includes green algae and land plants.

gross photosynthetic productivity   The efficiency with which all the plants in a given area use the light energy available to them to produce sugars.

gross primary productivity   In an ecosystem, the total amount of carbon fixed by photosynthesis, including that used for cellular respiration, over a given time period. Compare with **net primary productivity.**

ground meristem   The middle layer of a young plant embryo. Gives rise to the ground tissue system.

ground tissue system   In plants, all the tissues beneath the outer protective layers of epidermis and cork except for vascular tissue. Also called simply *ground tissue.*

groundwater   Any water below the land surface.

growth factor   Any of a large number of signaling molecules that are secreted by certain cells and that stimulate other cells to divide or to differentiate.

growth hormone (GH)   A peptide hormone, produced and secreted by the mammalian anterior pituitary, that promotes lengthening of the long bones in children and muscle growth, tissue repair, and lactation in adults. Also called *somatotropin.*

GTP   See **guanosine triphosphate.**

guanosine triphosphate (GTP)   A molecule consisting of guanine, a sugar, and three phosphate groups. Can be hydrolyzed to release free energy. Commonly used in many cellular reactions; also functions in signal transduction in association with G proteins.

guard cell   One of two specialized, crescent-shaped cells forming the border of a plant stoma. Guard cells can change shape to open or close the stoma. See also **stoma.**

gustation   The perception of taste.

guttation   Excretion of water droplets from plant leaves in the early morning, caused by root pressure.

gymnosperm   A vascular plant that makes seeds but does not produce flowers. The gymnosperms include four lineages of green plants (cycads, ginkgoes, conifers, and gnetophytes). Compare with **angiosperm.**

habitat destruction   Human-caused destruction of a natural habitat with replacement by an urban, suburban, or agricultural landscape.

habitat fragmentation   The breakup of a large region of a habitat into many smaller regions, separated from others by a different type of habitat.

Hadley cell   An atmospheric cycle of large-scale air movement in which warm equatorial air rises, moves north or south, and then descends at approximately 30°N or 30°S latitude.

hair cell   A pressure-detecting sensory cell, found in the cochlea, that has tiny "hairs" (stereocilia) jutting from its surface.

hairpin   A secondary structure in RNA consisting of a stable loop formed by hydrogen bonding between purine and pyrimidine bases on the same strand.

halophile   A bacterium or archaean that thrives in high-salt environments.

halophyte   A plant that thrives in salty habitats.

Hamilton's rule   The proposition that an allele for altruistic behavior will be favored by natural selection only if $Br > C$, where B = the fitness benefit to the recipient, C = the fitness cost to the actor, and r = the coefficient of relatedness between recipient and actor.

haploid   (1) Having one set of chromosomes ($1n$). (2) A cell or an individual organism with one set of chromosomes. Compare with **diploid.**

haploid number   The number of distinct chromosome sets in a cell. Symbolized as n.

haplotype   The set of alleles found on a single chromosome.

Hardy-Weinberg principle   A principle of population genetics stating that genotype frequencies in a large population do not change from generation to generation in the absence of evolutionary processes (e.g., mutation, migration, genetic drift, random mating, and selection).

heart   A muscular pump that circulates blood throughout the body.

heart murmur   A distinctive sound caused by backflow of blood through a defective heart valve.

heartwood   The older xylem in the center of an older stem or root, containing protective compounds and no longer functioning in water transport.

heat   Thermal energy that is transferred from an object at higher temperature to one at lower temperature.

heat of vaporization   The energy required to vaporize 1 gram of a liquid into a gas.

heavy chain   The larger of the two types of polypeptide chains in an antibody molecule; composed of a variable (V) region, which contributes to the antigen-binding site, and a constant (C) region. Differences in heavy-chain constant regions determine the different classes of immunoglobulins (IgA, IgE, etc.). Compare with **light chain.**

helicase   An enzyme that catalyzes the breaking of hydrogen bonds between nucleotides of DNA, "unzipping" a double-stranded DNA molecule.

helix-turn-helix motif   A motif seen in many repressor proteins in prokaryotes, consisting of two α-helices connected by a short stretch of amino acids that form a turn.

helper T cell   An effector T cell that secretes cytokines and in other ways promotes the activation of other lymphocytes. Is descended from an activated CD4⁺ T cell that has interacted with antigen presented by dendritic cells, macrophages, or B cells.

hemimetabolous metamorphosis   A type of metamorphosis in which the animal increases in size from one stage to the next, but does not dramatically change its body form. Also called *incomplete metamorphosis.*

hemocoel   A body cavity, present in arthropods and some molluscs, containing a pool of circulatory fluid (hemolymph) bathing the internal organs.

hemoglobin   An oxygen-binding protein consisting of four polypeptide subunits, each containing an oxygen-binding heme group. The major oxygen carrier in mammalian blood.

hemolymph   The circulatory fluid of animals with open circulatory systems (e.g., insects) in which the fluid is not confined to blood vessels.

hemophilia   A human disease, caused by an X-linked recessive allele, that is characterized by defects in the blood-clotting system.

herbaceous   Referring to a plant that is not woody.

herbivore   (adjective: herbivorous) An animal that eats primarily plants and rarely or never eats meat. Compare with **carnivore** and **omnivore.**

herbivory   The practice of eating plant tissues.

heredity   The transmission of traits from parents to offspring via genetic information.

heritable   Referring to traits that can be transmitted from one generation to the next.

hermaphrodite   An organism that produces both male and female gametes.

heterokaryotic   Describing a cell or fungal mycelium containing two or more nuclei that are genetically distinct.

heterospory   (adjective: heterosporous) In seed plants, the production of two distinct types of spore-producing structures and thus two distinct types of spores: microspores, which become the male gametophyte, and megaspores, which become the female gametophyte. Compare with **homospory.**

heterotherm   An animal whose body temperature varies markedly with environmental conditions. Compare with **homeotherm.**

heterotroph   Any organism that cannot synthesize reduced organic compounds from inorganic sources and that must obtain them by eating other organisms. Some bacteria, some archaea, and virtually all fungi and animals are heterotrophs. Also called *consumer.* Compare with **autotroph.**

heterozygote advantage   A pattern of natural selection that favors heterozygous individuals compared with homozygotes. Tends to maintain genetic variation in a population. Also called *heterozygote superiority.*

heterozygous   Having two different alleles of a certain gene.

hexose   A monosaccharide (simple sugar) containing six carbon atoms.

hibernation   An energy-conserving physiological state, marked by a decrease in metabolic rate, body temperature, and activity, that lasts for a prolonged period (weeks to months). Occurs in some animals in response to winter cold and scarcity of food. Compare with **torpor.**

hindbrain   One of the three main regions of the vertebrate brain; includes the cerebellum and medulla oblongata. Compare with **forebrain** and **midbrain.**

histamine   A molecule released from mast cells during an inflammatory response that causes blood vessels to dilate and become more permeable.

histone   One of several positively charged (basic) proteins associated with DNA in the chromatin of eukaryotic cells.

histone acetyl transferase (HAT)   In eukaryotes, one of a class of enzymes that loosen chromatin structure by adding acetyl groups to histone proteins.

histone deacetylase (HDAC)   In eukaryotes, one of a class of enzymes that recondense chromatin by removing acetyl groups from histone proteins.

HIV See human immunodeficiency virus (HIV).

holoenzyme A multipart enzyme consisting of a core enzyme (containing the active site for catalysis) along with other required proteins.

holometabolous metamorphosis A type of metamorphosis in which the animal completely changes its form. Also called *complete metamorphosis*.

homeosis Replacement of one body part by another normally found elsewhere in the body as the result of mutation in certain developmentally important genes (homeotic genes).

homeostasis (adjective: homeostatic) The array of relatively stable chemical and physical conditions in an animal's cells, tissues, and organs. May be achieved by the body's passively matching the conditions of a stable external environment (conformational homeostasis) or by active physiological processes (regulatory homeostasis) triggered by variations in the external or internal environment.

homeotherm An animal that has a constant or relatively constant body temperature. Compare with **heterotherm.**

homeotic gene Any gene that specifies a particular location within an embryo, leading to the development of structures appropriate for that location. Mutations in homeotic genes cause the development of extra body parts or body parts in the wrong places.

hominids Members of the family Hominidae, which includes humans and extinct related forms; chimpanzees, gorillas, and orangutans. Distinguished by large body size, no tail, and an exceptionally large brain. Also called *great apes.*

hominins Humans and extinct related forms; species in the lineage that branched off from chimpanzees and eventually led to humans.

homologous chromosomes In a diploid organism, chromosomes that are similar in size, shape, and gene content. Also called *homologs.*

homology (adjective: homologous) Similarity among organisms of different species due to their inheritance from a common ancestor. Features that exhibit such similarity (e.g., DNA sequences, proteins, body parts) are said to be homologous. Compare with **homoplasy.**

homoplasy Similarity among organisms of different species due to convergent evolution. Compare with **homology.**

homospory (adjective: homosporous) In seedless vascular plants, the production of just one type of spore. Compare with **heterospory.**

homozygous Having two identical alleles of a certain gene.

hormone Any of numerous different signaling molecules that circulate throughout the body in blood or other body fluids and can trigger characteristic responses in distant target cells at very low concentrations.

hormone-response element A specific sequence in DNA to which a steroid hormone-receptor complex can bind and affect gene transcription.

host An individual or a species in or on which a parasite lives.

host cell A cell that has been invaded by an organism such as a parasite or a virus.

Hox **genes** A class of homeotic genes found in several animal phyla, including vertebrates, that are expressed in a distinctive pattern along the anterior-posterior axis in early embryos and control formation of segment-specific structures.

human Any member of the genus *Homo,* which includes modern humans (*Homo sapiens*) and several extinct species.

human chorionic gonadotropin (hCG) A glycoprotein hormone produced by the human placenta from about week 3 to week 14 of pregnancy. Maintains the corpus luteum, which produces hormones that preserve the uterine lining.

Human Genome Project The multinational research project that sequenced the human genome.

human immunodeficiency virus (HIV) A retrovirus that causes AIDS (acquired immune deficiency syndrome) in humans.

humoral (immune) response The type of immune response that involves generation of antibody-secreting plasma cells from activated B cells. Defends against extracellular pathogens. Compare with **cell-mediated (immune) response.**

humus The completely decayed organic matter in soils.

Huntington's disease A degenerative brain disease of humans caused by an autosomal dominant allele.

hybrid The offspring of parents from two different strains, populations, or species.

hybrid zone A geographic area where interbreeding occurs between two species, sometimes producing fertile hybrid offspring.

hydrocarbon An organic molecule that contains only hydrogen and carbon atoms.

hydrogen bond A weak interaction between two molecules or different parts of the same molecule resulting from the attraction between a hydrogen atom with a partial positive charge and another atom (usually O or N) with a partial negative charge. Compare with **covalent bond** and **ionic bond.**

hydrogen ion (H^+) A single proton with a charge of 1+; typically, one that is dissolved in solution or that is being transferred from one atom to another in a chemical reaction.

hydrolysis A chemical reaction in which a molecule is split into smaller molecules by reacting with water. In biology, most hydrolysis reactions involve the splitting of polymers into monomers. Compare with **condensation reaction.**

hydrophilic Interacting readily with water. Hydrophilic compounds are typically polar compounds containing charged or electronegative atoms. Compare with **hydrophobic.**

hydrophobic Not interacting readily with water. Hydrophobic compounds are typically nonpolar compounds that lack charged or electronegative atoms and often contain many $C-C$ and $C-H$ bonds. Compare with **hydrophilic.**

hydroponic growth Growth of plants in liquid cultures instead of soil.

hydrostatic skeleton A system of body support involving fluid-filled compartments that can change in shape but cannot easily be compressed.

hydroxide ion (OH^-) An oxygen atom and a hydrogen atom joined by a single covalent bond and carrying a negative charge; formed by dissociation of water.

hyperpolarization Change in membrane potential from its resting negative state to an even more negative state; a normal phase in an action potential. Compare with **depolarization.**

hypersensitive response In plants, the rapid death of a cell that has been infected by a pathogen, thereby reducing the potential for infection to spread throughout a plant. Compare with **systemic acquired resistance.**

hypertension Abnormally high blood pressure.

hypertonic Comparative term designating a solution that has a greater solute concentration, and therefore a lower water concentration, than another solution. Compare with **hypotonic** and **isotonic.**

hypha (plural: hyphae) One of the strands of a fungal mycelium (the meshlike body of a fungus). Also found in some protists.

hypocotyl The stem of a very young plant; the region between the cotyledon (embryonic leaf) and the radicle (embryonic root).

hypothalamic-pituitary axis The functional interaction of the hypothalamus and the pituitary gland, which are anatomically distinct but work together to regulate most of the other endocrine glands in the body.

hypothalamus A part of the brain that functions in maintaining the body's internal physiological state by regulating the autonomic nervous system, endocrine system, body temperature, water balance, and appetite.

hypothesis A proposed explanation for a phenomenon or for a set of observations.

hypotonic Comparative term designating a solution that has a lower solute concentration, and therefore a higher water concentration, than another solution. Compare with **hypertonic** and **isotonic.**

immigration The migration of individuals into a particular population from other populations. Compare with **emigration.**

immune system In vertebrates, the system whose primary function is to defend the body against pathogens. Includes several types of cells (e.g., lymphocytes and macrophages) and several organs where they develop or reside (e.g., lymph nodes and thymus).

immunity (adjective: immune) State of being protected against infection by disease-causing pathogens either by relatively nonspecific mechanisms (innate immunity) or by specific mechanisms triggered by exposure to a particular antigen (acquired immunity).

immunization The conferring of immunity to a particular disease by artificial means.

immunoglobulin (Ig) Any of the class of proteins that function as antibodies.

immunological memory The ability of the immune system to "remember" an antigen and mount a rapid, effective response to a pathogen encountered years or decades earlier.

imperfect flower A flower that contains male parts (stamens) *or* female parts (carpels) but not both. Compare with **perfect flower.**

implantation The process by which an embryo buries itself in the uterine wall and forms a placenta. Occurs in mammals and a few other vertebrates.

imprinting A type of rapid, irreversible learning in which a young animal learns to recognize the individual caring for it. Occurs in birds and mammals.

inbreeding Mating between closely related individuals. Increases homozygosity of a population and often leads to a decline in the average fitness (inbreeding depression).

incomplete dominance An inheritance pattern in which the heterozygote phenotype is a blend or combination of both homozygote phenotypes.

incomplete metamorphosis See **hemimetabolous metamorphosis.**

independent assortment, principle of The concept that each pair of hereditary elements (alleles of the same gene) behaves independently of other genes during meiosis. One of Mendel's two principles of genetics.

indeterminate growth A pattern of growth in which an individual continues to increase its overall body size throughout its life.

indicator plate A laboratory technique for detecting mutant cells by growing them on agar plates containing a compound that when metabolized by wild-type cells yields a colored product.

induced fit Change in the shape of the active site of an enzyme, as the result of the initial weak binding of a substrate, so that it binds substrate more tightly.

inducer A small molecule that triggers transcription of a specific gene, often by binding to and inactivating a repressor protein.

inducible defense A defensive trait that is manifested only in response to the presence of a predator or pathogen. Compare with **constitutive defense.**

induction (1) The process by which one embryonic cell, or group of cells, alters the differentiation of neighboring cells. (2) Positive control of gene expression by a regulatory protein that binds to DNA and triggers transcription of a specific gene(s).

infection thread An invagination of the membrane of a root hair through which beneficial nitrogen-fixing bacteria enter the roots of their host plants (legumes).

inflammatory response An aspect of the innate immune response, seen in most cases of infection or tissue injury, in which the affected tissue becomes swollen, red, warm, and painful.

inhibition In ecological succession, the phenomenon in which early-arriving species make conditions less favorable for the establishment of certain later-arriving species. Compare with **facilitation** and **tolerance.**

inhibitory postsynaptic potential (IPSP) A change in membrane potential, usually hyperpolarization, at a neuron dendrite that makes an action potential less likely.

initiation (1) In an enzyme-catalyzed reaction, the stage during which enzymes orient reactants precisely as they bind at specific locations within the enzyme's active site. (2) In DNA transcription, the stage during which RNA polymerase and other proteins assemble at the promoter sequence. (3) In RNA translation, the stage during which a complex consisting of a ribosome, a mRNA molecule, and an aminoacyl tRNA corresponding to the start codon is formed.

initiation factors A class of proteins that assist ribosomes in binding to a messenger RNA molecule to begin translation.

innate behavior Behavior that is inherited genetically, does not have to be learned, and is typical of a species.

innate immune response See **innate immunity.**

innate immunity A set of nonspecific defenses against pathogens that exist before exposure to an antigen and involves mast cells, neutrophils, and macrophages; typically results in an inflammatory response. Compare with **acquired immunity.**

inner cell mass (ICM) A cluster of cells in the interior of a mammalian blastocyst that undergo gastrulation and eventually develop into the embryo.

inner ear The innermost portion of the mammalian ear, consisting of a fluid-filled system of tubes that includes the cochlea (which receives sound vibrations from the middle ear) and the semicircular canals (which function in balance).

insects The largest lineage of arthropods distinguished by having three body regions (a head, thorax, and abdomen), walking legs, and in most species one or two pairs of wings.

in situ hybridization A technique for detecting specific DNAs and mRNAs in cells and tissues by use of labeled probes. Can be used to determine where and when particular genes are expressed in embryos.

insulin A peptide hormone produced by the pancreas in response to high levels of glucose (or amino acids) in blood. Enables cells to absorb glucose and coordinates synthesis of fats, proteins, and glycogen. Compare with **glucagon.**

integral membrane protein Any membrane protein that spans the entire lipid bilayer. Also called *transmembrane protein.* Compare with **peripheral membrane protein.**

integration In the nervous system, processing of information from many sources.

integrator A component of an animal's nervous system that functions as part of a homeostatic system by evaluating sensory information and triggering appropriate responses. See **effector** and **sensor.**

integrin Any of a class of cell-surface proteins that bind to fibronectins and other proteins in the extracellular matrix, thus holding cells in place.

intercalated disc A specialized junction between adjacent heart muscle cells that contains gap junctions, allowing electrical signals to pass between the cells.

intermediate disturbance hypothesis The hypothesis that moderate ecological disturbance is associated with higher species diversity than either low or high disturbance.

intermediate filament A long fiber, about 10 nm in diameter, composed of one of various proteins (e.g., keratins, lamins); one of the three types of cytoskeletal fibers. Form networks that help maintain cell shape and hold the nucleus in place. Compare with **actin filament** and **microtubule.**

intermediate host The host species in which a parasite reproduces asexually. Compare with **definitive host.**

interneuron A neuron that passes signals from one neuron to another. Compare with **motor neuron** and **sensory neuron.**

internode The section of a plant stem between two nodes (sites where leaves attach).

interphase The portion of the cell cycle between one mitotic (M) phase and the next. Includes the G_1 phase, S phase, and G_2 phase.

interspecific competition Competition between members of different species for the same limited resource. Compare with **intraspecific competition.**

interstitial fluid The plasma-like fluid found in the region (interstitial space) between cells.

intertidal zone The region between the low-tide and high-tide marks on a seashore.

intraspecific competition Competition between members of the same species for the same limited resource. Compare with **interspecific competition.**

intrinsic rate of increase (r_{max}) The rate at which a population will grow under optimal conditions (i.e., when birthrates are as high as possible and death rates are as low as possible). Compare with **finite rate of increase.**

intron A region of a eukaryotic gene that is transcribed into RNA but is later removed, so it is not translated into a peptide or protein. Compare with **exon.**

invasive species An exotic (nonnative) species that, upon introduction to a new area, spreads rapidly and competes successfully with native species.

inversion A mutation in which a segment of a chromosome breaks from the rest of the chromosome, flips, and rejoins with the opposite orientation as before.

invertebrates A paraphyletic group composed of animals without a backbone; includes about 95 percent of all animal species. Compare with **vertebrates.**

ion An atom or a molecule that has lost or gained electrons and thus carries an electric charge, either positive (cation) or negative (anion), respectively.

ion channel A type of channel protein that allows certain ions to diffuse across a plasma membrane down an electochemical gradient.

ionic bond A chemical bond that is formed when an electron is completely transferred from one atom to another so that the atoms remain associated due to their opposite electric charges. Compare with **covalent bond** and **hydrogen bond.**

iris A ring of pigmented muscle just below the cornea in the vertebrate eye that contracts or expands to control the amount of light entering the eye through the pupil.

islets of Langerhans Clusters of cells in the pancreas that secrete insulin and glucagon directly into the blood.

isomer A molecule that has the same molecular formula as another molecule but differs from it in three-dimensional structure.

isotonic Comparative term designating a solution that has the same solute concentration and water concentration than another solution. Compare with **hypertonic** and **hypotonic.**

isotope Any of several forms of an element that have the same number of protons but differ in the number of neutrons.

joint A place where two components (bones, cartilages, etc.) of a skeleton meet. May be movable (an articulated joint) or immovable (e.g., skull sutures).

juvenile hormone An insect hormone that prevents larvae from metamorphosing into adults.

karyogamy Fusion of two haploid nuclei to form a diploid nucleus. Occurs in many fungi, and in animals and plants during fertilization of gametes.

karyotype The distinctive appearance of all of the chromosomes in an individual, including the number of chromosomes and their length and banding patterns (after staining with dyes).

keystone species A species that has an exceptionally great impact on the other species in its ecosystem relative to its abundance.

kidney In terrestrial vertebrates, one of a paired organ situated at the back of the abdominal cavity that filters the blood, produces urine, and secretes several hormones.

kilocalorie (kcal) A unit of energy often used to measure the energy content of food. Also called *Calorie.*

kinesin Any one of a class of motor proteins that use the chemical energy of ATP to transport vesicles, particles, or chromosomes along microtubules.

kinetic energy The energy of motion. Compare with **potential energy.**

kinetochore A protein structure at the centromere where spindle fibers attach to the sister chromatids of a replicated chromosome. Contains motor proteins that move a chromosome along a microtubule.

kingdom In Linnaeus' system, a taxonomic category above the phylum level and below the domain level.

kinocilium (plural: kinocilia) A single cilium that juts from the surface of many hair cells and functions in detection of sound or pressure.

kin selection A form of natural selection that favors traits that increase survival or reproduction of an individual's kin at the expense of the individual.

Klinefelter syndrome A syndrome seen in humans who have an XXY karyotype. People with this syndrome have male sex organs, may have some female traits, and are sterile.

knock-out mutant A mutant allele that does not function at all, or an organism homozygous for such a mutation. Also called *null mutant* or *loss-of-function mutant.*

Koch's postulates Four criteria used to determine whether a suspected infectious agent causes a particular disease.

Krebs cycle A series of chemical reactions, occurring in mitochondria, in which acetyl CoA is oxidized to CO_2, producing ATP and reduced compounds for the electron transport chain. Also called *citric acid cycle.*

labia major (plural: labium majus) One of two outer folds of skin that protect the labia minora, clitoris, and vaginal opening of female mammals.

labia minora (plural: labium minus) One of two inner folds of skin that protect the opening of the urethra and vagina.

labor The strong muscular contractions of the uterus that expel the fetus during birth.

lactation (verb: lactate) Production of milk from mammary glands of mammals.

lacteal A small lymphatic vessel extending into the center of a villus in the small intestine. Receives chylomicrons containing fat absorbed from food.

lactic acid fermentation Catabolic pathway in which pyruvate produced by glycolysis is converted to lactic acid in the absence of oxygen.

lagging strand In DNA replication, the strand of new DNA that is synthesized discontinuously

in a series of short pieces that are later joined together. Also called *discontinuous strand.* Compare with **leading strand.**

lamellae (singular: lamella) Any set of parallel platelike structures (e.g., the crescent-shaped flaps on the gill filaments of fish gills that serve to increase surface area for gas exchange).

large intestine The distal portion of the digestive tract consisting of the cecum, colon, and rectum. Its primary function is to compact the wastes delivered from the small intestine and absorb enough water to form feces.

larva (plural: larvae) An immature stage of a species in which the immature and adult stages have different body forms.

late endosome A membrane-bound vesicle that arises from an early endosome and develops into a lysosome.

lateral bud A bud that forms in the angle between a leaf and a stem and may develop into a lateral (side) branch. Also called *axillary bud.*

lateral gene transfer Transfer of DNA between two different species, especially distantly related species. Commonly occurs among bacteria and archaea via plasmid exchange; also can occur in eukaryotes via viruses and some other mechanisms.

lateral meristem A layer of undifferentiated plant cells found in older stems and roots that is responsible for secondary growth. Also called *cambium* or *secondary meristem.* Compare with **apical meristem.**

lateral root A plant root extending from another, older root.

leaching Loss of nutrients from soil via percolating water.

leading strand In DNA replication, the strand of new DNA that is synthesized in one continuous piece, with nucleotides added to the 3′ end of the growing molecule. Also called *continuous strand.* Compare with **lagging strand.**

leaf The main photosynthetic organ of vascular plants.

leak channel Potassium channel that allows potassium ions to leak out of a neuron in its resting state.

learning An enduring change in an individual's behavior that results from specific experience(s).

leghemoglobin An iron-containing protein similar to hemoglobin. Found in root nodules of legume plants where it binds oxygen, preventing it from poisoning a bacterial enzyme needed for nitrogen fixation.

legumes Members of the pea plant family that form symbiotic associations with nitrogen-fixing bacteria in their roots.

lens A transparent, crystalline structure that focuses incoming light onto a retina or other light-sensing apparatus of an eye.

leptin A hormone produced and secreted by fat cells (adipocytes) that acts to stabilize fat tissue mass in part by inhibiting appetite and increasing energy expenditure.

leukocytes Several types of blood cells, including neutrophils, macrophages, and lymphocytes, that circulate in blood and lymph and function in defense against pathogens. Also called *white blood cells.*

lichen A symbiotic association of a fungus and a photosynthetic alga.

life cycle The sequence of developmental events and phases that occurs during the life span of an organism, from fertilization to offspring production.

life history The sequence of events in an individual's life from birth to reproduction to death, including how an individual allocates resources to growth, reproduction, and activities or structures that are related to survival.

life table A data set that summarizes the probability that an individual in a certain population will survive and reproduce in any given year over the course of its lifetime.

ligand Any molecule that binds to a specific site on a receptor molecule.

ligand-gated channel An ion channel that opens or closes in response to binding by a certain molecule. Compare with **voltage-gated channel.**

light chain The smaller of the two types of polypeptide chains in an antibody molecule; composed of a variable (V) region, which contributes to the antigen-binding site, and a constant (C) region. Compare with **heavy chain.**

light-dependent reactions In photosynthesis, the set of reactions, occurring in photosystem I and II, that use the energy of sunlight to split water, producing ATP, NADPH, and oxygen.

light-independent reactions See **Calvin cycle.**

lignin A substance found in the secondary cell walls of some plants that is exceptionally stiff and strong. Most abundant in woody plant parts.

limiting nutrient Any essential nutrient whose scarcity in the environment significantly reduces growth and reproduction of organisms.

limnetic zone Open water (not near shore) that receives enough light to support photosynthesis.

lineage See **monophyletic group.**

LINEs (long interspersed nuclear elements) The most abundant class of transposable elements in human genomes; can create copies of itself and insert them elsewhere in the genome. Compare with **SINEs.**

linkage In genetics, a physical association between two genes because they are on the same chromosome; the inheritance patterns resulting from this association.

linkage map See **genetic map.**

lipase Any enzyme that can break down fat molecules into fatty acids and monoglycerides.

lipid Any organic subtance that does not dissolve in water, but dissolves well in nonpolar organic solvents. Lipids include fats, oils, phospholipids, and waxes.

lipid bilayer The basic structural element of all cellular membranes consisting of a two-layer sheet of phospholipid molecules whose hydrophobic tails are oriented toward the inside and hydrophilic heads, toward the outside. Also called *phospholipid bilayer.*

littoral zone Shallow water near shore that receives enough sunlight to support photosynthesis. May be marine or freshwater; often flowering plants are present.

liver A large, complex organ of vertebrates that performs many functions including storage of glycogen, processing and conversion of food and wastes, and production of bile.

locomotion Movement of an organism under its own power.

locus   (plural: loci) A gene's physical location on a chromosome.

logistic population growth   The density-dependent decrease in growth rate as population size approaches the carrying capacity. Compare with **exponential population growth.**

long-day plant   A plant that blooms in response to short nights (usually in late spring or early summer in the northern hemisphere). Compare with **day-neutral** and **short-day plant.**

long interspersed nuclear elements   See **LINEs.**

loop of Henle   In the vertebrate kidney, a long U-shaped loop in a nephron that extends into the medulla. Functions as a countercurrent exchanger to set up an osmotic gradient that allows reabsorption of water from a subsequent portion of the nephron.

loose connective tissue   A type of connective tissue consisting of fibrous proteins in a soft matrix. Often functions as padding for organs.

lophophore   A specialized feeding structure found in some lophotrochozoans and used in filter feeding.

lophotrochozoans   A major lineage of protostomes (Lophotrochozoa) that grow by extending the size of their skeletons rather than by molting. Many phyla have a specialized feeding structure (lophophore) and/or ciliated larvae (trochophore). Includes rotifers, flatworms, segmented worms, and molluscs. Compare with **ecdysozoans.**

loss-of-function mutant   See **knock-out mutant.**

lumen   The interior space of any hollow structure (e.g., the rough ER) or organ (e.g., the stomach).

lung   Any respiratory organ used for gas exchange between blood and air.

luteal phase   The second major phase of a menstrual cycle, after ovulation, when the progesterone levels are high and the body is preparing for a possible pregnancy.

luteinizing hormone (LH)   A peptide hormone, produced and secreted by the anterior pituitary, that stimulates estrogen production, ovulation, and formation of the corpus luteum in females and testosterone production in males.

lymph   The mixture of fluid and white blood cells that circulates through the ducts and lymph nodes of the lymphatic system in vertebrates.

lymphatic system   In vertebrates, a body-wide network of thin-walled ducts (or vessels) and lymph nodes, separate from the circulatory system. Collects excess fluid from body tissues and returns it to the blood; also functions as part of the immune system.

lymph node   One of numerous small oval structures through which lymph moves in the lymphatic system. Filter the lymph and screen it for pathogens and other antigens. Major sites of lymphocyte activation.

lymphocytes   Two types of leukocyte—B cells and T cells—that circulate through the bloodstream and lymphatic system and that are responsible for the development of acquired immunity.

lysogenic cycle   A type of viral replication in which a viral genome enters a host cell, is inserted into the host's chromosome, and is replicated whenever the host cell divides. When activated, the viral DNA enters the lytic cycle, leading to production of new virus particles. Also called *lysogeny* or *latent growth.* Compare with **lytic cycle.**

lysosome   A small organelle in an animal cell containing acids and enzymes that catalyze hydrolysis reactions and can digest large molecules. Compare with **vacuole.**

lysozyme   An enzyme that functions in innate immunity by digesting bacterial cell walls. Occurs in saliva, tears, mucus, and egg white.

lytic cycle   A type of viral replication in which a viral genome enters a host cell, new virus particles (virions) are made using host enzymes and eventually burst out of the cell, killing it. Also called *replicative growth.* Compare with **lysogenic cycle.**

macromolecule   Any very large organic molecule, usually made up of smaller molecules (monomers) joined together into a polymer. The main biological macromolecules are proteins, nucleic acids, and polysaccharides.

macronutrient   Any element (e.g., carbon, oxygen, nitrogen) that is required in large quantities for normal growth, reproduction, and maintenance of a living organism. Compare with **micronutrient.**

macrophage   A type of leukocyte, capable of moving through body tissues, that engulfs and digests pathogens and other foreign particles; also secretes cytokines and presents foreign antigens to $CD4^+$ T cells.

MADS box   A DNA sequence that codes for a DNA-binding motif in proteins; present in floral organ identity genes in plants. Functionally similar sequences are found in some fungal and animal genes.

major histocompatibility protein   See **MHC protein.**

malaria   A human disease caused by four species of the protist *Plasmodium* and passed to humans by mosquitoes.

malignant tumor   A tumor that is actively growing and disrupting local tissues and/or is spreading to other organs. Cancer consists of one or more malignant tumors. Compare with **benign tumor.**

Malpighian tubules   A major excretory organ of insects, consisting of blind-ended tubes that extend from the gut into the hemocoel. Filter hemolymph to form pre-urine and then send it to the hindgut for further processing.

mammals   One of the two lineages of amniotes (vertebrates that produce amniotic eggs) distinguished by hair (or fur) and mammary glands. Includes the monotremes (platypuses), marsupials, and eutherians (placental mammals).

mammary glands   Specialized exocrine glands that produce and secrete milk for nursing offspring. A diagnostic feature of mammals.

mandibles   Any mouthpart used in chewing. In vertebrates, the lower jaw. In insects, crustaceans, and myriapods, the first pair of mouthparts.

mantle   One of the three main parts of the mollusc body; the thick outer tissue that protects the visceral mass and may secrete a calcium carbonate shell.

Marfan syndrome   A human syndrome involving increased height, long limbs and fingers, an abnormally shaped chest, and heart disorders. Probably caused by mutation in one pleiotropic gene.

marsh   A wetland that lacks trees and usually has a slow but steady rate of water flow.

marsupials   A lineage of mammals (Marsupiala) that nourish their young in an abdominal pouch after a very short period of development in the uterus.

mass extinction   The extinction of a large number of diverse evolutionary groups during a relatively short period of geologic time (about 1 million years). May occur due to sudden and extraordinary environmental changes. Compare with **background extinction.**

mass feeder   An animal that takes chunks of food into its mouth.

mass number   The total number of protons and neutrons in an atom.

mast cell   A type of leukocyte that is stationary (embedded in tissue) and helps trigger the inflammatory response to infection or injury, including secretion of histamine. Particularly important in allergic responses and defense against parasites.

maternal chromosome   A chromosome inherited from the mother.

mechanoreceptor   A sensory cell or organ specialized for detecting distortions caused by touch or pressure. One example is hair cells in the cochlea.

medium   A liquid or solid in which cells can grow in vitro.

medulla   The innermost part of an organ (e.g., kidney or adrenal gland).

medulla oblongata   In vertebrates, a region of the brain stem that along with the cerebellum forms the hindbrain.

medusa   (plural: medusae) The free-floating stage in the life cycle of some cnidarians (e.g., jellyfish). Compare with **polyp.**

megapascal (MPa)   A unit of pressure (force per unit area), equivalent to 1 million pascals (Pa).

megaspore   In seed plants, a haploid (*n*) spore that is produced in a megasporangium by meiosis of a diploid (2*n*) megasporocyte; develops into a female gametophyte. Compare with **microspore.**

meiosis   In sexually reproducing organisms, a special two-stage type of cell division in which one diploid (2*n*) parent cell produces four haploid (*n*) reproductive cells (gametes); results in halving of the chromosome number. Also called *reduction division.*

meiosis I   The first cell division of meiosis, in which synapsis and crossing over occur, and homologous chromosomes are separated from each other, producing daughter cells with half as many chromosomes (each composed of two sister chromatids) as the parent cell.

meiosis II   The second cell division of meiosis, in which sister chromatids are separated from each other. Similar to mitosis.

meiotic map   See **genetic map.**

membrane potential   A difference in electric charge across a cell membrane; a form of potential energy. Also called *membrane voltage.*

memory   Retention of learned information.

memory cells   A type of lymphocyte responsible for maintenance of immunity for years or decades after an infection. Descended from a B cell or T cell activated during a previous infection.

meniscus   (plural: menisci) The concave boundary layer formed at most air-water interfaces due to surface tension.

menstrual cycle   A female reproductive cycle seen in Old World monkeys and apes (including humans) in which the uterine lining is shed (menstruation) if no pregnancy occurs. Compare with **estrous cycle.**

menstruation   The periodic shedding of the uterine lining through the vagina that occurs in females of Old World monkeys and apes, including humans.

meristem   (adjective: meristematic) In plants, a group of undifferentiated cells that can develop into various adult tissues throughout the life of a plant.

mesoderm   The middle of the three basic cell layers in most animal embryos; gives rise to muscles, bones, blood, and some internal organs (kidney, spleen, etc.). Compare with **ectoderm** and **endoderm.**

mesoglea   A gelatinous material, containing scattered ectodermal cells, that is located between the ectoderm and endoderm of cnidarians (e.g., jellyfish, corals, and anemones).

mesophyll cell   A type of cell, found near the surfaces of plant leaves, that is specialized for the light-dependent reactions of photosynthesis.

Mesozoic era   The period of geologic time, from 250 million to 65 million years ago, during which gymnosperms were the dominant plants and dinosaurs the dominant vertebrates. Ended with extinction of the dinosaurs. Also called *Age of Reptiles.*

messenger RNA (mRNA)   An RNA molecule that carries encoded information, transcribed from DNA, that specifies the amino acid sequence of a polypeptide.

meta-analysis   A comparative analysis of the results of many smaller, previously published studies.

metabolic pathway   An ordered series of chemical reactions that build up or break down a particular molecule. Often, each reaction is catalyzed by a different enzyme.

metabolic rate   The total energy use by all the cells of an individual. For aerobic organisms, often measured as the amount of oxygen consumed per hour.

metabolic water   The water that is produced as a by-product of cellular respiration.

metabolism   All the chemical reactions occurring in a living cell or organism.

metallothioneins   Small plant proteins that bind to and prevent excess metal ions from acting as toxins.

metamorphosis   Transition from one developmental stage to another, such as from the larval to the adult form of an animal.

metaphase   A stage in mitosis or meiosis during which chromosomes line up in the middle of the cell.

metaphase plate   The plane along which chromosomes line up during metaphase of mitosis or meiosis; not an actual structure.

metastasis   The spread of cancerous cells from their site of origin to distant sites in the body where they may establish additional tumors.

methanogen   A prokaryote that produces methane (CH_4) as a by-product of cellular respiration.

methanotroph   An organism that uses methane (CH_4) as its primary electron donor and source of carbon.

methylation   The addition of a methyl ($-CH_3$) group to a molecule.

MHC protein   One of a large set of mammalian cell-surface glycoproteins involved in marking cells as self and in antigen presentation to T cells. Also called *MHC molecule.* See **Class I** and **Class II MHC protein.**

microbe   Any microscopic organism, including bacteria, archaea, and various tiny eukaryotes.

microbiology   The field of study concerned with microscopic organisms.

microfilament   See **actin filament.**

micrograph   A photograph of an image produced by a microscope.

micronutrient   Any element (e.g., iron, molybdenum, magnesium) that is required in very small quantities for normal growth, reproduction, and maintenance of a living organism. Compare with **macronutrient.**

micropyle   The tiny pore in a plant ovule through which the pollen tube reaches the embryo sac.

microRNA (miRNA)   A small, single-stranded RNA associated with proteins in an RNA-induced silencing complex. Can bind to complementary sequences in mRNA molecules, allowing the associated proteins to degrade the bound mRNA or inhibit its translation. See **RNA interference.**

microsatellite   A noncoding stretch of eukaryotic DNA consisting of a repeating sequence 1- to 5-base pair long. Also called *simple sequence repeat.*

microspore   In seed plants, a haploid (n) spore that is produced in a microsporangium by meiosis of a diploid ($2n$) microsporocyte; develops into a male gametophyte. Compare with **megaspore.**

microtubule   A long, tubular fiber, about 2.5 nm in diameter, formed by polymerization of tubulin protein dimers; one of the three types of cytoskeletal fibers. Involved in cell movement and transport of materials within the cell. Compare with **actin filament** and **intermediate filament.**

microtubule-organizing center (MTOC)   General term for any structure (e.g., centrosome and basal body) that organizes microtubules in cells.

microvilli   (singular:microvillus) Tiny protrusions from the surface of an epithelial cell that increase the surface area for absorption of substances.

midbrain   One of the three main regions of the vertebrate brain; includes sensory integrating and relay centers. Compare with **forebrain** and **hindbrain.**

middle ear   The air-filled middle portion of the mammalian ear, which contains three small bones (ossicles) that transmit and amplify sound from the tympanic membrane to the inner ear. Is connected to the throat via the eustachian tube.

middle lamella   The layer of gelatinous pectins between the primary cell walls of adjacent plant cells. Helps hold the cells together.

migration   (1) In ecology, a cyclical movement of large numbers of organisms from one geographic location or habitat to another. (2) In population genetics, movement of individuals from one population to another.

millivolt (mV)   A unit of voltage equal to 1/1000 of a volt.

mimicry   A phenomenon in which one species has evolved (or learns) to look or sound like another species. See **Batesian mimicry** and **Müllerian mimicry.**

mineralocorticoids   A class of steroid hormones, produced and secreted by the adrenal cortex, that regulate electrolyte levels and the overall volume of body fluids. Aldosterone is the principal one in humans. Compare with **glucocorticoids.**

minisatellite   A noncoding stretch of eukaryotic DNA consisting of a repeating sequence that is 6 to 500 base pairs long. Also called *variable number tandem repeat (VNTR).*

mismatch repair   The process by which mismatched base pairs in DNA are fixed.

missense mutation   A point mutation (change in a single base pair) that causes a change in the amino acid sequence of a protein. Also called *replacement mutation.*

mitochondrial matrix   Central compartment of a mitochondrion, which is lined by the inner membrane; contains the enzymes and substrates of the Krebs cycle and mitochondrial DNA.

mitochondrion   (plural: mitochondria) A eukaryotic organelle that is bounded by a double membrane and is the site of aerobic respiration.

mitosis   In eukaryotic cells, the process of nuclear division that results in two daughter nuclei genetically identical to the parent nucleus. Subsequent cytokinesis (division of the cytoplasm) yields two daughter cells.

mitosis-promoting factor (MPF)   A complex of a cyclin and cyclin-dependent kinase that phosphorylates a number of specific proteins needed to initiate mitosis in eukaryotic cells.

mitotic (M) phase   The phase of the cell cycle during which cell division occurs. Includes mitosis and cytokinesis.

mitotic spindle   Temporary structure, composed largely of microtubules, that is involved in the movement of chromosomes to the equatorial plate and then to opposite sides of the cell during mitosis and meiosis.

model organism   An organism selected for intensive scientific study based on features that make it easy to work with (e.g., body size, life span), in the hope that findings will apply to other species.

molarity   A common unit of solute concentration equal to the number of moles of a dissolved solute in 1 liter of solution.

mole   The amount of a substance that contains 6.022×10^{23} of its elemental entities (e.g., atoms, ions, or molecules). This number of molecules of a compound will have a mass equal to the molecular weight of that compound expressed in grams.

molecular chaperone   A protein that facilitates the three-dimensional folding of newly synthesized proteins, usually by an ATP-dependent mechanism.

molecular clock   The hypothesis that certain types of mutations tend to reach fixation in populations at a steady rate over large spans of time. As a result, comparisons of DNA sequences can be used to infer the timing of evolutionary divergences.

molecular formula   A notation that indicates only the numbers and types of atoms in a molecule, such as H_2O for the water molecule. Compare with **structural formula.**

molecular weight   The sum of the mass numbers of all of the atoms in a molecule; roughly, the total number of protons and neutrons in the molecule.

molecule   A combination of two or more atoms held together by covalent bonds.

molluscs   Members of the phylum Mollusca. Distinguished by a body plan with three main parts: a muscular foot, a visceral mass, and a mantle. Include bivalves (clams, oysters), gastropods (snails, slugs), chitons, and cephalopods (squid, octopuses). Molluscs belong to the lophotrochozoan branch of the protostomes.

molting A method of body growth, used by ecdysozoans, that involves the shedding of an external protective cuticle or skeleton, expansion of the soft body, and growth of a new external layer.

monocot Any plant that has a single cotyledon (embryonic leaf) upon germination. Monocots form a monophyletic group. Also called a monocotyledonous plant. Compare with **dicot**.

monoecious Describing an angiosperm species that has both male and female reproductive structures on each plant. Compare with **dioecious**.

monohybrid cross A mating between two parents that are both heterozygous for a given gene.

monomer A small molecule that can covalently bind to other similar molecules to form a larger macromolecule. Compare with **polymer**.

monophyletic group An evolutionary unit that includes an ancestral population and all of its descendants but no others. Also called a *clade* or *lineage*. Compare with **paraphyletic group**.

monosaccharide A small carbohydrate, such as glucose, that has the molecular formula $(CH_2O)_n$ and cannot be hydrolyzed to form any smaller carbohydrates. Also called *simple sugar*. Compare with **disaccharide** and **polysaccharide**.

monosomy Having only one copy of a particular type of chromosome.

monotremes A lineage of mammals (Monotremata) that lay eggs and then nourish the young with milk. Includes just three living species: the platypus and two species of echidna.

morphological species concept The definition of a species as a population or group of populations that have measurably different anatomical features from other groups. Also called *morphospecies concept*. Compare with **biological** and **phylogenetic species concept**.

morphology The shape and appearance of an organism's body and its component parts.

motif In molecular biology, a domain (a section of a protein with a distinctive tertiary structure) found in many different proteins and often having specific functional properties.

motor neuron A nerve cell that carries signals from the central nervous system (brain and spinal cord) to an effector, such as a muscle or gland. Compare with **interneuron** and **sensory neuron**.

motor protein A class of proteins whose major function is to convert the chemical energy of ATP into motion. Includes dynein, kinesin, and myosin.

MPF See **mitosis-promoting factor**.

mRNA. See **messenger RNA**.

mucigel A slimy substance secreted by plant root caps that eases passage of the growing root through the soil.

mucosal-associated lymphoid tissue (MALT) Collective term for lymphocytes and other leukocytes associated with skin cells and with mucus-secreting epithelial tissues in the gut and respiratory tract. Plays important role in preventing entry of pathogens into the body.

mucus (adjective: mucous) A slimy mixture of glycoproteins (called mucins) and water that is secreted in many animal organs for lubrication.

Müllerian inhibitory substance A peptide hormone secreted by the embryonic testis that causes regression (withering away) of the female reproductive ducts.

Müllerian mimicry A type of mimicry in which two (or more) harmful species resemble each other. Compare with **Batesian mimicry**.

multicellularity The state of being composed of many cells that adhere to each other and do not all express the same genes with the result that some cells have specialized functions.

multiple fruit A fruit (e.g., pineapple) that develops from many separate flowers and thus many carpels. Compare with **aggregate** and **simple fruit**.

multiple sclerosis (MS) A human autoimmune disease caused by the immune system attacking the myelin sheaths that insulate nerve axons.

muscle fiber A single muscle cell.

muscle tissue An animal tissue consisting of bundles of long, thin contractile cells (muscle fibers).

mutagen Any physical or chemical agent that increases the rate of mutation.

mutant An individual that carries a mutation, particularly a new or rare mutation.

mutation Any change in the hereditary material of an organism (DNA in most organisms, RNA in some viruses).

mutualism (adjective: mutualistic) A symbiotic relationship between two organisms (mutualists) that benefits both. Compare with **commensalism** and **parasitism**.

mycelium (plural: mycelia) A mass of underground filaments (hyphae) that form the body of a fungus. Also found in some protists and bacteria.

mycorrhiza (plural: mycorrhizae) A mutualistic association between certain fungi and most vascular plants, sometimes visible as nodules or nets in or around plant roots.

myelin sheath Multiple layers of myelin, derived from the cell membranes of certain glial cells, that is wrapped around the axon of a neuron, providing electrical insulation.

myofibril Long, slender structure composed of contractile proteins organized into repeating units (sarcomeres) in vertebrate heart muscle and striated muscle.

myosin Any one of a class of motor proteins that use the chemical energy of ATP to move along actin filaments in muscle contraction, cytokinesis, and vesicle transport.

myriapods A lineage of arthropods with long segmented trunks, each segment bearing one or two pairs of legs. Includes millipedes and centipedes.

NAD⁺/NADH Oxidized and reduced forms, respectively, of nicotinamide adenine dinucleotide. A nonprotein electron carrier that functions in many of the redox reactions of metabolism.

NADP⁺/NADPH Oxidized and reduced forms, respectively, of nicotinamide adenine dinucleotide phosphate. A nonprotein electron carrier that is reduced during the light-dependent reactions in photosynthesis and extensively used in biosynthetic reactions.

natural experiment A situation in which groups to be compared are created by an unplanned, natural change in conditions rather than by manipulation of conditions by researchers.

natural selection The process by which individuals with certain heritable traits tend to produce more surviving offspring than do individuals without those traits, often leading to a change in the genetic makeup of the population. A major mechanism of evolution.

nauplius A distinct planktonic larval stage seen in many crustaceans.

Neanderthal A recently extinct European species of hominid, *Homo neanderthalensis*, closely related to but distinct from modern humans.

nectar The sugary fluid produced by flowers to attract and reward pollinating animals.

negative feedback A self-limiting, corrective response in which a deviation in some variable (e.g., body temperature, blood pH, concentration of some compound) triggers responses aimed at returning the variable to normal. Compare with **positive feedback**.

negative pressure ventilation Ventilation of the lungs that is accomplished by "pulling" air into the lungs by expansion of the rib cage. Compare with **positive pressure ventilation**.

negative-sense virus A virus whose genome contains sequences complementary to those in the mRNA required to produce viral proteins. Compare with **ambisense virus** and **positive-sense virus**.

nematodes. See **roundworms**.

nephron One of the tiny tubes within the vertebrate kidney that filter blood and concentrate salts to produce urine. Also called *renal tubule*.

neritic zone Shallow marine waters beyond the intertidal zone, extending down to about 200 meters, where the continental shelf ends.

nerve A long, tough strand of nervous tissue typically containing thousands of axons wrapped in connective tissue; carries impulses between the central nervous system and some other part of the body.

nerve cord A bundle of nerves extending from the brain along the dorsal (back) side of a chordate animal, with cerebrospinal fluid inside a hollow central channel. One of the defining features of chordates.

nervous tissue An animal tissue consisting of nerve cells (neurons) and various supporting cells.

net primary productivity (NPP) In an ecosystem, the total amount of carbon fixed by photosynthesis over a given time period minus the amount oxidized during cellular respiration. Compare with **gross primary productivity**.

net reproductive rate (R_0) The growth rate of a population per generation; equivalent to the average number of female offspring that each female produces over her lifetime.

neural Relating to nerve cells (neurons) and the nervous system.

neural tube A folded tube of ectoderm that forms along the dorsal side of a young vertebrate embryo and that will give rise to the brain and spinal cord.

neuroendocrine Referring to nerve cells (neurons) that release hormones into the blood or to such hormones themselves.

neuron A cell that is specialized for the transmission of nerve impulses. Typically has dendrites, a cell body, and a long axon that forms synapses with other neurons. Also called *nerve cell*.

neurosecretory cell A nerve cell (neuron) that produces and secretes hormones into the bloodstream. Principally found in the hypothalamus. Also called *neuroendocrine cell*.

neurotoxin Any substance that specifically destroys or blocks the normal functioning of neurons.

neurotransmitter A molecule that transmits electrical signals from one neuron to another or from a neuron to a muscle or gland. Examples are acetylcholine, dopamine, serotonin, and norepinephrine.

neutral In genetics, referring to any mutation or mutant allele that has no effect on an individual's fitness.

neutrophil A type of leukocyte, capable of moving through body tissues, that engulfs and digests pathogens and other foreign particles; also secretes various compounds that attack bacteria and fungi.

niche The particular set of habitat requirements of a certain species and the role that species plays in its ecosystem.

niche differentiation The change in resource use by competing species that occurs as the result of character displacement.

nicotinamide adenine dinucleotide See NAD⁺/NADH.

nicotinamide adenine dinucleotide phosphate See NADP⁺/NADPH.

nitrogen cycle, global The movement of nitrogen among terrestrial ecosystems, the oceans, and the atmosphere.

nitrogen fixation The incorporation of atmospheric nitrogen (N_2) into forms such as ammonia (NH_3) or nitrate (NO_3^-), which can be used to make many organic compounds. Occurs in only a few lineages of bacteria and archaea.

nociceptor A sensory cell or organ specialized to detect tissue damage, usually producing the sensation of pain.

node (1) In animals, any small thickening (e.g., a lymph node). (2) In plants, the part of a stem where leaves or leaf buds are attached. (3) In a phylogenetic tree, the point where two branches diverge, representing the point in time when an ancestral group split into two or more descendant groups. Also called *fork*.

node of Ranvier One of the periodic unmyelinated sections of a neuron's axon at which an action potential can be regenerated.

Nod factors Molecules produced by nitrogen-fixing bacteria that help them recognize and bind to roots of legumes.

nodule Lumplike structure on roots of legume plants that contain symbiotic nitrogen-fixing bacteria.

noncyclic electron flow See Z scheme.

nondisjunction An error that can occur during meiosis or mitosis in which one daughter cell receives two copies of a particular chromosome, and the other daughter cell receives none.

nonpolar covalent bond A covalent bond in which electrons are equally shared between two atoms of the same or similar electronegativity. Compare with **polar covalent bond**.

non-sister chromatids The chromatids of a particular type of chromosome (after replication) with respect to the chromatids of its homologous chromosome. Crossing over occurs between non-

sister chromatids. Compare with **sister chromatids**.

non-template strand The strand of DNA that is not transcribed during synthesis of RNA. Its sequence corresponds to that of the mRNA produced from the other strand. Also called *coding strand*.

nonvascular plants See **bryophytes**.

norepinephrine A catecholamine used as a neurotransmitter in the sympathetic nervous system. Also is produced by the adrenal medulla and functions as a hormone that triggers rapid responses relating to the fight-or-flight response.

Northern blotting A technique for identifying specific RNAs separated by gel electrophoresis by transferring them to filter paper and hybridizing with a labeled DNA probe complementary to the RNA of interest. Compare with **Southern blotting** and **Western blotting**.

notochord A long, gelatinous, supportive rod down the back of a chordate embryo, below the developing spinal cord. Replaced by vertebrae in most adult vertebrates. A defining feature of chordates.

nuclear envelope The double-layered membrane enclosing the nucleus of a eukaryotic cell.

nuclear lamina A lattice-like sheet of fibrous nuclear lamins, which are one type of intermediate filaments. Lines the inner membrane of the nuclear envelope, stiffening the envelope and helping organize the chromosomes.

nuclear localization signal (NLS) A short amino acid sequence that marks a protein for delivery to the nucleus.

nuclear pore An opening in the nuclear envelope that connects the inside of the nucleus with the cytoplasm and through which molecules such as mRNA and some proteins can pass.

nuclear pore complex A large complex of dozens of proteins lining a nuclear pore, defining its shape and transporting substances through the pore.

nuclease Any enzyme that can break down RNA or DNA molecules.

nucleic acid A macromolecule composed of nucleotide monomers. Generally used by cells to store or transmit hereditary information. Includes ribonucleic acid and deoxyribonucleic acid.

nucleic acid hybridization Base pairing between a single-stranded nucleic acid and a complementary sequence in a different nucleic acid (e.g., a labeled probe). Is used experimentally in Southern blotting and Northern blotting.

nucleoid In prokaryotic cells, a dense, centrally located region that contains DNA but is not surrounded by a membrane.

nucleolus In eukaryotic cells, specialized structure in the nucleus where ribosomal RNA processing occurs and ribosomal subunits are assembled.

nucleosome A repeating, bead-like unit of eukaryotic chromatin, consisting of about 200 nucleotides of DNA wrapped twice around eight histone proteins.

nucleotide A molecule consisting of a five-carbon sugar (ribose or deoxyribose), a phosphate group, and one of several nitrogen-containing bases. DNA and RNA are polymers of nucleotides containing deoxyribose (deoxyribonucleotides) and ribose (ribonucleotides), respectively. Equivalent to a nucleoside plus one phosphate group.

nucleotide excision repair The process of removing a damaged region in one strand of DNA and correctly replacing it using the undamaged strand as a template.

nucleus (1) The center of an atom, containing protons and neutrons. (2) In eukaryotic cells, the large organelle containing the chromosomes and surrounded by a double membrane. (3) A discrete clump of neuron cell bodies in the brain, usually sharing a distinct function.

null hypothesis A hypothesis that specifies what the results of an experiment will be if the main hypothesis being tested is wrong. Often states that there will be no difference between experimental groups.

null mutant See **knock-out mutant**.

nutrient A substance that an organism requires for normal growth, maintenance, or reproduction.

nutritional balance A state in which an organism is taking in enough nutrients to maintain normal health and activity.

nymph The juvenile form of an animal that undergoes hemimetabolous (incomplete) metamorphosis; resembles a miniature version of the adult form.

oceanic zone The waters of the open ocean beyond the continental shelf.

oil A fat that is liquid at room temperature.

Okazaki fragment Short segment of DNA produced during replication of the 5′ to 3′ template strand. Many Okazaki fragments make up the lagging strand in newly synthesized DNA.

olfaction The perception of odors.

olfactory bulb A bulb-shaped projection of the brain just above the nose. Receives and interprets odor information from the nose.

oligodendrocyte A type of glial cell that wraps around axons of some neurons in the central nervous system, forming a myelin sheath that provides electrical insulation. Compare with **Schwann cell**.

oligopeptide A chain composed of fewer than 50 amino acids linked together by peptide bonds. Often referred to simply as *peptide*.

ommatidium (plural: ommatidia) A light-sensing column in an arthropod's compound eye.

omnivore (adjective: omnivorous) An animal whose diet regularly includes both meat and plants. Compare with **carnivore** and **herbivore**.

oncogene Any gene whose protein product stimulates cell division at all times and thus promotes cancer development. Often is a mutated form of a gene involved in regulating the cell cycle. See **proto-oncogene**.

one-gene, one-enzyme hypothesis The hypothesis that each gene is responsible for making one (and only one) protein, in most cases an enzyme that catalyzes a specific reaction. Many exceptions to this hypothesis are now known.

oocyte A cell in the ovary that can undergoes meiosis to produce an ovum.

oogenesis The production of egg cells (ova).

oogonia (singular: oogonia) The diploid cells in an ovary that can divide by mitosis to create more oogonia and primary oocytes, which can undergo meiosis.

open circulatory system A circulatory system in which the circulating fluid (hemolymph) is not confined to blood vessels. Compare with **closed circulatory system**.

open reading frame (ORF) Any DNA sequence, ranging in length from several hundred to thousands of base pairs long, that is flanked by a start codon and a stop codon. ORFs identified by computer analysis of DNA may be functional genes, especially if they have other features characteristic of genes (e.g., promoter sequence).

operator In prokaryotic DNA, a binding site for a repressor protein; located near the start of an operon.

operculum The stiff flap of tissue that covers the gills of teleost fishes.

operon A region of prokaryotic DNA that codes for a series of functionally related genes and is transcribed from a single promoter into a polycistronic mRNA.

opsin A transmembrane protein that is covalently linked to retinal, the light-detecting pigment in rod and cone cells.

optical isomer A molecule that shares the same molecular formula as another molecule but differs in the arrangement of atoms or groups around a carbon atom; left-handed or right-handed form of a molecule. Compare with **geometric** and **structural isomer.**

optimal foraging The concept that animals forage in a way that maximizes the amount of usable energy they take in, given the costs of finding and ingesting their food and the risk of being eaten while they're at it.

orbital The spherical region around an atomic nucleus in which an electron is present most of the time.

order In Linnaeus' system, a taxonomic category above the family level and below the class level.

ORF See **open reading frame.**

organ A group of tissues organized into a functional and structural unit.

organelle Any discrete, membrane-bound structure within a cell (e.g., mitochondrion) that has a characteristic structure and functions.

organic For a compound, containing carbon and hydrogen and usually containing carbon-carbon bonds. Organic compounds are widely used by living organisms.

organism Any living entity that contains one or more cells.

organogenesis A stage of embryonic development, just after gastrulation in vertebrate embryos, during which major organs develop from the three embryonic germ layers.

orientation A deliberate movement that results in a change in position relative to some external cue, such as toward the Sun or away from a sound.

origin of replication The site on a chromosome at which DNA replication begins.

osmoconformer An animal that does not actively regulate the osmolarity of its tissues but conforms to the osmolarity of the surrounding environment.

osmolarity The concentration of dissolved substances in a solution, measured in moles per liter.

osmoregulation The process by which a living organism controls the concentration of water and salts in its body.

osmoregulator An animal that actively regulates the osmolarity of its tissues.

osmosis Diffusion of water across a selectively permeable membrane from a region of high water concentration (low solute concentration) to a region of low water concentration (high solute concentration).

osmotic potential See **solute potential.**

ossicle One of three small bones, in the middle ear of mammals, that transmit and amplify sound from the tympanic membrane to the inner ear.

ouabain A plant toxin that poisons the sodium-potassium pumps of animals.

outcrossing Reproduction by fusion of the gametes of different individuals, rather than self-fertilization. Typically refers to plants.

outer ear The outermost portion of the mammalian ear, consisting of the pinna (ear flap) and the ear canal. Funnels sound to the tympanic membrane.

outgroup A taxon that is closely related to a particular monophyletic group but is not part of it.

out-of-Africa hypothesis The hypothesis that modern humans (*Homo sapiens*) evolved in Africa and spread to other continents, replacing other *Homo* species without interbreeding with them.

oval window A membrane separating the fluid-filled cochlea from the air-filled middle ear through which sound vibrations pass from the middle ear to the inner ear in mammals.

ovary The egg-producing organ of a female animal, or the seed-producing structure in the female part of a flower.

oviduct See **fallopian tube.**

oviparous Producing eggs that are laid outside the body where they develop and hatch. Compare with **ovoviviparous** and **viviparous.**

ovoviviparous Producing eggs that are retained inside the body until they are ready to hatch. Compare with **oviparous** and **viviparous.**

ovulation The release of an ovum from an ovary of a female vertebrate. In humans, an ovarian follicle releases an egg at the end of the follicular phase of the menstrual cycle.

ovule In flowering plants, the structure inside an ovary that contains the female gametophyte and eventually (if fertilized) becomes a seed.

ovum (plural: ova) See **egg.**

oxidation The loss of electrons from an atom during a redox reaction, either by donation of an electron to another atom or by the shared electrons in covalent bonds moving farther from the atomic nucleus.

oxidative phosphorylation Production of ATP molecules from the redox reactions of an electron transport chain.

oxygen-hemoglobin equilibrium curve The graphical depiction of the percentage of hemoglobin in the blood that will bind to oxygen at various partial pressures of oxygen.

oxygenic Referring to any process or reaction that produces oxygen. Photosynthesis in plants, algae, and cyanobacteria, which involves photosystem II, is oxygenic. Compare with **anoxygenic.**

oxytocin A peptide hormone, secreted by the posterior pituitary, that triggers labor and milk production in females and that stimulates pair bonding, parental care, and affiliative behavior in both sexes.

p53 A tumor-suppressor protein (molecular weight of 53 kilodaltons) that responds to DNA damage by stopping the cell cycle and/or triggering apoptosis. Encoded by the *p53* gene.

pacemaker cell A specialized cardiac muscle cell in the sinoatrial (SA) node of the vertebrate heart that has an inherent rhythm and can generate an electrical impulse that spreads to other heart cells.

paleontology The study of organisms that lived in the distant past.

Paleozoic era The period of geologic time, from 543 million to 250 million years ago, during which fungi, land plants, and animals first appeared and diversified. Began with the Cambrian explosion and ended with the extinction of many invertebrates and vertebrates.

pancreas A large gland in vertebrates that has both exocrine and endocrine functions. Secretes digestive enzymes into a duct connected to the intestine and several hormones (notably, insulin and glucagon) into the bloodstream.

pandemic The spread of an infectious disease in a short time period over a wide geographic area and affecting a very high proportion of the population. Compare with **epidemic.**

parabiosis An experimental technique for determining whether a certain physiological phenomenon is regulated by a hormone, by surgically uniting two individuals so that hormones can pass between them.

parabronchus (plural: parabronchi) One of the many tiny parallel air tubes that run through a bird's lung where gas exchange occurs.

paracrine Relating to a chemical signal that is released by one cell and affects neighboring cells.

paraphyletic group An evolutionary unit that includes an ancestral population and *some* but not all of its descendants. Paraphyletic groups are not meaningful units in evolution. Compare with **monophyletic group.**

parapodia (singular: parapodium) Appendages found in some annelids from which bristle-like structures (chaetae) extend.

parasite An organism that lives on or in a host species and that damages its host.

parasitism (adjective: parasitic) A symbiotic term relationship between two organisms that is beneficial to one organism (the parasite) but detrimental to the other (the host). Compare with **commensalism** and **mutualism.**

parasitoid An organism that has a parasitic larval stage and a free-living adult stage. Most parasitoids are insects that lay eggs in the bodies of other insects.

parasympathetic nervous system The part of the autonomic nervous system that stimulates functions for conserving or restoring energy, such as reduced heart rate and increased digestion. Compare with **sympathetic nervous system.**

parathyroid glands Four small glands, located near or embedded in the thyroid gland of vertebrates, that secrete parathyroid hormone.

parathyroid hormone (PTH) A peptide hormone, secreted from the parathyroid glands, that increases blood calcium by promoting Ca^{2+} release from bones, Ca^{2+} uptake in the intestines, and Ca^{2+} reabsorption in the kidneys.

parenchyma cell In plants, a general type of cell with a relatively thin primary cell wall. These cells, found in leaves, the centers of stems and roots, and fruits, are involved in photosynthesis, starch storage, and new growth. Compare with **collenchyma cell** and **sclerenchyma cell.**

parental care Any action by which an animal expends energy or assumes risks to benefit its offspring (e.g., nest-building, feeding of young, defense).

parental generation The adult organisms used in the first experimental cross in a formal breeding experiment.

parietal cell A cell in the stomach lining that secretes hydrochloric acid.

parsimony The logical principle that the most likely explanation of a phenomenon is the most economical or simplest. When applied to comparison of alternative phylogenetic trees, it suggests that the one requiring the fewest evolutionary changes is most likely to be correct.

parthenogenesis Development of offspring from unfertilized eggs; a type of asexual reproduction.

partial pressure The pressure of one particular gas in a mixture; the contribution of that gas to the overall pressure.

pascal (Pa) A unit of pressure (force per unit area).

passive transport Diffusion of a substance across a plasma membrane or organelle membrane. When this occurs with the assistance of membrane proteins, it is called facilitated diffusion.

patch clamping A technique for studying the electrical currents that flow through individual ion channels by sucking a tiny patch of membrane to the hollow tip of a microelectrode.

paternal chromosome A chromosome inherited from the father.

pathogen (adjective: pathogenic) Any entity capable of causing disease, such as a microbe, virus, or prion.

pattern formation The series of events that determines the spatial organization of an embryo, including alignment of the major body axes and orientation of the limbs.

pattern-recognition receptor One of a class of membrane proteins on leukocytes that bind to molecules on the surface of many bacteria. Part of the innate immune response.

PCR See **polymerase chain reaction.**

peat Semidecayed organic matter that accumulates in moist, low-oxygen environments such as bogs.

pectin A gelatinous polysaccharide found in the primary cell wall and middle lamella of plant cells. Attracts and holds water, forming a gel that helps keep the cell wall moist.

pedigree A family tree of parents and offspring, showing inheritance of particular traits of interest.

pellet Any solid material that collects at the bottom of a test tube below a layer of liquid (the supernatant) during centrifugation.

penis The copulatory organ of male mammals, used to insert sperm into a female.

pentose A monosaccharide (simple sugar) containing five carbon atoms.

PEP carboxylase An enzyme that catalyzes addition of CO_2 to phosphoenol pyruvate, a three-carbon compound, forming a four-carbon organic acid. Found in mesophyll cells of plants that perform C_4 photosynthesis.

pepsin A protein-digesting enzyme present in the stomach.

pepsinogen The precursor of the digestive enzyme pepsin. Is secreted from cells in the

stomach lining and converted to pepsin by the acidic environment of the stomach lumen.

peptide See **oligopeptide.**

peptide bond The covalent bond (C−N) formed by a condensation reaction between two amino acids; links the residues in peptides and proteins.

peptidoglycan A complex structural polysaccharide found in bacterial cell walls.

per capita rate of increase (r) The growth rate of a population, expressed per individual. Calculated as the per capita birthrate minus the per capita death rate. Also called *per capita growth rate.*

perennial Describing a plant whose life cycle normally lasts for **more than one year.** Compare with **annual.**

perfect flower A flower that contains both male parts (stamens) and female parts (carpels). Compare with **imperfect flower.**

perforation In plants, a small hole in the primary and secondary cell walls of vessel elements that allow passage of water.

pericarp The part of a fruit, formed from the ovary wall, that surrounds the seeds and protects them. Corresponds to the flesh of most edible fruits and the hard shells of most nuts.

pericycle In plant roots, a layer of cells that give rise to lateral roots.

periderm The outermost portion of bark consisting of cork cambium, cork cells, and a third small layer of cells (phelloderm).

peripheral membrane protein Any membrane protein that does not span the entire lipid bilayer and associates with only one side of the bilayer. Compare with **integral membrane protein.**

peripheral nervous system (PNS) All the components of the nervous system that are outside the central nervous system (the brain and spinal cord). Includes the somatic nervous system and the autonomic nervous system.

peristalsis Rhythmic waves of muscular contraction that push food along the digestive tract.

permafrost A permanently frozen layer of icy soil found in most tundra and some taiga.

permeability The tendency of a structure, such as a membrane, to allow a given substance to diffuse across it.

peroxisome An organelle found in most eukaryotic cells that contains enzymes for oxidizing fatty acids and other compounds including many toxins, rendering them harmless. See **glyoxisome.**

petal One of the leaflike organs arranged around the reproductive organs of a flower. Often colored and scented to attract pollinators.

petiole The stalk of a leaf.

pH A measure of the concentration of protons in a solution and thus of acidity or alkalinity. Defined as the negative of the base-10 logarithm of the proton concentration: $pH = -\log[H^+]$.

phagocytosis Uptake by a cell of small particles or cells by pinching off the plasma membrane to form small membrane-bound vesicles; one type of endocytosis.

pharyngeal gill slits A set of parallel openings from the throat through the neck to the outside. A diagnostic trait of chordates.

phelloderm In the stems of woody plants, a thin layer of cells located between the outer cork cells and inner cork cambium.

pharyngeal jaw A secondary jaw in the back of the mouth, found in some fishes. Derived from modified gill arches.

phenetic approach A method for constructing a phylogenetic tree by computing a statistic that summarizes the overall similarity among populations, based on the available data. Compare with **cladistic approach.**

phenotype The detectable physical and physiological traits of an individual, which are determined its genetic makeup. Also the specific trait associated with a particular allele. Compare with **genotype.**

phenotypic plasticity Within-species variation in phenotype that is due to differences in environmental conditions. Occurs more commonly in plants than animals.

pheophytin In photosystem II, a molecule that accepts excited electrons from a reaction center chlorophyll and passes them to an electron transport chain.

pheromone A chemical signal, released by one individual into the external environment, that can trigger responses in a different individual.

phloem A plant vascular tissue that conducts sugars; contains sieve-tube members and companion cells. Primary phloem develops from the procambium of apical meristems; secondary phloem, from the vascular cambium of lateral meristems. Compare with **xylem.**

phonotaxis Orientation toward or away from sound.

phosphodiester linkage Chemical linkage between adjacent **nucleotide residues** in DNA and RNA. Forms when the phosphate group of one nucleotide condenses with the hydroxyl group on the sugar of another nucleotide. Also known as *phosphodiester bond.*

phosphofructokinase The enzyme that catalyzes synthesis of fructose-1,6-bisphosphate from fructose-6-phosphate, a key reaction (step 3) in glycolysis.

phospholipid A class of lipid having a hydrophilic head (a phosphate group) and a hydrophobic tail (one or more fatty acids). Major components of the plasma membrane and organelle membranes.

phosphorylase An enzyme that breaks down glycogen by catalyzing hydrolysis of the α-glycosidic linkages between the glucose residues.

phosphorylation (verb: phosphorylate) The addition of a phosphate group to a molecule.

phosphorylation cascade A series of enzyme-catalyzed phosphorylation reactions commonly used in signal transduction pathways to amplify and convey a signal inward from the plasma membrane.

photic zone In an aquatic habitat, water that is shallow enough to receive some sunlight (whether or not it is enough to support photosynthesis). Compare with **aphotic zone.**

photon A discrete packet of light energy; a particle of light.

photoperiodism Any response by an organism to the relative lengths of day and night (i.e., photoperiod).

photophosporylation Production of ATP molecules using the energy released as light-excited electrons flow through an electron transport chain during photosynthesis. Involves generation of a proton-motive force during electron transport and its use to drive ATP synthesis.

photoreceptor A molecule, a cell, or an organ that is specialized to detect light.

photorespiration A series of light-driven chemical reactions that consumes oxygen and releases carbon dioxide, basically reversing photosynthesis. Usually occurs when there are high O_2 and low CO_2 concentrations inside plant cells, often in bright, hot, dry environments when stomata must be kept closed.

photoreversibility A change in conformation that occurs in certain plant pigments when they are exposed to their preferred wavelengths of light and that triggers responses by the plant.

photosynthesis The complex biological process that converts the energy of light into chemical energy stored in glucose and other organic molecules. Occurs in plants, algae, and some bacteria.

photosystem One of two types of units, consisting of a central reaction center surrounded by antenna complexes, that is responsible for the light-dependent reactions of photosynthesis.

photosystem I A photosystem that contains a pair of P700 chlorophyll molecules and uses absorbed light energy to produce NADPH.

photosystem II A photosystem that contains a pair of P680 chlorophyll molecules and uses absorbed light energy to split water into protons and oxygen and to produce ATP.

phototaxis Orientation toward or away from light.

phototroph An organism that produces ATP through photosynthesis.

phototropins A class of plant photoreceptors that detect blue light and initiate phototropic responses.

phototropism Growth or movement of an organism in a particular direction in response to light.

phylogenetic species concept The definition of a species as the smallest monophyletic group in a phylogenetic tree. Compare with **biological** and **morphological species concept.**

phylogenetic tree A diagram that depicts the evolutionary history of a group of species and the relationships among them.

phylogeny The evolutionary history of a group of organisms.

phylum (plural: phyla) In Linnaeus' system, a taxonomic category above the class level and below the kingdom level. In plants, sometimes called a *division.*

physical map A map of a chromosome that shows the number of base pairs between various genetic markers. Compare with **genetic map.**

physiology The study of how an organism's body functions.

phytoalexin Any small compound produced by a plant to combat an infection (usually a fungal infection).

phytochrome A specialized plant photoreceptor that exists in two shapes depending on the ratio of red to far-red light and is involved in the timing certain physiological processes, such as flowering, stem elongation, and germination.

phytoplankton Small drifting aquatic organisms (plankton) that are photosynthetic.

phytoremediation The use of plants to clean contaminated soils.

pigment Any molecule that absorbs certain wavelengths of visible light and reflects or transmits other wavelengths.

piloting A type of navigation in which animals use familiar landmarks to find their way.

pinocytosis Uptake by a cell of extracellular fluid by pinching off the plasma membrane to form small membrane-bound vesicles; one type of endocytosis.

pioneering species Those species that appear first in recently disturbed areas.

pit In plants, a small hole in the secondary cell walls of tracheids that allow passage of water.

pitch The sensation produced by a particular frequency of sound. Low frequencies are perceived as low pitches; high frequencies, as high pitches.

pituitary gland A small gland directly under the brain that is physically and functionally connected to the hypothalamus. Produces and secretes an array of hormones that affect many other glands and organs.

placenta A structure that forms in the pregnant uterus from maternal and fetal tissues. Exchanges nutrients and wastes between mother and fetus, anchors the fetus to the uterine wall, and produces some hormones. Occurs in most mammals and in a few other vertebrates.

placental mammals See **eutherians.**

plankton Any small organism that drifts near the surface of oceans or lakes and swims little if at all.

Plantae The monophyletic group that includes red, green, and glaucophyte algae and land plants.

plant-defense hypothesis The hypothesis that rates of herbivory are limited by plant defenses such as toxins and spines.

plantlet A small plant, particularly one that forms on a parent plant via asexual reproduction and drops, becoming an independent individual.

plasma cell An effector B cell, which produces large quantities of antibodies. Is descended from an activated B cell that has interacted with antigen.

plasma membrane A membrane that surrounds a cell, separating it from the external environment and selectively regulating passage of molecules and ions into and out of the cell. Also called *cell membrane.*

plasmid A small, usually circular, supercoiled DNA molecule independent of the cell's main chromosome(s) in prokaryotes and some eukaryotes.

plasmodesmata (singular: plasmodesma) Physical connection between two plant cells, consisting of gaps in the cell walls through which the two cells' plasma membranes, cytoplasm, and smooth ER can connect directly. Functionally similar to gap junctions in animal cells.

plasmogamy Fusion of the cytoplasm of two individuals. Occurs in many fungi.

plastid One of a family of plant organelles, bounded by a double membrane, that includes chloroplasts; chromoplasts, which house pigment-containing vacuoles; and leucoplasts, which store oils, starch, or proteins.

plastocyanin A small protein that shuttles electrons from photosystem II to photosystem I during photosynthesis.

plastoquinone (PQ) A nonprotein electron carrier in the chloroplast electron transport chain. Receives excited electrons from pheophytin and passes them to more electronegative molecules in the chain. Also carries protons to the lumen side of the thylakoid membrane, generating a proton-motive force.

platelet A small membrane-bound cell fragment in vertebrate blood that functions in blood clotting. Derived from large cells in the bone marrow.

pleiotropy (adjective: pleiotropic) The ability of a single gene to affect more than one phenotypic trait.

ploidy The number of complete chromosome sets present. *Haploid* refers to a ploidy of 1; *diploid,* a ploidy of 2; *triploid,* a ploidy of 3; and *tetraploid,* a ploidy of 4.

podium (plural: podia) See **tube foot.**

point mutation A mutation that results in a change in a single nucleotide pair in a DNA molecule.

polar (1) Asymmetrical or unidirectional. (2) Carrying a partial positive charge on one side of a molecule and a partial negative charge on the other. Polar molecules are generally hydrophilic.

polar bodies The tiny, nonfunctional cells produced during meiosis of a primary oocyte, due to most of the cytoplasm going to the ovum.

polar covalent bond A covalent bond in which electrons are shared unequally between atoms differing in electronegativity, resulting in the more electronegative atom having a partial negative charge and the other atom, a partial positive charge. Compare with **nonpolar covalent bond.**

polar nuclei In flowering plants, the nuclei in the female gametophyte that fuse with one sperm nucleus to produce the endosperm. Most species have two.

pollen grain In seed plants, a male gametophyte enclosed within a protective coat.

pollen tube In flowering plants, a structure that grows out of a pollen grain after it reaches the stigma, extends down the style, and through which two sperm cells are delivered to the ovule.

pollination The process by which pollen reaches the carpel of a flower (in flowering plants) or reaches the ovule directly (in conifers and their relatives).

poly(A) tail In eukaryotes, a sequence of 100–250 adenine nucleotides added to the 3′ end of newly transcribed messenger RNA molecules.

polycistronic mRNA An mRNA molecule that contains more than one protein-coding segment, each with its own start and stop codons and each coding for a different protein. Common in prokaryotes.

polygenic inheritance The inheritance patterns that result when many genes influence one trait.

polymer Any long molecule composed of small repeating units (monomers) bonded together. The main biological polymers are proteins, nucleic acids, and polysaccharides.

polymerase chain reaction (PCR) A laboratory technique for rapidly generating millions of identical copies of a specific stretch of DNA by incubating the original DNA sequence of interest with primers, nucleotides, and DNA polymerase.

polymerization (verb: polymerize) The process by which many identical or similar small molecules (monomers) are covalently bonded to form a large molecule (polymer).

polymorphism (adjective: polymorphic) (1) The occurrence of more than one allele at a certain genetic locus in a population. (2) The occurrence of more than two distinct phenotypes of a trait in a population.

polyp The immotile (sessile) stage in the life cycle of some cnidarians (e.g., jellyfish). Compare with **medusa**.

polypeptide A chain of 50 or more amino acids linked together by peptide bonds. Compare with **oligopeptide** and **protein**.

polyploidy (adjective: polyploid) The state of having more than two full sets of chromosomes.

polyribosome A structure consisting of one messenger RNA molecule along with many attached ribosomes and their growing peptide strands.

polysaccharide A linear or branched polymer consisting of many monosaccharides joined by glycosidic linkages. Carbohydrate polymers with relatively few residues often are called *oligosaccharides*.

polyspermy Fertilization of an egg by multiple sperm.

poor-nutrition hypothesis The hypothesis that herbivore populations are limited by the poor nutritional content of plants, especially low nitrogen.

population A group of individuals of the same species living in the same geographic area at the same time.

population cycle A regular fluctuation in size exhibited by certain populations.

population density The number of individuals of a population per unit area.

population dynamics Changes in the size and other characteristics of populations through time.

population ecology The study of how and why the number of individuals in a population changes over time.

population viability analysis (PVA) A method of estimating the likelihood that a population will avoid extinction for a given time period.

positive feedback A physiological mechanism in which a change in some variable stimulates a response that increases the change. Relatively rare in organisms but is important in generation of the action potential. Compare with **negative feedback**.

positive pressure ventilation Ventilation of the lungs that is accomplished by "pushing" air into the lungs by positive pressure in the mouth. Compare with **negative pressure ventilation**.

positive-sense virus A virus whose genome contains the same sequences as the mRNA required to produce viral proteins. Compare with **ambisense virus** and **negative-sense virus**.

posterior Toward an animal's tail and away from its head. The opposite of anterior.

posterior pituitary The part of the pituitary gland that contains the ends of hypothalamic neurosecretory cells and from which oxytocin and antidiuretic hormone are secreted. Compare with **anterior pituitary**.

postsynaptic neuron A neuron that receives signals, usually via neurotransmitters, from another neuron at a synapse. Muscle and gland cells also may receive signals from presynaptic neurons.

post-translational control Regulation of gene expression by modification of proteins (e.g., addition of a phosphate group or sugar residues) after translation.

postzygotic isolation Reproductive isolation resulting from mechanisms that operate after mating of individuals of two different species occurs. The most common mechanisms are the death of hybrid embryos or reduced fitness of hybrids.

potential energy Energy stored in matter as a result of its position or molecular arrangement. Compare with **kinetic energy**.

prebiotic soup A hypothetical solution of sugars, amino acids, nitrogenous bases, and other building blocks of larger molecules that may have formed in shallow waters or deep-ocean vents of ancient Earth and given rise to larger biological molecules.

Precambrian era The interval between the formation of the Earth, about 4.6 billion years ago, and the appearance of most animal groups about 543 million years ago. Unicellular organisms were dominant for most of this era, and oxygen was virtually absent for the first 2 billion years.

predation The killing and eating of one organism (the prey) by another (the predator).

predator Any organism that kills other organisms for food.

prediction A measurable or observable result of an experiment based on a particular hypothesis. A correct prediction provides support for the hypothesis being tested.

pressure-flow hypothesis The hypothesis that sugar movement through phloem tissue is due to differences in the turgor pressure of phloem sap.

pressure potential (ψ_P) A component of the potential energy of water caused by physical pressures on a solution. In plant cells, it equals the wall pressure plus turgor pressure. Compare with **solute potential** (ψ_S).

presynaptic neuron A neuron that transmits signals, usually by releasing neurotransmitters, to another neuron or to an effector cell at a synapse.

pre-urine See **filtrate**.

prezygotic isolation Reproductive isolation resulting from any one of several mechanisms that prevent individuals of two different species from mating.

primary cell wall The outermost layer of a plant cell wall, made of cellulose fibers and gelatinous polysaccharides, that defines the shape of the cell and withstands the turgor pressure of the plasma membrane.

primary consumer An herbivore; an organism that eats plants, algae, or other primary producers. Compare with **secondary consumer**.

primary decomposer A decomposer (detritivore) that consumes detritus from plants.

primary growth In plants, an increase in the length of stems and roots due to the activity of apical meristems. Compare with **secondary growth**.

primary immune response An acquired immune response to a pathogen that the immune system has not encountered before. Compare with **secondary immune response**.

primary oocyte The large diploid cell in an ovarian follicle that can initiate meiosis to produce a haploid ovum.

primary producer Any organism that creates its own food by photosynthesis or from reduced inorganic compounds and that is a food source for other species in its ecosystem. Also called *autotroph*.

primary RNA transcript In eukaryotes, a newly transcribed messenger RNA molecule that has not yet been processed (i.e., it has not received a 5′ cap or poly(A) tail, and still contains introns).

primary spermatocyte A diploid cell in the testis that can initiate meiosis I to produce two secondary spermatocytes.

primary structure The sequence of amino acids in a peptide or protein; also the sequence of nucleotides in a nucleic acid. Compare with **secondary, tertiary,** and **quaternary structure**.

primary succession The gradual colonization of a habitat of bare rock or gravel, usually after an environmental disturbance that removes all soil and previous organisms. Compare with **secondary succession**.

primase An enzyme that synthesizes a short stretch of RNA to use as a primer during DNA replication.

primates The lineage of mammals that includes prosimians (lemurs, lorises, etc.), monkeys, and great apes (including humans).

primer A short, single-stranded RNA molecule that base pairs with the 5′ end of a DNA template strand and is elongated by DNA polymerase during DNA replication.

prion An infectious form of a protein that is thought to cause disease by inducing the normal form to assume an abnormal three-dimensional structure. Cause of spongiform encephalopathies, such as mad cow disease.

probe A radioactively or chemically labeled single-stranded fragment of a known DNA or RNA sequence that can bind to and thus detect its complementary sequence in a sample containing many different sequences.

proboscis A long, narrow feeding appendage through which food can be obtained.

procambium A group of cells in the center of a young plant embryo that gives rise to the vascular tissue.

product Any of the final materials formed in a chemical reaction.

productivity The total amount of carbon fixed by photosynthesis per unit area per year.

progesterone A steroid hormone produced in the ovaries and secreted by the corpus luteum after ovulation; causes the uterine lining to thicken.

programmed cell death See **apoptosis**.

prokaryote A member of the domain Bacteria or Archaea; a unicellular organism lacking a nucleus and containing relatively few organelles or cytoskeletal components. Compare with **eukaryote**.

prolactin A peptide hormone, produced and secreted by the anterior pituitary, that promotes milk production in female mammals and has a variety of effects on parental behavior and seasonal reproduction in other vertebrates.

prometaphase A stage in mitosis or meiosis during which the nuclear envelope breaks down and spindle fibers attach to chromatids.

promoter A short nucleotide sequence in DNA that binds RNA polymerase, enabling transcription to begin. In prokaryotic DNA, a single promoter often is associated with several contiguous genes. In eukaryotic DNA, each gene generally has its own promoter.

promoter-proximal elements In eukaryotes, regulatory sequences in DNA that are close to a promoter and that can bind regulatory transcription factors.

proofreading The process by which a DNA polymerase recognizes and removes a wrong base added during DNA replication and then continues synthesis.

prophase The first stage in mitosis or meiosis during which chromosomes become visible and the mitotic spindle forms. Synapsis and crossing over occur during prophase of meiosis I.

prosimians One of the two major lineages of primates, including lemurs, tarsiers, pottos, and lorises. Compare with **anthropoids.**

prostate gland A gland in male mammals that surrounds the base of the urethra and secretes a fluid that is a component of semen.

protease An enzyme that can degrade proteins by cleaving the peptide bonds between amino acid residues.

protein A macromolecule consisting of one or more polypeptide chains composed of 50 or more amino acids linked together. Each protein has a unique sequence of amino acids and, in its native state, a characteristic three-dimensional shape.

protein kinase An enzyme that catalyzes the addition of a phosphate group to another protein, typically activating or inactivating the substrate protein.

proteome The complete set of proteins produced by a particular cell type.

proteomics The systematic study of the interactions, localization, functions, regulation, and other features of the full protein set (proteome) in a particular cell type.

protist Any eukaryote that is not a green plant, animal, or fungus. Protists are a diverse paraphyletic group. Most are unicellular, but some are multicellular or form aggregations called colonies.

protoderm The exterior layer of a young plant embryo that gives rise to the epidermis.

proton-motive force The combined effect of a proton gradient and an electric potential gradient across a membrane, which can drive protons across the membrane. Used by mitochondria and chloroplasts to power ATP synthesis via the mechanism of chemiosmosis.

proton pump A membrane protein that can hydrolyze ATP to power active transport of protons (H^+ ions) across a plasma membrane against an electrochemical gradient. Also called H^+ ATPase.

proto-oncogene Any gene that normally encourages cell division in a regulated manner, typically by triggering specific phases in the cell cycle. Mutation may convert it into an oncogene.

protostomes A major lineage of animals that share a pattern of embryological development, including spiral cleavage, formation of the mouth earlier than the anus, and formation of the coelom by splitting of a block of mesoderm. Includes arthropods, mollusks, and annelids. Compare with **deuterostomes.**

proximal tubule In the vertebrate kidney, the convoluted section of a nephron into which filtrate moves from Bowman's capsule. Involved in the largely unregulated reabsorption of electrolytes, nutrients, and water. Compare with **distal tubule.**

proximate causation In biology, the immediate, mechanistic cause of a phenomenon (how it happens), as opposed to why it evolved. Also called *proximate explanation*. Compare with **ultimate causation.**

pseudocoelomate An animal with an internal fluid-filled body cavity (coelom) that is lined with endoderm and mesoderm layers.

pseudogene A DNA sequence that closely resembles a functional gene but is not transcribed. Thought to have arisen by duplication of the functional gene followed by inactivation due to a mutation.

pseudopodium (plural: pseudopodia) A temporary bulge-like extension of certain cells used in cell crawling and ingestion of food.

puberty The various physical and emotional changes that an immature animal undergoes leading to reproductive maturity. Also the period when such changes occur.

pulmonary artery A short, thick-walled artery that carries oxygen-poor blood from the heart to the lungs.

pulmonary circulation The part of the circulatory system that sends oxygen-poor blood to the lungs. Is separate from the rest of the circulatory system (the systemic circulation) in mammals and birds.

pulmonary vein A short, thin-walled vein that carries oxygen-rich blood from the lungs to the heart.

pulse-chase experiment A type of experiment in which a population of cells or molecules at a particular moment in time is marked by means of a labeled molecule and then their fate is followed over time.

pump Any membrane protein that can hydrolyze ATP to power active transport of a specific ion or small molecule across a plasma membrane against its electrochemical gradient. See **proton pump.**

Punnett square A diagram that depicts the genotypes and phenotypes that should appear in offspring of a certain cross.

pupa (plural: pupae) A metamorphosing insect that is enclosed in a protective case.

pupation A developmental stage of many insects, in which the body metamorphoses from the larval form to the adult form while enclosed in a protective case.

pupil The hole in the center of the iris through which light enters a vertebrate or cephalopod eye.

pure line In animal or plant breeding, a strain of individuals that produce offspring identical to themselves when self-pollinated or crossed to another member of the same population. Pure lines are homozygous for most, if not all, genetic loci.

purines A class of small, nitrogen-containing, double-ringed bases (guanine, adenine) found in nucleotides. Compare with **pyrimidines.**

pyrimidines A class of small, nitrogen-containing, single-ringed bases (cytosine, uracil, thymine) found in nucleotides. Compare with **purines.**

pyruvate dehydrogenase A large enzyme complex, located in the inner mitochondrial membrane, that is responsible for conversion of pyruvate to acetyl CoA during cellular respiration.

quantitative trait A heritable feature that exhibits phenotypic variation along a smooth, continuous scale of measurement (e.g., human height), rather than the distinct forms characteristic of discrete traits.

quaternary structure The overall three-dimensional shape of a protein containing two or more polypeptide chains (subunits); determined by the number, relative positions, and interactions of the subunits. Compare with **primary, secondary,** and **tertiary structure.**

radial cleavage The pattern of embryonic cleavage seen in protostomes, in which cells divide at right angles to each other to form tiers. Compare with **spiral cleavage.**

radial symmetry An animal body pattern in which there are least two planes of symmetry. Typically, the body is in the form of a cylinder or disk, with body parts radiating from a central hub. Compare with **bilateral symmetry.**

radiation Transfer of heat between two bodies that are not in direct physical contact. More generally, the emission of electromagnetic energy of any wavelength.

radicle The root of a plant embryo.

radula A rasping feeding appendage in gastropods (snails, slugs).

rain shadow The dry region on the side of a mountain range away from the prevailing wind.

range The geographic distribution of a species.

Ras protein A type of G protein that is activated by binding of signaling molecules to receptor tyrosine kinases and then initiates a phosphorylation cascade, culminating in a cell response.

Rb protein A tumor-suppressor protein that helps regulate progression of a cell from the G_1 phase to the S phase of the cell cycle. Defects in Rb protein are found in many types of cancer.

reactant Any of the starting materials in a chemical reaction.

reaction center Centrally located component of a photosystem containing proteins and a pair of specialized chlorophyll molecules. Is surrounded by antenna complexes and receives excited electrons from them.

reactive oxygen intermediates (ROIs) Highly reactive oxygen-containing compounds that are used in plant and animal cells to kill infected cells and for other purposes.

reading frame The division of a sequence of DNA or RNA into a particular series of three-nucleotide codons. There are three possible reading frames for any sequence.

realized niche The ecological niche that a species occupies in the presence of competitors. Compare with **fundamental niche.**

receptor-mediated endocytosis Uptake by a cell of certain extracellular macromolecules, bound to specific receptors in the plasma membrane, by pinching off the membrane to form small membrane-bound vesicles.

receptor tyrosine kinase (RTK) Any of a class of cell-surface signal receptors that undergo phosphorylation after binding a signaling molecule. The activated, phosphorylated receptor then triggers a signal-transduction pathway inside the cell.

recessive Referring to an allele whose phenotypic effect is observed only in homozygous individuals. Compare with **dominant.**

reciprocal altruism Altruistic behavior that is exchanged between a pair of individuals at different points in time (i.e., sometimes individual A helps individual B, and sometimes B helps A).

reciprocal cross A breeding experiment in which the mother's and father's phenotypes are the reverse of that examined in a previous breeding experiment.

recombinant Possessing a new combination of alleles. May refer to a single chromosome or DNA molecule, or to an entire organism.

recombinant DNA technology A variety of techniques for isolating specific DNA fragments and introducing them into different regions of DNA and/or a different host organism.

recombination See **genetic recombination.**

rectal gland A salt-excreting gland in the digestive system of sharks, skates, and rays.

rectum The last portion of the digestive tract where feces are held until they are expelled.

red blood cell A hemoglobin-containing cell that circulates in the blood and delivers oxygen from the lungs to the tissues. Also called *erythrocyte.*

redox reaction Any chemical reaction that involves the transfer of one or more electrons from one reactant to another. Also called *reduction-oxidation reaction.*

reduction An atom's gain of electrons during a redox reaction, either by acceptance of an electron from another atom or by the electrons in covalent bonds moving closer to the atomic nucleus.

reduction-oxidation reaction See **redox reaction.**

reflex An involuntary response to environmental stimulation. May involve the brain (e.g., conditioned reflex) or not (e.g., spinal reflex).

refractory No longer responding to stimuli that previously elicited a response. For example, the tendency of voltage-gated sodium channels to remain closed immediately after an action potential.

regeneration Growth of a new body part to replace a lost body part.

regulatory cascade In embryonic development, a progressive series of interactions among genes and/or cytoplasmic determinants that organizes the body plan of the embryo.

regulatory sequence, DNA Any segment of DNA that is involved in controlling transcription of a specific gene by binding certain proteins.

regulatory site A site on an enzyme to which a regulatory molecule can bind and affect the enzyme's activity; separate from the active site where catalysis occurs.

regulatory transcription factor General term for proteins that bind to DNA regulatory sequences (eukaryotic enhancers, silencers, and promoter-proximal elements), but not to the promoter itself, leading to an increase or decrease in transcription of specific genes. Compare with **basal transcription factor.**

reinforcement In evolutionary biology, the natural selection for traits that prevent interbreeding between recently diverged species.

release factors Proteins that can trigger termination of RNA translation when a ribosome reaches a stop codon.

releaser See **sign stimulus.**

renal corpuscle In the vertebrate kidney, the ball-like structure at the beginning of a nephron,

consisting of a glomerulus and the surrounding Bowman's capsule. Acts as a filtration device.

replacement mutation See **missense mutation.**

replacement rate The number of offspring each female must produce over her entire life to "replace" herself and her mate, resulting in zero population growth. The actual number is slightly more than 2 because some offspring die before reproducing.

replica plating A method of identifying bacterial colonies that have certain mutations by transferring cells from each colony on a master plate to a second (replica) plate and observing their growth when exposed to different conditions.

replication fork The Y-shaped site at which a double-stranded molecule of DNA is separated into two single strands for replication.

repolarization Return to a normal membrane potential after it has changed; a normal phase in an action potential.

repressor Any regulatory protein that inhibits transcription.

reproduction The ability of an organism to make an exact or nearly exact copy of itself.

reptiles One of the two lineages of amniotes (vertebrates that produce amniotic eggs) distinguished by adaptations for reproduction on land. Includes turtles, snakes and lizards, crocodiles and alligators, and birds. Except for birds, all are ectotherms.

reservoir In biogeochemical cycles, a location in the environment where elements are stored for a time.

residues In a polymer, the individual units derived from the monomers that covalently bind to form the polymer. Proteins contain amino acid residues; nucleic acids, nucleotide residues; and polysaccharides, sugar residues.

resilience, community A measure of how quickly a community recovers following a disturbance.

resistance, community A measure of how much a community is affected by a disturbance.

resistance (R) loci Genes in plants encoding proteins involved in sensing the presence of pathogens and mounting a defensive response. Compare with **avirulence (avr) loci.**

respiration See **cellular respiration.**

respiratory distress syndrome A syndrome in which premature infants can suffocate due to insufficient surfactant in their lungs.

respiratory system The collection of cells, tissues, and organs responsible for gas exchange between an animal and its environment.

resting potential The membrane potential of a cell in its resting, or normal, state.

restriction endonucleases Bacterial enzymes that cut DNA at a specific base-pair sequence (restriction site). Also called *restriction enzymes.*

restriction site A specific short sequence in a DNA molecule that is recognized and cleaved by a particular restriction endonuclease. Also called *restriction endonuclease recognition site.*

retina A thin layer of light-sensitive cells (rods and cones) and neurons at the back of a camera-type eye, such as that of cephalopods and vertebrates.

retinal A light-absorbing pigment, derived from vitamin A, that is linked to the protein opsin in rods and cones of the vertebrate eye.

retrovirus A virus with an RNA genome that reproduces by transcribing its RNA into a DNA sequence and then inserting that DNA into the host's genome for replication.

reverse transcriptase A enzyme of retroviruses (RNA viruses) that can synthesize double-stranded DNA from a single-stranded RNA template.

Rh factor A protein present on the surface of red blood cells in some but not all humans. Clumping may occur when blood from an Rh$^+$ individual is mixed with that from an Rh$^-$ individual.

rhizobia (singular: rhizobium) Members of the bacterial genus *Rhizobia;* nitrogen-fixing bacteria that live in root nodules of members of the pea family (legumes).

rhizoid The hairlike structure that anchors a bryophyte (nonvascular plant) to the substrate.

rhizome A modified stem that runs horizontally underground and produces new plants at the nodes (a form of asexual reproduction). Compare with **stolon.**

rhodopsin A transmembrane complex that is instrumental in detection of light by rods and cones of the vertebrate eye. Is composed of the transmembrane protein opsin covalently linked to retinal, a light-absorbing pigment.

ribonucleic acid (RNA) A nucleic acid composed of ribonucleotides that usually is single stranded and functions as structural components of ribosomes (rRNA), transporters of amino acids (tRNA), and translators of the message of the DNA code (mRNA).

ribonucleotide See **nucleotide.**

ribosomal RNA (rRNA) A RNA molecule that forms part of the structure of a ribosome.

ribosome A large complex structure that synthesizes proteins by using the genetic information encoded in messenger RNA strands. Consists of two subunits, each composed of ribosomal RNA and proteins.

ribosome binding site In a bacterial mRNA molecule, the sequence just upstream of the start codon to which a ribosome binds to initiate translation. Also called the *Shine-Dalgarno sequence.*

ribozyme Any RNA molecule that can act as a catalyst, that is, speed up a chemical reaction.

ribulose bisphosphate (RuBP) A five-carbon compound that combines with CO_2 in the first step of the Calvin cycle during photosynthesis.

RNA See **ribonucleic acid.**

RNA interference (RNAi) Degradation of an mRNA molecule or inhibition of its translation following its binding by a short RNA (microRNA) whose sequence is complementary to a portion of the mRNA.

RNA polymerase One of a class of enzymes that catalyze synthesis of RNA from ribonucleotides using a DNA template. Also called *RNA pol.*

RNA processing In eukaryotes, the changes that a primary RNA transcript undergoes in the nucleus to become a mature mRNA molecule, which is exported to the cytoplasm. Includes the addition of a 5′ cap and poly(A) tail and splicing to remove introns.

RNA replicase A viral enzyme that can synthesize RNA from an RNA template.

rod cell A photoreceptor cell with a rod-shaped outer portion that is particularly sensitive to dim light, but not used to distinguish colors. Also called simply *rod*. Compare with **cone cell.**

root   (1) An underground part of a plant that anchors the plant and absorbs water and nutrients. (2) In a phylogenetic tree, the bottom, most ancient node.

root apical meristem (RAM)   A group of undifferentiated plant cells at the tip of a plant root that can differentiate into mature root tissue.

root cap   A small group of cells that covers and protects the tip of a plant root. Senses gravity and determines the direction of root growth.

root hair   A long, thin outgrowth of the epidermal cells of plant roots, providing increased surface area for absorption of water and nutrients.

root pressure   Positive (upward) pressure of xylem sap in the vascular tissue of roots. Is generated during the night as a result of the accumulation of ions from the soil and subsequent osmotic movement of water into the xylem.

root system   The belowground part of a plant.

rotifers   Members of the phylum Rotifera. Distinguished by a cluster of cilia (corona) at the anterior end and a pseudocoelom. Rotifers belong to the lophotrochozoan branch of the protostomes.

rough endoplasmic reticulum (rough ER)   The portion of the endoplasmic reticulum that is dotted with ribosomes. Involved in synthesis of plasma membrane proteins, secreted proteins, and proteins localized to the ER, Golgi apparatus, and lysosomes. Compare with **smooth endoplasmic reticulum.**

roundworms   Members of the phylum Nematoda. Distinguished by an unsegmented body with a pseudocoelom and no appendages. Roundworms belong to the ecdysozoan branch of the protostomes. Also called *nematodes*.

rRNA   See **ribosomal RNA.**

rubisco   The enzyme that catalyzes the first step of the Calvin cycle during photosynthesis: the addition of a molecule of CO_2 to ribulose bisphosphate. Also called *ribulose 1,5-bisphosphate carboxylase/oxygenase*.

ruminants   A group of hoofed mammals (e.g., cattle, sheep, deer) that have a four-chambered stomach specialized for digestion of plant cellulose. Ruminants regurgitate the cud, a mixture of partially digested food and cellulose-digesting bacteria, from the largest chamber (the rumen) for further chewing.

sac fungus   A member of a monophyletic lineage of fungi (Ascomycota) that produce large, often cup-shaped reproductive structures that contain asci. Also called *cup fungus* and *ascomycete*.

salicylic acid   A compound produced by plants that may play a role in systemic acquired resistance (SAR) against pathogens; a component of aspirin.

salivary glands   Vertebrate glands that secrete saliva (a mixture of water, mucus-forming glycoproteins, and digestive enzymes) into the mouth.

sampling error   The accidental selection of a nonrepresentative sample from some larger population, due to chance.

saprophyte   An organism that feeds primarily on dead plant material.

sapwood   The younger xylem in the outer layer of wood of a stem or root, functioning primarily in water transport.

sarcomere   The repeating contractile unit of a skeletal muscle cell; the portion of a myofibril located between adjacent Z disks.

sarcoplasmic reticulum   Sheets of smooth endoplasmic reticulum in a muscle cell. Contains high concentrations of calcium, which can be released into the cytoplasm to trigger contraction.

saturated   Referring to fats and fatty acids in which all the carbon-carbon bonds are single bonds. Such fats have relatively high melting points. Compare with **unsaturated.**

scanning electron microscope (SEM)   A microscope that produces images of the surfaces of objects by reflecting electrons from a specimen coated with a layer of metal atoms. Compare with **transmission electron microscope.**

scarify   To scrape, rasp, cut, or otherwise damage the coat of a seed. Necessary in some species to trigger germination.

Schwann cell   A type of glial cell that wraps around axons of some neurons outside the brain and spinal cord, forming a myelin sheath that provides electrical insulation. Compare with **oligodendrocyte.**

sclereid   In plants, a type of sclerenchyma cell that usually functions in protection, such as in seed coats and nutshells. Compare with **fiber.**

sclerenchyma cell   In plants, a cell that has a thick secondary cell wall and provides support; typically contains the tough structural polymer lignin and usually is dead at maturity. Includes fibers and sclereids. Compare with **collenchyma cell** and **parenchyma cell.**

screen   See **genetic screen.**

scrotum   A sac of skin, containing the testes, suspended just outside the abdominal body cavity of many male mammals.

secondary cell wall   The inner layer of a plant cell wall formed by certain cells as they mature. Provides support or protection.

secondary consumer   A carnivore; an organism that eats herbivores. Compare with **primary consumer.**

secondary growth   In plants, an increase in the width of stems and roots due to the activity of lateral meristems. Compare with **primary growth.**

secondary immune response   The acquired immune response to a pathogen that the immune system has encountered before. Compare with **primary immune response.**

secondary production   The total amount of new body tissue produced by animals that eat plants. May involve growth and/or reproduction.

secondary spermatocyte   A cell produced by meiosis I of a primary spermatocyte in the testis. Can undergo meiosis II to produce spermatids.

secondary structure   In proteins, localized folding of a polypeptide chain into regular structures (e.g., α-helix and β-pleated sheet) stabilized by hydrogen bonding between atoms of the backbone. In nucleic acids, elements of structure (e.g., helices and hairpins) stabilized by hydrogen bonding and other interactions between complementary bases. Compare with **primary, tertiary,** and **quaternary structure.**

secondary succession   Gradual colonization of a habitat after an environmental disturbance (e.g., fire, windstorm, logging) that removes some or all previous organisms but leaves the soil intact. Compare with **primary succession.**

second law of thermodynamics   The principle of physics that the entropy of the universe or any closed system increases during any spontaneous process.

second-male advantage   The reproductive advantage of a male who mates with a female last, after other males have mated with her.

second messenger   A nonprotein signaling molecule produced or activated inside a cell in response to stimulation at the cell surface. Commonly used to relay the message of a hormone or other extracellular signaling molecule.

secretin   A peptide hormone produced by cells in the small intestine in response to the arrival of food from the stomach. Stimulates secretion of bicarbonate (HCO_3^-) from the pancreas.

sedimentary rock   A type of rock formed by gradual accumulation of sediment, as in riverbeds and on the ocean floor. Most fossils are found in sedimentary rocks.

seed   A plant reproductive structure consisting of an embryo, associated nutritive tissue (endosperm), and an outer protective layer (seed coat). In angiosperms, develops from the fertilized ovule of a flower.

seed coat   A protective layer around a seed that encases both the embryo and the endosperm.

seed plants   Members of several phyla of green plants that have vascular tissue and make seeds. Include **angiosperms** and **gymnosperms.**

seedless vascular plants   Members of several phyla of green plants that have vascular tissue but do not make seeds. Include horsetails, ferns, lycophytes, and whisk ferns.

segment   A well-defined region of the body along the anterior-posterior body axis, containing similar structures as other, nearby segments.

segmentation   Division of the body or a part of it into a series of similar structures; exemplified by the body segments of insects and worms and by the somites of vertebrates.

segmentation genes   A group of genes that affect body segmentation in embryonic development. Includes gap genes, pair-rule genes, and segment polarity genes.

segmented worms   See **annelids.**

segregation, principle of   The concept that each pair of hereditary elements (alleles of the same gene) separate from each other during the formation of offspring (i.e., during meiosis). One of Mendel's two principles of genetics.

selective adhesion   The tendency of cells of one tissue type to adhere to other cells of the same type.

selectively permeable membrane   Any membrane across which some solutes can move more readily than others.

selective permeability   The property of a membrane that allows some substances to diffuse across it much more readily than other substances.

self   Property of a molecule or cell such that immune system cells do not attack it, due to certain molecular similarities to other body cells.

self-fertilization   In plants, the fusion of two gametes from the same individual to form a diploid offspring. Also called *selfing*.

self-incompatible   Incapable of self-fertilization.

semen   The combination of sperm and accessory fluids that is released by male mammals and reptiles during ejaculation.

semiconservative replication   The mechanism of replication used by cells to copy DNA. Results in each daughter DNA molecule containing one old strand and one new strand.

seminal vesicles   In male mammals, paired reproductive glands that secrete a sugar-containing fluid into semen, which provides energy for sperm movement. In other vertebrates and invertebrates, often stores sperm.

senescence   The process of aging.

sensitive period   A short time period in a young animal's life during which learning of certain critical behaviors can occur. Also called the *critical period*.

sensor   Any cell, organ, or structure with which an animal can sense some aspect of the external or internal environment. Usually functions, along with an integrator and effector, as part of a homeostatic system.

sensory neuron   A nerve cell that carries signals from sensory receptors to the central nervous system. Compare with **interneuron** and **motor neuron**.

sepal   One of the protective leaflike organs enclosing a flower bud and later supporting the blooming flower.

septum   (plural: septa) Any wall-like structure. In fungi, septa divide the filaments (hyphae) of mycelia into cell-like compartments.

serotonin   A neurotransmitter involved in many brain functions, including sleep, pleasure, and mood.

serum   The liquid that remains when cells and clot material are removed from clotted blood. Contains water, dissolved gases, growth factors, nutrients, and other soluble substances. Compare with **plasma.**

sessile   Permanently attached to a substrate; not capable of moving to another location.

set point   A normal or target value for a regulated internal variable, such as body heat or blood pH.

severe combined immunodeficiency disease (SCID)   A human disease characterized by an extremely high vulnerability to infectious disease. Caused by a genetic defect in the immune system.

sex chromosome   Any chromosome carrying genes involved in determining the sex of an individual. Compare with **autosome.**

sex-linked inheritance   Inheritance patterns observed in genes carried on sex chromosomes, so females and males have different numbers of alleles of a gene and may pass its trait only to one sex of offspring. Also called *sex-linkage*.

sexual dimorphism   Any trait that differs between males and females.

sexual reproduction   Any form of reproduction in which genes from two parents are combined via fusion of gametes, producing offspring that are genetically distinct from both parents. Compare with **asexual reproduction.**

sexual selection   A pattern of natural selection that favors individuals with traits that increase their ability to obtain mates. Acts more strongly on males than females.

shell   A hard protective outer structure. In protists, also called a *test*.

Shine-Dalgarno sequence   See **ribosome binding sequence.**

shoot apical meristem (SAM)   A group of undifferentiated plant cells at the tip of a plant stem that can differentiate into mature shoot tissues.

shoot system   The aboveground part of a plant comprising stems, leaves, and flowers (in angiosperms).

short-day plant   A plant that blooms in response to long nights (usually in late summer or fall in the northern hemisphere). Compare with **day-neutral** and **long-day plant.**

short interspersed nuclear elements   See SINES.

shotgun sequencing   A method of sequencing genomes that is based on breaking the genome into small pieces, sequencing each piece separately, and then figuring out how the pieces are connected.

sieve plate   In plants, a pore-containing structure at one end of a sieve-tube member in phloem.

sieve-tube member   In plants, an elongated sugar-conducting cell in phloem that has sieve plates at both ends, allowing sap to flow to adjacent cells.

signal   In behavioral ecology, any information-containing behavior.

signal receptor   Any cellular protein that binds to a particular signaling molecule (e.g., a hormone or neurotransmitter) and triggers a response by the cell. Receptors for water-soluble signals are transmembrane proteins in the plasma membrane; those for many lipid-soluble signals (e.g., steroid hormones) are located inside the cell.

signal recognition particle (SRP)   A RNA-protein complex that binds to the ER signal sequence in a polypeptide as it emerges from a ribosome and transports the ribosome-polypeptide complex to the ER membrane where synthesis of the polypeptide is completed.

signal transducers and activators of transcription (STATs)   In mammals, a group of regulatory transcription factors that, upon phosphorylation, can activate transcription of certain genes.

signal transduction   The process by which a stimulus (e.g., a hormone, a neurotransmitter, or sensory information) outside a cell is amplified and converted into a response by the cell. Usually involves a specific sequence of molecular events, or signal transduction pathway.

signal transduction cascade   See **phosphorylation cascade.**

sign stimulus   A simple stimulus that elicits an invariant, stereotyped behavioral response (fixed action pattern) from an animal. Also called a *releaser*.

silencer   A regulatory sequence in eukaryotic DNA to which repressor proteins can bind, inhibiting transcription of certain genes.

silent mutation   A mutation that does not detectably affect the phenotype of the organism.

simple eye   An eye with only one light-collecting apparatus (e.g., one lens), as in vertebrates. Compare with **compound eye.**

simple fruit   A fruit (e.g., apricot) that develops from a single flower that has a single carpel or several fused carpels. Compare with **aggregate** and **multiple fruit.**

simple sequence repeat   See **microsatellite.**

SINEs (short interspersed nuclear elements)   The second most abundant class of transposable elements in human genomes; can create copies of itself and insert them elsewhere in the genome. Compare with **LINEs.**

single nucleotide polymorphism (SNP)   A site on a chromosome where individuals in a population have different nucleotides. Can be used as a genetic marker to help track the inheritance of nearby genes.

single-strand DNA-binding proteins (SSBPs)   A class of proteins that attach to separated strands of DNA during replication or transcription, preventing them from re-forming a double helix.

sink   Any tissue, site, or location where an element or a molecule is consumed or taken out of circulation (e.g., in plants, a tissue where sugar exits the phloem). Compare with **source.**

sinoatrial (SA) node   A cluster of cardiac muscle cells, in the right atrium of the vertebrate heart, that initiates the heartbeat and determines the heart rate. Compare with **atrioventricular (AV) node.**

siphon   A tubelike appendage of many molluscs, that is often used for feeding or propulsion.

sister chromatids   The paired strands of a recently replicated chromosome, which are connected at the centromere and eventually separate during anaphase of mitosis and meiosis II. Compare with **non-sister chromatids.**

sister groups   Closely related taxa, which occupy adjacent branches in a phylogenetic tree. Also called *sister taxa*.

skeletal muscle   The muscle tissue attached to the bones of the vertebrate skeleton. Consists of long, unbranched muscle fibers with a characteristic striped (striated) appearance; controlled voluntarily. Also called *striated muscle*. Compare with **cardiac** and **smooth muscle.**

sliding-filament model   The hypothesis that thin (actin) filaments and thick (myosin) filaments slide past each other, thereby shortening the sarcomere. Shortening of all the sarcomeres in a myofibril results in contraction of the entire myofibril.

slug   (1) A member of a certain lineage of terrestrial gastropods, closely related to snails but lacking a shell. (2) A mobile aggregation of cells of a cellular slime mold.

small intestine   The portion of the digestive tract between the stomach and the large intestine. The site of the final stages of digestion and of most nutrient absorption.

small nuclear ribonucleoproteins   See **snRNPs.**

smooth endoplasmic reticulum (smooth ER)   The portion of the endoplasmic reticulum that does not have ribosomes attached to it. Involved in synthesis and secretion of lipids. Compare with **rough endoplasmic reticulum.**

smooth muscle   The unstriated muscle tissue that lines the intestine, blood vessels, and some other organs. Consists of tapered, unbranched cells that can sustain long contractions. Not voluntarily controlled. Compare with **cardiac** and **skeletal muscle.**

snRNPs (small nuclear ribonucleoproteins)   Complexes of proteins and small RNA molecules that function in splicing (removal of introns from primary RNA transcripts) as components of spliceosomes.

sodium-potassium pump   A transmembrane protein that uses the energy of ATP to move sodium ions out of the cell and potassium ions in. Also called Na^+/K^+-ATPase.

solute   Any substance that is dissolved in a liquid.

solute potential (ψ_S)   A component of the potential energy of water caused by a difference in solute concentrations at two locations. Also called *osmotic potential*. Compare with **pressure potential (ψ_P)**.

solution   A liquid containing one or more dissolved solids or gases in a homogeneous mixture.

solvent   Any liquid in which one or more solids or gases can dissolve.

soma   See **cell body**.

somatic cell   Any type of cell in a multicellular organism except eggs, sperm, and their precursor cells. Also called *body cells*.

somatic nervous system   The part of the peripheral nervous system (outside the brain and spinal cord) that controls skeletal muscles and is under voluntary control. Compare with **autonomic nervous system.**

somatostatin   A hormone secreted by the pancreas and hypothalamus that inhibits the release of several other hormones.

somites   Paired blocks of mesoderm on both sides of the developing spinal cord in a vertebrate embryo. Give rise to muscle tissue, vertebrae, ribs, limbs, etc.

soredium   (plural: soredia) In lichens, a small reproductive structure that consists of fungal hyphae surrounding a cluster of green algae.

source   Any tissue, site, or location where a substance is produced or enters circulation (e.g., in plants, the tissue where sugar enters the phloem). Compare with **sink.**

Southern blotting   A technique for identifying specific DNA fragments separated by gel electrophoresis by transferring them to filter paper, separating the strands, and hybridizing with a labeled probe complementary to the fragment of interest. Compare with **Northern** and **Western blotting.**

speciation   The evolution of two or more distinct species from a single ancestral species.

species   A distinct, identifiable group of populations that is thought to be evolutionarily independent of other populations and whose members can interbreed. Generally distinct from other species in appearance, behavior, habitat, ecology, genetic characteristics, etc.

species–area relationship   The mathematical relationship between the area of a certain habitat and the number of species that it can support.

species diversity   The variety and relative abundance of the species present in a given ecological community.

species richness   The number of species present in a given ecological community.

specific heat   The amount of energy required to raise the temperature of 1 gram of a substance by 1°C; a measure of the capacity of a substance to absorb energy.

spectral karyotyping   See **chromosome painting.**

spectrophotometer   An instrument used to measure the wavelengths of light that are absorbed by a particular pigment.

sperm   A mature male gamete; smaller and more mobile than the female gamete.

spermatid   An immature sperm cell.

spermatogenesis   The production of sperm. Occurs continuously in a testis.

spermatogonia   (singular: spermatogonium) The diploid cells in a testis that can give rise to primary spermatocytes.

spermatophore   A gelatinous package of sperm cells that is produced by males of species that have internal fertilization without copulation.

sphincter   A muscular valve that can close off a tube, as in a blood vessel or a part of the digestive tract.

spicule   Stiff spike of silica or calcium carbonate found in the body of many sponges.

spindle fibers   Groups of microtubules that attach to chromosomes and push and pull them during mitosis and meiosis.

spiracle   In insects, a small opening that connects air-filled tracheae to the external environment, allowing for gas exchange.

spiral cleavage   The pattern of embryonic cleavage seen in deuterostomes, in which cells divide at oblique angles to form a spiral coil of cells. Compare with **radial cleavage.**

spleen   A dark red organ, found near the stomach of most vertebrates, that filters blood, stores extra red blood cells in case of emergency, and plays a role in immunity.

spliceosome   In eukaryotes, a large, complex assembly of snRNPs (small nuclear ribonucleoproteins) that catalyzes removal of introns from primary RNA transcripts.

splicing   The process by which introns are removed from primary RNA transcripts and the remaining exons are connected together.

sporangium   (plural: sporangia) A spore-producing structure found in seed plants, some protists, and some fungi (e.g., chytrids).

spore   (1) In bacteria, a dormant form that generally is resistant to extreme conditions. (2) In eukaryotes, a single cell produced by mitosis or meiosis (not by fusion of gametes) that is capable of developing into an adult organism.

sporophyte   In organisms undergoing alternation of generations, the multicellular diploid form that arises from two fused gametes and produces haploid spores. Compare with **gametophyte.**

sporopollenin   A watertight material that encases spores and pollen of modern land plants.

stabilizing selection   A pattern of natural selection that favors phenotypes near the middle of the range of phenotypic variation. Reduces overall genetic variation in a population. Compare with **disruptive selection.**

stamen   The male reproductive structure of a flower. Consists of an anther, in which pollen grains are produced, and a filament, which supports the anther. Compare with **carpel.**

standing defense   See **constitutive defense.**

stapes   The last of three small bones (ossicles) in the middle ear of vertebrates. Receives vibrations from the tympanic membrane and by vibrating against the oval window passes them to the cochlea.

starch   A mixture of two storage polysaccharides, amylose and amylopectin, both formed from α-glucose monomers. Amylopectin is branched, and amylose is unbranched. The major form of stored carbohydrate in plants.

start codon   The AUG triplet in mRNA at which protein synthesis begins; codes for the amino acid methionine.

statocyst   A sensory organ of many arthropods that detects the animal's orientation in space (i.e., whether the animal is flipped upside down).

statolith   A tiny stone or dense particle found in specialized gravity-sensing organs in some animals such as lobsters.

statolith hypothesis   The hypothesis that amyloplasts (dense, starch-storing plant organelles) serve as statoliths in gravity detection by plants.

STATs   See **signal transducers and activators of transcription.**

stem cell   Any relatively undifferentiated cell that can divide to produce daughter cells identical to itself or more specialized daughter cells, which differentiate further into specific cell types.

stereocilium   (plural: stereocilia) One of many stiff outgrowths from the surface of a hair cell that are involved in detection of sound by terrestrial vertebrates or of waterborne vibrations by fishes.

steroid   A class of lipid with a characteristic four-ring structure.

steroid-hormone receptor   One of a family of intracellular receptors that bind to various steroid hormones, forming a hormone-receptor complex that acts as a regulatory transcription factor and activates transcription of specific target genes.

sticky end   The short, single-stranded ends of a DNA molecule cut by a restriction endonuclease. Tend to form hydrogen bonds with other sticky ends that have complementary sequences.

stigma   The moist tip at the end of a flower carpel to which pollen grains adhere.

stolon   A modified stem that runs horizontally over the soil surface and produces new plants at the nodes (a form of asexual reproduction). Compare with **rhizome.**

stoma   (plural: stomata) Generally, a pore or opening. In plants, a microscopic pore on the surface of a leaf or stem through which gas exchange occurs.

stomach   A tough, muscular pouch in the vertebrate digestive tract between the esophagus and small intestine. Physically breaks up food and begins digestion of proteins.

stop codon   One of three mRNA triplets (UAG, UGA, or UAA) that cause termination of protein synthesis. Also called a *termination codon*.

strain   A population of genetically similar or identical individuals.

stream   A body of water that moves constantly in one direction.

striated muscle   See **skeletal muscle.**

stroma   The fluid matrix of a chloroplast in which the thylakoids are embedded. Site where the Calvin cycle reactions occur.

structural formula   A two-dimensional notation in which the chemical symbols for the constituent atoms are joined by straight lines representing single (—), double (=), or triple (≡) covalent bonds. Compare with **molecular formula.**

structural gene   A stretch of DNA that codes for a functional protein or functional RNA molecule, not including any regulatory sequences (e.g., a promoter, enhancer).

structural homology   Similarities in organismal structures (e.g., limbs, shells, flowers) that are due to inheritance from a common ancestor.

structural isomer A molecule that shares the same molecular formula as another molecule but differs in the order in which covalently bonded atoms are attached. Compare with **geometric** and **optical isomer.**

style The slender stalk of a flower carpel connecting the stigma and the ovary.

subspecies A population that has distinctive traits and some genetic differences relative to other populations of the same species but that is not distinct enough to be classified as a separate species.

substrate (1) A reactant that interacts with an enzyme in a chemical reaction. (2) A surface on which a cell or organism sits.

substrate-level phosphorylation Production of ATP by transfer of a phosphate group from an intermediate substrate directly to ADP. Occurs in glycolysis and in the Krebs cycle.

succession In ecology, the gradual colonization of a habitat after an environmental disturbance (e.g., fire, flood), usually by a series of species. See **primary** and **secondary succession.**

sucrose A disaccharide formed from glucose and fructose. One of the two main products of photosynthesis.

sulfate reducer A prokaryote that produces hydrogen sulfide (H_2S) as a by-product of cellular respiration.

summation The additive effect of different postsynaptic potentials at a nerve or muscle cell, such that several subthreshold stimulations can cause an action potential.

supernatant The liquid above a layer of solid particles (the pellet) in a tube after centrifugation.

surface tension The cohesive force that causes molecules at the surface of a liquid to stick together, thereby resisting deformation of the liquid's surface and minimizing its surface area.

surfactant A mixture of phospholipids and proteins produced by lung cells that reduces surface tension, allowing the lungs to expand more.

survivorship curve A graph depicting the percentage of a population that survives to different ages.

suspension culture A population of cells grown in a flask containing a liquid nutrient medium. The flask is rotated continuously to keep the cells suspended and the nutrients mixed.

suspension feeder Any organism that obtains food by filtering small particles or small organisms out of water or air. Also called *filter feeder.*

sustainability The planned use of environmental resources at a rate no faster than the rate at which they are naturally replaced.

sustainable agriculture Agricultural techniques that are designed to maintain long-term soil quality and productivity.

swim bladder A gas-filled organ of many ray-finned fishes that regulates buoyancy.

swamp A wetland that has a steady rate of water flow and is dominated by trees and shrubs.

symbiosis (adjective: symbiotic) Any close and prolonged physical relationship between individuals of two different species. See **commensalism, mutualism,** and **parasitism.**

symmetric competition Ecological competition between two species in which both suffer similar declines in fitness. Compare with **asymmetric competition.**

sympathetic nervous system The part of the autonomic nervous system that stimulates fight-or-flight responses, such as increased heart rate, increased blood pressure, and decreased digestion. Compare with **parasympathetic nervous system.**

sympatric speciation The divergence of populations living within the same geographic area into different species as the result of their genetic (not physical) isolation. Compare with **allopatric speciation.**

sympatry Condition in which two or more populations live in the same geographic area, or close enough to permit interbreeding. Compare with **allopatry.**

symplast In plant roots, a continuous pathway through which water can flow through the cytoplasm of adjacent cells that are connected by plasmodesmata. Compare with **apoplast.**

symporter A carrier protein that allows an ion to diffuse down an electrochemical gradient, using the energy of that process to transport a different substance in the same direction *against* its concentration gradient. Compare with **antiporter.**

synapomorphy A shared, derived trait found in two or more taxa that is present in their most recent common ancestor but is missing in more distant ancestors. Useful for inferring evolutionary relationships.

synapse The interface between two neurons or between a neuron and an effector cell.

synapsis The physical pairing of two homologous chromosomes during prophase I of meiosis. Crossing over occurs during synapsis.

synaptic cleft The space between two communicating nerve cells (or between a neuron and effector cell) at a synapse, across which neurotransmitters diffuse.

synaptic plasticity Long-term changes in the responsiveness or physical structure of a synapse that can occur after particular stimulation patterns. Thought to be the basis of learning and memory.

synaptic vesicle A small neurotransmitter-containing vesicle at the end of an axon that releases neurotransmitter into the synaptic cleft by exocytosis.

synaptonemal complex A network of proteins that holds non-sister chromatids together during synapsis in meiosis I.

synthesis (S) phase The phase of the cell cycle during which DNA is synthesized and chromosomes are replicated.

system In biology, a more complex organization resulting from the combination of various components, such as a group of organs that work together to perform a physiological function.

systemic acquired resistance (SAR) A slow, widespread response of plants to a localized infection that protects healthy tissue from invasion by pathogens. Compare with **hypersensitive response.**

systemic circulation The part of the circulatory system that sends oxygen-rich blood from the lungs out to the rest of the body. Is separate from the pulmonary circulation in mammals and birds.

systemin A peptide hormone, produced by plant cells damaged by herbivores, that initiates a protective response in undamaged cells.

systole The portion of the heartbeat cycle during which the heart muscles are contracting. Compare with **diastole.**

systolic blood pressure The force exerted by blood against artery walls during contraction of the heart's left ventricle. Compare with **diastolic blood pressure.**

taiga A vast forest biome throughout subarctic regions, consisting primarily of short conifer trees. Characterized by intensely cold winters, short summers, and high annual variation in temperature.

taproot A large vertical main root of a plant.

taste bud Sensory structure, found chiefly in the mammalian tongue, containing spindle-shaped cells that respond to chemical stimuli.

TATA-binding protein (TBP) A protein that binds to the TATA box in eukaryotic promoters and is a component of the basal transcription complex.

TATA box A short DNA sequence in many eukaryotic promoters about 30 base pairs upstream from the transcription start site.

taxis Movement toward or away from some external cue.

taxon (plural: taxa) Any named group of organisms at any level of a classification system.

taxonomy The branch of biology concerned with the classification and naming of organisms.

TBP See **TATA-binding protein.**

T cell A type of leukocyte that matures in the thymus and, with B cells, is responsible for acquired immunity. Involved in activation of B cells ($CD4^+$ helper T cells) and destruction of infected cells ($CD8^+$ cytotoxic T cells). Also called *T lymphocytes.*

T-cell receptor (TCR) A transmembrane protein found on T cells that can bind to antigens displayed on the surfaces of other cells. Composed of two polypeptides called the alpha chain and beta chain. See **antigen presentation.**

tectorial membrane A membrane in the vertebrate cochlea that takes part in the transduction of sound by bending the stereocilia of hair cells in response to sonic vibrations.

telomerase An enzyme that replicates the ends of chromosome (telomeres) by catalyzing DNA synthesis from an RNA template that is part of the enzyme.

telomere The region at the end of a linear chromosome.

telophase The final stage in mitosis or meiosis during which sister chromatids (replicated chromosomes in meiosis I) separate and new nuclear envelopes begin to form around each set of daughter chromosomes.

temperate Having a climate with pronounced annual fluctuations in temperature (i.e., warm summers and cold winters) but typically neither as hot as the tropics nor as cold as the poles.

temperature A measurement of thermal energy present in an object or substance, reflecting how much the constituent molecules are moving.

template strand (1) The strand of DNA that is transcribed by RNA polymerase to create RNA. (2) An original strand of RNA used to make a complementary strand of RNA.

tendon A band of tough, fibrous connective tissue that connects a muscle to a bone.

tentacle A long, thin, muscular appendage of gastropod molluscs.

termination (1) In enzyme-catalyzed reactions, the final stage in which the enzyme returns to its original conformation and products are released. (2) In DNA transcription, the dissociation of RNA polymerase from DNA when it reaches a termination signal sequence. (3) In RNA translation, the dissociation of a ribosome from mRNA when it reaches a stop codon.

territory An area that is actively defended by an animal from others of its species.

tertiary structure The overall three-dimensional shape of a single polypeptide chain, resulting from multiple interactions among the amino acid side chains and the peptide backbone. Compare with **primary, secondary,** and **quaternary structure.**

test A hard protective outer structure seen in some protists. Also called a *shell.*

testcross The breeding of an individual of unknown genotype with an individual having only recessive alleles for the traits of interest in order to infer the unknown genotype from the phenotypic ratios seen in offspring.

testis (plural: testes) The sperm-producing organ of a male animal.

testosterone A steroid hormone, produced and secreted by the testes, that stimulates sperm production and various male traits and reproductive behaviors.

tetrapod Any member of the taxon Tetrapoda, which includes all vertebrates with two pairs of limbs (amphibians, mammals, birds, and other reptiles).

theory A proposed explanation for a broad class of phenomena or observations.

thermal energy The kinetic energy of molecular motion.

thermocline A gradient (cline) in environmental temperature across a large geographic area.

thermophile A bacterium or archaean that thrives in very hot environments.

thermoreceptor A sensory cell or an organ specialized for detection of changes in temperature.

thermoregulation Regulation of body temperature.

thick filament A filament composed of bundles of the motor protein myosin; anchored to the center of the sarcomere. Compare with **thin filament.**

thigmotropism Growth or movement of an organism in response to contact with a solid object.

thin filament A filament composed of two coiled chains of actin and associated regulatory proteins; anchored at the Z disk of the sarcomere. Compare with **thick filament.**

thorn A modified plant stem shaped as a sharp protective structure. Helps protect a plant against feeding by herbivores.

threshold potential The membrane potential that will trigger an action potential in a neuron or other excitable cell. Also called simply *threshold.*

thylakoid A flattened, membrane-bound vesicle inside a plant chloroplast that functions in converting light energy to chemical energy. A stack of thylakoids is a granum.

thymus An organ, located in the anterior chest or neck of vertebrates, in which immature T cells generated in the bone marrow undergo maturation.

thyroid gland A gland in the neck that releases thyroid hormone (which increases metabolic rate) and calcitonin (which lowers blood calcium).

thyroid-stimulating hormone (TSH) A peptide hormone, produced and secreted by the anterior pituitary, that stimulates release of thyroid hormones from the thyroid gland.

thyroxine (T_4) A peptide hormone containing four iodine atoms that is produced and secreted by the thyroid gland. Acts primarily to increase cellular metabolism. In mammals, T_4 is converted to the more active hormone triiodothyronine (T_3) in the liver.

tight junction A type of cell-cell attachment structure that links the plasma membranes of adjacent animal cells, forming a barrier that restricts movement of substances in the space between the cells. Most abundant in epithelia (e.g., the intestinal lining). Compare with **desmosome** and **gap junction.**

tip The end of a branch on a phylogenetic tree. Represents a specific species or larger taxon that has not (yet) produced descendants—either a group living today or a group that ended in extinction. Also called *terminal node.*

Ti plasmid A plasmid carried by *Agrobacterium* (a bacterium that infects plants) that can integrate into a plant cell's chromosomes and induce formation of a gall.

tissue A group of similar cells that function as a unit, such as muscle tissue or epithelial tissue.

tolerance In ecological succession, the phenomenon in which early-arriving species do not affect the probability that subsequent species will become established. Compare with **facilitation** and **inhibition.**

tonoplast The membrane surrounding a plant vacuole.

top-down control hypothesis The hypothesis that the size of herbivore populations is limited by predation or disease rather than by limited or toxic nutritional resources.

topoisomerase An enzyme that cuts and rejoins DNA downstream of the replication fork, to ease the twisting that would otherwise occur as the DNA "unzips."

torpor An energy-conserving physiological state, marked by a decrease in metabolic rate, body temperature, and activity, that lasts for a short period (overnight to a few days or weeks). Occurs in some small mammals when the ambient temperature drops significantly. Compare with **hibernation.**

totipotent Capable of dividing and developing to form a complete, mature organism.

trachea (plural: tracheae) (1) In insects, one of the small air-filled tubes that extend throughout the body and function in gas exchange. (2) In terrestrial vertebrates, the airway connecting the larynx to the bronchi. Also called *windpipe.*

tracheid In vascular plants, a long, thin water-conducting cell that has gaps in its secondary cell wall, allowing water movement between adjacent cells. Compare with **vessel element.**

trade-off In evolutionary biology, an inescapable compromise between two traits that cannot be optimized simultaneously. Also called *fitness trade-off.*

trait Any heritable characteristic of an individual.

transcription The process by which RNA is made from a DNA template.

transcriptional control Regulation of gene expression by various mechanisms that change the rate at which genes are transcribed to form messenger RNA. In negative control, binding of a regulatory protein to DNA represses transcription; in positive control binding of a regulatory protein to DNA promotes transcription.

transcriptome The complete set of genes transcribed in a particular cell.

transduction Conversion of information from one mode to another. For example, the process by which a stimulus outside a cell is converted into a response by the cell.

transfer cell In land plants, a cell that transfers nutrients from a parent plant to a developing plant seed.

transfer RNA (tRNA) One of a class of RNA molecules that have an anticodon at one end and an amino acid binding site at the other. Each tRNA picks up a specific amino acid and binds to the corresponding codon in messenger RNA during translation.

transformation (1) Incorporation of external DNA into the genome. Occurs naturally in some bacteria; can be induced in the laboratory by certain processes. (2) Conversion of a normal cell to a cancerous one.

transgenic Referring to an individual plant or animal whose genome contains DNA introduced from another individual, either from the same or a different species.

transitional form A fossil species or population with traits that are intermediate between older and younger species.

transition state A high-energy intermediate state of the reactants during a chemical reaction that must be achieved for the reaction to proceed. Compare with **activation energy.**

transition state facilitation The second stage (after initiation) in enzyme-catalyzed reactions, in which the enzyme enables formation of the transition state.

translation The process by which proteins and peptides are synthesized from messenger RNA.

translational control Regulation of gene expression by various mechanisms that alter the life span of messenger RNA or the efficiency of translation.

translocation (1) In plants, the movement of sugars and other organic nutrients through the phloem by bulk flow. (2) A type of mutation in which a piece of a chromosome moves to a non-homologous chromosome. (3) The process by which a ribosome moves down a messenger RNA molecule during translation.

transmembrane protein Any membrane protein that spans the entire lipid bilayer. Also called *integral membrane protein.*

transmission The passage or transfer (1) of a disease from one individual to another or (2) of electrical impulses from one neuron to another.

transmission electron microscope (TEM) A microscope that forms an image from electrons that pass through a specimen. Compare with **scanning electron microscope.**

transpiration Water loss from aboveground plant parts. Occurs primarily through stomata.

transporter See **carrier protein.**

transport protein Collective term for any membrane protein that enables a specific ion or small molecule to cross a plasma membrane. Includes carrier proteins and channel proteins, which carry out passive transport (facilitated diffusion), and pumps, which carry out active transport.

transposable elements Any of several kinds of DNA sequences that are capable of moving themselves, or copies of themselves, to other locations in the genome. Include LINEs and SINEs.

tree of life A diagram depicting the genealogical relationships of all living organisms on Earth, with a single ancestral species at the base.

triacylglycerol See **fat.**

trichome A hairlike appendage that grows from epidermal cells of some plants. Trichomes exhibit a variety of shapes, sizes, and functions depending on species.

triglyceride See **fat.**

triiodothyronine (T_3) A peptide hormone containing three iodine atoms that is produced and secreted by the thyroid gland. Acts primarily to increase cellular metabolism. In mammals, T_3 has a stronger effect than does the related hormone thyroxine (T_4).

triose A monosaccharide (simple sugar) containing three carbon atoms.

triplet code A code in which a "word" of three letters encodes one piece of information. The genetic code is a triplet code because a codon is three nucleotides long and encodes one amino acid.

triploblast (adjective: triploblastic) An animal whose body develops from three basic embryonic cell layers: ectoderm, mesoderm, and endoderm. Compare with **diploblast.**

trisomy The state of having three copies of one particular type of chromosome.

tRNA See **transfer RNA.**

trochophore A larva with a ring of cilia around its middle that is found in some lophotrochozoans.

trophic level A feeding level in an ecosystem.

tropomyosin A regulatory protein present in thin (actin) filaments that blocks the myosin-binding sites on these filaments, thereby preventing muscle contraction.

troponin A regulatory protein, present in thin (actin) filaments, that can move tropomyosin off the myosin-binding sites on these filaments, thereby triggering muscle contraction. Activated by high intracellular calcium.

true navigation The type of navigation by which an animal can reach a specific point on Earth's surface.

trypsin A protein-digesting enzyme present in the small intestine that activates several other protein-digesting enzymes.

trypsinogen The precursor of protein-digesting enzyme trypsin. Secreted by the pancreas and activated by the intestinal enzyme enterokinase.

T tubules Membranous tubes that extend into the interior of muscle cells. Propagate action potentials throughout a muscle cell and trigger release of calcium from the sarcoplasmic reticulum.

tube foot One of the many small, mobile, fluid-filled extensions of the water vascular system of echinoderms; the part extending outside the body is called a podium. Used in locomotion and feeding.

tuber A modified plant rhizome that functions in storage of carbohydrates.

tumor A mass of cells formed by uncontrolled cell division. Can be benign or malignant.

tumor suppressor A gene (e.g., *p53* and *Rb*) or the protein it encodes that prevents cell division, particularly when the cell has DNA damage. Mutated forms are associated with cancer.

tundra The treeless biome in polar and alpine regions, characterized by short, slow-growing vegetation, permafrost, and a climate of long, intensely cold winters and very short summers.

turgid Swollen and firm as a result of high internal pressure (e.g., a plant cell containing enough water for the cytoplasm to press against the cell wall). Compare with **flaccid.**

turgor pressure The outward pressure exerted by the fluid contents of a plant cell against its cell wall.

Turner syndrome A human genetic disorder caused by the presence of only one X chromosome and no Y chromosome ("XO"). Individuals with this condition are female but sterile.

turnover In lake ecology, the complete mixing of upper and lower layers of water that occurs each spring and fall in temperate-zone lakes.

twofold rotational symmetry See **dyad symmetry.**

tympanic membrane The membrane separating the middle ear from the outer ear in terrestrial vertebrates, or similar structures in insects. Also called the *eardrum.*

ubiquinone See **coenzyme Q.**

ultimate causation In biology, the reason that a trait or phenomenon is thought to have evolved; the adaptive advantage of that trait. Also called *ultimate explanation.* Compare with **proximate causation.**

umami The taste of glutamate, responsible for the "meaty" taste of most proteins and of monosodium glutamate.

umbilical cord The cord that connects a developing mammalian embryo or fetus to the placenta and through which the embryo or fetus receives oxygen and nutrients.

unequal crossover An error in crossing over during meiosis I in which the two non-sister chromatids match up at different sites. Results in gene duplication in one chromatid and gene loss in the other.

unsaturated Referring to fats and fatty acids in which at least one carbon-carbon bond is a double bond. Double bonds produce kinks in the fatty acid chains and decrease the compound's melting point. Compare with **saturated.**

upstream In genetics, opposite to the direction in which RNA polymerase moves along a DNA strand. Compare with **downstream.**

urea A water-soluble excretory product of mammals and sharks. Used to remove from the body excess nitrogen derived from the breakdown of amino acids. Compare with **uric acid.**

ureter In vertebrates, a tube that transports urine from one kidney to the bladder.

urethra The tube that drains urine from the bladder to the outside environment. In male vertebrates, also used for passage of sperm during ejaculation.

uric acid A whitish excretory product of birds, reptiles, and terrestrial arthropods. Used to remove from the body excess nitrogen derived from the breakdown of amino acids. Compare with **urea.**

urochordates One of the three major chordate lineages (Urochordata), comprising sessile, filter-feeding animals that have a polysaccharide ex-oskeleton (tunic) and two siphons through which water enters and leaves; also called tunicates or sea squirts. Compare with **cephalochordates** and **vertebrates.**

uterus The organ in which developing embryos are housed in those vertebrates that give live birth. Common in most mammals and in some lizards, sharks, and other vertebrates.

vaccination The introduction into an individual of weakened, killed, or altered pathogens to stimulate development of acquired immunity against those pathogens.

vaccine A preparation designed to stimulate an immune response against a particular pathogen without causing illness. Vaccines consist of inactivated (killed) pathogens, live but weakened (attenuated) pathogens, or portions of a viral capsid (subunit vaccine).

vacuole A large organelle in plant and fungal cells that usually is used for bulk storage of water, pigments, oils, or other substances. Some vacuoles contain enzymes and have a digestive function similar to lysosomes in animal cells.

vagina The birth canal of female mammals; a muscular tube that extends from the uterus through the pelvis to the exterior.

valence The number of unpaired electrons in the outermost electron shell of an atom; determines how many covalent bonds the atom can form.

valence electron An electron in the outermost electron shell, the valence shell, of an atom. Valence electrons tend to be involved in chemical bonding.

valves In circulatory systems, flaps of tissue that prevent backward flow of blood, particularly in veins and between the chambers of the heart.

van der Waals interactions A weak electrical attraction between two hydrophobic side chains. Often contributes to tertiary structure in proteins.

variable number tandem repeat See **minisatellite.**

variable (V) region The portion of an antibody's light chains or heavy chains that has a highly variable amino acid sequence and forms part of the antigen-binding site. Compare with **constant (C) region.**

vasa recta In the vertebrate kidney, a network of blood vessels that runs alongside the loop of Henle of a nephron. Functions in reabsorption of water and solutes from the filtrate.

vascular bundle A cluster of xylem and phloem strands in a plant stem.

vascular cambium One of two types of lateral meristem, consisting of a ring of undifferentiated plant cells inside the cork cambium of woody plants; produces secondary xylem (wood) and secondary phloem. Compare with **cork cambium.**

vascular tissue system In plants, any tissue that is involved in conducting water or solutes from one part of a plant to another. Also called simply *vascular tissue.* See **phloem** and **xylem.**

vas deferens A pair of muscular tubes that store and transport semen from the epididymis to the ejaculatory duct. In nonhuman animals, called the *ductus deferens.*

vector A biting insect or other organism that transfers pathogens from one species to another. See also **cloning vector.**

vegetative organs The nonreproductive parts of a plant including roots, leaves, and stems.

vein   Any blood vessel that carries blood (oxygenated or not) under relatively low pressure from the tissues toward the heart. Compare with **artery.**

vena cava   (plural: vena cavae) A large vein that returns oxygen-poor blood to the heart.

ventral   Toward an animal's belly and away from its back. The opposite of dorsal.

ventricle   (1) A thick-walled chamber of the heart that receives blood from an atrium and pumps it to the body or to the lungs. (2) One of several small fluid-filled chambers in the vertebrate brain.

vertebra   (plural: vertebrae) One of the cartilaginous or bony elements that form the spine of vertebrate animals.

vertebrates   One of the three major chordate lineages (Vertebrata), comprising animals with a dorsal column of cartilaginous or bony structures (vertebrae) and a skull enclosing the brain. Includes fishes, amphibians, mammals, reptiles, and birds. Compare with **cephalochordates** and **urochordates.**

vessel element   In vascular plants, a short, wide water-conducting cell that has gaps through both the primary and secondary cell walls, allowing unimpeded passage of water between adjacent cells. Compare with **tracheid.**

vestigial trait   Any rudimentary structure of unknown or minimal function that is homologous to functioning structures in other species. Vestigial traits are thought to reflect evolutionary history.

vicariance   The physical splitting of a population into smaller, isolated populations by a geographic barrier.

villi   (singular: villus) Small, fingerlike projections (1) of the lining of the small intestine or (2) of the fetal portion of the placenta adjacent to maternal arteries. Function to increase the surface area available for absorption of nutrients and gas exchange (in the placenta).

virion   A single mature virus particle.

virulence   The ability of a pathogen or parasite to cause disease and death.

virulent   Referring to pathogens that can cause severe disease in susceptible hosts.

virus   A tiny intracellular parasite that uses host cell enzymes to replicate; consists of a DNA or RNA genome enclosed within a protein shell (capsid). In enveloped viruses, the capsid is surrounded by a phospholipid bilayer derived from the host cell plasma membrane, whereas nonenveloped viruses lack this protective covering.

visceral mass   One of the three main parts of the mollusc body; contains most of the internal organs and external gill.

visible light   The range of wavelengths of electromagnetic radiation that humans can see, from about 400 to 700 nanometers.

vitamin   An organic micronutrient that usually functions as a coenzyme.

vitelline envelope   A fibrous sheet of glycoproteins that surrounds mature egg cells in many vertebrates. Surrounded by a thick gelatinous matrix (the jelly layer) in some species. In mammals, called the *zona pellucida.*

viviparous   Producing live young (instead of eggs) that develop within the body of the mother before birth. Compare with **oviparous** and **ovoviviparous.**

volt (V)   A unit of electrical potential (voltage).

voltage   Potential energy created by a separation of electric charges between two points. Also called *electrical potential.*

voltage clamping   A technique for imposing a constant membrane potential on a cell. Widely used to investigate ion channels.

voltage-gated channel   An ion channel that opens or closes in response to changes in membrane voltage. Compare with **ligand-gated channel.**

wall pressure   The inward pressure exerted by a cell wall against the fluid contents of a plant cell.

water cycle, global   The movement of water among terrestrial ecosystems, the oceans, and the atmosphere.

water potential (ψ)   The potential energy of water in a certain environment compared with the potential energy of pure water at room temperature and atmospheric pressure. In living organisms, ψ equals the solute potential (ψ_S) plus the pressure potential (ψ_P).

water potential gradient   A difference in water potential in one region compared with that in another region. Determines the direction that water moves, always from regions of higher water potential to regions of lower water potential.

watershed   The area drained by a single stream or river.

water table   The upper limit of the underground layer of soil that is saturated with water.

water vascular system   In echinoderms, a system of fluid-filled tubes and chambers that functions as a hydrostatic skeleton.

Watson-Crick pairing   See **complementary base-pairing.**

wavelength   The distance between two successive crests in any regular wave, such as light waves, sound waves, or waves in water.

wax   A class of lipid with extremely long hydrocarbon tails, usually combinations of long-chain alcohols with fatty acids. Harder and less greasy than fats.

weather   The specific short-term atmospheric conditions of temperature, moisture, sunlight, and wind in a certain area.

weathering   The gradual wearing down of large rocks by rain, running water, and wind; one of the processes that transform rocks into soil.

weed   Any plant that is adapted for growth in disturbed soils.

Western blotting   A technique for identifying specific proteins separated by gel electrophoresis by transferring them to filter paper and exposing them to a labeled antibody that binds to the protein of interest. Compare with **Northern blotting** and **Southern blotting.**

wetland   A shallow-water habitat where the soil is saturated with water for at least part of the year.

white blood cells   See **leukocytes.**

wild type   The most common phenotype seen in a population; especially the most common phenotype in wild populations compared with inbred strains of the same species.

wildlife corridor   Strips of wildlife habitat connecting populations that otherwise would be isolated by man-made development.

wilt   To lose turgor pressure in a plant tissue.

wobble hypothesis   The hypothesis that some tRNA molecules can pair with more than one mRNA codon, tolerating some variation in the third base, as long as the first and second bases are correctly matched.

wood   Xylem resulting from secondary growth. Also called *secondary xylem.*

xeroderma pigmentosum   A human disease characterized by extreme sensitivity to ultraviolet light. Caused by an autosomal recessive allele that results in a defective DNA repair system.

X-linked inheritance   Inheritance patterns for genes located on the mammalian X chromosome. Also called *X-linkage.*

X-ray crystallography   A technique for determining the three-dimensional structure of large molecules, including proteins and nucleic acids, by analysis of the diffraction patterns produced by X-rays beamed at crystals of the molecule.

xylem   A plant vascular tissue that conducts water and ions; contains tracheids and/or vessel elements. Primary xylem develops from the procambium of apical meristems; secondary xylem, or wood, from the vascular cambium of lateral meristems. Compare with **phloem.**

xylem sap   The watery fluid found in the xylem of plants.

yeast   Any fungus growing as a single-celled form. Also, a specific lineage of ascomycetes.

Y-linked inheritance   Inheritance patterns for genes located on the mammalian Y chromosome. Also called *Y-linkage.*

yolk   The nutrient-rich cytoplasm inside an egg cell; used as food for the growing embryo.

Z disk   The structure that forms each end of a sarcomere. Contains a protein that binds tightly to actin, thereby anchoring thin filaments.

zero population growth (ZPG)   A state of stable population size due to fertility staying at the replacement rate for at least one generation.

zona pellucida   The gelatinous layer around a mammalian egg cell. In other vertebrates, called the *vitelline envelope.*

zone of (cellular) division   In plant roots, a group of apical meristematic cells just behind the root cap where cells are actively dividing.

zone of (cellular) elongation   In plant roots, a group of young cells, located behind the apical meristem, that are increasing in length.

zone of (cellular) maturation   In plant roots, a group of plant cells, located several centimeters behind the root cap, that are differentiating into mature tissues.

Z scheme   Path of electron flow in which electrons pass from photosystem II to photosystem I and ultimately to $NADP^+$ during the light-dependent reactions of photosynthesis. Also called *noncyclic electron flow.*

zygomycete   A member of a paraphyletic group of fungi (Zygomycota) characterized by a durable, thick-walled zygosporangium that forms from yoked hyphae of different mating types.

zygosporangium (plural: zygosporangia)   The spore-producing structure in fungi that are members of the Zygomycota.

zygote   The diploid cell formed by the union of two haploid gametes; a fertilized egg. Capable of undergoing embryological development to form an adult.

Credits

Chapter 15

Opener Halaska, Jacob Photolibrary.com

Chapter 16

Opener Joseph G. Gall. **16.05** Fred Stevens/Mykoweb. **16.06** Bert W. O'-Malley, M.D. Bert W. O'Malley, M.D. **16.09** Oscar Miller/Science Photo Library/Photo Researchers, Inc. **16.15a** Venkitaraman Ramakrishnan. **16.16** E.V. Kiseleva/Donald Fawcett/Visuals Unlimited. **16.19b** Bill Longcore /Photo Researchers, Inc. **16.21b** Bill Longcore /Photo Researchers, Inc.

Chapter 17

Opener EM Unit, VLA/Science Photo Library/Photo Researchers, Inc. **applying concepts** Michael Gabridge/Visuals Unlimited

Chapter 18

Opener Dr. Timothy J. Richmond Nature Magazine DC. **18.02a** Ada Olins/Don Fawcett/Photo Researchers, Inc. **18.04** Victoria E. Foe **18.04** Barbara Hamkalo

Chapter 19

Opener Tsien Laboratory **19.14** Baylor Coll. of Medicine/Peter Arnold, Inc. **19.17a** Brad Mogen/Visuals Unlimited

Chapter 20

Opener Sanger Institute Welcome Trust Medical Photographic Library. **20.07 and 20.07b** Reprinted by Permission from Macmillan Publishers Ltd. From Fig 4 in "Nature Medicine", 11 October 1, 2005 "Genetic Fingerprinting" by Alec J. Jeffreys pp1035-1039. The first application of DNA fingerprinting—an immigration case. **20.12** Patrick O. Brown, M.D.

Chapter 21

Opener Dr. Yorgos Nikas/Photo Researchers, Inc. **21.01** Chin-Sang. **21.02a-b** Elsevier Science Ltd. **21.04** Roslin Institute Empics. **21.05a** Richard Hutchings/Photo Researchers, Inc. **21.05b** Institute for Laboratory Animal Research ILAR/The National Academies. **21.06a** F. Rudolf Turner. **21.06b** Gary Grumbling. **21.07** Prof. Dr. Christiane Nusslein-Volhard. **21.09** Wolfgang Driever. **21.10a** Stephen Paddock. **21.10b** Stephen J. Small. **21.10c** Stephen Paddock. **21.12L** David Scharf/Photo Researchers, Inc. **21.12M** David Scharf/Peter Arnold, Inc. **21.12R** (c)Science VU/Dr. F.R. Turner/Visuals Unlimited. **21.14** Anthony Bannister/NHPA/Photo Researchers, Inc. **21.15a-b** Natural History Magazine

Chapter 22

Opener Yorgos Nikas/Photo Researchers, Inc. **22.02** Holger Jastrow. **22.04.** Gregory Ochocki/Photo Researchers, Inc. **22.07a** Michael Whitaker/Science Photo Library/Photo Researchers, Inc. **22.07bRML** Victor D. Vacquier. **22.14a** Kathryn W. Tosney. **22.14b** Gary C. Schoenwolf

Chapter 23

Opener Dr. John Runions/Oxford Brookes University **23.01** Holt Studios International/Photo Researchers, Inc. **23.03a** Ray Evert. **23.03c** Cabisco/Visuals Unlimited **23.04b** Biodisc/Visuals Unlimited **23.08b** Ken Wagner/Phototake NYC **23.10a-c** Neelima Sinha **23.12** John L.

Chapter 24

Opener Kim Taylor/NPL/Minden Pictures **24.02a** Bettmann/Corbis **24.02b** National Portrait Gallery, London **24.03a** Ken Lucas/Visuals Unlimited **24.03b** T. A. Wiewandt/DRK Photo **24.03c** Knut Finstermeier/Dr. Svante Paabo **24.04aR** Robert Lubeck/Animals Animals Earth Scenes **24.05aL** CMCD/Getty Images Inc. **24.05aR** Vincent Zuber/Custom Medical Stock Photo, Inc. **24.05bL** Mary Beth Angelo/Photo Researchers, Inc. **24.05bR** Custom Medical Stock Photo, Inc. **24.06aBL** George D. Lepp/Photo Researchers, Inc. **24.06aBR** Mickey Gibson Animals Animals/Earth Scenes **24.06aTL** Marie Read Animals Animals/Earth Scenes **24.06aTR** Tui De Roy Bruce Coleman Inc. **24.08L** Michael K. Richardson Springer-Verlag GmbH & Co KG **24.08M** Ronan O'Rahilly Springer-Verlag GmbH & Co KG **24.08R** National Museum of Health and Medicine **24.12** Frans Lanting/Minden Pictures **24.17aTL** From Fig 1C in Abzanhov et al. 2004. Science 305: 1462-1465 **24.17aTR** From Fig 1C in Abzanhov et al. 2004. Science 305: 1462-1465 **24.17aBL,BR** The Royal Society of London Reproduced by permission from K. Petren et al., A phylogeny of Darwin's finches based on microsatellite DNA length variation. Proceedings of the Royal Society of London, Series B, 266:321-329 (1999), p. 327, fig. 3. Copyright (c) 1999 Royal Society of London. Image courtesy of Kenneth Petren, University of Cincinnati. **24.17bL** From Figure 3D in Abzanhov et al. 2004. Science 305: 1462-1465 **24.17bR** From Fig 3E in Abzanhov et al. 2004. Science 305: 1462-1465

Chapter 25

Opener Phil Savoie/Minden Pictures **25.07a** Max Westby **25.08a** James Hughes/Visuals Unlimited **25.12a** Cyril Laubscher/Dorling Kindersley Media Library **25.13** Otorohanga Zoological Society Inc. **25.14a** Marc Moritsch/National Geographic Image Collection **25.15aT** Linda Mitchell Photography **25.15bB** R. & A. Simpson VIREO/The Academy of Natural Sciences **25.15bT** B. Schorre VIREO/The Academy of Natural Sciences **25.15cB** Van Os, Joseph/Getty Images Inc. **25.15cT** Jeremy Woodhouse/Getty Images, Inc

Chapter 26

Opener Wayne Lankinen/DRK Photo **26.09** Jordan Rofkar **26.12LMR** Jason Rick Loren H. Rieseberg

Chapter 27

Opener Pete Oxford/Minden Pictures **27.06a** William L. Crepet/American Journal of Botany **27.06b** Martin Land/Science Photo Library/Photo Researchers, Inc. **27.06c** Monte Hieb & Harrison Hieb/Monte Hieb/Geocraft **27.06d** John Gerlach/DRK Photo **27.09b** Nature Magazine **27.09b** Shuhai Xiao/Nature Magazine **27.09cB** Ken Lucas/Visuals Unlimited **27.09cT** Simon Conway Morris Simon Conway Morris **27.09dB** Ed Reschke/Peter Arnold, Inc. **27.09dT** B.J. Miller Biological Photo Service **27.12a** left, middle Jonathan B. Losos and Kevin de Quieroz/Prof. Jonathan B. Losos **27.13a** Colin Keates/Dorling Kindersley Media Library **27.13b** Tim Fitzharris/Minden Pictures **27.13c** Don P. Northup www.africancichlidphotos.com **27.13d** Joe Tucciarone/Photo Researchers, Inc. **27.15** David Hardy/Photo Researchers, Inc. **27.16bLR** Glen A. Izett/U.S. Geological Survey, Denver **27.16cR** Peter H. Schultz

Chapter 28

Opener Beth Donidow/Visuals Unlimited **28.03** Science VU/Visuals Unlimited **28.04** J. Robert Waaland/Biological Photo Service **28.06** Richard L. Carlton/Visuals Unlimited **28.08** Yul Roh American Association for the Advancement of Science **28.10** Kwangshin Kim/Photo Researchers, Inc. **28.10b** Todd Bannor/Custom Medical Stock Photo, Inc. **28.10c** Carolina Biological/Visuals Unlimited **28.13aL** CNRI/Science Photo Library/Photo Researchers, Inc. **28.13aR** Dr. Heide Schulz American Association for the Advancement of Science **28.13bL** Gary Gaugler/Visuals Unlimited **28.13bR** David M. Phillips/Visuals Unlimited **28.13cL** Linda Stannard University of Cape Town/Science Photo Library/Photo Researchers, Inc. **28.13c** Richard W. Castenholz **28.14a** R. Hubert, Microbiology Program, Dept. of Animal Science, Iowa State University **28.18** J. C. Revy/Phototake NYC **28.19** Centers for Disease Control & Prevention **28.20** S. Amano, S. Miyadoh and T. Shomura (The Society for Actinomycetes Japan) **28.21** R. Calentine/Visuals Unlimited **28.22** National Cancer Institute/SPL/Photo Researchers, Inc. **28.23a** Yves. V. Brun **28.23b** Prof. Hans Reichenbach **28.26** Kenneth M. Stedman **28.26** Corale Brierley/Visuals Unlimited **28.27** Marli Miller **28.27** Bonnie K. Baxter **28.29** Roland Birke/Tierbild Okapia/Photo Researchers, Inc.

Chapter 29

Opener Steve Gschmeissner/Photo Researchers, Inc. **29.02a** Norman T. Nicoll/Natural Visions **29.02b** Gregory Ochocki/Photo Researchers, Inc. **29.02c** Tom and Therisa Stack/Tom Stack & Associates, Inc. **29.02r** John Anderson/Animals Animals Earth Scenes **29.03a** Michael Abbey/Photo Researchers, Inc. **29.03b** Wim van Egmond/Visuals Unlimited **29.03c** Cabisco/Visuals Unlimited **29.03d** Sherman Thomson/Visuals Unlimited **29.03e** National Institute for Environmental Studies, Japan http://www.nies.go.jp/biology/mcc/home.htm **29.03f** Dr. Sc. Yuuji Tsuki **29.04** Sanford Berry/Visuals Unlimited **29.04** J. Robert Waaland/Biological Photo Service **29.07** Biophoto Associates/Photo Researchers, Inc. **29.13a** Andrew Syred/Science Photo Library/Photo Researchers, Inc. **29.13b** David M. Phillips/Visuals Unlimited **29.13c** Andrew Syred/Science Photo Library/Photo Researchers, Inc. **29.14a** Biophoto Associates/Photo Researchers, Inc. **29.14b** Bruce Coleman Inc. **29.15** M.I. Walker/Photo Researchers, Inc. **29.18** D. Fleetham/Animals Animals Earth Scenes **29.23** M. Abbey/Photo Researchers, Inc. **29.24** Carolina Biological/Visuals Unlimited **29.26** Michael Abbey/Visuals Unlimited **29.28** Dr. Philip J. Spagnuolo **29.30** Wim van Egmond/Visuals Unlimited **29.32** Phototake NYC **29.33** Andrew Syred/SPL/Photo Researchers, Inc. **29.34** Joyce Photographics/Photo Researchers, Inc. **29.36** Peter Parks/Image Quest 3-D **29.38** Robert De Goursey/Visuals Unlimited **29.40** Scott Camazine/Photo Researchers, Inc.

Chapter 30

Opener Darrell Gulin/DRK Photo **30.01a** Tom Myers/Photo Researchers, Inc. **30.01b** D. P. Burnside/Photo Researchers, Inc. **30.03a** Runk/Schoenberger/Grant Heilman Photography, Inc. **30.05** Holt Studios Int./Photo

Researchers, Inc. **30.06L** Linda Graham **30.06R** Lee W. Wilcox **30.07a/1** K.G. Vock/Okapia/Photo Researchers, Inc. **30.07a/2-3** Lee W. Wilcox **30.07b-1** Alvin E. Staffan/Photo Researchers, Inc. **30.07b/2** Biophoto Associates/Photo Researchers, Inc. **30.07b/3** Milton Rand/Tom Stack & Associates, Inc. **30.07b/4** Rod Planck/Photo Researchers, Inc. **30.07c-bM** Larry Lefever/Grant Heilman Photography, Inc. **30.07cBR** Stephen J. Krasemann/DRK Photo **30.07cTL** Walter H. Hodge/Peter Arnold, Inc. **30.07cTM** Degginger, E R/Animals Animals Earth Scenes **30.07cTR** ©ENLIGHTENED IMAGES/Animals Animals Earth Scenes **30.08/1** Paul K. Strother **30.08/2** Robert & Linda Mitchell Photography **30.08/3** Thomas A. Wiewandt/DRK Photo **30.08/4** David L. Dilcher **30.10** Lee W. Wilcox **30.13a-b** Lee W. Wilcox **30.14** David T. Webb **30.18a-bLR** Lee W. Wilcox **30.23a** James L. Castner **30.23b** Gerald C. Kelley/Photo Researchers, Inc. **30.23c** Gerry Ellis/GLOBIO **30.25a** John Cancalosi/Peter Arnold, Inc. **30.25b** Holt Studios Int.(Nigel Cattlin)/Photo Researchers, Inc. **30.27b/1** Robert & Linda Mitchell Photography **30.27b/2** Runk/Schoenberger/Grant Heilman Photography, Inc. **30.27b/3** Don Farrall/Getty Images, Inc. **30.27b/4** Kim Heacox Photography/DRK Photo **30.27T/2** Ed Reschke/Peter Arnold, Inc. **30.27T/3** Pat O'Hara/Corbis-Bettmann **30.27T/4** Robert & Linda Mitchell Photography **30.29a** Carolina Biological Supply Company/Phototake NYC **30.20bBL** C.C. Lockwood/Animals Animals Earth Scenes **30.20bTL** Runk/Schoenberger/Grant Heilman Photography, Inc. **30.20bR** Bill Banaszewski/Visuals Unlimited **30.29a-c** Lee W. Wilcox **30.31** Wim van Egmond **30.32** Andrew J. Martinez/Photo Researchers, Inc. **30.33** John Clayton/NIWA National Institute of Water and Atmospheric Research **30.35a-b** Robert and Beth Plowes Photograhy/Proteapix **30.36** Ed Reschke/Peter Arnold, Inc. **30.37** Visuals Unlimited **30.39** Lee W. Wilcox **30.40** Karen S. Renzaglia **30.41** Michael Clayton **30.42aL** David T. Webb **30.42aR** David Cavagnaro/DRK Photo **30.42b** Lee W. Wilcox **30.44** David T. Webb **30.45a** Biophoto Associates/Photo Researchers, Inc. **30.45b** Martin Land/Science Photo Library/Photo Researchers, Inc. **30.46** NICK GREEN/Photolibrary.com **30.47a-b** Lee W. Wilcox **30.49a** Brian Johnston **30.49bLR** Lee W. Wilcox. **30.48** Peter Chadwick/Dorling Kindersley Media Library

Chapter 31

Opener Steve Austin/CORBIS **31.01L** Mycorrhizal Applications, Inc. **31.01R** Jim Deacon Institute of Cell and Molecular Biology, The University of Edinburgh **31.03** John Cang Photography **31.04a** David M. Dennis /Tom Stack & Associates, Inc. **31.04b** Poehlmann/Phototake NYC **31.05a** George Musil/Visuals Unlimited **31.05b** Tony Brain/Science Photo Library/Photo Researchers, Inc. **31.06b** Biophoto Associates/Photo Researchers, Inc. **31.11a** Stan Flegler Visuals Unlimited **31.11b** David T. Webb **31.12** Michael P. Gadomski/Photo Researchers, Inc. **31.15** James A. Wubah **31.16** J. I. Ronny Larsson, Professor of Zoology **31.17** Biophoto Associates/Photo Researchers, Inc. **31.18** Jim Deacon Institute of Cell and Molecular Biology, The University of Edinburgh **31.19** Jeff Lepore/Photo Researchers, Inc. **31.20b** Pat O'Hara/DRK Photo **31.21** Stephen Sharnoff/Visuals Unlimited

Chapter 32

Opener David B. Fleetham/SeaPics.com **32.01** Perennou/Nuridsa /Photo Researchers, Inc. **32.01b** Mark W. Moffett/Minden Pictures **32.02a** Harry Rogers/Photo Researchers, Inc. **32.02b** Glenn Oliver/Visuals Unlimited **32.02cCM** Fallows/SeaPics.com **32.03a** Walker England/Photo Researchers, Inc. **32.03b** Ken Lucas/Visuals Unlimited **32.03c** David J. Wrobel/Visuals Unlimited **32.04** Roland Birke /Tierbild Okapia/Photo Researchers, Inc. **32.05a** Jeffrey L. Rotman/CORBIS **32.05b** Pitchal Frederic/Corbis Sygma **32.05c** J. Alex Halderman **32.09bB** L. Newman & A. Flowers/Photo Researchers, Inc. **32.09bT** Roger Steene/Image Quest 3-D **32.13a** J.P. Ferrero/Jacana/Photo Researchers, Inc. **32.12a** Flip Nicklin/Minden Pictures **32.12a** Flip Nicklin/Minden Pictures **32.12b** inset Marc Chamberlain/SeaPics.com **32.12b** Hiroya Minakuchi/Minden Pictures **32.13b** Steve Earley/Animals Animals Earth Scenes **32.14a** Pete Oxford/Minden Pictures **32.14b** Heidi & Hans-Jurgen Koch/Minden Pictures **32.15a** Daniel Sambraus/Photo Researchers, Inc. **32.15b** Anthony Bannister/NHPA/Photo Researchers, Inc. **32.15b** Josef Ramsauer & Prof. Dr. Robert Patzner **32.16a** Satoshi Kuribayashi/DRK Photo **32.16a** Satoshi Kuribayashi/OSF/DRK Photo **32.16b** Jim Brandenburg/Minden Pictures **32.17a** Andrew Syred/Photo Researchers, Inc. **32.17b** Eye of Science/Photo Researchers, Inc. **32.18a** Peter Parks/Image Quest 3-D **32.19a** Roland Seitre/Peter Arnold, Inc. **32.19b** Jeff Foott/DRK Photo **32.19c** Chamberlain, MC/DRK Photo **32.19d** Jim Greenfield/Image Quest 3-D **32.18b** Armin Maywald/Foto Natura/Minden Pictures **32.20bLBMBR** Proceedings of the National Academy of Sciences **32.22a** Doug Perrine/DRK Photo **32.22b** Dan Suzio/Photo Researchers, Inc. **32.21a** Karl-Heinz Marschner Holger de Groot/Biologis **32.21b** D. Parer & E. Parer-Cook/Auscape/Minden Pictures

32.23aTM Phototake NYC **32.23aB** Sinclair Stammers/Science Photo Library/Photo Researchers, Inc. **32.23bT** Jose B. Ruiz/Nature Picture Library **32.B** Jose B. Ruiz/Nature Picture Library **32.26** Andrew J. Martinez/Photo Researchers, Inc. **32.27a-b** Holger de Groot/Biologis **32.28** Andrew J. Martinez/Photo Researchers, Inc. **32.29** Matthew D. Hooge

Chapter 33

Opener Barbara Strnadova/Photo Researchers, Inc. **33.01a** Jean Claude Revy/Phototake NYC **33.01b** Sinclair Stammers/Photo Researchers, Inc. **33.04a** Peter Parks/Image Quest 3-D **33.05** Bill Beatty/Visuals Unlimited **33.04b** Carmel McDougall **33.09a** Mark W. Moffett/Minden Pictures **33.09b** Phototake NYC **33.09c** Dennis Kunkel Phototake NYC **33.10b** M.C. Chamberlain/DRK Photo **33.10c** Stephen Dalton/Photo Researchers, Inc. **33.12** Jan Van Arkel/Minden Pictures **33.13a** Newman & Flowers/Photo Researchers, Inc. **33.13b** Manfred Kage/Peter Arnold, Inc. **33.13c** Oliver Meckes/Photo Researchers, Inc. **33.14a** Peter Batson/Image Quest 3-D **33.14b** Alexa Bely **33.14c** Martin Dohrn/Photo Researchers, Inc. **33.15a** Degginger, E R/Animals Animals Earth Scenes **32.16b** Norbert Wu/Peter Arnold, Inc. **33.16a** Roger Steen/Image Quest Marine **33.17** Kjell B. Sandved/Photo Researchers, Inc. **33.18** Fred Bavendam/Peter Arnold, Inc. **33.20a** Michael Fogden/DRK Photo **33.20b** Diane Nelson/Visuals Unlimited **33.21** Luis M. de la Maza/Phototake NYC. **33.22** Mark Smith/Photo Researchers, Inc. **33.23a** Gerard Blondeau/Phototake NYC **T33.1/1** Jay Cossey **T33.1/2** Gustav W. Verderber **T33.1/3** Joachim Lippi **T33.1/4** Trounce Wikimedia Commons **T33.1/5** J. Gall/Photo Researchers, Inc. **T33.1/6** Nature's Images/Photo Researchers, Inc. **T33.1/7** Dr. J. K. Lindsey **T33.1/8** Scott Camazine/Photo Researchers, Inc. **33.24a** E. R. Degginger/Photo Researchers, Inc. **33.24b** M C Chamberlain/DRK Photo

Chapter 34

Opener Paul Nicklen/National Geographic Image Collection **34.02b** Thurston Lacalli **34.02a** Kaj R. Svensson/Photo Researchers, Inc. **34.03b** John D. Cunningham/Visuals Unlimited **34.09a** Gerald & Buff Corsi/Visuals Unlimited **34.09b** Greg Rouse/Scripps Institution of Oceanography Explorations **34.11** Rudie Kuiter/OceanwideImages.com **34.18a** James D. Watt/SeaPics.com **34.18b** S. Boon Fu/Peter Arnold, Inc. **34.20** James D. Watt/Image Quest 3-D **34.21R** Tom Stack/Tom Stack & Associates, Inc. **34.21** Southeastern Regional Taxonomic Center (SERTC)/South Carolina Department of Natural Resources **34.23a** Sue Daly/Nature Picture Library **34.23b** David Wrobel/SeaPics.com **34.25a** Tom McHugh/Photo Researchers, Inc. **34.25b** Tom Stack/Tom Stack & Associates, Inc. **34.24** Heather Angel/Natural Visions **34.26a** Fred McConnaughey/Photo Researchers, Inc. **34.26b** Jeff Jaskolski/SeaPics.com **34.27** Amar & Isabelle Guillen/SeaPics.com **34.28** Peter Scoones/Getty Images Inc **34.29a** Michael Fogden/DRK Photo **34.29b** R.Andrew Odum/Peter Arnold, Inc. **34.31** Tom McHugh/Photo Researchers, Inc. **34.32** Tom Vezo/Peter Arnold, Inc. **34.33** Gerry Ellis/Minden Pictures **34.35** Gerald and Buff Corsi/Visuals Unlimited **34.36** George Grall/National Geographic Image Collection **34.37** Doug Perrine/DRK Photo **34.38** Gerald and Buff Corsi/Visuals Unlimited **34.39** M.S. Mayilvahnan **34.39b** Tom McHugh/Photo Researchers, Inc.

Chapter 35

Opener Jed Fuhrman **35.02** NIBSC/Science Photo Library/Photo Researchers, Inc. **35.06a** Omikron/Photo Researchers, Inc. **35.06b** Biophoto Associates/Photo Researchers, Inc. **35.06c** K.G. Murti/Visuals Unlimited **35.06d** Oliver Meckes/E.O.S./Max-Planck-Institut-Tubingen/Photo Researchers, Inc. **35.13a-b** Abbott Laboratories **35.15aR** NIBSC/Science Photo Library/Photo Researchers, Inc. **35.15bR** James L. Van Etten **35.17** Hans Gelderblom/Eye of Science/Meckes/Ottawa/Photo Researchers, Inc. **35.18** Philip Leder **35.19** Nigel Cattlin/Holt Studios/Photo Researchers, Inc. **35.20** Lowell Georgia/Photo Researchers, Inc. **35.21** David Parker/SPL/Photo Researchers, Inc.

Chapter 36

Opener Tom McHugh/Photo Researchers, Inc. **36.04a** Tim Hauf Photography/Visuals Unlimited **36.04bLMR** John E. Weaver University of Nebraska-Lincoln Libraries **36.05** Phil Stoffer U.S. Geological Survey **36.06a** Wally Eberhart/Visuals Unlimited **36.06b** Wolfgang Poelzer/Water Frame/Peter Arnold, Inc. **36.08a-b** Gerald D. Carr **36.08c** Frans Lanting/Minden Pictures **36.09R** Alan Majchrowicz/Peter Arnold, Inc. **36.10e** Will Cook/Charles W. Cook **36.10a** Ken Wagner/Phototake NYC **36.10b** Kenneth W. Fink/Photo Researchers, Inc. **36.10c** Lee W. Wilcox **36.10d** Bill Beatty/Visuals Unlimited **36.12a-d** Lee W. Wilcox **36.14a** Lee W. Wilcox **36.14b** M. Harvey/DRK Photo **36.14d-f** Lee W. Wilcox **36.14c** Doug Wechsler/Animals Animals Earth Scenes **36.14e** TORSTEN BREHM/Nature Picture Library **36.15b** Walker/Photo Researchers, Inc. **36.15a** Ed Reschke/Peter Arnold, Inc. **36.18b** Lee W. Wilcox **36.18b** Ed Reschke/Peter Arnold,

Inc. **36.20RL** Lee W. Wilcox **36.22b** Lee W. Wilcox **36.21** Andrew Syred/Science Photo Library/Photo Researchers, Inc. **36.22a** Dr. Brad Mogen/Visuals Unlimited/Getty Images **36.23LR** Lee W. Wilcox **36-24a-c** Lee W. Wilcox **36.25a** Biophoto Associates/Science Source/Photo Researchers, Inc. **36.25b** Lee W. Wilcox **36.26a** Bruce Iverson/Bruce Iverson, Photomicrography **36.26b** G. Shih and R. Kessel/Visuals Unlimited **36.26c** Richard Kessel & Dr. Gene Shih Visuals Unlimited **36.27LMR** Lee W. Wilcox **36.28a-b** Michael Clayton **36.29c** Adam Hart-Davis/Science Photo Library/Photo Researchers, Inc. **36.29b** James W. Richardson/Visuals Unlimited **36.29a** Lee W. Wilcox;

Chapter 37

Opener David Nunuk/Science Photo Library/Photo Researchers, Inc. **37.02** Dr. Ulf Mehlig/Wikipedia Commons **37.02R** Alan Watson "Peter Arnold, Inc." **37.05a-b** David T. Webb **37.07LR** Lee W. Wilcox **37.08a** Lee W. Wilcox **37.09** Larry Lefever/Grant Heilman Photography, Inc. **37.11M** G. Shih and R. Kessel/Visuals Unlimited **37.11B** Lee W. Wilcox **37.11T** Ken Wagner/Phototake NYC **37.16a** John D. Cunningham/Visuals Unlimited **37.16a** Lee W. Wilcox **37.16b** Lee W. Wilcox **37.19R** Biophoto Associates/Photo Researchers, Inc. **36.18L** Jean Claude Revy/Phototake NYC **37.21** Martin H. Zimmerman/ Harvard Forest **37.24** Michael R. Sussman/American Society of Plant Biologists

Chapter 38

Opener Angelo Cavalli/Getty Images **38.03** Emanuel Epstein **38.03a** Photo Researchers, Inc. **38.02b** Nigel Cattlin/Photo Researchers, Inc. **38.02c** Photo Researchers, Inc. **38.06a** The Institute of Texan Cultures **38.06b** Thony Belizaire/Getty Images, Inc. **38.09R** Dennis Drenner/Visuals Unlimited **38.13b** Eduardo Blumwald **38.14** Hugh Spencer/Photo Researchers, Inc. **38.14** Andrew Syred/SPL/Photo Researchers, Inc. **38.15** E.H. Newcomb & S.R. Tandon/Biological Photo Service **38.17a** Richard Thom/Visuals Unlimited **38.17b** Gerry Ellis/GLOBIO **38.19** Frank Greenaway/Dorling Kindersley Media Library **38.18** Carlos Munoz-Yague Eurelios/Photographic Press Agency **38.20** Ken W. Davis /Tom Stack & Associates, Inc.

Chapter 39

Opener Lee W. Wilcox **39.04a-b** Malcolm B. Wilkins **39.12LR** Malcolm B. Wilkins **39.16a-b** Runk/Schoenberger/Grant Heilman Photography, Inc. **39.17a** American Society of Plant Biologists **39.20** Thomas Bjorkman **39.22L** Carolina Biological Supply Company/Phototake NYC **39.21** Donald Specker/Animals Animals Earth Scenes **39.23a-b** Lee W. Wilcox **39.27** Malcolm B. Wilkins **39.30b** American Society of Plant Biologists **39.32** Adel Kader **39.39** Nigel Cattlin/Holt Studios International /Photo Researchers, Inc.

Chapter 40

Opener Brian Johnston **40.01aL** Wayne P. Armstrong **40.01aR** Mitsuhiko Imamori Nature Production **40.01bL** Jerome Wexler/Photo Researchers, Inc. **40.01bR** Danny Ellinger Minden Pictures **40.03a** Dan Suzio/Photo Researchers, Inc. **40.03b-c** Jerome Wexler/Photo Researchers, Inc. **40.04bB** Rod Planck/Photo Researchers, Inc. **40.04bTL** John Gerlach/DRK Photo **40.04bTR** Tom & Therisa Stack/Tom Stack & Associates, Inc. **40.07a** D. Cavagnaro/DRK Photo **40.06a-b** Leonard Lessin/Photo Researchers, Inc. **40.07bTB** Lee W. Wilcox **40.13** "Candace Galen, University of Missouri. American Journal of Botany. 2003; 90:724-729. Sunny-side up: flower heliotropism as a source of parental environmental effects on pollen quality and performance in the snow buttercup, Ranunculus adoneus (Ranunculaceae) by Candace Galen, and Maureen L. Stanton. Reproduced by permission of the publisher. **40.18** Lee W. Wilcox **40.17TL** R.J. Erwin "Photo Researchers, Inc." **40.17TM** Mark Stouffer Animals Animals/Earth Scenes **40.17TR** Gerard Lacz/Animals Animals Earth Scenes **40.17ML** Tom Edwards/Visuals Unlimited **40.17MM** David Stuckel/Visuals Unlimited **40.17MR** Gregory K. Scott/Photo Researchers, Inc **40.17BL** Fritz Prenzel/Peter Arnold, Inc. **40.17BM** David M. Schleser/Nature's Images, Inc/Photo Researchers, Inc. **40.17BR** Sylvan Wittwer/Visuals Unlimited

Chapter 41

Opener Rob Nunnington/Minden Pictures **41.02** The Royal Society of London **41.03a** Educational Images/Custom Medical Stock Photo **41.03bT** Carolina Biological Supply Company/Phototake NYC **41.03bB** Carolina Biological Supply Company/Phototake NYC **41.03c** Carolina Biological Supply Company/Phototake NYC **41.04** Ed Reschke/Peter Arnold, Inc. **41.05a** Innerspace Imaging/Photo Researchers, Inc. **41.05b** Ed Reschke/Peter Arnold, Inc. **41.05c** G.W. Willis/Animals Animals Earth Scenes **41.06a** Nina Zanetti, Pearson Benjamin Cummings **41.10** Tom Stewart/CORBIS **41.11** Natalie Fobes/Getty Images Inc. **41.13a** Fred Hossler/Visuals Unlimited **41.13b** Oliver Meckes & Nicole Ottawa/Photo Researchers, Inc. **41.13c** P.M. Motta, A. Caggiati, G. Macchiarelli/Science Photo Library/Photo Researchers, Inc. **41.18a** Kent, Breck P./Animals Animals Earth Scenes **41.18b** Mendez, Raymond A/Animals Animals Earth Scenes **41.18c** Wayne Lynch/DRK Photo **41.19b** Editrice Kurtis S.r.l. **41.19a** Ed Reschke/Peter Arnold, Inc.;

Chapter 42

Opener Frans Lanting/Minden Pictures **42.13a** Fred Hossler/Visuals Unlimited

Chapter 43

Opener Jonathan Blair/Corbis Bettmann **43.03** Mark Smith/Photo Researchers, Inc. **43.04bLMR** Karel F. Liem **43.09a** Carolina Biological Supply Company/Phototake NYC **43.11a** Ed Reschke Peter Arnold, Inc.

Chapter 44

Opener Tony Freeman/PhotoEdit Inc. **44.06a** Walter E. Harvey/Photo Researchers, Inc. **44.06b** John D. Cunningham/Visuals Unlimited **44.22a** Lennart Nilsson Albert Bonniers Forlag AB **44.22b** Carolina Biological Supply/Visuals Unlimited

Chapter 45

Opener Marcus E. Raichle **45.02** Ronald F. Mervis, PhD, Neurostructural Research Labs and The Center for Aging and Brain Repair, University of South Florida College of Medicine **45.12a** C. Raines/Visuals Unlimited **45.14** Dennis Kunkel/Visuals Unlimited **45.16a** Oliver Meckes & Nicole Ottawa/Photo Researchers, Inc. **45.23a** Norbert Wu/Peter Arnold, Inc. **45.24b** Elsevier Science Ltd.

Chapter 46

Opener Stanley Breeden/DRK Photo **46.03a** Carole M. Hackney **46.07a** David Scharf/Peter Arnold, Inc. **46.10aR** Don Fawcett/T. Kwwabara/Photo Researchers, Inc. **46.13RL** David Quillin **46.14a** Michael and Patricia Fogden **46.14bLR** John A. Pearce **46.19B** James E. Dennis/Phototake NYC **46.19T** James E. Dennis/Phototake NYC **T46.01L** Brian Eyden/Science Photo Library/Photo Researchers, Inc. **T46.01M** Innerspace Imaging/Science Photo Library/Photo Researchers, Inc. **T46.01R** Brian Eyden/Science Photo Library/Photo Researchers, Inc.

Chapter 47

Opener Ralph A. Clevenger/Corbis **47.08R** Bernard Castelein/Nature Picture Library **47.08M** Duncan McEwan/Nature Picture Library **47.08L** Stephen Dalton/Photo Researchers, Inc. **47.09R** Perennou Nuridsany/Photo Researchers, Inc. **47.09M** Hans Pfletschinger/Peter Arnold Inc. **47.09L** BIOS/Heras Joël/Peter Arnold Inc. **47.11** Jordan Rehm/D. Scott Weigle

Chapter 48

Opener Leszczynski, Zigmund/Animals Animals/Earth Scenes **48.01a** Nova Scientific Corp/Oxford Scientific Films Ltd. **48.01b** Tom Adams/Visuals Unlimited **48.01c** Charles J. Cole **48.02a** Oxford Scientific Films/Animals Animals Earth Scenes **48.04TB** Reprinted by permission from Catherine S.C. Price et al., Sperm competition between Drosophila males involves both development and incapacitation. Nature 400:449-452 (1999), figs. 2 and 3. Copyright (c) 1999 Macmillan Magazines Limited. Image courtesy of Jerry A. Coyne, University of Chicago/Nature Magazine **48.05a** G. & C. Merker/B. Tomberlin/Visuals Unlimited **48.06a** Brent D. Opell Brent D. Opell **48.10B** From Nature "A stimulatory phalloid organ in a weaver bird" M. Winterbottom, T. Burke & T.R. Birkhead/Nature/ 6 May 1999, 399: 28 Fig 1b. Photo courtesy T.R. Birkhead **48.10T** From Nature "A stimulatory phalloid organ in a weaver bird" M. Winterbottom, T. Burke & T.R. Birkhead/Nature/6 May 1999, 399: 28 Fig 1a. Photo courtesy T.R. Birkhead **48.15a-c** Mitsuaki Iwago/Minden Pictures **48.16a** Claude Edelmann /Petit Format/Photo Researchers, Inc. **48.16b** Lennart Nilsson Albert Bonniers Forlag AB **48.16c** Petit Format/Nestle/Science Source/Photo Researchers, Inc. **48.18aLR** American Association for the Advancement of Science

Chapter 49

Opener Dennis Kunkel/CNRI Phototake NYC **49.01B** Fred Hossler/Visuals Unlimited **49.02a** Photo Researchers Inc **49.02b-c** Steve Gschmeissner/Photo Researchers, Inc. **49.05b** Steve Gschmeissner/SPL/Photo Researchers, Inc. **49.05a** David M. Phillips/Visuals Unlimited

Chapter 50

Opener Stephen J. Krasemann/DRK Photo **50.01a-d** Natalie B. Fobes Photography **50.05b** Gerry Ellis/Minden Pictures **50.05c** Tim Fitzharris/Minden Pictures **50.05a** Gerry Ellis/Minden Pictures **50.06** George Gerster/Photo Researchers, Inc. **50.07** Michael Collier **50.10a-c** Edward Kinsman/Photo Researchers, Inc. **50.12** Hans Reinhard/OKAPIA/Photo Researchers, Inc. **50.14** C.K. Lorenz/Photo Researchers, Inc. **50.16** © Judd Patterson **50.18** Tom Edwards/Animals Animals/Earth Scenes **50.20** Francis Lepine/Animals Animals Earth Scenes **50.22** Tom Bean/DRK Photo **50.27** Sandy Scheltema/Courtesy of The Age **50.29 inset** William H. Mullins/Photo Researchers, Inc. **50.29** Raymond Gehman/Corbis Bettmann **50.30b** Michael Fogden/DRK Photo **50.31** Ric Ergenbright/Corbis **50.32** Mark Newman/Photo Researchers, Inc. **50.32** Scott T. Smith Corbis Bettmann

Chapter 51

Opener Thomas Mangelsen/Minden Pictures **51.01** A. Flowers & L. Newman/Photo Researchers, Inc. **51.03a** © Callie de Wet **51.04** © Gary Woodburn/http://pbase.com/woody **51.06** Fred McConnaughey/Photo Researchers, Inc. **51.07** Nina Leen/Getty Images/Time Life Pictures **51.09a** Gavin Hunt **51.10** Photo courtesy of the Behavioural Ecology Research Group, Department of Zoology, Oxford University, Alexander Weir **51.11** Matt Meadows/Peter Arnold, Inc. **51.12a** Manuel Leal, Department of Biological Sciences, Union College **51.18a** Ken Lucas/Visuals Unlimited **51.18b** Paul A. Zahl/Photo Researchers, Inc. **51.18c** Ken Preston-Mafham/Premaphotos Wildlife **51.19** Bryan D. Neff **51.20** Tom Vezo/Peter Arnold, Inc.

Chapter 52

Opener National Geographic Image Collection **52.01** Robert Maier Animals Animals/Earth Scenes **52.06** Mark Trabue/USDA/NRCS **52.10LR** Marko Nieminen Ilkka Hanski **52.14a** Paal Hermansen /NHPA/Photo Researchers, Inc. **52.19a** BIOS (F. Mercay) Peter Arnold, Inc.

Chapter 53

Opener Ergenbright Scott Tuason/Image Quest Marine **53.01** PREMAPHOTOS/Nature Picture Library **53-01** Christian Ziegler **53.02a** G.R.Dick Roberts Photo Library/The Natural Sciences Image Library (NSIL) **53.02b** Fred Bavendam/Minden Pictures **53.02c** Tom Till/DRK Photo **53.02d** D. Cavagnaro/Visuals Unlimited **53.02e** Stephen J. Krasemann/DRK Photo **53.02f** Hal Beral/Visuals Unlimited **53.08LR** Jeff Podos **53.08MB** B.R. Grant **53.08MT** From Fig 1D in Science 14 July 2006 Vol. 313. no. 5784, pp. 224 – 226. Evolution of Character Displacement in Darwin's Finches Peter R. Grant* and B. Rosemary Grant. [Photos are by the authors] **53.09aL** Dave B. Fleetham/Visuals Unlimited **53.09aM** Chris Newbert/Minden Pictures **53.09aR** J. Sneesby/B. Wilkins/Getty Images Inc **53.10aL** Andrew J. Martinez Photo Researchers, Inc. **53.10aR** Zig Leszczynski/Animals Animals Earth Scenes **53.12a** Thomas G. Whitham/Northern Arizona University **53.15a** Mark Moffett/Minden Pictures **53.15b** Doug Perrine/SeaPics.com **53.18b** Thomas Kitchin/Tom Stack & Associates, Inc. **53.18a** Nancy Sefton/Photo Researchers, Inc. **53.20a** David Kjaer/Nature Picture Library **53.20b** Tony C. Caprio/Tony C. Caprio **53.21/1** Breck P. Kent/Animals Animals Earth Scenes **53.21/2** G. Carleton Ray/Photo Researchers, Inc. **53.21/3** Michael P. Gadomski/Animals Animals/Earth Scenes **53.21/4** James P. Jackson/Photo Researchers, Inc. **53.21/5** Bruce Heinemann/Getty Images, Inc. **53.21/6** Michael P. Gadomski/Photo Researchers, Inc. **53.22/1-4** Glacier Bay National Park Photo Glacier Bay National Park and Preserve **53.22/3** Christopher L. Fastie

Chapter 54

Opener Lonnie G. Thompson Byrd Polar Research Center The Ohio State University **54.02** MODIS NASA **54.11b** Randall J. Schaetzl, Michigan State University **54.11a** Richard Hartnup/Wikipedia Comons **54.12a** U.S. Forest Service, Northern Research Station

Chapter 55

Opener Micheline Pelletier/Corbis Bettmann **55.01b** Institute of Hydrobiology, Wuhan, China/The New York Times **55.01a** Tom & Pat Lesson/DRK Photo **55.02** Edward S. Ross, California Academy of Sciences **55.03a** Carr Clifton/Minden Pictures **55.03b** Kevin W. Fitz All Species Photography & Sound **55.04b** from: Fig1C in Nature Vol 436 August 18, 2005 page 1016 "Global hotspots of species richness are not congruent with endemism or threat". C. David L. Orne et al. **55.04a** from Fig 1A in Nature Vol 436 August 18, 2005 page 1016 "Global hotspots of species richness are not congruent with endemism or threat". C. David L. Orne et al. **55.05B** Fred Bruemmer/DRK Photo **55.07a** John Mitchell/Photo Researchers, Inc. **55.07b** Gary Braasch Gary Braasch Photography **55.07c** David Dennis/Animals Animals Earth Scenes **55.09** Jim Wark/Airphoto **55.08** Martin Harvey/Alamy **55.10a** Jacques Jangoux/Peter Arnold, Inc. **55.10bLR** NASA/Goddard Space Flight Center

Backmatter

BS7.02 Reproduced by permission from J.P. Ferris et al., Synthesis of long prebiotic oligomers on mineral surfaces. Nature 381:59-61 (1996), Fig. 2. Copyright (c) 1996 Macmillan Magazines Limited. Image courtesy of James P. Ferris, Rensselaer Polytechnic Institute **BS8.01aL** Photo courtesy of Peter M. O'Day, Juan Bacigalupo, Joan E. Haab, and Cecilia Vergara. The Journal of Neuroscience, October 1, 2000. 20(19): 7193-7198, Fig. 1C. (c) 2004 by the Society for Neuroscience **BS8-01bR** Dennis Kunkel/Phototake NYC **BS8.02aTL** Michael W. Davidson/Florida State University/Molecular Expressions **BS8.02bTR** Michael W. Davidson/Florida State University/Molecular Expressions; **BS8.03B** Rosalind Franklin/Photo Researchers, Inc.

ILLUSTRATION CREDITS

1.12 Data from J.J. Tewksbury and G.P. Nabhan. Seed dispersal direct deterrence by capsaicin in chilies. 2001. *Nature* 412:403–404, Fig. 1a. Used with permission from Macmillan Publishers Ltd.

3.25a Data from N.N. Nawani and B.P. Kapadnis. 2001. *Journal of Applied Microbiology* 90:803–808, Fig. 3. Data also from N.N. Nawani et al. 2002. *Journal of Applied Microbiology* 93:965–975, Fig. 7. Blackwell Publishing. **3.25b** Data from T. Hansen et al. 2002. *FEMS Microbiology Letters* 216:249–253, Fig. 1. Blackwell Publishing.

6.25a PDB Document Object Identifier (DOI):10.2210/pdb1J4N/pdb. **6.25b** PDB DOI:10.2210/pdb1ORS/pdb and PDB DOI:10.2210/pdb1ORQ/pdb.

7.4 PDB DOI:10.2210/pdb2AW4/pdb and PDB DOI:10.2210/pdb2AVY/pdb.

8.10b After *Molecular Biology of the Cell*, 4th edition, Fig. 19.5. © 2002 by Bruce Alberts et al. With permission from Garland Science/Taylor & Francis, LLC.

9.2 PDB DOI:10.2210/pdb1ERK/pdb and PDB DOI:10.2210/pdb2ERK/pdb. **9.15** PDB DOI:10.2210/pdb1PFK/pdb. **9.25** Modified from H. Wang and G. Oster. Energy transduction in the F_1 motor of ATP synthase. 1998. *Nature* 396:279–282, Fig. 1. With permission from Macmillan Publishers Ltd.

11.16 American Cancer Society's *Cancer Facts and Figures–2003.* Reprinted with permission.

16.2 Data from S. Murakami et al. 2002. *Science.* 296:1285–1290. Fig. 2a. Used with permission from AAAS. **16.15** PDB DOI:10.2210/pdb2J00/pdb and PDB DOI:10.2210/pdb2J01/pdb.

17.18b PDB DOI:10.2210/pdb2ORI/pdb.

18.2b After B. Dorigo et al. 2004. *Science* 306:1571–1573, Fig. 1. With permission from AAAS. **18.15** PDB DOI:10.2210/pdb1TSR/pdb.

20.1 Data from "Public collections of DNA and RNA reach 100 gigabases." 2005. EMBL-EBI Press Release, Hixton. **20.10** After *Molecular Cell Biology*, 5/e by Harvey Lodish, et al. © 1986, 1990, 1995, 2000, 2004 by W. H. Freeman and Company. Used with permission.

21.13 After S.B. Carroll. Homeotic genes and the evolution of arthropods and chordates (Review). 1995. Nature 376: 479–485. With permission from Macmillan Publishers Ltd.

24.4b *Pakicetus* based on J.G.M. Thewissen et al. 2001. *Nature* 413:277–281, Fig. 2. *Rhodocetus* based on P.D. Gingerich et al. 2001. *Science* 293:2239–2242, Fig. 3. **24.13** Data from P.T. Boag and P.R. Grant. 1981. *Science* 214:82–85. With permission from AAAS. **24.14** Modified from P.R. Grant and B.R. Grant. 2001. *Science* 296:707–711, Fig. 1. With permission from AAAS.

25.6 Data from S. Freeman and J. Herron, *Evolutionary Analysis*, 3e, Figs. 6.15a and 6.15c. © 2003 by Prentice-Hall, Inc. **25.7b** Data from W.E. Kerr and S. Wright. 1954. *Evolution* 8:172–177. **25.11** Data from M.O. Johnston. 1992. *Evolution* 46:688–702.

28.2 After G.L. Armstrong et al. 1999. *JAMA* 281:61–66, Fig. 1. © American Medical Association. All rights reserved. **28.5** Data from M.T. Madigan and J.M. Martinko. 2006. *Brock Biology of Microorganisms,* 11e, Fig. 5.9.

34.12 After E.B. Daeschler et al. 2006. *Nature* 440:757–763, Fig. 6. Also after N.H. Shubin et al. 2006. *Nature* 440:764–771, Fig. 4. With permission from Macmillan Publishers Ltd.

43.17 Data from L.O. Schulz et al. 2006. *Diabetes Care* 29:1866–1871. © 2006 American Diabetes Association. Used with permission from The American Diabetes Association.

48.4 Data from C.S.C. Price et al. Sperm competition between *Drosophila* males involves both displacement and incapacitation. 1999. *Nature* 400:449–452, Fig. 3c. Used with permission from Macmillan Publishers Ltd.

49.6a PDB DOI:10.2210/pdb1IGT/pdb. **49.6b** PDB DOI:10.2210/pdb1TCR/pdb. **49.7** PDB DOI:10.2210/pdb2VIR/pdb.

50.28 Data from A.K. Knapp et al. 2002. *Science* 298:2202–2205, Figs. 1, 2a, and 3b.

51.22 Data from J.L. Hoogland. 1983. *Animal Behavior* 31:472–479.

52.3 Modified from C.K. Ghalambor and T.E. Martin. 2001. *Science* 292:494–497, Fig. 1c. Used with permission from AAAS. **52.7b** Data from G.F. Gause. 1934. *The Struggle for Existence*. New York: Hafner Press. **52.9** Data from M. Dodd et al. 1995. *Journal of Ecology* 83:277–285. **52.14b** Data from T. Valverde and J. Silvertown. Variation in the demography of a woodland understorey herb (*Primula vulgaris*) along the forest regeneration cycle: Projection matrix analysis. 1998. *Journal of Ecology* 86:545–562. Blackwell Publishing.

53.4 Data from G.F. Gause. 1934. *The Struggle for Existence*. New York: Hafner Press. **53.18** Data from R.T. Paine. Intertidal community structure: Experimental studies on the relationship between a dominant competitor and its principal predator. 1974. *Oecologia* 15:93–120. With kind permission of Springer Science Business Media. **53.19** Data from R.T. Paine. 1974. *Oecologia* 15:93–120.

54.4 Data from A. Tsuda et al. 2003. *Science* 300:958–961. **54.18** Data from *BP Statistical Review of World Energy 2007*, pp. 6–21. **54.19a** Data from G. Beaugrand et al. 2002. *Science* 296:1692–1694. **54.19b** Data from N.L. Bradley et al. 1999. *Proceedings of the National Academy of Science* 96:9701–9704.

55.4a,b C.D.L. Orme et al. Global hotspots of species richness are not congruent with endemism or threat. 2005. *Nature* 436:1016–1019. Reprinted by permission from Macmillan Publishers Ltd. **55.4c** N. Myers et al. Biodiversity hotspots for conservation priorities. 2000. *Nature* 403:853–858, Fig. 1. Reprinted by permission from Macmillan Publishers Ltd. **55.6** O. Venter et al. 2006. *BioScience* 56:903–910, Fig. 2. Reproduced with permission of American Institute of Biological Sciences via Copyright Clearance Center. **55.12** Data from J.T. Hogg et al. 2006. *Proceedings of the Royal Society B: Biological Sciences* 273:1491–1499, Fig.1. © Proceedings of the Royal Society B: Biological Sciences. Additional unpublished from J.T. Hogg. **55.16** Data from E.I. Damschen et al. 2006. *Science* 313:1284–1286, Fig. 2b. Use with permission from AAAS.

Index

Boldface page numbers indicate a glossary entry; page numbers followed by an *f* indicate a figure; those followed by *t* indicate a table.

SUMMARY TABLES

BIOSKILLS

Dear Student:

As the author of *Biological Science*, **Third Edition**, I set out to write a book that would be an effective learning tool for you. I would appreciate hearing about your experience with this textbook and its multimedia support, and I invite your comments and suggestions for improvements!

Many thanks,

Scott Freeman

Grading System **A**=excellent/very useful, **B**=good/useful, **C**=moderately helpful, **D**=poor/very little help, **F**=not helpful

1. Please grade the following study aids in *Biological Science*, Third Edition:

(circle one)

Gold highlighting of Key Concepts and important information within the narrative Grade: **A B C D F**
Comments _____

"If you understand... you should be able to..." in blue typeface within the narrative Grade: **A B C D F**
Comments _____

Check Your Understanding boxes Grade: **A B C D F**
Comments _____

Figure caption questions and exercises Grade: **A B C D F**
Comments _____

Summary tables Grade: **A B C D F**
Comments _____

Overall Comments _____

2. Did you use your *MasteringBiology*™ subscription? Yes [   ] No [   ] If yes, please grade the following online resources:

Web Animations Grade: **A B C D F**
Comments _____

BioFlix animations and tutorials Grade: **A B C D F**
Comments _____

Instructor-assigned tutorials Grade: **A B C D F**
Comments _____

Chapter quizzes Grade: **A B C D F**
Comments _____

Answer keys for textbook questions Grade: **A B C D F**
Comments _____

3. Please list three things you liked most about the textbook and/or online resources:

Comments _____

4. Please describe two or three topics that were difficult for you to understand, and please include page numbers.

Comments _____

5. Do you have any other suggestions for improving future editions of *Biological Science*?
Please provide page numbers where appropriate.

Comments _____

School: _____

Your Name (Optional): _____

Email: _____

Date: _____

May Benjamin Cummings have permission to quote your comments in promotions for *Biological Science*?   ☐ Yes   ☐ No